Partially Integrable Evolution Equations in Physics

NATO ASI Series

Advanced Science Institutes Series

A Series presenting the results of activities sponsored by the NATO Science Committee, which aims at the dissemination of advanced scientific and technological knowledge, with a view to strengthening links between scientific communities.

The Series is published by an international board of publishers in conjunction with the NATO Scientific Affairs Division

A Life Sciences **B Physics**	Plenum Publishing Corporation London and New York
C Mathematical and Physical Sciences **D Behavioural and Social Sciences** **E Applied Sciences**	Kluwer Academic Publishers Dordrecht, Boston and London
F Computer and Systems Sciences **G Ecological Sciences** **H Cell Biology**	Springer-Verlag Berlin, Heidelberg, New York, London, Paris and Tokyo

Series C: Mathematical and Physical Sciences - Vol. 310

Partially Integrable Evolution Equations in Physics

edited by

Robert Conte

and

Nino Boccara

Service de physique du solide et de résonance magnétique,
Centre d'etudes nucléaires de Saclay,
Gif-sur-Yvette, France

Kluwer Academic Publishers

Dordrecht / Boston / London

Published in cooperation with NATO Scientific Affairs Division

Proceedings of the NATO Advanced Study Institute on
Partially Integrable Nonlinear Evolution Equations and Their Physical Applications
Les Houches, France
March 21–30, 1989

Library of Congress Cataloging in Publication Data

NATO Advanced Study Institute on Partially Integrable Nonlinear Evolution Equations and Their Physical Applications (1989 : Les Houches, Haute-Savoie, France)
Partially integrable evolution equations in physics / edited by Robert Conte and Nino Boccara.
p. cm. -- (NATO ASI series C, Mathematical and physical sciences ; 310)
"Proceedings of the NATO Advanced Study Institute on Partially Integrable Nonlinear Evolution Equations and Their Applications, Les Houches, France, 21-30 March 1989."

1. Evolution equations, Nonlinear--Numerical solutions--Congresses. 2. Differential equations, Nonlinear--Numerical solutions--Congresses. 3. Mathematical physics--Congresses. I. Conte, Robert, 1943- . II. Boccara, Nino. III. Title. IV. Series: NATO ASI series. Series C, Mathematical and physical sciences ; no. 310.
QC20.7.E88N38 1989
530.1'55353--dc20 90-34913
CIP

ISBN-13: 978-94-010-6754-6 e-ISBN-13: 978-94-009-0591-7
DOI: 10.1007/978-94-009-0591-7

Published by Kluwer Academic Publishers,
P.O. Box 17, 3300 AA Dordrecht, The Netherlands.

Kluwer Academic Publishers incorporates the publishing programmes of
D. Reidel, Martinus Nijhoff, Dr W. Junk and MTP Press.

Sold and distributed in the U.S.A. and Canada
by Kluwer Academic Publishers,
101 Philip Drive, Norwell, MA 02061, U.S.A.

In all other countries, sold and distributed
by Kluwer Academic Publishers Group,
P.O. Box 322, 3300 AH Dordrecht, The Netherlands.

Printed on acid-free paper

TABLE OF CONTENTS

IV. MATHEMATICAL METHODS

CONTRIBUTED PAPERS

PREFACE

In the many physical phenomena ruled by partial differential equations, two extreme fields are currently overcrowded due to recent considerable developments:

1) the field of completely integrable equations, whose recent advances are the inverse spectral transform, the recursion operator, underlying Hamiltonian structures, Lax pairs, etc

2) the field of dynamical systems, often built as models of observed physical phenomena: turbulence, intermittency, Poincaré sections, transition to chaos, etc.

In between there is a very large region where systems are neither integrable nor nonintegrable, but *partially integrable*, and people working in the latter domain often know methods from either 1) or 2).

Due to the growing interest in partially integrable systems, we decided to organize a meeting for physicists active or about to undertake research in this field, and we thought that an appropriate form would be a school. Indeed, some of the above mentioned methods are often adaptable outside their original domain and therefore worth to be taught in an interdisciplinary school.

One of the main concerns was to keep a correct balance between physics and mathematics, and this is reflected in the list of courses.

In the first section, a series of courses deals with the various kinds of waves encountered in physical systems. They present a review of experimental manifestations of solitary waves in solid state physics, condensed matter, fluid dynamics and nonlinear optics, and introduce different phenomenological models in order to describe what is observed. Some exact solutions are also presented, namely for the Boltzmann equation and for the Davey-Stewartson equation, i.e. the bidimensional generalization of nonlinear Schrödinger equation.

Next, three courses concern the instabilities arising from the existence or creation of defects in convection phenomena or more generally in any system modelizable by a Ginzgurg-Landau type partial differential equation. Simple topological considerations are often sufficient to describe the experiments at least qualitatively.

We have then gathered in a third section all the mathematical aspects of partial integrability; they represent a subset of the general, powerful methods dealing with integrability in general. This subset is made of the main methods able to predict constructively if a given partial differential equation is integrable or not. As everybody knows, the word "integrable" has many acceptions, and only a few of them are presented here, mainly the ones arising from singularity analysis like in Kowalevsky and Painlevé methods. It would be quite useful to clarify in the future the interrelations between these different kinds of integrability.

The fourth section contains mathematical methods other than the ones of singularity analysis. It is mainly related to group theory and differential geometry. Many of these methods, with sometimes the help of computer algebra, can provide physically useful information like conservation laws or exact particular solutions, often of solitary wave type. A promising approach is the one of inertial manifolds, the correct mathematical description of a finite number of degrees of freedom in an infinite dimensional dynamical system such as a partial differential equation.

Finally, many participants presented results of their own research, and these contributions are gathered at the end of the volume.

This meeting had a special scientific flavor since, besides the many students in the neighborhood of their thesis and the lecturers, there were also a significant amount of senior researchers; the result was that courses became sometimes very alive, so to say.

The school was funded for the most part by a grant from NATO as an Advanced Study Institute, and we are much indebted to Dr L. V. da Cunha and Dr G. Venturi for their help and advice in organizing this session. Other grants came from Direction des recherches, études et techniques (France), National Science Foundation (USA) and Special fund for scientists from Greece, Portugal and Turkey (NATO), and we thank all these organizations for their support of scientific research.

The wonderful location of Centre de physique in Les Houches was most appreciated by all participants, although some of them would have preferred more snow. The efficient help of the administrative staff of the Centre, as well as the good food and wine of the new chef Michel, were essential for the friendly atmosphere of this meeting.

We simply hope that this book will be useful to all physicists who want to use one of the many available mathematical methods.

Robert Conte

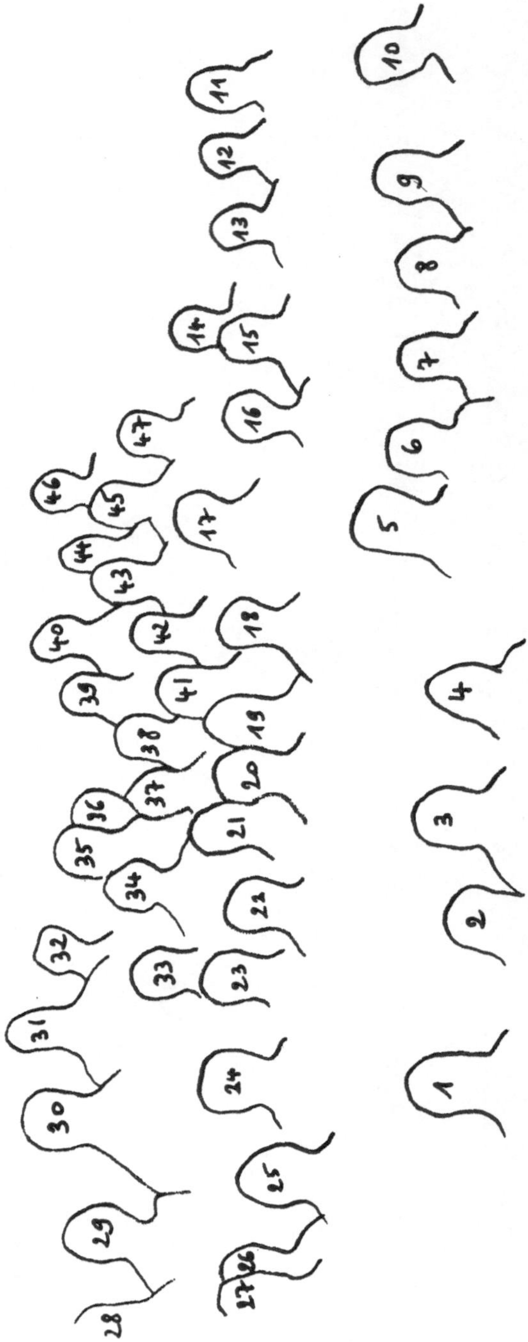

1
2
3
4
5
6
7
8
9
10
11
12
13
14
15
16
17
18
19
20
21
22
23
24
25
26
27
28
29
30
31
32
33
34
35
36
37
38
39
40
41
42
43
44
45
46
47

1. Franklin Lambert (VUB)
2. José M. Cerveró (Salamanca)
3. Frank Cariello (Columbia)
4. Mrs Cariello
5. Silvana de Lillo (Perugia)
6. Yavuz Nutku (Tübitak)
7. Radha Balakrishnan (Madras)
8. Michel Remoissenet (Dijon)
9. Olivier Larroche (Limeil)
10. Robert Conte (Saclay)
11. Anibal Rodriguez (Madrid)
12. Hasan Gümral (Tübitak)
13. Dominique Dehin (Liège)
14. Yvette Kosmann-Schwarzbach (Lille)
15. Véronique Hussin (Montréal)
16. Hugues Chaté (Saclay)
17. Richard Boesch (Boston)
18. Piotr Biler (Wroclaw)
19. Jacek Adam Tuszyński (Edmonton)
20. Willy Hereman (Madison)
21. Peter A. Clarkson (Exeter)
22. S. C. Mishra (Delhi)
23. Frank B. Estabrook (Jet Propulsion Lab)
24. Henri Cornille (Saclay)
25. Micheline Musette (VUB)
26. Liu Qichao (Beijing)
27. P. C. Dash (Bhubaneswar)
28. Pavel Winternitz (Montréal)
29. Mrs Estabrook
30. Jarmo Hietarinta (Turku)
31. Basil Grammaticos (Paris VII)
32. Jonathan Swinton (London)
33. Tom Kuusela (Turku)
34. Alexander V. Mikhailov (Landau Institute)
35. Jean-Claude Saut (Orsay)
36. Jean-Pierre Gazeau (Paris VII)
37. Peter J. Olver (Minneapolis)
38. Jean-Michel Ghidaglia (Orsay)
39. Wojciech Zakrzewski (Durham)
40. David J. Kaup (Clarkson)
41. Ralph Willox (VUB)
42. Martin D. Kruskal (Princeton)
43. John D. Gibbon (London)
44. Paolo Maria Santini (Roma)
45. John Weiss (Arlington)
46. Saadet Erbay (Tübitak)
47. Michel Grundland (Montréal)

Equations d'évolution non-linéaires partiellement intégrables et leurs applications physiques
Partially Integrable Nonlinear Evolution Equations and their Physical Applications

Les Houches 21-30 March 1989 - List of the 65 participants

Balakrishnan	Radha	The Institute of Mathematical Sciences	C.I.T. Campus	IND -Madras 600113
Bartuccelli	Michele	Imperial College, Mathematics Department		GB -London SW7 2BZ
Bessis	Daniel	CEN Saclay, Physique théorique		F -91191 Gif-sur-Yvette Cedex
Biler	Piotr	U. Wroclawski, Instytut matematyczny	Pl. Grunwaldzki 2/4	PL -50384 Wroclaw
Bishop	Alan	Los Alamos National Lab, CNLS	MS-B262	USA-Los Alamos, NM 87545
Boesch	Richard	Boston University, Department of Physics	590 Commonwealth Ave	USA-Boston, Ma 02215
Brenig	Léon	U. libre de Bruxelles, Chimie physique II	Campus Plaine, CP 231	B -1050 Bruxelles
Carati	Daniel	U. libre de Bruxelles, Chimie physique II	Campus Plaine, CP 231	B -1050 Bruxelles
Cariello	Frank	Columbia U., Department of Applied Physics		USA-New York, NY 10027
Cerveró	José M.	U. de Salamanca, Departamento de Fisica Teorica		E -37008 Salamanca
Chaté	Hugues	CEN Saclay, Physique du solide et résonance magnétique		F -91191 Gif-sur-Yvette Cedex
Clarkson	Peter A.	U. of Exeter, Department of Mathematics	North Park Road	GB -Exeter EX4 4QE
Conte	Robert	CEN Saclay, Physique du solide et résonance magnétique		F -91191 Gif-sur-Yvette Cedex
Cornille	Henri	CEN Saclay, Physique théorique		F -91191 Gif-sur-Yvette Cedex
Coullet	Pierre	U. de Nice, Physique théorique	Parc Valrose	F -06034 Nice Cedex
Csizmazia	A.		30, rue Léon Boquet	F -94100 Saint-Maur-des-Fossés
Dash	P. C.	Orissa U. of Agriculture and Technology, Dept of Physics		IND -Bhubaneswar 751003
Dehin	Dominique	U. de Liège, Physique théorique et math.	Au Sart Tilman B5	B -4000 Liège 1
Elmer	Franz-Josef	U. Basel, Institut für Physik	Klingerbergstrasse 82	CH -4056 Basel
Erbay	Saadet	Tübitak-TBAE, Dept of Applied Mathematics	PO Box 74	TR -41401 Gebze-Kocaeli
Estabrook	Frank B.	Jet Propulsion Laboratory, Mail code 169-327	4800 Oak Grove Dr.	USA-Pasadena, Ca 91109
Fordy	Allan P.	U. of Leeds, Center for nonlinear studies		GB -Leeds LS2 9JT
Fournier	Jean-Daniel	Observatoire de Nice	BP 139	F -06003 Nice Cedex
Gümral	Hasan	Tübitak-TBAE	PO Box 74	TR -41470 Gebze-Kocaeli
Gazeau	Jean-Pierre	U. Paris VII, Physique théorique et math.	Tour centrale	F -75251 Paris Cedex 05
Ghidaglia	Jean-Michel	U. Paris-sud, Laboratoire d'analyse numérique	Bâtiment 425	F -91405 Orsay Cedex
Gibbon	John D.	Imperial College, Mathematics Department		GB -London SW7 2BZ
Goriely	Alain	U. libre de Bruxelles, Chimie physique II	Campus Plaine, CP 231	B -1050 Bruxelles
Grammaticos	Basil	U. Paris VII, Physique nucléaire	Tour 24-14,5ième étage	F -75251 Paris Cedex 05
Grundland	Michel	U. de Montréal, Centre de recherches math.	CP 6128 Succ. A	CDN-Montréal, Québec H3C 3J7

Hereman	Willy	U. of Wisconsin, Department of Mathematics	610 Walnut street	USA-Madison, Wi 53705
Hietarinta	Jarmo	U. of Turku, Department of Physical Sciences		SF -20500 Turku 50
Hussin	Véronique	U. de Montréal, Centre de recherches math.	CP 6128 Succ. A	CDN-Montréal, Québec H3C 3J7
Karpman	Vladimir	IZMIRAN, Academic City	Troitsk	SU -Moscow Region 142092
Kaup	David J.	Clarkson U., Department of Mathematics		USA-Potsdam, NY 13676
Kosmann-Schw.	Yvette	USTL (Lille), UFR de math		F -59655 Villeneuve d'Ascq Cedex
Kruskal	Martin D.	Princeton University, Mathematics	Fine Hall	USA-Princeton, NJ 08544
Kuusela	Tom	U. of Turku, Department of Physical Sciences		SF -20500 Turku 50
Lambert	Franklin	Vrije U. Brussel, Theoretische Natuurkunde	Pleinlaan 2	B -1050 Bruxelles
Larroche	Olivier	Centre d'études de Limeil-Valenton	BP 27	F -94190 Villeneuve Saint-Georges
de Lillo	Silvana	U. di Perugia, Dipartimento di Fisica	Via Elce di Sotto	I -06100 Perugia
Liu	Qichao	Beijing Normal U., Physics Department		CHI -Beijing 100088
Mikhailov	Alexander V.	Landau Institute, Academy of Sciences		SU -Moscow 117940
Mishra	S. C.	U. of Delhi, Dept of Physics and Astrophysics		IND -Delhi 110007
Musette	Micheline	Vrije U. Brussel, Theoretische Natuurkunde	Pleinlaan 2	B -1050 Bruxelles
Newell	Alan C.	U. of Arizona, Department of Mathematics		USA-Tucson, Az 85721
Nutku	Yavuz	Tübitak-TBAE	PO Box 74	TR -41470 Gebze-Kocaeli
Olver	Peter J.	U. of Minnesota, School of Mathematics	206 Church st SE	USA-Minneapolis, Mn 55455
Peyrard	Michel	U. de Dijon, Optique du réseau cristallin	6, boulevard Gabriel	F -21100 Dijon
Pomeau	Yves	U. of Arizona, Department of Mathematics	Bldg 89	USA-Tucson, Az 85721
Rajeswari	N.	Nallamathu Gounder Mahalingam College		IND -Pollachi 642001, Tamilnadu
Ramani	Alfred	Ecole polytechnique, Centre de physique théorique		F -91128 Palaiseau Cedex
Remoissenet	Michel	U. de Dijon, Optique du réseau cristallin	6, boulevard Gabriel	F -21100 Dijon
Ribotta	Roland	U. Paris-sud, Physique des solides		F -91405 Orsay Cedex
Rodriguez	Anibal	U. Complutense de Madrid, Matematica Aplicada		E -28040 Madrid
Santini	Paolo Maria	U. di Roma I, Istituto di Fisica	Piazzale Aldo Moro 2	I -00185 Roma
Saut	Jean-Claude	U. Paris-sud, Mathématiques	Bâtiment 425	F -91405 Orsay Cedex
Scheurer	Bruno	Centre d'études de Limeil-Valenton	BP 27	F -94190 Villeneuve-Saint-Georges
Siggia	Eric	Cornell University, Department of Physics		USA-Ithaca, NY 14853
Swinton	Jonathan	Imperial College		GB -London SW7 2BZ
Tuszyński	Jacek Adam	U. of Alberta, Department of Physics		CDN-Edmonton, Alberta T6G 2J1
Weiss	John		6, Lockeland Ave	USA-Arlington, Ma 02174
Willox	Ralph	Vrije U. Brussel, Theoretische Natuurkunde	Pleinlaan 2	B -1050 Bruxelles
Winternitz	Pavel	U. de Montréal, Centre de recherches math.	CP 6128 Succ. A	CDN-Montréal, Québec H3C 3J7
Zakrzewski	Wojciech	U. of Durham, Dept of Mathematical Science	Science Site,South rd	GB -Durham DH1 3LE

Equations d'évolution non-linéaires partiellement intégrables et leurs applications physiques
Partially Integrable Nonlinear Evolution Equations and their Physical Applications

Les Houches 21-30 March 1989 - List of 32 first named contributors

Balakrishnan	Radha	The Institute of Mathematical Sciences	C.I.T. Campus	IND -Madras 600113
Bessis	Daniel	CEN Saclay, Physique théorique		F -91191 Gif-sur-Yvette Cedex
Bishop	Alan	Los Alamos National Lab, CNLS	MS-B262	USA-Los Alamos, NM 87545
Brenig	Léon	U. libre de Bruxelles, Chimie physique II	Campus Plaine, CP 231	B -1050 Bruxelles
Cerveró	José M.	U. de Salamanca, Departamento de Fisica Teorica		E -37008 Salamanca
Clarkson	Peter A.	U. of Exeter, Department of Mathematics	North Park Road	GB -Exeter EX4 4QE
Conte	Robert	CEN Saclay, Physique du solide et résonance magnétique		F -91191 Gif-sur-Yvette Cedex
Cornille	Henri	CEN Saclay, Physique théorique		F -91191 Gif-sur-Yvette Cedex
Coullet	Pierre	U. de Nice, Physique théorique	Parc Valrose	F -06034 Nice Cedex
Elmer	Franz-Josef	U. Basel, Institut für Physik	Klingerbergstrasse 82	CH -4056 Basel
Erbay	Saadet	Tübitak-TBAE, Dept of Applied Mathematics	PO Box 74	TR -41401 Gebze-Kocaeli
Estabrook	Frank B.	Jet Propulsion Laboratory, Mail code 169-327	4800 Oak Grove Dr.	USA-Pasadena, Ca 91109
Ghidaglia	Jean-Michel	U. Paris-sud, Laboratoire d'analyse numérique	Bâtiment 425	F -91405 Orsay Cedex
Grundland	Michel	U. de Montréal, Centre de recherches math.	CP 6128 Succ. A	CDN-Montréal, Québec H3C 3J7
Hereman	Willy	U. of Wisconsin, Department of Mathematics	610 Walnut street	USA-Madison, Wi 53705
Hietarinta	Jarmo	U. of Turku, Department of Physical Sciences		SF -20500 Turku 50
Karpman	Vladimir	IZMIRAN, Academic City	Troitsk	SU -Moscow Region 142092
Kosmann-Schw.	Yvette	USTL (Lille), UFR de math		F -59655 Villeneuve d'Ascq Cedex
Kruskal	Martin D.	Princeton University, Mathematics	Fine Hall	USA-Princeton, NJ 08544
Kuusela	Tom	U. of Turku, Department of Physical Sciences		SF -20500 Turku 50
Lambert	Franklin	Vrije U. Brussel, Theoretische Natuurkunde	Pleinlaan 2	B -1050 Bruxelles
Mikhailov	Alexander V.	Landau Institute, Academy of Sciences		SU -Moscow 117940
Mishra	S. C.	U. of Delhi, Dept of Physics and Astrophysics		IND -Delhi 110007
Newell	Alan C.	U. of Arizona, Department of Mathematics		USA-Tucson, Az 85721
Pomeau	Yves	U. of Arizona, Department of Mathematics	Bldg 89	USA-Tucson, Az 85721
Remoissenet	Michel	U. de Dijon, Optique du réseau cristallin	6, boulevard Gabriel	F -21100 Dijon
Ribotta	Roland	U. Paris-sud, Physique des solides		F -91405 Orsay Cedex
Santini	Paolo Maria	U. di Roma I, Istituto di Fisica	Piazzale Aldo Moro 2	I -00185 Roma
Saut	Jean-Claude	U. Paris-sud, Mathématiques	Bâtiment 425	F -91405 Orsay Cedex
Tuszyński	Jacek Adam	U. of Alberta, Department of Physics		CDN-Edmonton, Alberta T6G 2J1
Weiss	John		6, Lockeland Ave	USA-Arlington, Ma 02174
Winternitz	Pavel	U. de Montréal, Centre de recherches math.	CP 6128 Succ. A	CDN-Montréal, Québec H3C 3J7

COMPETING INTERACTIONS AND COMPLEXITY IN CONDENSED MATTER

A. R. BISHOP
Theoretical Division and Center for Nonlinear Studies
Los Alamos National Laboratory
Los Alamos, NM 87545, USA

ABSTRACT. Some major themes of space-time complexity in condensed matter contexts are reviewed. They are illustrated through model and physical systems, analysed by both analytical and numerical techniques.

1. Introduction

Two of the main directions that can be clearly discerned in current dynamical systems research are: (1) A return to the reality of *spatially extended* dynamical systems studied by a variety of novel techniques, including neural networks, cellular automata and coupled map lattices, as well as direct numerical integration of partial differential equations (pde's) and coupled ordinary differential equations (ode's); and (2) Recognition of the central role played by *competing interactions* (both of length and time scales) in pattern formation and complex dynamics.

Here we will introduce examples of, and approaches to, these issues in *condensed matter* contexts. Complexity in both time and space are important in condensed matter for device performance, response and transport properties, etc. However, condensed matter also provides excellent vehicles to study general issues in dynamical systems. This is because of the availablity of controlled materials, small scale experiments, and sophisticated probes of both time and space [1,2].

Broadly, three classes of problems might be distinguished -- these are separated mostly by historical developments and happily they are growing together as we focus increasingly on the real materials provided by nature. Nevertheless, it is easier to appreciate previous literature by separating:

(a) *Structural Complexity* in *classical* equilibrium Hamiltonian systems with competing (incommensurate) interactions or periods -- the analog of temporal problems with two or more incommensurate frequencies. The competing length scales lead to a variety of nontrivial, spatially inhomogeneous ground states and transitions, and there are by now many physical observations in widely varying experiments [3]. The inhomogeneous ground states often have the form of superlattice structures (e.g. of soliton arrays) similar to texturing in some metallurgical contexts [4].

R. Conte and N. Boccara (eds.), Partially Integrable Evolution Equations in Physics, 1–38.

Perhaps the biggest open question here is to understand *dynamics*, particularly large scale flow in such problems –– this is usually "glassy" and "hysteretic." Indeed much of materials science concerns the roles of "defects" in controlling *both* strength and flow.

(b) *Nonlinear pde's and coupled ode's*, again mostly classical, including effects of *external* forcing which naturally introduces competitions for length and time scales. Issues here include [1] the coexistence of coherence and chaos, the use of nonlinear spectral methods [5] to identify collective excitations, the importance of generalized homoclinic orbits [6], and the qualitatively new effects that can be induced by noise and disorder. Again, there are now many natural as well as more contrived experimental studies which are closely mimicked by such pde's and their equivalents (cellular automata, etc.) [1].

(c) *Quantum Hamiltonians* can also arise naturally in solid state and statistical physics. These offer some new approaches to probing the poorly understood area of "quantum chaos," i.e., quantum behavior of integrable and nonintegrable models with interesting (e.g. chaotic) classical limits. Recent examples here include quantum spins [7] and exciton-phonon coupled systems [8]. The focus is on gaining control of nonintegrability and $\hbar$ as "tuning" parameters, and studying wave-function structure and evolution as well as energy level distributions.

We should also emphasize the probable interconnectedness of the classes above. Indeed in a number of cases formal *mappings* can be identified between time-dependent and equivalent Hamiltonian systems (in a higher spatial dimension) or quantum models. In this way the central role played by competing interactions is revealed. Thus, for example, the inhomogeneous "ground states" referred to in (a) contain the character of "intermittency" observed in appropriate regimes of (b). The reader is referred to [9] for more details.

By way of illustrating some of the above phenomena, techniques and outstanding issues, we mainly focus here on the *sine-Gordon* (SG) and closely related cubic nonlinear Schrödinger (NLS) equations in the presence of various kinds of perturbations. SG has been a template of nonlinear pde's for many years because: (1) In its pristine (1 + 1)-dimensional form it is strictly integrable. This means that elegant analytic techniques (e.g. nonlinear spectral analysis [5]) for isolating true "solitions" are available, which has given insight into ideas of "collective coordinates" and "particle-like" solutions of wider applicability. Further, we are able to identify the deviation from integrability as a controlled parameter; and, equally importantly, (2) SG represents a *class* of nonlinear Klein-Gordon equations which arise naturally in many branches of physics [10]. Nature is often quite closely modeled (on many spatial scales) by coupled nonlinear oscillators, and solitons or solitary waves can control transport, statistical mechanics, localization, radiation absorption frequencies, etc.

As a simple example consider the problem of "current oscillations" in the dynamics of near-commensurate systems [11]. The 1 + 1 dimensional driven, damped SG equation,

$$\ddot{\psi}(x,t) + \alpha\,\dot{\psi}(x,t) - \psi''(x,t) + sin\,\psi(x,t) = \Gamma\ , \tag{1}$$

has been used as a model for, e.g. charge-density-waves or a vortex lattice in a superconductivity film whose thickness is periodically modulated in one-direction. In eq. (1) the overdot is $\partial/\partial t$, the prime is $\partial/\partial x$, α is a damping parameter, and Γ is a DC-driving field. The boundary conditions may be, e.g., periodic or periodic mod

(2π) to allow for a fixed average density of kinks (dicommensurations), n_k. If $n_k = 0$ the problem is said to be commensurate while $n_k \neq 0$ measures the deviation from commensurability. The observable "current" $< \dot{\psi} > = L^{-1} \int dx\ \dot{\psi}(x,t)$. Physically, the commensurability is between two characteristic length scales. Thus in the charge dnesity wave case $\psi(x,t)$ corresponds to the phase field of the order parameter (a periodic lattice and charge density distortion induced by electron-lattice coupling) and the charge density wave with a commensurability wavelength Ma/N (with M, N reduced integers and a the lattice constants) has the form $\sim \cos(2\pi Nx/Ma + \phi)$. This charge density can couple to the component of the lattice potential with periodicity a/N and produce an interaction energy $\sim \cos(M\phi)$: thus $\psi = M\phi$ in eq. (1). In the incommensurate case the charge density wave has the form $\sim \cos[(2\pi N/Ma + \delta q)\, x + \phi]$ with the interaction energy $\sim \cos(M\, \phi + M\, \delta q\, x)$; thus $\psi(x) = M\, \phi + M\, \delta q\, x$ and δq determines the boundary condition in eq. (1). In the case of the vortex lattice $\psi(x,t)$ is a center of mass field, the pinning force $\sim \sin \psi$ is due to the thickness modulation, and eq. (1) corresponds to deviation of the vortex average spacing (determined by an applied magnetic field) from the period of the thickness modulation. Further details and references concerning both of the above problems can be found in [11].

There are now *numerous* numerical and analytical studies of the SG equation in different dimensions and under a variety of physically relevant perturbations. We record here only a few representative source references [1,5,12].

The remainder of this report describes a selected sequence of problems in more detail. Section 2 is devoted to the (1+1)-dimensional SG with spatially uniform AC-driving and damping. Section 3 considers the same problem but with DC-driving instead. Section 4 introduces a discrete SG system but with nonconvex interparticle interactions, admitting an internal competition of length scales. Section 5 returns to a *single* particle problem but, by including quantum effects, raises questions of whether the scaling and statistical approaches to "chaos" developed for classical problems can be useful in the quantum domain. Section 6 contains a brief summary and some concluding remarks.

Although, we are concerned here with problems motivated by condensed matter, the dynamical systems issues are of course much more general. This report should be read in conjunction with those of e.g., Coullet, Ghidhalgia, Newell, Ribotta and Pomeau.

2. Sine-Gordon Equation with Damping and AC-Driving: A Quasi-Periodic Route to Chaos in a Near-Integrable PDE

Solutions of nonlinear evolution equations often exhibit rich patterns in space and time which may have both coherent and chaotic components. In both dissipation-dominated and near-conservative cases, the solutions reside in an infinite dimensional phase space but may approach attractors which are low-dimensional. Hence, the mathematical techniques developed in recent dynamical systems theory are potentially relevant. In particular, these techniques explain how motion near a low dimensional attractor of a deterministic system can act chaotically, and they offer means to characterize the nature of the attractor quantitatively. On the other hand, techniques from modern nonlinear partial differential equations provide coordinates for the attractors. These coordinates capture coherent spatial patterns of the solutions. It is certainly natural to try to combine insight from these two approaches.

We focus in this section on near integrable nonlinear wave equations. The underlying integrability provides a wide selection of solutions for potential nonlinear resonances, and it offers sufficient structure for the possibility of analytical coordinates for the attractors. Thus, near integrability provides the analytical tools for a precise description of near conservative phenomena, some of which appear more generic in near conservative cases than the integrable methods might suggest.

For near-integrable problems a strategy is to find a nonlinear resonance, and then to study the system in a neighborhood of this resonance. Here we study, as an example, the damped, AC-driven sine-Gordon equation under periodic boundary conditions [5]:

$$\phi_{tt} - \phi_{xx} + sin\phi = \varepsilon[-\alpha\phi_t + \Gamma sin(\omega t)], \tag{2.1}$$

$$\phi(x+L,t) = \phi(x,t), \tag{2.1b}$$

$$\phi(x,t=0) = \phi_{in}(x), \tag{2.1c}$$

$$\phi_t(x,t=0) = v_{in}(x). \tag{2.1d}$$

Here $0 < \varepsilon \ll 1$, and the control parameters, are α (the strength of the dissipation), Γ (the amplitude of the ac driver), ω (the frequency of the AC driver), L (the spatial period), and the initial data (ϕ_{in}, v_{in}).

For the purposes of illustration we further specialize here to the case where the frequency ω is near, but less than, unity,

$$0 \ll \omega \lesssim 1. \tag{2.2}$$

This choice places us in a "nonlinear (cubic) Schrödinger (NLS) regime"; that is, when $\omega \lesssim 1$, one can use singular perturbation methods to approximate a class of equations, which includes the sine-Gordon equation (1.1), by a NLS equation. Elsewhere [13], we have studied lower driving frequencies (e.g. $\omega \simeq 0.6$) for which the NLS approximation is not valid; in these regions of parameter space, the chaotic attractors are dominated by "breather" to "kink-antikink" transitions (see 2B below). Here, in the NLS regime, we will see that the attractors are dominated by similar, but distinct, collective-mode transitions, namely breather-radiation interactions.

Classical dynamical systems diagnostics applied to the results of careful numerical experiments have identified (in a particular parameter range) intermittency between quasi-periodic and chaotic states. We have used soliton modes to begin an effective coordinatization of the attractors, capturing both their temporal and spatial structures. This is done through a nonlinear spectral transform which permits several new conclusions. It : (1) confirms that even the chaotic attractors can be well described by a few soliton modes; (2) establishes the existence of homoclinic orbits in the underlying integrable problem; (3) measures the presence of homoclinic crossings in the perturbed system; and (4) shows the importance of soliton interactions in the transitions between metastable parts of the attractor. Thus, far more precise information is now available about the nature of the onset of chaos for this near integrable example than can ever be expected for pde's with less structure.

First we summarize the results of extensive numerical experiments [5]. The global picture of transitions in the NLS regime is depicted in the schematic diagram of fig. 1. Before describing the space-time structures in each region of this diagram, we first set the stage for these experiments. The parameters in the system (2.1) were chosen as: $\varepsilon = 0.1$, $\alpha = 0.4$, and $L = 24$. The remaining parameters are then varied: ω is varied below, but near, unity, and Γ is varied near 0. The initial conditions $\phi_{in}(x)$, $v_{in}(x)$ are taken to be a whole-line sine-Gordon breather localized inside the period L, and extended periodically to the whole line.

In the "quasi-periodic" regime III of fig. 1 there are also windows of subharmonic locking rather than true quasi-periodicity –– a familiar situation in two-frequency dynamical systems (e.g. circle maps). In fact around $\omega \sim 0.9$ the quasi-periodicity is suppressed completely (see fig. 1) and an unusual transition from regime II to IV occurs directly.

Pertinent questions about the bifurcation sequencies in fig. 1 include: what is the origin of the second frequency in the quasi-periodic region?; how does this second independent frequency correlate with the increased spatial structure?; do these spatial structures have a meaningful, quantitative interpretation in terms of the exact sine-Gordon theory?; does the dynamics in these regions (including region IV) admit to a perturbation analysis of the integrable sine-Gordon equation?

To be specific we focus on parameter values $L = 24$, $\varepsilon = 0.1$, $\varepsilon\alpha = 0.04$ and $\omega = 0.87$, with initial data a SG breather with frequency parameter $\omega_{br} = 0.77$. The control parameter for the experiment is Γ, the amplitude of the AC driver –– ($\varepsilon\Gamma$) ranges over (0.0, 0.116). For purposes of orientation, this experiment is represented by a line in fig. 1.

The attractor in this bifurcation experiment may be represented as Γ increases by the symbolic sequence

$$\begin{pmatrix} FLAT, \\ PERIODIC \end{pmatrix} \rightarrow \begin{pmatrix} PERIOD\ 1, \\ PERIODIC \end{pmatrix}$$

$$\rightarrow \begin{pmatrix} \sim PERIODIC\frac{1}{2} \\ QUASI\!-\!PERIODIC \end{pmatrix} \rightarrow \begin{pmatrix} \sim PERIODIC\frac{1}{2} \\ CHAOTIC \end{pmatrix},$$

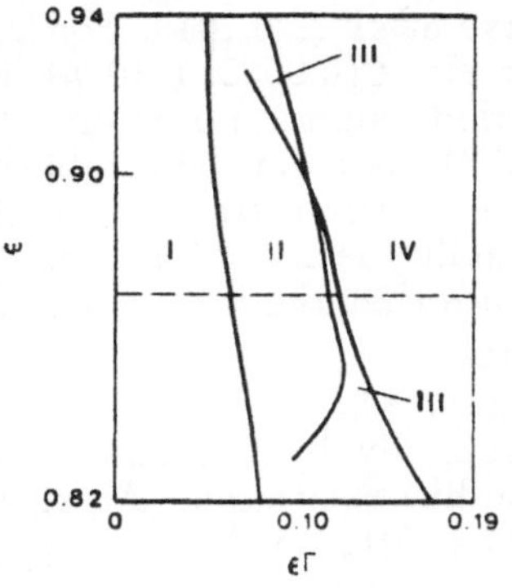

Figure 1. Semi-schematic bifurcation diagram for the ac-driven damped SG equation in the NLS regime. Other parameter values are $\varepsilon\alpha = 0.04$ and L = 24. The final attractors are shown as functions of driving frequency (ω) and strength ($\varepsilon\Gamma$) and labeled as: I, period, locked to the driver (x-independent); II, period, locked to the driver (one localized breather); III, quasiperiodic (weak period L/2 state superimposed on II); and IV, chaotic, with intermittent laminar regions (competition between two breathers and anharmonic L/2 radiation).

which we abbreviate by

$$(F, P) \to (P1, P) \to (\sim P\frac{1}{2}, QP) \to (\sim P\frac{1}{2}, C).$$

For small driving amplitudes ($0 < \varepsilon\Gamma < 0.0585$) the periodic spatial structure of the initial breather decays as a transient, and the attractor is an x-independent flat state with no spatial structure, which is periodic in time. This state is locked to the AC driver with its temporal period. The existence and stability of this locked state can be established with classical mathematical analysis, both with and without the NLS approximation. For example, in the NLS approximation, the field ϕ is represented as

$$\phi(x,t) = \sqrt{6\varepsilon}[A(\varepsilon t, \sqrt{\varepsilon}x)e^{it} + c.c.], \tag{2.3}$$

where the complex amplitude $A(T, X)$ satisfies the driven, damped NLS equation

$$-2iA_T + A_{XX} + 3AA^*A = i\alpha\, A + \Gamma^{\varepsilon} e^{-i(1-\omega)T}, \tag{2.4}$$

where $\Gamma^{\varepsilon} = \Gamma/2\sqrt{6\varepsilon}$. A flat locked state is represented by a solution of this equation in the form

$$A(X,T) = Ce^{-i(1-\tilde{\omega})T/2}, \tag{2.5}$$

where the complex constant C satisfies

$$[-(1-\tilde{\omega}) + 3CC^*]C = i\alpha\, C + \Gamma^{\varepsilon}. \tag{2.6}$$

A typical hysteresis diagram depicting the solutions of this algebraic equation, together with the stability of the locked state (2.5) to arbitrary perturbations with spatial period L, is shown in ref. [5]. Note in particular that the flat state on the lower branch is stable to all perturbations, while the state on the upper branch can become unstable, at large enough Γ values, to spatially dependent perturbations. These instabilities are long wavelength modes; the short wavelength modes are always stable. As the amplitude of the flat locked state increases, the most unstable mode changes from a flat ($\kappa = 0$) state, to a $\kappa_1 = 1(2\pi/L)$ state, on to a $\kappa_2 = 2(2\pi/L)$ state, etc.

As we increase Γ, the amplitude of the flat attractor rises along the lower branch, until it reaches a "knee" in the hysteresis curve at $\varepsilon\Gamma \simeq 0.1015$. For large values of Γ, the attractor must change. In fact, for $\varepsilon\Gamma \in (0.0585, 0.1015)$, one stable attractor is a single excitation within each period, superimposed on a flat background. The existence and stability of this (P1,P) state can be established by mathematical analysis. Γ is further increased, all of the attractors develop instabilities at shorter wavelength. The first such instability is a $\kappa = 2$ mode, with spatial wavenumber $\kappa_2 = 2(2\pi/L)$. Thus, we anticipate that the locked breather state will become unstable, as Γ is increased, to P$\frac{1}{2}$ states.

As Γ is further increased beyond $\varepsilon\Gamma = 0.1015$ the attractor becomes quasi-periodic in time and appears to be characterized spatially by the same single breather, but now accompanied by a $\kappa = 2$ radiation-like excitation. As $\varepsilon\Gamma$ approaches 0.1053 from below, the upper threshold of the ($\sim$ P$\frac{1}{2}$, QP) attractor, the amplitude in the $\kappa = 2$ radiation visibly increases and initial transient times increase. The long transients, because of their large $\kappa = 2$ component, often appear as approximate P$\frac{1}{2}$ spatial structures in the sense that two coherent excitations are present in the spatial period. Indeed, at large dissipation, a locked two-breather state, periodic in time, is achieved, instead of a quasi-periodic state —– energy is permanently transferred from $\kappa = 1$ to $\kappa = 2$.

For $\varepsilon\Gamma \geq 0.1053$, the P$\frac{1}{2}$ tendency in the above transient persists for all time and is visibly enhanced: at times the attractor appears as two localized breathers per spatial period, and and at other times as extended $\kappa = 2$ anharmonic radiation. The overall temporal behavior is now chaotic, with intermittent visitation of a small number of distinct metastable parts of the attractor. Broadly speaking the attractors consist of at least two metastable parts: (1) striking "laminar" regimes, which are essentially the same as the quasi-periodic attractors ($\sim$ P$\frac{1}{2}$, QP) of the prechaotic regime, which occur at smaller $\varepsilon\Gamma$ values; and (2) intermittent chaotic bursts. However, upon closer inspection, these chaotic bursts reveal substructure characterized by a dynamical energy exchange between (a) predominantly $\kappa = 2$ radiation, and (b) two-breather states per period. In addition some of the dynamics and transitions between metastable parts of the attractor are accompanied by *relative* motion of the coherent components of the $\sim$ P$\frac{1}{2}$ structure. As the threshold $\varepsilon\Gamma = 0.1053$ is approached from above, the fraction of time spent in laminar regions increases.

These features of the bifurcation sequence are substantiated by the use of many standard diagnostics from dynamical systems theory. Namely: (i) time series of spatially averaged quantities ($H = \frac{1}{2}\,\phi_t^2 + \frac{1}{2}\,\phi_x^2 + 1$ - cos ϕ, and displacement ϕ); (ii) phase planes for $\mathbf{P}(t) = (\phi(x_1,t), \phi(x_2,t))$, with x_1 and x_2 arbitrarily chosen points on the chain; (iii) Poincaré sections (using variables as in (ii) to define a plane in a three-dimensional phase space); (iv) temporal power spectra, $S(x_1,\omega)$, at $x_1 = 0$ (the center of the chain); (v) leading Lyapunov exponent (computed from two initially neighboring trajectories with a distance norm $n = \int_0^L(\phi_t^2 + \phi_x^2)dx$); (vi) the correlation dimension computed according to the algorithm of Grassberger and Procaccia. Complete details may be found in [5]. As an example we show simple time series in fig. 2. Note in particular: (1) in the quasi-periodic cases, the modulation of the time signal by the second (lower) frequency and the growth of the amplitude of this modulation with increasing Γ; (2) in the chaotic case (fig. 2a), temporal intermittency characterized by the presence of laminar regions separated by (chaotic) bursts: (3) the quasi-periodic nature of these laminar regions (cf. figs. 2b) and the linear growth with time of the *modulational* amplitude; (4) the very similar <H> values (or modulational amplitudes) at which all laminar regions are exited.

These and other [5] conventional dynamical systems diagnostics yield temporal information. For a partial differential equation such temporal data should be correlated with spatial information. One possibility is a linear spectral analysis in κ-, as well as ω-, space. For near integrable nonlinear pde's such as that discussed here, more insight is gained from a ***nonlinear spectral analysis***, correlated with space-time profiles.

First, we briefly describe the nonlinear transform. For more complete detail, see refs. [14]. The method may be summarized as follows: at time t; we take the (numerically) generated spatial profile $\{\phi(x,t)|\forall x \in [0,L]\}$ and numericallly perform a spectral transform to obtain $\{\hat{\phi}(\lambda,t)|\forall\lambda\}$. This transform maps the field

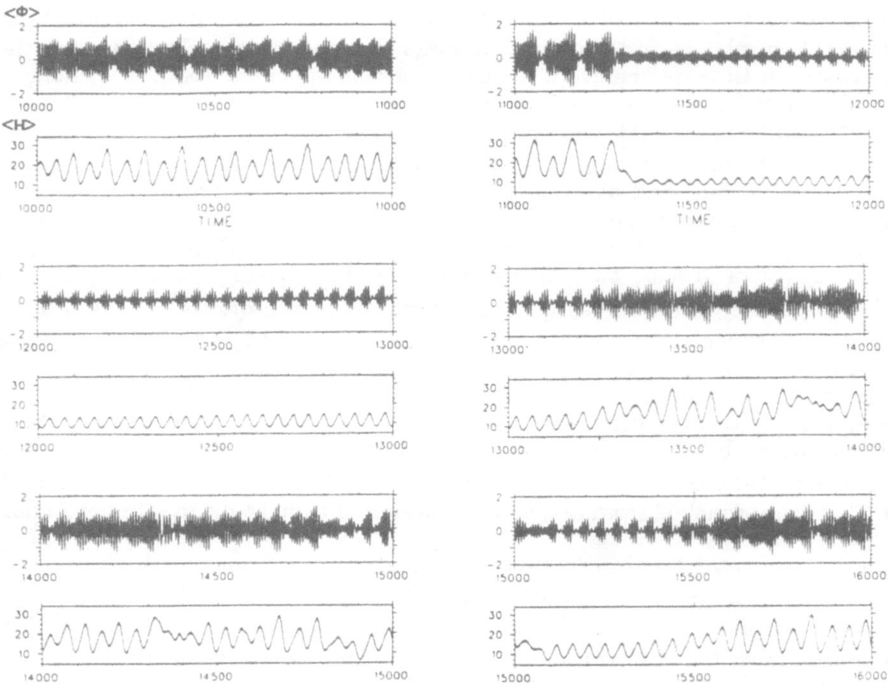

Figure 2. Time series for spatially-averaged SG energy ($\langle H \rangle$; $H = \frac{1}{2}\ \phi_t^2\ +\ \frac{1}{2}\phi_x^2\ +\ 1 - cos\phi$) and field ($\langle \phi \rangle$): *above*: $\varepsilon\Gamma = 0.1055$ (intermittent chaos); *below*: top left: $\varepsilon\Gamma = 0.101$ (periodic); Lower left: 0.102 (weakly quasi-periodic); 0.1055 (top right: small, and lower right: large, amplitude "quasi-periodic" laminar intervals from chaotic case (above)).

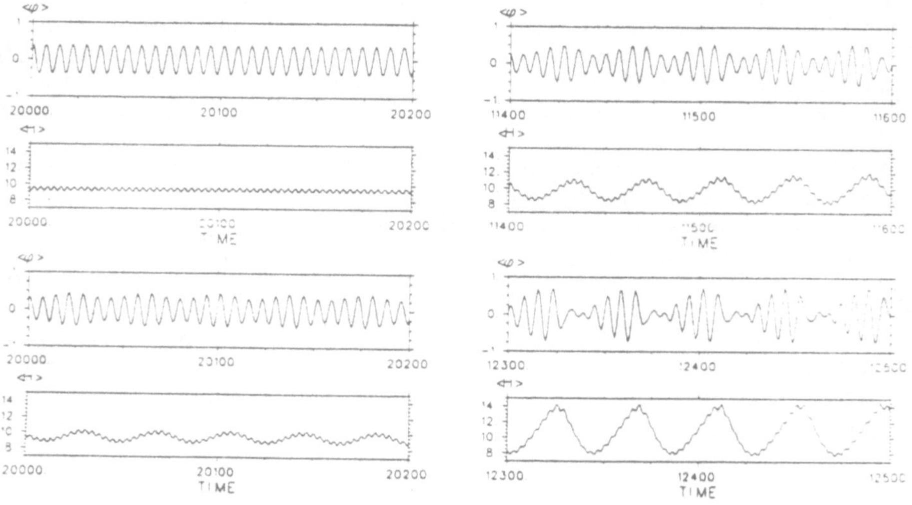

from its spatial representation *onto a representation on a basis of solitons*, in terms of which the unperturbed sine-Gordon equation is exactly separable. Thus the spectral representation $\hat{\phi}(\lambda, t)$ measures precisely the number, types and physical characteristics of the solitons (the localized coherent states) which are present in the wave at time t. If the temporal dynamics were the perfect sine-Gordon equation, these spectral properties would be invariant in time. However, because of the perturbations, they change with t and must be measured successively at t increases.

This transformation from the "spatial representation" $\{ \phi(x)\}$ to the spectral representation $\{\hat{\phi}(\lambda)\}$ is defined through a linear, non-self-adjoint eigenvalue problem, with spectral parameter λ lying in the complex plane. For periodic ϕ, the spectrum of this eigenvalue problem is entirely continuous spectrum residing on curves in the complex λ-plane. The real axis is always spectrum, and complex spines of spectrum attached to the real axis are associated with excited "radiation-like" modes in $\phi(x)$. In particular a spine near $\lambda^2 = \frac{1}{16}$ indicates the presence of a long wavelength mode; while spines near $\lambda^2 \simeq 0$ and $\lambda^2 \simeq \infty$ correspond to modes with high spatial wave number κ. Curves of spectrum in the complex λ plane which are not tied to the real axis are associated with soliton wavetrains in ϕ. These coherent excitations come in two types, namely kink (and/or antikink) trains and breather trains. Kink trains are associated with bands of spectrum on the imaginary λ axis; breather trains are associated with a pair of bands in the first and second quadrants, and their complex conjugates. In the NLS regime, where displacements of ϕ are small (compared with 2π), only breather trains are accessible.

Given $\phi(x)$, the spectrum of the linear eigenvalue problem is actually determined through the "Floquet discriminant" $\Delta(\lambda, \phi)$, an analytic function of both ϕ and the spectral parameter λ. The spectrum comprises those curves in the complex λ plane for which $\Delta(\lambda)$ is real and $-2 \leq \Delta(\lambda) \leq 2$. It is these curves of spectrum which are depicted in the figures shown below. Several additional remarks are necessary in order to interpret these figures:

i) Typically, bands of spectrum off the real axis are very short. If one band were to degenerate to a point in the upper half λ plane (which cannot happen for periodic potentials), a soliton would be present in ϕ. An isolated point on the imaginary axis $\lambda = i\nu$ would indicate a kink (or antikink) which would evolve under sine-Gordon dynamics with velocity $(1 - 16\nu^2)/(1 + 16\nu^2)$. An isolated point in the first quadrant would indicate a breather with sine-Gordon velocity $(1 - 16|\lambda|^2)/(1 + 16|\lambda|^2)$ and an internal breather frequency $\nu = \cos[\tan^{-1}(\lambda_I/\lambda_R)]$. Because the bands of spectrum are so short, these facts about solitons are useful qualitative approximations even in the periodic case.

ii) Excitations with spectrum inside the circle of radius $\frac{1}{4}$ travel to the right, while those with spectrum outside this circle travel to the left.

iii) Symmetries of ϕ imply symmetries in its spectrum. Since ϕ is real, its spectrum comes in quartets: if λ belongs to the spectrum, so do $-\lambda$ and $\pm\lambda^*$. Thus, it is sufficient to examine the closed first quadrant in the complex λ plane. In addition, if ϕ and ϕ_t are chosen to be even functions of x (here about $x = 0$) this additional symmetry induces another symmetry in the spectrum about the circle of radius $\frac{1}{4}$: namely, if λ belongs to the spectrum so does $1/(16\lambda)$.

iv) In these figures, "" denotes a band of spectrum where $-2 \leq \Delta(\lambda) \leq 2$. Points λ_p, where $\Delta(\lambda_p) = +2$, are denoted "⊞"; points λ_0, where $\Delta(\lambda_0) = 0$, are denoted "0"; points λ_a, where $\Delta(\lambda_a) = -2$, are denoted "$\diamond$". Note also that the associated spatial profiles $\phi(x)$ and $\phi_t(x)$ are included as inserts in the spectral plots, with the solid lines representing ϕ and the dotted lines ϕ_t.

We now return to the bifurcation sequence at $\omega = .87$. First consider $\Gamma = 0.101$ which is in the (P1,P) regime of one excitation in each spatial period, locked temporally to the AC driver. Fig. 3 shows the nonlinear spectrum of the locked state at three instants in time. It detects one breather at rest in the laboratory frame, riding over a flat ($\kappa = 0$) background. The curve of spectrum associated to this breather is located on the circle of radius $\frac{1}{4}$, at an angle of $\sim 30°$ with the real axis. No radiation modes, other than $\kappa = 0$, are visible, although numerical data shows that low κ modes are very weakly excited, increasingly so as $\varepsilon\Gamma$ approaches the quasi-periodic threshold ($\varepsilon\Gamma \simeq 0.1015$). This spectrum is not t-independent, as it is for completely integrable SG dynamics, but it executes $O(\varepsilon)$ fluctuations. For instance the breather spectrum oscillates periodically with the driving frequency between $\sim 29°$ and $\sim 32°$. This small periodic fluctuation in the spectrum is consistent with the small temporal oscillation of <H>, fig. 2.

As $\varepsilon\Gamma$ is increased into the quasi-periodic regime to a value of 0.104, the P$\frac{1}{2}$ character accompanying the single breather is clearly visible in the spatial profiles (see fig. 4a). In the nonlinear transform of these profiles, the $\kappa = 2$ mode is now visibly excited, as well as weaker modes of shorter wavelength. In addition, a particularly interesting new feature is present in the nonlinear spectra for this case. Namely, at times the spine of spectrum associated to the breather merges with that spine attached to the real axis at $\lambda = \frac{1}{4}$ (and associated with the background) and at later times (fig. 4b) becomes a "cross" of spectrum. This cross corresponds to the presence of "periodic" extended $\kappa = 1$ anharmonic radiation mode in the profile. Oscillation in the spectrum between the "cross" and "breather" curves occurs regularly in t, with the period of the *slow* underlying modulation. Thus, these spectral measurements show that the state oscillates (on a slow scale) between that of a predominantly coherent breather (fig. 4a) and that of a predominantly extended $\kappa = 1$ anharmonic radiation (fig. 4b), each accompanied by $\kappa = 2$ radiation. The transitions between these coherent breather and extended radiation states are correlated with the oscillations in <H>, which are now of larger amplitude (see fig. 2): the maxima of <H> correspond to the breather configurations, while the minima of <H> correspond to extended states.

The breather configuration and the extended configuration are separated by a state which is a *homoclinic orbit* under pure sine-Gordon dynamics. It is analogous to the separatrix in the single pendulum's phase space in that it has infinite temporal period. As discussed in refs. [6,15,21] homoclinic crossings play a central role in the chaotic regime just as for the single pendulum. Here the homoclinic states separate in phase space states with distinct spatial structures -- in the present case breather from radiation, and in other cases (see below) kink-antikink from breather.

Detailed nonlinear spectral analysis of the chaotic regime in fig. 1 can be found in Ref. [5]. A major conclusion is that, just as for the quasi-periodic precursor regime above, even the chaotic attractor (at least near the chaotic threshold) can also be described by a few ($\sim$ 3-4) nonlinear modes. The basic scenario quantified by nonlinear spectral analysis is captured in the space-time profiles shown in fig. 5. In the chaotic (intermittency) regime, energy is nonlinearly transferred from the κ=1 breather to the nonlinear κ=2 mode which grows into two independent

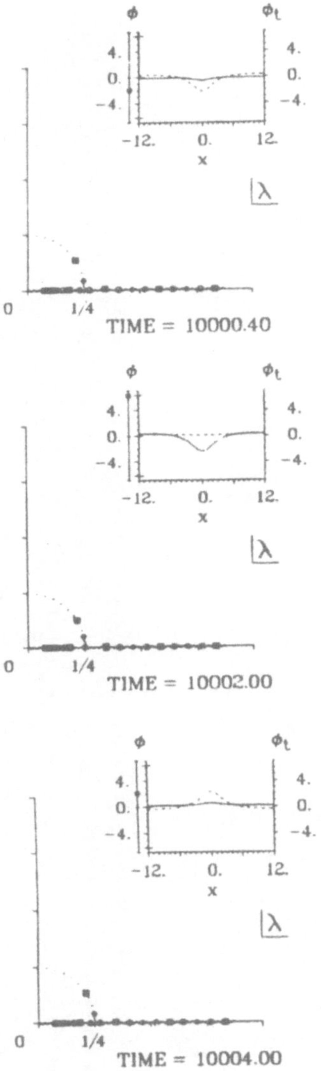

Figure 3. Nonlinear spectral transform of the (P1,P) attractor at $\varepsilon\Gamma = 0.101$, through approximately one half of the temporal period.

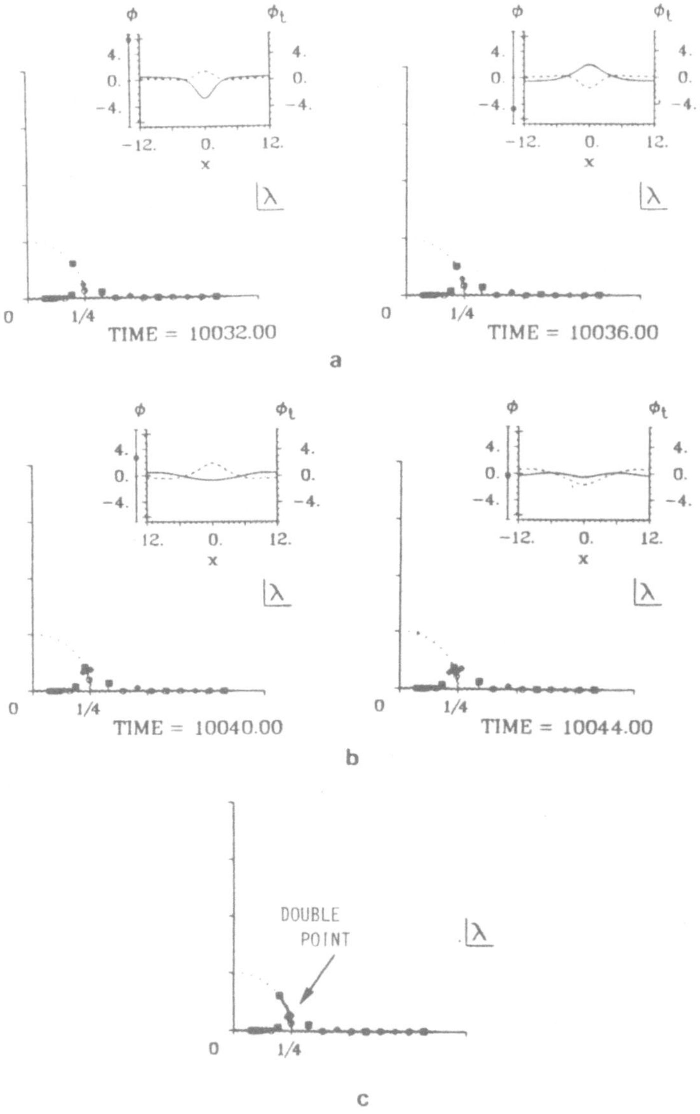

Figure 4. Nonlinear spectral transform of the ($\sim P\frac{1}{2}$, QP) attractor at $\varepsilon\Gamma$ 0.104. Note the two states (a) and (b), which represent, respectively, a localized breather state and an extended $\kappa = 1$ radiation state. These states are separated by a state which is a homoclinic orbit in the unperturbed system. Fig. 4c is a schematic representation of the spectral transform of this homoclinic state. At lower driving strengths such homoclinic states are not crossed; here they are crossed regularly; in the chaotic regime such homoclinic states are crossed aperiodically.

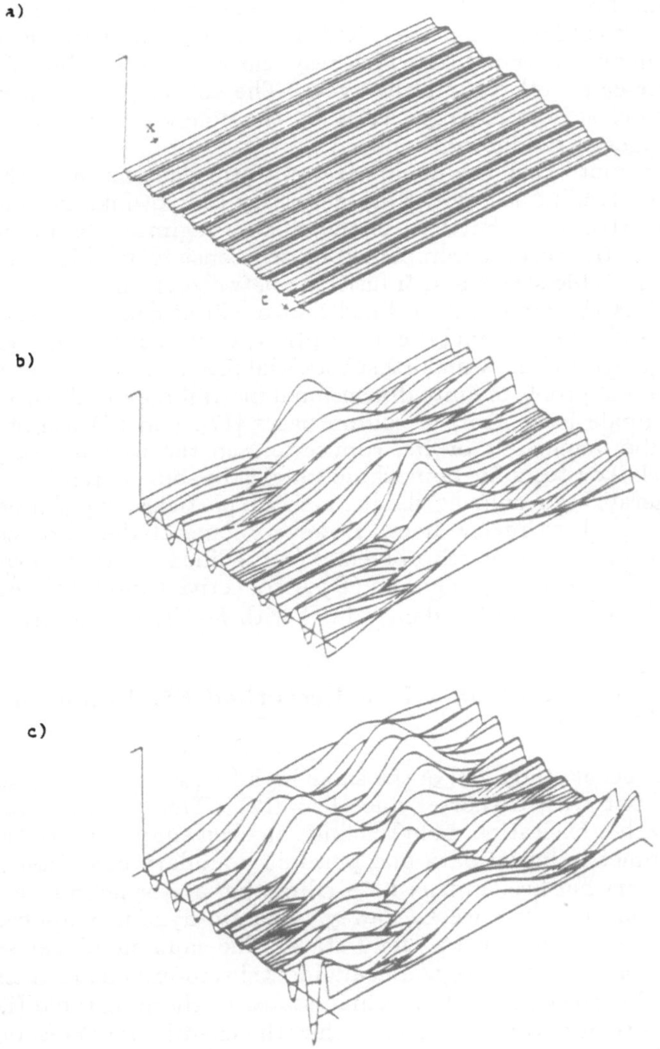

Figure 5. Space-time profiles at $\omega = 0.87$: (a) $\varepsilon\Gamma = 0.050$ (flat in space, periodic in time); (b) $\varepsilon\Gamma = 0.105$ (quasi-periodic); (c) $\varepsilon\Gamma = 0.110$ (chaotic). Note the second hump centered at $x = \pm L/2$ in the quasi-periodic case and the changes in spatial symmetry in the chaotic case.

and oppositely moving breathers. These breathers bounce and interact irregularly (through the periodic boundary conditions) and eventually annihilate each other. This results in κ=1 nonlinear radiation which resynchronizes into a breather plus κ=2 radiation and (irregularly) repeats the cycle. The collective nonlinear mode formation and dynamics is isolated very clearly by the nonlinear spectral analysis applied at various stages of the intermittency.

It is important to note that the nonlinear transform demonstrates that *both* coherent and extended (radiation) modes are necessary to coordinatize the quasi-periodic and chaotic attractors. Further, in the chaotic regimes, the motion and interaction of coherent structures, accompanied by self-consistent radiation modes, describe both the metastable states and transitions between them.

An ultimate test of the approach outlined above will of course be the use of "soliton collective coordinates" in analytic descriptions of transitions to and qualities of "chaos." This program is now making substantial progress in several respects, including: (i) extension of proofs of finite dimensional inertial manifolds (previously developed for linear mode bases [16]) to soliton bases [17]; and (ii) use of soliton coordinates to describe homoclinic orbits, connections in the presence of perturbations, and e.g., Melnikov tests for homoclinic tangles in some extended systems [18,19] —— again generalizing results available for single particles [20]. These works are not yet mature enough to review here. However, some of the ideas can usefully be introduced through two model problems: (A) A linear mode truncation of damped SG with *high* frequency driving; and (B) A collective coordinate reduction for single breather-K$\bar{\text{K}}$ breakup in the damped SG with *low* frequency driving.

2(A). Correlations Between Chaos in a Perturbed SG Equation and a Truncated Model System

Here we summarize a recent study [21] of the same high frequency driving, damped SG system as above but with a shorter line (L = 12). This length is found to support only *one* breather excitation and thus finds an even simpler route to chaos. Specifically, with symmetric initial data and periodic boundary conditions, a discrete (bimodel) symmetry has been imposed —— a breather may synchronize equally well at its seed position or a distance L/2 away. Chaotic dynamics appears as a fluttering between these locations via a "flat" state. The same nonlinear spectral analysis as above supports this picture and again identifies intermediate states that are O(1) unstable and correspond to homoclinic states in the integrable (i.e. unperturbed) limit. It is then natural to surmise that the chaotic dynamics on these attractors is due to the perturbation of the homoclinic configurations. The purpose of this section is to introduce a model dynamical system that mimics such behavior and which has proven analytically tractable. The model is derived by a low-order *linear* mode truncation of the nonlinear Schrödinger equation. Such a truncation is generally dangerous. However its regimes of validity can be monitored here by the parallel use of the nonlinear spectral transform on the full pde.

Considering eqs.(2.1) again we use $\varepsilon\alpha = 0.04$ and $\omega = 0.87$ but, as mentioned above, L = 12. As expected the system resonates with a *low* amplitude breather-like spatial mode accompanying a k = 0 flat motion. In this low amplitude regime we can easily derive a perturbed nonlinear Schrödinger envelope as a good approximation: ($\omega \equiv 1 - \varepsilon\tilde{\omega}$). Writing

$$\phi = 2(\varepsilon\tilde{\omega})^{1/2}[B(X,T)e^{iwt}tc.c.] + 0(\varepsilon) \tag{2.7}$$

$$X = (2\varepsilon\tilde{\omega})^{1/2}\alpha \;\; ; T \;=\; \varepsilon\tilde{\omega}t \; ,$$

the slowly varying envelope B(X,T) satisfies

$$iB_T + B_{XX} + (|B|^2 - 1)B = i\tilde{\alpha}B + \tilde{\Gamma} \; . \tag{2.8}$$

Note that this amplitude equation has preserved an integrable unperturbed limit (the cubic NLS) and has factored out one frequency (that of the driver, ω). Thus steady solutions of (2.8) correspond to frequency locked solutions of (2.1), while periodic flows of (2.8) which are incommensurate with ω, correspond to quasi-periodic perturbed SG solutions.

Based on the numerical observation of a low amplitude breather on a flat background, we examine a further severe mode truncation of (2.8):

$$B(X,T) = c(T) + b(T)cos(kX) \; , \tag{2.9}$$

$$k = 2\pi/L_X \quad , \qquad L_X = L(2\tilde{\omega})^{1/2} \; .$$

Inserting (2.9) into (2.8) and retaining cubic terms in the complex Fourier amplitudes c(T),b(T) yields

$$ic_T + (|\,c|^2 + \frac{1}{2}|\,b|^2 - 1)c + \frac{1}{2}(cb^* + c^*b)b = i\tilde{\alpha}c + i\tilde{\Gamma} \tag{2.10}$$

$$-ib_T + (|\,c|^2 + \frac{3}{4}|\,b|^2 - (1{+}k^2))b + (cb^* + bc^*)c = i\tilde{\alpha}b \; .$$

This 4-dimensional dynamical system is of course not expected to be in *quantitative* agreement with the full p.d.e. and the effect of including an additional mode is discussed in Ref. [21]. However: (i) it has guided intuition on the nonlinear soliton mode basis in the presence of the same perturbations, both in terms of geometric phase space structure and connections, including implementation of Melnikov tests [18,22]; and (ii) the ansatz (2.9) has proven capable of modeling several of the features observed in the full p.d.e., including the chaotic fluttering of a weakly unstable breather, via an intermediate flat O(1) unstable state, mentioned above. Specifically, the 2-mode ansatz admits the symmetry (c,b) $\rightarrow$ (c,-b), corresponding to a translation of ϕ by L/2, and b = 0 is an invariant subspace which corresponds to the flat intermediate structure. In the nonlinear spectral language, the ansatz is robust enough to capture all three key spectral configurations of the "gap" state, the "cross" state, and the intermediate state with complex double points and associated homonclinic character.

Very importantly, the *unperturbed* limit of the truncated system is an integrable Hamiltonian system, with two real independent integrals:

$$I \;\equiv |\,c|^2 + \frac{1}{2}|b|^2 \tag{2.11}$$

$$H \equiv \frac{1}{2}|c|^2 + |b|^2|c|^2 + \frac{3}{16}|b|^4 - \frac{1}{2}(1+k^2)|b|^2$$
$$- |c|^2 + \frac{1}{4}(b^2c^{*2} + b^{*2}c^2) .$$

Complete analytical analysis of fixed point structure and stability are possible as well as explicit solution by quadrature [21]. Furthermore, a direct correspondence can be established between the ode fixed points and SG solutions. For instance, the ring of fixed points (c,b) = $(e^{i\phi}, o)$ $(\phi \in [o, 2\pi])$ in the b = o invariant subspace corresponds to $\phi \sim 2(\varepsilon\tilde{\omega})^{1/2}[c\, e^{i\omega t} + c * e^{-i\omega t}]$, describing the flat (pendulum) solution frequency-locked to the driver. The fixed points are readily found to be O(1) unstable, reflecting the same O(1) instability of the flat SG solution. Moreover, the orbits homoclinic to the ring reflect the SG solutions which are homoclinic to the pendulum solution with frequency $\omega = 0.87$. In fact identifying the integral

$$h \equiv H - (\frac{1}{2}I^2 - I) \tag{2.12}$$

we can easily find an effective oscillator equation for $Z \equiv B^2$ ($b = B\, e^{i\beta}$):

$$\frac{1}{2}Z_T^2 - \frac{Z^2}{32}(Z - 8k^2)(7Z - 8(2-k^2)) = h = 0 \tag{2.13}$$

with infinite period solution: $O \rightarrow [\frac{8}{7}(2-k^2)]^{1/2} \rightarrow O$.

Other fixed points of the ode system can similarly be assigned in the full pde. A complete bifurcation and stability picture for the *perturbed* ode system has been determined numerically and many of the key elements can be derived analytically or by straightforward perturbation theory [21].

The correspondence of homoclinic crossings in the ode and pde has been checked in several ways. In the perturbed pde these crossings are measured by following the exact nonlinear SG spectrum of ϕ at each time step and identifying transitions from "gap" to "cross" states (c.f. fig. 4). In the ode system we have followed h = H - $(\frac{1}{2}I^2 - I)$ and checked for zero crossings. As a final check we have combined these two diagnostics. We take c(T), b(T) during the flow generating h, reconstruct the perturbed SG solution according to the approximation (2.7), and then compute the SG spectral components of ϕ. The question is whether h passing through zero corresponds to the SG field ϕ passing through a homoclinic spectral configuration? As shown in [21] the agreement is indeed rather good except very near to the homoclinic structure ($h \simeq 0$), as might be expected due to the approximation by linear Fourier modes.

Recent analysis of the truncated model suggests [22] that a homoclinic loop (rather than a tangle) is responsible for chaos in this system. Interestingly it appears that the geometric structure of the model problem can indeed by lifted to soliton variables and applied to the full perturbed SG and NLS pde's.

Finally, we note that the model problem introduced here is in fact a particular case (with somewhat different perturbations) of a general class of problems introduced by Holm et al. [23], motivated by polarization dynamics in nonlinear optical beams. Detailed discussions of reductions to finite-dimensional inertial manifolds, homoclinic crossings, chaos and Arnol'd diffusion have also been given by these authors.

2(B). A Mode Reduction for the Breather -- Kink-Antikink Transition and Associated Chaotic Dynamics

We now turn to an equally drastic approximation for chaos via a second homoclinic crossing in the SG equation, namely the breakup of a breather into a counterpropagating kink-antikink pair. This scenario has also been explored in great detail numerically [12,13] although usually not associated with an underlying homoclinic structure. It is especially relevant with lower frequency AC-driving since this induces resonant breathers with large amplitude $\lesssim 2\pi$. Although the collective coordinate scheme we introduce [24] is much oversimplified, it turns out to be surprisingly successful. It will also serve to introduce the basic ideas of homoclinic orbits and Melnikov's criterion for the onset of complexity. In general pde's pose problems of coexisting homoclinic structures and connections, and far more general phase space geometry than in an ode. However the basic ingredients in chaos really are one-dimensional in the present problem and captured by breather/kink-antikink collective coordinates.

Specifically, we consider here a one-dimensional SG system in which a single breather excitation is present. When this system is perturbed by spatially-uniform DC and AC driving, and space-independent dissipation, the breather can break into a kink-antikink (K$\bar{\text{K}}$) pair, which can then recombine into a breather soliton. This process may occur repeatedly, with a frequency that is not necessarily related to that of the driver. To understand this competition, consider the breather and K$\bar{\text{K}}$ solutions to the unperturbed SG equation:

Breather;

$$\phi_B(x,t) = 4\, tan^{-1}\left\{\frac{tan\theta\ sin\ [\nu(t)]}{cosh\ [2k(x-x_0)]}\right\}\ , \tag{2.14}$$

$$\nu(t) = \nu_0 + t\ cos\theta\ ,$$

$$k = \frac{1}{2}\ sin\theta\ ,$$

$$E_B = 16\ sin\theta\ ,$$

K$\bar{\text{K}}$;

$$\phi_{K\bar{K}}(x,t) = 4\, tan^{-1}\left\{\frac{sinh[u(t-t_0)\sqrt{1-u^2}}{u\ cosh[(x-x_0)/\sqrt{1-u^2}]}\right\}\ , \tag{2.15}$$

$$E_{K\bar{K}} = \frac{16}{\sqrt{1-u^2}} > 16, \quad u \geq 0\ .$$

Here, ν_0, x_0 and $0 \geq \theta < 2\pi$ are constants, u is the t $\rightarrow \infty$ velocity of the kink and -u is the t $\rightarrow \infty$ velocity of the antikink. From Eqs. (2.14) and (2.15), we see that the breather's internal frequency, ω_B, is cosθ. The breather soliton can be viewed as the bound state of a kink-antikink pair, with threshold binding energy ΔE = 16(1 - sinθ), and where u $\rightarrow$ 1/(i tanθ) and $t_0 \rightarrow -\nu_0/\cos\theta$. Now, if the system supports a breather excitation, and is externally driven by γ sinωt, and damped by $\alpha\partial\phi_B/\partial$t,

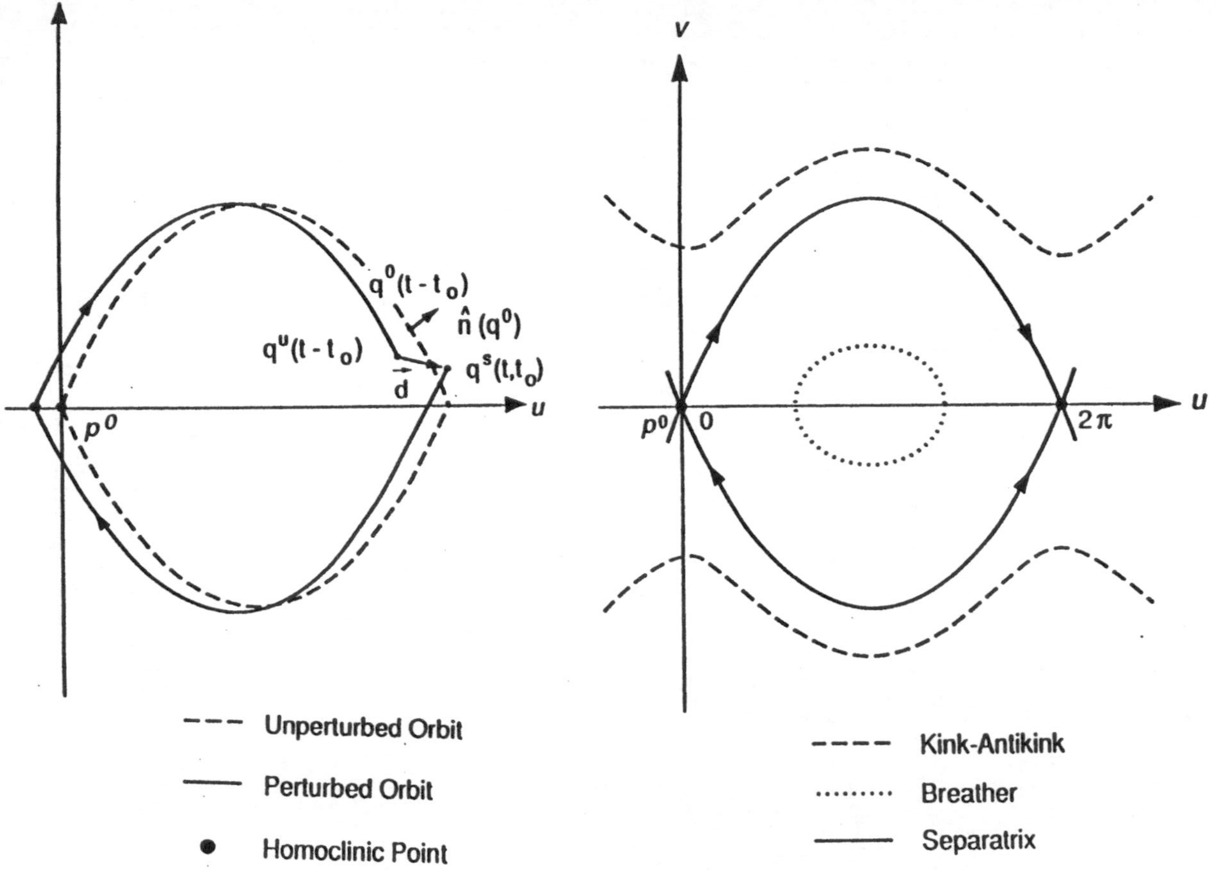

Figure 6. Schematic Poincaré map for the system under consideration. The closed orbits correspond to breather motion, while the unbounded orbits correspond to kink-antikink-pair motion. The separatrix then corresponds to a breather whose amplitude just attains 2π, or a kink-antikink pair with zero relative velocity, so that they may not escape each other. In this figure, the points 0 and 2π are identified; (b) A sketch of the perturbed separatrix orbit. The Melnikov distance, $d(t_o)$, is the projection of $\vec{d}(t_o)$ on $\hat{n}(q^o)$, the normal to the unperturbed orbit at $t = t_o$ (see text).

then the rate of change of the system's energy is $\dot{H} \approx \gamma$ sinωt $\int$ dx$(\partial\phi_B/\partial t)$ - $\alpha \int$ dx$(\partial\phi_B/\partial t)^2$, so that $\dot{H}$ can be either positive or negative. When $\dot{H}$ is positive, enough energy can be fed into the system that the breather attains E = 16, and breaks into a K$\bar{K}$ pair. Likewise, if $\dot{H}$ is negative, the K$\bar{K}$ pair may lose enough energy to recombine into a breather.

The specific equation under consideration is eq. (2.1) which we write as

$$\phi_{tt}(x,t;\varepsilon) - \phi_{xx}(x,t;\varepsilon) + sin\ [\phi(x,t;\varepsilon)] = \varepsilon F(t,\phi_t)\ ; \qquad (2.16)$$

where εF(t,ϕ_t) is the perturbation,

$$\varepsilon F(t,\phi_t)\ = \eta(t) - \alpha\phi_t(x,t;\varepsilon)\ ; \qquad (2.17)$$

$$\eta(t)\ = \Gamma + \gamma sin\omega$$

$$\varepsilon\ = max\{\alpha,\Gamma,\gamma\}\ .$$

Here, the variables x and t denotes space and time, respectively, while subscripts x and t denote partial derivatives w.r.t. these variables. Also, ε is the overall strength of the perturbation, Γ represents the DC driving, γ sinωt denotes the AC driving and $\alpha\Phi_t$(x,t;ε) is the damping.

In order to investigate this pattern competition in Eq. (2.16), we make a severe mode truncation to the breather collective coordinates, which leads to a three-dimensional set of ode's which may then be analyzed by Melnikov's method [20]. Salerno has shown [25] that this procedure yields a system whose separatrix dynamics is equivalent to the separatrix dynamics of the full pde. Several interesting predictions have been obtained [24] for space-time instabilities in this system. In particular, the perturbation can result in transverse crossings of the stable and unstable orbits to the homoclinic point, which corresponds to intermittent binding and unbinding of the K$\bar{K}$ pair. This is indeed the basic nature of low-dimensional chaos, and has been observed numerically [13] for the AC-driven, underdamped sine-Gordon chain when the frequency of the driver is low.

Complete details of this analysis can be found in ref. [24]. Here we only sketch the basic steps.

First we must determine equations of motion for the collective coordinates. This can be achieved within several perturbative approaches. One such approach is to take the ansatz that the solution to the perturbed SG equation has the same form as the unperturbed breather, Eq. (2.14), except that now we allow θ and ν_0 to be functions of time, so that

$$\nu(t) = \nu_0(t) + \int^t dt' cos\theta(t')\ . \qquad (2.18)$$

By making this approximation, we are assuming that the perturbation is sufficiently weak that its main effect on the system is to continuously alter the phase and the frequency of the breather. Since the breather energy is intimately related to its frequency ($E_B = 16\ \sqrt{1-\omega_B^2}$), we see that this is valid when $\omega_B \ll 1$. Next, we introduce the breather collective coordinates, (u(t;ε), v(t;ε)), where

$$u(t;\varepsilon) = \phi_B(x = x_0, t; \varepsilon) = 4tan^{-1}A(t) \tag{2.19}$$

and

$$v(t;\varepsilon) = \left.\frac{\partial \phi_B}{\partial t}\right|_{x=x_0} = \frac{4\dot{A}(t)}{1 + A(t)^2} , \tag{2.20}$$

where A(t) = tanθ sin ν(t).

In terms of A(t) and its time derivatives we have

$$\dot{u} = v$$
$$\dot{v} = 4(\ddot{A}\ cos^2(u/4) - 2(v/4)^2\ tan(u/4)) . \tag{2.21}$$

To determine the equations of motion for the breather collective modes, Ä must be found in terms of u(t;ε), v(t;ε). After substantial algebra this may be obtained to consistent perturbative order, and, introducing

$$Z = tan(u/4), \qquad W = \dot{Z} ,$$

(2.21) can be written as the following autonomous dynamical system

$$\vec{\chi} = \vec{f}(\vec{\chi}) + \varepsilon\vec{g}(\vec{\chi}, \psi) ,$$
$$\dot{\psi} = \omega \tag{2.22}$$

with

$$\vec{\chi} = \begin{pmatrix} z \\ w \end{pmatrix} ;$$
$$\vec{f}(\chi) = \begin{pmatrix} w \\ -\frac{z(l - w^2)}{1 + z^2} \end{pmatrix}$$
$$\varepsilon\vec{g}(\vec{\chi}); t) = \begin{pmatrix} 0 \\ \tilde{\eta}(t)g_\eta(z,w) - \alpha g_\alpha(z,w) \end{pmatrix} ,$$

where

$$g_\eta(z,w) = \frac{1}{(1 + sw^2)(w^2 + z^2)(1 + s)\sqrt{1 + z^2}} \times \left[\left(w^2 + z^2\right)z^2 - s\left\{ z^4 + w^2\left[z^2 + \left(1 - w^2\right)\left(1 + z^2\right)\right] + 2w^2z^2s^2\left(1 + z^2\right)^2 + w^2s^3\left(1 + 2z^2\right)\left(1 + z^2\right)^2 \right] \right.$$

and

$$
\begin{aligned}
g_\alpha(z,w) &= \frac{w}{(1+sw^2)(w^2+z^2)(1+z^2)} \\
&\times \left[\left(2+z^2\right)\left(w^2+z^2\right)z^2\right. \\
&- s\left\{z^4 - w^2\left[\left(2+z^2\right)z^4 - \left(1-w^2\right)\left(1+z^2\right)\right]\right\} \\
&\left. + w^2 s^2 \left(1+2z^2\right)\left(1+z^2\right)\right] .
\end{aligned}
$$

In order to study the onset of irregular behavior in this system, we consider the Poincaré map, $P_\varepsilon^{t_o}$ where $t_o \in [0,2\pi/\omega)$, fig. 6. The distance, $d(t_o)$, between the perturbed stable and unstable orbits to the homoclinic point is given by Eq. (4.5.11) in Ref. [20]:

$$
d(t_o) = \frac{1}{\left| f\left(\vec{q}^{\,o}(t_o)\right)\right|} M(t_o) , \tag{2.23}
$$

where

$$
\left| f\left(\vec{q}^{\,o}(t_o)\right)\right| = \left[w^2 + \frac{z^2(1-w)^2}{(1+z^2)^2}\right]^{1/2} ,
$$

and $M(t_o)$ is the Melnikov function

$$
\begin{aligned}
M(t_o) = \int_{-\infty}^{\infty} dt\ \vec{f}\left(\vec{q}^{\,o}(t-t_o)\right) \wedge \varepsilon\vec{g}\left(q^o(t-t_o);t\right) \\
exp\left[-\int_{t_o}^{t} dt' Tr Df\left(\vec{q}^{\,o}(t')\right)\right] .
\end{aligned}
$$

In the definition of $d(t_o)$, q^o is the unperturbed separatrix orbit, $\vec{f} \wedge \varepsilon\vec{g}$ is the exterior product of $\vec{f}$ and $\varepsilon\vec{g}(\vec{f} \wedge \varepsilon\vec{g} = \sum \varepsilon_{ij}\, f_i \varepsilon g_j = f_1 g_2 - f_2 g_1$, with ε_{ij} the Levi-Civita tensor), and $Tr\{Df(q^o(t'))\}$ is the trace of the Jacobian matrix of $\vec{f}$ evaluated on the separatrix:

$$
Tr\{Df(\vec{x})\} = 0 + \frac{2wz}{1+z^2} .
$$

The separatrix orbit is easily determined from Eq. (2.22). From the second comoponent,

$$
\dot{w} = z = -\frac{z(1-w^2)}{1+z^2} ,
$$

we see that z changes sign when $w = \dot{z} = \pm 1$, so that, on the separatrix,

$$
z = \pm(t - t_o)
$$

and

$$w = \pm 1 \ . \tag{2.24}$$

The upper (resp. lower) sign corresponds to the upper (lower) orbit in fig. 6.
When Eqs. (2.22) and (2.24) are substituted into Eqs. (2.23), we get

$$d(t_o) = \int_{-\infty}^{\infty} \frac{dt}{(1+s^2)} \left(1+t^2\right)^{-7/2}$$

$$\left\{\tilde{\eta}(t+t_o)\left[\left(1+s\right)\left(1+t^2\right)t^2 + \right.\right.$$

$$\left. 2t^2\left(1+t^2\right)s^2 + \left(1+t^2\right)^2\left(1-2t^2\right)s^3\right]$$

$$-\alpha\frac{1+s}{\sqrt{1+t^2}}\left[t^2\left(2+t^2\right)\left(1+t^2\right)\right.$$

$$\left.\left. - st^4\left(1-t^2\right) + s^2\left(1+t^2\right)\left(1+2t^2\right)\right]\right\} , \tag{2.25}$$

where $\tilde{\eta}(t + t_o) = \frac{\pi}{4} \{\Gamma + \gamma \sin[\omega(t + t_o)]\}$. Expanding $\sin[\omega(t + t_o)]$ and using the fact that s is even in t, Eq. (2.25) has the form

$$d(t_o) = M(t_o) = \frac{\pi}{4}[\Gamma\, m_\Gamma + \gamma\, p(\omega)\ \sin \omega t_o] - \alpha m_\alpha \ , \tag{2.26}$$

where m_Γ and m_α are numerical constants, and p is a function of ω only [24]. The zeros of the Melnikov function then yield the parameter values for which the stable and unstable manifolds to the homoclinic point intersect, so that "complicated" breather/kink-antikink behavior may occur whenever

$$\frac{\gamma}{\alpha} \geq \left(\frac{\gamma}{\alpha}\right)_c = \frac{1}{p(\omega)}\left(\frac{4m_\alpha}{\pi} - \frac{\Gamma m_\Gamma}{\alpha}\right) . \tag{2.27}$$

Since our primitive soliton mode truncation has excluded the possibility of exciting radiation modes, the breather in the full pde may destabilize at a lower driving strength. In fact, the competition between a single breather state and a two-breather state, mediated by a radiative state has been observed for periodic SG. Notice that the threshold for breather breakup is a monotonic increasing function of the driving frequency. This can be seen by realizing that we are looking at motion near the separatrix, so that $\omega \approx 0$ is close to the frequency of the breather, and energy is efficiently fed into the system. When the driving frequency is nearly one, we are far off resonance, and the amplitude of the driving must be large.

From Eq. (2.27) we see that the effect of adding DC driving to the system is always to decrease the threshold for breather breakup (m_Γ is positive). This is readily understood: The addition of spatially homogeneous, time-independent driving to the SG equation may be viewed as the addition of a term of the form $-\Gamma\phi$ to the potential, $1 - \cos\phi$. When $V(\phi) = 1 - \cos\phi - \Gamma\phi$ is plotted versus ϕ, we see that small amplitude motions correspond to breather states, while large ϕ solutions correspond to kink solitons. Since the DC contribution decreases the potential, thereby making large-amplitude solutions possible at lower energies, addition of DC driving will enhance the soliton conversion.

Surprisingly the numerical values of $(\gamma/\alpha)_c$ calculated by this method agree with previous numerical work quite well. For example, with $\alpha = 0.2$, $\omega = 0.6$ the threshold for intermittent chaos is $\gamma \simeq 0.9$ [13], i.e. $\gamma/\alpha \simeq 4.5$. The Melnikov criterion developed here predicts $(\gamma/\alpha)_c = 4.4$.

3. Sine-Gordon Equation with Damping and DC-Driving: A Model Transverse Instabilities on Propagating Interfaces

We now turn to spatially uniform DC driving instead of the AC-driving of the previous section. This may appear much less interesting. However, especially with *weak* damping the pattern formation, hysteresis and complex dynamics is in fact extremely rich.

The phenomena basically described by this equation include zero-field steps in Josephson junctions, spontaneous nucleation of kink-antikink pairs in stressed materials, and transverse structures on propagating domain-wall interfaces. The last example includes such systems as charge-density-wave materials, magnets, ferroelectrics, and many other phenomena appearing in the presence of a nonlinear, periodic potential. While we stress the wide applicability of our results, we shall present this section in the language of Josephson junctions for concreteness.

The structure of zero-field steps (ZFS's) in overlap-geometry Josephson junctions has been the subject of many theoretical and experimental studies. With recent advances in materials technology, it has become possible to manufacture these junctions in other configurations -- notably, in an annular geometry. From a theoretical point of view, this geometry has important advantages, since an analysis of periodic SG can make use of rigorous analytical results, such as inverse-scattering theory. This situation gives us the opportunity to study issues that are central to dynamical-systems theory for spatially-extended systems (such as space-time complexity, pattern formation, pattern competition, and mode conversion) from both a well-established theoretical framework, and in a controlled physical system.

The results that we present here [26] are for the one-dimensional, DC-driven SG equation with dissipation, viz.

$$\phi_{xx}(x,t) - \phi_{tt}(x,t) - sin[\phi(x,t)] = \varepsilon\phi_t(x,t) - \Gamma_0 \,, \tag{3.1}$$

with periodic boundary conditions,

$$\phi(x,t) = \phi(x+L,t)$$

and

$$\phi_\xi(x,t) = \phi_\xi(x+L,t) \,.$$

Here, $\phi_\xi = \partial\phi/\mathrm{d}\xi$, and $\xi = x$ or t. In Eqs. (3.1), x is the spatial variable, normalized to the Josephson penetration depth, λ_J [$\lambda_J = \hbar/2\mu_0 eJd$, where $\hbar$ is Planck's constant, μ_0 is the magnetic permeability of the vacuum, e is the electron charge, J is the maximum pair-current density and d is the magnetic thickness of the insulating layer]. Also, t is the time variable, in units of the inverse plasma frequency, $1/\omega_J$ [$1/\omega_J = (\hbar C/2Je)^{1/2}$, where C is the capacitance per unit area of the junction]. On the right-hand side of Eq. (3.1), Γ_0 is the DC-driving, and $\varepsilon\phi_t$ represents the dissipation (this is the only form of damping considered here). In Eq. (3.1), L is the length of the system. For our numerical studies, we take ε=0.1, $0 \le \Gamma_0 \le 1.0$, $L = 24$, and consider a fairly discrete system, with 76 lattice sites.

As we shall see below, this system possesses a rich variety of multisoliton wave-trains and transitions between them, including the same breather-K$\bar{\text{K}}$ breakup discussed in section (2B). First we summarize the results of a numerical study [26]. Fig. 7 presents data in the form of current ($=\Gamma_0$) versus voltage ($=\ll \phi_t \gg$, in our normalized units, where the double angular brackets denote averages over space and time). This picture was generated by starting with a stationary, large-amplitude, spatially-random profile, with $\Gamma_0 = 1.0$. We then permitted the system to evolve in time according to Eqs. (3.1) until it reached a t $\rightarrow \infty$ attractor; the homogeneous-spatial profile. The driving, Γ_0, was then "adiabatically" reduced to zero, to generate the backbone of the curve. Each of the steps was then developed by starting with the state at which an abrupt drop in voltage occurred, and then increasing the driving "adiabatically" until the voltage abruptly jumped to the power-balance regime –– in this portion of the I-V curve, the energy input by the DC driving matches the time-averaged dissipation. (The arrows on the curves in this figure indicate the direction in which Γ_0 was changed to trace out the steps.)

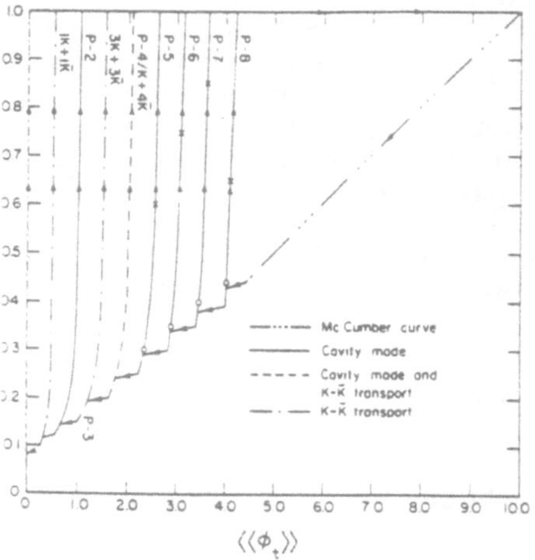

Figure 7. A summary of numerical results. The arrows indicate the direction in which the driving, Γ_0, was adiabatically changed. The x's on the periodic steps (labeled P-N) denote where the periodic symmetry breaks to another P-N state (see Fig. 8) when Γ_0 is increased, and the o's indicate where the symmetry is regained as Γ_0 is decreased. Notice that for $0.15 \le \Gamma_0 \le 0.18$ on the 3K-3$\bar{\text{K}}$ step, the spatial pattern is periodic.

From Fig. 7, we see that the system displays a number of interesting features. The most prominent of these is the existence of two different transport mechanisms. In the high-voltage steps immediately below the power balance, or McCumber curve, the transport occurs through cavity modes (which are standing waves of phase-locked, multibreather wave trains), where current flows because a standing wave is superimposed on a running average (see Fig. 8). In the language of solitons, this running average is due to the fact that during its oscillation, $\phi(x,t)$ crosses a potential barrier at each instant that its spatial profile is flat. On these steps, the amplitude and wavelength increase as the step number decreases. In the region of low voltage, the standing waves have developed a full 2π amplitude, and have increased the number of active modes by delocking the phases of the breathers. Then sustained K$\bar{\text{K}}$ pairs are nucleated, thereby giving transport by the *transverse* motion of kinks. [Notice that the imposition of periodic boundary conditions in Eq. (3.1) necessitates the creation of *pairs* of kinks and antikinks.] On these steps, the number of K$\bar{\text{K}}$ collisions cannot be sustained as the current is lowered further). The final K$\bar{\text{K}}$ is stable above $\Gamma_0 \cong 0.08$, while below this critical driving strength, dissipation dominates, and no current flows. In the low-voltage regime, there does exist one additional phase-locked state with spatial period equal to two. This mode is also unusual in that the other states appearing in Fig. 7 may be accessed by evolving a random-spatial profile according to Eqs. (3.1) with an appropriate Γ_0, but the period-two state cannot. These observations lead us to believe that the length of our system is nearly consistent with an exact period-two solution to the SG equation, subject to periodic boundary conditions.

This system also possesses a regime where the two types of transport yield the same *I-V* characteristic, *viz.*, the period-four (P-4) and 4K-4$\bar{\text{K}}$ curve, cf. Figs. 7, 9(a), and 9(b). On this step, the P-4 structure was obtained by adiabatically decreasing Γ_0, while the 4K-4$\bar{\text{K}}$ state was generated by setting $\Gamma_0 = 0.30$ and using an initial state with a random spatial profile. We found that these coexisting solutions to Eqs. (3.1) are mutually stable with respect to spatially homogeneous, AC driving and point-impurity-type perturbations with strengths up to $0.25\Gamma_0$. Since we observed that both types of transport mechanisms display the same *I-V* characteristics, it is necessary to measure some other quantity, such as the power spectrum, in order to distinguish between them.

From the recipe for generating Fig. 7, it is evident that Eqs. (3.1) display hysteresis: After the spatial pattern abruptly changes with a small change in Γ_0, changing Γ_0 in the opposite direction does not recover the previous state. Instead, the system remains in the state with the new spatial pattern, and a new current-voltage branch is traced out. In addition to this large-scale hysteresis, there is an additional irreversibility that occurs *within* the higher steps. On the high-voltage, cavity-mode steps, the simple structure like that in Fig. 8(b) is replaced by another where the amplitude locking is broken, so that it contains interleaved breathers of different sizes, as in Figs. 8(c) and 8(d). This symmetry breaking occurs at high driving for increasing Γ_0, and persists until the spatial pattern jumps to the spatially homogeneous state (the McCumber curve). However, if the driving is decreased after the symmetry is broken, the lower symmetry state lasts until the driving is reduced to near, but before, the onset of the step (see Fig. 7, where the x's represent approximately where the symmetry breaks with increasing Γ_0, and the o's represent where the symmetry is regained with decreasing Γ_0). We believe that this breaking of the spatial pattern is caused by the multibreather-wave-train

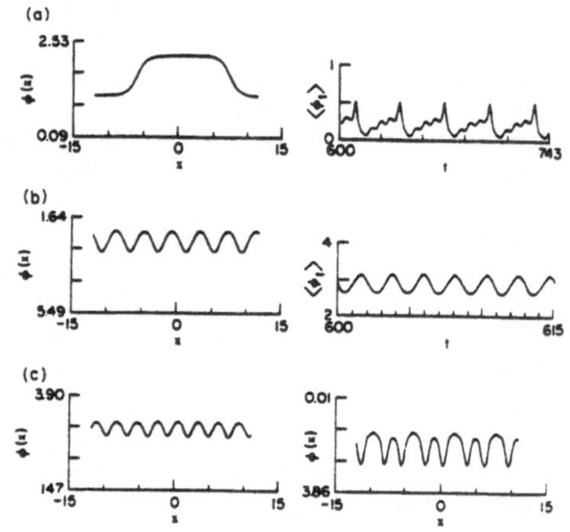

Figure 8. In each of the ϕ vs x pictures, the value of ϕ is given *modulo* 2π. (In each case, the difference between the top and the bottom of the graphs is 15 units): (a) The 1K + 1$\bar{K}$ state. Here, transport is due to the transverse motion of the kinks, and results in the lowest step in Fig. 7; (b) The P-6 state. Here, transport is due to the running average of the phase-locked wave train, as can be seen in the $\langle \phi_t \rangle$ vs time plot on the right; (c) and (d) On the left-hand side are the P-6 and P-8 spatial states that are stable for low driving. On the right-hand side are the symmetry-broken states obtained when the driving is increased beyond 0.75 and 0.65, respectively.

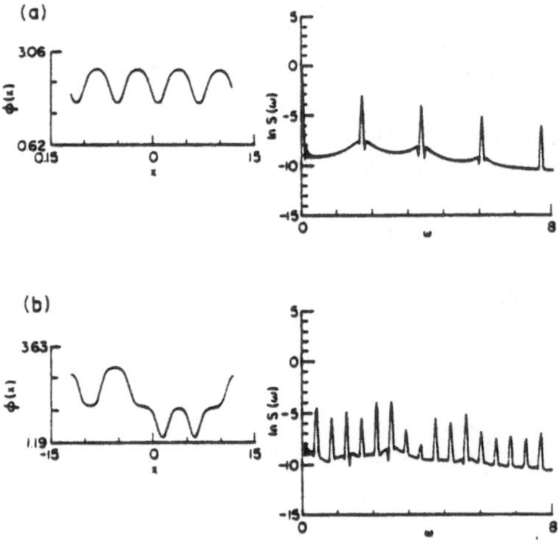

Figure 9. (a) On the left-hand side is a snapshot of the spatial profile of the P-4 state, on the right-hand side is the power spectrum of ϕ averaged over x for this state; (b) A snapshot of the spatial profile for the 4K-4$\bar{K}$ state. The additional peaks in the power spectrum are due to K-$\bar{K}$ collisions. In each of the ϕ vs x pictures, the values are given modulo 2π, with a difference of 15 units between the bottom and top of the graphs.

nature of these cavity-mode states, which results in a *competition* of length scales. Since the amplitude and width of a breather are related to its internal frequency, and the applied current drives this frequency, the system eventually reaches a point where it is energetically favorable for it to create a second type of breather to relieve the internal stress.

It is helpful to contrast these ZFS structures with those that occur in overlap-geometry Josephson junctions. As we mentioned earlier, the imposition of periodic boundary conditions guarantees that only K-$\bar{\text{K}}$ pairs will be excited, so that only even steps are generated from the uniform spatial state. The patterns that we observe, cavity modes and fluxon-antifluxon pairs, are simpler than the symmetric and asymmetric modes that are obtained with Neumann boundary conditions.

Analytic treatments of this system are possible in principle both in terms of linear stability analysis and in a soliton basis –– indeed the system is sufficiently simple that it should be a valuable testing ground for nonlinear collective mode reductions in the presence of perturbations.

Linear stability of the spatially homogeneous rotating state to a period-N, cavity mode state is furthest developed to date –– at least for onset conditions; the nonlinear saturation of period-N linear modes requires fully a nonlinear mode basis. Linear stability analysis starts with the spatially uniform solution to the unperturbed SG equation and adds a perturbation of the form $y(t)f(x)$, where $f(x)$ is periodic. When this solution is substituted into Eq. (3.1) and the result is linearized, we get

$$f''(x)y(t) - f(x)\ddot{y}(t) - f(x)y(t)\left[1 - 2sn^2\left[\frac{t}{\alpha};\alpha\right]\right] \\ = \varepsilon f(x)\dot{y}(t) + [\varepsilon\dot{\phi}_0(t) - \Gamma_0] \ , \tag{3.2}$$

where $\text{sn}(u, \alpha)$ is a Jacobi elliptic function, α is the modulus of this function, and

$$\dot{\phi}_0(t) = \frac{2}{\alpha}\left[1 - \alpha^2 sn^2\left[\frac{t}{\alpha};\alpha\right]\right]^{1/2} .$$

Next, write $f(x)$ as

$$f(x) = \sum_{n=-\infty}^{\infty} f_n e^{inkx} \ , \tag{3.3}$$

where $k = 2\pi N/L$, and N is the spatial periodicity of the final state. When we multiply through by exp($-imkx$) and integrate over x, we get (with $f_m \neq 0$)

$$\ddot{y}(t) + \varepsilon\dot{y}(t) + (m^2k^2 + 1)y(t) - 2y(t)sn^2\left[\frac{t}{\alpha};\alpha\right] \\ = [\Gamma_0 - \varepsilon\dot{\phi}_0(t)]\delta_{m,0} \ .$$

The resulting ode's can be solved perturbatively by expanding $\text{sn}^2\left[\frac{1}{2}\ t;\ \alpha\right]$, as explained in [26]. This approach is extremely successful quantitatively in predicting the transition points between uniform and period-N or period-N and period-(N$-$1) states.

Attempts to use a full soliton mode basis are in progress [27]. These exploit the integrability of the unperturbed SG equation, as discussed in section 2. The nonlinear spectral transform can be used to characterize the various branches in terms of their $K\bar{K}$, breather and radiation components. Various analytic perturbation schemes can then be used to describe the time evolution of, e.g., Θ-function representations [27] under our perturbations. Correspondingly, we have also used the nonlinear spectral scheme described in section 2 to numerically follow the various soliton components —— this involves some subtlety of interpretation to separate a rotating background from additional dynamic spatial structure.

Finally, we mention that this (1+1)-dimensional SG system can be used to approximately model *wall* dynamics in (2+1)-dimensional, discrete SG problems. This is accomplished by describing the fluctuations relative to the wall center of mass, as described in ref. [28]. Correspondinly wall motion in (N+1)-dimensional SG can be reduced approximately to a (N–1)+1)-dimensional system. In this way the phenomena described in this section model the propagation of walls by *transverse* pattern formation —— e.g. transverse $K\bar{K}$ nucleation, as observed in dislocation motion in crystals. Analysis of transverse patterns on propagating interfaces is important in many fields (e.g. cellular textures on flame fronts) and is likely to receive further attention experimentally in condensed matter settings —— for instance domain wall motion in hard magnetic materials.

4. The (1+1)-Dimensional SG Model with Nonconvex Interparticle Interactions

As we mentioned in the Introduction, solid state physics in the last 15 years has rediscovered the SG equation in the context of commensurate-incommensurate phase transitions now observed experimentally in very many kinds of materials [3]. In these situations, several length scales are in competition and lead to intrinsically inhomogeneous ground states (frequently, ordered or irregular arrays of kink-solitons).

Most of the theoretical studies to date have concentrated on cases where the interparticle interactions are *convex*, as in the *discrete* version of the SG equation (2.1). A typical case is where there are *two* competing lengths: the lattice constant between an array of particles (e.g. atoms) and a periodic potential in which they sit with a periodicity which is incommensurate with the lattice constant.

Recently, motivated by physical concerns, the class of problems has been extended to include *nonconvex* interactions between particles, leading to additional, and qualitatively different, competing length scales [2,9]. Nonconvex interparticle interactions also arise effectively in a Ginzburg-Landau free-energy functional for the strains (~ differences in displacements between neighboring particles) in materials undergoing elastic phase transitions. The model described below may, for example, be useful in descriptions of twin boundary dynamics in martensitic materials [29]. Here the substrate potential models, e.g., the parent phase and other terms describe an expansion of the elastic free energy as a function of the strain and strain gradients.

One model which has been studied in some detail [29] is

$$H = \sum_n \frac{1}{2}\, \dot{u}_n^2 + \alpha(u_{n+1} - u_n)^4 - \beta(u_{n+1} - u_n)^2$$

$$+\frac{1}{2}\,\gamma(u_{n+1}-2u_n+u_{n-1})^2-cos u_n\ . \qquad (4.1)$$

Here, there is a competition between the substrate periodicity 2π and the natural lattice constants $2\pi \pm \ell_0$, $\ell_o^2 = \beta/2\alpha$. Further generalizations are given in [29]. It is easy to see that for $\gamma \ll 1$ the system will behave as for $\gamma = 0$ with $\beta \rightarrow \beta - 2\gamma$. The ground state is then *dimerized* (i.e. a periodic line of "short" and "long" lattice constants) for $\beta - 2\gamma > \frac{1}{8}$. On the other hand, for large values of α, β, γ (i.e. when the substrate is weak), it can be shown [29] that a long-period superlattice (a "kink-antikink" or "twinning" [30] pattern) is formed and stabilized by the substrate competition. Roughly these patterns are a periodic array of N short lattice constants, followed by M long ones, with the (N+M) unit cell periodically repeated. N and M are integers and may be equal.

The detailed ground state phase diagram has been obtained [29] by a combination of ansatz, numerically exact transfer integral techniques, and numerical relaxation of the equations of motion following from (4.1) The results (for $\ell_o \ll 2\pi$, so that the dominant nonlinearity is in the interaction potential and SG "solitons" are not important) are summarized in fig. 10. The notation in this figure is given in the figure caption. We draw attention to two interesting features: (1) The transition from the uniform to long-period ground states contains at worst a triple point (three coexisting phase) and possibly a Lifshitz point character; (2) Transitions between long-period ground states are typically first-order, not continuous. These results distinguish model (4.1) from, e.g., the Frenkel-Kontorova model and are a direct consequence of the interparticle *nonconvexity*. In particular the interactions between defects ("solitons") within each long-period pattern (see fig. 10) is attractive leading to the nucleative first order character. This is in contrast with the convex interaction case, where the interaction is repulsive, giving rise to a soliton lattice (incommensurate structure) with a *continuous* transition in the soliton spacing. The soliton nucleation transition mechanism pre-empts an alternative "soft mode" (or "phonon-mediated") scenario in which the period of the patterns could change homogeneously and continuously: this can happen in principle because of the competing first- and second-neighbor gradient terms. The exceptions to this rule are transitions from that *uniform* ground state. Linearizing we find that $\omega(q) \sim -q^2 + q^4$: when $-q^2$ is not stabilized by q^4, a transition can occur. Specifically [29], linear stability shows two regimes: (a) $\gamma < 1/16$, where the homogeneous state becomes unstable and bifurcates into a dimerized one along the line $\beta = 2\gamma + \frac{1}{8}$; and (b) $\gamma > 1/16$, where the homogeneous state becomes unstable along the curve $\beta = \alpha^{1/2}$ and bifurcates into modulated states with wavenumber $q_c = (2\pi)^{-1}\cos^{-1}(\frac{1}{2}\gamma^{-1/2} - 1)$ —— an infinite number (incomplete "Devil's staircase") of both commensurate and incommensurate phases. As stated above, a corresponding linear stability analysis of the modulated phases themselves shows that the actual (nucleation driven) phase boundaries always lie *inside* the domains of linear stability. The strong hysteresis accompanying the transitions can be clearly seen in numerical relaxations studies as parameters are adiabatically varied.

This complex ground state structure and first order transitions naturally leads to highly hysteretic, multiple time scale dynamics. Indeed, this model will be an excellent template for analyzing "glassy" transport and response in competing interaction problems, as discussed in section 1. Dynamic studies to date [29] include:

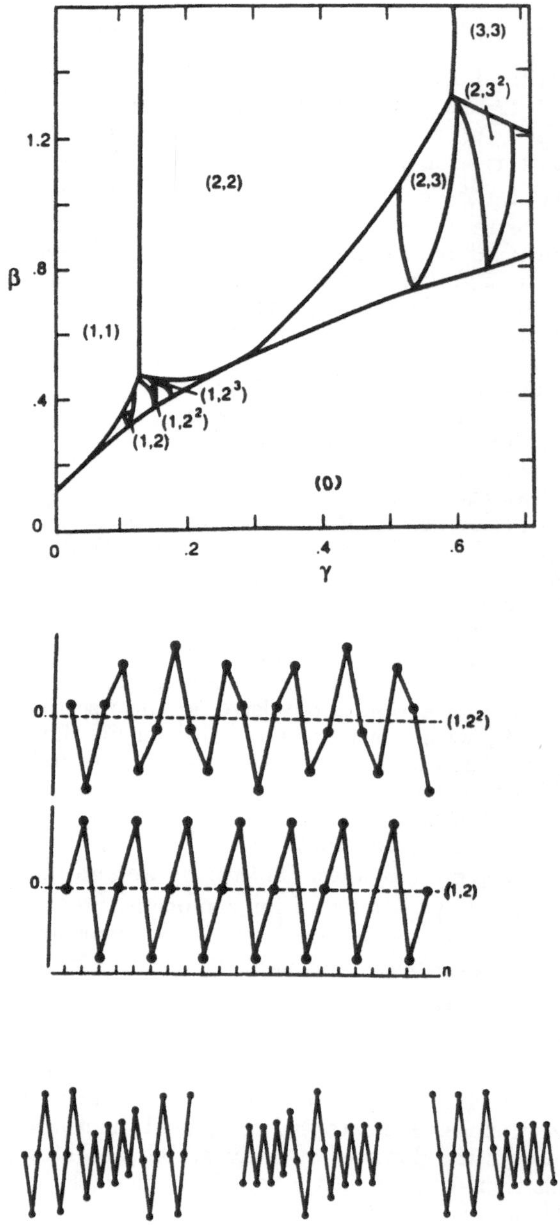

Figure 10. (After Floria and Marianer [29]): (a) The ground state phase diagram obtained by the method of effective potentials; (b) The notation (N^p, M^q) is illustrated; and (c) examples (β=1.0) of a (1,1) soliton in the (2,2) ground state, a (2,2) "soliton" in the (1,1) ground state, and a (1,1) to (2,2) "interface" configuration.

(1) analysis of the optic and acoustic phonons accompanying the homogeneous ground state; (2) relaxation of "hot" (thermalized) initial conditions, showing the importance of pinning effects which lock defects randomly in the true ground state; and (3) effects of a DC-field and damping. Point (3) is extremely interesting and appears to contain the ingredients of "1/f" noise and a relationship between time *and* space characterizations of this phenomenon. Indeed this "noise" occurs through the motion of local defects, a notion which is now rising even in theories of fully developed turbulence [31]. Comparison with the DC-driven SG system of section 3 is instructive. There, linear stability analysis of the *uniform* rotating state identified periodic wave-train unstable modes, which saturated into breather and $K\bar{K}$ wave-trains. Here, the rotating state is itself inhomogeneous because of the internal ground state complexity [29]. Linear stability analysis then shows that this state destabilizes through *local* modes being excited and saturating in the nonlinearity as the rotation rate decreases and allows the complex landscape of metastable pinned configurations to be felt. These modes contain explicit frequency- and spatial-scale relationships. In this sense the rigidly rotating state is "critically unstable" [32] as the rotation rate is decreased to a critical value and is destabilized by a low density of defects (with respect to the rotating pattern) being excited and moving slowly relative to the average rotation rate –– they then control long-wavelength and long-time character in a *connected* way, giving rise to 1/f behavior. We can expect that the "critical" properties will be very interesting, as discussed in other competing interaction systems with critical transport in the presence of DC-driving [11].

5. Quantum Dynamics of a Pulsed Spin System

In this section we introduce some recent approaches to *quantum* mechanics of systems with interesting (e.g. chaotic) classical limits. This field (often colloquially referred to as "quantum chaos") is associated with its own considerable history and literature. Our intention in the model problem introduced below is quite limited. We wish to emphasize in the context of this Institute that: (a) Competing interactions are equally important in quantum problems as in classical ones; (b) It is very important to have control of models which contain an integrable limit and where deviations from integrability can be "tuned," as for classical models; and (c) It is natural to try to characterize such quantum problems by scaling and "fractal" measures, analagous to the successful approaches developed for classical limits. Furthermore, this includes studying wave-function and dynamics as well as more traditional investigations of energy level distributions.

Considering quantum *dynamics*, we can expect that the coherent structure of wavefunctions is greatly affected by quantum interference, leading to the suppresion of anomalous diffusion features characteristic of chaos and eventually to the vanishing of Kolmogorov-Sinai entropy and of other characteristic exponents. If we examine a semiclassical regime, however, new phenomena can appear, not present in either the classical or quantum limit. In particular, semiclassical wavefunction patterns may exhibit several distinctive behaviors in a recurrent time-regime beyond the crossover time t_c at which the classical and quantum correspondence breaks down. (In the case of hyperbolic fixed points of the corresponding classical motion, $t_c \sim$ (Lyapunov exponent) $^{-1}$x $\ln(\hbar^{-1})$. (See, e.g. [33].) But these behaviors have not so far been examined systematically; most previous studies on the dynamics of wavefunctions have concentrated on the time regime $t < t_c$. For a study of long-time and semiclassical behaviors, quantum-large spin systems are especially advantageous because the finite dimensionality of their Hilbert space requires no artificial truncation of energy matrices. Further, for spin echo experiments in electron spin resonance, for instance, an assembly of spin 1/2 systems behaves coherently and effectively constitutes a single large quantum-spin.

Here, we report [7] on the long-time behavior of wavefunctions in a periodically-pulsed large quantum-spin system whose classical limit exhibits a transition from predominantly regular orbits to global chaos as the pulse strength is increased. The effect of dissipation is omitted in the present treatment. We shall attempt to characterize wavefunction patterns in terms of the singularity spectra $f(\alpha)$, which has proven very useful recently in quantifying mulifractal aspects of chaotic systems [34].

The quantum dynamics for our spin system with $\vec{S} = (S^x, S^y, S^z)$ is described by $ih\dot{\Psi} = H\,\Psi$ where $H = H_0 + \sum_{n=-\infty}^{\infty} V\,\delta\,(t - 2\pi n)$ with $H_0 = A(S^z)^2$ and $V = -\mu B S x$. A (>0) and μB (>0) represent a planar anisotropy and pulse strength, respectively. (Here we have chosen a convenient model Hamiltonian. One may make other choices, e.g., $H_0 = AS^z$ and $V = -\mu BS^x \cos(\omega t)$, without changing the qualitative features of the results below.) We solve the above Schrödinger equation by rewriting it immediately in a matrix form; a set of eigenstates of S^z is chosen as basis kets. Then, the wavefunction Ψ just after the n-th pulse is given by $\Psi\,(2\pi n + 0) = \sum_{m=-S}^{S} C_m\,(2\pi n + 0) \mid m >$ with $\vec{C}(2\pi n + 0) = \sum_{\alpha} \exp\,(-2\pi i n E_\alpha/\hbar) \cdot [\vec{X}_\alpha + \cdot \vec{C}_\alpha(+0]\vec{X}_\alpha$. Here $\{E_\alpha\}$, $\{\vec{X}_\alpha\}$ are quasi-energies and quasi-eignefunctions for the one-period propagator represented by the unitary matrix:
$\hat{U} = \exp\,(-\,(i\hbar)\,\hat{V}\,\exp\,(-\,(i/\hbar)\,2\pi\hat{H}_0)$. $\hat{V}$ and $\hat{H}_0$ are matrices for V and H_0, respectively. The probability density function is given in terms of SU(2S + 1) coherent state representations as $P_n(\theta,\phi) = [(2S + 1)/4\pi] \mid < \theta, \phi \mid 2\pi n + 0 > \mid^2$, where the first factor on the right-hand side is due to the normalization over the surface of a unit radius sphere. In the following, A = 1.0 sets the energy unit. Further, we employ S = 128 and choose $\hbar = 1\,\sqrt{S(S+1)}$ so that the observable spin magnitude maintains the scaled value for the classical spin vector, i.e., $\hat{S}^{\,2} = S\,(S + 1)\,\hbar^2 = 1$.

In fig. 11, very early stages (n = 1,2,3) of the temporal evolution of initially (n = 0) localized wavepackets are shown. For a weak pulse ($\mu\,\tilde{B} \equiv \mu B/A = 0.01$), $P_n(\theta,\phi)$ shows a simple unidirectional diffusion (see figs. 11(a) - (c)) corresponding to regular behaviors in classical dynamics. [Note that investigation of classical dynamics indicates the presence of two characteristic fields $\mu\tilde{B}_1 \cong 0.1$ and $\mu\tilde{B}_2 \cong 0.5$, where the fraction of chaotic trajectories increases strongly and the last KAM torus disappears, respectively.] However, for a strong pulse ($\mu\tilde{B} = 1.0$), remarkably isotropic and irregular diffusions begin after the period of "classical" stretching-and folding-type diffusion. Fig. 11 also indicates $t_c = 0(1)$ for both the $\mu\tilde{B} = 0.01$ and 1.0 cases. The above results resemble a quantized version of

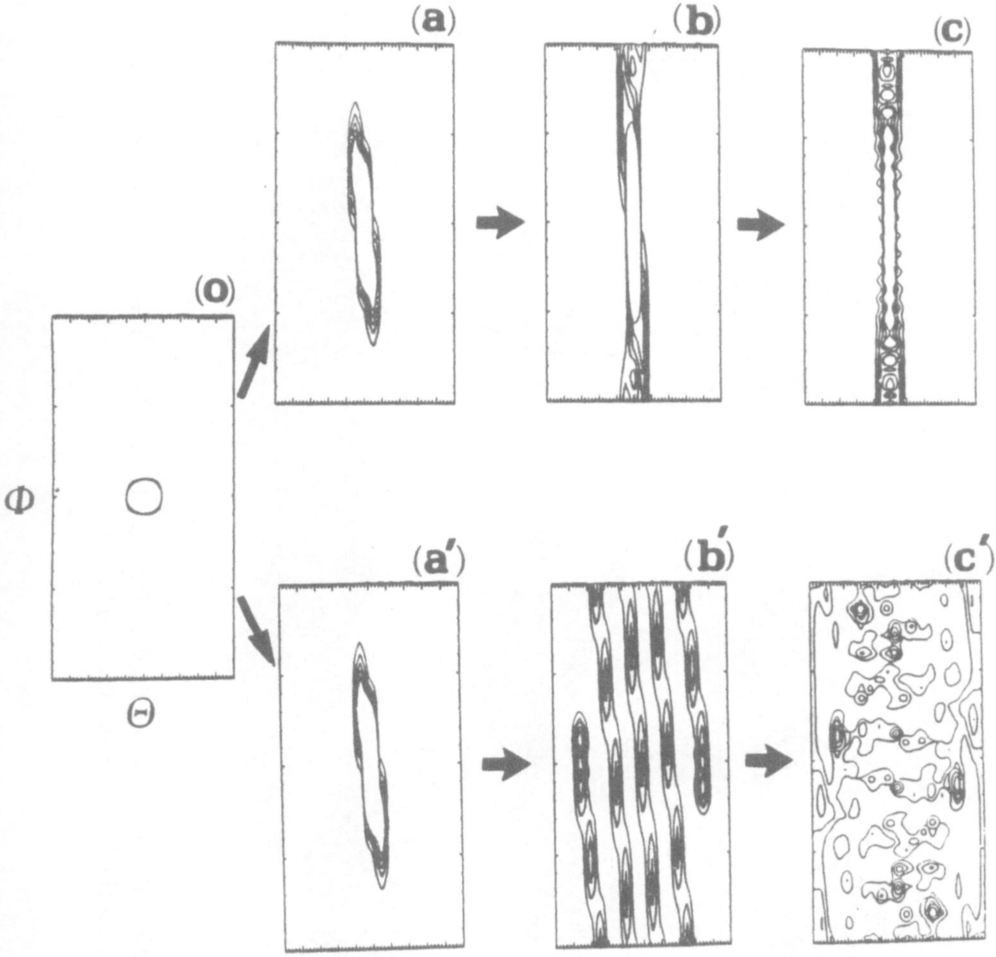

Figure 11. Contour map for very early stages of $P_n(\theta,\phi)$: (0) Initial (n = 0) wave packet; (a) - (c) time evolution for $\mu\tilde{B} = 0.01$; (a′ - (c′) time evolution for $\mu\tilde{B} = 1.0$. From the left, n = 1, 2 and 3.

abstract dynamical systems (e.g., C- or K-systems), in which wavefunctions have been reported [35] to exhibit highly irregular patterns after stretching and folding. Also, certain eigenstates were found with anomalous localization lengths.

We now proceed to examine $P_n(\theta,\phi)$ in large n regions (n = 70,90,110) (see figs. 12). While the exact classical-quantum correspondence has been lost in this time region, these figures clearly maintain some images of the underlying classical dynamics: figs. 12(a) - 12(c), 12(a′) - 12(c′) and 12(a″) - 12(c″) retain signatures of nonergodicity at $\mu\tilde{B} = 0.01$, of partial ergodicity at $\mu\tilde{B} = 0.2$, and of complete ergodicity at $\mu\tilde{B} = 1.0$, respectively. In fact, localized regular structures with large

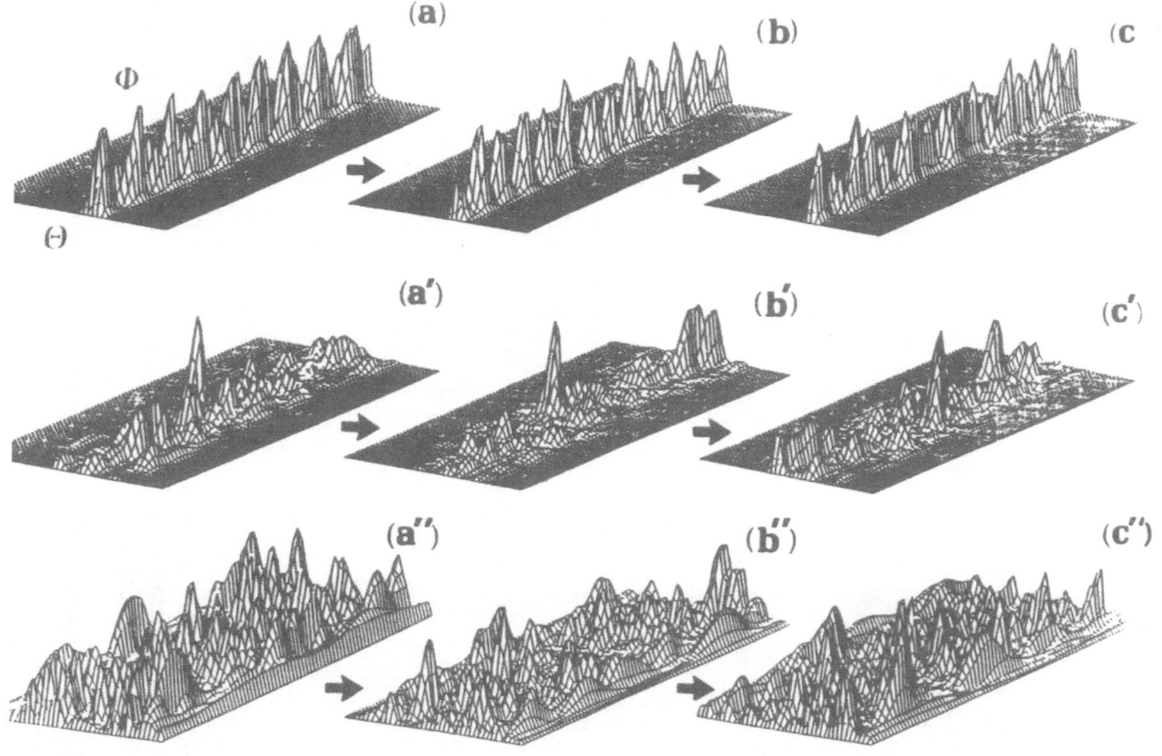

Figure 12. Time evolution of 3D profiles of $P_n(\theta, \phi)$ for $n \gg 1$: (a) - (c) $\mu\tilde{B} = 0.01$; (a′) - (c′)$\mu\tilde{B} = 0.2$; (a″) - (c″)$\mu\tilde{B} = 1.0$. From the left, n = 70, 90 and 110.

amplitude keep a quasi-periodic oscillation for $\mu\tilde{B} = 0.01$ and fine structures with small amplitudes continue to occupy the global phase space for $\mu\tilde{B} = 1.0$. For $\mu\tilde{B} = 0.2$, fine structures continue to occupy a limited portion of phase space.

We now try to quantify $P_n(\theta, \phi)$ in terms of mulitfractals [34]. Since $P_n(\theta, \phi)$ is already normalized to unity in the θ - ϕ plane, the calculation of the singularity spectra $f(\alpha)$ is straightforward: for a linear scale ℓ, we consider the square ℓ x ℓ mesh $A_i(\ell)$ around the position (θ, ϕ) and calculate $P_{n,i}(\ell) = \int_{(\theta,\phi)\subset A_i(\ell)} P_n(\theta, \phi) \sin\theta \, d\theta \, d\phi$. Summing $P_{n,i}{}^q(\ell)$ over all meshes, we obtain the partition function $\Gamma(q,\ell)$. The scaling property of $\Gamma(q,\ell)$ is then examined by changing ℓ according to $\ell = \ell_0 x 2^m$ (m = 0,1,2,...) with $\ell_0 = 0(S^{-1/2})$. The scaling exponents τ_q thus obtained are used to find $f(\alpha)$. It should be noted, however, that our numerical data $P_n(\theta, \phi)$ are reliable only to the order of 10^{-5}. Using them as inputs, we can obtain wide scaling regions for $\Gamma(q,\ell)$ in the case $q \geq 0$, but it is difficult to explore sufficiently wide scaling regions in the case $q < 0$. So our reports of $f(\alpha)$ below will be limited to $q \geq 0$. This restriction does not prevent us from studying the general tendencies of fluctuations of singularities or local dimensions α. Fig. 13 represents $f(\alpha)$ with $q \geq 0$ for several $\mu\tilde{B}$ values at a fixed time n = 90. We find that fluctuations of α for $\mu\tilde{B} = 0.01$ and 1.0 fall into a narrow range and those for $\mu\tilde{B} = 0.2$ extend over a much wider range. The large fluctuation in the latter case signifies the inhomogeneous distribution of measure $P_n(\theta, \phi)$ in fig. 12(b′), which reflects the coexistence of classical KAM orbits and localized chaos in a transitional region leading to global chaos. This large fluctuation is reminiscent of the critical fluctuations at an equilibrium phase transition. The relatively small fluctuation for $\mu\tilde{B} = 1.0$ signifies the uniform distribution of measures in Figs. 12(b″).

Using our data for $f(\alpha)$ with $q \geq 0$, we now estimate the effective range of fluctuations $\alpha^*_{min} \leq \alpha \leq \alpha^*_{max}$: α^*_{max} and α^*_{min} denote the value at which $f(\alpha)$ takes the maximum (i.e., fractal dimension) and the value at which $f(\alpha)$ takes 2/3 times its maximum (an arbitrary choice). For $\mu\tilde{B} = 0.2$ at n = 90, for example, $\alpha^*_{max} = 1.98 \pm 0.01$, $\alpha^*_{min} = 1.35 \pm 0.01$. The features in fig. 13, which have now been quantified, are found to persist throughout the temporal evolution. Careful examinations indicate: (1) the effective range of α shows distinctive temporal variations for $\mu\tilde{B} = 0.2$; (2) on the other hand, it remains almost unchanged for $\mu\tilde{B} = 1.0$ (despite the absence of dissipation in the present system), which reflects a well-organized ergodicity in this case.

The mixing and ergodic features of classical chaos have helped to establish relationships with the formalism of equilibrium statistical mechanics. In the field of quantum chaos, most of the literal definitions of classical chaos lose their significance. Nonetheless, we have still found here complicated behaviors in the quantum mechanical treatment of chaotic systems. We believe that the characterization given here will be a vehicle for more profound understanding of these complexities. Summarizing, despite the complete absence of classical and quantum correspondence, the long-time behavior of semiclassical wavefunctions maintains the ergodic and nonergodic features possessed by the underlying classical dynamics. The enhanced fluctuation of their local dimensions in a transitional region leading to global chaos persists throughout the time evolution. This is reminiscent of critical fluctuations at an equilibrium phase transition.

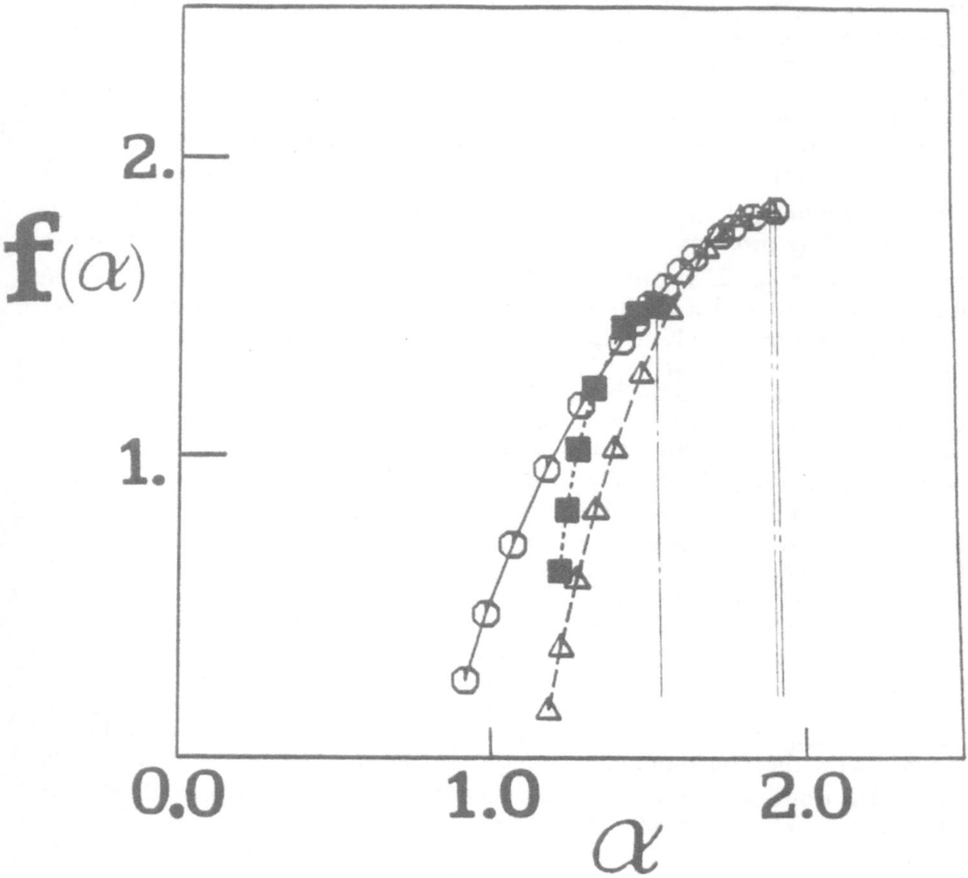

Figure 13. f(α) in q$\geq$0 regime at n = 90. Squares, circles and triangles correspond to $\mu\tilde{\mathrm{B}}$ = 0.01, 0.2, and 1.0, respectively.

6. Summary

In summary, studies of complexity in models motivated by condensed matter and materials science have focused attention on a number of issues of wide importance in dynamical systems: (i) The importance of understanding complexity in extended (space-time) systems, so that *interrelations* of pattern formation and complicated dynamics can be included; (ii) The fundamental role played by *competing interactions*, including both length scales and frequencies; (iii) *Mappings* between time-dependent, nonequilibrium systems (pde's, neural networks, coupled map lattice, cellular automata), and to effective, equilibrium Hamiltonian systems in higher spatial dimensions and with competing length scales. (Excellent examples are found in the liquid crystal convection cell experiments of Ribotta (see these proceedings)); (iv) The value of *nonlinear* collective (e.g. soliton-like) *mode reduction* as a basis for describing low-dimensional attractors in many degree-of-freedom systems; (v) The role of generalized (space-time) *homoclinic orbits* for describing the geometry of both temporal sensitivities and spatial instabilities; and (vi) The important role played by "defects" in controlling the large scale flow (transport) of competing interaction systems.

From a theoretical perspective, a number of techniques have been developed and tested on specific models with which to describe aspects seen in real or numerical experiments: e.g. perturbation theory around linear and nonlinear states [37]; perturbation theory in a collective coordinate or soliton basis [38]; energy balance criteria for nonlinear modes [39]; algorithms for computing ground states in the presence of competing interactions [29], etc. Much development remains necessary, however, in terms of, e.g., incorporating collective coordinates in Renormalization Group or Fokker-Planck descriptions, controlling estimates of center manifold dimensionality, describing energy transfer between nonlinear modes, analyzing frequency and phase "pulling" of coupled nonlinear oscillator, understanding coexisting homoclinic orbits and their connections, including effects of noise and disorder in nonlinear systems [40], describing dynamics in competing interaction systems, etc.

Finally, while the precision, elegance and variety of condensed matter experiments which probe "complexity" in time and space have increased dramatically, it should be remembered that a plethora of phenomena in mainstream *metallurgy* share similar features [2]. We expect that a great deal of attention will soon focus on microscopic modeling of texturing and of nontrivial dynamics in material science, and on the bridge to condensed matter, where we have gained some understanding via simpler, controlled systems.

We gratefully acknowledge close collaborations over the last several years with many colleagues, including S. Aubry, J. Ariyasu, G. Forest, P. Lomdahl, S. Marianer, A. Mazor, D. McLaughlin, K. Nakamura and E. Overman, II.

REFERENCES

1. "Spatio-temporal coherence and chaos in physical systems," eds. A. Bishop, G. Gruner, B. Nicolaenko (North-Holland, 1986).
2. "Competing interactions and microstructures: statics and dynamics," eds. R. LeSar, A. Bishop, R. Heffner (Springer-Verlag, 1988).
3. P. Bak, Rep. Phys. **45**, 587 (1982).
4. See, e.g., articles in Ref. 2.
5. E. A. Overman, II, D. W. McLaughlin and A. R. Bishop, Physica D **19**, 1 (1986).
6. A. R. Bishop, D. W. McLaughlin, M. G. Forest and E. A. Overman, Jr., Phys. Lett. **127A**, 335 (1988).
7. K. Nakamura, A. R. Bishop et al, Phys. Rev. Lett. **54**, 861 (1985); Phys. Rev. Lett. **57**, 5 (1986); Phys. Lett. **117A**, 459 (1986); Phys. Rev. B, in press (1989).
8. A. C. Scott et al, preprint (1989).
9. R. Eykholt, A. Bishop, P. Lomdahl and E. Domany, Physica **23D**, 102 (1986).
10. A. R. Bishop, J. A. Krumhansl and S. E. Trullinger, Physica **1D**, 1 (1980).

11. A. R. Bishop, B. Horovitz and P. S. Lomdahl, Phys. Rev. B **38**, 4853 (1988).
12. A. R. Bishop and P. S. Lomdahl, Physica **18D**, 54 (1986).
13. A. R. Bishop et al, Phys. Rev. Lett. **50**, 1095 (1983).
14. See, M. G. Forest and D. W. McLaughlin, J. Math. Phys. **73**, 1248 (1982).
15. N. M. Ercolani, M. G. Forest, and D. W. McLaughlin, Physica D (in press).
16. C. R. Doering, J. D. Gibbon, D. D. Holm and B. Nicolaenko, Nonlinearity **1**, 179 (1988).
17. D. W. McLaughlin, private communication.
18. N. M. Ercolani, M. G. Forest and D. W. McLaughlin, preprint (1988).
19. J. C. Ariyasu and A. R. Bishop, Phys. Rev. B **35**, 3207 (1987).
20. e.g. J. Guckenheimer and P. Holmes, "Nonlinear Oscillations, Dynamical Systems, and Bifurcations of Vector Fields," (Springer-Verlag 1983).
21. A. R. Bishop et al, Phys. Lett A (in press) (1989); J. Math. Phys. (in press).
22. G. Kovacic and S. Wiggins, preprint (1988).
23. D. David, D. D. Holm and M. V. Tratnik, Phys. Lett. **137A**, 355 (1989); and Physics Reports (in press).
24. J. C. Ariyasu and A. R. Bishop, Phys. Rev. A **39**, 6409 (1989).
25. M. Salerno, Phys. Lett A (in press) (1989).
26. See ref. 24.
27. R. Flesch, M. G. Forest and A. Sinha, in preparation.
28. J. Pouget, S. Aubry, A. Bishop and P. Lomdahl, Phys. Rev. B **35**, 9500 (1989).
29. S. Aubry, K. Fesser and A. R. Bishop, J. Phys. A **18**, 3157 (1985); S. Marianer and A. R. Bishop, Phys. Rev. B **37**, 9893 (1988); S. Marianer and M. Floria, Phys. Rev. B **38**, 12054 (1988); A. R. Bishop, S. Marianer and M. Floria, Phys. Rev. B (in press).
30. G. Barsch, B. Horovitz and J. Krumhansl, Phys. Rev. Lett. **59**, 1251 (1987).
31. See P. Coullet and L. Gil, these proceedings and references therein.
32. e.g. P. Bak and C. Tang, Physics Today **42**, S27 (1989).
33. M. V. Berry and N. L. Balazs, J. Phys. **A12**, 625 (1979).
34. e.g. T. Halsey et al, Phys. Rev. **A33**, 1141 (1986).
35. N. L. Balazs and A. Voros, Europhys. Lett. **4**, 1089 (1987), and Annals of Phys. (in press).
36. R. Ribotta and A. Joets, these proceedings.
37. e.g. A. Mazor and A. R. Bishop, Physica **27D**, 269 (1987); D. W. McLaughlin, A. Pearlstein and G. Terrones, preprint (1988).
38. N. M. Ercolani, M. G. Forest and D. W. McLaughlin, Lectures in Appl. Math. **23**, 3 (1985); and preprint (1988).
39. P. S. Lomdahl and M. R. Samuelson, Phys. Rev. A **34**, 664 (1986).
40. "Disorder and Nonlinearity," eds. A. R. Bishop, D. K. Campbell and S. Pnevmatikos (Springer-Verlag 1989).

EXACT SOLUTIONS OF THE BOLTZMANN EQUATION

H. CORNILLE
Service de Physique Théorique de Saclay
Laboratoire de l'Institut de Recherche Fondamentale
du Commissariat à l'Energie Atomique
91191 Gif-sur-Yvette Cedex, France

ABSTRACT We review the exact solutions of the discrete and continuous Boltzmann Equations.

For the discrete B.E. the velocity $\vec{V}$ can only take discrete values $\vec{V}_i$. The discrete models equations are nonlinear but they include the linear conservation laws. The simplest solutions are the exponential similarity shock waves *and the solutions found in* 1+1, 2+1, 2D, 3D dimensions *are sums of similarity waves*. For a well-defined class of (1+1)-dimensional rational functions it is proved that they are the only possible solutions if two linear conservation laws are present. Different solutions are presented: with nonuniform Maxwellians, shock wave with or without specular reflection, periodic, semiperiodic.

The conservation laws being not directly included into the nonlinear continuous B.E. the exact solutions are different. For the Kac model we construct the even velocity BKW (Bobylev-Krook-Wu) solution, its odd velocity partner and an inhomogeneous solution, while for the homogeneous B.E. we derive the BKW solution. For the inhomogeneous B.E. a time-dependent harmonic potential, which vanishes at infinite time, is introduced, but the equilibrium state is an absolute Maxwellian. This is *the only known exact inhomogeneous example relaxing toward the true equilibrium state* and not toward the trivial vacuum state

R. Conte and N. Boccara (eds.), Partially Integrable Evolution Equations in Physics, 39–81.

CONTENTS

Keywords : Exact solutions, Boltzmann Equation, Kinetic Theory, Discrete Kinetic Theory, Homogeneous BE, Inhomogeneous BE, Similarity Solutions, Multidimensional Solutions.

1. INTRODUCTION TO THE BOLTZMANN EQUATION

We start with a system of N particles (equal mass m) and for each particle define a phase point $\vec{Y}_r = \vec{X}_r, \vec{V}_r$ with $\vec{X}_r$ for the space and $\vec{V}_r$ for the velocity. Let $Y = \{\vec{Y}_r\} r = 1, \ldots N$ be the associated $6N$-dimensional phase space and $f_N(Y,t)$ the N-particle distribution function. From the conservation of probability in phase space $\int f_N dY$ =const or $\frac{df_N}{dt} = 0$ we obtain the Liouville equation with $\vec{F}_r$ for the interparticle forces

$$\frac{\partial f_N}{dt} + \sum_1^N \vec{V}_r \cdot \frac{\partial f_N}{\partial \vec{X}_r} + \frac{1}{m} \vec{F}_r \cdot \frac{\partial f_N}{\partial \vec{V}_r} = 0 \tag{1.1}$$

Integrating $\int f_N d\vec{Y}_{r+1}, \ldots d\vec{Y}_N$ we define the 1-particle distribution function $f_1(\vec{Y}_1, t)$ for $r = 1$, the 2-particle $f_2(\vec{Y}_1, \vec{Y}_2, t)$ for $r = 2, \ldots$ From (1.1) we deduce the so-called BGKY hierarchy of linear equations (not closed) between f_1 and f_2, f_3 and $f_4, \ldots$ and for instance obtain for (f_1, f_2)

$$(\partial_t + \vec{V}_1 . \partial_{\vec{X}_1}) f_1 = (N-1) \int d\vec{Y}_2 \vec{F}_{12} \cdot \frac{\partial f_2}{\partial \vec{V}_1} \tag{1.2}$$

These equation being linear, their evolution in time requires the knowledge of all the initial data f_r at $t = 0$, $r = 1, \ldots N$. Further they satisfy classical mechanics and are reversible with respect to t. Is it possible to "deduce" from them the irreversible Boltzmann equation for f_1 alone ? A lot of assumptions are considered:

(i) The density of particles is sufficiently low (dilute gas) so that only binary collisions occur, (ii) the spatial dependence is sufficiently slow so that the collisions between particles are localised in the space, (iii) the interparticle forces are sufficiently short so that (i) is meaningful (iv) the 2-particle distribution is without correlations (molecular chaos)

$$f_2(\vec{Y}_1, \vec{Y}_2, t) = f_1(\vec{Y}_1, t) f_1(\vec{Y}_2, t) \tag{1.3}$$

which can be correct only for pre-collisional particles.

The Boltzmann gas limit (BGL) requires that $N \longrightarrow \infty$, $m \longrightarrow 0$, $Nm \longrightarrow$ const, $N\sigma^2$ =(mean free path)$^{-1}$ $\longrightarrow$const (σ being a parameter for the range of the interparticle forces), $N\sigma^3$(volume)$\longrightarrow 0$. The modern derivation is due to Grad who introduces "truncated distribution" $f_1^\sigma, f_2^\sigma, \ldots$ with for the integration the restriction that in f_1 no particles are within σ of particle 1... We have to begin with spherical molecules, rewrite the Liouville equation for $f_1^\sigma, f_2^\sigma, \ldots$ apply both BGL and molecular chaos and finally extend to other interparticles forces. Let us put $f_1(\vec{V}_1, \vec{X}_1 t) = f(\vec{V}, \vec{X}, t) > 0$ and the Boltzmann equation becomes:

$$\mathcal{L} f(\vec{V}, \vec{X}, t) = \text{Col } f, \qquad \mathcal{L} = \partial_t + \vec{V} . \partial_{\vec{X}} + \partial_{\vec{V}} . \vec{A}_F \tag{1.4}$$

if an external force $\vec{A}_F$ is present. Colf f, the nonlinear part, corresponds to elastic binary collisions with momentum and energy conservations: $\vec{V}+\vec{V}_1 = \vec{V}'+\vec{V}_1'$, $\vec{V}^2+\vec{V}_1^2 = \vec{V}'^2+\vec{V}_1'^2$. Let $\vec{g} = \vec{V}-\vec{V}_1$, $\vec{g}' = \vec{V}'-\vec{V}_1'$, $|\vec{g}|=|\vec{g}'|$ be the relative velocities, withθ the angle between $\vec{g}$ and $\vec{g}'$ and $d\Omega$ the solid angle $d\Omega = \sin\theta d\theta d\varepsilon$. We define $I(|\vec{g}|,\theta)$ for the interaction between the particles and Col f becomes a gain term minus a loss term.

$$4\pi \text{ Col } f(\vec{V},\vec{X},t) = \int d\Omega d\vec{V}_1 \, |\vec{g}| \, I(|\vec{g}|,\theta)[f(\vec{V}_1')f(\vec{V}') - f(\vec{V})f(\vec{V}_1)] \qquad (1.5)$$

or a balance between particles which appear and those which disappear due to binary collisions.

For the interaction we can consider hard spheres or interparticles forces with inverse power law ν :$|\vec{g}I|=|\vec{g}|^{1-4/(\nu-1)}$ multiplied by a function of θ. For $\nu = 5$, the "Maxwell particles", with $\sigma(\theta)$ for the cross-section, the BE becomes

$$4\pi \text{ Col } f = \int d\vec{V}_1 d\Omega \sigma(\theta)(f(\vec{V}_1')f(\vec{V}') - f(\vec{V})f(\vec{V}_1)) \qquad (1.6)$$

and is the only one for which exact solutions are known.

In order to discuss the *symmetry properties of* Col f, let $\phi(\vec{V})$ be an arbitrary function, and $I(\phi) = \int \text{Col } f \, \phi(\vec{V})d\vec{V}$. With interchanges of the integration variables we obtain

$$4I(\phi) = \int [\phi(\vec{V}) + \phi(\vec{V}_1) - \phi(\vec{V}') - \phi(\vec{V}_1')][f_1'f' - ff_1] \, |\vec{g}| \, Id\Omega d\vec{V} d\vec{V}_1 \qquad (1.7)$$

ϕ is called a summational Invariant if $I(\phi) = 0$ which is trivial if $\phi = 1, \vec{V}, \vec{V}^2$. As an application we obtain the linear macroscopic conservation laws (mass, momentum, energy)

$$\int \mathcal{L}f \begin{pmatrix} 1 \\ \vec{V} \\ \vec{V}^2 \end{pmatrix} d\vec{V} = \int \text{Col } f \begin{pmatrix} 1 \\ \vec{V} \\ \vec{V}^2 \end{pmatrix} d\vec{V} = \begin{pmatrix} 0 \\ 0 \\ 0 \end{pmatrix} \qquad (1.8)$$

For instance if $\rho(\vec{X},t) = \int fd\vec{V}$ and $\left\langle \vec{V} \right\rangle = \rho^{-1} \int f\vec{V}d\vec{V}$ are the mass density and the mean velocity then we deduce the mass conservation law $\rho_t + \partial_{\vec{X}} \cdot \rho \left\langle \vec{V} \right\rangle = 0$ and similarly for the momentum and energy conservation laws. *The importent point is that the macroscopic quantities* $\rho, \left\langle \vec{V} \right\rangle \rho, \ldots$ *depend on* $\vec{X},t$ *and satisfy linear relations while the microscopic distribution function* f *depends also on* $\vec{V}$ *and satisfies the nonlinear BE.*

Let us show the H-Theorem for the homogeneous $f(\vec{V},t)$ with $f_t = \text{Col } f$. We define $H = \int d\vec{V} f\log f$, $X = f(\vec{V}')f(\vec{V}_1')/f(\vec{V})f(\vec{V}_1) > 0$ and apply (1.7)

$$\begin{aligned} H_t &= \int d\vec{V}(1+\log f)f_t = \int d\vec{V} \text{ Col } f(1+\log f) \\ &= -\frac{1}{4}\int d\vec{V}d\vec{V}_1 \, |\vec{g}| \, If(\vec{V})f(\vec{V}_1)(X-1)\log X \leq 0 \end{aligned} \qquad (1.9)$$

The entropy $(-H)$ increases up to the equilibrium state $t = \infty$ where the equality occurs with Col $f = 0$ or $f(\vec{V})f(\vec{V}_1) = f(\vec{V}')f(\vec{V}_1')$. Let $\vec{C} = \vec{V} - \left\langle \vec{V} \right\rangle$ be the so-called peculiar velocity, T with $3T\rho = \int f\vec{C}^2 d\vec{V}$ be the temperature, we obtain the absolute Maxwellian

$$f_{AM} = \rho \, e^{-\vec{C}^2/2RT}(2\pi RT)^{3/2}$$

with ρ, T being constants.

References: Grad H. "Handbuch der Physik" Springer-Verlag, Berlin 1958; Harris S. "Introduction to the Theory of the B.E." Holt, Rinehart and Winston, Inc 1971; Truesdell C. and Muncaster R.G. "Fundamentals of Maxwell's Kinetic Theory" Academic Press, New York 1980; Cercignani C. "Theory and application of the B.E.", Scottish Academic Press, Edinburgh and London 1975; "Lecture Notes in Physics" 35, 1975, Lanford OE, p.1; Studies in Statistical Mechanics Vol.X, "The B.E.", 1983, Editors Montroll E.W. and Lebowitz J.L.; Hoare M.R. "Advances in Chemical Physics" LVI, Ed. Prigogine and Rice, p.2, 1984; The more recent and complete monograph on the homogeneous B.E. with Maxwell molecules is due to Bobylev S. Sov. Sci. Rev. C - Math. Phys. Vol.7, 1988.

2. EXACT SOLUTIONS OF THE DISCRETE BOLTZMANN MODELS

The continuous Boltzmann equation being very difficult to study, physicists and mathematicians have proposed a lot of discrete models. For these models, the velocity $\vec{V}$ can only take a finite number of discrete values $\vec{V}_i$, $i = 1, 2, ...$ and to each $\vec{V}_i$ is associated a density N_i. The exact solutions are mainly known for the models with one speed $|\vec{V}_i| = 1$ for which mass and energy conservation laws cannot be distinguished. In general we discuss the Boltzmann type of collisions which means binary collision. However, in order to eliminate spurious conservation laws, appearing in some planar velocity models, nonbinary collision terms have been introduced.

The most popular model, which is at the origin of the discrete kinetic theory, is the cubic Broadwell model. It is the three dimensional generalization of the square planar model. We will insist on these two models.

For the discrete models, space and time are continuous variables. However a great success has been obtained with lattice gas models for which space and time are also discretised. We do not discuss these models because, up to now, no exact solutions are known.

The collision rule for these models is that only particles with opposite velocities can collide for binary collisions. For p couples of opposite velocities $\vec{V}_{2j-1} + \vec{V}_{2j} = 0$, then, for the collision term associated to (N_{2j-1}, N_{2j}), $N_{2j-1}N_{2j}$ is a lost term while $N_{2k-1}N_{2k}$, $k \neq j$, are gain terms.

Physically relevant solutions must at least be positive and satisfy the H*-theorem.* In general (except for the $2\vec{V}_i$ models), the H-theorem is automatically satisfied and it remains to study the positivity. On the other hand the physically interesting solutions are the multidimensional ones with at least two different spatial coordinates. Unfortunately the positivity proof becomes more and more complicated in higher dimensions. For pedagogical reason we present here complete positivity proofs only for the unidimensional similarity solutions and the (1+1)-dimensional solutions.

In the presentation of these models (section 2.1) we insist on the fact that linear conservation laws are included into the discrete models equations. In 2.2 we present the exponential similarity waves, in 2.3 simple examples of (1+1)-dimensional solutions, in 2.4 we explain why the exact (1+1)-dimensional solutions are sums of similarity waves and in 2.5 present the multidimensional solutions.

2.1. Discrete models

2.1.a. models on a line $x_1 \in R^1$

We begin with the $2\vec{V}_i$ models with $\vec{V}_1$ lying along the positive x-axis and $\vec{V}_1 + \vec{V}_2 = 0$. We write down, for the associated densities N_i, both the nonlinear equations

$$\ell_i N_i = \mathrm{Col}N_i, \quad i = 1,2, \qquad \ell_1 = \partial_t + \partial_x, \qquad \ell_2 = \partial_t - \partial_x \tag{2.1}$$

and the two linear mass M and momentum J conservation laws.

$$M_t + J_x = \sum_1^2 \mathrm{Col}N_i = 0, \qquad M = N_1 + N_2, \qquad J = N_1 - N_2 \tag{2.2}$$

$$J_t + (N_1 + N_2)_x = \mathrm{Col}N_1 - \mathrm{Col}N_2 = 0 \tag{2.3}$$

The solution being $\mathrm{Col}N_1 = \mathrm{Col}N_2 = 0$ is impossible. So we require that only the mass conservation (2.2) is satisfied while the momentum conservation (2.3) is violated. For an arbitrary quadratic collision term we obtain 2 equations with 1 conservation law.

$$\ell_1 N_1 = -\ell_2 N_2 = aN_1^2 + bN_1N_2 + cN_2^2 \tag{2.4}$$

For $a + b + c = 0$ $a \leq 0, b \geq 0$ we have the Illner model, for $b = 0$, $a = c = 1$ the Carleman model and $a = 0$, $c = -b = 1$ the McKean one.

We go on, adding $\vec{V}_3 = 0$ and an associated density N_3, with the $3\vec{V}_i$ models. The nonlinear equations (2.1) become

$$\ell_i N_i = \mathrm{Col}N_i, \qquad i = 1,2,3, \qquad \ell_3 = \partial_t \tag{2.5}$$

and we write down the linear mass, momentum and energy E conservation laws.

$$M_t + J_x = \Sigma \mathrm{Col}N_i = 0, \qquad M = \Sigma N_i, \qquad J = N_1 - N_2 \tag{2.6}$$

$$J_t + (N_1 + N_2)_x = \mathrm{Col}N_1 - \mathrm{Col}N_2 = 0 \tag{2.7}$$

$$E_t + (N_1 - N_2)_x = \mathrm{Col}N_1 + \mathrm{Col}N_2 = 0, \qquad E = N_1 + N_2 \tag{2.8}$$

leading to $\mathrm{Col}N_i = 0$, $i = 1,2,3$. We give up energy conservation and maintain mass and momentum conservation laws. For (N_1, N_2) the loss term is N_1N_2, and the gain term N_3^2 . Microscopically the momentum conservation $\vec{V}_1 + \vec{V}_2 = 2\vec{V}_3 = 0$ is satisfied but not energy: $\vec{V}_1^2 + \vec{V}_2^2 \neq 2\vec{V}_3^2$. This $3\vec{V}_i$ model is only a mathematical one.

$$N_{1t} + N_{1x} = N_{2t} - N_{2x} = -N_{3t}/2 = N_3^2 - N_1N_2 \tag{2.9}$$

2.1.b. planar models $(x_1, x_2) \in R^2$

We begin with the simplest model with $\vec{V}_1, \vec{V}_3$ lying respectively along the positive x_1, x_2 axes, and $\vec{V}_1 + \vec{V}_2 = \vec{V}_3 + \vec{V}_4 = 0$. The associated densities satisfy the nonlinear equations $\ell_i N_i = \mathrm{Col}N_i$, $i = 1,...4$, $\ell_{2j-1} = \partial_t + \partial_{x_j}$, $\ell_{2j} = \partial_t - \partial_{x_j}$, $j = 1,2$, the mass M and momentum $\vec{J} = \sum_1^2 J_j \vec{i}_{x_j}$ conservation laws being

$$M_t + \partial_{\vec{x}}.\vec{J} = \sum_1^4 \text{Col}N_i = 0, \quad M = \Sigma N_i, \quad J_j = N_{2j-1} - N_{2j} \tag{2.10}$$
$$\partial_t J_j + \partial_{x_j}(N_{2j-1} + N_{2j}) = \text{Col}N_{2j-1} - \text{Col}N_{2j}, \quad j = 1,2$$

The conditions $\text{Col}N_{2j-1} = \text{Col}N_{2j}$ for the conservation laws can be satisfied. N_1N_2 being a lost term term and N_3N_4 a gain term for the couple (N_1, N_2) we obtain 4 equations

$$\ell_1 N_1 = \ell_2 N_2 = -\ell_3 N_3 = -\ell_4 N_4 = N_3 N_4 - N_1 N_2 \tag{2.11}$$

with 3 linear conservation laws equations.

We go on with the regular polygonial $2p\vec{V}_i$ model with $\vec{V}_1$ along the positive x_1−axis, $\vec{V}_{2j-1} + \vec{V}_{2j} = 0$, the angle between $\vec{V}_{2j-1}$ and $\vec{V}_1$ being $(j-1)\pi/p$, $j = 1, 2, ...p$. The densities N_i satisfy

$$\ell_{2j-1}N_{2j-1} = \ell_{2j}N_{2j} = -(p-1)N_{2j-1}N_{2j} + \sum_{k \neq j} N_{2k-1}N_{2k} = \text{Col}N_{2j-1},\ j = 1, ...p \tag{2.12}$$

with $\ell_i = \partial_t + \vec{V}_i.\partial_{\vec{x}}$ and $\Sigma\text{Col}_i = 0$. Among the $2p$ equations $p+1$ are linear while only three physical conservation laws (mass and momentum) can exist. This means that with the binary collisions alone there exist $p-2$ spurious conservation laws: 0 for $p = 2$, 1 for $p = 3, ...$ The mathematical escape to eliminate these unphysical conservation laws is the introduction of multiparticle collisions, for instance ternary collisions for the $p = 3$ hexagonal model, ternary, cubic,... for higher p value models.

2.1.c. 3-dimensional models $(x_1, x_2, x_3) \in R^3$

The simplest model is the cubic $6\vec{V}_i$ Broadwell model with $\vec{V}_{2j-1}$, $j = 1, 2, 3$ lying along the positive x_j−axis and $\vec{V}_{2j-1} + \vec{V}_{2j} = 0$. We write down the nonlinear equations and the four conservation laws for the mass M and the momentum $\vec{J} = \Sigma J_j \vec{i}_{x_j}$:

$$\ell_i N_i = \text{Col}N_i, \quad i = 1, ...6, \quad \ell_{2j-1} = \partial_t + \partial_{x_j}, \quad \ell_{2j} = \partial_t - \partial_{x_j}, \quad j = 1,2,3 \tag{2.13}$$

$$M_t + \partial_{\vec{x}}.\vec{J} = \Sigma\text{Col}N_i = 0, \quad M = \Sigma N_i, \quad J_j = N_{2j-1} - N_{2j} \tag{2.14}$$

$$\partial_t J_j + \partial_x(N_{2j-1} + N_{2j}) = \text{Col}N_{2j-1} - \text{Col}N_{2j} = 0 \quad j = 1,2,3 \tag{2.15}$$

which are satisfied if $\text{Col}N_{2j-1} = \text{Col}N_{2j}$ and $\text{Col}N_1 + \text{Col}N_3 + \text{Col}N_5 = 0$. Taking into account for $\text{Col}N_{2j-1}$ both detailed balance and the fact that $N_{2j-1}N_{2j}$ is a lost term while $N_{2k-1}N_{2k}$ $k \neq j$ are gain terms we obtain 6 equations

$$\ell_1 N_1 = \ell_2 N_2 = N_5 N_6 + N_3 N_4 - 2N_1 N_2 \tag{2.16}$$
$$\ell_3 N_3 = \ell_4 N_4 = N_5 N_6 + N_1 N_2 - 2N_3 N_4 \tag{2.17}$$
$$\ell_5 N_5 = \ell_6 N_6 = -\ell_1 N_1 - \ell_3 N_3 \tag{2.18}$$

which contain 4 conservation laws.

There exist other 3-dimensional models, for instance the $14\vec{V}_i$ Cabannes model...

2.1.d. hypercubic p-dimensional models $(x_1, ... x_p) \in R^p$

It is clear that the $2p\vec{V}_i = 6\vec{V}_i$ Broadwell model is the $p = 3$-dimensional generalization of the $2p\vec{V}_i = 4\vec{V}_i$ planar model. Both models have neither spurious conservation laws nor violation of conservation laws. We introduce a $p = 4, 5, ...$ generalization in a p-dimensional space. Let $\vec{V}_{2j-1}$ lying along the positive x_j axis and $\vec{V}_{2j-1} + \vec{V}_{2j} = 0$. The $2p$ equations for the $2p$ associated N_i satisfy

$$(\partial_t + \partial_{x_i})N_{2i-1} = (\partial_t - \partial_{x_i})N_{2i} = -(p-1)N_{2i-1}N_{2i} + \sum_{k \neq i} N_{2k-1}N_{2k} \tag{2.19}$$

and contain $(p+1)$ linear relations which correspond to the $p+1$ mass and momentum conservation laws.

In conclusion, contrary to the continuous B.E. the discrete models equations include the linear conservation laws. This will be the clue for the understanding of the exact (1+1), (2+1), 2D, 3D dimensional solutions

References

Broadwell J.E., Phys. Fluids 7, 1243, 1964; Harris S., Phys. Fluids 9, 1328, 1966, Hardy J. and Pomeau Y., J.M.P. 13, 1042, 1972; McKean H., Comm. Pure Appl. Math. 28, 435, 1975; Gatignol R., "Lecture Notes in Physics" 36, 1975; Tartar L., Seminaire Goulaouic-Schwartz, 1975-1976; Cabannes M., J.Méc. Théor. Appl. 17, 1, 1978, Mech. Res. Comm. 10, 317, 1983; Illner R., Math. Math. Appl. Sci. 1, 187, 1979, Trans. Theo. Stat. Phys. 13, 431, 1984; Beale T., Com. Math. Phys. 102, 217, 1985; Ruijqrook Th. and Wu T.T., Physica A113, 401 (1982); Frisch U., D'Humières O., Hasslacher B., Lallemand P., Pomeau Y., Rivet J.P. Complex Systems 1, 649, 1987; Platkowski T. and Illner R., SIAM Review 30, 213, 1988; Cercignani C., Illner R., Shinbrot M., Comm. Math. Phys. 114, 687, 1988 with references to the Broadwell model; Longo E. and Monaco R., Proceed XVI Rar. Gas Dynam. Pasadena 1988, Piechor K., Arch. Mechan. 4, 1988; d'Humières D. "Bibliography on Lattice Gas and Related Topics", to appear in the proceedings of the Les Houches workshop, February-March 1989.

2.2. Exponential Similarity waves

Let V, W, Z be independent densities satisfying the (1+1)-dimensional equations

$$V_t + V_x = -\alpha_1 W_t - \beta_1 W_x = -\alpha_2 Z_t - \beta_2 Z_x = aV^2 + cW^2 + bVW + dZ^2 \tag{2.20}$$

for different discrete models: (i) $2\vec{V}_i$ models $\alpha_1 = 1$, $\beta_1 = -1$, $Z = 0$, for instance the Illner model $a + b + c = 0$, $a \leq 0$, $c \geq 0$, the Carleman model $-a = c = 1$, the McKean model $a = 0$, $c = 1$; (ii) $3\vec{V}_i$ models $\alpha_1 = -1$, $\beta_1 = 1$, $\alpha_2 = 1/2$, $\beta_2 = 0$, $a = c = 0$, $d = -b > 0$; (iii) $4\vec{V}_i$ model $\alpha_1 = -1$, $\beta_1 = 1$, $\alpha_2 = 1$, $\beta_2 = 0$, $a = c = 0$, $d = -b > 0$ and the Broadwell model with the only change $\alpha_2 = 2$.

We seek exponential type similarity solutions with the variable $\varphi = \gamma x + \rho t$. Then with V_φ, W_φ, Z_φ being proportional we can rewrite the densities as

$$V = v_0 + v/D, \qquad W = w_0 + w/D, \qquad Z = z_0 + z/D, \qquad D = D(\varphi) \tag{2.21}$$

and (2.21) becomes a Riccati equation for D

$$\begin{gathered} eD_\varphi + a_0 + a_1 D + a_2 D^2 = 0, \qquad e = v(\rho + \gamma) = -w(\alpha_1\rho + \beta_1\gamma) = -z(\alpha_2\rho + \beta_2\gamma) \\ a_0 = av^2 + cw^2 + bvw + dz^2, \qquad a_1 = 2avv_0 + 2cww_0 + 2dzz_0 + b(vw_0 + wv_0) \\ a_2 = av_0^2 + cw_0^2 + bv_0w_0 + dz_0^2 \end{gathered} \tag{2.22}$$

which is integrable: (i) If $a_2 = 0$, D is a constant plus an exponential. Choosing $e = -a_1 = a_0$ we find D of the form

$$D = 1 + de^\varphi \qquad d > 0 \tag{2.23}$$

(ii) If $a_2 \neq 0$, putting $D = e\ E_\varphi/Ea_2$ we find a second order differential equation $e^2 E_{\varphi\varphi} +$ $a_1 e E_\varphi +$ $a_0 a_2 E = 0$ with two exponential $\exp(\mu_\pm \varphi)$ solutions, $\mu_\pm = (-a_1 \pm \sqrt{A})\ /2e$ if $A = a_1^2 - 4a_0a_2 \neq 0$. If $A = 0$, the two independent solutions are $\exp - a_1\varphi/2e$ and $\varphi\exp - a_1\varphi/2e$ but for D the exponential disappears and it remains only rational (linear in φ) power type solutions that we exclude. If $A \neq 0$ then D^{-1} can be written $c_1 + c_2$ $/(1 + c_3 \exp(\mu_+ - \mu_-)\varphi)$, c_i being constants, which are of the exponential type with a rescaling of the arbitrary φ.

2.2.a. Exponential Similarity waves for the $2\vec{V}_i$ models

We call $N_\pm$ the densities associated to $\vec{V}_\pm$, $\vec{V}_+ + \vec{V}_- = 0$, $\vec{V}_+$ lying along the positive x-axis and $H = \Sigma N_\pm \log N_\pm$

$$(\eta\partial_t + \partial_x)N_\eta = \mathrm{Col}N_+, \qquad \mathrm{Col}N_+ = aN_+^2 + bN_+N_- + cN_-^2, \qquad \eta = \pm \tag{2.24}$$

$$H_t + \partial_x(N_+\log N_+ - N_-\log N_-) = \mathrm{Col}_H = (\log N_+/N_-)\mathrm{Col}N_+ \tag{2.25}$$

These models satisfy mass conservation $\Sigma(\partial_t + \eta\partial_x)N_\eta = 0$, $\eta = \pm 1$ but violate momentum conservation. However, due to their simplicity, they represent pedagogical examples for testing physical properties of their solutions. Physically relevant densities must be positive $N_\pm > 0$ and satisfy the H-Theorem or $\mathrm{Col}_H \leq 0$.

For the Illner models $a+b+c=0$, $a \le 0$, $c \ge 0$, the H-theorem is automatically satisfied once the $N_\pm$ are positive

$$\mathrm{Col}_{N_+} = (N_- - N_+)(cN_- - aN_+), \qquad a \le 0, \ c \ge 0 \tag{2.26}$$

Furthermore, satisfying detailed balance, they are more physical (in ColN_+, the negative sign for N_+^2 and positive one for N_-^2 can be interpreted as a loss term for N_+ and a gain term for N_-) than the other mathematical (2.24) models.

We consider similarity waves of the (2.23) type and write down the positivity constraints.

$$N_\pm = n_{0\pm} + n_\pm/D, \ \ D = 1+u, \ \ u = e^{\rho t + \gamma x}; \ \ N_\pm \ge 0 \ \ if \ \ n_{0\pm} \ge 0, \ \ n_{0\pm} + n_\pm \ge 0 \tag{2.27}$$

We substitute (2.27) into (2.24) and obtain

$$\begin{gathered} an_{0+}^2 + bn_{0-}n_{0+} + cn_{0-}^2 = 0 \\ n_+(\rho+\gamma) = n_-(-\rho+\gamma) = -2an_+n_{0+} - b(n_+n_{0-} + n_-n_{0+}) - 2cn_-n_{0-} \\ = an_+^2 + bn_+n_- + cn_-^2, \quad an_{0+}^2 + bn_{0-}n_{0+} + cn_{0-}^2 = 0 \end{gathered} \tag{2.28}$$

The frequency ρ, the wave number γ or the sound speed $s = \rho/\gamma$ are deduced from the $n_\pm$

$$s(n_- + n_+) = n_- - n_+, \quad \rho 2n_+n_- = (n_- - n_+)(an_+^2 + bn_+n_- + cn_-^2) \tag{2.29}$$

From the $n_{0\pm}$ alone relation and those mixing $n_\pm$ and $n_{0\pm}$ we find two possible classes

$$n_{0+} = \lambda_\eta n_{0-}, \quad n_{0+} + n_+ = \lambda_{-\eta}(n_{0-} + n_-), \quad 2a\lambda_\eta = -b + \eta\sqrt{b^2 - 4ac}, \quad \eta = \pm 1 \tag{2.30}$$

The positivity conditions (2.27) can be satisfied only if $\lambda_\pm > 0$ or only if $ac > 0$, $ab < 0$, $b^2 > 4ac$. For the Illner $ac \le 0$ models, $\lambda_\pm = 1$, c/a only $ac = 0$ (which includes the McKean model but excludes the Carleman one) can lead to $N_\pm > 0$. The solutions depend on two arbitrary parameters chosen among the four $n_{0\pm}, n_\pm$. If further we wish that the similarity waves can be interpreted as shock waves then the shock speed s must have $|s| < 1$.

We begin with the $ac \neq 0$, $ac > 0$, $ab < 0$ models (not of the Illner type) satisfying positivity. For the H-Theorem we directly check whether or not $\mathrm{Col}_H \le 0$, hence the signs of $N_+ - N_-$ and of ColN_+ must be opposite

$$\begin{gathered} (-1 + N_+/N_-)DN_- = (\lambda_{-\eta} - 1)(n_{0-} + n_-) + un_{0-}(\lambda_\eta - 1) \\ D^2 a\mathrm{Col}N_+ = -a^2u(n_+ - \lambda_- n_-)(n_+ - \lambda_+ n_-) = -a^2u(\lambda_+ - \lambda_-)^2(n_- + n_{0-})n_{0-} < 0 \end{gathered} \tag{2.31}$$

So ColN_+ having the $-a$ sign, $N_+ - N_-$ must have the a sign.

Class 1: If $0 < a < c$, $2\sqrt{ac} < -b \leq a + c$; $n_{0+} > 0$ arbitrary, $n_+ > 0$ arbitrary, and the choices $\lambda_- n_{0-} = n_{0+}$, $\lambda_+(n_{0-} + n_-) = n_{0+} + n_+$ then both $N_\pm > 0$ and $\mathrm{Col}_H \leq 0$. Further $\mid s \mid < 1$ can exist. First $\lambda_\pm$ are real, $0 < \lambda_- < \lambda_+$, leading to $n_{0-} > 0$, $n_{0-} + n_- > 0$ and $N_\pm > 0$. Second $\mathrm{Col}N_+ < 0$ it remains to show $N_+ > N_-$ or (see (2.31)) $\lambda_- \geq 1$ or $\sqrt{b^2 - 4ac} \leq -2a - b$. From $b^2 > 4ac > 4a^2$ we find $-b > 2a$ and $\lambda_- \geq 1$ becomes $a + b + c \geq 0$. Third for $\mid s \mid < 1$, from (2.30) with $n_+ > 0$ we must have $n_- > 0$ or $n_- \lambda_+ = n_+ + n_{0+}(1 - \lambda^+/\lambda^-) > 0$. It is an easy exercise to find another class.

Class 1bis: If $a < c < 0$, $2\sqrt{ac} < b \leq -a - c$; $n_{0+} > 0$ arbitrary, $n_+ > 0$ arbitrary and the choices $\lambda_+ n_{0-} = n_{0+}$, $\lambda_-(n_{0-} + n_-) = n_{0+} + n_+$ then both $N_\pm > 0$ and $\mathrm{Col}_H \leq 0$. Further $\mid s \mid < 1$ if $n_- \lambda_- = n_+ + n_{0+}(1 - \lambda_-/\lambda_+) > 0$. (For the proof notice that $0 < \lambda_+ < \lambda_- \leq 1$ and $\mid s \mid < 1$ if $n_- < 0$).

Now we look at the models $ac = 0$: either $a = 0$, $c \neq 0$ or $c = 0$, $a \neq 0$ and disregard the case $a = c = 0$, $b \neq 0$ which leads to the completely integrable Ruijgrok-Wu model. If $a = 0$, $b \neq 0$, then (2.28) for $n_{0\pm}$ alone leads to two classes of solutions depending whether $n_{0-} = 0$ or $bn_{0+} + cn_{0-} = 0$. Substituting into the (2.28) relation mixing $n_{0\pm}$, $n_\pm$ we reconstruct all parameters as fonctions of two arbitrary ones. Contrary to above, here the Illner models $a = 0$, $b = -c$ and $c = 0$, $b = -a$ are included.

Class 2: If $a = 0$, $c \geq -b > 0$, $n_{0-} > 0$ arbitrary, such that $n_+ + n_{0+} > 0$, $n_{0+} = -cn_{0-}/b > 0$, $n_- + n_{0-} = 0$, then $N_\pm \geq 0$, $\mathrm{Col}_H \leq 0$ and $\mid s \mid < 1$ can exist.
Due to $n_{0\pm} > 0$, $n_{0\pm} + n_\pm \geq 0$ then $N_\pm \geq 0$. Further we find

$$\begin{aligned} D^2 \mathrm{Col}N_+ &= un_{0-}b(n_+ + n_{0+}) < 0 \\ n_{0-}u(N_+/N_- - 1) &= n_{0+} + n_+ - n_{0-}(1 + c/b)u > 0 \end{aligned}$$

leading to $\mathrm{Col}_H \leq 0$. Due to $n_- < 0$, we get $\mid s \mid < 1$ if $n_+ < 0$.

Class 2bis: If $a = 0$, $c \geq -b > 0$, $n_{0+} > 0$ arbitrary, $n_- > 0$ arbitrary, $n_{0-} = 0$, $n_+ + n_{0+} = -cn_-/b > 0$, then $N_\pm \geq 0$, $\mathrm{Col}_H \leq 0$ and further $\mid s \mid < 1$ can exist.
$N_\pm \geq 0$ is trivial, for $\mathrm{Col}N_+ < 0$ and $N_+ \geq N_-$ we use

$$D^2 \mathrm{Col}N_+ = un_- n_{0+} b < 0, \quad -1 + N_+/N_- = -1 - c/b + un_{0+}/n_- > 0$$

We find $\mid s \mid < 1$ from $n_- > 0$, chosen such that $n_+ > 0$.

The models $c = 0$, $a \leq -b < 0$ lead to two other classes.

Class 3: If $c = 0$, $-a \geq b > 0$, $n_{0+} > 0$ arbitrary, n_- arbitrary such that $n_- + n_{0-} > 0$, $n_+ + n_{0+} = 0$, $n_{0-} = -an_0^+/b > 0$, then $N_\pm \geq 0$, $\mathrm{Col}_H \leq 0$, $\mid s \mid < 1$ can exist.

$$\begin{aligned} D^2 \mathrm{Col}N_+ &= bun_{0+}(n_- + n_{0-}) > 0, \\ D(N_+ - N_-) &= n_{0+}u(1 + a/b) - n_{0-} - n_- < 0 \end{aligned}$$

For $|s|<1$, due to $n_+<0$, we choose the arbitrary n_- such that $-n_{0+}<n_-<0$.

Class 3bis: If $c=0$, $-a\geq b>0$, $n_{0-}>0$ arbitrary, $n_+>0$ arbitrary, $n_{0+}=0$, $n_-+n_{0-}=-an_+/b>0$, then $N_\pm>0$, $\mathrm{Col}_H\leq 0$ and further $|s|<1$ can exist

$$D^2\mathrm{Col}N_+=bun_+n_{0-}>0,\quad D(N_+-N_-)=n_+(1+a/b)-n_{0-}u\leq 0$$

For $|s|<1$ we choose the arbitrary parameters such that $n_-=-an_+/\,b-n_{0-}>0$.

For the Classes 2, 2bis, 3, 3bis we note that when $|x|\longrightarrow\infty$, one of the shock limits $n_{0\pm}$ or $n_{0\pm}+n_\pm$ is always zero, leading to the interpretation of an infinite Mach schock.

Coming back to the Illner model (2.26) (only $ac=0$ leads to $N_\pm\geq 0$), adding linear terms so that the collision term (2.24) becomes:

$$\mathrm{Col}N_+=(N_--N_+)(cN_--aN_++d)\ ,\quad a\leq 0,\ \ c\geq 0,\ \ d>0,\ \ \mathrm{Col}H\leq 0,\ \ \forall N_\pm>0\ ,$$

we check the positivity for (2.27). In $\mathrm{Col}N_+$, again the constant term is zero while the coefficients of D^{-1}, D^{-2} are opposite. The first, $n_{0\pm}$ alone, relation $(n_{0-}-n_{0+})\,(cn_{0-}-an_{0+}+d)=0$ having a positive second factor gives $n_{0-}=n_{0+}$. The second relation, mixing $n_{0\pm}$ and $n_\pm$, being $(n_--n_+)\,(c(n_{0-}+n_-)-a(n_{0+}+n_+)+d)=0$, with a positive second factor leads to $n_-=n_+$. Hence the similarity solution being $N_+=N_-$, with a vanishing collision term, is impossible.

2.2.b. Similarity Solutions for models with more than $2\vec{V}_i$

The following (1+1)-dimensional equations for three independent densities N_i :

$$N_{1t}+N_{1x}=N_{2t}-N_{2x}=-bN_{3t}=\mathrm{Col}N_1=\tau(N_3^2-N_1N_2),\quad b=1/2,1,2 \qquad (2.32)$$

correspond to different models: (i) $3\vec{V}_i$ model with $\vec{V}_1$ lying along the positive $x=x_1$ axis, $\vec{V}_2+\vec{V}_1=0$, $\vec{V}_3=0$, and $b=\frac{1}{2}$, $\tau=1$; (ii) $4\vec{V}_i$ model with $\vec{V}_1,\vec{V}_3$ lying along the $x=x_1$ and x_2 axes, $\vec{V}_1+\vec{V}_2=\vec{V}_3+\vec{V}_4=0$, $N_4=N_3$ and $b=1$, $\tau=1$; (iii) $6\vec{V}_i$ Broadwell model with $\vec{V}_1,\vec{V}_3,\vec{V}_5$ lying along the positive $x=x_1$ and x_2,x_3 axes, $\vec{V}_1+\vec{V}_2=\vec{V}_3+\vec{V}_4=\vec{V}_5+\vec{V}_6=0$, $N_6=N_5=N_4=N_3$ and $b=2$, $\tau=2$.
Both the $H-$Theorem $H_t+\partial_x(...)=(\log N_1N_2/N_3^2)\ \mathrm{Col}N_1\leq 0$, the mass $M=\Sigma N_i$ and momentum $J=N_1-N_2$ conservation laws $M_t+J_x=0$, $J_t+(N_1+N_2)_x=0$ are satisfied. Rescaling the N_i we choose $\tau=1$.

We want to construct positive similarity waves.

$$N_i=n_{0i}+n_i/D,\qquad D=1+e^{pt+\gamma x}\qquad \text{with}\quad n_{0i}\geq 0,\ \ \Sigma_i=n_{0i}+n_i\geq 0 \qquad (2.33)$$

For the relations to be satisfied by the parameters, we write down that in (2.32) the constant term is zero while the coefficients of D^{-1}, D^{-2} are opposite

$$n_{03} = \sqrt{n_{01}n_{02}}$$
$$n_1(\rho+\gamma) = n_2(\rho-\gamma) = -bn_3\rho = n_3^2 - n_1n_2 = -2n_{03}n_3 + n_{01}n_{02} + n_{02}n_1 \tag{2.34a}$$

There exist three arbitrary parameters chosen to be $n_{0i} \geq 0$, $i = 1,2$, $y = n_2/n_1 < 0$, from which we easily construct the others: we define $\tilde{y} = 1 + y^2 + 2y(1 - 2/b^2) > 0$ for $y < 0$, $b \in [1/2, 2]$ and find

$$n_1 = -(1+y)(4n_{03}y/b + (1+y))(n_{01}y + n_{02}))/yy^2, \quad n_2 = n_1y, \quad n_3 = -2yn_1/b(1+y)$$
$$2\rho = -n_1\tilde{y}/(1+y), \quad s = -\rho/\gamma = (1+y)/(1-y), \quad |\, s \,| < 1 \quad \text{for} \quad y < 0 \tag{2.34b}$$

For a study of the positivity we write down Σ_i in terms of the arbitrary parameters

$$\Sigma_1 = -[2y\sqrt{n_{01}}/b + (1+y)\sqrt{n_{02}}]^2/y\tilde{y} \geq 0, \quad \Sigma_2 = -y[\sqrt{n_{01}}(1+y) + 2\sqrt{n_{02}}/b]^2/\tilde{y} \geq 0$$
$$\Sigma_3 = [2y\sqrt{n_{01}}/b + (1+y)\sqrt{n_{02}}][\sqrt{n_{01}}(1+y) + 2\sqrt{n_{02}}/b]/\tilde{y} \tag{2.35}$$

and obtain two classes of physically acceptable shock waves.

Class 1: If $n_{0i} > 0$, $i = 1,2$ and either $0 > y > -b\sqrt{n_{02}}/(b\sqrt{n_{02}} + 2\sqrt{n_{01}})$ or $y < -1-(2/b)$ $\sqrt{n_{02}}/n_{01}$ then $N_i \geq 0$.

Class 2: If either $n_{01} = n_{03} = 0$, $n_{02} > 0$, $-1 < y < 0$ or $n_{02} = n_{03} = 0$, $n_{01} > 0$, $y < -1$ then $N_i \leq 0$. We notice that $|\, s \,| < 1$ and there exist solutions with one shock limit equal to zero, like in an infinite Mach shock.

2.2.c. Similarity Solutions for the hexagonal model

We consider the hexagonal model with $\vec{V}_0$ along the positive x axis, $\vec{V}_1$ with an angle of $\pi/3$ and $\vec{V}_2$ an angle of $2\pi/3$ while $\vec{V}_0 + \vec{V}_3 = \vec{V}_1 + \vec{V}_4 = \vec{V}_2 + \vec{V}_5 = 0$. In one spatial dimension, due to $N_5 = N_1$, $N_4 = N_2$, we only have four independent densities N_i which satisfy for binary collisions alone:

$$\ell_0 N_0 = \ell_3 N_3 = -2\ell_1 N_1 = -2\ell_2 N_2 = 2\text{Col}_B = 2(N_1N_2 - N_0N_3)$$
$$\ell_0 = \partial_t + \partial_x, \quad \ell_3 = \partial_t - \partial_x, \quad \ell_1 = \partial_t + \partial_x/2, \quad \ell_2 = \partial_t - \partial_x/2 \tag{2.36}$$

Positive similarity waves of the type (2.33) have been constructed. Now we introduce ternary collisions for which two velocities are opposite. The collision term becomes $\text{Col}_T = (1 + \tau M)\ \text{Col}_B$ with $M = \Sigma N_i$ being the total mass. The similarity waves have square-root branch points and a class of positive solutions satisfying the $H-$Theorem has been obtained. However for Col_B as well as for Col_T exist three linear relations and only two conservation laws. In order to eliminate this spurious conservation law we introduce another ternary collisions term $\text{Col}_{T'}$ with $2\pi/3$ angles among the incoming and among the outgoing velocities. The equations become:

$$\ell_0 N_0 + \ell_3 N_3 = 2\ \text{Col}_T, \qquad \ell_0 N_0 - \ell_3 N_3 = 2\mu\ \text{Col}_{T'}$$
$$\ell_3 N_3 - 4\ell_1 N_1 = 3\ell_0 N_0, \qquad \ell_0 N_0 - 4\ell_2 N_2 = 3\ell_3 N_3$$
$$\text{Col}_T = (1 + \tau M)(N_1N_2 - N_0N_3), \qquad \text{Col}_T = N_3N_1^2 - N_0N_2^2 \tag{2.37}$$

We obtain that the $H-$Theorem:

$$H_t + \partial_x(...) = \mathrm{Col}_T \log(N_1N_2/N_0N_3) + \mu \mathrm{Col}_{T'} \log(N_3N_1^2/N_0N_2^2)$$

is satisfied if

$$1 + \tau M \geq 0, \qquad \mu > 0 \tag{2.38}$$

The similarity waves have square root branch points

$$\begin{array}{llll} N_i = n_{0i} + n_i/\sqrt{D}, & M = m_0 + m/\sqrt{D}, & D = 1 + e^{\gamma(x+st)}, & n_{0i} + n_i \geq 0 \\ n_{0i} \geq 0, & m_0 = n_{00} + n_{03} + 2(n_{01} + n_{02}) & m = n_0 + n_3 + 2(n_1 + n_2) & \end{array} \tag{2.39}$$

We define scaled parameters $\bar{n}_i = n_i/n_0$ for $i \neq 0$ and assume:

$$n_0 = -\varphi, \quad \varphi = \pm 1, \quad n_{00} = 1, \quad n_{01} = \bar{n}_1, \quad n_{02} = \bar{n}_1^3, \quad n_{03} = \bar{n}_1^4 \tag{2.40}$$

Our first goal is to show that all parameters are functions of $\bar{n}_1$. In the first two nonlinear (2.37) relations the ℓ.h.s. have $D^{-1/2}$, $D^{-3/2}$ terms but not D^{-1}. Hence in Col_T, $\mathrm{Col}_{T'}$, the D^{-1} terms are zero

$$\tau(\varphi m - m_0) = 1, \quad \bar{n}_1^6 - \bar{n}_2^2 = 2\bar{n}_1^2(\bar{n}_1\bar{n}_2 - \bar{n}_3) \tag{2.41}$$

giving for the nonlinear (2.37) relations

$$\begin{array}{c} \gamma n_0(1 + s + \bar{n}_3(s-1)) - 4\tau m(\bar{n}_1\bar{n}_2 - \bar{n}_3) = 0 \\ \gamma n_0(1 + s - \bar{n}_3(s-1)) - 4\mu(\bar{n}_3\bar{n}_1^2 - \bar{n}_2^2) = 0 \end{array} \tag{2.42}$$

The two linear (2.37) relations give two other relations

$$s(\bar{n}_3 - 3 - 4\bar{n}_1) = (\bar{n}_3 + 3 + 2\bar{n}_1), \bar{n}_3(8 + 9\bar{n}_1 + \bar{n}_2) + 8\bar{n}_1\bar{n}_2 + 9\bar{n}_2 + \bar{n}_1 = 0 \tag{2.43}$$

We eliminate $\bar{n}_3$ between the (2.41-43) relations and find a cubic relation $f(\bar{n}_2, \bar{n}_1) = 0$ for $\bar{n}_2$ with $\bar{n}_1-$dependent coefficients

$$f = \bar{n}_2^3 + (\bar{n}_2^2 - \bar{n}_1^6)(9\bar{n}_1 + 8) + (\bar{n}_2^2 + 1)2\bar{n}_1^3 + \bar{n}_2(18(\bar{n}_1^2 + \bar{n}_1^4) + 32\bar{n}_1^2 - \bar{n}_1^6) \tag{2.44}$$

From arbitrary $\bar{n}_1$ values, the algebraic construction of the solution can be performed. First (2.40) gives n_{0i}, $i = 1, 2, 3$. Second $\bar{n}_2$ and $\bar{n}_3, s$ are found from (2.44) and the (2.43) relations. Third we obtain $n_i = n_0\bar{n}_i$, m, m_0 are deduced from (2.39) and τ with (2.41). Finally (2.42) gives both μ and γ.

If we choose $\bar{n}_1 > 0$ then the positivity constraints $n_{0i} > 0$, $m_0 > 0$ are satisfied. It remains to find an $\bar{n}_i$ interval for which the other positivity constraints $n_{0i} + n_i \geq 0$,

$1+\tau M \geq 0$, $\mu > 0$ are satisfied. Let us assume that we have proved:

$$|\bar{n}_2| < \bar{n}_1^3, \quad |\bar{n}_3| < \bar{n}_1^4, \quad \varphi m < 0, \quad \bar{n}_2 < 0, \quad \bar{n}_3 < 0 \tag{2.45}$$

First due to $n_{01}+n_1 = \bar{n}_1(1-\varphi) \geq 0$, $n_{02}+n_2 = \bar{n}_1^3 - \varphi\bar{n}_2$, $n_{03}+n_3 = \bar{n}_1^4 - \varphi\bar{n}_3$ we see that the positivity constraints $n_{0i}+n_i \geq 0$, $m_0+m > 0$ or $N_i \geq 0$, $M \geq 0$ are satisfied. Second we find $\tau = 1/(\varphi m - m_0) \leq 0$, $1+\tau m_0 = \tau\varphi m > 0$, $\tau < 0$, $1+\tau(m+m_0) = \tau m(1+\varphi) \geq 0$ and the first condition $1+\tau M \geq 0$ for the H-Theorem is satisfied. We rewrite μ

$$\mu = \tau m\varphi(\bar{n}_1\bar{n}_2 - \bar{n}_3)/(\bar{n}_2^2 - \bar{n}_3\bar{n}_1^2)$$

Due to the assumption $|\bar{n}_2| < \bar{n}_1^3$ it follows that the r.h.s. in the (2.41) relation is positive, $\bar{n}_1\bar{n}_2 - \bar{n}_3 > 0$ and μ has the $\tau m\varphi > 0$ sign. The second $\mu > 0$ condition for the $H-$theorem is satisfied. It remains to prove the assumptions (2.45) for an $\bar{n}_1$ interval. If

$$1/3 < \bar{n}_1 \leq 1/2 \tag{2.46}$$

we prove that (2.45) follows. First we remark that the roots of the cubic (2.44) equation are negative. From $f(0,\bar{n}_1) > 0$, $f(\bar{n}_2 = -8, \bar{n}_1) > 0$, $f(\bar{n}_2 = -\bar{n}_1^3, \bar{n}_1) < 0$ it follows that one $\bar{n}_2$ root exists between $-\bar{n}_1^3$ and 0 and we choose this root for the solution. From the second (2.41) relation we find $\bar{n}_3 < 0$ and $\bar{n}_3 + \bar{n}_1^4 = [(\bar{n}_1^3 + \bar{n}_2)/\bar{n}_1]^2/2 > 0$.

In conclusion when ternary collisions are included we still find physically relevant similarity solutions. In the sequel we always assume binary collisions.

References

Kolodner I., Ann. Math. Pure Appl. 63, 11, 1963, Broadwell J.E., Phys. Fluids 7, 1243, 1964; Wick J. Math. Meth. Appl. Sci., 6, 515, 1984; Platkowski T., J. Méc. Théor. Appl. 4, 555, 1985; Dukek G. and Nonnenmacher T.F., Physica A135, 167, 1986; Cornille H., J.M.P. 28, 1567, 1987, 29, 1667, 1988, J. Phys. A20, 1973, 1987, J. Stat. Phys. 48, 789, 1987, "Lecture Notes in Mathematics" Springer-Verlag Ed. Toscani G-1989, Let. Mat. Phys. 1989; Zanette D., Physica A153, 612, 1988; "Discrete Kinetic Theory, Lattice Gas,..." Ed. Monaco R. World Scientific, Singapore Publishing 1989, Boffi V., Dukek G. and Spiga G., p.315, Platkowski T., p.248.

2.3. Simple examples of (1+1)-dimensional solutions

We again choose the simple (1+1)-dimensional models with three independent densities N_i : (i) $3\vec{V}_i$ with $b = \frac{1}{2}$, (ii) $4\vec{V}_i$ with $b = 1$ and $N_4 = N_3$, (iii) $6\vec{V}_i$ Broadwell model with $b = 2$ and $N_6 = N_5 = N_4 = N_3$

$$N_{1t} + N_{1x} = N_{2t} - N_{2x} = -bN_{3t} = N_3^2 - N_1N_2 \tag{2.47}$$

2.3.a. Shock waves

We start with an ansatz which is a sum of similarity waves

$$N_i = n_{0i} + \sum_{j=1}^{J} n_{ji}/D_j, \quad D_j = 1 + u_j, \quad u_j = d_j e_j^{\gamma} x + \rho_j t, \quad d_j > 0 \tag{2.48}$$

that we substitute into (2.47). We first obtain the relations for each similarity wave $n_{03} = \eta\sqrt{n_{01}n_{02}}$, $\eta^2 = 1$, choose $\eta > 0$ and

$$n_{j1}(\rho_j+\gamma_j) = n_{j2}(\rho_j-\gamma_j) = -bn_{j3}\rho_j = n_{j3}^2 - n_{j1}n_{j2} = -2n_{03}n_{j3} + n_{01}n_{j2} + n_{02}n_{j1} \tag{2.49}$$

Second it is clear that we cannot sum an arbitrary number of similarity waves. In order that the sum in (2.48) be a solution then other constraints arise from the vanishing of $D_p^{-1}D_q^{-1}$, $p \neq q$.

$$2n_{p3}n_{q3} = n_{p1}n_{q2} + n_{p2}n_{q1} \tag{2.50}$$

What is the maximal number of components ? For each component we introduce scaled parameters $y_j = n_{j2}/n_{j1}$, $\bar{n}_{3j} = n_{3j}/n_{j1}$ and obtain from (2.49-50)

$$\begin{aligned} &\bar{n}_{j3} = -2y_j/b(1+y_j), \quad 2\bar{n}_{p3}\bar{n}_{q3} = y_p + y_q \\ &f(y_p, y_q) = y_p^2 + B_q y_p + y_q = 0, \quad B_q(1+y_q) = 1 + y_q^2 + 2y_q(1-4/b^2) \end{aligned} \tag{2.51}$$

Can we have a superposition of three similarity waves with y_1, y_2, y_3 different satisfying $f = 0$? From $f(y_p, y_q) = f(y_p, y_r) = 0$ we deduce that

$$y_p(y_p + y_q + y_r) + 1 + y_p + y_r + 2y_p(1 - 4/b^2) = 0 \tag{2.52}$$

must hold for the three y_1, y_2, y_3 or $y_1 + y_2 + y_3 = -1 + 8/b^2$. Substituting into (2.52) we find $y_1 + y_2 + y_3 = -1$ or $1/b^2 = 0$ which is impossible.

In conclusion we cannot have more than two similarity waves and so $J = 2$ in (2.48). There exist thirteen parameters, ten relations and three arbitrary parameters chosen to be

$$n_{0i} \geq 0 \qquad i = 1,2 \qquad -1 < y_1 < 0 \tag{2.53}$$

Only the y_2 sign is unknown. From the quadratic $f(y_2, y_1) = 0$ equation (2.51) we see that the two roots $y_2^{\pm}$ are real with $y_2^+ y_2^- = y_1 < 0$. We choose the negative root y_2^- :

$$2y_2 = -B_1 - \sqrt{B_1^2 - 4y_1} < 0, \qquad B_1(1 + y_1) = 1 + y_1^2 + 2y_1(1 - 4/b^2)$$

To prove that $y_2 < -1$ is equivalent to $\sqrt{B_1^2 - 4y_1} > 2 - B_1$ or to $(B_1 - 1 - y_1)(1 + y_1) \equiv -8y_1/b^2 > 0$ which is true. We define $\tilde{y}_j = 1 + y_j^2 + 2y_j(1 - 2/b^2) > 0$ for $y_j < 0$ and can apply the formalism of subsection 2b

$$\begin{array}{ll} n_{j1} = -(1 + y_j)(4n_{03}y_j/b + (1 + y_j)(n_{01}y_j + n_{02}))/y_j\tilde{y}_j, & n_{ji} = \bar{n}_{ji}n_{j1} \\ 2\rho_j = -n_{j1}\tilde{y}_j/(1 + y_j), \quad s_j = \rho_j/\gamma_j = (1 + y_j)/(y_j - 1), & \mid s \mid < 1 \text{ for } y_j < 0 \end{array} \tag{2.54}$$

With the change $y \longrightarrow y_j$, $j = 1,2$ the expressions Σ_i written down in (2.35) become $\Sigma_{ji} = n_{0i} + n_{ji}$. We have $\Sigma_{j1} \geq 0$, $\Sigma_{j2} \geq 0$, and for the last one $\Sigma_{j3} \geq 0$ if

$$X_2 = -2/b(1 + y_2) < \sqrt{\frac{n_{01}}{n_{02}}} < -b(1 + y_1)/2y_1 = X_1 \tag{2.55}$$

We must verify that $X_2 < X_1$ or $y_1 > b^2(1 + y_1)(1 + y_2)/4 = 2y_1y_2/(y_1 + y_2)$ or $y_2 < y_1$ which is true for $-1 < y_1 < 0$, $y_2 < -1$.

The first difference with the similarity case of section 2b is that Σ_{ji} are not necessarily the shock limits when $\mid x \mid \longrightarrow \infty$. This is true if $\gamma_1\gamma_2 < 0$. On the contrary if $\gamma_1\gamma_2 > 0$ the shock limits are n_{0i} and $n_{0i} + n_{1i} + n_{2i}$. We must prove that in the (X_1, X_2) interval, there exists a sub-interval for which $\gamma_1\gamma_2 < 0$. We write $\gamma_1\gamma_2$:

$$4\gamma_1\gamma_2 = \prod_j (y_j - 1)n_{02}^2 X_1 X_2, \qquad X_j = n_{01}/n_{02} + 4\sqrt{\frac{n_{01}}{n_{02}}}/b(1 + y_j) + 1/y_j, \quad j = 1,2,$$

and $\gamma_1\gamma_2 < 0$ if $X_1X_2 < 0$. Let us consider the real roots $X_j^{\pm}$, $X_j^+ > X_j^-$ of $X_j = 0$

$$X_2^{\pm} = -2(1 \pm \sqrt{C_2})/b(1+y_2), \quad X_1^{\pm} = -2(1 \mp \sqrt{C_1})/b(1+y_1), \quad C_i = 1 - b^2(1+y_i)^2/4y_i$$

with $C_i > 0$ and $X_i^+ > 0$. Let us assume that the following inequalities hold:

$$X_1^+ < X_2 < X_1 < X_2^+ \tag{2.56}$$

then for $\sqrt{n_{01}/n_{02}} \in [X_1, X_2]$ we have both $\gamma_1\gamma_2 < 0$ and $\Sigma_{ji} \leq 0$. The first assumption $X_1 < X_2^+$ is equivalent to $b^2(1+y_1)(1+y_2)/4y_1 < 1+\sqrt{C_2}$ or, using $(1+y_1)(1+y_2)(y_1+y_2) = 8y_1y_2/b^2$, equivalent to $1 - 2y_1/(y_1 + y_2) < \sqrt{C_2}$ or $4y_1y_2 > [b(1 + y_2)(y_1 + y_2)/2]^2/y_2$, which holds because the r.h.s. is positive while the ℓ.h.s. is negative.

The second assumption $X_1^+ < X_2$ is equivalent to $\sqrt{C_1} < 1 - (1 + y_1)/(1 + y_2)$ or $(1 + y_1)(1 + y_2)(b^2/4y_1 + 1/(1 + y_2)^2) < 2$ or $1 + y_2^2 + y_2(2 - 4/b^2) > 0$ which is true. In conclusion if the three arbitrary parameters satisfy $-1 < y_1 < 0$, $X_1 < \sqrt{n_{01}/n_{02}} < X_2$, choosing $n_{03} = \sqrt{n_{01}n_{02}}$ and the $y_2 < -1$ root then both $\gamma_1\gamma_2 < 0$ and the shock limits $\Sigma_{ji} \leq 0$.

The second difference with the similarity case it that non-negative shock limits are not sufficient to guarantee the positivity for all x values. We must add constraints on the $d_j > 0$ arbitrary parameters in $D_j = 1 + u_j$ and sometimes this is not sufficient. Positivity can be discussed at $t = 0$ for $u_j = d_j e^{\gamma_j x}$ because the Boltzmann models equations carry out positivity for $t > 0$.

More generally (omitting the subscript i) let us discuss the positivity of $N = n_0 + \Sigma n_j / D_j$ when $\gamma_1 \gamma_2 < 0$ and the two shock limits are $n_0 + n_j \geq 0, j = 1, 2$

$$N D_1 D_2 = \Sigma(n_0 + n_j)u_i + n_0 + \Sigma n_j + n_0 u_1 u_2, \qquad i \neq j \tag{2.57}$$

(i) If $n_0 + n_j > 0$, $j = 1, 2$ then $N > 0$ is possible. In that case we cannot have both $n_0 < 0$ and $n_0 + n_1 + n_2 < 0$. Firstly if $n_0 < 0$ but $n_0 + n_1 + n_2 > 0$, and $\gamma_k > 0$, $\gamma_\ell < 0$ let us compare $n_0 u_k u_\ell < 0$ and $(n_0 + n_\ell)u_k > 0$ for $x \geq 0$, $(n_0 + n_k)u_\ell > 0$ for $x \leq 0$. We find

$$N > 0 \quad \text{if} \quad d_j > -(n_0 + n_j)/n_0, \quad j = 1, 2$$

Secondly if $n_0 > 0$ but $n_0 + n_1 + n_2 < 0$, $\gamma_k > 0$, $\gamma_\ell < 0$, we compare $n_0 + n_1 + n_2 < 0$ with $(n_0 + n_\ell)\, u_k > 0$ for $x \geq 0$ and $(n_0 + n_k)u_\ell > 0$ for $x \leq 0$. We find

$$N > 0 \quad \text{if} \quad d_j > -(n_0 + n_1 + n_2)/(n_0 + n_i), \quad i \neq j$$

(ii) If both $n_0 + n_j = 0$ then, except if $\gamma_1 + \gamma_2 = 0$, the positivity is violated. We find $N D_1 D_2 = n_0(u_1 u_2 - 1)$ with two different signs when $|\, x \,| \longrightarrow \infty$ if $\gamma_1 + \gamma_2 \neq 0$. On the contrary if $\gamma_1 + \gamma_2 = 0$, then $N > 0$ either if $n_0 > 0$, $d_1 d_2 > 1$ or $n_0 < 0$, $d_1 d_2 < 1$.

(iii) If $n_0 + n_2 = 0$, $n_0 + n_1 > 0$, we cannot always concluded with these informations alone. We have $N D_1 D_2 = (n_0 + n_1)u_2 + n_1 + n_0 u_1 u_2$ and both n_0, n_1 cannot be negative. If $n_1 < 0$, $n_0 > 0$ and $\gamma_2(\gamma_1 + \gamma_2) > 0$, then always for one of the two $|\, x \,| \longrightarrow \infty$ limits, we have $N D_1 D_2 \longrightarrow n_1 < 0$ and positively is violated. On the contrary if $\gamma_2 < 0$, $\gamma_1 + \gamma_2 > 0$, in these limits n_1 is dominated by a positive term and we find

$$N > 0 \quad \text{if} \quad d_2 > -n_1/(n_0 + n_1), \qquad d_1 d_2 > -n_1/n_0 \qquad \text{and} \qquad \gamma_2(\gamma_1 + \gamma_2) < 0$$

Similarly if $n_1 > 0$, $n_0 < 0$ and $\gamma_2(\gamma_1 + \gamma_2) < 0$ then one of the two asymptotic limits is negative while if $\gamma_2(\gamma_1 + \gamma_2) > 0$ we can control positivity with the d_j.As a result of this discussion we must be careful when we seek infinite Mach shock with one shock limit equal to zero. However we can always define an "almost infinite" shock with a limit $n_0 + n_j = \varepsilon$, ε arbitrarily small but fixed, and apply the (i) results.

In conclusion for all $3\vec{V}_i$, $4\vec{V}_i$ and Broadwell models, a class of physically acceptable (1+1)-dimensional shock waves exist.

As an illustration we choose the $b = 1/2$, $3\vec{V}_i$ model with $y_1 = -0.1$, $n_{01} = 1$, $n_{02} = 0.616$ for the arbitrary parameters and deduce: $y_2 = -4.47$, $u_{03} = 0.785$, $\gamma_1 = -0.998$, $\gamma_2 = 2.58$, $\rho_1 = 0.82$, $\rho_2 = 1.6$, $s_1 = -0.81$, $s_2 = 0.63$, $| s_i |< 1$; $u_{1i} = -0.61, 0.061, -0.27$, $u_{2i} = 0.13, -0.61, -0.7$, $i = 1,2,3$. We can verify that the Σ_{ji} are strictly positive.

2.3.b. Periodic waves

We again consider the three (2.47) models $b = \frac{1}{2}, 1, 2$. Our ansatz is a sum of complex conjugate similarity waves

$$N_i = n_{0i} + 2\mathrm{Re}\ n_i/D, \qquad D = 1 + de^{\rho t + \gamma x}, \qquad \rho = \rho_R + i\rho_I, \quad \gamma = i\gamma_I, \qquad d = d_R + id_I$$

which are periodic in x but either propagating $\rho_I \neq 0$ or nonpropagating $\rho_I = 0$ with t. We get from the const. D^{-1}, $| D |^{-2}$ relations

$$n_{03} = \varphi\sqrt{n_{01}n_{02}}, \qquad \varphi = \pm 1, \qquad 2 | n_3 |^2 = n_1 n_2^* + n_2 n_1^* \tag{2.58}$$

$$n_1(\rho + \gamma) = n_2(\rho - \gamma) = -bn_3\rho = n_3^2 - n_1 n_2 = -2n_{03}n_3 + n_{01}n_2 + n_{02}n_1 \tag{2.59a}$$

we rewrite (2.59a): $\gamma(n_2 + n_1) = \rho(n_2 - n_1)$ and

$$n_3 = -2n_1n_2/b(n_1+n_2) = -b\rho + b(n_1+n_2)/2, \ n_3(b\rho - 2n_{03}) + n_{01}n_2 + n_{02}n_1 = 0 \tag{2.59b}$$

We show that only the $b = 2$ Broadwell model can have nonpropagating $\rho_I = 0$ waves. From $\rho = \rho_R$, $\gamma = i\gamma_I$, (2.59a-b) and the complex conjugate relations we find

$$n_3/n_3^* = n_1/n_2^* = n_2/n_1^* = (2\rho + n_1 + n_2)/(2\rho + n_1^* + n_2^*) \tag{2.60}$$

Hence $n_2 = n_1^*$, n_3 is real and $n_{02} = n_{01}$, $n_{03} = \varphi n_{01}$ in the last (2.59b) relation. Writing $n_i = n_{iR} + in_{iI}$, the second (2.58) relation becomes $n_3^2 = n_{1R}^2 - n_{1I}^2$, $bn_{1R}n_3 = - | n_1 |^2$ or $n_{1I}^4 + n_{1R}^2 n_{1I}^2\,(b^2+2) + n_{1R}^4(1-b^2) = 0$ and $n_{1I}^2/n_{1R}^2 > 0$ cannot exist for $b \leq 1$. Only $b = 2$ is possible with $n_{1I}^2 = (-3+2\sqrt{3})n_{1R}^2$. We obtain the solutions with $\varphi = \pm 1$, $n_2 = n_1^*$ and $n_{01} = n_{02} = \varphi n_{03}$ as the arbitrary parameter

$$\begin{aligned} &n_{1R} = -n_{01}(2+\sqrt{3})(1+\varphi(\sqrt{3}-1))/\sqrt{3}, \quad n_{1I} = -n_{01}(1+\varphi(\sqrt{3}-1))\sqrt{1+2/\sqrt{3}}, \\ &n_3 = n_{1R}(1-\sqrt{3}), \quad \rho_R = n_{01}(\varphi + (1+\sqrt{3})/2), \quad \gamma_I = -\rho_R\sqrt{-3+2\sqrt{3}} \end{aligned} \tag{2.61}$$

Two classes of positive solutions correspond to $n_{01}\varphi > 0$:

(i) If $n_{01} > 0$ and $\varphi = 1$ then $\rho_R > 0$ and $N_i \longrightarrow n_{0i} = n_{01} > 0$ when $t \longrightarrow \infty$. From $| D |>| 1 - | d | e^{\rho_R t} |$, choosing $| d |$ sufficiently large, we see that the contribution of the periodic term becomes arbitrarily small compared to $n_{01} > 0$ and $N_i > 0$.

(ii) If $n_{01} < 0$ and $\varphi = -1$ then $\rho_R < 0$ and $N_i \longrightarrow n_{0i} + 2n_{iR} = n_{01}(1 - 2/\sqrt{3}) > 0$ when $t \longrightarrow \infty$. From $| D |>| 1- | d | e^{\rho_R t} |$, choosing $| d |$ arbitrarily small, we see that the contribution of the periodic term becomes arbitrarily close to its asymptotic ($t \longrightarrow \infty$) value $2n_{iR}$ and $N_i > 0$.

We go on with the more physically interesting propagating $\rho_I \neq 0$ *waves.* We introduce the parameter $y = n_2/n_1 =| y | e^{i\theta}$ and intermediate parameters: $\bar{n}_3 = n_3/n_1$, $\bar{\rho} = \rho/n_1$, $\bar{\gamma} = \gamma/n_1$. First we substitute $\bar{n}_3 = -2y/b(1+y)$ into the second (2.58) relation

$$4\cos\theta = -\Lambda + \sqrt{\Lambda^2 + 32/b^2} > 0, \qquad \Lambda =| y | + | y |^{-1} \geq 2 \tag{2.62}$$

The condition $\cos\theta < 1$ satisfied for $b = 2$ gives $\Lambda > 2$ for $b = 1$ and $\Lambda > 14$ for $b = 1/2$, which confirms that in the $\rho_I = 0$ case, $| y |= 1$, $\Lambda = 2$, only $b = 2$ remains. We still have the two classes $n_{03} = \varphi\sqrt{n_{01}n_{02}}$ with $\varphi = \pm 1$. Second we determine the intermediate parameters $2\bar{n}_{3I} = -b\cos\theta\sin\theta$, $\bar{n}_{3R}/\bar{n}_{3I} = -\bar{\gamma}_I/\bar{\gamma}_R = (\cos\theta + | y |)/\sin\theta$, $\bar{n}_{3R} < 0$, $2\bar{\rho}_R = -\sin^2\theta < 0$, $2\bar{\rho}_I = \sin\theta(\cos\theta - | y |)$. From $\gamma_R = 0$ we deduce $\bar{\gamma}_I/\bar{\gamma}_R = n_{1R}/n_{1I} = -\bar{n}_{3R}/\bar{n}_{3I}$ or $n_{3I} = 0$ and $n_3 = n_{3R}$ is real like in the above case but $n_2 \neq n_1^*$. We deduce some useful signs relations

$$\begin{aligned} &n_3/n_{1R} = (\bar{n}_{3R}^2 + \bar{n}_{3I}^2)/\bar{n}_{3R} < 0, \quad \rho_R/n_{1R} = -| y | \sin^2\theta/(| y | + \cos\theta) < 0 \\ &n_3 b = (2n_{03}\bar{n}_{3R} - n_{01} | y | \cos\theta - n_{02})/\bar{\rho}_R > 0 \quad \text{if} \quad n_{0i} > 0 \end{aligned} \tag{2.63}$$

For instance if we construct solutions with $n_{0i} > 0$, then $n_3 > 0$, $n_{1R} < 0$, $\rho_R > 0$ and, like in the above (i) class for $\rho_I = 0$, these solutions are positive. For the determination of the n_{0i} we start with $n_{02} + n_{01}y - 2n_{03}\bar{n}_3 + bn_3\bar{\rho} = 0$ whose real (last (2.63) relation) and imaginary parts allow us to eliminate n_3 :

$$n_{02}A + n_{01} = 2n_{03}B, \quad A = (\cos\theta - | y |)/ | y | (1 - | y | \cos\theta), \quad A + 2B/b + 1 = 0 \tag{2.64a}$$

Choosing $\varphi = 1$ or $n_{03} = \sqrt{n_{02}n_{01}}$ then from the quadratic $\sqrt{n_{01}/n_{02}}$ equation

$$n_{01}/n_{02} - 2B\sqrt{n_{01}/n_{02}} + A = 0, \qquad \sqrt{n_{01}/n_{02}} = B \pm \sqrt{B^2 - A}\,. \tag{2.64b}$$

We choose $n_{0i} = 1$, $| y |$ arbitrary and we can reconstruct all other parameters: n_{02} from (2.64b) and so n_{03}, then n_3, n_{1R}, ρ_R, n_{1I} from (2.63), θ and y from (2.62). Finally $n_2 = yn_1$, $\rho = n_1\bar{\rho}$, $\gamma = n_1\bar{\gamma}$. If we choose $\varphi = -1$, we still obtain a quadratic equation for $\sqrt{n_{01}/n_{02}}$ and from n_{01} and $| y |$ construct another class of solutions.

We prove that for the $b = 2$ Broadwell model, there exist $N_i > 0$ for all arbitrary $| y |$ values, $n_{01} > 0$ and the choice $\eta = 1$ or $n_{03} = \sqrt{n_{01}n_{02}} > 0$. As stated above if the n_{0i} are positive then $\rho_R > 0$ and $N_i > 0$ exist. It is sufficient that in (2.64b) positive $\sqrt{n_{01}/n_{02}}$ roots exist. We notice that $\cos\theta <| y |$ is equivalent to $\sqrt{\Lambda^2 + 8} < 4 | y | + \Lambda$ or $| y |^2 > 0$. Similarly $\cos\theta <| y |^{-1}$ is equivalent to $\sqrt{\Lambda^2 + 8} < \Lambda + 4/ | y |$ or $1/ | y |^2 > 0$.

Consequently $\cos\theta < \inf(|y|, 1/|y|)$ and $A < 0$ in (2.64a-b). Hence the roots of (2.64b) exist and always one $\sqrt{n_{01}/n_{02}} > 0$. Then n_{02} is positive as is n_{03}. For the other $b = 1/2$, 1 models we have not found positive solution.

As an illustration we give a numerical example for $b = 2, u_{01} = 1, |y| = 0.5$ and deduce: $n_{02} = 0.11$, $n_{03} = 0.33$, $n_1 = -23.7 + i27.4$, $n_2 = -16.8 - i6.8$, $n_3 = 14.5$, $\rho = 13.0 - i10.3$, $\gamma = i12.8$.

2.3.c. Nonuniform Stationary Solutions and associated (1+1)-dimensional solutions

For the multidimensional models, in order to obtain (1+1)-dimensional solutions we can require that these solutions are on some particular line. Let us consider the Broadwell model with solutions such that $x_3 = x_2 = x_1 = x$ and assume $N_5 = N_1$, $N_6 = N_2$:

$$N_{1t} + N_{1x} = N_{2t} - N_{2x} = -(N_{3t} + N_{3x})/2 = (N_{4x} - N_{4t})/2 = N_3N_4 - N_1N_2 \quad (2.65)$$

First we determine stationary solutions $N_i^s = n_{0i} + n_i/D$, $D = 1 + e^{\gamma x}$ that we substitute into (2.65). We obtain $n_{04} = n_{01}n_{02}/n_{03}$ and

$$n_1 = -n_2 = -n_3/2 = n_4/2 = -\gamma/3, \quad n_{02} - n_{01} + 2(n_{04} - n_{03}) = \gamma \quad (2.66)$$

$$N_5^s = N_1^s = n_{01} + n_1/D, \quad N_2^s + N_1^s = N_6^s + N_5^s = n_{01} + n_{02},$$
$$N_3^s = n_{03} - 2n_1/D, \quad N_4^s + N_3^s = n_{03} + n_{04} \quad (2.67)$$

For the positivity we must have $n_{0i} > 0$, $n_{0i} + n_i > 0$. For the arbitrary parameters we choose $n_{01} = 1$, $n_{02} = n_{03} > 0$ and obtain $n_{04} = 1$, $3n_1 = n_{02} - 1$, $n_1 + n_{01} = n_3 + n_{03} = (2 + n_{02})/3$, $n_2 + n_{02} = n_4 + n_{04} = (1 + 2n_{02})/3$.

Second we add a time-dependent term $N_i = N_i^s + p_i/\Delta$, $\Delta = 1 + de^{\rho t}$ substitute into (2.65) and obtain:

$$p_1 = p_2 = -p_3/2 = -p_4/2 = \rho/3, \quad \rho = n_{01} + n_{02} + 2n_{03} + 2n_{04} > 0 \quad (2.68)$$

Due to $\rho > 0$, the N_i relax towards N_i^s when $t \longrightarrow \infty$. Choosing $d > 0$ sufficiently large, then p_i/Δ is arbitrarily small. It follows that the positivity of N_i^s guarantees the positivity of N_i. Finally the total mass $M = \Sigma N_i = 2n_{01} + 2n_{02} + n_{03} + n_{04}$ is a constant.

Third adding $\delta_i e^{\mu t}$ to N_i^s and linearizing we find $\delta_1 = \delta_2 = -\delta_3/2 = -\delta_4/2$ and $\mu = -2n_{03} - 2n_{04} - n_{01} - n_{02}$ is a negative eigenvalue.

2.3.d. Specular Reflection boundary Condition

We go on with the $b = 1/2, 1, 2$ models (2.47) and assume, for shock waves, a specular reflection boundary condition at a wall $x = 0$ or $N_1(x = 0, t) = N_2(x = 0, t)$. We start with an ansatz

$$\begin{pmatrix} N_1 \\ N_2 \\ N_3 \end{pmatrix} = n_{01} \begin{pmatrix} 1 \\ 1 \\ \eta \end{pmatrix} + \begin{pmatrix} n_1 & n_2 \\ n_2 & n_1 \\ n_3 & n_3 \end{pmatrix} \begin{pmatrix} \frac{1}{D_+} \\ \frac{1}{D_-} \end{pmatrix}, \quad D_\pm = 1 + d\, e^{\rho t \pm \gamma x}, \quad \eta^2 = 1 \tag{2.69}$$

which satisfies such boundary conditions. Substituting into (2.47) we find five relations

$$n_1(\rho+\gamma) = n_2(\rho-\gamma) = -bn_3\rho = n_3^2 - n_1 n_2 = n_{01}(n_1 + n_2 - 2\eta n_3),$$

$$n_1^2 + n_2^2 = 2n_3^2 \tag{2.70}$$

and six parameters $n_i, \rho, \gamma, n_{01}$. We choose $n_{01} > 0$ as the arbitrary parameter, define $y = n_2/n_1$, $\bar{n}_3 = n_3/n_1 = -2y/b(1+y)$, $B = \sqrt{1+8/b^2}$ and from $2\bar{n}_3^2 = 1 + y^2$ obtain a product of two factors

$$\prod_{\kappa=-1,1} (y^2 + (1+\kappa B)y + 1) = 0, \qquad y = y_\pm = -(1+\kappa B) \pm \sqrt{B^2 - 3 + 2\kappa B},$$

For $\kappa = -1$, the solutions associated to $b = 2$ (y complex), $b = 1$ ($y = 1$), $b = \frac{1}{2}$ (y real leading to a violation of positivity) are not acceptable. In the following we restrict our study to $\kappa = 1$ with

$$b = 2, \quad 2y_\pm = -1 - \sqrt{3} \pm \sqrt{2\sqrt{3}} < 0, \quad b = 1, \quad y_\pm = -2 \pm \sqrt{3} < 0,$$

$$b = \frac{1}{2}, \quad 2y_\pm = -1 - \sqrt{33} \pm \sqrt{30 + 2\sqrt{33}} < 0 \tag{2.71}$$

From the mixed n_{01}, n_i relation we find n_1/n_{01}

$$n_1/n_{01} = b(1+y)E_\eta/y, \qquad E_\eta = (b(1-B) + 4\eta))/(4 + b^2(B-1))$$

from which we deduce $n_2 = yn_1$ and $n_3 = n_1\bar{n}_3 = -2n_{01}E_\eta$. For the frequency, the wave number and the shock speed we find:

$$2\rho b/n_{01} = 4\eta + b(1-B), \qquad \rho(y-1) = \gamma(y+1), \qquad |\, s \,|=|\, \rho/\gamma \,|< 1$$

Introducing the total mass $M = N_1 + N_2 + 2bN_3$, we write down N_i and M with n_{01} as a scaling parameter and $y = y_\pm$ as numbers

$$N_1/n_{01} = 1 + b(1+y)E_\eta(1/yD_+ + 1/D_-), \qquad N_2 = N_1(D_+ \longleftrightarrow D_-)$$

$$N_3/n_{01} = \eta - 2E_\eta(1/D_+ + 1/D_-), \quad M/n_{01} = 2(1+b_\eta) - b(3+B)E_\eta(1/D_+ + 1/D_-)$$

n_{01} has only two relevant values ± 1, but $n_{01} = -1$ leads to violation of positivity and so we assume $n_{01} > 0$. Furthermore the solutions for n_{01} and η fixed are invariant in the exchange $y_+ \longleftrightarrow y_- = 1/y_+$ and we find: $y \longleftrightarrow 1/y$, $(1+y)/y \longleftrightarrow 1+y$, $n_1 \longleftrightarrow n_2$, $n_3 \longleftrightarrow n_3$, $\rho \longleftrightarrow \rho$, $\gamma \longleftrightarrow -\gamma$, $D_+ \longleftrightarrow D_- \Longrightarrow N_i \longleftrightarrow N_i$. Finally there exist two independent solutions, with $n_{01} = 1$ and $\eta = \pm 1$. Due to the fact that the wave numbers are opposite in $D_\pm$, the shock limits, when $| x | \longrightarrow \infty$, are $\Sigma_1 = n_{01} + n_1$, $\Sigma_2 = n_{01} + n_2$, $\Sigma_3 = \eta n_{01} + n_3$. We define $\bar{\Sigma}_i = \bar{\Sigma}_i \Lambda$ with $\Lambda = 4 + b^2(B-1) > 0$, $n_{01} = 1$ and find:

$$\bar{\Sigma}_1 = (b + 2\eta + b/y)^2 > 0, \quad \bar{\Sigma}_2 = \bar{\Sigma}_1(y \longrightarrow 1/y) > 0, \quad \bar{\Sigma}_3 = -4\eta + b(B-1)(2+2b\eta) > 0$$

$$\bar{\Sigma}_3 : \eta = \pm 1, \quad b = 2, \ 1, \ \frac{1}{2} \longrightarrow 4, 6, 4 + 3(\sqrt{33}-1)/4; \quad \longrightarrow 4(2\sqrt{3}-3), 2, (5\sqrt{33}-11)/4$$

These shock limits being strictly positive, we can manage the arbitrary d parameter of $D_\pm$ so that $N_i > 0$.

As an illustration in fig.1 we give two numerical examples for the $b = 1/2$, $3\vec{V}_i$ model with $y = y_+ = -0.15167$, $12E_\eta = 6 + 15\eta - (2+\eta)\sqrt{33}$ and $n_{01} = 1$:
(i) $\eta = 1$, $n_3 = (\sqrt{33}-7)/2 = -0.62$, $n_1 = (7-\sqrt{33})(1+1/y)/8 = -0.87$, $n_2 = y n_1 = 0.133$, $\rho = (9-\sqrt{33})/2 = 1.62$, $s = \rho/\gamma = (1-\sqrt{33})/\sqrt{30+2\sqrt{33}} = -0.736$;
(ii) $\eta = -1$, $n_3 = (\sqrt{33}+9)/6 = 2.45$, $n_1 = -(9+\sqrt{33})(1+1/y)/24 = 3.4$, $n_2 = -0.52$, $\rho = -(7+\sqrt{33})/2 = -6.37$, $s = -0.736$. For the $N_i(x,t) > 0$ and $M(x,t)$ profiles we choose $d = 0.5$. We observe the displacement of the frontshock when the time is growing.

References

For the Carleman model: Bobylev A.V., Math. Congress Warsaw 1983 (Book Abstract B29); Wick J. Math. Methods Appl. Sci. 6, 515, 1984, a result due to K. Piechor is quoted in Platkowski T. Méc. Théo Appl. 4, 555, 1985. For the Illner model and other models with more than $2\vec{V}_i$: Cornille H., J.M.P. 28, 1567, 1987, J. Phys. A20, 1973, 1987, Phys. Lett. A125, 253, 1987, J.Stat.Phys. 48, 7, 89, 1987, CRAS 304, 1091, 1987, "Inverse Problems" Ed. Sabatier PC p.481, World Scientific 1987, J.M.P. 29, 1667, 1988, Let. Math. Phys. 16, 245, 1988; Cabannes H. and Tiem D.H., CRAS 304, 29, 1987, Complex Systems 1, 574, 1987.

2.4. Exact (1+1)-dimensional solutions as sums of similarity waves

We recall that the exponential similarity waves are of the type $N_i = n_{0i} + n_i/D$, $D = 1 + de^{\rho t + \vec{\gamma}.\vec{x}}$. All (1+1)-dimensional and multidimensional exact solutions which have been found are sums of similarity waves

$$N_i = n_{0i} + \sum_{j=1}^{J} n_{ji}/D_j, \qquad D_j = 1 + u_j, \qquad u_j = d_j \exp(\rho_j t + \vec{\gamma}_j.\vec{x}) \tag{2.72}$$

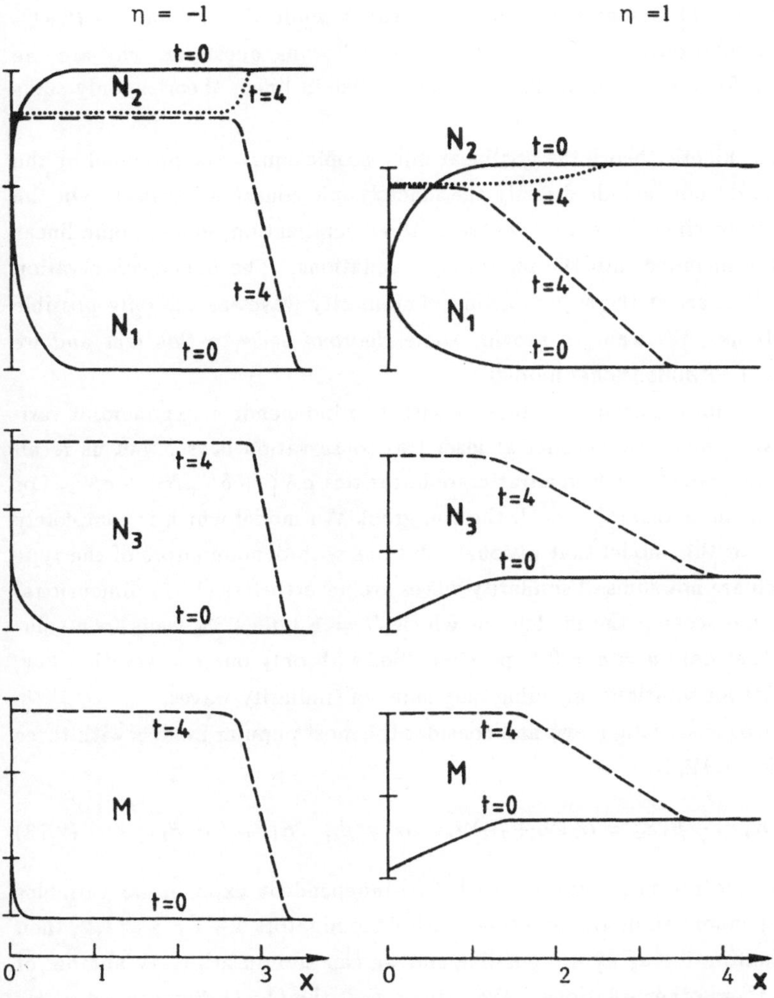

fig.1: (1+1)-dimensional solution with specular reflection for the $3\vec{v}_i$ model

with $J = 2$ for the (1+1)-dimensional and 2D solutions while $J = 3$ for the (2+1)-dimensional and 3D solutions. We try to answer the following question: why are the exact solutions found for the nonlinear discrete models, like in linear theories, only sums of similarity waves ?

In the continuous kinetic theory the nonlinear microscopic equations, provided by the Boltzmann equation, do not include directly the macroscopic conservation laws. On the contrary for the discrete kinetic theory, because of the discretization, macroscopic linear conservation laws are included into the microscopic equations. The linear conservation laws, acting like a filter, select the superposition of similarity waves as the only possible exact rational solutions. *We want to provide some rigorous basis to this fact and we restrict our study to 1+1 dimensional models.*

Let us consider rational densities, solutions with two independent exponential variables $\rho_i\gamma_j \neq \rho_j\gamma_i$. Why must we consider at least two conservation laws ? Let us recall the results for the $2\vec{V}_i$ models with quadratic nonlinearities $aN_+^2 + bN_+N_- + cN_-^2$. For $a = c = 0$ there exists an associated model, the Ruijgrook-Wu model which is completely soluble. This means for this model that rational solutions with denominators of the type $D = 1 + \sum_1^q u_j$ which are not sums of similarity waves are nevertheless (1+1)-dimensional solutions. Furthermore seeking the models for which $D = 1 + u_1 + u_2$ leads to a solution, it was shown that only $a = c = 0$ is possible. So, with only one conservation law, counterexamples exist for solutions including only sums of similarity waves.

So we assume two conservation laws and consider the most popular models with three independent densities $\mathcal{V}, \mathcal{W}, \mathcal{Z}$.

$$-\mathcal{V}_t - \mathcal{V}_x = \alpha_1\mathcal{Z}_t + \beta_1\mathcal{Z}_x = \alpha_2\mathcal{W}_t + \beta_2\mathcal{W}_x \qquad \alpha_i \neq \beta_i, \quad \alpha_1\beta_2 \neq \alpha_2\beta_1 \tag{2.73}$$

We look at the possible rational solutions with two independent exponential variables u_1, u_2. For two independent similarity solutions with denominators $1 + u_j$, $j = 1, 2$, then others with denominators $1 + u_1^m u_2^n$ are possible and we can add an arbitrary number of such solutions. Are there other solutions ? We assume that the (1+1)-dimensional u_1, u_2 rational solutions reduce to the similarity waves $1/(1 + u_i)$ when $u_j \neq 0$ and $u_i = 0$ and that the three densities have a common denominator $D(u_1, u_2)$. Starting with a fourth order D of the type

$$D = 1 + \Sigma u_i(1 + d_{3i}u_iu_j + d_{4i}u_i^2u_j) + du_1u_2 + d_4u_1^2u_2^2, \qquad i = 1, 2, \quad j \neq i$$

it has been proved that only factorized $D = (1 + u_1)(1 + u_i)(1 + d_4u_1u_2)$ leading to sums of two or three similarity waves are possible. Here, for simplicity, we reproduce the proof only for D of the second order

$$\mathcal{V} = v_{00} + V/D, \quad \mathcal{W} = w_{00} + W/D, \quad \mathcal{Z} = z_{00} + Z/D, \quad D = 1 + \Sigma u_i + d u_1 u_2 \quad (2.74)$$

References

Cornille H., J.M.P. 28, 1567, 1987 (with a result obtained in collaboration with T.T. Wu in Appendix), J. Stat. Phys. 48, 789, 1987, "Mathematical Aspects of Fluids and Plasma Dynamics", Lecture Notes in Mathematics, Springer-Verlag, Ed. Toscani G., 1989, Book in honor of A. Martin, Ed. Sciulli, Springer, 1990.

2.5. Multidimensional exact solutions

These solutions have been obtained for the p-dimensional hypercubic model with $2p$ densities, including for $p = 2$ the $4\vec{V}_i$ model and for $p = 3$ the $6\vec{V}_i$ Broadwell model

$$N_{2i-1t} + N_{2i-1x_i} = N_{2it} - N_{2ix_i} = -(p-1)N_{2i-1}N_{2i} + \sum_{k\neq i} N_{2k-1}N_{2k}, \; i = 1,2,...p \quad (2.75)$$

We sketch briefly some results, the main difficulty being the proof of $N_i > 0$.

2.5.a. (2+1)-dimensional shock waves

The solutions are the superposition of three similarity waves: $N_i = n_{0i} +$ $\Sigma n_{ji}/D_j$, $D_j = 1 + d_j \exp(\vec{\gamma}_j.\vec{x} + \rho_j t)$, $d_j > 0$, $j = 1,2,3$. We call $\vec{\gamma}_j.\vec{x} = y_j$. In a two-dimensional y_1, y_2 space necessarily $y_3 = \tau_1 y_1 + \tau_2 y_2$ and the asymptotic shock limits become plateaus in the y_1, y_2 plane: (i) Only four plateaus if $\tau_1\tau_2 = 0$, for instance if $\tau_2 = 0$, $\tau_1 > 0$ they are n_{0i}, $n_{0i} + n_{2i}$, $n_{0i} + n_{1i} + n_{3i}$, $\sum_0^3 n_{ki}$, (ii) 6 plateaus if $\tau_1\tau_2 \neq 0$, for instance if $\tau_1 > 0$, $\tau_2 > 0$ they are n_{0i}, $n_{0i} + n_{ji}$, $n_{0i} + n_{ji} + n_{3i}$, $j = 1,2, \Sigma n_{ki}$. It has been proved that if these asymptotic limits are positive then we can find conditions on the d_j so that $N_i > 0 \; \forall y_1, y_2$, $t \geq 0$.

We explicit the simplest solution for the Broadwell model. We start with an ansatz $N_i(x_1, x_2 + x_3, t): N_5 = N_3$, $N_6 = N_4$ and

$$\begin{pmatrix} N_1 \\ N_2 \\ N_3 \\ N_4 \end{pmatrix} = \begin{pmatrix} n_{01} \\ n_{02} \\ n_{03} \\ n_{03} \end{pmatrix} + \begin{pmatrix} n_1 & n_1 & n_{31} \\ n_1 & n_1 & n_{32} \\ n_3 & n_4 & n_{33} \\ n_4 & n_3 & n_{33} \end{pmatrix} \begin{pmatrix} 1/D_+ \\ 1/D_- \\ 1/D_3 \end{pmatrix}, \quad D_\pm = 1 + d\, e^{\rho t \pm \gamma(x_2+x_3)}$$

$$D_3 = 1 + d_3 e_3^{\rho} t + \gamma_3 x_1, \; (N_{1t} + N_{1x_1})/2 = (N_{2t} - N_{2x_1})/2 \quad (2.76)$$

$$= -N_{3t} - N_{3x_2} = -N_{4t} + N_{4x_2} = N_3 N_4 - N_1 N_2$$

Satisfying at a wall $x_2 + x_3 = 0$ a specular reflection boundary condition $N_3 = N_4 = N_5 = N_6$. Firstly we write the relations for the two first components

$$\rho n_1/2 = -n_3(\rho+\gamma) = n_4(-\rho+\gamma) = n_3 n_4 - n_1^2 = n_1(n_{01}+n_{02}) - n_{03}(n_3+n_4)$$
$$n_{03} = \sqrt{n_{01}n_{02}}, \quad n_3^2 + n_4^2 = 2n_1^2$$

Let $n_{01} = 1$ ($n_{02} = n_{03}^2$) and $n_{03} > 0$ be the arbitrary parameters. We introduce $\bar{n}_i = n_i/n_1$, $\mathcal{S} = \bar{n}_3 + \bar{n}_4$, $\mathcal{P} = \bar{n}_3\bar{n}_4$, $A_1 = \mathcal{P} - 1 < 0$ which are numbers and reconstruct the two components from n_{03}.

$$4\mathcal{S} = -1+\sqrt{33}, \quad 2\bar{n}_3 = -1+\sqrt{1+4\mathcal{S}} > 0, \quad 4\mathcal{P}+\mathcal{S} = 0, \quad n_1 A_1 = 1 + n_{03}^2 - n_{03}\mathcal{S},$$
$$n_i = \bar{n}_i n_1, \quad i = 3,4, \quad \rho = 2(-n_1 + n_3 n_4/n_1), \quad \gamma(n_4+n_3) = \rho(n_4 - n_3)$$

Secondly the third component has relations $n_1(n_{31}+n_{32}) = n_{33}(n_3+n_4)$ and

$$n_{31}(\rho_3+\gamma_3)/2 = n_{32}(\rho_3-\gamma_3)/2 = -n_{33}\rho_3 = n_{33}^2 - n_{31}n_{32} = n_{01}n_{32} + n_{02}n_{31} - 2n_{03}n_{33}$$

We introduce numbers $z = n_{32}/n_{31} = -1 + (-\mathcal{S} + \sqrt{2+7\mathcal{S}/2})/2 < 0$, $\bar{n}_{33} = n_{33}/n_{31} = (z+1)/\mathcal{S} = -z/(1+z) > 0$, $A_3 = \bar{n}_{33}^2 - z > 0$. We reconstruct the parameters from n_{03} : $n_3 A_3 = z + n_{03}^2 - 2n_{03}\bar{n}_{33}$, $n_{32} = zn_{31}$, $n_{33} = n_{31}\bar{n}_{33}$, $\rho_3 = n_{31}n_{32}/n_{33} - n_{33}$, $\gamma_3(n_{32}+n_{31}) = \rho_3(n_{32}-n_{31})$.

For the positivity of the N_i it is sufficient to check the positivity of the eight shock limits $\Sigma_i^2 = n_{0i} + n_i = n_{01}+n_1$, $n_{02}+n_1$, $n_{03}+n_3$, $n_{03}+n_4$ and $\Sigma_i^3 = \Sigma_i^2 + n_{3i}$. For this solution, as well as for all other found (2+1)-dimensional solutions, we always obtain Σ_i in a factorized form: $n_{03}\Sigma_i = \Omega_i(n_{03} - n_{02}B_k)(n_{03} - n_{02}B_{k'})$. For the positivity of Σ_i it is sufficient to check the sign of Ω_i and of each of the two factors. This leads to some n_{02}/n_{03} interval in which $\Sigma_i = 0$. It remains to verify that the intersection of these intervals when i is varying is not empty. For the Σ_i^2 we define $\bar{\Sigma}_i^2 = \Sigma_i^2/\Omega_i^2$ and obtain

$$\begin{aligned} \bar{\Sigma}_1^2 &= (n_{03}-\bar{n}_3)(1-n_{03}/\bar{n}_4) & \bar{\Sigma}_2^2 &= (n_{03}-1/\bar{n}_4)(1-\bar{n}_3 n_{03}) \\ \bar{\Sigma}_3^2 &= (n_{03}-\bar{n}_3)(1-n_{03}\bar{n}_3) & \bar{\Sigma}_4^2 &= (n_{03}-1/\bar{n}_4)(1-n_{03}/\bar{n}_4) \end{aligned} \tag{2.77}$$

Due to $\bar{n}_3 > 0$, $\bar{n}_4 < 0$, $\Omega_2^2 = \Omega_1^2 = -\bar{n}_4/A_1 < 0$, $\Omega_3^2 = -1/A_1 > 0$, $\Omega_4^2 = \bar{n}_4\Omega_2^2 > 0$ we find that $\Sigma_i^2 > 0$ if 0.715$= 1/\bar{n}_3 < n_{03} < \bar{n}_3 = 1.398$. Similarly for the Σ_i^3 let us define $\bar{\Sigma}_i^3 = (1 - A_3A_1)\,\Sigma_i^3/\Omega_i^3$, $A_3A_1 < 0$ and obtain

$$\begin{aligned} \bar{\Sigma}_1^3 &= (n_{03}-B_3)(1-B_4 n_{03}) & \bar{\Sigma}_2^3 &= (n_{03}-\tilde{B}_4)(1-\tilde{B}_3 n_{03}) \\ \bar{\Sigma}_4^3 &= (n_{03}-B_3)(1-\tilde{B}_3 n_{03}) & \bar{\Sigma}_3^3 &= (n_{03}-\tilde{B}_4)(1-B_4 n_{03}) \end{aligned} \tag{2.78}$$

Due to $z < 0$, $\Omega_1^3 = \bar{n}_{33}A_1 + \bar{n}_4 A_3 < 0$, $\Omega_2^3 = \bar{n}_4 A_3 + z\bar{n}_{33}A_1 > 0$, $\Omega_3^3 = -3z/4 > 0$, $\Omega_4^3 = A\bar{n}_{33}^2 + \bar{n}_4^2 A_3 < 0$, $B_4 = (A_1 + A_3)/\,\Omega_1^3 > 0$, $B_3 = \Omega_3^3/\Omega_1^3 < 0$, $\tilde{B}_3 = \Omega_3^3/\Omega_2^3 > 0$, $\tilde{B}_4 = (\bar{n}_{13}A_3 + \bar{n}_{33}zA_1)/\Omega_3^3 > 0$, $1/\tilde{B}_3 < 1/B_4 < \tilde{B}_4$ we see that $\Sigma_i^3 > 0$ if $1/B_4 = 1.308 < n_{03} < \tilde{B}_4 = 4.37$.

In conclusion the shock limits Σ_i^2, Σ_i^3 are positive and lead to positive densities only if $1/B_4 = 1.308 < n_{03} < \bar{n}_3 = 1.398$. This simple solution illustrates the difficulty of the positivity problem. Here with an arbitrary parameter n_{03} we show that only in a small interval (1.31,1.4) can positive solutions exist. This simple solution is positive for all hypercubic models except for $p = 2$.

For the $p = 2$ square model, other types of positive (2+1)-dimensional shock waves were found. For the general Broadwell model with six different densities and six asymptotic plateaus for each density, this means that we must find in the arbitrary parameter space the intersection of thirty six subdomains leading to the positivity property.

Let us introduce the total mass $M = \Sigma N_i$

$$\begin{aligned} M &= m_0 + mQ + m_3/(1+\bar{d}_3 e^x), \qquad m_0 = 1 + n_{02} + 4n_{03}, \\ m &= 2(n_1 + n_3 + n_4), \quad m_3 = n_{31} + n_{32} + 4n_{33}, \quad Q = 1/D^+ + 1/D^- \end{aligned} \tag{2.79}$$

$D_\pm = 1 + \bar{d}e^{\pm y}$, $\bar{d} = de^{\rho_1 t}$, $\bar{d}_3 = d_3 e^{\rho_3 t}$, $y = \gamma(x_2 + x_3)$, $x = \gamma_3 x_1$.
We note that M is even in y and obtain for the $x(y)$ equidensity lines $M(x, y, t)$ =const.

$$x = -\log\bar{d} + \log(m_3/(m_0 + mQ - M) - 1)$$

We choose an example with $n_{03} = 1.329$, $d = 0.95$, $d_3 = 500$ leading to $n_i = -0.918$, -1.285, 0.1948, $n_{3i} = -0.0542$, 0.0191, -0.0295, $i = 1,2,3$. $\gamma = -3.$, $\gamma_3 = -0.14$, $\rho = 2.4$, $\rho_3 = 0.065$, $m_0 = 8.088$, $m = -4.017$, $m_3 = -0.153$. In fig.2 we present the M equidensity lines (with arrays for decreasing values) at $t = 0$ and 5. At $t = 0$ we observe the asymptotic plateaus $m_0 + m + m_3 < m_0 + m$ along parallels to the x-axis, we see the shock due to the third $x-$dependent similarity component. For positive time sufficiently large, the asymptotic plateaus become the smallest equidensity lines in both up and downdomains. The highest plateau is the Maxwellian m_0 and we observe a bump around the x-axis which spreads out with increasing time.

2.5.b. 2D, 3D, pD dimensional solutions

For the $p = 2$, $4\vec{V}_i$ model we start with a 2D dimensional ansatz

$$\begin{pmatrix} N_1 \\ N_2 \\ N_3 \\ N_4 \end{pmatrix} = \begin{pmatrix} n_{01} \\ n_{02} \\ n_{01} \\ n_{02} \end{pmatrix} + \begin{pmatrix} n_1 & n_3 \\ n_1 & n_4 \\ n_3 & n_1 \\ n_4 & n_1 \end{pmatrix} \begin{pmatrix} 1/D_2 \\ 1/D_1 \end{pmatrix}, \quad D_k = 1 + de^{\rho t + \gamma x_k}, \ k = 1,2, \ d > 0 \tag{2.80}$$

that we generalize for the p-dimensional hypercubic model into:

$$\begin{aligned} N_{2q-1} &= n_{01} + n_1 \sum_{k \neq q} 1/D_k + n_{2p-1}/D_q, \quad D_k = 1 + d\, e^{\rho t + \gamma x_k} \\ N_{2q} &= n_{02} + n_1 \sum_{k \neq q} 1/D_k + n_{2p}/D_q, \quad q = 1,2,...p \ \ d > 0 \end{aligned} \tag{2.81}$$

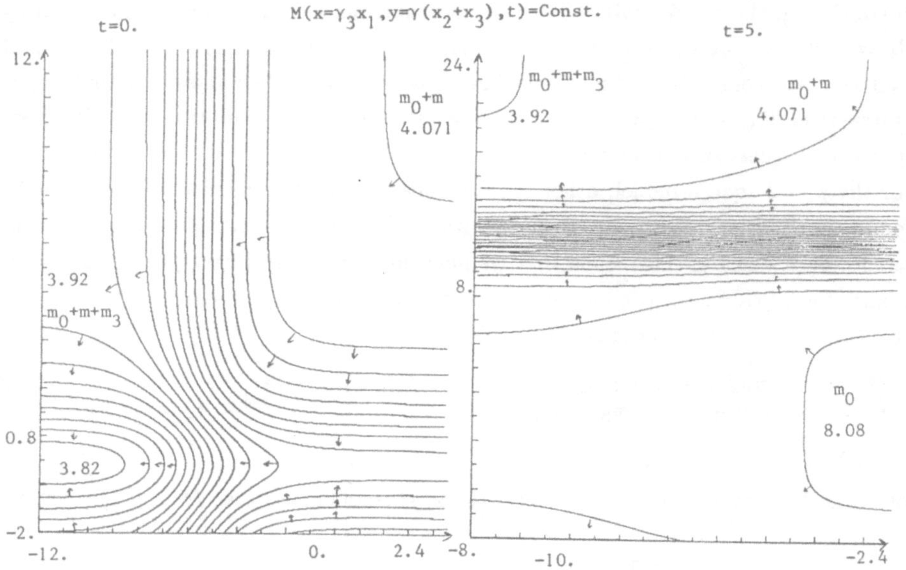

fig.2: (2+1)-dimensional shock waves for the Broadwell model

The conditions $N_1 = N_3$, $N_2 = N_4$ for a specular reflection at an hyperplane $x_1 = x_2$ are satisfied. The coefficients of D_q^{-1}, D_q^{-2} in the discrete hypercubic model equations give

$$\begin{array}{l} \rho n_1 = n_{2p-1}n_{2p} - n_1^2 = n_1(n_{01} + n_{02}) - n_{2p-1}n_{02} - n_{2p}n_{01} \\ n_{2p-1}(\rho + \gamma) = n_{2p}(\rho - \gamma) = -(p-1)n_1\rho \end{array} \tag{2.82}$$

while the vanishing of $(D_k D_{k'})^{-1}$ gives a new relation $n_{2p-1} + n_{2p} = 2n_1$ for $p > 2$ that we still assume for $p = 2$.

For the construction of the solutions we define $\bar{n}_i = n_i/n_1$ $i = 2p-1, 2p$, choose $\bar{n}_{2p-1} = 1 + \sqrt{p}$, $\bar{n}_{2p} = 1 - \sqrt{p}$. We assume for the arbitrary parameters $n_{02} > n_{01} > 0$ and obtain for the others: $\sqrt{p}n_1 = n_{02} - n_{01} > 0$,

$$\begin{array}{l} n_{2p} = (1/\sqrt{p} - 1)(n_{02} - n_{01}) < 0, \quad n_{2p-1} = (1 + 1/\sqrt{p})(n_{02} - n_{01}) > 0, \\ \rho = \sqrt{p}(n_{01} - n_{02}) < 0, \quad \gamma = -\sqrt{p} \;\; \rho > 0, \quad | \, c = \gamma/\rho \, | = 1/\sqrt{p} < 1 \end{array}$$

The N_i with i odd are sums of positive terms while for i even they are also positive: $n_{02} + n_{2p}/D_j \geq (n_{02} + n_{01}(\sqrt{p} - 1))/\sqrt{p} > 0$.

We introduce the total mass $M = \Sigma N_i$

$$\begin{array}{l} M = m_0 + m \sum_1^p 1/D_j, \quad m_0 = p(n_{01} + n_{02}) > 0, \quad m = 2\sqrt{p}(n_{02} - n_{01}) > 0 \\ D_j = 1 + d \exp(n_{02} - n_{01})(x_j - t/\sqrt{p}) \quad j = 1, 2, ...p \end{array} \tag{2.83}$$

The time dependence corresponds to a translation

$$M(x_1, x_2, ...x_p; t) = M(x_1 + ct, x_2 + ct, ...x_p + ct; 0) \tag{2.84}$$

In the p-dimensional space there exist 2^p sectors corresponding to the different signs of the p coordinates $x_1, ...x_p$. In these sectors the asymptotic limits are p-dimensional manifolds which are the extension of the 2-dimensional plateaus of the $4V_i$ model.

For the equidensity manifolds M =const=C we can write down x_p as a function of $x_1, ...x_{p-1}$, in particular for $p = 3$:

$$x_3 = \gamma^{-1} \log \left[\bar{d}^{-1} \left(-1 + m/(C - m_0 - m \sum_1^2 (1 + \bar{d} \exp \gamma x_i)^{-1} \right) \right], \quad \bar{d} = de^{\rho t} \tag{2.85}$$

In fig. 3 we present such equidensity lines in section inside the wall $x_2 = x_1$ or parallel to the wall $x_2 = x_1 + 3$ where we observe two or three shocks connecting two, three or four plateaus.

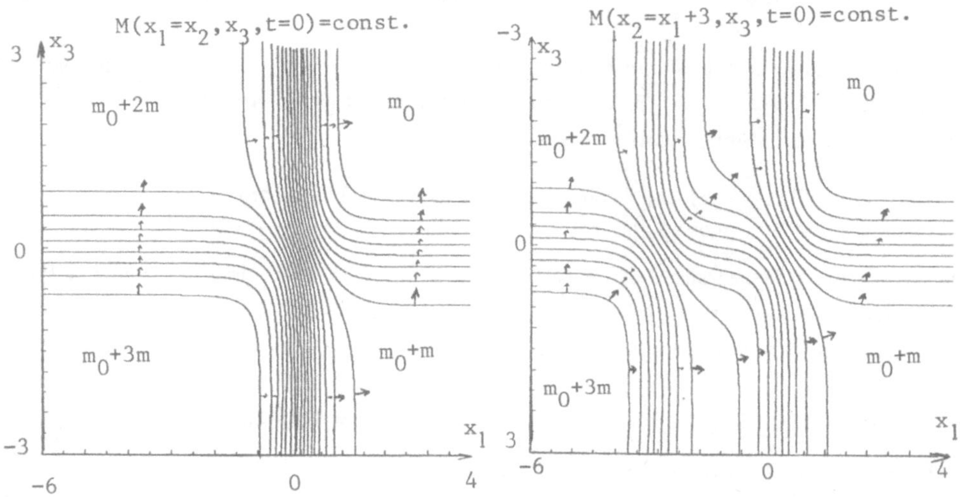

fig.3 3D Solutions for the Broadwell Model

2.5.c. Nonuniform Maxwellians for the hypercubic models

We first construct the nonuniform Maxwellians N_i^M and later add a time-dependent component

$$N_i^M = n_{0i} + \sum_{j=1}^{2} n_{ji}/(1+d_j \exp\vec{\gamma}_j.\vec{x}), \ \vec{\gamma}_j.\vec{x} = \sum_{1}^{p} \gamma_{ji}x_i, \ N_i = N_i^M + q_i/(1+d_3 e^{\rho t}) \quad (2.86)$$

For the Maxwellians we obtain $n_{ji} + n_{ji+1} = 0$, i odd, and the wave numbers are deduced from the n_{ji} : $n_{ji}\gamma_{ji} = (p-1)n_{ji}^2 - \Sigma n_{jk}^2$, i and k odd and the problem is reduced to the determination of the n_{0i} and of the n_{ji}, i odd. Putting $n_{m,n}^0 = n_m^0 - n_n^0$ they satisfy the relations

$$\begin{aligned} &n_{0i}n_{0i+1} = n_{0k}n_{0k+1}, \ i \text{ odd}, \ k \text{ odd}, \quad n_{1i}n_{2i} = n_{11}n_{21}, \ i \text{ odd} \\ &(p-1)n_{ji}^2 - \sum_k n_{jk}^2 = (p-1)n_{i+1,i}^0 n_{ji} - \sum n_{k+1,k}^0 n_{jk}, \ i \text{ and } k \text{ odd } k \neq i \end{aligned} \quad (2.87)$$

With substractions, the last (2.87) relations become $n_{ji}^2 - n_{j1}^2 = n_{ji}n_{i+1,i}^0$, i odd. We divide both sides by n_{j1}, substract the $j = 1, 2$ relations, take into account the n_{ji} alone (2.87) relations and assuming $n_{1i}/n_{2i} \neq n_{11}/n_{21}$ obtain the relation

$$n_{0i} + n_{1i} + n_{2i} = n_{0i+1} \qquad i \text{ odd} \quad i \leq 2p-1 \quad (2.88)$$

For the construction of the solution we start with the $p+2$ arbitrary parameters n_{0i}, i odd, n_{02}, n_{11} and first obtain n_{0i} i even, $n_{02i} = n_{01}n_{02}/n_{02i-1}$ and $n_{21} = n_{02} - n_{01} - n_{11}$. Second in the (2.87) relation we obtain quadratic relations for the n_{ji} which are deduced from the arbitrary parameters

$$\begin{aligned} &n_{1i} = z_{i\pm}, \ n_{2i} = z_{i\mp}, \ 2z_{i\pm} = n_{0i+1} - n_{0i} \pm \sqrt{\Delta_{i+1,i}} \\ &\Delta_{i+1,i} = (n_{0i+1} - n_{0i})^2 - 4n_{11}n_{21} \qquad i \quad \text{odd} \end{aligned} \quad (2.89)$$

Third $n_{ji+1} = -n_{ji}$, i odd, we know all n_{0i}, n_{ji} and deduce the wave vector components γ_{jq}. We notice the relation $N_i^M + N_{i+1}^M = n_{0i} + n_{0i+1}$, i odd.

For $N_i^M > 0$ is is sufficient that the asymptotic limits be positive

$$n_{0i} > 0, \ n_{0i} + n_{1i} + n_{2i} > 0, \ n_{0i} + n_{ji} > 0 \,(\text{or } -n_{0i} < n_{ji} < n_{0i+1} \ i \text{ odd}), \ j = 1,2 \quad (2.90)$$

We prove that it is sufficient that the arbitrary parameters satisfy the inequalities

$$n_{0i} > 0 \qquad i \text{ odd and } i = 2, \qquad 0 < n_{11} < n_{02} < n_{01} \quad (2.91)$$

The two first inequalities relations in (2.90) are trivially satisfied as well as the last inequalities for $i = 1$. For other odd i values we rewrite (2.89)

$$n_{0i+1} - z_{i\pm} = n_{0i} + z_{i\mp} = (n_{0i} + n_{0i+1} \mp \sqrt{\Delta_{i+1,i}})/2 \tag{2.89'}$$

and the last inequalities $n_{ji}+n_{0i} > 0$ are satisfied if the r.h.s. is positive or $\sqrt{\Delta_{i+1,i}} < n_{0i}+n_{0i+1}$. Due to $n_{21} < 0$ we see that $z_{i\pm}$ are real. For the ultimate result which guarantees the positivity of the reconstructed densities N_i we write down a set of inequalities $-n_{11}n_{21} = n_{11}(n_{11} + n_{01} - n_{02}) < n_{11}n_{01} < n_{02}n_{01}$ or $\Delta_{i+1,i} < (n_{0i} + n_{0i+1})^2$.

Adding a third component to N_i^M, it can be shown that this component is only time-dependent. We define $n^+_{\ell,m} = n_{0\ell} + n_{0m}$, obtain $3p - 1$ relations for the $2p + 1$ parameters q_i, ρ :

$$q_i = q_{i+1}, \sum_k q_k = 0,\ \rho q_i = -(p-1)q_i^2 + \sum_{k\neq i} q_k^2 = (p-1)q_i n^+_{i,i+1} - \sum_{k\neq i} q_k n^+_{k,k+1},\ i \text{ odd},\ k \text{ odd} \tag{2.92}$$

requiring $p - 2$ new relations for the $p + 2$ arbitrary parameters of N_i^M. We give a simple solution valid for $p > 2$:

$$\begin{aligned} &q_i = q, \quad i = 1,2,...2p-2, \quad q_{2p-1} = q_{2p} = -q(p-1) \\ &\rho = qp(p-2) = n^+_{1,2} + (p-1)n^+_{2p-1,2p}, \quad n_{0i} = n_{01}, \quad i = 1,3,...2p-3 \end{aligned} \tag{2.93}$$

Due to $\rho > 0$ we see that the tird component goes to zero when $t \longrightarrow \infty$ and choosing d sufficiently large, it becomes arbitrarily small so that $N_i > 0$ is a consequence of $N_i^M > 0$. The four arbitrary parameters for N_i being $n_{01} > 0$, $n_{02} > 0$, $n_{2p-1} > 0$ $n_{02p-1} > 0$, then the sufficient positivity conditions (2.91) lead to $N_i^M > 0$ and $N_i > 0$.

Linearizing around N_i^M and assuming time-dependent perturbations $N_i = N_i^M + \delta_i \exp(\mu t)$, we want to verify for the simplest $p = 2$ and 3 models that $\mu > 0$ eigenvalues do not exist. We find $\delta_i = \delta_{i+1}$, $\Sigma\delta_i = 0$, i odd and putting $s_i = n^+_{2i-1,2i}$ $S_1 = \Sigma s_i$, $S_2 = \Sigma s_i s_j$ we obtain: (i) $p = 2$, $\mu = -S_1 < 0$; (ii) $p = 3$, $\mu^2 + 2\mu S_1 + S_2 = 0$, $2(S_1^2 - S_2) = \Sigma(s_i - s_j)^2 > 0$ and the two real μ roots are negative.

2.5.d. Semiperiodic Solutions

These solutions represent periodic waves, submitted to a strong perturbation and rebuilt after the shock. Although a whole class of solutions has been constructed, here we present a simple one for which positivity is easily checked. For the Broadwell model we start with an ansatz $N_5 = N_3$, $N_6 = N_4$ and for the 4 other N_i :

$$N_i = N_i + n_{3i}/D_3,\ P_i = n_{0i} + 2\mathrm{Re}(n_i/D),\ P_{i+1} = n_{0i+1} + 2\mathrm{Re}(n_i^*/D)\ i = 1,3$$

$$D = 1 + d\, e^{\rho t + i\vec{\gamma}.\vec{x}}, \vec{\gamma}.\vec{x} = \gamma_1 x_1 + \gamma_2(x_2 + x_3),\ D_3 = 1 + d_3 e^{\rho_3 t + \vec{\gamma}_3.\vec{x}},\ \vec{\gamma}_3\vec{x} = \gamma_{31}x_1 + \gamma_{32}(x_2 + x_3)$$

and obtain for the periodic part P_i a solution associated to $n_{1I} = n_{1R}\sqrt{1.1}$

$n_{01} = 1,\ n_{03} = \frac{1}{2},\ n_{02} = (105 + 2\sqrt{814})/68,\ n_{04} = 2n_{02},\ \rho = 5(407 + 5\sqrt{814})/714 > 0,$
$n_1 = -8\rho(1 + i\sqrt{11/10})/15,\ n_3 = \rho(1 - i\sqrt{37/5})/15,$

$$D = 1 + d\ \exp\rho(t - x_1\sqrt{11/10} - (x_2 + x_1)\sqrt{37/5})$$

which satisfies the relations:

$$n_1(\rho + i\gamma_1)/2 = n_i^*(\rho - i\gamma_1)/2 = \mid n_3 \mid^2 - \mid n_1 \mid^2 = n_{01}n_1^* + n_{02}n_1 - n_{03}n_3^* - n_{04}n_3 =$$
$$-n_3(\rho + i\gamma_2) = n_3^*(-\rho + i\gamma_i),\ n_{01}n_{02} = n_{03}n_{04},\ \mathrm{Re}(n_1^2 - n_3^2) = 0$$

For the third shock wave component, we introduce $\bar{n}_{3i} = n_{3i}/n_{31}$ and obtain the parameter values

$$84\bar{n}_{32} = -101 + \sqrt{3145},\ 21\begin{pmatrix}\bar{n}_{34}\\ \bar{n}_{33}\end{pmatrix} = (17 - \sqrt{3145})\begin{pmatrix}1\\ 1\end{pmatrix} + \sqrt{11/2}(\sqrt{85} - 5\sqrt{37})\begin{pmatrix}1\\ -1\end{pmatrix},$$

$$n_{31} = (n_{02}(1 - 2\bar{n}_{33}) + \bar{n}_{32} - \bar{n}_{34}/2)/3\bar{n}_{32} = -0.9667 < 0,\ n_{3i} = \bar{n}_{3i}n_{31} = 0.517 > 0,$$

$$-0.489 < 0, 4.08 > 0,\ i = 2,3,4;\ \rho_3/n_{31} = (-17 + \sqrt{3145})/24,\ \rho_3/\gamma_{32} = \sqrt{34/55} < 1,$$

$$-\rho_3/\gamma_{31} = \sqrt{17/185} < 1,\quad \rho_3 = -1.349$$

which satisfy the relations

$$n_{31}(\rho_3 + \gamma_{31})/2 = n_{32}(\rho_3 - \gamma_{31})/2 = n_{33}n_{34} - n_{31}n_{32} = n_{01}n_{32} + n_{02}n_{31} - n_{03}n_{34}$$
$$n_{04}n_{33} = -n_{33}(\rho_3 + \gamma_{32}) = -n_{34}(\rho_3 - \gamma_{32}),\ n_3n_{34} + n_3^*n_{33} = n_1n_{32} + n_1^*n_{31}$$

For the positivity the important point is $\rho > 0$ and $\lim N_i = n_{0i} > 0$ when $t \longrightarrow \infty$. Choosing d large we can have $\mid P_i - n_{0i} \mid$ very small and the positivity of the N_i is reduced to the positivity of the shock wave component $n_{0i} + n_{3i} > 0$. Due to $n_{32} > 0$, $n_{34} > 0$ we only have to check $n_{01} + n_{31} = 0.03 > 0$, $n_{03} + n_{33} = 0.01 > 0$.

For the physical discussion it is convenient to introduce two new spatial coordinates $y2\pi = \vec{\gamma}.\vec{x}$, $x = \vec{\gamma}_3.\vec{x}$. We define the total mass $M = M^p + m_3/D_3$, $M^p = m_0 + 2\mathrm{Re}(m/D)$, $m_0 = 13.91$, $m = -3.08$, $m_3 = 6.74$ and can study the equidensity lines M =const. As an application in fig. 4 we choose $d = 12(1 + i)$, $d_3 = 1$ and $t = 0, 0.5, 4$. The periodic part has equidensity lines parallel to the x-axis while the shock strip is around $\vec{\gamma}_3.\vec{x} = 0$. At fixed t, the shock strip has moved towards $\vec{\gamma}_3.\vec{x} = -\rho_3 t$ and for t large M becomes equal to a similarity shock wave $M = m_0 + m_3/D_3$ with two shock limits.

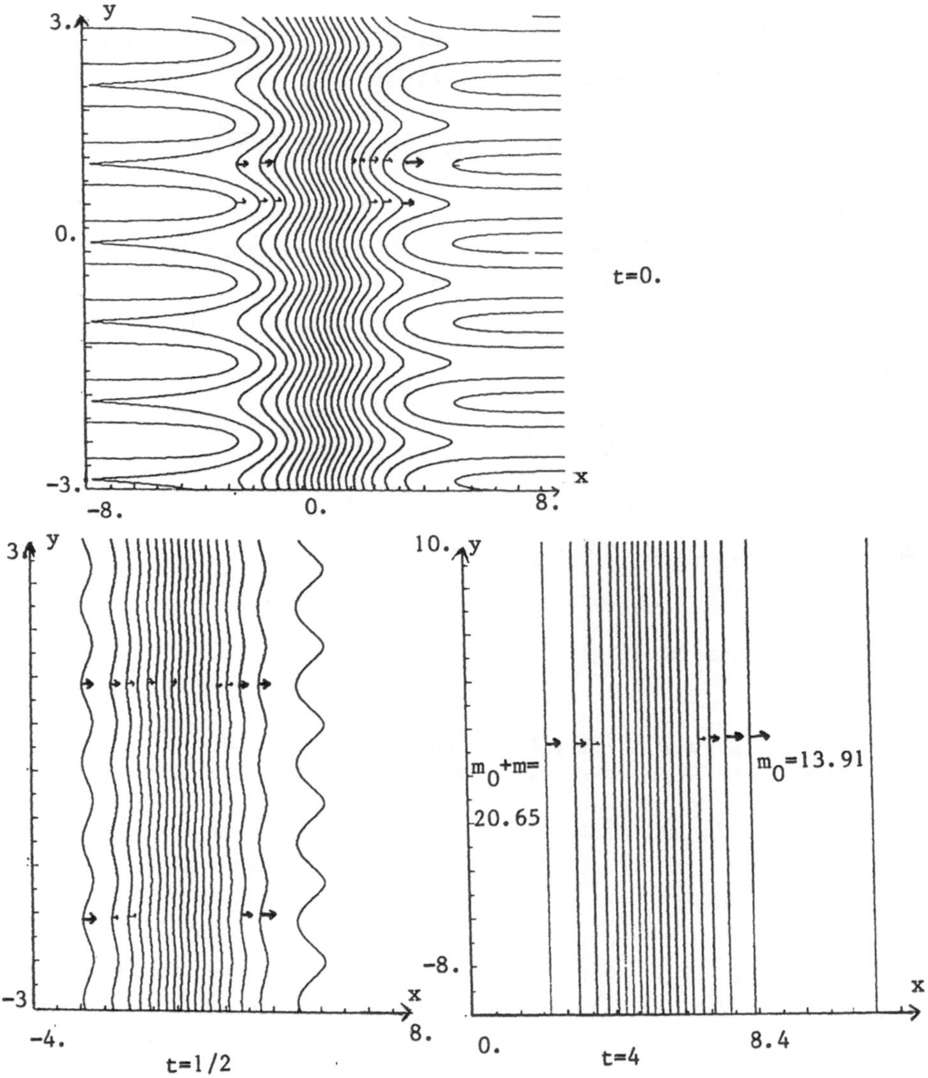

fig.4: Semiperiodic waves for the Broadwell Model

$M(\vec{\gamma}.\vec{x}=2\pi y,\ \vec{\gamma}_3.\vec{x}=x,\ t)=\text{const.}$

References

Cornille H., J. Phys. A20, L1063, 1987, J. Stat. Phys. 52, 897, 1988, "Topics on Inverse Problems" Ed. Sabatier PC. World Scientific, Singapore 1988, p.101, J.M.P. 30, 789, 1989, TTSP, 18, 33, 1989; XVI Int. Sympos. Rarefied Gas Dynamics, Prog. Astron. Aeron. Ed. Muntz vol.118, AIAA p.131, 1989, "Kinetic Theory, Lattice Gas..." Ed. Monaco World Scientific, Singapore 1989, p.83, J. Phys. A22, 4787, 1989.

3. Exact Solutions of the Continuous B.E.

There exist different methods for the determination of exact homogeneous solutions:
(i) Start with ansatz which are gaussians with time-dependent width multiplied by even $\vec{V}^2$ polynomials. This is a direct method that we shall use for the Bobylev-Krook-Wu (BKW) solution.
(ii) Transform the BE, through generalized Laplace Transforms, into nonintegrable NLPDE. For instance $f \longrightarrow H$, $V^2 \longrightarrow y$

$$(1+\partial_t)(1+\partial_y)H = H^2 \tag{3.1}$$

The exponential similarity and (1+1)-dimensional solutions have denominators of the type $(1+u)^2, (1+u_1+u_2)^2$, $u_i = \exp(\rho_i t + \gamma_i y)$. Apart the BKW, no new positive solutions have been obtained for the BE. However their study was interesting for a class of nonintegrable NLPDE with factorized linear operators and power type nonlinearities. It was found that still the denominators were of the type $(1+u_1+u_2)$ to some power.
(iii) use of Laguerre series, Fourier transforms,...

For the construction of inhomogeneous solutions it is convenient to consider the Nikolski Transform which connects homogeneous and inhomogeneous solutions.

3.1. Kac model in 1+1+1 (velocity v, space x, time t) dimensions

For this simple model the mass and energy conservation laws are satisfied but not the momentum one

$$\begin{gathered}(\partial_t + v\partial_x)f(v,x,t) = \int_{-1}^{+1} \sigma(\theta) \int_{-\infty}^{+\infty} (f(v' = v\cos\theta - w\sin\theta)f(w' = v\sin\theta + w\cos\theta) \\ -f(v)f(w))dwd\theta \\ \sigma(\theta) = \sigma(-\theta) > 0, \quad v'^2 + w'^2 = v^2 + w^2, \quad v' + w' \neq v + w\end{gathered} \tag{3.2}$$

3.1.a. Even velocity $\mathbf{f = f^+(v^2)}$ or one-dimensional BKW solution

If the ansatz polynomial is of degree $2n_+$, we find $v^{2n_+ + 2}$ and v^{4n_+} for the highest ℓ.h.s and r.h.s monomials, hence $n_+ = 1$

$$f^+\sqrt{2\pi}\exp(b(t)v^2/2) = \alpha_0(t) + \alpha_2(t)v^2/2$$

We substitute into (3.2), define the moment $\sigma_2 = \int \sigma \sin^2\theta\cos^2\theta > 0$, use $\int_{-\infty}^{+\infty} e^{-bu^2/2} u^{2p} = \Gamma(p+1)(2/p)^{p+\frac{1}{2}}$ and from the monomials const.,v^2, v^4 terms obtain three relations: $-b_t b^{1/2} = \alpha_2\sigma_2$, $4b^{3/2}\alpha_{0t} = 3\alpha_2^2\sigma_2$, $\alpha_{2t} - b_t\alpha_0 = -3\sigma_2\alpha_2^2 b^{-3/2}$. Two of them can be replaced by the two conservation laws $\int f dv = \int f v^2 dv = 1$ or $\alpha_2 = (b-1)b^{3/2}$, $\alpha_0 = (3-b)b^{1/2}/2$.

This leads to the integrable equation $b_t = b(1-b)\sigma_2$ for b and finally to the solution found by Ernst

$$f^+ = f^+_{BKW} = \left(e^{-v^2 b/2}/2\sqrt{2\pi b^{-1}}\right)(3-b+b(b-1)v^2) \xrightarrow[t \longrightarrow \infty]{} f_{AM} = e^{-v^2/2}/\sqrt{2\pi}$$
$$b^{-1} = 1 - ce^{-\sigma_2 t}, \quad 0 < c < 2/3, \quad f^+ > 0. \tag{3.3}$$

3.1.b. Odd partner $f^-(v)$ to f^+_{BKW}

For a sum $f = f^+(v^2) + f^-(v)$ the Kac model has two equations: one even part with f^+_{BKW} solution and an odd part $(\theta, w \longrightarrow -\theta, -w$ then $v', w' \longrightarrow v', -w')$.

$$f_t = \int_{-\pi}^{+\pi} d\theta\sigma(\theta) \int_{-\infty}^{+\infty} dw(f^+(w')f^-(v') - f^-(v)f^+(w)) \tag{3.4}$$

mixing $f^{\pm}$. For the positivity of the sum f^- must increase less than f^+, hence $f^- 2\sqrt{\pi} = \alpha_1 v\exp - bv^2/2$. We substitute into (3.4), define new moments $\tau_m = \int \sigma(\theta)\cos\theta^m d\theta$, find from the two v and v^3 monomials $-b_t = b(b-1)(\tau_1 - \tau_3)$, $\partial_t \log\alpha_1 = \tau_1 - \tau_0 + 3/2 \log b$ and obtain the complete solution with b in (3.3)

$$f(v,t) = f^+_{BKW} + ve^{-bv^2/2} db^{3/2} e^{(\tau_1 - \tau_0)t}/\sqrt{2\pi} \xrightarrow[t \longrightarrow 1]{} e^{-v^2/2}/\sqrt{2\pi}, \; \tau_0 > \tau_1,$$
$$\sigma_2 - \tau_1 + \tau_3 = 0, \quad f > 0 \;\text{ if }\; d^2 \le 2c(2-3c)/(1-c) \tag{3.5}$$

3.1.c. Inhomogeneous $f(v,x,t) = f^+ + f^-$ solution

We have two coupled equations for $f^{\pm}(v,x,t)$

$$f_t^+ + vf_x^- = \int\int \sigma(\theta)\left[f^+(w')f^+(v') - f^+(w)f^+(v)\right] dw d\theta \tag{3.6a}$$

$$f_t^- + vf_x^+ = \int\int \sigma(\theta)\left[f^+(w')f^-(v') - f^-(v)f^+(w)\right] dw d\theta \tag{3.6b}$$

Still choosing as ansatz the product of a gaussian with $b(t,x)$ by polynomials and requiring positivity for the highest monomials we find a v^2 even and v odd polynomials but $b_x = 0$

$$\sqrt{2\pi} f^+ e^{b(t)v^2/2} = \alpha_0(t,x) + \alpha_2(t,x)v^2/2, \quad \sqrt{2\pi} f^- e^{b(t)v^2/2} = \alpha_1(t,x)v/2$$

In (3.6a) vf_x^- enters only in the v^2 term so that we have the two previous const. and v^4 relations $\alpha_2\sigma_2 = -b^{1/2}b_t$, $4\alpha_{0t} = \sigma_2\alpha_2^2 b^{-5/2}$. We find that α_2 and α_0 are only t-dependent. At this stage $f^+(v,t)$, $\int fv^i dv = N_i(t)$, $i = 0,2$ are x-independent. What about f^-? Let us define the momentum $\mathcal{J}_0 = \int fvdv = \alpha_1 b^{-3/2}/\sqrt{2}$ and apply the mass conservation law $N_{0t} + \mathcal{J}_{0x} = 0$. We see that $\mathcal{J}_0$ and α_1 are linear in x so that $f^- = (x - x_0)\tilde{f}^-(v,t)$.

Coming back to $f_t + vf_x = \text{Col}f$, multiplying by $1, v, v^2, v^3$ and integrating over v we obtain two linear conservation laws and two nonlinear relations. We define $\mathcal{J}_i = \int v^{1+i} f dv = (x - x_0)\tilde{\mathcal{J}}_i(t)$, $i = 0, 2$ and find a system of (1+1)-dimensional NLPDE which are integrable.

$$N_{it} + \tilde{\mathcal{J}}_i = 0, \; i = 0,2, \; \tilde{\mathcal{J}}_{0t} = (\tau_1 - \tau_0)N_0\tilde{\mathcal{J}}_0, \; \tilde{\mathcal{J}}_{2t} = (\tau_3 - \tau_0)N_0\tilde{\mathcal{J}}_2 + 3\tilde{\mathcal{J}}_0 N_2(\tau_1 - \tau_3)$$

We obtain two classes of solutions relaxing: (i) towards Maxwellians and $N_0(t) \longrightarrow$ Const., (ii) towards vacuum, $N_0(t) \longrightarrow 0$. We present the (i) class. From the density $N_0 = (1-y)/(1+y)$, $y = c_1 e^{-c_2 t}$, $c_2 > 0$ we get the energy N_2 and then $\mathcal{J}_0$, $\mathcal{J}_2$.

$$[3(\tau_1 - \tau_3) + (\tau_3 - \tau_0)N_0\partial_t]N_2 = N_{2tt}, \quad \mathcal{J}_i = (x_0 - x)N_{it}$$

This gives the only known inhomogeneous exact solution (without external force) which relaxes toward an Absolute Maxwellian:

$$\begin{cases} \sqrt{2\pi} f = b^{3/2} e^{-bv_2/2}[b^{-1}(N_0 + (1/2\sigma_2)\partial_t \log b + v(x_0 - x)N_{0t} - (v^2/2\sigma_2)\partial_t \log b] \\ ((2\partial_t N_0 - N_0\partial_t)\nu\sigma_2 - \partial^2_{t^2})b^{-1}(t) = 0, \; \sigma_2 = \tau_1 - \tau_3 = \mu(\tau_0 - \tau_1) = c_2\mu/\nu, \; 0 < \mu < 2 \end{cases} \tag{3.7}$$

The positivity is satisfied for $| x - x_0 | <$ const$_1$ (exp const$_2 t$), interval going to infinity when $t \longrightarrow \infty$. The density $N_0 \longrightarrow 1$ when $t \longrightarrow \infty$ with either an expansion $c_1 < 0$ or a contraction $c_1 > 0$. When $t \longrightarrow \infty$, $b \longrightarrow 1$, $b_t \longrightarrow 0$, $N_{0t} \longrightarrow 0$ and $f \longrightarrow (\sqrt{2\pi})^{-1}$ exp-$v^2/2$ an absolute Maxwellian.

References

Kac M., "Fundation of Kinetic Theory", 3rd Berkeley Symp. on Math. Stat. and Prob. vol. III, 1956; Uhlenbeck G.L. and Ford C.W. "Lectures in Statistical Mechanics", Ed. Kac M., Am. Math. Soc. Providence p.99, 1963; Henin F., Bul. Ac. Roy Bel. LX, 721, 1974; Ernst M., Phys. Lett. A61, 320, 1979; Cornille H., J.Phys. A17, L234, 2355, 1984, JMP 26, 1203, 1985, J.Phys. A18, 1209, 1985, TTSP 16, 795, 1987.

3.2. Three-dimensional homogeneous BE and BKW solution

We rewrite the BE for pseudo-Maxwell molecules with a cross-section $\sigma(\theta)$

$$4\pi f(\vec{v}, t) = \int d\vec{w} \sin\theta d\theta d\varepsilon \sigma(\theta)[f(\vec{w}')f(\vec{v}') - f(\vec{w})f(\vec{v})]$$
$$\begin{pmatrix} v'^2 \\ w'^2 \end{pmatrix} = \begin{pmatrix} v^2 \\ w^2 \end{pmatrix} + [(w^2 - v^2)\sin^2\theta/2 + | \vec{v} || \vec{w} | \sin\theta\cos\varepsilon\sin(\vec{v}, \vec{w})] \begin{pmatrix} 1 \\ -1 \end{pmatrix} \tag{3.8}$$

3.2.a. Ansatz gaussian multiplied by an even polynomial

The balance ℓ.h.s versus r.h.s still gives v^{2n+2}, v^{4n} and so

$$(2\pi b^{-1})^{3/2} f(v^2,t) = \exp(-v^2 b(t)/2)[\alpha_0(t) + \alpha_2(t)v^2/2]$$

Like for the Kac model we have three relations coming from const, v^2, v^4 and we replace two of them by the mass and energy conservation laws. Due to $f_{AM}(2\pi)^{3/2} = \exp{-v^2/2}$ we find $N_2 = \int f v^2 d\vec{v} = 3N_0 = 3\int f d\vec{v} = 3$ or $1 = \alpha_0/b + 5\alpha_2/2b^2 = \alpha_0 + 3\alpha_0/2b$ or $\alpha_2 = b(b-1)$, $2\alpha_0 = 5 - 3b$. For the v^4 contribution we note that only $4v'^2 w'^2 \simeq v^4 \sin^2\theta + \mathcal{O}(v^4)$ contributes in the r.h.s.

$$\pi^{-3}\alpha_2^2/16\ \mathrm{e}^{-bv^2/2} \int_0^{\pi} d\theta \int_0^{\infty} dw\ \mathrm{e}^{-bw^2} w^2 \sigma(\theta) \sin^3\theta$$

We define a moment $8\sigma_2 = \int \sigma(\theta)\sin^3\theta d\theta$, obtain $-b_t = \sigma_2 b(b-1)$, deduce $b^{-1} = 1 - c\ \mathrm{e}^{-\sigma_2 t}$, $f > 0$ if $0 < c < 2/5$ and find

$$f_{BKW} = (2\pi b^{-1})^{-3/2} \mathrm{e}^{-v^2 b/2}[5 - 3b + v^2 b(b-1)]/2 \xrightarrow[t \longrightarrow \infty]{} f_{AM} \tag{3.9}$$

3.2.b. NLPDE associated to the B.E.

We assume $\sigma(\theta) \simeq (\sin^2\theta)^{q-1}$ and introduce a transform with the confluent hypergeometric function ${}_1F_1$. We define

$$\Gamma(3/2)H(p,t) = \int_0^{\infty} dv\ v_1^{1/2} F_1(q, 3/2, -pv) f(v,t) dv \tag{3.10}$$

which represents generating functionals of power moments of f, and obtain a class of NLPDE for integer q values

$$(\partial_t + 1) \prod_{m=0}^{q-1} (1 + \partial_y/(q+m))\, H(y = \log p/(p+1), t) = H^2 \tag{3.11}$$

which reduces to (3.1) when $q = 1$ and σ is a constant. For this $q = 1$ case exist nontrivial similarity and (1+1)-dimensional solutions

$$\begin{aligned} H_i &= 1/D,\ \ 6u/D,\ \ (1+2u)/D,\ \ i = 1,2,3,\ \ D = (1+u)^2,\ \ u = \exp(\rho t + \gamma y) \\ H_4 &= 6u_1u_2/\Delta,\ \ H_5 = u_2(2 + u_1 + u_2)/\Delta,\ \ \Delta = (1 + u_1 + u_2)^2,\ \ u_i = \exp(\rho_i t + \gamma_i y) \end{aligned} \tag{3.12}$$

Unfortunately only H_3 (with $u = p\mathrm{e}^{-t/6}/(p+1)$) does not violate some physical property of the BE. Using the Laguerre polynomials formula

$$\Gamma(3/2)p^n/(p+1)^{n+1} = \int {}_1F_1(1, 3/2, -pu) L_n^{1/2}(u) du$$

we go back to f and find for H_3 the BKW solution associated to $q = 1$. For the more general $q \neq 1$ class of NLPDE we find other solutions for instance $(1+(q+1)u)/(1+u)^{q+1}$, $(1+qu_2+u_1)/(1+u_1+u_2)^{q+1}, \ldots$ but the only physical solutions still correspond to BKW. For these NLPDE exist also nonrational solutions, for instance for (3.1) a Weirstrassian equianharmonic elliptic function

$$H = 1 - u^2\mathcal{P}(u = \mathrm{e}^{-t/6}p/(p+1), \quad g_2 = 0, \quad g_3 \neq 0), \quad \mathcal{P}_{uu} = \mathcal{P}^2 \tag{3.13}$$

$f > 0$ requires $H(u) > 0$ for $u > 0$ but H violates positivity an infinite number of times along $u > 0$.

References

Bobylev V., Sov. Phys. Dokl. 20, 823, 1976, Theor. Mat. Phys., 8, 279, 1984, and the recent monograph Sov. Sci. Rev. C. Math. Phys. 7, 111, 1988; Krook M. and Wu T.T., Phys. Rev. Lett. 16, 1107, 1976; Ernst M., Phys. Lett. A61, 320, 1979. For the NLPDE associated to the BE: Krook M. and Wu T.T. Phys. Fluids 20, 1989, 1979, (eq. (3.1)); Cornille H. and Gervois A., Phys. Lett. 83A, 251, 1981, Ernst M. and Hendriks E.M. Phys. Lett. 81A, 371, 1981, (eq. (3.10)), Cornille H., J. Stat. Phys. 45, 611, 1986; For a class of NLPDE sharing common properties with B.E.: Cornille H. and Gervois A., Physica 6D, 29, 1982, Cornille H., Phys. Let. A91, 211, 1982, J. Phys. A15, L529, 1982, JMP 25, 1335, 1984. For the "nonisotropic BE" see Hendriks E.M. and Nieuwenhuizen T.M., J. Stat. Phys. 29, 591, 1982; for the "extended BE" see Spiga G., ASMF, Univ. Modena 36, 355, 1988; Dukek G. and Spiga G., ZAMP, 39, 924, 1988,...

3.3. Inhomogeneous solutions relaxing toward $\mathbf{f_{AM}}$

We comeback to the inhomogeneous BE of section 1: $\mathcal{L}f = \mathrm{Col}f, \mathcal{L} = \partial_t + \vec{v}.\partial_{\vec{x}} + \partial_{\vec{v}}.\vec{A}$, $I = \mid \vec{g} \mid^{-4/(\nu-1)}$ with an external force $\vec{A}$ and an inverse power law ν for the interparticle force. Noticing that an exact inhomogeneous solution relaxing toward f_{AM} is missing we remark:

(i) There exists an inhomogeneous solution for the Kac model but this model violates momentum conservation law.

(ii) If we apply the Nikolski Transform (which deduces inhomogeneous solution from homogeneous one) to BKW we obtain the Bobylev-Muncaster solution:

$$f(2\pi(1-\phi))^{3/2} = \mathrm{e}^{-C^2t^2/2(1-\phi)}[1 + \phi(-3 + C^2t^2/(1-\phi))/2(1-\phi)] \tag{3.14}$$

with $\vec{c} = \vec{v} - \langle\vec{v}\rangle$ for the peculiar velocity $C^2 = \vec{c}^2$ and when $t \longrightarrow \infty$, $\phi \longrightarrow$ const. and $f \longrightarrow \exp\text{-}\vec{c}^2t^2 \longrightarrow 0$. The equilibrium state is the vacuum.

(iii) f is a local Maxwellian if Col$f = 0$ or

$$f_{Loc.Max}(2\pi T)^{3/2} = e^{-C^2/2T}, \quad \mathcal{L}(f_{LM}) = 0 \tag{3.15}$$

One century ago Boltzmann has obtained exact f_{LM} solutions with the introduction of an harmonic external potential. Can we extend these Boltzmann exact solutions when Col$f \neq 0$?

3.3.a. Nikolski Transform with external forces: $\mathcal{L} = \partial_t + \vec{\mathbf{v}}.\partial_{\vec{\mathbf{x}}} + \vec{\mathbf{A}}.\partial_{\vec{\mathbf{v}}}$

Let f be an inhomogeneous solution $\mathcal{L}f =$ Col$f$, F an homogeneous one $dF/d\tau =$ Col$F$ and assume

$$f(\vec{v}, \vec{x}, t) = f(\vec{\eta}(\vec{v}, \vec{x}, t), \tau(t)) \equiv F(\vec{\eta}, \tau), \quad \vec{\eta} = \gamma(\vec{v} - \vec{v}_0) \tag{3.16}$$

with γ and $\vec{v}_0$ unknown $\vec{x}, t-$dependent functions. We show that the following properties hold

$$\vec{\eta} = \vec{c}\, T^{-1/2}, \; \rho = T^{3/2}, \; \Sigma f_{\eta_j} \mathcal{L}(\eta_j) = 0, \; d\tau/dt = T^{2(1-1/(\nu-1))} \tag{3.17}$$

$\vec{c}$, ρ and T being the peculiar velocity, density and temperature associated to f while $\vec{\eta}$, τ are the homogeneous momentum and time. For the two first (3.17) relations we compare macroscopic quantities in both homogeneous and inhomogeneous formalisms. For F we have $N_0 = \int F d\vec{\eta} = 1$, $N_2 = \int F \vec{\eta}^2 d\vec{\eta} = 3$ and we choose $N_0 \langle \vec{\eta} \rangle = \int \vec{\eta} F d\vec{\eta} = 0$. We notice that $d\vec{\eta} = \gamma^3 d\vec{v}$ and obtain for f : $\rho = \int f d\vec{v} = \gamma^{-3}$, $\langle \vec{v} \rangle \rho = \int f \vec{v} d\vec{v} = \gamma^{-3} \int d\vec{\eta} f(\vec{\eta}/\gamma + \vec{v}_0) = \vec{v}_0 \rho$, hence $\vec{c} = \vec{v} - \vec{v}_0$. For the temperature $3T\rho = \int f \vec{c}^2 d\vec{v} = \gamma^{-5} N_2 = 3\gamma_2 \rho$ we deduce both $T = \gamma^{-2}$ and the two first (3.17) relations. For the two last ones we compare Col$f$ and Col$F$ which contain the interparticle forces $|\, \vec{g}\, |^{1/(4-\nu)}$, $|\, \vec{G}\, |^{1/(4-\nu)}$ with $\vec{g}$, $\vec{G}$ for the relative homogeneous and inhomogeneous velocities. Noticing $d\vec{v} = T^{3/2} d\vec{\eta}$, $|\, \vec{g}\, | = |\, \vec{G}\, |\; T^{1/2}$ we deduce Col$f = T^{2(1-1/(\nu-1))}$Col$F$. Then $\mathcal{L}f = \dot{\tau} f_\tau + \Sigma f_{\eta_j} \mathcal{L}(\eta_j) = T^{2(1-1/(\nu-1))}ColF$ is equivalent to the homogeneous $F_\tau =$Col$F$ if the two last (3.17) relations hold.

In conclusion to any homogeneous solution we can associate an inhomogeneous one $f(\vec{\eta} = \vec{c}T^{-1/2}, \tau(t))$ provided $\mathcal{L}(\eta_j) = 0$, $\tau = \int^t T^{2(1-1/(\nu-1))} dt'$. Furthermore, here $\rho = T^{3/2}$ for the density, doing a similar restriction into the Boltzmann study of $\mathcal{L}(f_{LM}) = 0$ gives in (3.15) the condition $\mathcal{L}(\vec{\eta}^2 = \Sigma \eta_j^2 = \vec{c}^2/T) = 0$. There is a closed analogy between the one-century ago Boltzmann calculation of the exact solutions for Col$f = 0$ and the present one for the determination of inhomogeneous solutions deduced from homogeneous ones.

The last problem is the determination of compatible forces $\vec{A}$, temperature T and mean velocity $\langle\vec{v}\rangle$ so that $\mathcal{L}(\eta_j) = 0$ or

$$(\partial_t + \Sigma v_i \partial_{x_i} + A_i \partial_{v_i})(T^{-1/2}(v_j - \langle\vec{v}\rangle_j)) = 0 \tag{3.18}$$

For this polynomial with const., v_j, $v_i v_j$ terms, we require that their coefficients vanish. We obtain

$$\vec{A} = a(t)\vec{x} + \vec{A}_0(t),\ \partial^2_{t^2} T^{-1/2} = a(t)T^{-1/2}$$

and choosing for simplicity $\vec{A}_0 = 0$ then $2\langle\vec{v}\rangle = -\vec{x}\partial_t \log T$. We find an harmonic potential and a linear differential equation for the temperature. If the external force is missing, or $a(t) = 0$, then $T \simeq t^{-2} \longrightarrow 0$, $\rho \simeq t^{-3} \longrightarrow 0$ and like in the Bobylev-Muncaster solution (3.14), we find a relaxation toward vacuum. On the contrary if $a(t) \neq 0$, $a(t)$ decreasing more than t^{-2}, then, like in the Schrödinger equation, we can define a "Jost solution"

$$T^{-1/2} = \text{const} + \int_t^\infty (t' - t)T^{-1/2}(t')a(t')dt' \ \text{ with } \ T(\infty) = \text{const.} \tag{3.19}$$

For instance exist explicit Bessel solutions: if $a(t) = ae^{-t/2}$ then $T^{-1/2} \equiv I_0\,(2ae^{-t/2})$ or $\mathcal{J}_0(2ae^{-t/2})$, $a < 1.44$. For these $a(t)\vec{x}$ forces, both the temperature and the density go to constants and the relaxation is not toward vacuum. All these results are independent of the interparticle power law force. However if we want to construct exact inhomogeneous solutions we need homogeneous ones and so we go back to the Maxwell molecules.

3.3.b. Maxwell Molecules

We start with the homogeneous F_{BKW} and construct the inhomogeneous solution associated with an harmonic potential $\vec{A} = a(t)\vec{x}$, $a(t) \longrightarrow 0$ when $t \longrightarrow \infty$ such that $\partial^2_{t^2} T^{-1/2} = a(t)T$ with $T(\infty)$ =const, $\vec{C} = \vec{v} + (\vec{x}/2)\partial_t \log T$, $\rho \simeq T^{3/2}$ and

$$f(\vec{v},\vec{x},t)(2\pi(1-\phi))^{3/2} = e^{-\vec{C}^2/2T(1-\phi)}[1 + \vec{\phi}(\vec{C}^2(T(1-\phi))^{-1} - 3)/2(1-\phi)], \tag{3.20}$$

$$\phi = d\exp -\sigma_2 \int_0^t T^{3/2}(t')dt' \xrightarrow[t \longrightarrow \infty]{} 0,\ 0 < d < 2/5,\ (2\pi)^{3/2} f \longrightarrow \exp -\vec{C}^2/2T(\infty)) = f_{AM}$$

This forces disappears at infinite time $(a(t) \longrightarrow 0)$ but, acting like a screen, the equilibrium state is an absolute Maxwellian. Up to now this is the only known exact inhomogeneous example relaxing toward the true equilibrium state and not toward the trivial vacuum state.

References

Boltzmann L. "Wissenschaptlich Abhandlunger" Edt. Hasenorl F., Leipzig Vol.II, p.83, 1909; Nikolskii A., Sov. Phys. Dokl. 8, 633, 1964; Bobylev V., Sov. Phys. Dokl. 20, 823, 1976; Muncaster R.C., Arch. Rat. Mech. Anal. 70, 115, 1979; Cornille H., J. Phys. A18, L839, 1985, JMP 27, 1373, 1986, J. Stat. Phys. 45, 611 (1986), TTSP 16, 155, 1987.

NONLINEAR INTERACTION BETWEEN SHORT AND LONG WAVES

V.I.Karpman

IZMIRAN, Troitsk, Academic City, Moscow Region 142092, USSR

Abstract

A unified Hamiltonian system of evolution equations describing ponderomotive interaction between a high- and low-frequency wave is derived and investigated. It is applied to the modulational instability and parametric decay of the high-frequency wave due to that interaction. We show that in the linear approximation they have almost identical dispersion relations, but the nonlinear equations for them are different: the parametric decay equations are the lowest order truncated system, obtained from the equations describing the modulational instability. The nonlinear developments of both instabilities are compared.

An approximate analytical theory of the cells arising as a result of the modulational instability is developed. The results obtained are in good agreement with numerical solutions of basic equations in (1 + 2) dimensions.

R. Conte and N. Boccara (eds.), Partially Integrable Evolution Equations in Physics, 83–159.

1. Introduction. Basic equations.

We consider a class of nonlinear equations describing the ponderomotive interaction between short (high-frequency) and long (low-frequency) waves. Despite of different appearance, they can be reduced to a unified Hamiltonian form. This permits to develop a general approach to their investigation and solution, which is the main purpose of this paper.

In all cases considered we assume that the high-frequency wave has the form

$$Re\left\{\Psi(\underset{\sim}{R},t)\exp[i(kz-\omega t)]\right\} \tag{1.1}$$

where $\Psi(\underset{\sim}{R},t)$ is a slow function of t and $\underset{\sim}{R}=\{x,y,z\}$, and the carrier wave vector $\underset{\sim}{k}$ is directed along the z-axis. The carrier frequency $\omega=\omega(\underset{\sim}{k})$ is defined by linear dispersion law in homogeneous media. Then the complex amplitude $\Psi(\underset{\sim}{R},t)$ satisfies nonlinear Schrödinger equation of the form (e.g.,/1/)

$$i\left(\frac{\partial\Psi}{\partial t}+v_g\frac{\partial\Psi}{\partial z}\right)+\frac{1}{2}\left(S_1\Delta_\perp\Psi+S_2\frac{\partial^2\Psi}{\partial z^2}\right)=\delta\omega\Psi, \tag{1.2}$$

where $\Delta_\perp=\partial^2/\partial x^2+\partial^2/\partial y^2$,

$$v_g=(\partial\omega/\partial k_z)_{k_\perp=0}$$

$$S_1=(\partial^2\omega/\partial k_\perp^2)_{k_\perp=0} \tag{1.3}$$

$$S_2=(\partial^2\omega/\partial k_z^2)_{k_\perp=0}$$

and $\delta\omega$ is the nonlinear frequency shift. In all systems considered below this quantity is proportional to the variation of medium density, i.e.

$$\delta\omega = c_1 \nu, \qquad \nu = (\rho - \rho_0)/\rho_0 \tag{1.4}$$

where $\rho(\underset{\sim}{R}, t)$ is the density and ρ_0 the unperturbed density of the medium and c_1 is one of the constants characterizing wave-medium interaction.

Equations (1.2) and (1.4) must be supplemented by medium equations, defining the density variation in the presence of the wave (1.1). Here we assume that they are of acoustic type, i.e. the interaction of the high-frequency wave with the medium produces acoustic type perturbations which represent the low-frequency, or long, waves. As typical illustrations, we consider below some important examples from plasma physics. It appears that the corresponding equations are of more general character. For instance, they can describe nonlinear wave dynamics in isotropic and anisotropic molecular systems.

In section 2 such equations are written in a unified Hamiltonian form, which is convenient for investigation of different nonlinear processes resulting from the interaction between short and long waves. Typical and important examples are modulational instability and parametric decay; they are considered in section 3. One of the consequences of the modulational instability is the formation of nonlinear cells (or lumps). In section 4 we consider a two-dimensional example of such processes, when the cells appear as a result of the two-mode initial perturbations of a plane high frequency wave.

1.1. Interaction of electromagnetic and ion-sound waves in isotropic collisionless plasma

From the Maxwell equations follows

$$\mathrm{rot}\,\mathrm{rot}\,\underset{\sim}{E} = -\frac{1}{c^2}\frac{\partial^2}{\partial t^2}\hat{\epsilon}\,\underset{\sim}{E} \tag{1.5}$$

where $\hat{\epsilon}$ is the dielectric permeability operator of an inhomogeneous and nonstationary plasma with frequency and space dispersion. Assuming the conditions

$$v_T/c \ll 1, \quad (k r_D)^2 \ll 1, \quad |\gamma| \ll 1 \tag{1.6}$$

where $v_T = (T_e/m_e)^{1/2}$ is the electron thermal velocity (indices e and i are related to electrons and ions), $\omega_p = (4\pi N_0 e^2/m_e)^{1/2}$ is the equilibrium plasma frequency, N_0 unperturbed electron density, $r_D = \omega_p/v_T$ is the Debye radius, and neglecting the terms of higher order than in (1.6), we may write $\hat{\epsilon} = \hat{\epsilon}^0 + \delta\hat{\epsilon}$ where $\delta\hat{\epsilon}$ is the part arising from inhomogeneity and nonstationarity of the plasma. The explicit expressions are

$$(\hat{\epsilon}^0 \underset{\sim}{E})_\alpha = \hat{\epsilon}^0_{\alpha\beta} E_\beta, \quad (\delta\hat{\epsilon}\underset{\sim}{E})_\alpha = \delta\hat{\epsilon}_{\alpha\beta} E_\beta,$$

$$\hat{\epsilon}^0_{\alpha\beta} = \epsilon^0_{\alpha\beta}(\hat{\omega}, \hat{k}), \quad \delta\hat{\epsilon}_{\alpha\beta} = \delta\epsilon_{\alpha\beta}(\hat{\omega}; t, \underset{\sim}{R}), \tag{1.7}$$

$$\hat{\omega} = i\partial/\partial t, \quad \hat{k}_\alpha = -i\partial/\partial x_\alpha,$$

$$\epsilon^{0}_{\alpha\beta}(\omega,k) = \delta_{\alpha\beta}\,(1-\omega_p^2/\omega^2 - k^2 v_T^2\,\omega_p^2/\omega^4) - 2k_\alpha k_\beta\, v_T^2\,\omega_p^2/\omega^4, \tag{1.8a}$$

$$\delta\epsilon_{\alpha\beta}(\omega;\underset{\sim}{R},t) = -(\omega_p^2/\omega^2)\,\vartheta(\underset{\sim}{R},t)\,\delta_{\alpha\beta} \tag{1.8b}$$

For $\underset{\sim}{E}(\underset{\sim}{R},t)=\underset{\sim}{E}^{0}(\underset{\sim}{R},t)\exp[i(\underset{\sim}{k}\underset{\sim}{R}-\omega t)]$ where $\underset{\sim}{E}^{0}(\underset{\sim}{R},t)$ is a slow function of t and R, one has /2-6/

$$\epsilon_{\alpha\beta}(\hat{\omega},\hat{k};\underset{\sim}{R},t)E_\beta(\underset{\sim}{R},t) = \Big\{\epsilon_{\alpha\beta}(\omega,k;\underset{\sim}{R},t) + \frac{i}{2}\Big[\frac{\partial^2\epsilon_{\alpha\beta}(\omega,\underset{\sim}{k};\underset{\sim}{R},t)}{\partial\omega\,\partial t} - \frac{\partial^2\epsilon_{\alpha\beta}(\omega,\underset{\sim}{k};\underset{\sim}{R},t)}{\partial k_\gamma\,\partial R_\gamma}\Big]+\dots\Big\}_{\substack{\omega=\hat{\omega}\\ k=\hat{k}}} E_\beta(\underset{\sim}{R},t) \tag{1.9}$$

With accepted accuracy, from (1.5) and (1.7)-(1.9) we have the following equation

$$(\nabla_\alpha\nabla_\beta - \delta_{\alpha\beta}\Delta)E_\beta = c^{-2}\Big[(\hat{\omega}^2-\omega_p^2)E_\alpha + v_T^2\,\omega_p^2\,\hat{\omega}^{-2}(\Delta E_\alpha + 2\nabla_\alpha\nabla_\beta E_\beta) - \omega_p^2\,\vartheta E_\alpha\Big] \tag{1.10}$$

When transforming the right hand side of eq. (1.10), we have used the general relation

$$F(\hat{\omega})\{E^{0}(t)e^{-i\omega t}\} = \Big[F(\omega)E^{0}(t) + i\frac{\partial F}{\partial\omega}\frac{\partial E^{0}}{\partial t} - \frac{1}{2}\frac{\partial^2 F(\omega)}{\partial\omega^2}\frac{\partial^2 E^{0}}{\partial t^2}+\dots\Big]e^{-i\omega t} \tag{1.11}$$

Now we assume $T_e \gg T_i$. Then the Landau damping can be neglected and plasma motion may be described by the hydrodynamic equations with account of the ponderomotive force from the field $\underset{\sim}{E}$. At $|\nu| \ll 1$ these equations can be linearized with respect to ν. As a result we arrive at the following equation

$$\nu_{tt} - c_s^2 \Delta \nu = \frac{\omega_p^2}{16\pi \rho_0 \omega^2} \Delta (E_\alpha^* E_\alpha) \tag{1.12}$$

where c_s is the ion sound velocity

$$c_s \approx (T_e/m_i)^{1/2}. \tag{1.13}$$

In the right hand side of (1.12) we have neglected the contribution of the space dispersion and also terms of the order of $\nu|E|^2$.

Equations (1.10) and (1.12) form a closed system. Its solutions describe both longitudinal and transverse waves. In the first case we assume

$$\underset{\sim}{E} = -\nabla \varphi(\underset{\sim}{R}, t)\, e^{-i\omega_p t}, \tag{1.14}$$

where $\varphi(\underset{\sim}{R}, t)$ is a slow function of t and

$$|\nabla \varphi|^2 r_D^2 / \varphi^2 \ll 1 \tag{1.15}$$

Then we arrive at the equation

$$2i\omega_p \partial \underset{\sim}{E}/\partial t + 3v_T^2 \Delta \underset{\sim}{E} = \omega_p^2 \nu E \tag{1.16}$$

Here we have used relation (1.11) where the terms with $\partial^2 E^\circ/\partial t^2$ have been neglected. The system of equations (1.14), (1.16) and (1.12) is known as Zakharov equations /7/.

Along with that, equations (1.10) and (1.12) have solutions describing the transverse field with

$$\operatorname{div} \underset{\sim}{E} = 0 \tag{1.17}$$

In this case eq. (1.10) takes the form

$$\partial^2 \underset{\sim}{E}/\partial t^2 - c^2[1+(v_T^2/c^2)\omega_p^2 \hat{\omega}^{-2}]\Delta \underset{\sim}{E} = \omega_p^2(1+\nu)\underset{\sim}{E} \tag{1.18}$$

Neglecting the terms of the order of

$$(v_T/c)^2 (\omega_p/\omega)^2 \tag{1.19}$$

where ω is the characteristic frequency of the field $\underset{\sim}{E}$, we have the equation

$$\partial^2 \underset{\sim}{E}/\partial t^2 - c^2 \Delta \underset{\sim}{E} = \omega_p^2 (1+\nu)\underset{\sim}{E} \; . \tag{1.20}$$

The system of equations (1.20) and (1.12) was considered in ref. /8/, and the waves described by it were called there electroacoustic. In one dimensional case it has the same form as the Zakharov equations. One dimensional soliton solutions of the system (1.20) and (1.12) were investigated in ref./8/ and subsequent papers. Similar to Zakharov solitons they are unstable under transverse perturbations.

Assume in (1.10)

$$\underset{\sim}{E} = \underset{\sim}{E}^{0}(\underset{\sim}{R}, t) \exp[i(kz - \omega t)], \tag{1.21}$$

where $\underset{\sim}{E}^{0}(\underset{\sim}{R}, t)$ is a slow function of t and R. Then $\underset{\sim}{E}^{0}$ satisfies a nonlinear Schrödinger equation. To derive it we suppose that

$$\underset{\sim}{E}^{0} = \underset{\sim}{E}^{0}(\lambda \underset{\sim}{R}, \lambda t, \lambda^2 t, \dots),$$
$$\underset{\sim}{E}^{0} = \underset{\sim}{E}^{00} + \lambda \underset{\sim}{E}^{01} + \lambda^2 \underset{\sim}{E}^{02} + \dots, \quad \nu \sim \lambda^2 \tag{1.22}$$

where $\lambda \ll 1$ (after calculations we put $\lambda = 1$). Then, in the zeroth approximation we obtain two branches

$$\text{(i)} \quad E_z^{00} = 0, \quad \omega^2 = \omega_p^2 + k^2 c^2, \tag{1.23}$$

$$\text{(ii)} \quad \underset{\sim}{E}_{\perp}^{00} = 0, \quad \omega = \omega_p [1 + (3/2)(k r_D)^2] \tag{1.24}$$

(the symbol $\perp$ means a direction in the plane perpendicular to the z -axis). Deriving (1.23) and (1.24), we also neglected the terms of order of (1.19) and $(k r_D)^4$.

Consider case (i). Here vector $\underset{\sim}{E}^{00}$ is perpendicular to the z -axis. In next approximations we obtain

$$E_z^{01} = (i/k)(\nabla_{\perp} \underset{\sim}{E}_{\perp}^{00}), \; E_z^{02} = -(1/k^2)\nabla_z (\nabla_{\perp} E_{\perp}^{00}) \tag{1.25a}$$

Without loss of generality we assume

$$\underset{\sim}{E}_{\perp}^{01} = \underset{\sim}{E}_{\perp}^{02} = 0 \tag{1.25b}$$

From (1.25) and (1.23), including terms of the order of λ^2, we obtain

$$div\, \underset{\sim}{E} \equiv div\left\{ \underset{\sim}{E}^{\circ} \exp(ikz) \right\} = 0 \tag{1.26}$$

Then, considering λ^2-approximation, we arrive at the following equation for $E^{\circ\circ}$

$$i\left(\frac{\partial \underset{\sim}{E}^{\circ\circ}}{\partial t} + v_g \frac{\partial \underset{\sim}{E}^{\circ\circ}}{\partial z}\right) + \left(S_1 \Delta_\perp \underset{\sim}{E}^{\circ\circ} + S_2 \nabla_z^2 \underset{\sim}{E}^{\circ\circ}\right) = (\omega_p^2/2\omega)\, \nu \underset{\sim}{E}^{\circ\circ} \tag{1.27a}$$

where v_g, S_1 and S_2 are defined by formulas (1.3) with account of (1.23), i.e.

$$v_g = c^2 k/\omega, \quad S_1 = c^2/\omega, \quad S_2 = c^2\omega_p^2/\omega^3 \tag{1.27b}$$

Now consider case (ii), when the vector $\underset{\sim}{E}^{\circ\circ}$ is parallel to the z-axis. Then

$$\underset{\sim}{E}_\perp^{01} = -(i/k)\nabla_\perp E_z^{\circ\circ}, \quad \underset{\sim}{E}_\perp^{02} = (1/k^2)\nabla_\perp \nabla_z E_z^{\circ\circ} \tag{1.28a}$$

Without loss of generality we assume

$$E_z^{01} = E_z^{02} = 0 \tag{1.28b}$$

Again, considering the second approximation with respect to λ, we obtain equation for $E^{\circ\circ}(\underset{\sim}{R}, t)$. It coinsides with (1.27a), where S_1 and S_2 are now obtained by substitution dispersion equation from (1.24) into (1.3).

This gives

$$v_g = 3\omega_p k r_D^2, \qquad S_1 = S_2 = 3\omega_p r_D^2 \qquad (1.29)$$

1.2. Interaction between electromagnetic and MHD waves in a magnetized plasma

There are several branches of electromagnetic escillations in a magnetized plasma which can represent high-frequency waves. Here we restrict ourselves by a sufficiently simple, but important case when the frequency ω is in the interval $\omega_{LH} \ll \omega < \omega_c$. Here $\omega_c = eB_0/cm_e$ is the electron hyrofrequency and $\omega_{LH} \approx (m_e/m_i)^{1/2}\omega_c$ the low hybrid frequency. Waves belonging to this branch are called whistlers. They have right polarization and the dispersion law

$$\omega \approx \omega_c k_z (k_z^2 + k_\perp^2)^{1/2} [(\omega_p/c)^2 + k_z^2 + k_\perp^2]^{-1}. \qquad (1.30)$$

Let us assume

$$B_0^2/4\pi \gg N(T_e + T_i), \qquad \omega_p^2 \gg \omega_c^2 . \qquad (1.31)$$

The first of conditions (1.31) means that we consider low-β plasma. Consitions (1.31) are often realized in laboratory plasmas. They also take place in a wide region of the near Earth plasma (ionosphere and magnetosphere) where whistlers play a fundamental role, similar to that of Langmuir waves in isotropic plasma. Collisional as well as collisionless damping are neglected. Assuming that the wave electric field is of the form (1.21) and using (1.22) we arrive at the Schrödinger equation (1.2) /9,10/, where

$$\Psi=(E_x-iE_y)/E_0,\quad E_0=[8u(1-u)]^{1/2}(\omega/\omega_p)B_0, \tag{1.32a}$$

$$v_g=2(1-u)v_\varphi,\quad S_1=(1-2u)v_\varphi/k,\quad S_2=2(1-u)(1-4u)v_\varphi/k,$$
$$v_\varphi=\omega/k=cu^{1/2}(1-u)^{1/2}\omega_c/\omega_p,\quad u=\omega/\omega_c, \tag{1.32b}$$

and

$$\delta\omega=-\omega[(1-u)\nu-b],\quad b=(B_z-B_0)/B_0 \tag{1.33}$$

(as in ref. /11,12/, we neglect the Doppler term in $\delta\omega$; for the processes which are investigated here, it is small).

Now let us consider equations for ν and b . Assuming that $T_e \gg T_i$, one can use MHD description of plasma, i.e. MHD equations supplemented by the ponderomotive force caused by the whistler wave field /9/. As was shown in refs. /11,12/, ν and b remain to be small even in a nonlinear stage. Therefore MHD equations can be linearized with respect to ν and b . (Nonlinearity, however, is contained in the ponderomotive force $f\sim\nabla|E|^2$ as well as in the right hand side of eq. (1.2).) Then the MHD equations can be transformed to the form /12/

$$\frac{\partial^2}{\partial t^2}(\nu - b) - c_s^2 \frac{\partial^2 \nu}{\partial z^2} = -c_A^2 \left(\frac{\partial^2}{\partial z^2} + \frac{2}{v_g} \frac{\partial^2}{\partial t \partial z} \right) |\Psi|^2, \quad (1.34a)$$

$$\frac{\partial^2 b}{\partial t^2} - c_A^2 \Delta b - c_s^2 \Delta_\perp \nu = \frac{u\, c_A^2}{1-u} \Delta_\perp |\Psi|^2 \quad (1.34b)$$

where $c_A = B_0 / (4\pi N_0 m_i)^{1/2}$ is the Alfven velocity and c_s ion sound velocity (1.13). The terms in the right hand sides of eqs. (1.34) describe the action of the ponderomotive force. It can be shown that these equations are equivalent to those used in refs. /9,13/.

Equations (1.2), (1.33) and (1.34) form a closed system. Without the right hand sides it describes two branches of MHD waves: fast and slow magnetosonic waves. The right hand sides describe their interaction to the considered above high-frequency waves (whistlers). Thus here we have typical interaction between high- and low-frequency waves.

An analysis of the system of equations (1.2), (1.33) and (1.34) shows that it describes a development of two types of nonlinear structures corresponding to fast and slow magnetosonic branches. They differ, first of all, by their velocities and, secondly, by the relations between ν and b. As has been shown in ref. /12/, for the structures of the first type

$$b \approx \nu \quad (1.35)$$

The same relation holds for the linear fast magnetosonic waves. Therefore, following ref. /12/, we use the term fast magnetic sound (FMS)branch for this nonlinear waves . Substituting (1.35) into (1.33) and (1.34b) and taking into account that

$$c_s^2 / c_A^2 \approx 4\pi N T_e / B_0^2 \ll 1 \qquad (1.36)$$

(cf.(1.31)) we obtain the following equations for a whistler- wave interacting with FMS branch

$$i(\Psi_t + v_g \Psi_z) + \frac{1}{2}(S_1 \Delta_\perp \Psi + S_2 \Psi_{zz}) = \omega u \nu \Psi, \qquad (1.37a)$$

$$\nu_{tt} - c_A^2 \Delta \nu = [u c_A^2 / (1-u)] \Delta_\perp |\Psi|^2 . \qquad (1.37b)$$

For the second low-frequency branch interacting with a whistler wave

$$b \ll \nu \qquad (1.38)$$

which is the same as for slow magnetic sound. Thus, it is called nonlinear slow magnetic sound (SMS) branch. Structures of this type propagate with very low velocities (smaller or of the order of c_s) /12/. Substituting (1.38) into (1.33) and (1.34a) we arrive at the following equations for a whistler-wave interacting with SMS branch

$$i(\Psi_t + v_g \Psi_z) + \frac{1}{2}(S_1 \Delta_\perp \Psi + S_2 \Psi_{zz}) = -\omega(1-u)\nu \Psi \qquad (1.39a)$$

$$\nu_{tt} - c_s^2 \nu_{zz} = -c_A^2 (\partial / \partial z)^2 |\Psi|^2 \qquad (1.39b)$$

(the second term in the right hand side of eq. (1.34a) is neglected because of small velocity of the SMS waves).

The additional assumption that z -derivatives are small in comparison to those in transverse directions (which is justified by final results) leads to equations derived in ref. /14/. Here, however, this will not be assumed.

Using MHD equations for the low-frequency waves we, thus, have neglected their dispersion. If it is taken into account, there appears the third, Alfven, branch interacting with the ψ -field /15/. Without dispersion this branch coincides with the usual Alfven oscillations, which does not interact with the high-frequency field when dispersion is neglected. Thus, they have not appeared above. Here this branch is not considered (some of its properties are described in ref. /15/).

The possibility of separation of the nonlinear FMS and SMS branches is conditioned by their different physical properties. For example, as the velocities of FMS disturbances are much larger than those of SMS disturbances, the initial perturbations decay rather rapidly into two separate pulses which are described by systems (1.37) and (1.39), respectively. As it is known, similar situation permits to use reduced equations in other physical cases (e.g. Korteveg-de Vries or Kadomtzev-Petviashvili equations).

It is convenient to itroduce dimensionless variables. Defining

$$t' = \omega t, \qquad \underset{\sim}{R}' = k \underset{\sim}{R}, \tag{1.40}$$

where ω and k are carrier frequency and wave number in (1.1), one can rewrite the obtained above equations in unified notations.

For instance, the system of equations describing the ponderomotive interaction of the linearly polarized transverse electromagnetic wave with the ion sound in isotropic plasma takes the form (the primes denoting the dimensionless variables (1.40) are omitted)

$$i\left(\Psi_t + V_1 \Psi_z\right) + \frac{1}{2}\left(s_1 \Delta_\perp \Psi + s_2 \Psi_{zz}\right) = c_1 \nu \Psi, \tag{1.41}$$

$$\nu_{tt} - V_2^2 \Delta \nu = c_2 \Delta |\Psi|^2, \tag{1.42}$$

where

$$\Psi = E/E_0, \quad E_0^2 = 16\pi N_0 T_e, \tag{1.43}$$

$$c_1 = \omega_p^2/2\omega^2, \quad c_2 = (c_s k \omega_p)^2/\omega^4, \tag{1.44}$$

$$V_1 = c^2 k^2 (\omega_p^2 + c^2 k^2)^{-1}, \quad V_2 = c_s k (\omega_p^2 + c^2 k^2)^{-1/2}, \tag{1.45a}$$

$$s_1 = c^2 k^2 (\omega_p^2 + c^2 k^2)^{-1}, \quad s_2 = (c k \omega_p)^2 (\omega_p^2 + c^2 k^2)^{-2}. \tag{1.45b}$$

Here a characteristic scale for k is ω_p/c. It is important that $V_1 \gg V_2$ even for small k (but $kc/\omega_p \gg c_s/c$).

The same equations (1.41)-(1.44) describe the system Langmuir - ion sound waves, but now

$$V_1 = 3(kr_D)^2 \omega_p/\omega \approx 3(kr_D)^2, \; V_2 = c_s k/\omega \approx (m_e/m_i)^{1/2} kr_D, \quad (1.46a)$$

$$s_1 = s_2 = 3(\omega_p/\omega)(kr_D)^2 \approx 3(kr_D)^2. \quad (1.46b)$$

In this case $V_1 \gg V_2$ at

$$3kr_D \gg (m_e/m_i)^{1/2}. \quad (1.47)$$

Similarly, the system (1.37) in dimensionless variables consists of eq. (1.41) and the equation

$$\nu_{tt} - V_2^2 \Delta \nu = c_2 \Delta_\perp |\psi|^2 \quad (1.48)$$

where ψ is defined in (1.32a) and

$$V_1 = 2(1-u), \quad V_2 = (m_e/m_i)^{1/2} u^{-1/2} (1-u)^{-1/2}, \quad (1.49a)$$

$$s_1 = 1-2u, \quad s_2 = 2(1-u)(1-4u), \quad (1.49b)$$

$$c_1 = u, \quad c_2 = (m_e/m_i)(1-u)^{-2}. \quad (1.49c)$$

In this case

$$V_1/V_2 = 2(m_i/m_e)^{1/2} u^{1/2} (1-u)^{3/2} \quad (1.49d)$$

At last, the system (1.49) in dimensionless variables consists of eq. (1.41) and the equation

$$\vartheta_{tt} - v_2^2 \vartheta_{zz} = c_2 (\partial^2/\partial z^2)|\psi|^2, \tag{1.50}$$

where

$$c_1 = -(1-u), \quad c_2 = -(c_A/v_\varphi)^2 = -(m_e/m_i)u^{-1}(1-u)^{-1} \tag{1.51a}$$

$$v_2 = c_s/v_\varphi = (m_e/m_i)^{1/2}(4\pi N_0 T_e/B_0^2)^{1/2} u^{-1/2}(1-u)^{-1/2} \tag{1.51b}$$

and v_1 , s_1 and s_2 are the same as in (1.49).
Then

$$v_1/v_2 = 2(m_i/m_e)^{1/2} u^{1/2}(1-u)^{3/2}(B_0^2/4\pi N_0 T_e)^{1/2}. \tag{1.51c}$$

It is assumed that u is not too close to the unity (in order to neglect whistler dispersion and cyclotron damping) and $u \gg (m_e/m_i)^{1/2}$ (because $\omega \gg \omega_{LH}$). Therefore for both branches, FMS and SMS, $v_1/v_2 \gg 1$.

It will be shown in the following, that the systems, consisting of eq. (1.41) together with one of the equations (1.42), (1.48) or (1.50), can be considered from a unified point of view.

2. Hamiltonian formalism. Conservation laws.

Different physical cases, discussed in the preceding section, can be considered from a unified point of view in terms of canonical variables of the Hamiltonian formalism.

Consider, at first, the system (1.41) and (1.42). Introducing the velocity potential $v = -\nabla \Phi$ and using the linearized continuity equation, we obtain

$$\partial \nu / \partial t = \Delta \Phi \tag{2.1}$$

From (2.1) and (1.42) follows

$$\partial \Phi / \partial t = v_2^2 \nu + c_2 \Psi^* \Psi \tag{2.2}$$

It is easy to see that the system (1.41), (2.1) and (2.2) can be written in the following Hamiltonian form

$$i \frac{\partial \Psi(\underset{\sim}{R}, t)}{\partial t} = \frac{\delta H}{\delta \Psi^*(\underset{\sim}{R})}, \tag{2.3}$$

$$\frac{c_1}{c_2} \frac{\partial \Phi(\underset{\sim}{R}, t)}{\partial t} = \frac{\delta H}{\delta \nu(\underset{\sim}{R})}, \quad \frac{c_1}{c_2} \frac{\partial \nu}{\partial t} = -\frac{\delta H}{\delta \Phi(\underset{\sim}{R})}, \tag{2.4}$$

where

$$H = H_1 + H_2 + H_{12}, \tag{2.5}$$

$$H_1 = \int dR \, \{(i v_1/2)(\Psi \Psi_z^* - \Psi^* \Psi_z)$$

$$+ (1/2)[s_1 (\nabla_\perp \Psi)(\nabla_\perp \Psi^*) + s_2 (\nabla_z \Psi)(\nabla_z \Psi^*)]\}, \tag{2.6a}$$

$$H_2 = (c_1/2c_2) \int d\underset{\sim}{R} \, [v_2^2 \nu^2 + (\nabla \Phi)^2], \tag{2.6b}$$

$$\mathcal{H}_{12} = c_1 \int d\underset{\sim}{R}\, \vartheta \Psi \Psi^* . \tag{2.6c}$$

From (2.3) and (2.4) we see that Ψ and Ψ^*, Φ and $(c_1/c_2)\vartheta$ are the canonically conjugate variables.

Other considered systems can also be written in the Hamiltonian form (2.3) and (2.4), but explicit expressions for the Hamiltonian $\mathcal{H}$ depend on the case /17,18/. In them Φ is not, generally speaking, the velocity potential, but an auxiliary function introduced to reduce the initial system to equations of the first order with respect to time t.

To derive a unified approach, it is simpler to use the Fourier representation. Introduce the notations

$$a(\underset{\sim}{P}, t) = (2\pi)^{-D/2} \int \Psi(\underset{\sim}{R}, t)\, e^{-i\underset{\sim}{P}\underset{\sim}{R}}\, d\underset{\sim}{R}, \tag{2.7a}$$

$$\tilde{\vartheta}(\underset{\sim}{P}, t) = (2\pi)^{-D/2} \int \vartheta(\underset{\sim}{R}, t)\, e^{-i\underset{\sim}{P}\underset{\sim}{R}}\, d\underset{\sim}{R}, \tag{2.7b}$$

$$\tilde{\Phi}(\underset{\sim}{P}, t) = (2\pi)^{-D/2} \int \Phi(\underset{\sim}{R}, t)\, e^{-i\underset{\sim}{P}\underset{\sim}{R}}\, d\underset{\sim}{R} \tag{2.7c}$$

where D is the dimension of space under consideration (D =1,2,3). Then equation (1.41), common for all systems considered, takes the form

$$i a_t(\underset{\sim}{p}, t) = \sigma(\underset{\sim}{p})\, a(\underset{\sim}{p}) + c_1 (2\pi)^{-D/2} \int \tilde{\nu}(\underset{\sim}{p} - \underset{\sim}{p}')\, a(\underset{\sim}{p}')\, d\underset{\sim}{p}' \qquad (2.8)$$

where

$$\sigma(\underset{\sim}{p}) = v_1 p_z + (1/2)(s_1 p_\perp^2 + s_2 p_z^2). \qquad (2.9)$$

Here $\sigma(\underset{\sim}{p})$ plays the role of the frequency of the $a(\underset{\sim}{p}, t)$ - field when the interaction is absent. Actually, it is the addition to the carrying frequency $\omega(\underset{\sim}{k})$, if we neglect the terms of order of p^3 in the total frequency

$$\omega(\underset{\sim}{k} + \underset{\sim}{p}) = \omega(\underset{\sim}{k}) + \sigma(\underset{\sim}{p}) + O(p^3) \qquad (2.10)$$

(the nonlinear Schrödinger equation (1.2) is valid just with such accuracy).

Equation for the ν -field in general case can be written as

$$\tilde{\nu}_{tt}(\underset{\sim}{p}) + \eta^2(\underset{\sim}{p})\nu(\underset{\sim}{p}) = -c_2 G(\underset{\sim}{p})(2\pi)^{-D/2} \int a^*(\underset{\sim}{p}' - \underset{\sim}{p})\, a(\underset{\sim}{p}')\, d\underset{\sim}{p}' \qquad (2.11)$$

where $\eta(\underset{\sim}{p})$ is the frequency of the $\nu(\underset{\sim}{p})$ -component without interaction and $G(\underset{\sim}{p})$ is a function depending on the type of interaction. For the considered above equations (1.42), (1.48) and (1.50) we have, respectively,

$$\eta(\underset{\sim}{p}) = v_2 p, \qquad G(\underset{\sim}{p}) = p^2, \qquad (2.12a)$$

$$\eta(\underset{\sim}{\rho}) = v_2 \rho, \quad G(\underset{\sim}{\rho}) = \rho_\perp^2 , \tag{2.12b}$$

$$\eta(\underset{\sim}{\rho}) = v_2 |\rho_z|, \quad G(\underset{\sim}{\rho}) = \rho_z^2 . \tag{2.12c}$$

The functions $\eta(\underset{\sim}{\rho})$ and $G(\underset{\sim}{\rho})$ satisfy the relations

$$\eta(-\underset{\sim}{\rho}) = \eta(\underset{\sim}{\rho}), \quad G(-\underset{\sim}{\rho}) = G(\underset{\sim}{\rho}) \tag{2.12d}$$

Now let us reduce eq. (2.11) to a system of two equations of the first order with respect to t by introducing auxiliary function $\tilde{\Phi}(\underset{\sim}{\rho}, t)$, assuming that it is the Fourier transform of a real function $\Phi(\underset{\sim}{R}, t)$ (see (2.7c)). Generalizing (2.1), we write

$$\tilde{\nu}_t = B(\underset{\sim}{\rho})\tilde{\Phi}(\underset{\sim}{\rho}, t) \tag{2.13a}$$

where $B(\underset{\sim}{\rho})$ is still unknown funktion. Substitution of (2.13a) into (2.11) gives

$$\tilde{\Phi}_t(\underset{\sim}{\rho}, t) = -[\eta^2(\underset{\sim}{\rho})/B(\underset{\sim}{\rho})]$$
$$- c_2[G(\underset{\sim}{\rho})/B(\underset{\sim}{\rho})](2\pi)^{-D/2}\int d\underset{\sim}{\rho}' a^*(\underset{\sim}{\rho}' - \underset{\sim}{\rho}) a(\underset{\sim}{\rho}'). \tag{2.13b}$$

From the reality of $\nu(\underset{\sim}{R}, t)$ and $\Phi(\underset{\sim}{R}, t)$ follows

$$\tilde{\nu}^*(\underset{\sim}{\rho}) = \tilde{\nu}(-\underset{\sim}{\rho}), \quad \tilde{\Phi}^*(\underset{\sim}{\rho}) = \tilde{\Phi}(-\underset{\sim}{\rho}). \tag{2.14}$$

The system (2.13) describes the ϑ -field interacting with the ψ -field. Now we require that the function $\Phi(\underset{\sim}{\rho}, t)$ and $(c_1/c_2)\,\vartheta(\underset{\sim}{\rho}, t)$ must be canonically conjugated functions satisfying equations (2.4). This determines $B(\underset{\sim}{\rho})$. In order to realize it, let us write eqs. (2.3) and (2.4) in the Fourier representation

$$i a_t(\underset{\sim}{\rho}, t) = \delta H / \delta a^*(\underset{\sim}{\rho}), \tag{2.15a}$$

$$(c_1/c_2)\,\tilde{\Phi}_t(\underset{\sim}{\rho}, t) = \delta H / \delta \tilde{\vartheta}(-\underset{\sim}{\rho}), \tag{2.15b}$$

$$(c_1/c_2)\,\tilde{\vartheta}_t(\underset{\sim}{\rho}, t) = -\delta H / \delta \tilde{\Phi}(-\underset{\sim}{\rho}) \tag{2.15c}$$

where H is defined by (2.5) and H_1 , H_{12} by (2.6a) and (2.6c). Then

$$H_1 = \int d\underset{\sim}{\rho}\, \sigma(\underset{\sim}{\rho})\, a^*(\underset{\sim}{\rho})\, a(\underset{\sim}{\rho}), \tag{2.16}$$

$$H_{12} = c_1 (2\pi)^{-D/2} \int d\underset{\sim}{\rho}'\, d\underset{\sim}{\rho}''\, \tilde{\vartheta}(\underset{\sim}{\rho}')\, a(\underset{\sim}{\rho}'')\, a^*(\underset{\sim}{\rho}' + \underset{\sim}{\rho}'') \tag{2.17}$$

and eq. (2.15a) in $(\underset{\sim}{\rho}, t)$ representation leads to (2.8). The term H_2 should be a quadratic form of $\tilde{\vartheta}$ and Φ which can be easily obtained from (2.13) and (2.15). The system (2.8) and (2.13) is compatible with eqs. (2.15) if

$$B(\underset{\sim}{\rho}) = -G(\underset{\sim}{\rho}), \tag{2.18}$$

and

$$\mathcal{H}_2 = \frac{c_1}{2c_2} \int d\underset{\sim}{p} \{ [\eta^2(\underset{\sim}{p}) / G(\underset{\sim}{p})] \tilde{\nu}(\underset{\sim}{p}) \tilde{\nu}(-\underset{\sim}{p}) + G(\underset{\sim}{p}) \tilde{\Phi}(\underset{\sim}{p}) \tilde{\Phi}(-\underset{\sim}{p}) \} \tag{2.19}$$

Now, following ref. /16/, let us perform transformation from $\tilde{\nu}(\underset{\sim}{p},t)$, $\tilde{\Phi}(\underset{\sim}{p},t)$ to "canonical" variables $b(\underset{\sim}{p},t)$ and $b^*(\underset{\sim}{p},t)$, choosing them in such a way that $\mathcal{H}_2$ would have a "diagonal" form

$$\mathcal{H}_2 = \int d\underset{\sim}{p}\, \eta(\underset{\sim}{p})\, b^*(\underset{\sim}{p})\, b(\underset{\sim}{p}). \tag{2.20}$$

Then

$$\tilde{\nu}(\underset{\sim}{p}) = -i \left[\frac{c_2 G(\underset{\sim}{p})}{2c_1\, \eta(\underset{\sim}{p})} \right]^{1/2} [b(\underset{\sim}{p}) - b^*(-\underset{\sim}{p})], \tag{2.21a}$$

$$\tilde{\Phi}(\underset{\sim}{p}) = \left[\frac{c_2}{2c_1} \; \frac{\eta(\underset{\sim}{p})}{G(\underset{\sim}{p})} \right]^{1/2} [b(\underset{\sim}{p}) + b^*(-\underset{\sim}{p})] \tag{2.21b}$$

and

$$\mathcal{H}_{12} = -i \int d\underset{\sim}{p}'\, d\underset{\sim}{p}''\, V(\underset{\sim}{p}' - \underset{\sim}{p}'')\, a^*(\underset{\sim}{p}')\, a(\underset{\sim}{p}'') \times [b(\underset{\sim}{p}' - \underset{\sim}{p}'') - b^*(\underset{\sim}{p}'' - \underset{\sim}{p}')] \tag{2.22}$$

where

$$V(\underset{\sim}{p}) = (2\pi)^{-D/2}\left[\frac{c_1 c_2 G(\underset{\sim}{p})}{2\eta(\underset{\sim}{p})}\right]^{1/2}. \qquad (2.23)$$

Now the Hamilton equations take the form

$$i\,a_t(\underset{\sim}{p},t) = \delta H/\delta a^*(\underset{\sim}{p}),$$
$$i\,b_t(p,t) = \delta H/\delta b^*(\underset{\sim}{p}) \qquad (2.24)$$

where H is defined by formulas (2.5), (2.16), (2.20) and (2.22). The functions $a(\underset{\sim}{p},t)$ and $b(\underset{\sim}{p},t)$ are canonical Hamiltonian variables conjugated with $a^*(\underset{\sim}{p},t)$ and $b^*(\underset{\sim}{p},t)$

Calculating the right hand sides in (2.24), we obtain the basic equations in the unified form /17/

$$a_t(\underset{\sim}{p},t) = -i\sigma(\underset{\sim}{p})\,a(\underset{\sim}{p},t)$$
$$+\int d\underset{\sim}{p}'\,V(\underset{\sim}{p}')\,a(\underset{\sim}{p}'+\underset{\sim}{p},t)[b^*(\underset{\sim}{p}',t)-b(-\underset{\sim}{p}',t)], \qquad (2.25a)$$

$$b_t(\underset{\sim}{p},t) = -i\eta(\underset{\sim}{p})\,b(\underset{\sim}{p},t) + V(\underset{\sim}{p})\int d\underset{\sim}{p}'\,a^*(\underset{\sim}{p}',t)\,a(\underset{\sim}{p}+\underset{\sim}{p}',t) \qquad (2.25b)$$

where $\sigma(\underset{\sim}{p})$, $\eta(\underset{\sim}{p})$ and $V(\underset{\sim}{p})$ satisfy the following general relations

$$\sigma(p_z,\underset{\sim}{p}_\perp) = \sigma(p_z,-\underset{\sim}{p}_\perp),\ \eta(\underset{\sim}{p}) = \eta(-\underset{\sim}{p}),\ V(\underset{\sim}{p}) = V(-\underset{\sim}{p}), \qquad (2.26b)$$

$$\sigma(0) = \eta(0) = V(0) = 0. \tag{2.26b}$$

If a solution of eqs. (2.25) is known, the function $\Psi(\underset{\sim}{R}, t)$ and $\vartheta(\underset{\sim}{R}, t)$ are restored by the formulas

$$\Psi(\underset{\sim}{R}, t) = (2\pi)^{-D/2} \int d\underset{\sim}{P}\, a(\underset{\sim}{P}, t)\, e^{i\underset{\sim}{P}\underset{\sim}{R}} \tag{2.27a}$$

$$\vartheta(\underset{\sim}{R}, t) = -i\left(\frac{c_2}{2c_1}\right)^{1/2} (2\pi)^{-D/2} \times \int d\underset{\sim}{P} \left[\frac{G(\underset{\sim}{P})}{\eta(\underset{\sim}{P})}\right]^{1/2} \left[b(\underset{\sim}{P}, t) - b^*(-\underset{\sim}{P}, t)\right] e^{i\underset{\sim}{P}\underset{\sim}{R}} \tag{2.27b}$$

As it is clear from the above analysis, $\Psi(\underset{\sim}{R}, t)$, $\vartheta(\underset{\sim}{R}, t)$ and auxiliary function $\Phi(\underset{\sim}{R}, t)$ satisfy the Hamilton equations (2.3) and (2.4) in all cases considered. The Hamiltonian parts H_1 and H_2 have the forms (2.6a) and (2.6c) in all cases as well, but the term H_2 in $(\underset{\sim}{R}, t)$ rep resentation depends on the case. For the equations (2.1) and (2.2), which correspond to (1.42), it is given by (2.6b). For the FMS waves, the equation (1.48) is reduced to the system

$$\vartheta_t = \Delta_\perp \Phi, \tag{2.28a}$$

$$\Delta_\perp\left[\left(\Phi_t - c_2|\Psi|^2\right) - v_2^2\,\vartheta\right] = v_2^2\,\vartheta_{zz}, \tag{2.28b}$$

which is easily derived from the above written Fourier representation. This corresponds to /18/

$$H_2 = (c_1/2c_2)\int d\underset{\sim}{R}\left\{v_2^2[\nu^2 + (\nabla_\perp \nabla_z \chi)^2] + \frac{1}{2}(\nabla_\perp \Phi)^2\right\} \quad (2.29)$$

where auxiliary function $\chi(\underset{\sim}{R}, t)$ satisfies the equation

$$\Delta_\perp \chi = \nu \quad (2.30)$$

χ may be considered as a functional of ν, defined by the equation

$$\chi(\underset{\sim}{R}) = \int K(\underset{\sim}{R} - \underset{\sim}{R}')\,\nu(\underset{\sim}{R}')\,dR' \quad (2.31)$$

where $K(\underset{\sim}{R} - \underset{\sim}{R}')$ is the Green function of the equation (2.30). From (2.31) follows

$$\delta\chi(\underset{\sim}{R})/\delta\nu(\underset{\sim}{R}') = K(\underset{\sim}{R} - \underset{\sim}{R}') \quad (2.32)$$

Then it is easy to check that eqs. (2.4) with (2.5), (2.6a), (2.6c), (2.29) and (2.32) give eqs. (2.28) which correspond to (1.48).

For the SMS waves, the equation (1.50) is reduced to the system

$$\nu_t = \Phi_z, \quad \Phi_t = c_2|\psi|^2 + v_2^2\nu \quad (2.33)$$

which corresponds to /17/

$$H_2 = (c_1/2c_2)\int d\underset{\sim}{R}\,[v_2^2 \vartheta^2 + (\nabla_z \Phi)^2] \tag{2.34}$$

Along with the Hamiltonians, let us also write the explicit expressions for Lagrangians. They can be divided in two parts:

$$\mathcal{L} = \int L\, dR \qquad L = L^{(1)} + L^{(2)} \tag{2.35}$$

where

$$L^{(1)} = \frac{i}{2}[\Psi^* \Psi_t - \Psi_t^* \Psi + v_1(\Psi^* \Psi_z - \Psi_z^* \Psi)]$$
$$- \frac{1}{2}[s_1 |\nabla_\perp \Psi|^2 + s_2 |\nabla_z \Psi|^2] + (c_1/c_2)\vartheta(\Phi_t - c_2 \Psi^* \Psi), \tag{2.36}$$

which is common for all three systems considered, and

$$L^{(2)} = -(c_1/2c_2)[v_2^2 \vartheta^2 + (\nabla \Phi)^2] \tag{2.37}$$

for the system (2.1) and (2.2),

$$L^{(2)} = -(c_1/2c_2)\{v_2^2[\vartheta^2 + (\nabla_\perp \nabla_z \chi)^2] + (\nabla_\perp \Phi)^2\} \tag{2.38}$$

for the system (2.28), and

$$L^{(2)} = -(c_1/2c_2)[v_2^2 \vartheta^2 + (\nabla_z \Phi)^2] \tag{2.39}$$

for the system (2.33).

Calculating $\mathcal{D}_t L$, where $\mathcal{D}_t$ means total derivatives with respect to t, and using the homogeneitity of L in time, we obtain, with account of the equations of motion, the following conservation law

$$\frac{\partial}{\partial t}\left(\frac{\partial L}{\partial \psi_t^*}\psi_t^* + \frac{\partial L}{\partial \psi_t}\psi_t + \frac{\partial L}{\partial \Phi_t}\Phi_t - L\right)$$

$$+\nabla_\alpha\left(\frac{\partial L}{\partial \nabla_\alpha \psi^*}\psi_t^* + \frac{\partial L}{\partial \nabla_\alpha \psi}\psi_t + \frac{\partial L}{\partial \nabla_\alpha \Phi}\Phi_t\right) \qquad (2.40)$$

$$+(c_1 v_2^2/c_2)[\nabla_z(\chi_z \partial_t) - \nabla_\perp(\chi_z \nabla_\perp \chi_{zt})] = 0$$

where $\chi(\mathbf{R},t)$ satisfies eq. (2.30) in the case (2.38) and vanishes in two other cases, (2.37) and (2.39). Using eqs. (2.35)-(2.39), it is easy to check that

$$\int\left(\frac{\partial L}{\partial \psi_t^*}\psi_t^* + \frac{\partial L}{\partial \psi_t}\psi_t + \frac{\partial L}{\partial \Phi_t}\Phi_t - L\right)d\mathbf{R} = H \qquad (2.41)$$

for all cases considered. Thus, as it should be expected, eq. (2.40) expresses the energy conservation.

In a similar way, calculating $\mathcal{D}_\alpha L$ ($\alpha = x, y, z$) and using the homogeneity of L in space, we obtain after some calculations

$$\frac{\partial}{\partial t}\left(\frac{\partial L}{\partial \psi_t^*}\nabla_\alpha\psi^* + \frac{\partial L}{\partial \psi_t}\nabla_\alpha\psi + \frac{\partial L}{\partial \Phi_t}\nabla_\alpha\Phi\right)$$

$$+\nabla_\beta\left(\frac{\partial L}{\partial \nabla_\beta \psi^*}\nabla_\alpha\psi^* + \frac{\partial L}{\partial \nabla_\beta \psi}\nabla_\alpha\psi + \frac{\partial L}{\partial \nabla_\beta \Phi}\nabla_\alpha\Phi - \delta_{\alpha\beta}L\right) \qquad (2.42)$$

$$+(c_1 v_2^2/c_2)[\nabla_z(\chi_z \nabla_\alpha \partial) - \nabla_\perp(\chi_z \nabla_\perp \nabla_\alpha \chi_z)] = 0$$

This equation expresses the conservation of the vector

$$P_\alpha = \int \left(\frac{\partial L}{\partial \psi_t^*} \nabla_\alpha \psi^* + \frac{\partial L}{\partial \psi_t} \nabla_\alpha \psi + \frac{\partial L}{\partial \Phi_t} \nabla_\alpha \Phi \right) d\underset{\sim}{R} . \quad (2.43)$$

It is natural to interpret the expression (2.43) as the momentum. Performing calculations, we have

$$P_\alpha = \int \left[(i/2)(\psi^* \nabla_\alpha \psi - \psi \nabla_\alpha \psi^*) + (c_1/c_2) \nu \nabla_\alpha \Phi \right] d\underset{\sim}{R} \quad (2.44)$$

for all cases considered.

Apart of them , there are two more conserving quantities

$$J_1 = \int |\psi|^2 d\underset{\sim}{R} , \qquad J_2 = \int \nu \, d\underset{\sim}{R} \quad (2.45)$$

3. Modulational instability of a "high-frequency" wave. Relationship between modulational instability and parametric decay.

As one of the important applications of the system (2.25), consider the modulational instability of a high-frequency wave.

The basic equations for the interacting ψ and ν fields are satisfied by the "unperturbed" solution

$$\psi = \psi_0 , \qquad \nu = 0 \quad (3.1)$$

describing a plane wave with the amplitude ψ_0 and propagating in the medium with constant density $\rho = \rho_0$. Without loss of generality we assume that ψ_0 is real. The modulational instability means that small initial perturbations of the background state (3.1) increase with time. Let

$$\Psi = \Psi_0 + \delta\Psi(\underset{\sim}{R}, t), \quad \vartheta = \vartheta(\underset{\sim}{R}, t) \tag{3.2}$$

Consider behaviourof the fluctuations $\delta\Psi(\underset{\sim}{R}, t)$ and $\vartheta(\underset{\sim}{R}, t)$, assuming that at least one of these functions is nonzero at t =0 and they are small enough at sufficiently small t . Following usual scheme, suppose that at t =0 only one modulation mode with the wave vector $\underset{\sim}{q}$ is excited. Then the solution of eqs. (2.25) can be taken in the form

$$a(\underset{\sim}{\rho}, t) = (2\pi)^{D/2}\Psi_0 \sum_{m=-\infty}^{\infty} A_m^*(t)\,\delta(\underset{\sim}{\rho} + m\underset{\sim}{q}) \tag{3.3a}$$

$$b(\underset{\sim}{\rho}, t) = (2\pi)^{D/2}\Psi_0 \sum_{n=-\infty}^{\infty} B_n(t)\,\delta(\underset{\sim}{\rho} - n\underset{\sim}{q}), \quad B_0(t) \equiv 0, \tag{3.3b}$$

where

$$A_0(0) = 1, \quad |A_1(0)|^2 + |B_1(0)|^2 \neq 0,$$
$$A_n(0) = B_n(0) = 0 \qquad (n \neq 0, 1) \tag{3.4}$$

Substituting (3.3) into (2.25), we obtain the following equations for $A_m(t)$ and $B_m(t)$ /19/

$$\frac{dA_m}{dt} = i\hat{\sigma}(-m\underset{\sim}{q})A_m + \gamma(\underset{\sim}{q})\sum_{n=-\infty}^{\infty} |n|^{1/2}\left(A_{m-n}B_n - A_{m+n}B_n^*\right) \tag{3.5a}$$

$$\frac{dB_n}{dt} = -i\eta(nq)B_n + \gamma(q)|n|^{1/2}\sum_{m=-\infty}^{\infty} A_m^* A_{m+n} \tag{3.5b}$$

where

$$\begin{aligned} \gamma(q) &= (2\pi)^{D/2}\, V(\underset{\sim}{q})\Psi_0 \\ &= (c_1 c_2/2)^{1/2}\, |G(\underset{\sim}{q})/\eta(\underset{\sim}{q})|^{1/2}\Psi_0 \end{aligned} \tag{3.6}$$

Eqs. (3.5) can be written in a Hamiltonian form

$$i\frac{\partial A_m}{\partial t} = -\frac{\partial h}{\partial A_m^*}, \qquad i\frac{\partial B_n}{\partial t} = \frac{\partial h}{\partial B_n^*} \tag{3.7}$$

where

$$h = h_1 + h_2 + h_{12}, \tag{3.8a}$$

$$h_1 = \sum_{m=-\infty}^{\infty} G(-mq)A_m^* A_m, \quad h_2 = \sum_{n=-\infty}^{\infty} \eta(nq)B_n^* B_n, \tag{3.8b}$$

$$h_{12} = -i\gamma\sum_{m,n=-\infty}^{\infty} |n|^{1/2} A_m^*\left(A_{m-n}B_n - A_{m+n}B_n^*\right). \tag{3.8c}$$

It is easy to check that the Hamiltonian h is real. From (3.7) follows that A_m and B_m are canonically conjugated with A_m^* and B_m^*.

Now let us assume that $A_0(t) \approx 1$ and $A_n(t), B_n(t)$

at $n \neq 0$ are small at sufficiently small t . Then, neglecting the second order quantities we obtain the following closed system of equations for $A_{\pm 1}(t)$ and $B_{\pm 1}(t)$

$$dA_1/dt = i\sigma(-\underset{\sim}{q})A_1 + \gamma(B_1 - B^*_{-1}) \tag{3.9a}$$

$$dB_1/dt = -i\eta(\underset{\sim}{q})B_1 + \gamma(A_1 + A^*_{-1}) \tag{3.9b}$$

$$dA^*_{-1}/dt = -i\sigma(\underset{\sim}{q})A^*_{-1} + \gamma(B^*_{-1} - B_1) \tag{3.9c}$$

$$dB^*_{-1}/dt = i\eta(\underset{\sim}{q})B^*_{-1} + \gamma(A^*_{-1} + A_1) \tag{3.9d}$$

Supposing that $A_1(t)$, $B_1(t)$, $A^*_{-1}(t)$ and $B^*_{-1}(t)$ are proportional to $exp(-i\Omega t)$, we obtain the following dispersion equation for the frequency Ω

$$[\Omega^2 - \eta^2(\underset{\sim}{q})][(\Omega - v_1 q_z)^2 - \tfrac{1}{4}(s_1 q_x^2 + s_2 q_z^2)^2] = 2\gamma^2(\underset{\sim}{q})\eta(\underset{\sim}{q})(s_1 q_x^2 + s_2 q_z^2) \tag{3.10}$$

Here we have taken q_y =0, without loss of generality. As has been shown in refs. /17,19/, at $v_1 \gg v_2$ (which will be assumed below)eq. (3.10) approximately coinsides with the dispersion equation of the modulational instability for a whistler wave,obtained in refs. /9,12/. Using a detailed analysis of ref. /12/, let us present the following summary of results conderning solutions of eq. (3.10) at $|\Psi_0|^2 \ll 1$. There are two types of unstable solutions of this equation.

The first type is of the form

$$\Omega(\underset{\sim}{q}) = \Omega_0(\underset{\sim}{q}) + i\,\Omega_1(\underset{\sim}{q}), \quad |\Omega_1| \ll \Omega_0 \tag{3.11a}$$

and the maximum growth rate is reached if $\Omega_0(q)$ is the common root of each bracket in the left hand side of (3.10), i.e.

$$\Omega_0(q) = \eta(q) = -\sigma(-q). \tag{3.11b}$$

Substituting (3.11a) into (3.10) and taking into account (3.11b), we have

$$\Omega_1(\underset{\sim}{q}) = \gamma(\underset{\sim}{q}). \tag{3.11c}$$

Thus, in this case (which is possible if the unperturbed amplitude Ψ_0 is sufficiently small, in order to fulfil $\gamma(\underset{\sim}{q}) \ll \eta(\underset{\sim}{q})$) the solution of (3.10) with maximum growth rate is

$$\Omega(\underset{\sim}{q}) = \eta(\underset{\sim}{q}) + i\,\gamma(\underset{\sim}{q}) \tag{3.12a}$$

and the vector $\underset{\sim}{q}$ must satisfy the equation

$$\sigma(-\underset{\sim}{q}) + \eta(\underset{\sim}{q}) = 0, \tag{3.12b}$$

which defines a curve in the plane (q_x , q_z). From the results of ref. /12/ follows that the region of instability is a rather narrow band along this curve (Fig. 1), at which there is the maximum growth rate. From (2.9) and (2.12) follows that at $v_1 \gg v_2$ the equation (3.12b) is satisfied only in the cases (2.12a) and (2.12b). The corresponding vectors q are almost perpendiqular to the z -axis:

$$q_z / q_x \sim v_2 / v_1 \ll 1$$

For the SMS branch, when eq. (2.12c) takes place, the equation (3.12b) can not be satisfied.

Taking into account (2.10) we can rewrite eq. (3.12b) as

$$\omega(\underset{\sim}{k}) - \omega(\underset{\sim}{k} - \underset{\sim}{q}) = \eta(\underset{\sim}{q}) \tag{3.14}$$

This is just the condition of the resonant decay of the initial short wave into that with the frequency $\omega(\underset{\sim}{k} - \underset{\sim}{q})$ and a long wave with the frequency $\eta(\underset{\sim}{q})$. Thus, in the case under discussion the maximum growth rate of the modulational instability is realized if the resonant decay condition (3.14) takes place. However, we shall see below that in the nonlinear stage the modulational instability does not reduce to the decay process.

The solutions of the second type appear when $v_1 q_z \gg \eta(\underset{\sim}{q})$. This takes place, for instance, if a whistler wave interacts with the SMS branch (when $\eta(\underset{\sim}{q}) = v_2 q_z$). In this case, as has been mentioned, the condition (3.12b) can not be satisfied. However, now there is an unstable branch with very low frequencies

$$\eta(\underset{\sim}{q}) \lesssim |\Omega| \ll v_1 q_z , \tag{3.15}$$

Therefore, we call this the low frequency modulational instability. In the region (3.15), eq. (3.10) can be written as

$$[\Omega^2 - \eta^2(\underset{\sim}{q})][2\Omega v_1 q_z - v_1^2 q_z^2 + (1/4)(s_1 q_x^2 + s_2 q_z^2)^2]$$
$$= 2\gamma^2(\underset{\sim}{q})\eta(\underset{\sim}{q})(s_1 q_x^2 + s_2 q_z^2) \tag{3.16}$$

(in the second bracket we have neglected Ω^2 which is of the second order of smallness). Observing that

$$(1/4)(s_1 q_x^2 + s_2 q_z^2)^2 = v_1^2 q_z^2 + \sigma(\underset{\sim}{q})\sigma(-\underset{\sim}{q})$$

we rewrite (3.16) in the form

$$2\Omega^3 v_1 q_z - 2\Omega\eta^2(\underset{\sim}{q}) v_1 q_z + [\Omega^2 - \eta^2(\underset{\sim}{q})]\sigma(\underset{\sim}{q})\sigma(-\underset{\sim}{q})$$
$$= 2\gamma^2(\underset{\sim}{q})\eta(\underset{\sim}{q})[v_1^2 q_z^2 + \sigma(\underset{\sim}{q})\sigma(-\underset{\sim}{q})] \tag{3.17}$$

This is cubic equation for Ω. A numerical investigation of (3.17) shows that the complex roots appear only if $\sigma(\underset{\sim}{q})\sigma(-\underset{\sim}{q})$ is small. Thus, in the unstable region the $\underset{\sim}{q}$-vector should satisfy the condition

$$v_1 q_z \approx \pm (1/2)(s_1 q_x^2 + s_2 q_z^2) \tag{3.18}$$

and equation (3.17) takes the form

$$\Omega^3 - \eta^2(\underset{\sim}{q})\Omega \pm 2\eta(\underset{\sim}{q})\gamma^2(\underset{\sim}{q}) = 0 \qquad (3.19)$$

This equation has complex roots at

$$3^{3/4}\gamma(\underset{\sim}{q}) > \eta(\underset{\sim}{q}) \qquad (3.20a)$$

From (3.6) and (3.20a) follows that the low-frequency modulational instability exists if, in addition to (3.18),

$$\Psi_0 > \Psi_{cr} = 3^{-3/4}(2\pi)^{-D/2}V^{-1}(\underset{\sim}{q})\eta(\underset{\sim}{q}) \qquad (3.20b)$$

If $\gamma(\underset{\sim}{q}) \gg \eta(\underset{\sim}{q})$ (or $\Psi_0 \gg \Psi_{cr}$), the second term in (3.19) can be neglected and the unstable root is

$$\Omega \approx [2\eta(\underset{\sim}{q})\gamma^2(\underset{\sim}{q})]^{1/3}\exp(i\pi/3) \qquad (3.21)$$

for any sign in (3.18). Numerical investigation of eq.(3.19) shows that expression (3.21) gives a correct order of Ω also in the case when Ψ_0/Ψ_{cr} is not so large. Thus, for the low-frequency modulational instability $Re\,\Omega \sim Im\,\Omega$, in contrast to (3.11a). However, the growth rates of high and low frequency instabilities can be comparable.

The band of the low-frequency instability is also rather narrow (Fig. 1). From (3.18) follows that

$$q_z/q_x \sim s_1 q_x/v_1 \lesssim q/k \ll 1 \qquad (3.22)$$

(The initial equations are valid if $q \ll k$; as far as we use normalization (1.40), $k = 1$.) Condition (3.18) can be written in the form

$$\omega(\underset{\sim}{k} \pm \underset{\sim}{q}) \approx \omega(\underset{\sim}{q}) \tag{3.23}$$

Actually, this relation is a result of the fact that the growth rate of the low frequency modulational instability is of the order of or greater than $\eta(\underset{\sim}{q})$. This situation is a characteristic feature of modified decay /20/.

Now, consider the nonlinear stage of the modulational instability for the one mode initial modulations (3.4). Introducing in system (3.5) the notations

$$\sigma(-m\underset{\sim}{q}) = \sigma_m(\underset{\sim}{q}), \quad \eta(m\underset{\sim}{q}) = \eta_m(q) \tag{3.24}$$

and new unknown functions α_m and β_m

$$\begin{aligned} \alpha_m(t) &= A_m(t)\exp(-i\sigma_m t) \\ \beta_m(t) &= B_m(t)\exp(i\eta_m t) \end{aligned} \tag{3.25}$$

we arrive at the system of equations

$$\begin{aligned} d\alpha_m/dt = \gamma \sum_{n=-\infty}^{\infty} |n|^{1/2} \{ &\alpha_{m-n}\beta_n \exp[i(\sigma_{m-n} - \sigma_m - \eta_n)t] \\ &- \alpha_{m+n}\beta_n^* \exp[i(\eta_n + \sigma_{m+n} - \sigma_m)t] \}, \end{aligned} \tag{3.26a}$$

$$d\beta_n/dt = \gamma|n|^{1/2}\sum_{m=-\infty}^{\infty} \alpha_m^* \alpha_{m+n} \exp[i(\eta_n + \sigma_{m+n} - \sigma_m)t], \tag{3.26b}$$

with initial conditions

$$\alpha_0(0) = 1, \quad |\alpha_1(0)|^2 + |\beta_1(0)|^2 \neq 0, \tag{3.27a}$$

$$\alpha_k(0) = \beta_k(0) = 0 \qquad (k \neq 0, 1). \tag{3.27b}$$

Expressing the Hamiltonian (3.8) through $\alpha_m(t)$ and $\beta_m(t)$, we have

$$h_1 = \sum_{m=-\infty}^{\infty} \sigma_m \alpha_m^* \alpha_m, \quad h_2 = \sum_{n=-\infty}^{\infty} \eta_n \beta_n^* \beta_n, \tag{3.28a}$$

$$h_{12} = 2\gamma \, \mathrm{Im} \sum_{m,n=-\infty}^{\infty} |n|^{1/2} \alpha_m^* \alpha_{m-n} \beta_n \times \exp[i(\sigma_{m-n} - \sigma_m - \eta_n)t] \tag{3.28b}$$

Then

$$h\{\alpha_m(t), \beta_m(t), t\} = \sigma_1|\alpha_1(0)|^2 + \eta_1|\beta_1(0)|^2 + 2\gamma \, \mathrm{Im}[\alpha_0(0)\alpha_1^*(0)\beta_1(0)]. \tag{3.29}$$

A number of important results, following from eqs. (3.26), can be derived by truncations of this system . We start with the truncated system which contains the same unknown functions as the linearized equations (3.9). In terms of $\alpha_n(t)$ and $\beta_n(t)$ the corresponding nonlinear system is

$$d\alpha_1/dt = \gamma\{\alpha_0\beta_1 \exp[-i(\sigma_1+\eta_1)t] - \alpha_0\beta^*_{-1}\exp[i(\eta_1-\sigma_1)t]\}, \quad (3.30a)$$

$$d\beta_1/dt = \gamma\{\alpha_0^*\alpha_1 \exp[i(\sigma_1+\eta_1)t] + \alpha_0\alpha^*_{-1}\exp[i(\eta_1-\sigma_{-1})t]\}, \quad (3.30b)$$

$$d\alpha_0/dt = -\gamma\{\alpha_1\beta_1^*\exp[i(\sigma_1+\eta_1)t] - \alpha_{-1}\beta_1\exp[i(\sigma_{-1}-\eta_1)t] + \alpha_{-1}\beta^*_{-1}\exp[i(\sigma_{-1}+\eta_1)t] - \alpha_1\beta_{-1}\exp[i(\sigma_1-\eta_1)t]\}, \quad (3.30c)$$

$$d\beta_{-1}/dt = \gamma\{\alpha_0^*\alpha_{-1}\exp[i(\sigma_{-1}+\eta_1)t] + \alpha_0\alpha_1^*\exp[i(\eta_1-\sigma_1)t]\}, \quad (3.30d)$$

$$d\alpha_{-1}/dt = -\gamma\{\alpha_0\beta_1^*\exp[i(\eta_1-\sigma_{-1})t] - \alpha_0\beta_{-1}\exp[-i(\sigma_{-1}+\eta_1)t]\} \quad (3.30e)$$

with the initial conditions (3.27).

Now let us assume that the vector $\underset{\sim}{q}$ satisfies eq. (3.12b) or, in notations (3.24),

$$\sigma_1(\underset{\sim}{q}) + \eta_1(\underset{\sim}{q}) = 0 \tag{3.31}$$

Then in eqs. (3.30a), (3.30b) and (3.30c) the terms appear which do not contain oscillating factors. If t is not too large, they play the main role. Retaining only such terms we arrive at the third order system

$$d\alpha_1/dt = \gamma\alpha_0\beta_1, \quad d\beta_1/dt = \gamma\alpha_0^*\alpha_1 \tag{3.32a}$$

$$d\alpha_0/dt = -\gamma\alpha_1\beta_1^* \tag{3.32b}$$

with the same initial conditions (3.27). The system (3.32) has two conserving quantities, I_1 and I_2 ,

$$I_1 = |\alpha_0(t)|^2 + |\alpha_1(t)|^2, \; I_2 = |\alpha_0(t)|^2 + |\beta_1(t)|^2 \tag{3.33}$$

Neglecting the change of the initial wave amplitude, i.e. substituting into (3.32a) $\alpha_0(t) = 1$, we obtain the linear system of equations which has the exponentially growing solution with the growth rate γ . It corresponds to the solution of system (3.9) with Ω given by (3.11). Eqs. (3.32) are the well known nonlinear equations describing the three wave resonant parametric interaction. Their general solution, regular at real t , is expressed through the elliptic Jacobi functions

$$\alpha_0(t) = A e^{i\varphi_0} sn[B\gamma(t-t_0)],$$
$$\alpha_1(t) = B e^{i\varphi_1} dn[B\gamma(t-t_0)], \tag{3.34}$$
$$\beta_1(t) = -A e^{i(\varphi_1-\varphi_0)} cn[B\gamma(t-t_0)]$$

where $A > 0$, $B > 0$ and the module of the functions in (3.34) is $k = A/B$.

The truncated systems of higher order than (3.30) contain α_n and β_n with $|n| \leq 2$ (the number of equations N =9), $|n| \leq 3$ (N =13), etc. They can be easily solved numerically. An analysis of numerical solutions shows /19/ that solutions of the systems with N =5,9,13 at $\tau < 10$ where $\tau = \gamma t$, and $\underset{\sim}{q}$ satisfies (3.31) are rather close to the solution of eq. (3.32). Solutions of the systems with N =9 and N =13 are close to each other, at least for $\tau <$ 30 (if $\psi_0 \leq$ 0.02). This is seen, e.g., from the Fig.2.

At the incomplete parametric resonance, when

$$\Delta\omega(\underset{\sim}{q}) \equiv \eta_1(\underset{\sim}{q}) + \sigma_1(\underset{\sim}{q}) \neq 0 \tag{3.35}$$

and $\Delta\omega(\underset{\sim}{q}) \ll \eta_1(\underset{\sim}{q})$, we obtain from (3.10) the following expression for the unstable branch (a more general case is considered in the Appendix A)

$$\Omega(\underset{\sim}{q}) = \eta(\underset{\sim}{q}) - \Delta\omega(\underset{\sim}{q})/2 + i\Gamma(\underset{\sim}{q}) \tag{3.36a}$$

$$\Gamma = [\gamma^2 - (\Delta\omega/2)^2]^{1/2}, \tag{3.36b}$$

From (3.36b) follows that the instability takes place at

$$|\Delta\omega| < 2\gamma. \tag{3.37}$$

Expressions (3.36) can also be obtained from the truncated system of the lowest order (N =3). Indeed, from (3.30) with (3.35) we obtain

$$\frac{d\alpha_1}{dt} = \gamma\alpha_0\beta_1 e^{-i\Delta\omega t}, \quad \frac{d\beta_1}{dt} = \gamma\alpha_0^*\alpha_1 e^{i\Delta\omega t}, \tag{3.38a}$$

$$d\alpha_0/dt = -\gamma\alpha_1\beta_1^* \exp(i\Delta\omega t) \tag{3.38b}$$

This system has again two integrals of motion (3.33). In the linear approximation ($\alpha_0(t) = 1$) from (3.38a) follows

$$\alpha_1(t) = \alpha_1^0 \exp(-ift),$$
$$\beta_1(t) = -i(f/\gamma)\alpha_1^0 \exp[i(\Delta\omega - f)t], \tag{3.39}$$

where $f = \Delta\omega/2 \pm i\Gamma$ with Γ from (3.36b). This gives again the equation (3.36a).

Thus, the high frequency modulational instability and parametric decay have almost identical dispersion relations, but the nonlinear equations for them are different: the parametric decay equations form the lowest (third) order truncated system, obtained from the equations describing the modulational instability.

On the other hand, for the low frequency modulational instability (when (3.15) takes place) the third order truncated system (3.38) does not give correct results. Indeed, in this case $\sigma_1(\underset{\sim}{q}) = 0$ and, therefore, $\Delta\omega = \eta_1(\underset{\sim}{q})$. Together with (3.37), this gives the instability threshold which differs from (3.20a) by a numerical factor. The correct threshold follows only from the fifth order system (3.9) which leads to the dispersion equation (3.10). Thus, for the low frequency modulational instability the correct lowest order truncated system has the form (3.30) which after linearization gives (3.9).

4. Two-mode initial conditions. The cells.

The solutions (3.3) of the system (2.25) have generalizations describing developments of the multi-mode initial modulations of the plane wave. The simplest of them contains two initial modes with the wave vectors, $\underset{\sim}{q}$ and $\bar{\underset{\sim}{q}}$, symmetrical with respect to the z-axis,

$$\underset{\sim}{q} = \{q_x, q_z\}, \quad \bar{\underset{\sim}{q}} = \{-q_x, q_z\} \tag{4.1}$$

In this case we look for the solution of the system (2.25) in the form

$$a(\underset{\sim}{\rho}, t) = (2\pi)^{D/2} \Psi_0 \sum_{\substack{m=-\infty \\ n=-\infty}}^{\infty} A^*_{mn}(t)\, \delta(\underset{\sim}{\rho} + m\underset{\sim}{q} + n\bar{\underset{\sim}{q}}), \tag{4.2a}$$

$$b(\underset{\sim}{p},t) = (2\pi)^{D/2}\Psi_0 \sum_{\substack{m=-\infty \\ n=-\infty}}^{\infty} B_{mn}(t)\,\delta(p - m\underset{\sim}{q} - n\underset{\sim}{\bar{q}}) \tag{4.2b}$$

$$B_{00}(t) \equiv 0 \tag{4.2c}$$

Substituting (4.2) into (2.25), we arrive at the infinite system of ordinary differential equations

$$dA_{mn}/dt = i\sigma_{mn}A_{mn} - \sum_{m',n'} \gamma_{m'n'} A_{m+m',n+n'} (B^*_{m'n'} - B_{-m',-n'}), \tag{4.3a}$$

$$dB_{mn}/dt = -i\eta_{mn}B_{mn} + \gamma_{mn} \sum_{m',n'} A^*_{m'n'} A_{m+m',n+n'}, \tag{4.3b}$$

where m' and n' are integer numbers,

$$-\infty < m', n' < \infty$$

and

$$\sigma_{mn} = \sigma(-m\underset{\sim}{q} - n\underset{\sim}{\bar{q}}), \quad \eta_{mn} = \eta(m\underset{\sim}{q} + n\underset{\sim}{\bar{q}}), \tag{4.4a}$$

$$\gamma_{mn} = (2\pi)^{D/2} V(m\underset{\sim}{q} + n\underset{\sim}{\bar{q}})\Psi_0 . \tag{4.4b}$$

Taking into account (2.26) and (4.4) we have

$$\sigma_{mn} = \sigma_{nm}, \quad \eta_{mn} = \eta_{nm}, \quad \gamma_{mn} = \gamma_{nm}, \qquad (4.5)$$
$$\sigma_{00} = \eta_{00} = \gamma_{00} = 0$$

The system (4.3) is Hamiltonian

$$i\,dA_{mn}/dt = -\partial h/\partial A^*_{mn}$$
$$i\,dB_{mn}/dt = \partial h/\partial B^*_{mn} \qquad (4.6)$$

where $h = h_1 + h_2 + h_{12}$,

$$h_1 = \sum_{m,n} \sigma_{mn} A^*_{mn} A_{mn}, \quad h_2 = \sum_{m,n} \eta_{mn} B^*_{mn} B_{mn} \qquad (4.7a)$$

$$h_{12} = -i \sum_{m,n} \sum_{m',n'} \gamma_{m'n'} A^*_{mn} \big(A_{m-m',\,n-n'} B_{m'n'}$$
$$- A_{m+m',\,n+n'} B^*_{m'n'} \big). \qquad (4.7b)$$

Thus, A_{mn} and B_{mn} are canonically conjugated to A^*_{mn} and B^*_{mn} .

We consider the initial conditions

$$\left.\begin{array}{l} A_{00}(0) = 1, \; A_{10}(0) = A_1(0), \; A^{(0)}_{01} = \rho A_1(0) \\ B_{10}(0) = B_1(0), \; B_{01}(0) = \rho B_1(0), \end{array}\right\} \qquad (4.8a)$$

$$|A_1(0)|^2 + |B_1(0)|^2 \neq 0 \qquad (4.8b)$$

$$A_{mn}(0) = A_{nm}(0) = 0 \quad (m,n < 0 \text{ or } m+n > 1) \tag{4.8c}$$

where ρ is an arbitrary complex number

$$\rho = |\rho| \exp(i\delta) \tag{4.9}$$

At $\rho = 0$ we arrive at the solutions corresponding to the single-mode initial modulations considered in the previous section. At $\rho = 1$ we have completely symmetrical two-mode initial modulations. From (4.5) follows that in this case

$$A_{mn}(t) = A_{nm}(t), \quad B_{mn}(t) = B_{nm}(t), \tag{4.10}$$

i.e. symmetrical initial modulations lead to the symmetry with respect to z-axis at any t. Such solutions may be considered as a two dimensional model of axially symmetric three dimensional forms. One can also consider more general, quasisymmetrical initial modulations, assuming that $|\rho| = 1$ but the phase δ is arbitrary. This is a simple model of three dimensional perturbations with the amplitude symmetry.

Consider now the modulational instability developing from the initial conditions (4.8). Assuming that $A_{00}(t) = 1$ and the rest $A_{mn}(t)$ are small quantities and linearizing eqs. (4.3), we obtain

$$dA_{mn}/dt = i\sigma_{mn} A_{mn} + \gamma_{mn}(B_{mn} - B^*_{-m,-n})$$

$$dB_{mn}/dt = -i\eta_{mn} B_{mn} + \gamma_{mn}(A_{mn} + A^*_{-m,-n}) \tag{4.11}$$

From the initial conditions (4.8) follows that system (4.11) has nonvanishing solutions if $|m|+|n|=1$. Introducing the notations

$$A_{00}(t)=A_0(t),\ A_{\pm 1,0}(t)=A_{\pm 1}(t),\ B_{\pm 1,0}(t)=B_{\pm 1}(t) \tag{4.12}$$

$$\sigma_{10}=\sigma_1\,,\quad \sigma_{-1,0}=\sigma_{-1}\,,\qquad \eta_{10}=\eta_{-1,0}=\eta_1 \tag{4.13}$$

and observing that $\gamma_{10}=\gamma$ where γ is defined in (3.6), we arrive at the system coinciding with (3.9) which gives the dispersion equation (3.10). Thus, for the two mode initial modulations with the wave vectors (4.1), the frequencies and growth rates of the modulational instability are the same as for one mode with the wave vector q . More general dispersion relations, following from (4.11), are derived in Appendix A.

Consider now the nonlinear stage of the instability with the initial conditions (4.8). Introducing in (4.3) the new unknown functions

$$\begin{aligned} \alpha_{mn}(t) &= A_{mn}(t)\exp(-i\sigma_{mn}t) \\ \beta_{mn}(t) &= B_{mn}(t)\exp(i\eta_{mn}t) \end{aligned} \tag{4.14}$$

we obtain the system of equations

$$\begin{aligned} d\alpha_{mn}/d\tau = -\gamma^{-1}\sum_{m',n'}\gamma_{m',n'}\,\alpha_{m+m',n+n'}\big[\beta^*_{m',n'} \\ \times\exp(i f^{m',n'}_{m,n}\tau)-\beta_{-m',-n'}\exp(i g^{m',n'}_{m,n}\tau)\big], \end{aligned} \tag{4.15a}$$

$$d\beta_{mn}/d\tau = (\gamma_{mn}/\gamma)\sum_{m';n'} \alpha^*_{m';n'}\,\alpha_{m+m';n+n'} \times \exp(i\,h^{m';n'}_{m,n}\,\tau) \qquad (4.15b)$$

where $\tau = \gamma t$, and $f^{m';n'}_{m,n}$, $g^{m';n'}_{m,n}$ and $h^{m';n'}_{m,n}$ are notations for the normalized combination frequencies

$$f^{m';n'}_{mn}(\underset{\sim}{q}) = \gamma^{-1}[\sigma_{m+m';n+n'}(\underset{\sim}{q}) - \sigma_{mn}(\underset{\sim}{q}) + \eta_{m';n'}(\underset{\sim}{q})], \qquad (4.16a)$$

$$g^{m';n'}_{mn}(\underset{\sim}{q}) = \gamma^{-1}[\sigma_{m+m';n+n'}(\underset{\sim}{q}) - \sigma_{mn}(\underset{\sim}{q}) - \eta_{m'n'}(\underset{\sim}{q})], \qquad (4.16b)$$

$$h^{m';n'}_{mn}(\underset{\sim}{q}) = \gamma^{-1}[\sigma_{m+m';n+n'}(\underset{\sim}{q}) - \sigma_{m';n'}(\underset{\sim}{q}) + \eta_{mn}(\underset{\sim}{q})]. \qquad (4.16c)$$

An investigation of the combination frequencies is given in Appendix B.

Consider simple truncated systems following from (4.15). Firstly, we retain in (4.15) only the terms with $|k| + |\ell| \leq 1$, i.e. those where at least one of the indices is zero and the other is zero or ± 1. It is easy to check that it has a solution with

$$\alpha_{01}(t) = \rho\,\alpha_{10}(t), \quad \alpha_{0,-1}(t) = \rho^*\alpha_{-1,0}(t)$$
$$\beta_{01}(t) = \rho\,\beta_{10}(t), \quad \beta_{0,-1}(t) = \rho^*\beta_{-1,0}(t) \qquad (4.17)$$

Evidently, this solution satisfies the initial conditions (4.8). Introducing the notations

$$\alpha_{00} = \alpha_0 , \quad \alpha_{\pm 1,0} = \alpha_{\pm 1} , \quad \beta_{\pm 1,0} = \beta_{\pm 1} , \tag{4.18}$$

we obtain the closed system of equations

$$d\alpha_1/d\tau = \alpha_0 \{ \beta_1 \exp[-i(\sigma_1 + \eta_1)\gamma^{-1}\tau] - \beta_{-1}^* \exp[i(\eta_1 - \sigma_1)\gamma^{-1}\tau] \} , \tag{4.19a}$$

$$d\beta_1/d\tau = \alpha_0^* \alpha_1 \exp[i(\sigma_1 + \eta_1)\gamma^{-1}\tau] + \alpha_0 \alpha_{-1}^* \exp[i(\eta_1 - \sigma_{-1})\gamma^{-1}\tau] , \tag{4.19b}$$

$$d\alpha_0/d\tau = -(1 + |\rho|^2) \{ \alpha_1 \beta_1^* \exp[i(\sigma_1 + \eta_1)\gamma^{-1}\tau] - \alpha_{-1} \beta_1 \exp[i(\sigma_{-1} - \eta_1)\gamma^{-1}\tau] - \alpha_1 \beta_{-1} \exp[i(\sigma_1 - \eta_1)\gamma^{-1}\tau] + \alpha_{-1} \beta_{-1}^* \exp[i(\sigma_{-1} + \eta_1)\gamma^{-1}\tau] \} , \tag{4.19c}$$

$$d\alpha_{-1}/d\tau = -\alpha_0 \{ \beta_1^* \exp[i(\eta_1 - \sigma_{-1})\gamma^{-1}\tau - \beta_{-1} \exp[-i(\eta_1 + \sigma_{-1})\gamma^{-1}\tau] \} , \tag{4.19d}$$

$$d\beta_{-1}/d\tau = \alpha_0 \alpha_1^* \exp[i(\eta_1 - \sigma_1)\gamma^{-1}\tau] + \alpha_0^* \alpha_{-1} \exp[i(\eta_1 + \sigma_{-1})\gamma^{-1}\tau] . \tag{4.19e}$$

Now assume that the vector $\tilde{q}$ satisfies the resonance condition (3.31). Then some terms in the eqs. (4.19) do not contain oscillating factors and one can expect that they play the main role if

$$\gamma \ll |\sigma_{mn}(q)|, \quad |\eta_{mn}(q)| \tag{4.20}$$

(which is a condition of smallness of Ψ_0). Neglecting oscillating terms in eqs. (4.19), we arrive at the third order system

$$d\alpha_1/dt = \gamma \alpha_0 \beta_1, \quad d\beta_1/dt = \gamma \alpha_0^* \alpha_1, \tag{4.21a}$$

$$d\alpha_0/dt = -\gamma(1+|\rho|^2)\alpha_1 \beta_1^* . \tag{4.21b}$$

The solution of eqs. (4.21) is

$$\begin{aligned} \alpha_0(t) &= Ae^{i\varphi_0} sn[(1+|\rho|^2)^{1/2} B\gamma(t+t_0)], \\ \alpha_1(t) &= Be^{i\varphi_1} dn[(1+|\rho|^2)^{1/2} B\gamma(t+t_0)], \\ \beta_1(t) &= -(1+|\rho|^2)^{-1/2} Ae^{i(\varphi_1-\varphi_0)} cn[(1+|\rho|^2)^{1/2} B\gamma(t+t_0)] \end{aligned} \tag{4.22}$$

where $A > 0$, $B > 0$ and the module of elliptic functions in (4.22) is

$$k = (A/B)(1+|\rho|^2)^{-1/2} . \tag{4.23}$$

These results are in rather good agreement with numerical

solutions of higher order truncated systems at sufficiently small Ψ_0 and not too large t. Consider, for example, the truncated system with $|k|+|\ell| \leq 2$, with $\rho=1$, consisting of 15 equations (because of (4.10)).

Its solution for the case (2.12b), which corresponds to the nonlinear interaction of a whistler with FMS waves, is shown in Fig.3. The parameters in the equations and initial conditions are the same as in refs. /12,22/ where this interaction has been investigated numerically both by means of the Schrödinger and modified MHD equations (1.34) /12/ and the reduced equations (1.41), (1.48) /22/. The corresponding values are

$$\left.\begin{array}{llll} v_1 = 1.2, & v_2 = 0.1, & s_1 = 0.2, & s_2 = -0.72 \\ c_1 = 0.4, & c_2 = (2/3)\times 10^{-2}, & & q = 0.19 \end{array}\right\} \quad (4.24a)$$

$$\left.\begin{array}{lll} \rho = 1, & \alpha_0(0) = 1, & \alpha_1(0) = 0.05 \\ \beta_1(0) = -0.00995. & & \end{array}\right\} \quad (4.24b)$$

As in the mentioned papers, the vector $\underset{\sim}{q}$ satisfies the resonant condition (3.31) which gives

$$q_x = 0.1890768, \qquad q_z = 0.0187075 \quad (4.24c)$$

The plots of $|\alpha_0(\tau)|$, $|\alpha_1(\tau)|$ and $|\beta_1(\tau)|$ at $\Psi_0 = 0.02$, shown in Fig.3a, are almost the same as those obtained from (4.21) at $\rho=1$ at least for $\tau < 22$. The functions $|\alpha_{-1}(\tau)|$ and $|\beta_{-1}(\tau)|$ are rather small (we use notations (4.18)). The phases

$$\psi_0(\tau)=\arg\alpha_0(\tau),\ \psi_1(\tau)=\arg\alpha_1(\tau),\ \psi_2(\tau)=\arg\beta_1(\tau), \quad (4.25)$$

shown in fig.3b, are in qualitative agreement with the results following from (4.22) where $\varphi_0 = \varphi_1$. Indeed, according to (4.22) and (4.24b), the phase ψ_0 jumps from zero to π at the moment τ when $\alpha_0(\tau) = 0$ and then, in a half of the period, again vanishes. The phase $\psi_1(\tau) = 0$, because $\alpha_1(\tau)$ is positive. The function $\beta_1(\tau)$, soon after $\tau = 0$, becomes to be positive and then $\psi_2(\tau) = 0$ up to the moment τ when $\beta_1(\tau) = 0$. At this τ the phase $\psi_2(\tau)$ jumps from zero to π , etc. Also, it follows from (4.19d) and (4.19e) that $\arg\alpha_{-1}(\tau)$ and $\arg\beta_{-1}(\tau)$ have fast oscillations, which is in agreement with fig.3b.

The results of the numerical solution of the same system at Ψ_0 =0.03 are shown in Fig.4. In this case the difference from (4.22) is larger, but the behaviour of $|\alpha_0(\tau)|$, $|\alpha_1(\tau)|$ and $|\beta_1(\tau)|$ is qualitatively the same. The functions $|\alpha_{-1}(\tau)|$ and $|\beta_{-1}(\tau)|$ are more random than at Ψ_0 =0.02. The plots of phases significantly differ from those in fig.3a at sufficiently large τ .

At larger values of Ψ_0 the higher mode instabilities are possible. They are considered in Appendix A where it is shown that for the parameters (4.24) they appear at $\Psi_0 > 0.08$. It should also be noted that a truncated system with $|k|+|l|\le N, N\geq 2$ may occasionally contain terms with small combination frequencies (4.16), despite of (4.20). This may be of importance at sufficiently large τ and ψ_0 (Appendix B).

Now, consider the structures of ψ and γ fields. From (2.27a), (4.2a) and (4.14) follows

$$\psi(R,t) = \psi_0 \sum_{m,n} \alpha^*_{mn}(t) \exp\{-i[(m\underset{\sim}{q} + n\underset{\sim}{\bar{q}})R + \sigma_{mn} t]\} \tag{4.26}$$

From this we can calculate $|\psi(\underset{\sim}{R},t)|^2$. To obtain simple analytical expressions, let us confine ourself to the largest terms, α_{00}, α_{10} and α_{01} (see, e.g., Figs.3,4). Assuming that ρ =1, we have $\alpha_{10} = \alpha_{01}$ and using (4.18) we obtain

$$\psi(\underset{\sim}{R},t) \approx \psi_0 \{\alpha_0^*(t) + 2\alpha_1^*(t) \exp[-i(q_z z + \sigma_1 t)] \times \cos(q_x x)\} \tag{4.27}$$

Experssion (4.27) describes a modulated wave which consists of cells propagating in z -direction with the velocity close to the group velocity v_g of the carrier wave. From (4.27) follows

$$|\psi(R,t)|^2 \approx \psi_0^2 \{|\alpha_0(t)|^2 + 4|\alpha_0(t)\alpha_1(t)| \cos y_1 \cos y_2 + 4|\alpha_1(t)|^2 \cos^2 y_2\} \tag{4.28}$$

where, taking into account (3.31),

$$y_1 = q_z z - \eta_1 t + \psi_1(t) - \psi_0(t), \tag{4.29}$$

$$y_2 = q_x x \tag{4.30}$$

with notations (4.25). The cell structure can be characterized by the contour maps $|\Psi(\underset{\sim}{R},t)|^2 = const$. It is similar to that one obtained in refs. /11,12,22/ by means of direct numerical solutions of the equations describing whistler-FMS waves interaction.

From (4.28) follows that maxima of the function $|\Psi(\underset{\sim}{R},t)|^2$ are reached at

$$y_1 = \pi m, \qquad y_2 = \pi n, \tag{4.31}$$

where m and n are integer numbers (positive or negative) and $m+n$ is an even. At these points

$$|\Psi|^2_{max} = \Psi_0^2 \{|\alpha_0(t)| + 2|\alpha_1(t)|\}^2. \tag{4.32}$$

Now consider the minima of $|\Psi|^2$. It is easy to find that for

$$|\alpha_0(t)| \geq 2|\alpha_1(t)|, \tag{4.33}$$

the minimum positions are defined again by equations (4.31), but $m+n$ are now odd numbers and

$$|\Psi|^2_{min} = \Psi_0^2 \{|\alpha_0(t)| - 2|\alpha_1(t)|\}^2. \tag{4.34}$$

If

$$|\alpha_0(t)| \leq 2|\alpha_1(t)|, \tag{4.35}$$

the minimum positions are defined by the equations

$$y_1 = \pi m, \quad \cos y_2 = \frac{1}{2}(-1)^{m+1}\left|\frac{\alpha_0(t)}{\alpha_1(t)}\right| \tag{4.36}$$

and, instead of (4.34),

$$|\psi|^2_{min} = 0 \tag{4.37}$$

Now let us investigate $\vartheta(\underset{\sim}{R}, t)$. From (2.27b), (4.2b) and (4.14) we have at $\rho = 1$

$$\vartheta(R,t) = (2c_2/c_1)^{1/2}\psi_0 \sum_{\substack{m=-\infty \\ n=-\infty}}^{\infty} [G(m\underset{\sim}{q} + n\underset{\sim}{\bar{q}})/\eta_{mn}(\underset{\sim}{q})]^{1/2} \tag{4.38}$$

$$\times|\beta_{mn}(t)|\sin[(m\underset{\sim}{q} + n\underset{\sim}{\bar{q}})\underset{\sim}{R} - \eta_{mn}(\underset{\sim}{q})t + \gamma_{mn}(t)]$$

where $\gamma_{mn} = \arg\beta_{mn}$. Applying this expression to the case (2.12b), we again confine ourselves by the main terms with $\beta_{10} = \beta_{01} = \beta_1$. Then

$$\vartheta(R,t) \approx -2\left(\frac{2c_2 q_\perp^2}{c_1 v_2 q}\right)^{1/2}\psi_0|\beta_1(t)|\cos X_1 \cos X_2, \tag{4.39}$$

where

$$X_1 = q_z z - \eta_1 t + \psi_2(t) + \pi/2, \quad X_2 = q_x x \tag{4.40}$$

It is evident that the points, where $\vartheta = \vartheta_{min}$ and $\vartheta = \vartheta_{max}$, are defined by the equations

$$X_1 = \pi n, \qquad X_2 = \pi m, \tag{4.41}$$

where $n + m$ is an even for the minima and odd for the maxima. From that follows

$$\vartheta_{max}(t) = 2(2c_2 q_\perp^2 / c_1 v_2 q)^{1/2} \Psi_0 |\beta_1(t)| \tag{4.42a}$$

$$\vartheta_{min}(t) = -\vartheta_{max}(t) \tag{4.42b}$$

Therefore, in the approximation considered, ϑ_{max} differs from ϑ_{min} only by sign and they change in time as $|\beta_1(t)|$.

The plots of expressions (4.32) and (4.34) with (4.37) are shown in Fig.5 and those of (4.42) in Fig.6, where the functions $|\alpha_0(t)|$, $|\alpha_1(t)|$ and $|\beta_1(t)|$ are found by solving equations (4.19) with the parameters (4.24) and Ψ_0 =0.02. The solutions of the truncated system with $|k| + |l| \leq 2$ give almost the same pictures. The Figs. 5 and 6 are in a good agreement with those obtained in numerical experiments of refs./22/. The same can be said about other properties of the cells at Ψ_0 =0.02, in particular, about the phase shifts between the positions of $|\Psi|^2_{max}$ and ϑ_{min} . Comparing (4.40) with (4.29) and taking into account the behaviour of the phases (4.25) which shown in Fig.3b, we arrive at approximately same conclusions as in numerical experiments /22/.

However, at sufficiently large τ or ψ_0 there are increasing discrepancies between the presented above results and numerical experiments. The latter show irreversibility and chaotic behaviour of $|\psi|^2$ and ν with time. At ψ_0 =0.03 they are observable after two nonlinear periods /22/. Theoretically, the irreversibility should follow from higher order truncated systems. An investigation of the physical reasons of these phenomena is of big interest. Some of possible mechanisms are considered in Appendixes A and B.

APPENDIX A

The unstable modes of higher orders

Investigating the linear stage of the modulational instability with the two mode initial conditions, we have taken them in the form (4.8). Therefore in the system (4.11) it was sufficient to assume that $|m|+|n|=1$. However, due to nonlinear effects, nonvanishing A_{mn} and B_{mn} with $|m|+|n|>1$ will appear with time. These modes may be unstable, which would lead to a noticeble deviation from a lower order approximation.

Consider, for example, the instability of the mode $q_{mn}=mq+n\bar{q}$ assuming that it developes in the time interval when $A_{00}(t)\approx 1$. Then the initial stage of this instability is described by the linear system of the fourth order, consisting of eqs. (4.11) and two more equations obtained from (4.11) by substitution $m\rightarrow -m$ and $n\rightarrow -n$. Looking for a solution where A_{mn}, B_{mn} and $A^*_{-m,-n}$, $B^*_{-m,-n}$ are proportional to $exp(-i\Omega t)$, we obtain the following dispersion equation

$$(\Omega^2-\eta^2_{mn})(\Omega-\sigma_{-m,-n})(\Omega+\sigma_{m,n}) = 2\gamma^2_{mn}\eta_{mn}(\sigma_{mn}+\sigma_{-m,-n}) \tag{A.1}$$

At m =1 and n =0 , from (A.1) follows eq. (3.10).

Assume that

$$|\Omega| \gg \gamma_{mn} \quad . \tag{A.2}$$

Then, generalizing (3.36a), one can look for the solution of eq. (A.1) of the form

$$\Omega = \eta_{mn} + \Omega' \tag{A.3}$$

where $|\Omega'| \ll \eta_{mn}$. Designate

$$\Delta\omega_{mn}(\underset{\sim}{q}) = \eta_{mn}(\underset{\sim}{q}) + \sigma_{mn}(\underset{\sim}{q}) \, . \tag{A.4}$$

From what follows it will be seen that the instability is possible if

$$|\Delta\omega_{mn}(\underset{\sim}{q})| \ll \eta_{mn}(\underset{\sim}{q}) \, . \tag{A.5}$$

Then from (A.1) one obtains

$$\Omega'_{mn} \approx -\Delta\omega_{mn}/2 \pm [(\Delta\omega_{mn}/2)^2 - \gamma^2_{mn}]^{1/2} \tag{A.6}$$

i.e. the instability takes place at

$$|\Delta\omega_{mn}(q)| < 2\gamma_{mn}(q) \, . \tag{A.7}$$

This justifies (A.5). From (A.7) at m =1 and n =0 follow formulae (3.36) and (3.37). Using (4.4b) and (2.23) we have in the case (2.12a)

$$\gamma_{mn}(q) = \Psi_0 (c_1 c_2 / 2 v_2)^{1/2} [(m-n) q_\perp^2 + (m+n) q_z^2]^{1/4}, \quad \text{(A.8a)}$$

and in the case (2.12b)

$$\gamma_{mn}(q) = \Psi_0 (c_1 c_2 / 2 v_2) \times |m-n| q_\perp [(m-n)^2 q_\perp^2 + (m+n)^2 q_z^2]^{-1/4}. \quad \text{(A.8b)}$$

Applying the criterion (A.7) to the nonlinear instability which appears when A_{mn} and B_{mn} arise due to the development of the initial perturbations (4.8), one should take into account that the vector $\tilde{q}$ satisfies the condition of the linear instability (3.31). This restricts possibilities of the origin of the nonlinear instability.

As an example, which can be compared with numerical solutions of the system (1.41) and (1.48) obtained in ref. /22/, consider parameters from (4.24). If Ψ_0 =0.02, the condition (A.7) is fulfilled for (m,n) =(1,0), (10,3), etc. The first pair corresponds to linear instability investigated in section 4. The modes of higher order can develope only at rather large t and the corresponding nonlinear instabilities would hardly appear because of the influence of lower modes. With increase of Ψ_0, the criterion (A.7) is satisfied by lower (m,n). For example, at $\Psi_0 > 0.08$ the lowest unstable mode appearing after (1,0) is (m,n) = =(2,0). It can develop fast enough.

APPENDIX B

INVESTIGATION OF THE COMBINATION FREQUENCIES (4.16)

If Ψ_0 is suffuciently small, the condition (4.20) takes place and so the combination frequencies (4.16) are, generally, large. However, there can be exceptions for some particular combinations of indices and one can expect that the corresponding terms in eqs. (4.15) are more important than others. Here we find such frequencies, assuming that the vector $\underset{\sim}{q}$ satisfies eq. (3.31) and $|q| \ll 1$. Then

$$|q_z| \ll |q_x| \sim |q| \ll 1 \tag{B.1}$$

and if, occasionally, the terms with q_z, q_x and q_x^2 cancel out in a particular combination frequency, it has the order of q_z^2 . To find such frequencies, we start with identities

$$\begin{aligned} &\sigma_{m+m', n+n'}(\underset{\sim}{q}) - \sigma_{m,n}(\underset{\sim}{q}) = -v_1 (m'+n') q_z \\ &+ (1/2)\{(m'-n')[2(m-n)+m'-n'] s_1 q_x^2 \\ &+ (m'+n')[2(m+n)+m'+n'] s_2 q_z^2 , \end{aligned} \tag{B.2}$$

and

$$\sigma_{m+m', n+n'}(\underset{\sim}{q}) - \sigma_{m';n'}(\underset{\sim}{q}) = -2v_1(m+n)q_z$$
$$+(1/2)\{(m-n)[(m-n)+2(m'-n')]s_1 q_x^2 \qquad \text{(B.3)}$$
$$+(m+n)[(m+n)+2(m'+n')]s_2 q_z^2\}$$

which can easily be obtained from (4.4a) and (2.9). Then, taking into account (3.31) and (B.2), we have

$$\gamma f_{m,n}^{m';n'} = -v_1(m'+n'-|m'-n'|)q_z$$
$$+(1/2)\{(m'-n')[2(m-n)+m'-n']s_1 q_x^2$$
$$+(m'+n')[2(m+n)+m'+n']s_2 q_z^2\}$$
$$-(1/2)|m'-n'|(s_1 q_x^2 + s_2 q_z^2) \qquad \text{(B.4)}$$
$$+[(m'-n')^2 q_x^2 + (m'+n')^2 q_z^2]^{1/2} - |m'-n'|q$$

The combination of two last terms in (B.4) is of the order of q_z^2. The first term vanishes if

$$n' = 0,\ m' \geq 0 \qquad \text{or} \qquad n' \geq 0,\ m' = 0 \tag{B.5}$$

Thus, if one of the conditions (B.5) takes place, the frequencies $f_{m,n}^{m',n'}$ contain only terms with q_x^2 and q_z^2. The corresponding expressions are

$$f_{m,n}^{m',0} = (m'/2\gamma)\{[2(m-n) + m' - 1]s_1 q_x^2 + [2(m+n) + m' - 1]s_2 q_z^2\}, \quad m' \geq 0, \tag{B.6a}$$

$$f_{m,n}^{0,n'} = (n'/2\gamma)\{[2(n-m) + n' - 1]s_1 q_x^2 + [2(m+n) + n' - 1]s_2 q_z^2\}, \quad n' \geq 0 \tag{B.6b}$$

Performing similar transformations for $g_{m,n}^{m',n'}$ and $h_{m,n}^{m',n'}$, with account of (B.2) and (B.3), we conclude that only the following expressions do not contain the first order terms with respect to q :

$$g_{m,n}^{m',0} = (m'/2\gamma)\{[2(m-n) + m' - 1]s_1 q_x^2 + [2(m+n) + m' - 1]s_2 q_z^2\}, \quad m' \leq 0; \tag{B.7a}$$

$$g_{m,n}^{0,n'} = (n'/2\gamma)\{[2(n-m)+n'-1]s_1 q_x^2$$

$$+[2(m+n)+n'-1]s_2 q_z^2\}, \quad n' \leq 0; \tag{B.7b}$$

and

$$h_{m,0}^{m',n'} = (m/2\gamma)\{[m+2(m'-n')-1]s_1 q_x^2$$

$$+[m+2(m'+n')-1]s_2 q_z^2\}, \quad m \geq 0; \tag{B.8a}$$

$$h_{0,n}^{m',n'} = (n/2\gamma)\{[n+2(n'-m')-1]s_1 q_x^2$$

$$+[n+2(m'+n')-1]s_2 q_z^2\}, \quad n \geq 0 \tag{B.8b}$$

As $q_z^2 \ll q_x^2$, these quantities become to be much smaller, if the terms with q_x^2 cancel out. The corresponding expressions and indices are

$$f_{mn}^{m',0} = 2m'ns_2 q_z^2/\gamma, \quad m' = 2(n-m)+1 \geq 0, \tag{B.9a}$$

$$f_{mn}^{0,n'} = 2n'ms_2 q_z^2/\gamma, \quad n' = 2(m-n)+1 \geq 0; \tag{B.9b}$$

$$g_{m,n}^{m',0} = 2m'ns_2q_z^2/\gamma, \quad m' = 2(n-m)+1 \le 0; \tag{B.9c}$$

$$g_{m,n}^{0,n'} = 2n'ms_2q_z^2/\gamma, \quad n' = 2(m-n)+1 \le 0; \tag{B.9d}$$

$$h_{m,0}^{m',n'} = 2n'ms_2q_z^2/\gamma, \quad n'-m' = (m-1)/2, \, m \ge 0; \tag{B.9e}$$

$$h_{0,n}^{m',n'} = 2m'ns_2q_z^2/\gamma, \quad m'-n' = (n-1)/2, \, n \ge 0 \tag{B.9f}$$

Expressions (B.9) are multiples of the frequency

$$f = 2s_2q_z^2\gamma^{-1} \tag{B.10}$$

which may be rather small if $q_z^2 \ll 1$. For the parameters (4.24) and ψ_0 =0.02 we have f =-0.50. For larger values of ψ_0, the frequency f changes inversely proportional to ψ_0. At some indices the expressions (B.9) identically vanish, which corresponds to the complete resonance (3.31). At small m, n and m', n', the frequencies (B.9) are of the same order of smallness as f. They play an important role in the truncated systems with

$$|k| + |l| \leq N \quad , \quad N \geq 2 \ .$$

References

1. Karpman,V.I. (1975), 'Nonlinear Waves in Dispersive Media,' Pergamon, Oxford.
2. Pitaevski,L.P. (1960), Zh. Eks. Teor. Fiz. 39, 1450 [Sov. Phys. JETP (1961), 12, 1008].
3. Kadomtsev,B.B. (1964), 'Plasma Turbulence', Academic Press, New York.
4. Kravtsov,Yu.A. and Stepanov,N.S. (1969), Zh. Eks. Teor. Fiz. 57, 1730 [Sov. Phys. JETP (1970), 30, 935].
5. Mikhailovski,A.B. (1977),' Theory of Plasma Instabilities', vol.2, Consultants Bureau.
6. Barash,Yu.S. and Karpman,V.I. (1983), Zh. Eks. Teor. Fiz. 85, 1962 [Sov. Phys. JETP 58, 1139].
7. Zakharov,V.E. (1972), Zh. Eks. Teor. Fiz. 62, 1745 [Sov. Phys. JETP, 35, 908].
8. Gurovich,V.T. and Karpman,V.I. (1969), Zh. Eks. Teor. Fiz. 56, 1952 [Sov. Phys. JETP, 29, 1048].
9. Karpman,V.I. and Washimi,H. (1977), J.Plasma Phys. 18, 173.
10. Karpman,V.I. and Kaufman,R.N., (1982), Physica Scripta, T2/1, 252, Errata: (1984), 29, 288.
11. Karpman,V.I. and Shagalov,A.G. (1984), Zh. Eks. Teor. Phys. 87, 422 [Sov. Phys. JETP 60, 242].
12. Karpman,V.I. and Shagalov,A.G. (1987), J.Plasma Phys. 38, 155; ibid (1988), 39, 1.
13. Stenflo,L., Yu,M.Y. and Shukla,P.K. (1986), J.Plasma Phys. 36, 447.

14. Karpman,V.I. and Stenflo,L. (1988), Phys. Lett. A 127, 99.

15. Karpman,V.I. and Shagalov,A.G. (1989), J.Plasma Phys., 41, 289.

16. Zakharov,V.E., Musher,S.L. and Rubenchik,A.M. (1985), Phys. Rep. 129, 285, and references therein.

17. Karpman,V.I. (1989), Phys. Lett. A, 136, 216.

18. Karpman,V.I., Rasmussen,J.J. and Turitsyn,S.K. (1989), Phys. Lett. A 139, 423.

19. Karpman,V.I. (1989), Phys. Lett. A., 136, 221.

20. Galeev,A.A. and Sagdeev,R.Z. (1973),'Voprosy Teorii Plasmy', Vol.7, Moscow.

21. Karpman,V.I. (1989), Phys. Lett. A 137, 379.

22. Karpman,V.I., Hansen,F.R., Huld,T., Lynov,J.P., Pecseli,H.L. and Rasmussen,J.J. (1989), "Nonlinear Evolution of the Modulational Instability of Whistler Waves". Preprint Risö National Lab., Denmark (to be published).

23. Davydov,A.S. (1979), Physica Scripta 20, 387.

24. Zakharov,V.E. and Shchur,L.N. (1981), Zh. Eksp. Teor. Fiz. 81, 2019.

25. Karpman,V.I. and Kaufman,R.N. (1981), Pis'ma Zh. Eksp. Teor. Fiz. 33, 266 [JETP Lett. 33, 252].

26. Karpman,V.I. and Shagalov,A.G. (1984), Zh. Eksp. Teor. Fiz. 87, 422 [Sov. Phys. JETP 60, 242].

Figure captions

Fig.1. Region of the modulational instability (hatched) for the parameters (4.24) and Ψ_0 =0.02

(1) "high frequency" modulational instability

(2) "low frequency" modulational instability

Fig.2. (a) Plots of $|\alpha_0|$ versus $\tau = \gamma t$ obtained from the truncated system of the N-th order with N =3,5,9,13.

(b) Plots of $I_1(\tau)$ and $I_2(\tau)$ from (3.33).

The parameters are given in (4.24) and Ψ_0 =0.02.

Fig.3. Numerical solution of the truncated system obtained from (4.3) at $|k|+|l| \leq 2$. The parameters are given in (4.24) and Ψ_0 =0.02.

(a) Plots of $|\alpha_{mn}|$ and $|\beta_{mn}|$ versus $\tau = \gamma t$.

(1) $|\alpha_0|$, (2) $|\alpha_1|$, (3) $|\beta_1|$,

(4) α_{-1} , (5) β_{-1} .

(b) Plots of $\arg \alpha_m(\tau)$ and $\arg \beta_m(\tau)$.

Numbering of curves is the same as in (a).

Fig.4. The same as Figure 3 for Ψ_0 =0.03.

Fig.5. (1) $|\Psi/\Psi_0|^2_{max}$ versus τ , (2) $|\Psi/\Psi_0|^2_{min}$ versus τ for Ψ_0 =0.02 and parameters (4.24).

Fig.6. (1) ϑ_{max} versus τ , (2) ϑ_{min} versus τ for Ψ_0 =0.02 and parameters (4.24).

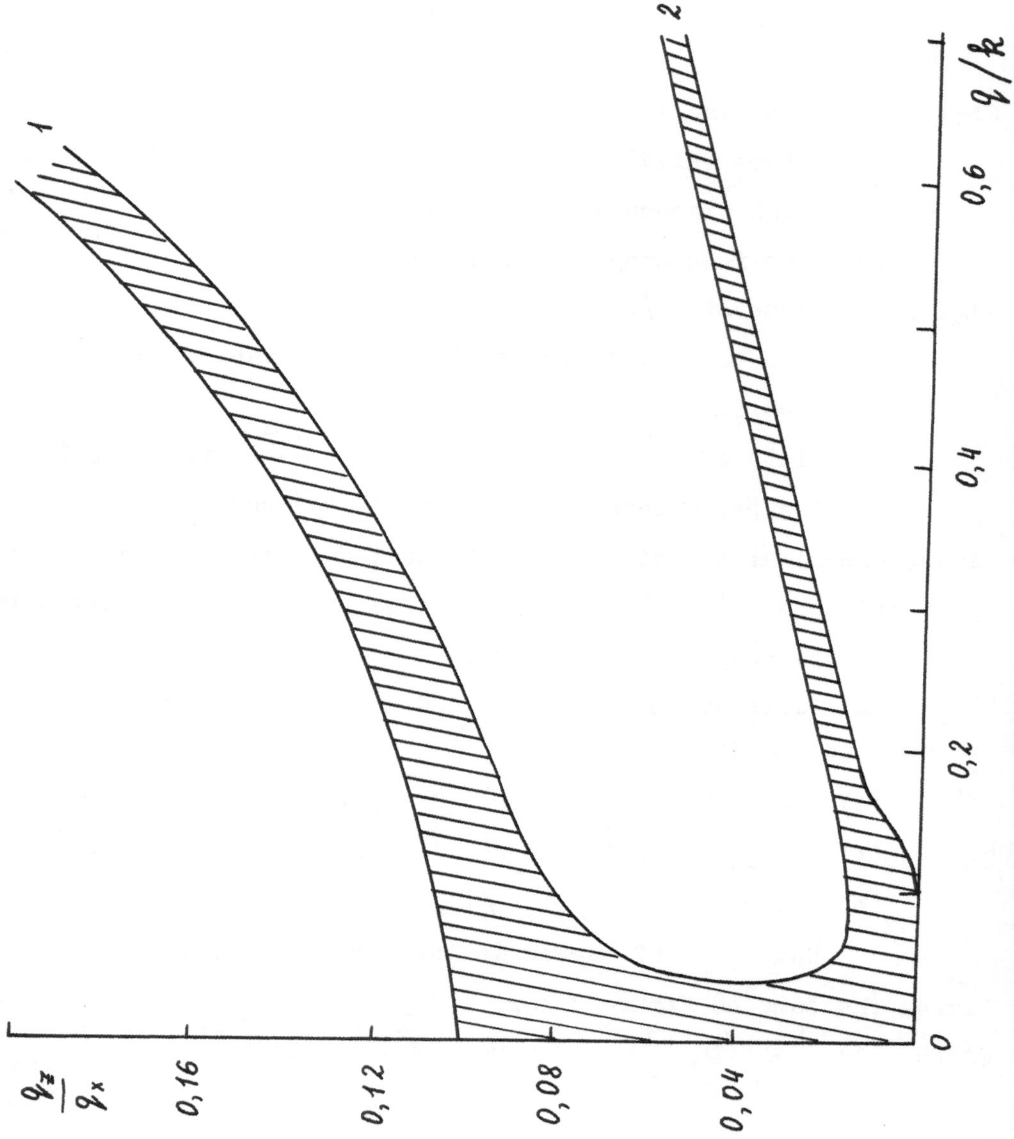

q_z/q_x
0,16
0,12
0,08
0,04
0
0,2
0,4
0,6
q/k
1
2

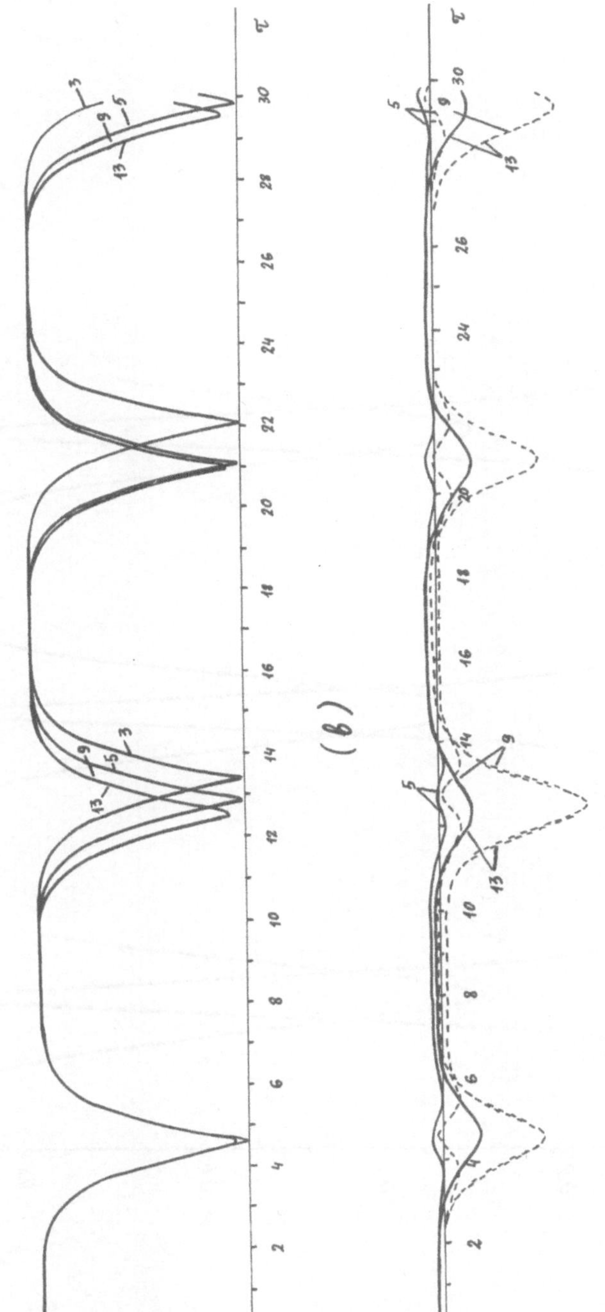
(a)
(b)
τ
1.0
0.8
0.6
0.4
0.2
0.0
1.04
1.02
1.00
0.98
0.96
0.94
0
2
4
6
8
10
12
14
16
18
20
22
24
26
28
30
3
5
9
13

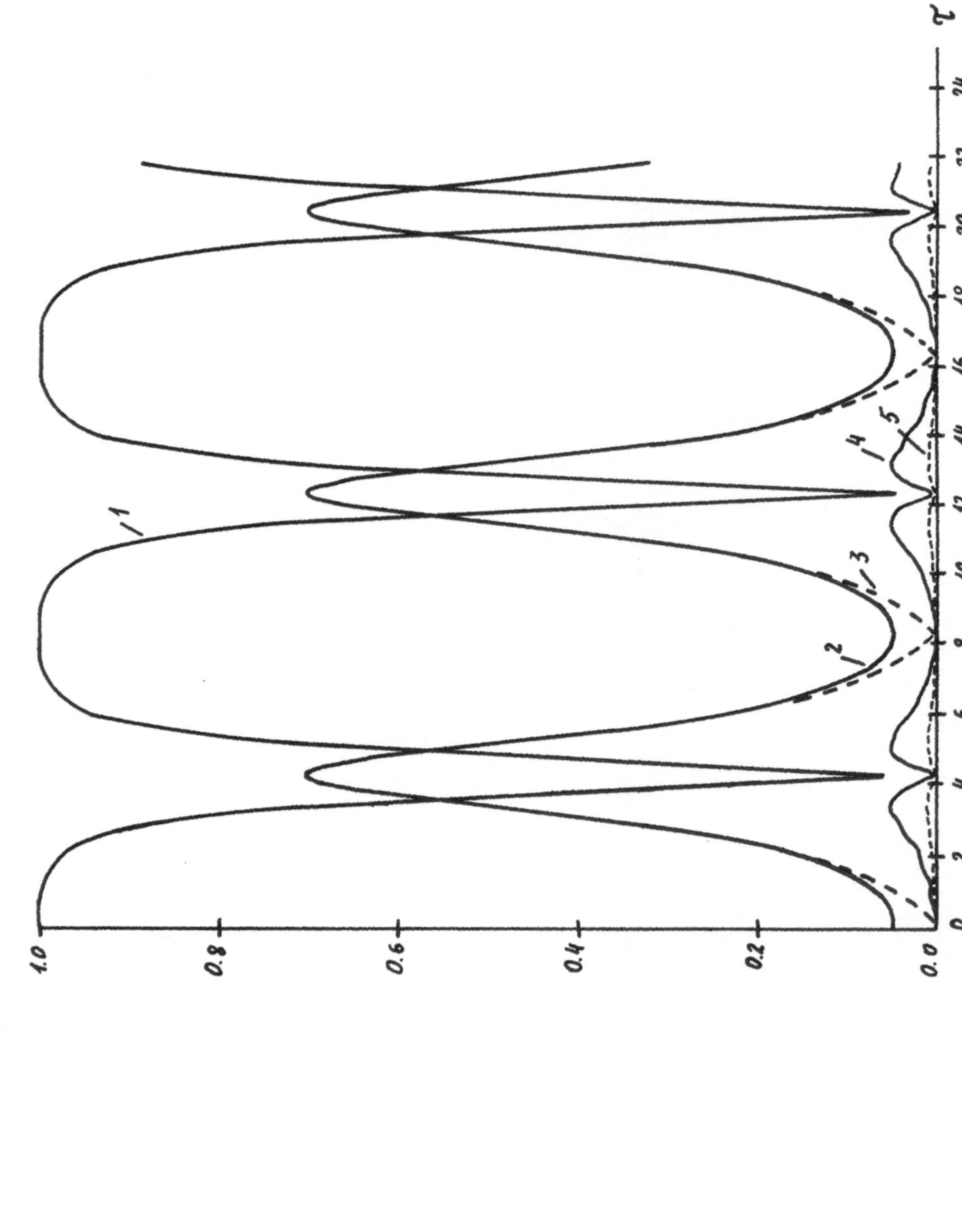

1.0
0.8
0.6
0.4
0.2
0.0
0
2
4
6
8
10
12
14
16
18
20
22
24
τ
1
2
3
4
5

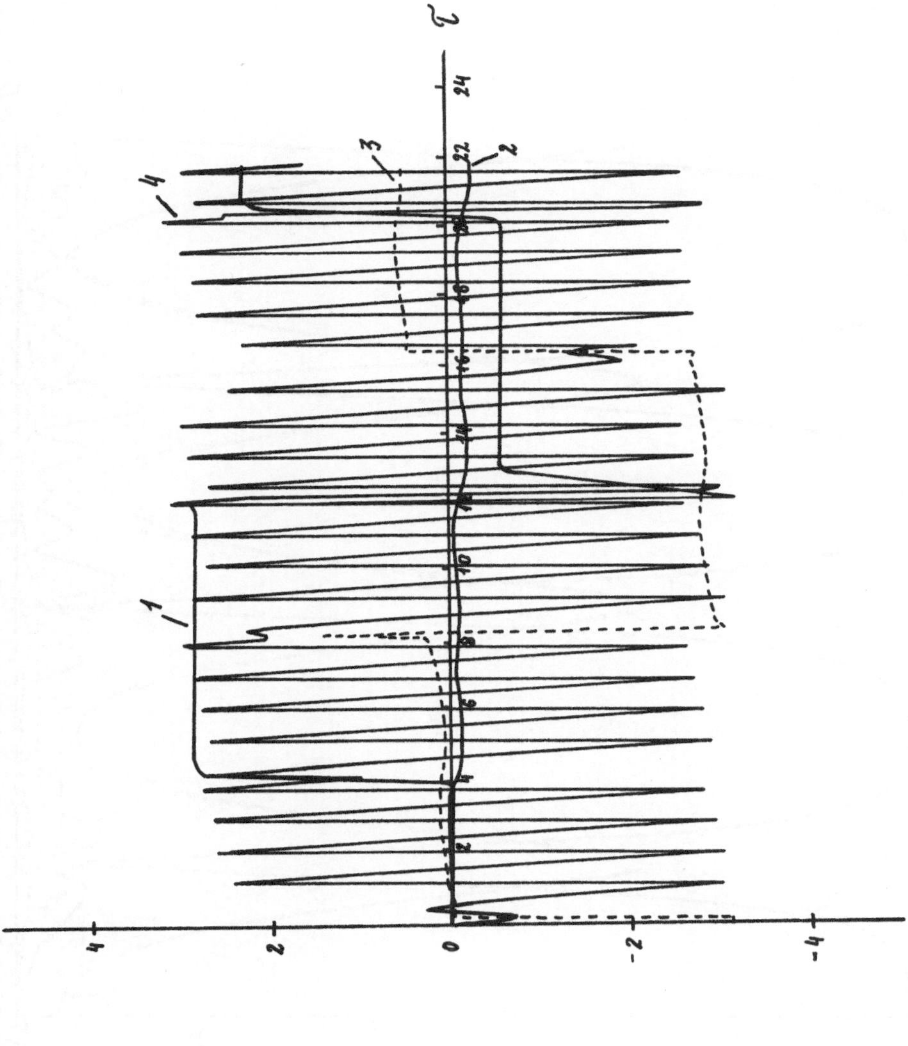
τ
24
22
4
2
0
-2
-4
1
2
3
4

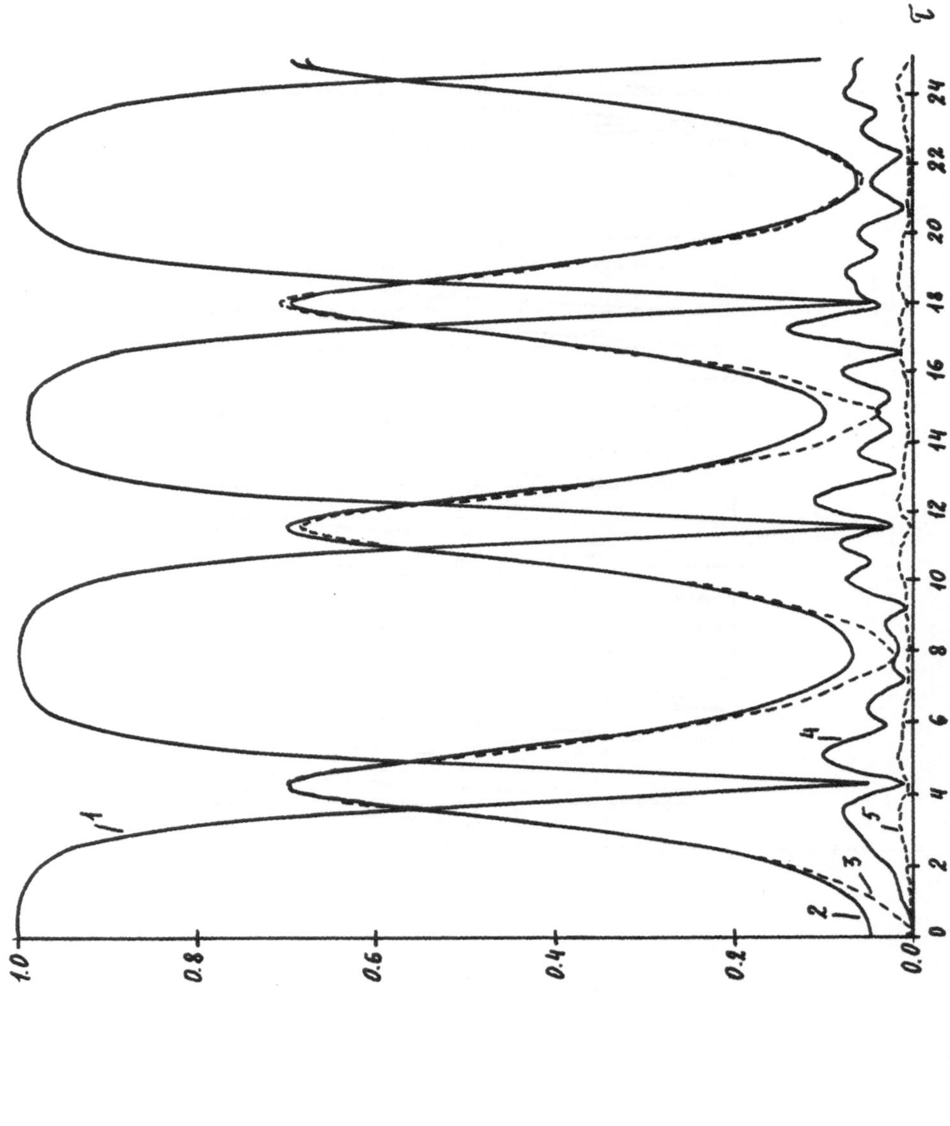
1.0
0.8
0.6
0.4
0.2
0.0
0
2
4
6
8
10
12
14
16
18
20
22
24
τ
1
2
3
4
5

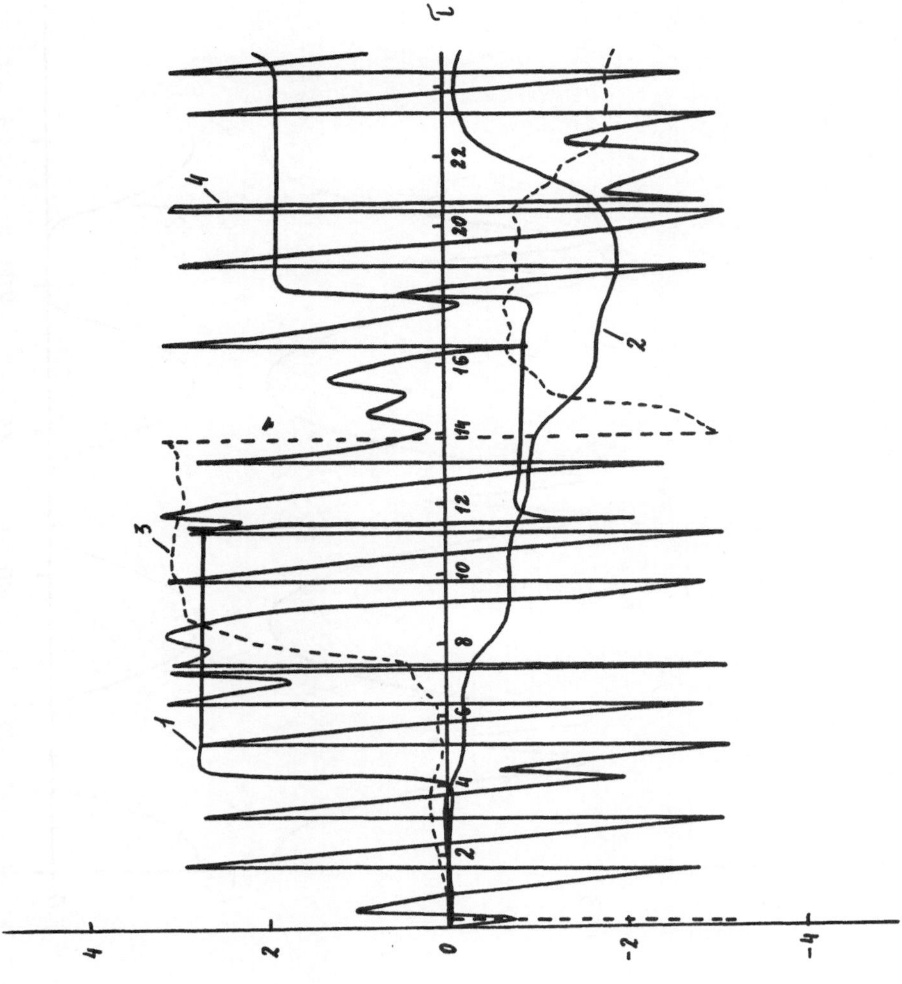
4
2
0
-2
-4
τ
2
4
6
8
10
12
14
16
20
22
1
2
3
4

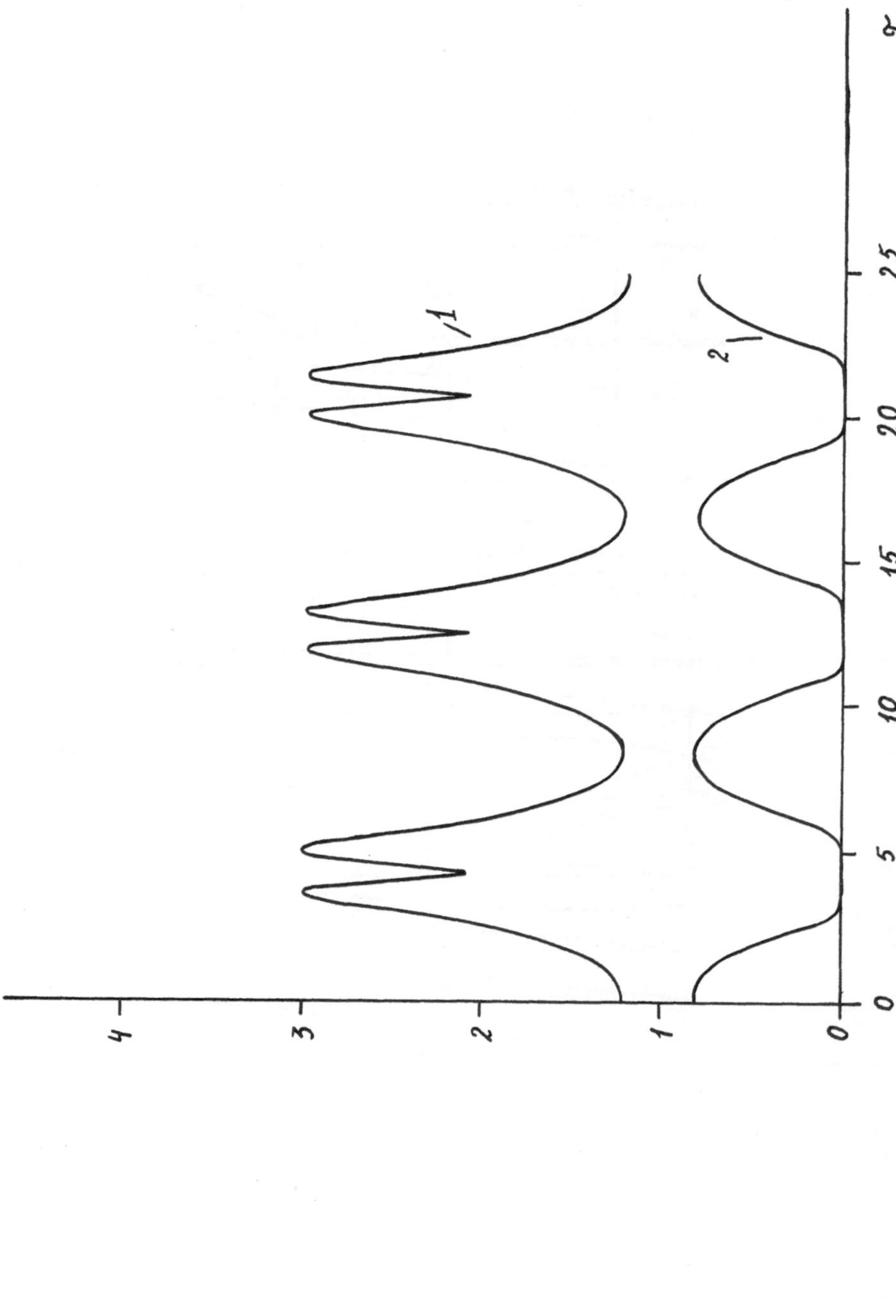
4
3
2
1
0
0
5
10
15
20
25
τ
1
2

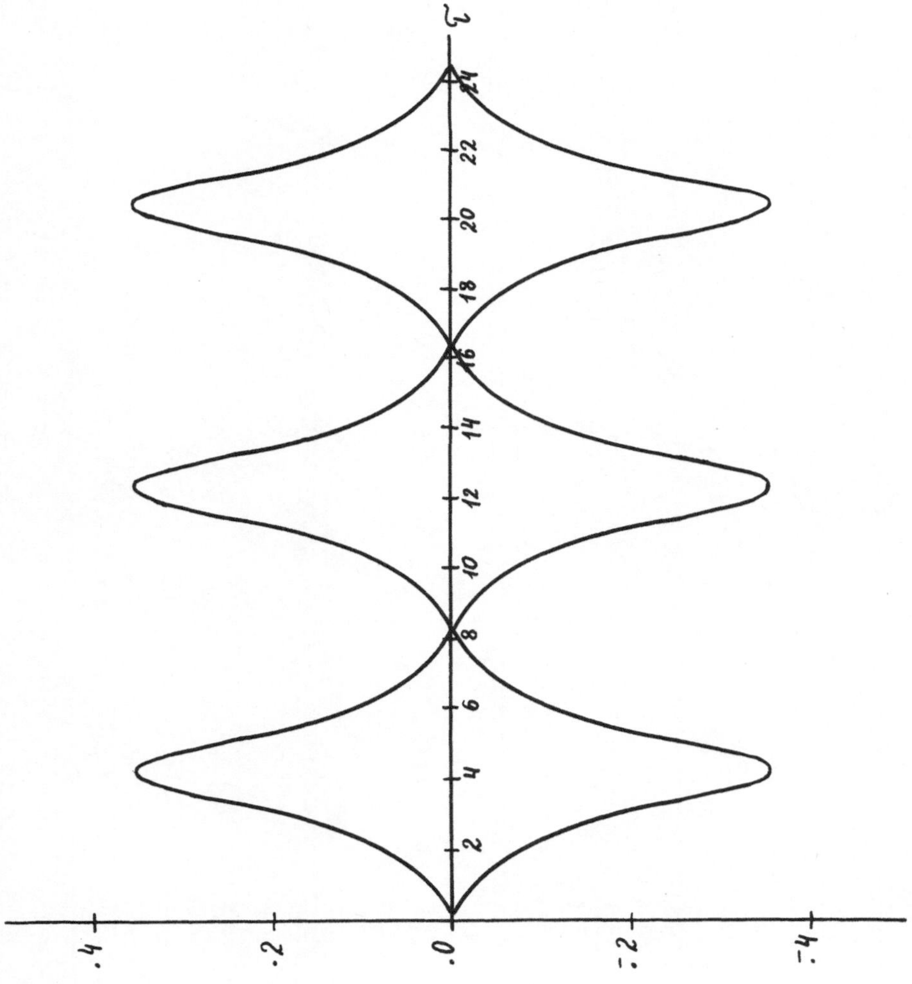

.4
.2
.0
-.2
-.4
2
4
6
8
10
12
14
16
18
20
22
24
ζ

NONLINEAR OPTICS

Alan C. Newell, J.V. Moloney
Department of Mathematics
University of Arizona
TUCSON, AZ 85721

ABSTRACT. Nonlinear optics is a relatively young subject having to await the development of the laser in order to achieve light intensities sufficient to cause an intensity-dependent refractive index. The potential of the field is enormous from the points of view of fundamental science and technology. The field is ideal for the applied mathematician. It uses all the methodology (modelling, analysis, experimental simulation) and exhibits all the challenges of the more traditional areas of applications, like continuum mechanics, but has the advantage that its equations of motion, while still difficult, are somewhat easier. The reason is that, in a wide variety of circumstances, one is dealing with sets of almost monochromatic, dispersive wavepackets with relatively weak nonlinearities, which ingredients are the recipe for equations of nonlinear Schrödinger type about which much is known. One also frequently finds situations in which three and four wave mixing processes (e.g. Raman and Brillouin scattering) are important. Again, many of the properties of the nonlinear envelope equations describing these interactions are familiar to the applied mathematician, but are still relatively new to the optical scientist.

Because of this form of the governing equations, soliton-like coherent pulses play central roles and many extremely interesting and challenging applications of the perturbation theory of soliton equations are encountered. Foremost among the examples of these applications is the use of coherent pulses in communication in optical fibres. Mollenauer and Stolen at Bell Laboratories have managed to overcome the loss and distortion problem with the use of Raman pumping to reinforce the signal at intervals of approximately 40 km without the use of electronic repeaters, and it is likely that solitons in light fibres will revolutionize communication within the next decennium. Other examples of the use of the soliton paradigm are nonlinear waveguides and optically bistable Fabry-Pérot and ring cavities for the design of ultra-fast (pico-second) switches and logic elements with the ultimate goal of designing an all-optical signaling system (optical computers) in which light beams are controlled by other light beams. It also turns out that nonlinear optics and in particular multistable cavities and lasers provides an excellent vehicle in which to study in detail examples of spatio-temporal chaos.

A review article which summarizes these lectures will be shortly published in Physica D.

R. Conte and N. Boccara (eds.), Partially Integrable Evolution Equations in Physics, 161.

THE PHASE DIFFUSION AND MEAN DRIFT EQUATIONS FOR CONVECTION AT FINITE RAYLEIGH NUMBERS IN LARGE CONTAINERS I.

Alan C. Newell, Thierry Passot[†], Mohammad Souli
Arizona Center for the Mathematical Sciences
University of Arizona
TUCSON, AZ 85721

ABSTRACT. We derive the phase diffusion and mean drift equations for the Oberbeck-Boussinesq equations in large aspect ratio containers. We are able to recover all the long wave instability boundaries (Eckhaus, zig-zag, skew-varicose) of straight parallel rolls found previously by Busse and his colleagues. We can calculate the wavenumber selected by curved patterns and find very close agreement with the dominant wavenumbers observed by Heutmaker and Gollub at Prandtl number 2.5 and by Steinberg, Ahlers and Cannell at Prandtl number 6.1. We find a new instability, the focus instability, which causes circular target patterns to destabilize and which, at sufficiently large Rayleigh numbers, plays a major role in the onset of time dependence. Further, we predict the values of the Rayleigh number at which the time dependent but spatially ordered patterns will become spatially disordered. The key difficulty in obtaining these equations is the fact that the phase diffusion equation appears as a solvability condition at order ε (the inverse aspect ratio) whereas the mean drift equation is the solvability condition at order ε^2. Therefore, we had to use extremely robust inversion methods to solve the singular equations at order ε and the techniques we use should prove to be invaluable in a wide range of similar situations. In addition to providing more details of comparisons between theory and experiment, we plan to make these techniques and the program to implement them available in paper II.

Reference: Submitted to Journal of Fluid Mechanics.

The authors of both papers are grateful for support under Air Force grant number AFOSR4962086C0130 and NSF grant number DMS 8703397. One of us (ACN) is also grateful for a Humboldt Fellowship and the hospitality he received at the University of Bayreuth.

[†] Observatoire de Nice, BP 139, 06003 Nice Cedex, France.

R. Conte and N. Boccara (eds.), Partially Integrable Evolution Equations in Physics, 163.

NONLINEAR EVOLUTION EQUATIONS, QUASI-SOLITONS AND THEIR EXPERIMENTAL MANIFESTATION

M. REMOISSENET
Laboratoire O. R. C
Université de Bourgogne
21000 Dijon France.

ABSTRACT. We review the typical experimental facts which characterize quasi-solitons in one-dimensional real systems, in connection with their modeling by nonlinear partial differential equations.We consider these nonlinear waves or excitations in two different domains of the real world : the macroworld and the microworld. In the macroworld we examine typical one-dimensional devices : the electrical networks, the Josephson transmission lines and the optical fibers, where the localized waves or pulses can be simply and coherently created , easily observed and manipulated on a macroscopic scale. In the microworld , we consider the magnetic chains and polymers, where the indirect experimental signatures of the localized nonlinear excitations are more subtle than for the nonlinear macrowaves. We finally discuss some open problems in the complex and important field of biological chains such as DNA.

Introduction.

Nature provides many examples of coherent nonlinear structures and waves which have been observed in various fields ranging from fluids, plasmas, solid state physics, chemistry and biology. Among these beautiful nonlinear phenomena, wave packets, localized waves and solitary waves which propagate without dispersing, are a continuous source of fascination for the physicist; they have been observed experimentally in various real systems. Completly integrable equations, with their remarkable soliton solutions, which arise naturally as models of these systems have proved to be very efficient to predict their typical dynamical properties. In fact, in the real world none of these integrable field equations is ever realized exactly, instead, they are obtained by a small parameter expansion : they correspond to the lowest nonlinear approximation. Today, the experiments are becoming more and more sensitive and accurate, allowing the observation of effects which could not be detected before. Meanwhile, the models can be improved : the higher order terms first ignored in the expansion can be taken into account, leading generally to near integrable or partially integrable equations with quasi-soliton solutions. By a

R. Conte and N. Boccara (eds.), Partially Integrable Evolution Equations in Physics, 165–221.

quasi-soliton, in the following we understand a solitary wave, i. e a localized nonlinear wave or excitation which has properties close to those of a soliton, although to day, the word soliton is often used by physicists where solitary waves are meant.
Although a considerable body of literature now exists that includes theoretical and numerical studies of localized nonlinear waves , the number of papers related to the experimental activity is growing rather slowly. Quasi one-dimensional systems are especially interesting because localized waves and excitations have been clearly identified experimentally.

This review is intended to provide an introduction to the typical experimental facts characterizing quasi-solitons in one-dimensional real systems, in connection with their modeling by nonlinear partial differential equations.We consider nonlinear waves or excitations in two different domains of the real world : the macroworld and the microworld.
The first three chapters deal with waves in the macroworld : we examine sucessively the following devices. The electrical networks, which are useful to simulate various physical systems and also have direct applications to modern electronic systems. The Josephson transmission lines which have proven to be successful for testing grounds of non linear wave theory and are quite attractive for information processing and storage. The optical fibers, where the explotation of the typical nonlinear effects has led to the demonstration of various quasi-soliton properties and which are very interesting guiding systems with potential applications.These systems provide the simplest examples of one-dimensional devices, where the localized waves or pulses can be simply and coherently created , easily observed and manipulated on a macroscopic scale.
In the microworld (chapter 4) , we consider the magnetic chains , where the localized nonlinear wave modes have a spatial extension ranging from less than a few microns to a few angströms. These systems, where moving domain walls can be viewed as quasi-solitons, are now currently used to carry out experiments on nonlinear excitations in condensed matter. In these chains the excitations are mainly created by thermal processes, sometimes by some external stimulus; their experimental manifestation is indirect, their observation is more subtle than for the nonlinear macrowaves , it needs the various and sophisticated experiments of condensed matter physics.In chapter 5, we finally discuss some open problems in the fascinating, important and complex field of biological chains such as DNA, where the developement of the models and experiments is still in its infancy.

During the past years a considerable number of papers has been published on the subject considered here and we have therefore restricted ourselves to an overview of the theory and experiments related to some specific quasi-one-dimensional systems. We apologize to those whose contributions to this field are not specifically cited.

1. Electrical transmission lines.

Nonlinear electrical transmission lines (Scott, 1970. Lonngreen and Scott, 1978. Peterson, 1984)) are perhaps the most simple one-dimensional experimental devices to observe and study the propagation and properties of nonlinear waves . They have received much attention since a pioneering work on a simulation line (Hirota and Suzuki, 1970, 1973) of the integrable Toda lattice (Toda, 1967,1970) . Some fundamental properties of nonlinear systems have been confirmed experimentally : the shock wave properties (Freeman et al, 1977. Watanabe et al, 1978. Yoshinaga et al, 1980), the generation of solitary waves and the soliton interaction (Watanabe, 1982. Noguchi, 1974. Jäger, 1978), the recurence phenomenon (Fukushima et al, 1980. Watanabe et al, 1980), the lattice properties (Nagashima et al, 1978. Kofane et al, 1988).

The transmission lines are efficient to simulate various physical systems like potential problems in quantum mechanics (Yazaki et al, 1985), waves in plasmas (Landt, 1985. Nejoh, 1985) and optical bistabilty in relation with solitons (Paulus et al, 1984).

The electrical transmission lines have also direct applications to engineering and modern electronic systems for : harmonic generation (Benson et al, 1965), pulse shaping (Yagi et al, 1977), pulse compression (Tan et al, 1988), data transmission and coding (Susuki et al, 1973a, 1973b. Chu et al, 1978).

Here, we consider elementary nondissipative and dissipative transmission lines which can be modeled by nonlinear partial differential equations (NLPDE) and present some experimental results on (pulse and envelope) solitary waves or quasi-solitons propagation. Then, we review shortly results on more complicated transmission lines.

1.1. SOLITARY WAVES ON AN ELEMENTARY TRANSMISSION LINE.

1.1.1. *Non dissipative transmission line.*

The existence of solitary waves is associated with the presence of specific type of dispersion (periodic structures, discretness) besides the nonlinearity of the electrical elements (capacitance with dielectrics or semiconductors, inductances with ferrites).Here, we consider a simple transmission line (Figure 1.1) where the nonlinearity is caused by the capacitance of a semiconductor junction (reversed bias diode), where the dispersion is due to the discrete structure of the components : the linear inductance L and the nonlinear capacitance C(V) , in each unit section.

In a first approximation we neglect the small dissipation and inhomogeineities introduced by these elements. V_n (t) is the voltage across the nth capacitance and I_n (t) is the current through the nth inductance. From the Kirchoff's law we get

$$L\frac{dI_n}{dt} = V_{n-1} - V_n \quad , \qquad \frac{dQ_n}{dt} = I_n - I_{n+1} \tag{1.1}$$

where $Q_n(V_n)$ denotes the charge stored in the nth capacitor. If $Q(V_n)$ is given by

$$Q(V_n) = Q(V_o)Ln(1+ V_n/F(V_o)\,) \qquad (1.2)$$

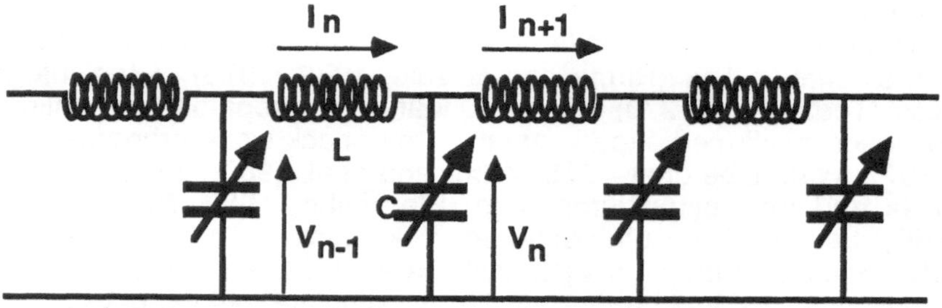

Figure 1.1. An electrical transmission line with a linear inductance and a nonlinear capacitance in each unit section.

The circuit equations (1.1) and (1.2) give

$$LQ(V_o)\frac{d^2}{dt^2}Ln\,(1+\frac{V_n}{F(V_o)}) = V_{n+1} + V_{n-1} - 2V_n \qquad (1.3)$$

Here $Q(V_o)$ and $F(V_o)$ are constant parameters depending of the DC bias voltage V_o , they are determined from the characteristic of the capacitor. In this special case equation (1.3) is just the Toda lattice equation. Therefore the experiment of the Toda lattice using a LC transmission line is realizable if the nonlinear charge satisfies exactly (Hirota and Susuki, 1970. Watanabe, 1982) equation (1.3). Usually it is satisfied only in a limited voltage range or it is not satisfied at all. Consequently instead of eq (1.2) one assumes

$$Q_n(V_n) = C_o\,(\,V_n - a\,V_n^2 + bV_n^3 + \ldots\ldots.\,) \qquad (1.4)$$

In the following, we first consider the case where $bV_n^3 << aV_n^2$. From eqs(1.1) and (1.4) we obtain a set of N nonlinear differential equations

$$\frac{d^2}{dt^2}(\,V_n - aV_n^2) = \frac{1}{LC_o}(V_{n+1} + V_{n-1} - 2V_n)\ , \quad (n = 1,2,....N), \qquad (1.5)$$

These equations cannot be solved analytically and we use the continuum approximation or long wavelength approximation :$V_n(\,t\,) \rightarrow V(\,n,t\,)$ and

$$V_{n\pm1} = V \pm \frac{\partial V}{\partial n} + \frac{1}{2}\frac{\partial^2 V}{\partial n^2} \pm \frac{1}{3!}\frac{\partial^3 V}{\partial n^3} + \frac{1}{4!}\frac{\partial^4 V}{\partial n^4} + \ldots\ldots\ldots \qquad (1.6)$$

Combining equations (1.5) and (1.6) yields a Boussinesq type equation

$$\frac{\partial^2 V}{\partial t^2} - a\frac{\partial^2 V^2}{\partial t^2} = \frac{1}{LC_o}\left(\frac{\partial^2 V}{\partial n^2} + \frac{1}{12}\frac{\partial^4 V}{\partial n^4}\right) \tag{1.7}$$

This equation is not integrable but looking for a solution of the form V(n - vt), one can get a solitary wave solution

$$V(n,t) = \frac{3}{2a}\frac{v^2 - v_o^2}{v^2}\,\mathrm{sech}^2\left[\frac{3(v^2-v_o^2)}{v_o^2}\right]^{1/2}(n - vt) \tag{1.8}$$

If we next introduce the expansion

$$V = \varepsilon V_1 + \varepsilon^2 V_2 + \ldots\ldots \tag{1.9}$$

and streched variables (coordinate system moving with the velocity v_o) :

$$X = \varepsilon^{1/2}(n - v_o t) \tag{1.10}$$

$$T = \varepsilon^{3/2} v_o t/24 \tag{1.11}$$

where $\varepsilon \ll 1$ and $v_o = (LC_o)^{-1/2}$.Subsituting eqs(1.9) to (1.11) in eq (1.7) , keeping terms of order ε^3 (Fukushima et al, 1980) and puttting $\alpha = 24a$, we obtain a KdV equation

$$V_{1T} + \alpha V_1 V_X + V_{1XXX} = 0 \tag{1.12}$$

In terrms of the original variables (n, t), the one soliton solution of eq (1.12) becomes

$$V = \varepsilon V_1(n,t) = \varepsilon\frac{3}{a}\left(\frac{u}{v_o} - 1\right)\mathrm{sech}^2\left\{\left[6\left(\frac{u}{v_o} - 1\right)\right]^{1/2}(n - ut)\right\} \tag{1.13}$$

where $u = v_o(1 + \varepsilon v/24)$ For small velocities (small amplitudes): $v \rightarrow u = v_o + \delta v_o$, ($\delta \ll 1$), it can be easily shown that the solitary wave solution (1.8) reduces to the soliton solution (1.13).This integrable KDV equation describes the weakly nonlinear and weakly dispersive waves . In the continuous limit it represents the zeroth order approximation of the physical system. To higher order one, a KDV type equation with additional derivative terms can be obtained (Yoshinaga et al, 1984).

For higher amplitudes pulses one must take into account the higher order nonlinear and dispersion terms in the expansion of the discrete equations (1.1) and one has additional terms in equation (1.7) . In this case the propagation of a solitary wave in a LC line was experimentally studied (Watanabe,1982). It was shown that the waveform approaches that of the Toda lattice soliton only when the amplitude is small, but gradually deviates from it when the amplitude is increased.

If in each unit cell of the previous elementary circuit (Figure 1.1) one has a shunt capacitor C_1 in parallel with the inductance L, the presence of this capacitance C_1 leads to an *additional dispersion* : *time dispersion.* If C_1 is sufficiently large and if one neglects higher order terms, in the continuous limit eq (1.7) (Fukushima et al, 1980. Jäger, 1982. Kolosick et al, 1974) becomes an improved Boussinesq equation (IBq) (Mankankov, 1978). The nonintegrable IBq was used to model a transmission line (Kolosick et al,1974) and the following properties of its solitary waves solutions were observed experimentally : an asymmetry in propagation of positive or negative signals ,a dependence of the number of solitary waves on the width of an exciting pulse, the recurrence phenomena, the non destructive collision phenomena. These results show that such solitary waves (quasi-solitons) are close to the ideal soliton properties.

1.1.2. *Experiments on pulse propagation.*

The transmission lines described in the litterature are constructed with a large number of sections : from 50 to 1000. The components are carefully selected and adjusted for minimal ohmic losses and reflection effects caused by inhomogeneities. the nonlinear capacitors, which consist of reversed biased diodes, are biased by the dc voltage V_0. At this bias one determines the quantities C_0 and a which characterise the nonlinear capacitor and the charge voltage relations [see eq(1.4)].

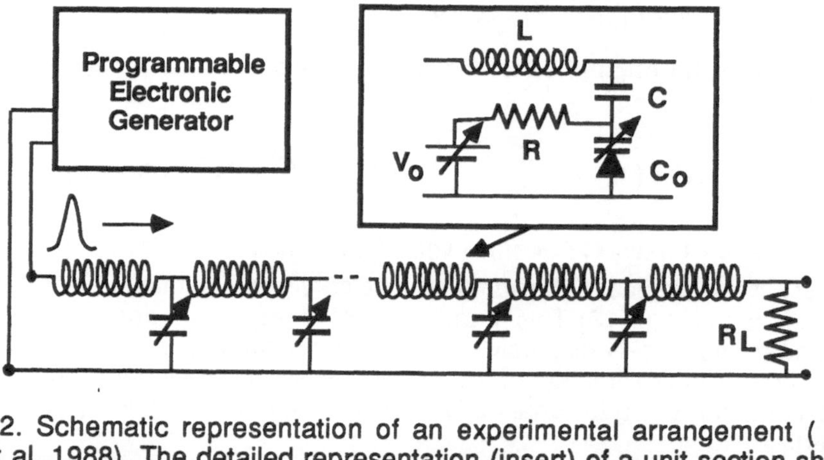

Figure 1.2. Schematic representation of an experimental arrangement (see Kofane et al, 1988). The detailed representation (insert) of a unit section shows that each diode with nonlinear capacitance C(V) is biased through a large resistance R. The linear capacitance C >> C(V) is used to block the DC bias current corresponding to V_0.

In a typical experiment (Figure 1.2), an initial pulse, with a given profile, is launched at one end of the transmission line which is terminated by a resistor adjusted to minimize the reflection effect at the output terminal.
The voltage waveforms are examined at various points of the line using an oscilloscope (analog or digital). In some experiments the profile of the initial wave at the input terminal of the line must very close to the shape predicted theoretically (sech-shaped pulse for example). For this purpose it is convenient (Kofane et al, 1988) to use a generator where the form, amplitude and width of the input signal can be carefully programmed and adjusted close to the calculated solitary wave solution; the parasitic oscillations, which constitute the tail, resulting from the pulse shaping are drastically reduced. The maximum duration of the propagation of a pulse along the line is limited by the number of sections. To study long term effects it is interesting to increase artificially its length : the pulse at the output terminal is amplified and launched once more at the input terminal, and so on (Kuusela et al, 1987).

1.1.3. *Dissipative transmission lines.*

Weak dissipation .

Real transmission lines are dissipative and let us first discuss the effects of weak losses on the solitary wave propagation. In a unit section of the transmission line (Figure 1.1) one adds a resistance R in serie with L and a conductance G in parallell with C(V), these dissipative elements are for exemple related to the finite losses of the inductor and the capacitor respectively. In the continuous limit one can model the dissipative transmission line by

$$V_{tt} - aV^2_{tt} + (R/L)(V_t - aV^2_t) + (G/C_o)V_t + (RG/LC_o)V = (1/LC_o)[V_{nn} + (1/12)V_{4n}] \tag{1.14}$$

As previously we use the reductive pertubation method [see eqs(1.9) to (1.11)] and the transformation :$R/L \to \varepsilon^{3/2}R/L$, $G/C_o \to \varepsilon^{3/2}G/C_o$. To $O(\varepsilon^3)$ we find a dissipative KdV equation

$$V_{1T} + \alpha V_1 V_X + V_{1XXX} = (12/\tau v_o)V_1 \tag{1.15}$$

where $1/\tau = R/L + G/C_o$. For $\tau \to \infty$ (R and G $\to$ 0), eq (1.15) reduces obviously to the standard KdV equation with the one soliton solution given (in terms of the new variables X and T) by eq (1.13). Equation (1.15) is non integrable, but under usual experimental conditions R and G are very small, which leads to $(12/\tau v_o) \ll 1$, so that a perturbational treatement is applicable. Karpman and Maslov (1978) developed a perturbational approach to the KdV equation on the basis of the inverse scattering method. According to them a soliton-like pulse

solution is obtained with a non soliton part which corresponds to the soliton tail. Nagashima and Amagishi (1979) have compared their experimental results with the perturbation theories especially for the damping time and the distorted shape of the soliton followed by a dip or bump on its tail. The agreement is not quite satisfactorily indicating the limit of the perturbation theory.
The perturbation theory for the inverse scattering transform was also applied (Kako, 1979) to study a dissipative transmission line which is equivalent to the Toda lattice in the limit of vanishing dissipation. The time evolution of a one soliton solution was calculated, it was found that local dissipation leads to a change in the amplitude and velocity of the soliton, the creation of small oscillations and the formation of a tail behind the initial soliton . Note that a recent numerical experiment on transmission line, simulating a strongly dissipative Toda lattice, shows that a head-on collision between quasi-solitons is purely elastic (T. Kuusela et al, 1989).

Large dissipation.

Let us now consider a transmission line (Ostrowski et al, 1972. Tagaki, 1983) where in each unit cell (Figure 1.1), one has a resitance R is in serie with the nonlinear capacitor C (V_n). In this case, assuming that Q_n (V_n) is given by eq (1.6) and using the continuum approximation [see eq(1.6)], we have the following partial differential equation

$$V_{tt} - aV^2_{tt} - (R/L)[V_{nn} + (1/12) V_{4n}]_t - (R/L)[V^2_{nn} + (1/12) V^2_{4n}]_t$$
$$= (1/ LC_o)[V_{nn} + (1/ 12) V_{4n}] \qquad (1.16)$$

Using the transformation $X = \varepsilon^{1/2} (n - v_o t)$, $T = \varepsilon^{3/2} v_o t$ and neglecting the higher order derivatives with respect to X, we have

$$V_T + VV_X - (RC_o/2)v_o V_{XX} = 0 \qquad (1.17)$$

This nonlinear diffusion equation is the so called Burgers equation (see Dodd et al, 1984) which appears in many domains of physics.

1.1.4. *Modulated waves.*

Untill now, we have considered pulses propagation, but modulated waves can also exist. Let us consider an LC electrical transmission line (Figure 1.1) in the continuous limit, where instead of (1.4 the charge voltage relation can be approximated by : $Q(V) = C_o [V - |V|^{1/2} V]$,(Fukushima et al, 1980. Fukushima, 1983). Inserting Q (V) in eq (1.5), in the continuum approximation, neglecting terms V_{4n} and V_{4ntt} yields

$$(V - a'|V|^{1/2}V)_{tt} = (1/ LC_o)[V_{nn} + C_1 V_{nn\,tt}] \qquad (1.18)$$

The voltage waveform is assumed to be sinusoidal with frequency ω and wave number k : $V = V(n, t)\, e^{i(\omega t - kn)} + cc$, where cc denotes the complex conjugate and where the amplitude $V(n, t)$ is assumed to vary slowly compared to the sinusoidal phase term. Inserting the expression of V in (1.18), one gets a nonlinear dispersion relation which can now be used [see Karpman and Krushkal (1969) and chapter 3 on optical fibers] to construct the equation for the wave envelope propagation using the Taylor expansion of ω around the carrier frequency ω_o the wavenumber k_o and the amplitude $|V_o|^{1/2}$

$$\omega-\omega_o = (\frac{\partial\omega}{\partial k})(k-k_o) + \frac{1}{2}(\frac{\partial^2\omega}{\partial k^2})(k-k_o)^2 + (\frac{\partial\omega}{\partial |V|^{1/2}})(|V|^{1/2}-|V_o|^{1/2}) + ... \quad (1.19)$$

We now replace $(k - k_o)$ by a spatial operator and $(\omega - \omega_o)$ by a temporal operator : $(k - k_o) \rightarrow -i\,\partial/\partial x$ $\quad (\omega - \omega_o) \rightarrow i\partial/\partial t$, we substitute these operators in (1.19) and let the resulting expression operate on V, we get

$$-i\,[\frac{\partial V}{\partial t} + (\frac{\partial\omega}{\partial k})\frac{\partial V}{\partial n}] - \frac{1}{2}(\frac{\partial^2\omega}{\partial k^2})\frac{\partial^2 V}{\partial n^2} + (\frac{\partial\omega}{\partial |V|^{1/2}})(\,|V|^{1/2}-|V_o|^{1/2})V = 0, \quad (1.20)$$

Introducing the transformation (see chapter 3 : optical fibers) : $z = n - v_g t$, $T = t$ with $v_g = \partial\omega/\partial k$, from equation (1.20) with $V_o = 0$, one obtains an equation of the nonlinear Schrödinger (NLS) type

$$i\,\frac{\partial V}{\partial T} - \frac{1}{2}\,(\frac{\partial v_g}{\partial k})\,\frac{\partial^2 V}{\partial z^2} + (\frac{\partial\omega}{\partial |V|^{1/2}})\,|V|^{1/2}V = 0 \quad (1.21)$$

Here the coefficient ($\partial v_g/\partial k$) is the Group Velocity Dispersion (GVD) and the coefficient ($\partial\omega/\partial|V|^{1/2}$) can be calculated from the nonlinear dispersion relation . Equation (1.21) can be more rigourously obtained (Fukushima et al, 1980) by using the reductive perturbation method (Taniuti and Yajima, 1969) [see also chapter 3]. Fukushima et al (1980) have found envelope pulse solutions for eq(1.21) and compared them with the waveforms they measured experimentally ,a good agreement was obtained .

1.2. OTHER TRANSMISSION LINES.

Besides the elementary transmission line considered above, other transmission lines can be realized by a proper choice of the elements . For example, instead of using a nonlinear capacitance one can use (Yagi and Noguchi, 1977) a nonlinear inductor (such as a gyromagnetic ferrite core). A typical unit section of the transmission line is represented on Figure 1.3, where : C is a linear capacitor and $\Phi(I)$ is the flux in the

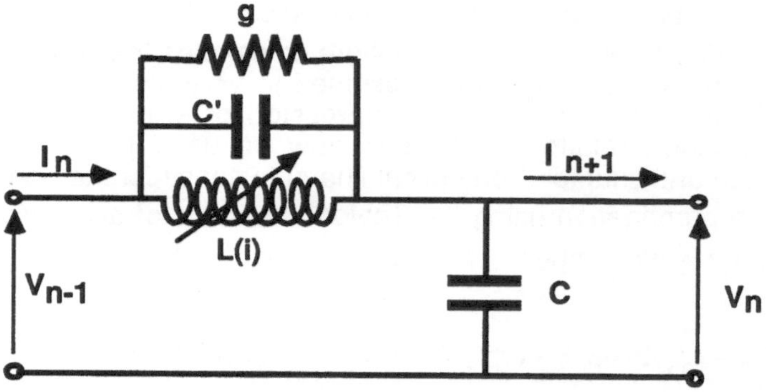

Figure 1.3. Typical section of a transmission line with a nonlinear inductance L(i) with its equivalent conductance g and capacitance C'.

nonlinear inductance L(I) is approximated by :$\Phi(I) = L_o (1 - aI^2) I$, with $L_o = C^{te}$, C' is a stray capacitance and g^{-1} a parallel resistance. From the discrete circuit equations, in the continuum approximation, neglecting higher order terms, one obtains the following equation describing the change of current i (n, t) in the nonlinear inductance

$$C\frac{\partial^2\Phi(i)}{\partial t^2} = [\frac{\partial^2 i}{\partial n^2} + \frac{1}{12}\frac{\partial^4 i}{\partial n^4}] + C'\frac{\partial^4\Phi(i)}{\partial n^2\partial t^2} + g\frac{\partial^3\Phi(i)}{\partial n^2\partial t} \tag{1.22}$$

For C'=0 and g =0, eq (1.22) reduces to a Boussinesq type equation, similar to eq(1.7) which has solitary wave solutions. Using the following transformation

$$T = \varepsilon^{1/2}(t - n/v_o)\ ,\ z = \varepsilon^{3/2}n/2v_o\ ,\ G \rightarrow \varepsilon^{1/2}g\ ,\ i \rightarrow \varepsilon^{1/2}i \tag{1.23}$$

where $\varepsilon \ll 1$, and substituting eqs (1.23) in eq (1.22) leads to a modidied Korteweg de Vries - Burgers equation

$$i_z - ai^3{}_t - (v_o^2 + C')\, i_{ttt} - g i_{tt} = 0 \tag{1.24}$$

In the dissipationless limit g = 0, eq(1.24) obviously reduces to the modified KdV equation which has soliton solutions. The pulse shaping and pulse transmission along this line was studied (Yagi and Noguchi, 1977) experimentally and analyzed in terms of the solutions of the KdV equation . Indeed the physics of the line is better described by eq (1.24) or (1.22).

Envelope modes.

Let us assume that in the above model equation (1.22), the dispersion due to the capacitance C' (time dispersion) is larger than the dispersion due to the capacitance C (spatial dispersion). We have C'>>C, and looking for modulated waves, like for eq (1.24), we can derive (Yagi and Noguchi, 1976) a nonlinear dispersion relation : using the same procedure than in § 1.1.4 we can obtain a NLS equation for the slowly varying envelope I(n,t) :

$$iI_t + P\, I_{nn} + Q|I|^2 I = 0 \qquad (1.25)$$

where $P = \frac{1}{2}\left(\frac{\partial v_g}{\partial k}\right)$ and $Q = \left(\frac{\partial \omega}{\partial |I|^2}\right)$. Equation (1.25) has been used to study the modulational instability on the nonlinear inductive line. This modulational instability was first studied by Benjamin and Feir (1967) in the context of deep water waves. Namely, it occurs when PQ > 0, i.e a plane wave is unstable and one has envelope soliton solutions ; when PQ < 0, a plane wave is stable and one has dark or hole soliton solutions (see also § 3.2.2). Other kinds of transmission lines were also constructed (Sakai et al, 1976. Kawata et al, 1977) to study modulational instability and dark soliton formation (Muroya et al, 1982).

Nonlinear wave interactions.

Nonlinear transmission lines can be constructed where two or more different wave modes can interact and exchange energy. Using the discrete transmission line represented on Figure 1.4, the interaction between a pair of waves with equal group velocities, was investigated theoretically (Yoshinaga et al, 1981) .

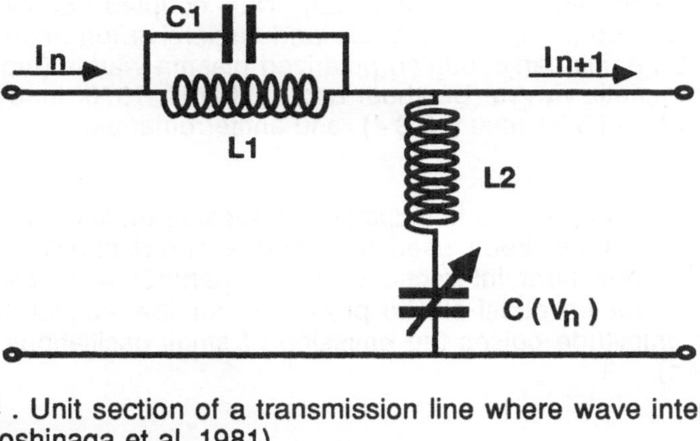

Figure 1.4 . Unit section of a transmission line where wave interactions can be studied (Yoshinaga et al, 1981).

If the nonlinear capacitance is assumed to be : $C(V_n) = C_o (1 + 2\mu V_n + 3\nu V_n^2)$, for the transmission line described on Figure 1. 4, one obtains the following differential equation for V_n

$$\left[\frac{d^4}{dt^4} + (1+\alpha)\frac{d^2}{dt^2} + \alpha\right] (V_{n-1} + V_{n+1} - 2V_n) - \beta\frac{d^2V_n}{dt^2} - \beta\frac{d^2}{dt^2}(\mu V_n^2 + \nu V_n^3)$$

$$+ \left(\frac{d^4}{dt^4} + \alpha\frac{d^2}{dt^2}\right)\left[\mu(V^2_{n-1} + V^2_{n+1} - 2V^2_n) + \nu(V^3_{n-1} + V^3_{n+1} - 2V^3_n) \right] = 0$$

(1.26)

where $\alpha = C_oL_2/ C_1L_1$, $\beta = C_o/C_1$ and the time is measured in units of $(C_oL_2)^{1/2}$. From (1.26) it can be shown that wo kinds of modes can exist : low frequency (LF) modes and High frequency (HF) modes. Using the reductive perturbation expansion (see chapter 4), it was shown that equation (1.26) can be reduced to a simple set of coupled NLS equations

$$i\frac{\partial A_1}{\partial T} + P_1\frac{\partial^2 A_1}{\partial s^2} = Q_1 | A_1|^2 A_1 + R_1 | A_2|^2 A_1 \qquad (1.27a)$$

$$i\frac{\partial A_2}{\partial T} + P_2\frac{\partial^2 A_1}{\partial s^2} = Q_2 | A_2|^2 A_2 + R_2 | A_1|^2 A_2 \qquad (1.27b)$$

where $T = \varepsilon^2 t$, $s = \varepsilon(n - v_g t)$, $v_{gi} = \partial\omega_i/\partial k_i$, $P_i = (1/2)(\partial v_{gi}/\partial k_i)$ with $i = 1, 2$. The nonlinear coefficients Q_i and R_i are functions of ω_i and k_i; A_1 and A_2 correspond to the slow varying envelopes of the LF and HF modes. To our knowledge, such modes have not been studied experimentally. NLS coupled equations similar two eqs (1.35) were obtained for Langmuir and dispersive ion acoustic waves (Spatschek, 1978), nonlinearly coupled polarized plasma waves (Inoue, 1976), coupled electromagnetic waves (Berkhoer and Zakharov, 1976. Inoue, 1977) in a dielectric, for optical fibers (see § 3.3.4) and antiferromagnetic chains (see § 4.4.1).

Coupled NLPDE also appear in bi-inductance transmission lines (Kofane et al, 1988), these lines have been used to simulate one-dimensional diatomic lattices with cubic nonlinear interactions. The experimental studies show a quasi-soliton behavior i.e quasi-elastic properties for low amplitude pulses, while for higher amplitude pulses the emission of small oscillations increases with the amplitude).

2. Josephson transmission lines.

The superconducting Josephson junctions (see the books : Barone and Paterno, 1982. Likharev, 1986 and the reviews : Scott, 1970, Parmentier, 1978. Lomdahl, 1985. Pedersen, 1988) have proven to be one of the most successful testing grounds for nonlinear wave theory and their use for information processing and storage is quite attractive. In the Josephson transmission line (long junction) the physical quantity of interest is a quantum of magnetic flux, or a fluxon, which has an approximate soliton behavior. It is a remarkably robust and stable object , which can be easily manipulated at high speed and stored electronically. Consequently it should be used as a basic bit in information processing systems. We now give an overview of the properties of the long junction or transmission line, which is presently well characterized in the context of the low-temperature superconductors. One expects modifications in the near future with the new high temperature superconductors.

2.1. MODIFIED SINE GORDON PICTURE OF THE LONG JUNCTION.

A *small Josephson junction* (Josephson,1960), consists of two layers of supraconducting metal separated by a thin dielectric barrier layer which is small enough to permit tunelling of supraconducting electron pairs or, equivalently, coupling of the wave functions (see Feynman, 1960) of the two superconductors To the lowest rate of approximation the electrical behavior is described by the two basic Josephson equations

$$j = j_o \sin \phi , \qquad \frac{\partial \phi}{\partial t} = \frac{2\pi}{\Phi_o} V \qquad (2.1a,b)$$

Here j(x,t) is the tunnelling supercurrent, j_o being its maximum value which depends on the material and the geometry of the junction. ϕ is the phase difference between the macroscopic quantum wave functions, V(x,t) is the voltage drop across the insulating barrier. Where Φ_o = h/2e = 2,064 10^{-15} Wb is the flux quantum (where h = 2πh is the Planck constant and e the electronic charge).

To the lowest order of approximation *the long Josephson junction* may be modelled (Swihart,1961.see also Pagano,1987) by the continuous electrical transmission line structure (Figure2.1) where L is the inductance per unit length, C the capacitance per unit length and $j_o \sin \phi$ the Josephson tunneling current per unit length. Dealing with real transmission lines we must take into account : losses, bias and junction irregularities which influence the motion of real fluxons. On the equivalent transmission line model the dissipative effects can be represented (Figure 2.1) by the following elements : a serie resistance R which represents the scattering of quasiparticles in the surface layers of the two superconductor, a parallel conductance G which is referred to as the quasiparticle tunnelling current (Josephson, 1964). With the presence of these losses, one also consider a uniformly distributed bias current j_B per unit length.

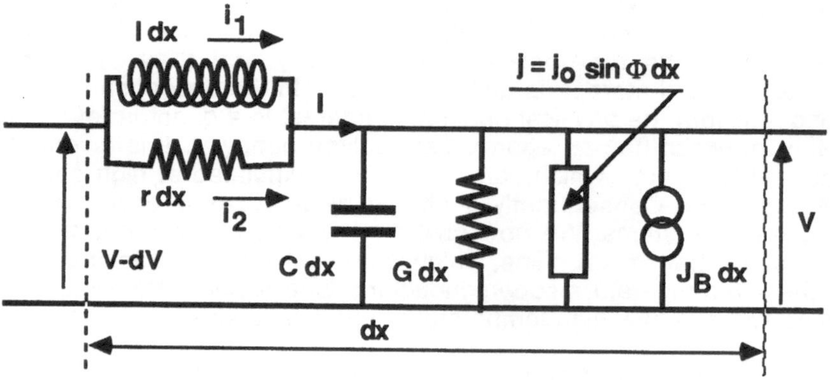

Figure 2.1. Equivalent electrical transmission line of the dissipative long Josephson junction.

Applying Kirschoff's laws to the model of figure (2.1) gives

$$\frac{\partial V}{\partial x} = -L\frac{\partial I_1}{\partial t} \tag{2.2a}$$

$$\frac{\partial I}{\partial x} = -RI_2 \tag{2.2b}$$

$$\frac{\partial}{\partial x}(I_1 + I_2) = -C\frac{\partial V}{\partial t} - GV - j_o \sin\phi + j_B \tag{2.2c}$$

where :$\lambda_J = (\Phi_o/2\pi\, j_o\, L)^{\frac{1}{2}}$ is the Josephson length and $\omega_J = (2\pi\, j_o/\Phi_o\, C)^{\frac{1}{2}}$ is the Josephson plasma frequency..From (2.2) , using (2.1b) and introducing the transformation $X = X/\lambda_J$, $Y = Y/\lambda_J$, $T = t\,\omega_J$, yields a modified SG equation.

$$\phi_{XX} - \phi_{TT} - \alpha\phi_T + \beta\,\phi_{XXT} = \sin\phi - \gamma \tag{2.3}$$

The constants α , β, and γ are defined by : $\alpha = G/\omega_J C$, $\beta = \omega_J/LR$, $\gamma = j_B/j_o$. When the coefficients α, β and γ are neglected, eq (2. 3) reduces to the Sine Gordon equation which has the well known (kink-antikink) soliton solutions propagating with velocity v :

$$\phi_{\pm}(X,T) = 4\,\mathrm{tg}^{-1}\left[\pm\frac{(X-vT)}{(1-v^2)^{1/2}}\right] \tag{2.4}$$

Physically the kink (antikink) correspond to a $\pm 2\pi$ jump in phase difference ϕ accross the insulating barrier separating the two superconductors (Figure 2.2). In other words it is a flux quantum vortex, a current loop which consists of a surface current and a tunneling supercurrent, connecting the two surface layers via the barrier. This current loop supports one quantum Φ_0 of magnetic flux and the soliton (antisoliton) is therefore often called a fluxon (antifluxon). The moving soliton described by eq (2.4) is accompanied by a voltage pulse and a current pulse. The shapes of these pulses can be calculated from eqs (2.2).
When α and β are sufficiently small, solutions of eq (2.3) can be calculated using a perturbation analysis (Mc Laughlin et al, 1978. Scott et al, 1978) which has proven quite sucssesful to describe experimental results in the limit of small bias current (Davidson, 1986).

The effect of the damping terms α and β was also analysed numerically (Lomdahl et al, 1982), when these terms are important they can lead to interesting properties of fluxons for digital applications (Nitta et al, 1984) .

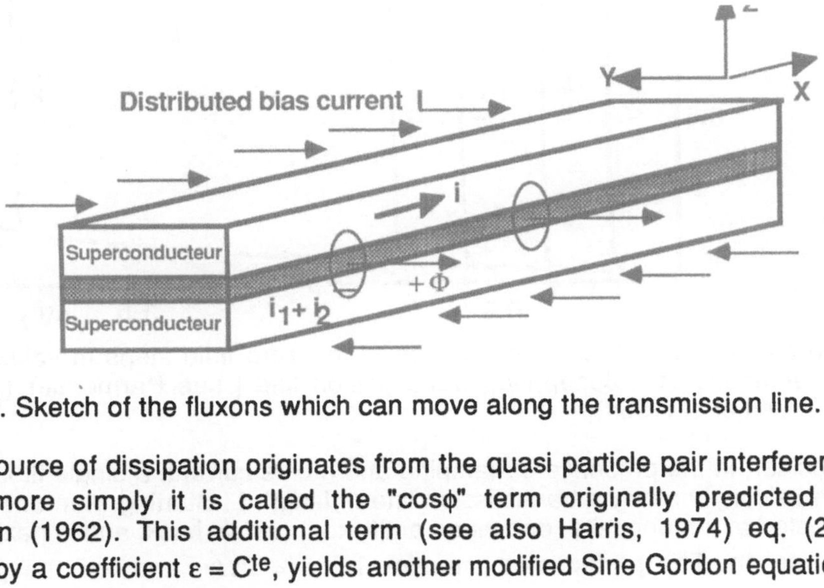

Figure 2.2. Sketch of the fluxons which can move along the transmission line.

Another source of dissipation originates from the quasi particle pair interference current, more simply it is called the "cosϕ" term originally predicted by Josephson (1962). This additional term (see also Harris, 1974) eq. (2.3) weighted by a coefficient $\varepsilon = C^{te}$, yields another modified Sine Gordon equation

$$\phi_{XX} - \phi_{TT} - \alpha(1 + \varepsilon \cos \phi)\, \phi_T + \beta\, \phi_{XXT} = \sin \phi - \gamma \qquad (2.5)$$

2.2. EXPERIMENTAL OBSERVATION OF FLUXONS.

2.2.1. *Indirect method.*

The modified Sine Gordon equation can be used to analyze the d.c current singularities which occur in the voltage-current characteristics of a long one-dimensional Josephson junction and are an experimental manifestation of fluxons motion in the line. To the propagating fluxon corresponds a voltage pulse which can be detected at either and of the transmission line. Current singularities in the current voltage characteristics were observed in the absence of external applied magnetic field, an explanation of their existence was proposed by Fulton and Dynes (1973) on the basis of fluxon motion inside the junction. There singularities known as zero field steps which are essentially a series of current spikes at approximately equidistant intervals of D.C. Voltage were first observed by Chen et al (1971). Their occurence at non zero voltages suggests an explanation in terms of fluxon motion.

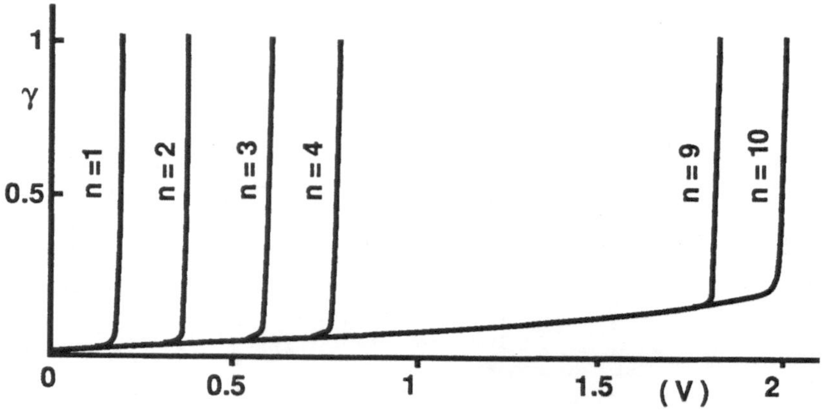

Figure 2.3. Schematic representation of the zero field steps in voltage current characteristic of a Josephson transmission line [see Parmentier (1978) and Pedersen et al (1984)].

Consider, in the presence of damping and a bias current, a single fluxon moving with velocity v along a transmission line of length L. Starting from one end, it is reflected at the opposite end as an antifluxon, comes back, and is next reflected, and so on. The period of the motion is $\tau = 2L/v$, and for one period the corresponding phase change is $\Delta\phi = 4\pi$. For N fluxons moving on the line the phase change will be $\Delta\phi_N = 4N\pi$. The mean voltage over one period τ or dc voltage V_{DC} is calculated (see for example Lomdahl, 1985) from eq (2.1b) and replacing t by its value one gets : $V_{DC} = \Phi_0 Nv_c/L$, the voltage at which the socalled Zero Field Steps occur. Here v_c is the limiting velocity of the fluxon in the presence of damping α and bias. For a given junction geometry this steady states fluxon velocity can be calculated from equation (2.11) where $\beta = 0$ and α and γ are treated as small perturbations, one gets : $v_c = [1 + (\frac{4\alpha}{\pi\gamma})^2]^{1/2}$. If the junction length is very large, the details at the boundaries play a minor role, and the voltage V_{dc} at the first step is given by the above value. For N = 1, one has the first Zero Field Step, for N = 2 one has the second one and so on. (Figure 2.3).

It was shown (Costabile and Parmentier, 1975) that if the loss term in eq. (2.5), (with $\beta = 0$), is quadratic rather than linear, it is possible to find exact analytical solutions for steady propagation. The model equation considered was

$$\phi_{XX} - \phi_{TT} - \Gamma |\phi_T| \phi_T = \sin\phi - \gamma \qquad (2.6)$$

The solutions of this equation were used to (Costabile and Parmentier, 1975) to calculate the average value of V i.e. the Zero Field Steps (Parmentier, 1978) .

Recently using eq (2.5) (with $\beta = 0$) it proved possible to calculate the shape of the zero field steps in transmission lines with different geometries : an inline junction, where the bias current is fed in the direction of propagating fluxons, an overlap junction, where the bias current is fed at right angles to fluxon velocity.The experimental findings agree very well with the results of this modified Sine Gordon model (Levring et al, 1982. Pedersen et al, 1984).

In the junctions geometries mentionned above the boundary effects and fluxon antifluxon interactions are present and are expected to be of importance. Consequently, a new experimental approach was to choose a geometry that has not boundaries at all : an annular or ring shaped junction which consists of an overlap junction folded back into itself (Davidson et al, 1986). This geometry is unique in that it allows for the study of undisturbed fluxon motion as well as fluxon.-antifluxon collisions, since there are no boundary effects. Moreover these are the closest experimental conditions under which the perturbation theory on its original form is expected to apply.

2.2.2. *Direct experimental observation of fluxons.*

In §2.2.1 we have considered the fluxon propagation properties for a Josephson Transmission line, which were mainly investigated through the analysis of Zero-Field Steps. Recently direct observation systems have been developed to study the propagation properties of the fluxons : the first direct measurement was recently performed (Matsuda et al, 1982, 1983. Matsuda, 1983. Nitta, 1984). In the experiments an input pulse is fed into one end of the transmission line and output voltage which appears across the terminal resistance at the output end is detected. A low-noise and fast rise time detector together with minicomputer signal processing were used to improve the signal/noise ratio and observe the fluxon. Observed waveforms are shown in figure 2. 4 this number of pulses varies according to changes in input pulse height.

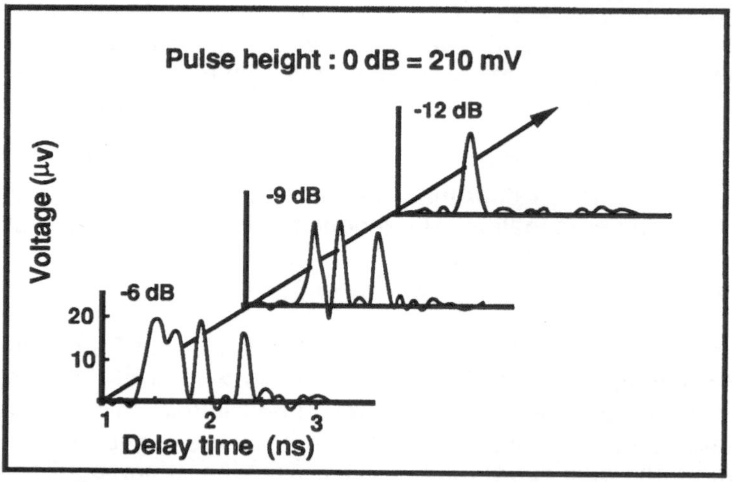

Figure 2.4. Sketch of waveforms profiles (see Matsuda et al, 1982) which change as a function of the input pulse height.

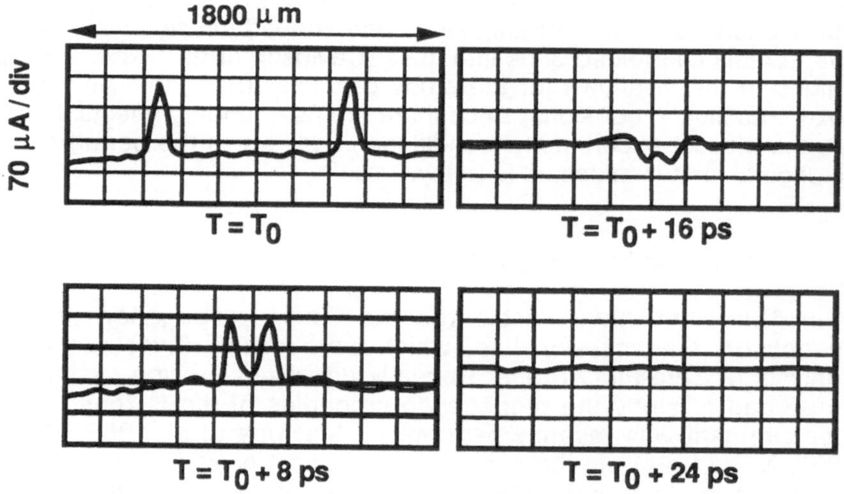

Figure 2.5. Schematic representation of a fluxon-antifluxon collision which results in a breather decay mode (see Fujimaki, 1987).

The experimental behavior of a fluxon-antifluxon collision was also investigated (Matsuda, 1986) using an improved direct measurement system based on a signal processing technique . The experiment showed an increase in propagation delay time following a collision. This result was qualitatively explained as the effect of dissipation [see eq (2. 3)] using numerical simulation and perturbation theory, where the origin of the increase in the delay time is explained in terms of a power balance equation (Scott et al, 1984) .

Very recently a direct experimental observation of a fluxon-antifluxon collision was reported on a discrete Josephson transmission line, i.e. a line composed of 31 discrete junctions (Fujimaki et al, 1987). By precise control of the collision of a fluxon and an antifluxon in the line the annhilation process, into a breather decay mode, was observed in time but also in space (Figure 2.5). This destructive collision can be well described in terms of the modified (dissipative) Sine Gordon equation.

3. Optical fibers.

3.1. PRELIMINARIES.

With lasers producing high intensity short duration optical pulses it is now possible to probe the interesting and potentially useful, nonlinear effects (Shen, 1985) in optical systems (Newell 1989, these proceedings and waveguides (Stegeman et al,1988. Firth et al, 1988). Among the guiding structures the optical fiber is an interesting (Gloge, 1979) and important device (Mollenauer and Stolen, 1982. Doran and Blow , 1983, 1987). Namely, in an optical transmission system using linear pulses, the bit rate of transmission is limited by the dispersive character of the material which causes the pulse to spread out , eventually overlap to such an extend that all the information is lost. To overcome this limitation Hasegawa and Tappert (1973a,1973b) proposed to compensate the dispersive effect by the nonlinear change (Kerr effect) of the refractive index ,or dielectric response, of the fiber material. When the frequency shift due to the Kerr effect is balanced with that due to the dispersion, the initial optical pulse may tend to form a nonlinear stable pulse called "optical soliton" which in fact is a quasi-soliton. This prediction and subsequent observation of optical solitons propagation (Mollenauer et al, 1980) in a single fiber has stimulated theoretical and experimental studies on nonlinear guided waves. Namely, the exploitation of the typical nonlinear effects in optical devices has led to the demonstration of various (quasi) soliton properties such as pulse compression (Mollenauer et al, 1983), the soliton laser (Mollenauer and Stolen, 1984), Raman reshaping (Hasegawa, 1983. Mollenauer et al, 1985). The consequences for future long distance high bit rate communications systems are the subject of much study (Mollenauer and Smith, 1988). In recent years, a considerable body of literature have been published on the subject and in the following we just give an overview on modeling of optical fibers by NLPDE and typical experiments.

3.2. OPTICAL FIBERS PROPERTIES AND APPROXIMATE MODELING.

3.2.1. *Typical properties.*

Generally an optical fiber is a cladded cylindrical waveguide which consists of a highly transparent (high index) dielectric core surrounded by a second layer of a second dielectric with a lower optical index. Optical fibers guide light by total internal reflection : this is achieved by a refractive index n(x,y) [or dielectric function ε(x,y)] variation decreasing in the radial direction from the center to the periphery (Figure 3.1).

In multimode fibers certain rays or modes can propagate at different velocities, these are in fact solutions of the wave equation for the cylindrical dielectric wave-guide. One can design the fiber in such a way that it transmits only one mode and attenuates all other modes by absorption or leakage. This is done by modifying the index profile at a sufficiently small index difference so that only one state is supported, one has a monomode fiber. While multimode fiber is cheaper and easier to produce and can be used with incoherent sources, single mode fiber offers much greater bit capacity and is of primary interest for optical communication systems.

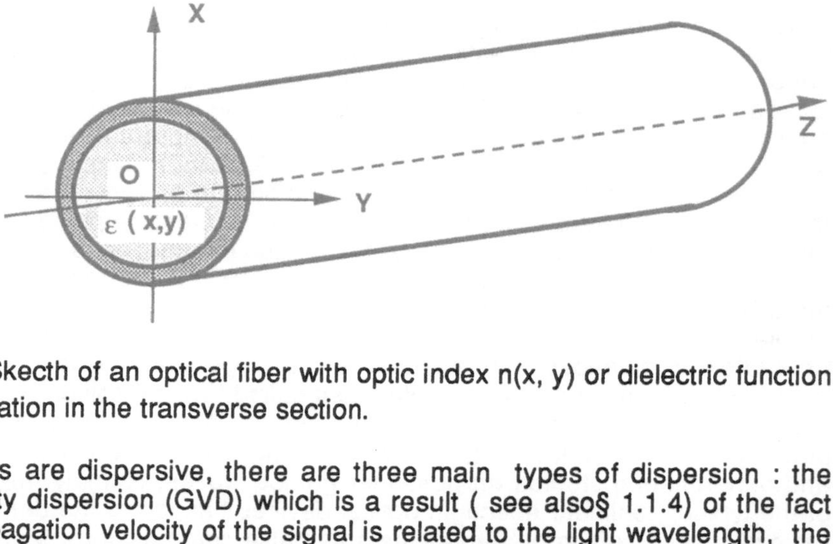

Figure 3.1. Skecth of an optical fiber with optic index n(x, y) or dielectric function ε (x,y) in variation in the transverse section.

Optical fibers are dispersive, there are three main types of dispersion : the group velocity dispersion (GVD) which is a result (see also§ 1.1.4) of the fact that the propagation velocity of the signal is related to the light wavelength, the modal dispersion which is inherent to multimode fibers, the waveguide or geometrical effects. Optical fibers exhibit losses up to 1 dB/km, although their quality has improved in recent years; this will be discussed in the following.

3.2.2. *NLS picture.*

Let us consider a monomode fiber where or a sake of simplicity we ignore the transverse (x, y) dependence of the electric field vector **E.** The one-dimensional wave equation for a linearly polarized optical wave obtained (Jain and Tzoar, 1978. 1987) from the Maxwell equations (Karpman, 1975) is given by

$$\frac{\partial^2 \mathbf{E}}{\partial z^2} - \frac{1}{c^2}\frac{\partial^2 \mathbf{D}_L}{\partial t^2} - \frac{2n_o n_2}{c^2}\frac{\partial^2}{\partial t^2}\mathbf{E}^2\mathbf{E} = 0 \tag{3.1}$$

Here D_L is the linear part of the electric displacement, n_o and n_2 are respectively the linear part and nonlinear part (intensity dependent) of the refactive index n, the Fourier transform of which is given by : $n(\omega) = n_o(\omega) + n_2 |E|^2$, $c = (\varepsilon_o \mu_o)^{-1/2}$ is the speed of light. We consider a monochromatic field amplitude

$$E(z,t) = \frac{1}{2}\left[A(z,t)\, e^{i(kz-\omega t)} + A^*(z,t)\, e^{-i(kz-\omega t)}\right] \tag{3.2}$$

with the z axis as the direction of propagation, A the slowly varying transverse component of the electric field and A^* its complex conjugate, k the wavenumber and ω the frequency.Inserting (3.2) in (3.1) one gets a nonlinear see dispersion relation : $c^2k^2/\omega^2 = n^2(\omega)$ which can be formally written $k = k(\omega, |A|^2)$. We now follow the procedure used for the electrical transmission lines i.e the dispersion relation is used (see § 1.1.4, on electrical transmission lines) to construct the (Karpman et al, 1973. Hasegawa et al, 1980) equation fo the wave envelope propagation in the z direction using the Taylor expansion of k around the carrier frequency ω_o, the wave number k_o and the amplitude $|A_o|^2 = 0$ .Describing the dynamic evolution of A by using a coordinate system which moves at the group velocity $v_g = \partial\omega/\partial k$, where $s = z$, $t' = t - z/v_g$, gives a NLS equation

$$i\frac{\partial A}{\partial s} + \frac{1}{2}k''\frac{\partial^2 A}{\partial t'} + Q|A|^2 A = 0 \tag{3.3}$$

where $Q = (\partial k/\partial |A|^2)$ and $k'' = \partial^2 k/\partial\omega^2$. By the transformation $q = Q(A)^{1/2}$, $t' = (k'')^{1/2}\tau$, it is convenient to normalize the NLS eq(3.14a) into

$$i\frac{\partial q}{\partial s} + \frac{1}{2}\frac{\partial^2 q}{\partial \tau^2} + |q|^2 q = 0 \tag{3.4}$$

Equation (3.4) support N soliton solutions (Zakharov and Shabat, 1972. Yajima and Satsuma, 1974). The relative signs of the dispersive term and nonlinear term are important (see chapter 1), for optical fibers Q is always positive, k" can be negative or positive (GVD > 0 or GVD < 0) . When their product is positive eq (3. 3) admits an envelope (bright) soliton solutions which for the dimensionless form (3.4) is given by

$$q = A \operatorname{sech}[A(\tau - vs)] \exp[-iv\tau + i(v^2 - A^2)s/2] \tag{3.5}$$

Moreover in the case a plane wave in unstable for modulation. Hasegawa and Tappert (1973) first modelled the propagation of a guided mode in a perfect nonlinear monomode fiber by the NLS equation. They find that the optical pulse is an envelope soliton solution of eq (3.4) and they predicted the stationnary transmission of the pulse at the anomalous dispersion (negative group dispersion : $\partial^2 k/\partial\omega^2 < 0$) regime.Then this prediction was successfully verified by the experiments of Mollenauer et al (1980). Using a monomode fiber they demonstrated a dispersionless transmission of an optical pulse (7ps width) with a peak power of about $P = 1W$ at a wavelength $\lambda = 1.45\mu m$ for a distance $d \cong 700$ m. This modulational instability is an example of the Benjamin Feir instability (§1.1.4) first discussed in the context of water waves . It was predicted (A. Hasegawa and Brinkman,1980. Hasegawa, 1984) and recently observed in optical fibers (Tai and al, 1986). For $k'' Q < 0$. ($k'' < 0$) eq (3.4) admits soliton solutions called dark solitons (Zhakarov and Shabat, 1973) because they are characterized by the absence of light (Tomlison et al, 1989) in a bright background. They were recently observed in optical fibers (Krökel et al, 1988).Equation (3.4) with the initial condition

$$q(s = 0,\tau') = A \operatorname{sech}(\tau') \tag{3.6}$$

was studied analytically (Satsuma and Yajima , 1974). The soliton is obtained whenever A is an integer N. For N = 1 [one soliton state equivalent to eq 3.6] the pulse remains forever like that at the imput, but for higher N, the solutions have period $s = \pi/2$, i.e. they pulsate with this period. For example, explicit expressions for $q(s,\tau')$ have been obtained for N = 1,2 :

$$q_1(s,\tau') = e^{-is/2}\operatorname{sech}(\tau') \tag{3.7a}$$

$$q_2(s,\tau') = 4e^{-is/2}\frac{[\cosh(3\tau')+3e^{-i4s}\cosh(\tau')]}{\cosh(4\tau)+4\cosh(2\tau')+3\cos(4s)} \tag{3.7b}$$

as illustrated on Figure 3.2 (upper part).

This pulse shape narrowing and splitting , with prediction based on the NLS equation was observed experimentally by Mollenauer et al (1980). The seven picosecond duration pulses from a mode locked color center laser were observed by autocorrelation on a monomode silica glass fiber at negative group velocity dispersion (Stolen et al, 1983). The pulse narrowing or compression represents an efficient and simple way to produce useful subpicosecond pulses (Mollenauer et al, 1983. Salin et al, 1986).

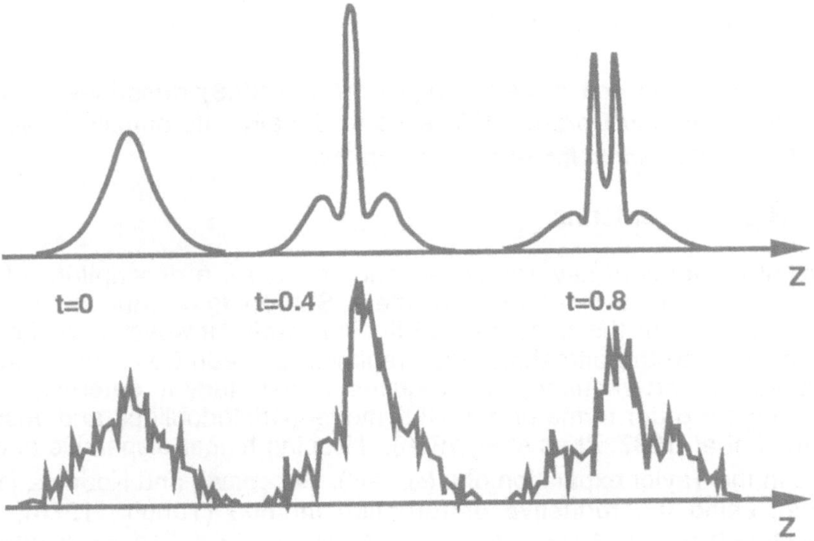

Figure 3.2. Predicted theoretical pulse evolution : narowing and splitting, and sketch of the real shapes of the pulse as observed by autocorrelation (see Mollenauer et al, 1980).

3. 3. MODIFIED NLS PICTURE.

3.3.1. *Effect of losses*

Although the quality of optical fibers has dramatically improved over recent years, the fibers do exhibit some attenuation resulting in an attenuation of the pulse transmission which can no longer be neglected for long distance telecommunications. In the linear regime of a monomode optical fiber, the loss is typically in the range 02-1.0 dB/km. There are, however special wavelengths or spectral windows where the attenuation is particularly low , for 1.3 μm $\le \lambda \le$ 1. At high powers a nonlinear loss term would result from stimulated processes such as Brillouin scattering (Cotter, 1982a 1982b). Linear absorptive or scattering loss simply adds a term $i\gamma q$ to eq (3.4) which becomes

$$i\frac{\partial q}{\partial s} + \frac{1}{2}\frac{\partial^2 q}{\partial \tau'^2} + |q|^2 q + i\gamma q = 0 \qquad (3.8)$$

The dispersive term is small, so that it can be treated as a perturbation. Following the perturbation method developed in (Kodama and Ablowitz, 1980. Hasegawa and Kodama, 1981) the damped pulse solution can be calculated :

$$q(\tau',s) \cong A \operatorname{sech}(A\tau')e^{i\sigma} + O(\gamma^2) \tag{3.9}$$

with $A = A_o\, e^{-2\gamma s}$ and $\sigma = (A^2{}_o/8\gamma)\,(1-e^{-4\gamma s})$. Eq(3.9) describes a pulse which spreads (Blow and Doran, 1985) exponentionally : its amplitude A decays as $\exp(-2\gamma s)$ and its width increases as $\exp(2\gamma s)$.

3.3.2. Higher order terms

For pulse widths of several picoseconds or more, a description of pulses in optical fibers in term of the Nonlinear Schrodinger equation seems to be reasonnable to fit the experimental findings well. However experiments in the high power and the ultrashort pulse regimes, as seen from the studies of pulse compression and frequency shift requires further study to determine the effects of the higher order terms on the NLS model (Cristodoulides and Joseph, 1985. Bourkoff et al, 1987. Zhas et al, 1988). Keeping higher dispersive and nonlinear terms in the Taylor expansion of k $(\omega, |A|^2)$, Hasegawa and Kodama (1982) have shown, using the reductive perturbation method (Tanuiti, 1974), that the propagation of the pulses in the fiber can be modelled by a modified NLS equation with the following form

$$i\frac{\partial q}{\partial s} + \frac{1}{2}\frac{\partial^2 q}{\partial \tau^2} + |q|^2 q = -i\left[\beta_1 \frac{\partial^3 q}{\partial \tau^3} + \beta_2 \frac{\partial}{\partial \tau}(|q|^2 q) + (\beta_3 + i\sqrt{}\ \sigma_3)\, q \frac{\partial}{\partial \tau}|q|^2\right] \tag{3.10}$$

where q is [see eq (3.8)] the complex envelope of the electric field and s and τ describes the direction of propagation and the time measured in the group velocity frame, respectively [see (3.3) and (3.4)]. In this equation the higher order terms : β_1 , β_2 and β_3 correspond to higher order dispersion. The term σ_3 is due to the induced Raman effect, it leads to the downshift of the carrier frequency of the pulse (Kodama, 1985. Hasegawa et al, 1987). The amount of the frequency shift is found to be proportional to the distance of propagation as well as the fourth power of the pulse (quasi-soliton) amplitude in agreement with the experiments on the propagation of subpicosecond pulses along monomode fiber (Mitschke et al, 1986. Beaud et al, 1987) . For 120 femtosecond pulses net frequency shifts as great as 10 % were observed .

3.3.3. *Zero dispersion*

In optical fibers at the so called zero dispersion point $\lambda \approx 1{,}27$ mm, the second order dispersion $k'' = \partial v_g/\partial k$ (GVD) is zero, but the next term k''' is not zero.

Consequently ignoring the higher orders terms other than the third order dispersion term in the expansion of $k(\omega, |A|^2)$ one gets

$$i\frac{\partial q}{\partial s} + i\,\beta_1\frac{\partial^3 q}{\partial q^3} + |q|^2 q = 0 \qquad (3.11)$$

This non integrable equation, with quasi-soliton solutions numerically investigated, which models the fiber properties at the zero dispersion point has been used to propose a new method for pulse compression. Because of the smaller dispersion present, the minimum power required to produce the quasi solitons is much smaller than at using other wavelength. For example, for a one picosecond pulse, only 240 mW is sufficient to produce a quasi soliton, as compared to 1600 mW required for a quasi-soliton at λ = 1,3 μm in the anomalous dispersion region (GVD < 0) [see Wai et al, 1986, 1987].

3.3.4. *Birefringent effects.*

The so called monomode fibers with nominal circular symetry about the fiber axis are not really simple mode, but are in fact bimodal due to the presence of birefringence (Kaminov, 1981) : they can propagate two which are normally assumed to be nearly degenerate modes with orthogonal polarizations. The electric field in the fiber is written

$$E(r,\theta,z,t) = A_1(z,t)R_1(r,\theta)\,e^{i(k_1-\omega t)} + A_2(z,t)R_2(r,\theta)e^{i(k_2-\omega t)} + c.c, \qquad (3.12)$$

Here A_1 and A_2 designate the slowly varying envelopes in each of the two polarizations and R_1 and R_2 the respective transverse linear mode modulations , k_1 and k_2 are the wave numbers, r, θ the polar coordinates in the fiber section. The equations for A_1 and A_2 may be determined (Menyuk, 1987a, 1987b) in a similar maner as in non birefringent fiber by expanding $k(\omega, |A|^2)$. One obtains two coupled NLS equation.

$$i\left(\frac{\partial A_1}{\partial z} + v_{g1}\frac{\partial A_1}{\partial t}\right) - \frac{1}{2}\left(\frac{\partial v_{g1}}{\partial k_1}\right)\frac{\partial^2 A_1}{\partial t^2} + \frac{\chi_{NL}}{2}\left(|A_1|^2 + \frac{2}{3}|A_2|^2\right)A_1 = 0$$

$$i\left(\frac{\partial A_2}{\partial z}\right) + v_{g2}\frac{\partial A_2}{\partial t}\Big) - \frac{1}{2}\left(\frac{\partial v_{g2}}{\partial k^2}\right)\frac{\partial A_2}{\partial t^2} + \frac{\chi_{NL}}{2}\left(|A_2|^2 + \frac{2}{3}|A_1|^2\right)A_2 = 0$$

(3.13a,b)

Here v_{g1} and v_{g2} are the respective group velocities of the modes, $\chi_{NL}$is the nonlinear coefficient (Kerr effect) which is assumed to be the same for the two modes (Wabnitz, 1988) . Such coupled equations are found in other domains of physics (see chap 1 and 4). Bound solitary waves solutions of these equations were analyzed (Tratruk and Sipe, 1988). As far as we know, until now no experiments have been reported which confirm the birefringent effects modeled by eqs (3.13). Very recently, the modulational instability of two copropagating beams in an optical fiber, was analyzed (Agrawal et al, 1989) in terms of coupled NLS identical to (3.13).

4. Magnetic chains.

Many quasi one-dimensional (1D) magnetic compounds (Birgeneau et al, 1978) have now been shown to provide a good support for the study of nonlinear excitations (Lovesey et al, 1984. Balucani et al,1987. Boucher et al,1987) . These compounds where moving domain walls can be viewed as soliton-like-excitations are now currently (Izyumov,1989) used to carry out accurate experiments on nonlinear excitations in condensed matter. The domain walls are highly localized nonlinear objects, they correspond to transition regions between two different but energetically degenerate ground states. In a ferromagnetic chain such walls connect for example neighboring spin up and spin down regions, whereas in an antiferromagnet they form the liaison between the two ground state configurations obtained by interchange of the two sublattices (Figures4.3, 4.6, 4.7) A and B.

Among the interesting 1D magnets one has the ferromagnetic compound $CsNiF_3$, the antiferromagnetic compound $(CH_3)_4$ NM_nCl_3 also called TMMC, and $CsCoCl_3$. The structure of these compounds is such that each chain is well separated from other parallel chains by large cations (Cs^+ or $(CH_3)_4$ N^+).

With spins S = 1 and S = 5/2 $CsNiF_3$ and TMMC are usually treated respectively as classical ferromagnetic and antiferromagnetic spin chains, i.e. quantum effects (Cieplak et al, 1980. Balakrisnan et al, 1985) in the spin dynamics can be neglected (Mikeska, 1978, 1980, 1981.) In these systems the nonlinear excitations (domain walls) have relatively large spatial extensions and the dynamics of the discrete spins can be approximated, in the continuum limit, by nonlinear partial differential equations.

On the other hand $CsCoCl_3$ is described by an antiferromagnetic quantum spin (S = 1/2) chain (Villain, 1975). Unlike $CsNiF_3$ and TMMC, in the Ising-like spin chain such as $CsCoCl_3$ (or $CsCoBr_3$) the widths of the quasisolitons are very narow and the continuum approximation cannot be used. These interesting systems (Boucher, 1988) and other new magnetic compounds (De Groot et al, 1986. Kopinga and De Jonge, 1987) will not be considered here.

In these magnetic chains, the importance of theoretically tractable models, describing experimentally accessible systems including the nonlinear excitations and their interactions, is clearly central to developing the understanding of nonlinear phenomena. Starting from model Hamiltonians, in the continuum limit, the spin dynamics of $CsNiF_3$ and TMMC can be described by complicated NLPDE . However, like for physical systems examined in the preceeding chapters, to lowest nonlinear order of approximation , the nonlinear excitations are frequently taken to be solutions of an integrable prototype equation which in this case is the Sine Gordon equation. The approximate validity of this model is limited to low or high external magnetic field as we shall see in the following. In the low amplitude limit the spin dynamics can be modeled by NLS-like equations, this point will be also shortly examined.

4.1. SPIN CHAINS.

4.1.2. *Spins chains in the classical approximation*

Adopting the Heisenberg model for $CsNiF_3$ and TMMC in the classical approximation the spins are treated as vectors $\mathbf{S}_n$. The model Hamiltonian is

$$H = -2J \sum_n \mathbf{S}_n \mathbf{S}_{n+1} + A \sum_n (S^z{}_n)^2 - g\,\mu_B\, H_x\, S^x{}_n \qquad (4.1)$$

Here, J s the exchange interaction between the spins on adjacent lattice sites n and n+1 (Tjon et al, 1977. Fogedby, 1980). One has : $J > 0$ for $CsNiF_3$ and $J < 0$ for TMMC. The second term represents the influence of crystalline anisotropy : A is the single ion anisotropy constant, the third term represents the Zeeman energy of the spins in an external symetry breaking magnetic field ,$H = H_x$, perpendicular to the chain axis $\mathbf{z}$. The quantities g and μ_B are respectively the Landé factor and the Bohr magneton. The dynamics of these classical spin vectors is described by the undamped Bloch equation :

$$h\frac{d\mathbf{S}_n}{dt} = \mathbf{S}_n \times [J(\mathbf{S}_{n-1} + \mathbf{S}_{n+1}) + g\,\mu_B\,\mathbf{H} - 2A\, S^z{}_n\, \mathbf{z}] \qquad (4.2)$$

4.2. FERROMAGNETIC CHAIN.

4.2.1. *Dynamics and equations of motion.*

Let us first consider the ferromagnetic chain. The spins are treated classically as having fixed length S and orientations parametrized by the spherical polar angles θ_n and φ_n :

$$S^x{}_n = S\cos\theta_n \cos\varphi_n\,,\ S^y{}_n = S\cos\theta_n \sin\varphi_n\,,\ S^z{}_n = S\sin\theta_n \qquad (4.3)$$

In these coordinates, using the continuum approximation, the equations of motion (4.2) become

$$\theta_T = Ja^2(\varphi_{zz}\cos\theta - 2\theta_z\,\varphi_z \sin\theta) - b\sin\varphi \qquad (4.4a)$$

$$\varphi_T \cos\theta = -Ja^2\theta_{zz} + 2A\sin\theta\cos\theta\left(1 - \frac{Ja^2}{2A}\varphi^2{}_z\right) + b\sin\theta\cos\varphi \qquad (4.4b)$$

where $T = (S/\hbar t)$ and $b = g\,\mu_B/S$. In the limit $\theta \ll 1$ and $b \ll 2A$, eqs (4.4) can be approximated by the Sine Gordon equation :

$$\varphi_{TT} - c_o^2\,\varphi_{zz} + \omega_o^2 \sin\varphi = 0 \qquad (4.5)$$

with : $\varphi_T = 2A\theta$. Here $c_o = (2AJ)^{1/2}a$ is the linear spin waves velocity and one has $\omega_o^2 = 2Ab$. The well known solutions of the SG eq. (4.5) are the kink-solitons, breathers and the small amplitude waves which correspond to spin waves or magnons. The single kink-soliton solution , which here (see chapter 2) corresponds to a magnetic domain wall, has the form

$$\varphi = 4\tan^{-1}\exp\left[\pm\frac{z - vt}{d\left(1 - \frac{v^2}{c_o^2}\right)^{1/2}}\right] \qquad (4.6)$$

where v is the propagation velocity along the z axis. here $d = c_o/\omega_o$ provides a measure of the soliton at rest (v = 0). Here in the case of $C_s\,Ni\,F_3$ one has "broad" solitons in the sense that the number ΔN_s of spins inside the soliton (wall) is large : ΔN_S = 10 to30 (Boucher et al, 1987). The ± sign in the exponent of (4.6) refers respectively to the kink-soliton and antikink-soliton solution. In the underlying ferromagnet a kink-soliton physically means a complete 2π turn of the spin in the xy plane (easy plane) as one proceeds along the chain axis z (Figure 4.1).

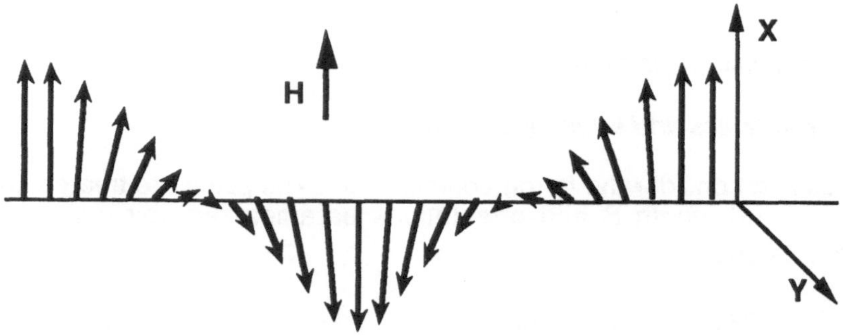

Figure 4.1. Representation of a ferromagnetic domain wall, soliton in the Sine-Gordon approximation, which corresponds to a 2π rotation of the spins in the xy plane.

4.2.2. *Experiments and validity of the Sine Gordon picture.*

The essential quantity that enters most important physical properties is the two spin correlation function at temperature T

$$s(z, t) = < S^a (0,0)\, S^a (z, t) > \qquad (4.7)$$

which gives the probability that given a spin at position 0 and time 0, then, a spin at position z, some time t later will also be pointing in direction a. Here < > denotes an equilibrium average at temperature T. One usually defines a transverse correlation function $s_T (z,t)$ and a longitudinal correlation function $s_L(z,t)$. Most peculiar features of magnetic chains manifest themselves in the space time Fourier transform of s(z,t) which is called *the spectral density or the dynamical structure factor* S(q,ω).This quantity is directly accessible experimentally with neutron scattering technique (see P A Lindgård, 1984), where one resolves both energy and momentum, and measures S(q,ω) directly. If one considers the dynamics of classical spin chains with Heisenberg exchange, at zero temperature the spins will be perfectly aligned. In this limit, the first excited states correspond to infinitely long lived spin waves or magnons which appear as poles in S(q,ω). Above zero temperature kinks (anti-kinks) are created, they destroy the long range order and induce a loss of correlation.Because of their translational invariance the quasisolitons can be anywhere on the chain and they scatter the magnetic neutrons giving a (central) peak near q =0. Because they are not vibrational modes but are translating, they give a (central) peak near ω = 0

The loss of order is reflected in a weakening of the Bragg peak intensity and the appearance of a more diffuse scattering pattern around the Bragg peak. In other terms the soliton contribution redistributes strength from the Bragg peak into a quasielastic contribution to the spectrum.

In the context of the Sine Gordon model, considering a gas of non interacting kink solitons, including magnons corrections, Mikeska (1978,1980) calculated the transverse and longitudinal part of S(q,ω) for the easy plane ferromagnet $CsNiF_3$. Originally this model was considered to give a striking evidence of S.G. kink-solitons.

Indeed, in $CsNiF_3$ a central peak was observed and fitted in terms of the approximate Sine Gordon approach (Steiner et al, 1980, 1983). Then it was realized that the Sine Gordon picture was reliable in a restricted domain of fields and temperature. Namely, if the external magnetic field is comparable to or greater than a small fraction of the anisotropy field, the spins are unstable towards rather large off plane excursions, the Sine Gordon model becomes inadequate. (Kumar 1982. Magyari and Thomas, 1982. Wysin et al, 1982, 1984).The instability is created by the soliton motion which also causes spins to go off plane, such that the critical magnetic field for instability rapidly decreases with increasing soliton velocity. The instability was predicted to occur at $H_c \approx 18$ K

Oerstedt (Kumar, 1982) but, in recent neutron experiments the out of plane component of the central peak versus field shows no indication of instability (Kakurai et al, 1986). Implication of these results suggested (Wysin et al, 1984) that nonlinear elementary excitations in these chains are breather modes rather than isolated quasi-solitons. For $CsNiF_3$ the expected "soliton phase diagram" i.e. temperature T versus external magnetic field H was discussed (Boucher et al, 1987) the soliton picture is expected to be approximately valid below the limiting value H_c of the external magnetic field.
Quasi-one dimensional magnetic systems can be used to test the possibility that nonlinear excitations give significant contributions to the thermodynamic properties and to verify that they may be described quite generally as thermal excitations present, in addition to the small amplitude linear modes : the magnons. For example, other experimental evidence for quasi-solitons was investigated using thermal probes (Ramirez et al, 1982) such as specific heat measurements.

4.3. ANTIFERROMAGNETIC CHAIN.

4.3..1. *Equations of motion and excitations.*

Let us now consider the antiferromagnetic chain modelled by the Hamiltonian (4.1) with J < 0.The even or odd lattice sites , are referred to as the A and B sublattices ; from the Bloch equation (4.2) one obtains six equations of motion for the spin components of each sublattice. Then, at low temperature, a slow variation of the classical spin vectors in space is expected and a continuum approximation which is adapted to these slow variations is performed in the following way. Again, the spins vectors are parametrized in spherical polar coordinates (Mikeska, 1980) in terms of four angles $\Theta, \Phi, \theta, \phi$:

$$\mathbf{S}_{A,B} = S[\ \sin(\Theta\pm\theta)\cos(\Phi\pm\phi),\ \sin(\Theta\pm\theta)\sin(\Phi\pm\phi),\ \cos(\Theta\pm\theta)\ ] \quad (4.8)$$

which are all functions of time and continuum position z on the chain. Angles Θ and Φ are considered large compared to the small angles θ and ϕ. If $\theta = \phi = 0$, then the spins S_A and S_B are exactly antiparallel and thus θ, ϕ describe small deviations from antiferromagnetic alignement which occur owing to small spatial variations and to small values of the local magnetization.
In order to take the continuum limit , we assume $\theta, \phi << 1$, as well as slow spatial variations in Θ and Φ. Assuming $2A/J<<4$, which is the case for most real materials, the small angles can be eliminated. Then, in terms of the fields introduced above, the equations which govern the dynamics are given (Wysin et al, 1986. Gouvea and Pires, 1986) by

$$(c_o^2\Phi_{zz} - \Phi_{tt})\sin\Theta = 2\ (\Phi_t\Theta_t - c_o^2\Phi_z\Theta_z)\cos\Theta - \omega_1^2\sin\Theta\ \sin\Phi\cos \\ + 2\omega_1\Theta_t\ \sin\Theta\cos\Phi \quad (4.9a)$$

$$(c_o^2\Theta_{zz} - \Theta_{tt})\csc\Theta = (c_o^2\Phi_z^2 - \Phi_t^2)\cos\Theta + (\omega_1^2\cos 2\Phi - 4\omega_2^2)\cos\Theta - 2\,\omega_1\Phi_t \sin\Theta\cos\Phi \qquad (4.9b)$$

where $c_o = 2JSa$, $\omega_1 = g\mu_B H$ and $\omega_2 = (8AJS)^{1/2}$. Measuring the space in units of the lattice constant a and the time in units of h/JS, putting $c_o = 2$, $\alpha = 2A/J$ and $\beta = g\mu_B H/JS$, eqs (4.9) are transformed into

$$(\Phi_{zz} - \frac{1}{4}\Phi_{tt})\sin\Theta = 2(\frac{1}{4}\Phi_t\Theta_t - \Phi_z\Theta_z)\cos\Theta - \frac{1}{4}\beta^2\sin\Theta\sin\Phi\cos\Phi + \frac{1}{2}\beta\Theta_t\sin\Theta\cos\Phi \qquad (4.10a)$$

$$(\Theta_{zz} - \frac{1}{4}\Theta_{tt})\csc\Theta = (\Phi_z^2 - \frac{1}{4}\Phi_t^2)\cos\Theta + (\frac{1}{4}\beta^2\cos 2\Phi - \alpha)\cos\Theta - \frac{1}{2}\beta\Phi_t\sin\Theta\cos\Phi \qquad (4.10b)$$

4.3.2. *Sine Gordon approximations.*

Analytical solutions of equations (4.10) have not been found and particular cases were examined which correspond to excitations in the xy and yz planes respectively i. e transverse (T) and longitudinal (L) kink-solitons . Specifically, it is assumed (Mikeska, 1980. Wysin et al, 1986) that $\Theta = \frac{\pi}{2} - \Theta_s$, where $\Theta_s \ll 1$. Then, linearizing eqs (4.10) in Θ_s yields to a Sine-Gordon equation for the variable $\psi = 2\Phi - \pi$, and a ψ dependant equation for Θ_s :

$$\psi_{tt} - 4\psi_{zz} + \beta^2\sin\psi = 0 \qquad (4.11a)$$

$$\Theta_{szz} - 4\Theta_{szz} = (4\psi_z^2 - \psi_t^2 + \beta^2\cos^2\psi - 4\alpha)\cos\Theta - 2\beta\psi_t\cos\psi \qquad (4.11b)$$

In the low field limit : $\beta^2 \ll 4\alpha$, ignoring the second derivatives of Θ_s in eqs (4.11), a kink-soliton solution can be obtained, which corresponds to a π rotation of the spins (Figure 4.2) in the xy plane, propagating at velocity v along the z axis :

$$\Phi = \frac{\pi}{2} + 2\tan^{-1}[\exp\frac{\beta\gamma}{2}(z - vt)] \quad , \Theta_s \approx -\frac{\beta^2\gamma v}{4\,\alpha}\operatorname{sech}^2(z - vt) \qquad (4.12a,b)$$

with $\gamma = (1 - \frac{1}{4}v^2)^{-1/2}$

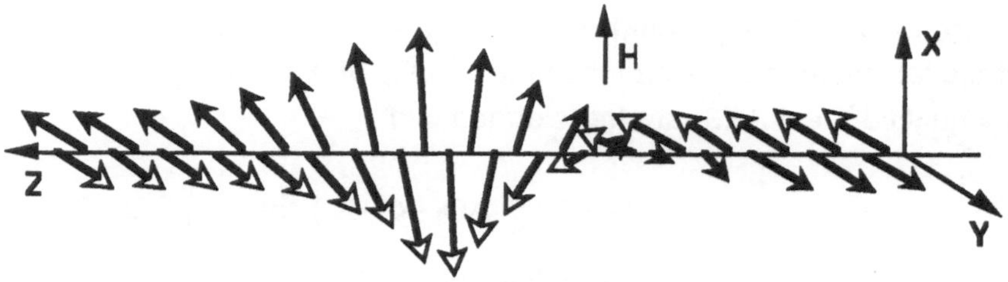

Figure 4.2. Representation of a xy domain wall (SG soliton) in an antiferromagnetic chain.

In the case of the antiferromagnet TMMC, the spatial extension of the kink-solitons is of the same order than for the ferromagnet $CsNiF_3$: ΔN_s=10-30 spins.By numerical integration of the equations of motion(Wysin et al, 1986) it was shown these kinks are stable over a wide range of fields and velocities.
By taking $\Phi = \pi/2$ and $\Theta_e = 2\Theta - \pi$, eq(4.10b) reduces to a Sine-Gordon equation in the variable Θ_e

$$\Theta_{e,tt} - 4\Theta_{e,zz} + 4a\sin\Theta_e = 0 \tag{4.13}$$

with kink-soliton solutions which now correspond to π rotations (Figure 4.3) in the yz plane :

$$\Theta_e = \frac{\pi}{2} + 2\tan^{-1}[\exp \gamma\alpha^{1/2}(z - vt)] \tag{4.14}$$

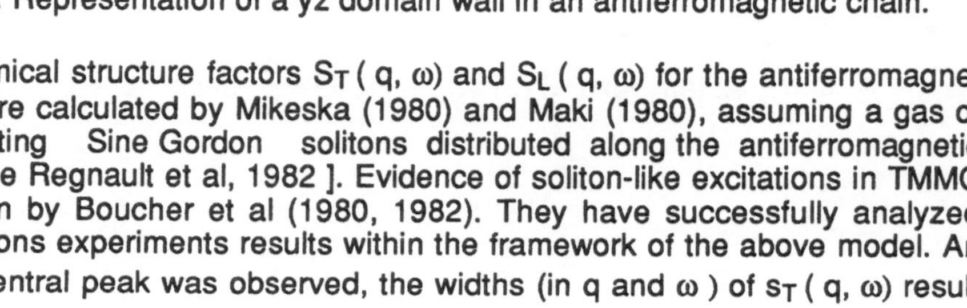

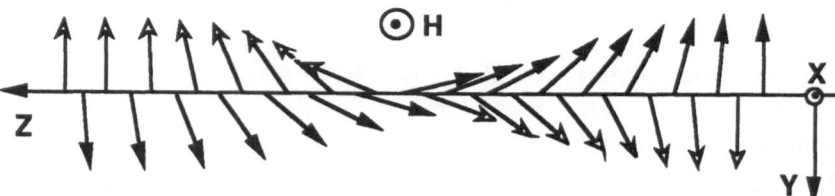

Figure 4.3. Representation of a yz domain wall in an antiferromagnetic chain.

The dynamical structure factors $S_T(q, \omega)$ and $S_L(q, \omega)$ for the antiferromagnet TMMC were calculated by Mikeska (1980) and Maki (1980), assuming a gas of noninteracting Sine Gordon solitons distributed along the antiferromagnetic chain [see Regnault et al, 1982]. Evidence of soliton-like excitations in TMMC was shown by Boucher et al (1980, 1982). They have successfully analyzed their neutrons experiments results within the framework of the above model. An intense central peak was observed, the widths (in q and ω) of $s_T(q, \omega)$ result

from the flipping of the spins associated with the sublattices each time a soliton passes by.
The spin dynamics can also be probed by the nuclear magnetic resonance technique (Boucher and Renard, 1980), it is a particularly interesting method when the neutron measurements are not possible. The spin lattice relaxation rate $1/T_1$, of nuclei located in the neighborhood of the chains is governed by the flipping of the magnetic sublattices. Clearly, as a quasi-soliton passes each point in the chain a spin flip occurs and the previous spin configuration is relaxed. One has a dominant contribution of S_T (q,ω) and can approximately write $1/T_1 \sim S_T(q, \omega)$.

4.3.3. *General regime : quasi-solitons.*

A phase diagram of the soliton-like excitations in TMMC was determined experimentally (Boucher et al, 1984) by both varying the external magnetic field H and the temperature T , from which the (Transverse and Longitudinal) soliton picture appears to be approximatively valid when the condition H/T >> 10kOe/K is satisfied. In the low field limit the soliton energy is proportional to H with a constant different of the constant predicted theoretically. This difference was first attributed to quantum effects (Maki, 1981. Mikeska, 1982) but it is now expected to be the due to the out of plane spin motions (Gouvea an Pires, 1986).
In the large field limit the (L) soliton energy is independant of H. In fact, as previously noticed the T and L kink- solitons are two limiting cases of a more complex solution of the full discrete evolution equations which describe the dynamics of the spins which was investigated numerically by Wysin et al (1986) . Their results show that the out of plane spin components play an important role in the behavior of the excitations. Precisely, in the full field range, two different zones can be defined depending on wether the field H is higher or lower than a critical field H_c. For v = 0, one has static T-kinks in the Sine-Gordon regime as calculated above, their energy is proportional to H. When v increases, the deviation from the xy plane increases and the T like excitations have their spins tilted out of this transverse plane. For $H = H_c$ all these soliton-like excitations become static and ones has also L excitations. For $H > H_c$, one has T excitations, and L excitations with their rest energy proportional to H_c .

The thermodynamical relevance of quasi-solitons in TMMC was tested experimentally by specific heat measurements (Borsa et al, 1983) and thermal conductivity measurements (Gronckel e(al, 1988).These last results were compared to a thermal conductivity model (Wysin and Kumar, 1987) where the contribution of magnons and solitons are taken into account.

4.4. EXCITATIONS IN THE LOW AMPLITUDE LIMIT.

In the weak amplitude limit for simple model Hamiltonians, the dynamics of the spins can be described in terms of the nonlinear Schrödinger equation to lowest nonlinear order of approximation (Laksmanan, 1977. Corones, 1977. Cieplak and Turski, 1980. Makankov, 1980). More precisely, in the case of the ferromagnetic chain , for $\varphi<<1$ and $\theta<<1$, eqs (4.6) reduce to the NLS equation (Remoissenet ,1986). The associated excitations, which are envelope modes, correspond to *nonlinear magnons or low amplitude breathers*, until now they have not been observed experimentally.

4.4.1. *Antiferromagnetic chain.*

For the antiferromagnetic chain weak nonlinear effects have been clearly identified experimentally very recently. In the following, we shall examine these effects and also the reduction of eqs (4.14) to coupled NLS equations .

Using eqs (4.9a) and (4.9b) we first redefine Φ as $\pi/2+\Phi$ and Θ as $(\pi/2-\Theta)$. Then, we consider weakly nonlinear oscillations around the ground state which is assumed to be zero i.e. in the small angle limit : $\Phi\rightarrow\delta\Phi$ and $\Theta\rightarrow\delta\Theta$ with the small parameter $\delta<<1$.Under these conditions eqs (4.9) which give the the spin dynamics reduce to

$$L\Phi=-\frac{2}{3}\omega^2{}_1\delta^2\Phi^3+2\varepsilon^2(\,c^2{}_0\Phi_z\Theta_z-\Phi_t\Theta_t\,)\Theta+\,2\omega_1\varepsilon\Theta_t\Phi \tag{4.15a}$$

$$L'\Theta=-\frac{2}{3}\omega^2{}_2\delta^2\Theta^3+\delta^2(\Phi^2{}_t-c^2{}_0\Phi^2{}_z)\Theta-\omega^2{}_1\delta^2\Phi^2\Theta-2\omega_1\delta\Phi_t\Phi \tag{4.15b}$$

with $\omega_1=g\mu_BH$, $\omega_2=S4AJ$, $c_o=2J\,S$ and where the linear operators L and L' are given by :

$$L=c^2{}_0\partial^2/\partial z^2-\partial^2/\partial t^2-\omega_1{}^2\quad,\;L'=c^2{}_0\partial^2/\partial z^2-\partial^2/\partial t^2-\omega_2{}^2 \tag{4.16a,b}$$

In equations (4.16) Φ and Θ are respectively the small in plane (IP) and out of plane (OP) angles (Figure 4.4). We now use the multiple scale expansion technique. Accordingly, we write

$$\frac{\partial}{\partial x}=\frac{\partial}{\partial x}+\varepsilon\frac{\partial}{\partial_1}+\varepsilon^2\frac{\partial}{\partial X_2}+.....\quad,\quad\frac{\partial}{\partial t}=\frac{\partial}{\partial t}+\varepsilon\frac{\partial}{\partial T_1}+\varepsilon^2\frac{\partial}{\partial T_2}+.... \tag{4.17a,b}$$

$$\Phi=\varepsilon\Phi_1+\varepsilon^2\Phi_2+\varepsilon^3\Phi_3+.....,\qquad\Theta=\varepsilon\Theta_1+\varepsilon^2\Theta_2+\varepsilon^3\Theta_3+.....,\tag{4.17c,d}$$

To lowest order in ε, one has

$$L\Phi_1=0\;\;,\;L'\Theta_1=0\;, \tag{4.18a,b}$$

yielding the solutions.

$$\Phi_1=A(X_1, X_2, T_1,T_2)e^{i\psi_1}+c.c, \quad \Theta_1=B(X_1, X_2, T_1,T_2)e^{i\psi_2}+c.c \qquad (4.19a,b)$$

with $\psi_1= k_1x - \Omega_1t$, $\psi_2= k_2x - \Omega_2t$, and where $\Omega^2{}_1= \omega^2+c^2{}_0k^2{}_1$, $\Omega^2{}_2= \omega^2+ c^2{}_0k^2{}_2$ are the linear dispersion relations in the long wavelength limit.

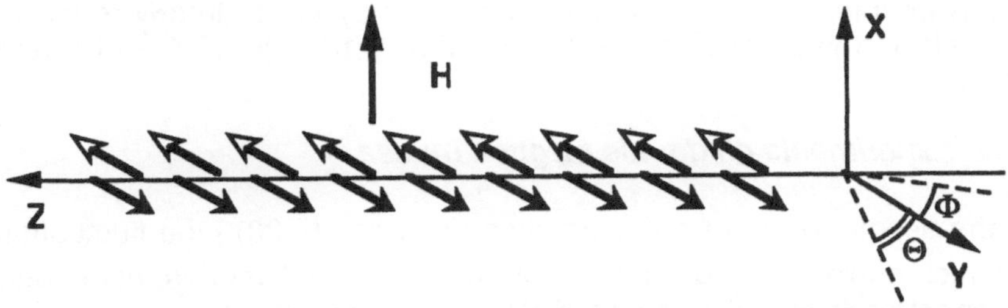

Figure 4.4. Representation of the spin direction with respect to the chain and the field axis and the associated IP and OP small angles φ and Θ.

The solutions (4.19) are now introduced in the second order equations and from the removal of the secular terms we get

$$\Phi_2= i[\, F^+ABe^{i(\psi_1+\psi_2)} - F^-AB^*e^{i(\psi_1-\psi_2)}\,] + C.C \qquad (4.20a)$$

$$\Theta_2= i\,[\, G^+ABe^{i[\psi_1(k_1)+\psi_1(k_2))} - G^-AB^{*i\,[\psi_1(k_1)+\psi_1(k_2)]}\,] + C.C \qquad (4.20b)$$

with

$$F^{\pm}= 2\omega_1\varepsilon\Omega_2/[\, c^2{}_0(k_1\pm k_2)^2-(\Omega_1 \pm \Omega_2)^2+\omega^2{}_1] \qquad (4.21c)$$

$$G^{\pm}=-2\omega_1\varepsilon\Omega_1/[\, c^2{}_0(k_1\pm k_2)^2-(\Omega_1(k_1) \pm \Omega_2(k_1))^2+\omega^2{}_2] \qquad (4.21d)$$

To $O(\varepsilon^3)$, with the transformation $s = X_1$, $\tau= T_2$, after removing the secular terms two coupled Nonlinear Schrödinger (NLS) equations are obtained

$$iA_\tau + pA_{ss} + q|A|^2A + r|B|^2A = 0 \qquad (4.22a)$$

$$iB_\tau + p'B_{ss} + q'|B|^2B + r'|A|^2B = 0 \qquad (4.22b)$$

where $\partial/\partial s = \partial/\partial(X_1- v_{g1}T_1) = \partial/\partial(X_1- v_{g2}T_1)$ with the group velocities $v_{g1}=\partial\Omega_1/\partial k_1$, $v_{g2}=\partial\Omega_2/\partial k_2$, and $\tau = T_2$. Here p, p', q, q', r and r' are functions (Remoissenet, 1989) of k_1, k_2, ω_1, ω_2 and c_o.

For the system (4.22) the complete integrability is established (Zhakarov and Schulman, 1982) for the specific parameter restrictions : p = p' ; q = q' = r = r' or p

= - p' ; q = q' = - r = - r' and soliton solutions can be calculated (Shadevan et al, 1986). In the present physical case these coefficients are functions of k_1, k_2, ω_1, ω_2 and the above constraints are not fullfilled : the equations (4.22) can have solitary wave solutions (Bhakta, 1987) depending on the signs of the coefficient a_o = (rp' - pq')/(rr' - qq') for A and of the coefficient b_o =(r'p - qp')/(rr' - qq') for B.NLS coupled equations similar to eqs (4.25) were already found for electrical transmission lines and optical fibers (see § 1.2 , § 3.3.4 and references cited therein).

4.4.2. *Experiments on double magnon modes.*

To the second order of approximation [see eqs (4.20)] the fluctuations in Φ and Θ, which correspond to new double magnon (DM) modes, give contributions to the spectral density, let us now discuss this problem.

In fact, conventional DM modes have been observed (Cowley et al, 1969) by neutron ineslastic scattering ad well-defined modes, which can be analysed as a function of both wavevector q and frequency ω. Modes of this type , are a direct consequence of linear magnon theory.

In the case of planar antiferromagnetic chains considered above, the spin-flop configuration (Figure4.4) describes the equilibrium spin orientations at low temperatures when a large magnetic field **H**, is applied perpendicular to the chain axis **z**. If **H** is parallel to **x**, a single magnon is a manifestation of small oscillations of the spins in the xy and yz planes. Such oscillations, which correspond to small changes of the in-plane (IP) and out of plane (OP) angles Φ and Θ, give rise to spin fluctuations which are predominantly in the x and z directions. These fluctuations, deemed IP and OP respectively, are *perpendicular* to the equilibrium spin direction. Unlike these new DM modes, the conventional DM modes are the result of the same angular oscillations of the spins, but correspond to the small changes which these oscillations induce in the spin-component *parallel* to the equilibrium spin direction : y.

From the above equations the IP and OP fluctuations are nonlinearly coupled leading to a mixing of their frequencies. Very recently, such predicted DM modes whith frequencies which are the sum and difference of IP and OP single magnon frequencies , were observed by polarized neutron scattering experiments on TMMC (Boucher et al, 1989). These results are interesting and promising, because they show that small nonlinear effects can be successfully probed by experiments. Nevertheless, the experimental investigation of the higher order nonlinear excitations, which correspond to the solutions of the coupled NLS equations (Remoissenet , 1989), remains an open problem.

5. Biopolymers.

The idea that the principles of physics could be applied to biology was initiated by Szent-Giörgyi (1941) then, Fröhlich (1968, 1969) suggested that elecrical vibrations, with frequencies in the microwave region, may be excited coherently in active biological materials through chemical (metabolic) processes.

Now, it is apparent that theoretical physics must play a major role in the modeling (Davydov, 1985. Scott, 1985a. Lomdahl et al, 1985. Balanowski et al, 1985. Krumhansl et al, 1983) and elucidating of mechanisms underlying the functionning of biological systems, that are complex and highly organized systems. Soliton-like excitations were indentified in a synthetic polymer (Heeger et al, 1988) like polyacetylene, and one expects that nonlinear excitations may play a role in the properties of biopolymers. Recent investigations indicate that this may indeed the case in certain structures, deoxyribonucleic acid (DNA) being the most promising but the most complex one. In fact, the complexity of DNA coupled to the refinement, variety and range of function that it performs with such admirable accuracy is a challenge for theoreticians . Although topologically different classes of quasi-solitons may arise according to the particular system under study, these coherent excitations may be involved (Fröhlich, 1968) in information and energy transfer (Baverstock et al, 1988) along the biomolecular chains.

Obviously, models involving nonlinear excitations in biology cannot explain every aspect of dynamics, but they are motivating exciting questions and experiments. Here,we first consider the Davydov model which is perhaps the earliest suggestion for a localized nonlinear excitation in such systems, in order to understand how metabolic energy is stored and transported in biological molecules. Then we summarize some important properties and phenomena in DNA and we review shortly the recent nonlinear models.

5.1. DAVYDOV MODEL.

Among the interesting peculiar motions in biological molecules, an interesting candidate is the amide I, or costretching, vibration in protein. A linear theory is not appropriate because if amide I vibrational energy is assumed to be localized on (one or) a few neighboring peptide groups at some time, it will rapidly spread out and distribute itself over the molecule. This dispersion due to dipole-dipole interaction implies a lifetime of the order of 10^{-12} seconds which is much to short for normal biological mechanisms.

For an α-helix protein, Davydov et al, proposed a nonlinear mechanism that might prevent energy dispersion, i.e. increase this lifetime. This mechanism involves the interaction of longitudinal sound waves (stretching of the hydrogen bonds) with localized amide I bond energy. The longitudinal sound waves act as a potential well to trap the bond energy and prevent its dispersion. This self trapped state often called the "Davydov soliton" is rather similar to a polaron and it can be shown that there is a relationship between the Davydov and Fröhlich models (Tuszynski, 1984). In solid state physics, the polaron is an example of self trapped state . In this case an electron (or a hole) moves through a crystal as a localized wave function rather than an extended Bloch state ; since the electron is localized, it polarizes the medium in its vicinity, thereby, lowering its energy which keeps it localized (Ziman, 1980). In other terms the polaron is a quasi particle arising from phonon dressing of an electron or a hole. In fact, for "Davydov soliton", localized amide I vibrational energy distorts the lattice in its vicinity (inducing longitudinal sound waves on the helix) lowering its energy and thereby trapping it.

The analytical theory of this quasi-soliton has been discussed in a number of papers by Davydov and his coworkers (see Davydov, 1985 and references therein, and Scott, 1982). In the original model one has two coupled discrete differential equations. They describe the time evolution of amide I vibrational energy coupled to the displacements of the hydrogen-bonded chain of peptide group. Specifically, in the continuum limit these equations can be reduced to the following equations.

$$i\frac{\partial a}{\partial t} + \frac{\partial^2 a}{\partial x^2} = - k_1 \chi \rho\, a \qquad (5.1a)$$

$$\frac{\partial^2 \rho}{\partial x^2} - \frac{1}{c^2}\frac{\partial^2 \rho}{\partial t^2} = k_2 \chi \frac{\partial^2 (|a|^2)}{\partial x^2} \qquad (5.1b)$$

Here a (x,t) is the probability amplitude of the excitation on the peptide), ρ (x,t) is the gradient of the longitudinal sound (displacements of the peptide group), χ is the nonlinear coupling parameter between amide I vibration and lattice distorsion which in other terms represents the charge in amide I energy per unit extension of an adjacent hydrogen bond, c is the longitudinal sound speed, and k_1 and k_2 are constants that are determined from the energetic parameters of the problem. Eqs(5.1) which are not completely integrable are called the "Zhakarov" equations because they first arose as model equations (in 3 dimensions) in the context of plasma physics to describe the interaction between ion sound and Langmuir waves (Zhakarov, 1972). Seeking travelling wave solutions of eqs (6.1) on the form of excitations that propagate along the chain with a velocity v : $\rho(x,t) = \rho(x-vt)$, allows to reduce these equations reduce to a NLS equation with soliton solutions. The work of Davydov was pursed by Hyman et al (1981) and J. Scott (1982) who predicted that Davydov solitons should appear under normal physological conditions. This conclusion was supported by the assignment to internal vibrations of the quasi-soliton, of a measured laser Raman spectrum of metabolically active cell (Lomsdahl et al, 1982).

Careri et al (1984) have also discussed the formation of Davydov-like solitons in acetanilide (ACN) which is organized in hydrogen bonded chains with a remarkable similarity to the chain structure of hydrogen-bonded peptide groups in α-helical proteins.

Although the results are promising, the signatures of quasi-solitons of the Davydov type in biopolymers are not completely clear (Barthès, 1989) and further experimental and theoretical (Cottingham et al, 1989. Wang et al, 1989) investigations are needed.

5.2. TYPICAL FEATURES OF DNA.

Among the fascinating features of DNA are its complicated static structure, dynamic structure , replication and transcription phenomena. Here, we summarize these properties, for details the interested reader is refered to the folowing excellent references in molecular biology (Freifelder, 1987. Saenger,1984. Dickerson, 1983).

5.2.1. *Some preliminaries on the DNA Structure.*

DNA fibers can exist in two fundamental forms : the B configuration which is stable under conditions of high humidity (about 92%) and the A configuration when the humidity is lower (about 66%).

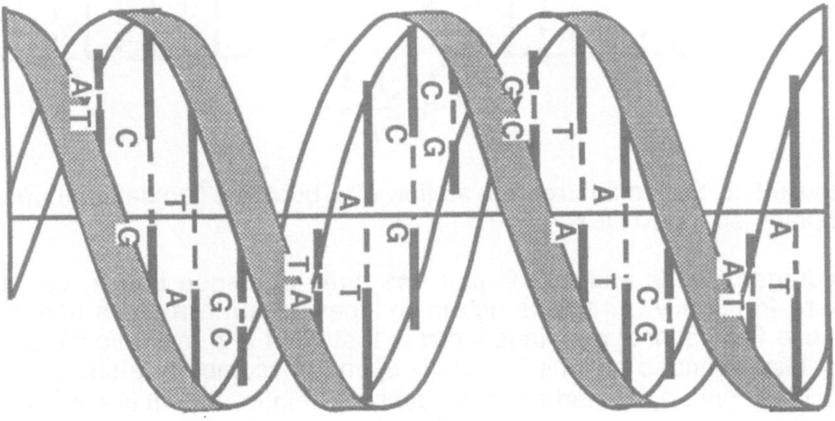

Figure 5.1. Schematic representation of the DNA double helix with A-T and G-C base pairs drawn symbolically as bars (see Saenger, 1984. Freifelder, 1987) between chains, the dotted lines represent the hdrogen bonds.

The DNA molecule is composed of two strands (Figure 5.1) wrapped helically around a central axis. In each strand there are chains of alternating sugar rings and phosphate groups (sugar- phosphate backbones) which follow a helical path at the outer edge of the molecule and the bases are in helical array in the central core. The bases of one strand and those of the other strand are held together by hydrogen bonds; these base pairs are of two kinds : the A (adenine) - T thymine) base pairs and the G (guanine)- C (cytosine) base pairs. Adenine is normally paired with thymine by *two* hydrogen bonds and guanine is paired with cytosine by *three* hydrogen bonds. The bases are stacked with their planes separated by a spacing a = 3.4 Angström. The helix has two external helical grooves : a deep wide one, called the major groove and a shalow narow one called the minor groove.

5.2.2. *Dynamic fluctuations in the structure.*

Important features at physiological temperature are the fluctuational openings or dynamic fluctuations in the structure, this important phenomena is of great interest to molecular biologists, it exposes the bases either individually or perhaps coordinately. Namely, due to thermal fluctuations DNA double helix experiences bending motions, torsional motions (Lewitt, 1982) and *base - pairs openings* also called *breathing* or *local melting* (Lilley, 1988). In the latter process hydrogen bonds (and also hydrophobic interactions) which are weak, are easily disrupted by the thermal motion . Since an A-T pair has two

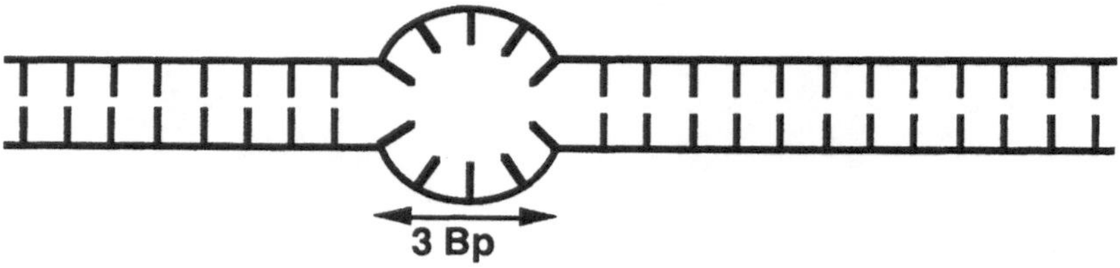

Figure 5.2. Schematic representation of a bubble (fluctuational opening) along doubled stranded DNA.

hydrogen bonds and a G-C pair has three hydrogen bonds, openings occurs more frequently in regions rich in A-T pairs than in regions rich in G-C pairs. These fluctuational openings indicate that *DNA is a dynamic structure*, in which doubled-stranded regions frequently open to become, for a short while (order of 10^{-12} seconds), singled-stranded bubbles (Figure 5.2). It is obvious that in such base unpairing, large amplitude (nonlinear) motions are involved. This process has to occur for the biological function of DNA, most specifically during the initiation of transcription when about 12 bases pairs are opened immediatly upstream of the point where RNA synthesis starts, as we shall discuss shortly in the following.

.2.2.3 DNA replication and transcription.

Two important and complex processes occur in DNA : *replication and transcription.*

The purpose of DNA *replication* is to create daughter DNA molecules that are identical to the parental molecule. The relevant genetic information which resides in the bases sequence of double-stranded DNA molecules is transferred from parent to progeny by a faithful replication of the parental DNA molecules.

The process by which RNA molecules are synthetized is called the *transcription.* In this process , the transfer of genetic information is accomplished from DNA to RNA molecules and then from RNA to protein molecules. It is catalyzed by enzymes called RNA polymerases .

In the first step of the transcription RNA polymerase moves rapidly along the DNA molecule and then binds to it. Binding occurs in particular selected sites called *promoters,* which are base sequences in which several interactions occur. More precisely, the RNA polymerase must recognize a specific DNA sequence, attach in a proper conformation, open the DNA strands in order to gain acces to the bases to be copied, and initiate synthesis of RNA. During synthesis one open strand of DNA acts as a template. Termination of synthesis occurs at specific sites (bases sequences) within the DNA molecule, then the RNA is released .The process of base recognition is facilitated by the presence of fluctuational openings or breathing bubbles mentionned above, which occur along DNA helix. That is, while base pairs are temporarily broken, the polymerizing enzyme can penetrate the helix and read the base sequence, as must be necessary for RNA synthesis.

It is established that studies involving nonlinear excitations are important and useful to approach the complex mechanisms considered above, which occur in the replication and transcription of messenger ribonucleic acid (RNA). Several models have been proposed suggesting that soliton-like excitations may play a role in the dynamical properties of DNA but the question is far from being settled. The preliminary step to understand these mechanisms is to study the nature, dynamics and formation of the openings; we now consider the recent models which deal with these problems.

5.3. OPEN PROBLEMS AND DNA MODELS.

In a complex system such as DNA, the degrees of freedom are numerous, coupled, and influenced by the environement in which the molecule sits, the problem of solving the equations of molecular motion is untractable. Rather, a more feasible route is to select and isolate some degrees of freedom which dominate a specific conformational change (Krumhansl et al, 1983). For example as discussed above, the Davydov model for α-helix protein has essentially two degrees of freedom : the displacement of the peptide unit and the quantized carbonyl vibrational mode.

One can consider two classes of nonlinear models which have been proposed to elucidate open problems related to the complex dynamical properties of DNA.

First, one has models where the nonlinearity is assumed to result mainly from some effective nonlinear substrate potentials (Sine-Gordon like or multistable potentials) with corresponding topological kink (anti-kink) excitations. Second, one has models where the nonlinear coupling between molecular units is assumed to play a dominant role, in this case the excitations are nontopological, as for example the Davydov solitons dicussed previously. Let us review successively both classes of models.

Englander et al. (1980) first suggested a model to explain the existence of transiently open states in DNA which has been as demonstrated by hydrogen exchange measurements. The ground state of the double helix was modeled as two linear chains of pendula, which represent the bases, connected by linear springs representing the sugar-phosphate backbones.This coupled-pendulum system which was first introduced by Scott (1969) to is an example mechanical model to simulate the propagation of (Sine-Gordon) fluxons on the Josephson transmission line (see chapter 4). The bonding to an opposite base is represented by the gravitational force (nonlinear Sine Gordon term). A soliton like excitation, which corresponds to an opening configuration of the base pairs, is viewed as a localized twist corresponding for exemple to a $\pm 2\pi$ rotation of the pendula (bases). This mobile open unit may diffuse along the double helix. From the data of kinetic and thermodynamic experiments Englander et al concluded that these open configurations may consist of mobile segments of 10 base pairs in length.

Later Yomosa (1983, 1984) proposed a model where the conformation and stability of DNA was mainly determined by the energy of hydrogen bonds between interstrand complementary base pairs, the stacking energy between intrastrand adjacent bases and the torsional energy of polynucleotide strands. Taking account on of these energy terms and of the kinetic energy of the bases rotations he derived, in the continuum approximation, coupled Sine-Gordon-like equations which describe the dynamics of the rotations of the bases

$$I\,\phi_{tt} + A\sin\phi + B\sin\phi\cos\phi' - Sa^2\phi_{zz} = 0 \qquad (5.2a)$$

$$I\,\phi'_{tt} + A\sin\phi + B\sin\phi'\cos\phi - Sa^2\phi'_{zz} = 0 \qquad (5.2b)$$

$$I\,\Phi_{tt} + 2A\sin(\Phi/2)\cos(\varphi/2) + B\sin\Phi - Sa^2\Phi_{zz} = 0 \qquad (5.2c)$$

$$I\,\varphi_{tt} + 2A\sin(\varphi/2)\cos(\Phi/2) + B\sin\varphi - Sa^2\varphi_{zz} = 0 \qquad (5.2d)$$

where A and B are constant parameters, I is the mean value of the moment of inertia of the bases for their rotations ϕ and ϕ'around the axes (Figure 5.3), with $\Phi = \phi + \phi'$ and $\varphi = \phi - \phi'$. The quantity a is the lattice spacing (distance between bases along the z axis).

In some particular cases soliton-like solutions were found and Yomosa estimated the velocity (8.3 10^3 cm/sec), width (9a), and number density of these excitations ($2.20^{-5}/a$) , he also calculated also the average number of bases in open states ($10^{-4}/a$) , where a is the lattice (bases) spacing given in § 5.2.1.

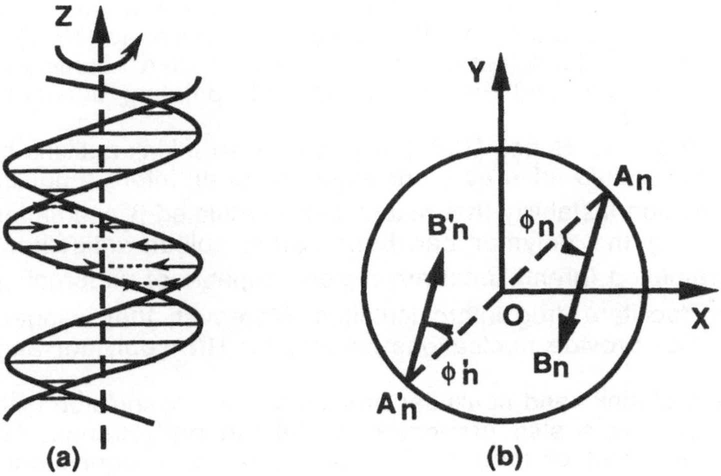

Figure 5.3. Schematic drawing of the base pairs B_n and $B'_n$and their rotation angles ϕ and ϕ'in the plane normal to helix axis.z (Yomosa 1983,1984).

Takeno and Homma (1984) improved the previous plane base rotator model of Yomosa by taking into account the discretness of the structure obtaining a generalized Frenkel-Kontorova system, where a model field equation constitutes a coupled pair of nonlinear differential difference equations which in a particular case could be decoupled to give a discrete version of the Double Sine-Gordon equation. When the intrastrand are base interactions are much larger than the inter-strand ones the continuum approximation can be used and the model system admits kink-solitons propagating along the helical axis. This model was generalized by Xiao et al (1987) who included the longitudinal vibrations (Mei et al, 1981. Kim et al, 1987) of DNA and found these have a weak influence on quasi-solitons properties.

Modeling DNA as a rod, these longitudinal acoustic waves were analyzed (Scott, 1985a,b) in terms of a Boussinesq equation and more recently using the Ostrovskii-Sutin equation (OSE) with a possible additional samping and forcing terms (Muto et al, 1988).

The plane base rotator model was also refined by Zhang (1987). In addition to the inter-strand interactions via the hydrogen bonds, he considered the predominant dipole-dipole interaction, the dipole-induced-dipole interaction, and the dispersion interaction between two bases in a complementary base pair. Again, the dynamics of the rotation of the bases was described by a set of coupled Double Sine-Gordon equations with quasi soliton solutions.

Instead of including the nonlinearity in an effective substrate potential for the base rotations, Krumhansl and Alexander (1983) have suggested the existence of soliton like excitations is associated to the dominant nonlinearity (multistability) of sugar puckering .Krumhansl and coworkers (1985) have made extensive computer simulation on the dynamics of DNA segments up to 200 Base pairs in length, and also study kink interpolating between A and B conformation of DNA.

Banerjee(1983) and Sobell (1984) have proposed a kink-antikink bound state structure which should arise as a consequence of an intrinsic nonlinear ribose inversion inversion instability that results in a modulated β alternation in sugar-puckering along the polymer backbone. Such soliton-antisoliton-like pairs contain β premelted (premeltons) are regions capable of undergoing breathing motions that facilitate drug intercalculation. Moreover, they suggested that β premeltons may provide nucleations centers for RNA polymer and promoter recognition.

The existence of kinks and nonlinear breathing mode associated with collective sugar puckering was also proposed by Takeno and Homma (1987). The potential energy part of their model Hamiltonian was equivalent to a spin Hamitonian (4.2) , without external field ,as described in §4.1

Del Giudice et al (1982) have discussed a model that involves the role of "Davydov like solitons " in the dynamics of DNA. Balanovski et al (1985) also presented a modified Davydov model, where the quasi-soliton formation is mainly due to three effects : the strength of the nonlinear interaction between certain molecular units, the energy release due to adenosine triphosphate (ATP) hydrolysis, the constant presence of an electromagnetic background arising from different components of the cell. Balanovski (1987) discussed the DNA properties and replication in terms of this model.

Very recently, Prohofsky (1988) considered that the nonlinear hydrogen stretch modes (10-120 cm^{-1}) should play a role in quasi-soli ton energy concentration. He used the Davydov model and the NLS equation to speculate on quasi-soliton formation and a mechanism for base pair melting in RNA polymerase transcription.

Two recent models have dealt with the thermalization of DNA. The statistical mechanics of a simple lattice model (Peyrard and Bishop, 1989) was proposed. It consits of two nonlinearly coupled chains. the corresponding nonlinear potential, which corresponds to the H bonds is approximated by a Morse potential. The results show that the denaturation "bubble" which is observed experimentally at the beginning of the denaturation process of DN A could be created by energy localization due to nonlinear effects. Another model reducing the DNA lattice to a Toda lattice was also considered recently by Muto et al (1988), to calculate the thermal equilibrium of solitons in DNA as a function of temperature and the number of base pairs. The results show that the number of thermally generated solitons is significant at physiological temperature.

Concluding remarks.

Thee has been little space in this short review to mention all the connections between the different systems we have considered. However, the following conclusions can be stated :
(i) in the macroworld, quasi-solitons are important just as well for applications, like the transport of information and energy, as for testing grounds of nonlinear wave theory : there is a good agreement between experiments and model equations.
(ii) in the microworld, the experimental signatures of nonlinear excitations are clear for kink-solitons which present topological properties. By contrast, one has no sufficient evidence of non topological quasi-solitons such as pulse, envelope modes and polarons.
(iii) the progress made in the study of the above systems and an interdisciplinary approach will surely benefit in elucidating the fascinating properties of a biopolymer such as DNA which present nonlinear features.

ACKNOWLEDGEMENTS.

I would like to thank A. Fujimaki, A. Matsuda, R. D. Parmentier and N. F. Pedersen for sending me reprints and preprints of their work on Josepshon transmission lines. I am greatful to M. Peyrard for critically reading the manuscript and helpful comments.

References.

CHAPTER 1.

Benjamin, T.B. and Feir, J.F. (1967). The desintegration of wavetrains on deep water, J Fluid Mech, 27,417-430.
Benson, F. A. and Last, D. J . (1965) Nonlinear transmission line harmonic generator. Proc IEEE, 112, 635-643.
Berkhoer, A. L and Zakharov, V. E (1976). Self excitation of waves with different polarizations in nonlinear media, Soviet Phys JETP, 31, 486-489.
Chu,P.L. and Whitebread, T. (1978) Applications of solitons to communication system. Electron Lett, 14, 531-532.
Dodd, R.K. Eilbeck, J.C. Gibbon, J. D. and Morris, H. C. (1984) Solitons and nonlinear wave equations, Academic Press, New York.
Freeman, R. H. and Karbowiack, A. E. (1977) An investigation of nonlinear transmission lines and shock waves. J Phys D : Appl Phys , 10, 633-643.
Fukushima, K. (1983) Modulated wavetrain in a nonlinear transmission line. J Phys Soc Japan, 52, 376-379.
Fukushima, K. Wadati, M. and Narahara, Y. (1980) Envelope soliton in a new nonlinear transmission line. J Phys Soc Japan, 49, 1593-1597.

Fukushima, K. Wadati, M. Kotera, T. Sawada, K. and Narahara, Y. (1980) Experimental and theoretical study of recurrence phenomena in nonlinear transmission line. J Phys Soc Japan, 48,1029-1034.
Gasch, A. Wedding, B. and Jäger, D. (1984) Multistability and soliton modes in nonlinear microwave resonators. Appl Phys Lett, 44, 1105-07.
Hirota, R. and K. Susuki, K. (1973) Theoretical and experimental studies of lattice solitons in nonlinear lumped networks. Proc IEEE , 61, 1483-367.
Hirota, R. and Susuki, K . (1970) Studies of lattice soliton by using electrical networks. J Phys Soc Japan, 28, 1366-1491.
Inoue, Y. (1976). Nonlinear coupling of polarized plasma waves, J Plasma Phys,16, 439-459.
Inoue, Y. (1977). Nonlinear interaction of dispersive waves with equal group velocity, J Phys Soc Japan, 16, 243-249.
Jäger, D. (1978) Soliton propagation along periodic transmission lines. Appl Phys, 16, 35-38.
Jäger, D.(1982) Experiments on KdV solitons. J Phys Soc Japan, 51,1686-1693.
Kako, F. (1979) Propagation of solitons in a dissipative transmission line. J Phys Soc Japan, 47, 1686-1692.
Karpman, V. I. and Maslov. (1978) E. M. Perturbation theory for solitons, Sov Phys JETP, 46, 281-291.
Karpman,V. I. and Krushkal, E.M. (1969). Modulated waves in nonlinear dispersive media. Sov Phys JETP, 28, 277-281.
Kawata, T. Sakai, J. and Inoue, H. (1977) Nonlinear dispersive waves and parametric interaction in the transmission line, 60, 339-346.
Kofane, T. Michaux, B. and Remoissenet, M. (1988) Theoretical and experimental studies of diatomic lattice solitons using an electrical transmission line. J Phys C : Solid State Phys, 21, 1395-1412.
Kolosick, J. A. Landt, D. L. Hsuan, H.C. and Lonngreen, K. E. (1974) Properties of solitary waves as observed on a nonlinear dispersive transmission line. Proc I EEE, 62, 578-581.
Kuusela, T and Hietarinta, J. (1989) Elastic scattering of solitary waves in the strongly dissipative Toda lattice, Phys Rev Lett, 62, 700-703.
Kuusela, T. Hietarinta, J . Kolko, K. and Larbro, R. (1987) Soliton experiments in a nonlinear electrical transmission line. Eur J Phys, 8, 27-33.
Landt, D. L. (1972) An experimental simulation of waves in plasmas. Am J Phys, 40, 1493-1497.
Longreen K. E. (1978) ' Observation of solitons on nonlinear dispersive transmission lines ' in Lonngreen, K. and Scott , A. C. (Eds) Solitons in action. Academic Press, New York, pp 127-152.
Mankankov, V.G. (1978) Dynamics of classical solitons. Physics Reports , 35, 2-128.
Muroya, K. Saitoh, N. and Watanabe, S (1982) Experiment on lattice soliton by nonlinear LC circuit, observation of a dark soliton, J Phys Soc Japan, 51, 1024-1029.

Nagashima, H. and Amagishi, H. (1979) Experiment on solitons in the dissipative Toda lattice using nonlinear transmission lines. J Phys Soc Japan, 47, 2021-2027.

Nagashima, H. and Amagishi, Y. (1978) Experiment on the Toda lattice using nonlinear transmission lines. J Phys Soc Japan, 45, 680-688.

Nejoh, Y. (1985) Envelope soliton of the electron plasma wave in a nonlinear transmission line. Phys Script, 31, 415-418.

Newell, A.C. (1985). Solitons in mathematics and physics, Soc for Ind and Appl Maths, Philadelphia.

Noguchi, A. (1974) Solitons in a nonlinear transmission line. Elec and Commun Japan. 57A, 9-13.

Ostrovskii, L. A and Papko, V. V. (1972) Solitary electromagenetic waves in nonlinear lines. Radiophysics, 15, 438-446.

Paulus, P. Wedding, B. Gasch, A. and Jäger, D. (1984) Bistability and solitons observed in a nonlinear ring resonator. Phys Lett, 102A, 89-92.

Peterson, G. E. (1984) Electrical transmission lines as models forsoliton propagation in materials : elementary aspect of video solitons. AT§TBell Lab Tech J , 63, 901-919.

Sakai, J. and Kawata, T. (1976) Nonlinear wave modulation in the transmission line. J Phys Soc Japan, 41, 1819-1820.

Scott, A. C. (1970) Active and nonlinear wave propagation in electronics. Wiley Interscience, New York.

Spatschek, K. H. (1978). Coupled localized electron-plasma waves and oscillatory ion-acoustic perturbations, Phys Fluids, 21, 1032-1035.

Suzuki, K. Hirota and Yoshikawa, K. (1973)The properties of phase modulated soliton trains. Jap J Appl Phys, 12, 361-365.

Suzuki, K. Hirota, R. and Yoshikawa, K. (1973) Amplitude modulated soliton trains and coding-decoding applications. Int J Electron, 34, 777.

Takagi, K. (1983) The power spectrum of a white noise passed through a nonlinear transmision line. Jpn J Appl Phys. 22,1466.

Tan, M. Su, C.Y. and Anklam, W.J. (1988) 7. electrical pulse compression on an inhomogeneous nonlinear transmission line. Electron Lett, 24, 213-215.

Taniuti, T. and Yajima, N. (1969) Perturbation method for a nonlinear wave modulation. J Math Phys, 10, 1369-.1372

Toda, M (1967) Vibrations of a chain with nonlinear interaction. J Phys Soc Japan, 22,431-436.

Toda, M. (1970) Waves in nonlinear lattice. Prog Theor Phys Japan Suppl. 45, 174-201.

Watanabe, S. (1982) Solitons in nonlinear transmission line. J Phys Soc Japan , 51, 1030-1036.

Watanabe, S. Miyakawa, M. and Muroya, K. (1980) Experiment on recurrence in nonlinear LC circuit. J Phys Soc Japan, 48, 825-831.

Watanabe, S. Miyakawa,M. and Toda, M. (1978) Asymptotic behavior of collisionless shock in nonlinear LC circuit. J Phys Soc Japan, 45, 2030.

Yagi, T and Noguchi, A. (1977) Gyromagnetic nonlinear element and its application as a pulse-shaping transmission line. Electron Letters, 13, 683-685.
Yagi, T. and Noguchi,A. (1976) Experimental studies on modulational instability by using nonlinear transmission lines. Elec and Commun in Japan, 59A, 1-6.
Yazaki, T and Fukushima, K. (1985) Experimental studies of potential problems in quantum mechanics. Am J Phys, 53, 1186-1191.
Yoshinaga, T. and Kakutani, T. (1980) Solitary and shock waves on a coupled transmission line. J Phys Soc Japan, 49, 2072-2074.
Yoshinaga, T. and Kakutani, T. (1984) Second order KDV soliton on a nonlinear transmission line, J Phys Soc Japan, 53, 85-92.
Yoshinaga, T.Sugimoto, N and Kakutani, T. (1981) Nonlinear wave interactions on a discrete transmission line; J. Phys Soc Japan, 50, 2122-2128.

CHAPTER 2.

Barone, A. and Paterno, G. (1982) Physics and application of the Josephson effect . Wiley, New York.
Chen, J. T. Finnegan, T. F. and Langenberg, D. N. (1971) Physica 55, 413.
Costabile, G and Parmentier, R. D. (1975) Analytic solution for fluxon propagation in Josephson junctions with bias and loss, in low Temperature Physics - LT 14, vol. 14, M. Krusius and M. Vuorio Eds, North Holland, Amsterdam, pp. 112-115.
Costabile, G. Parmentier, R. D. Savo, B. Mac Laughlin, D. W and Scott, A. C. (1978) Exact solutions in a long (but finite) Josephson junction. Appl. Phys. Lett. 32, 587-589.
Davidson, A. Ducholm, B. and Pedersen, N. F. (1986) Experiments on soliton motion in annular Josephson Junctions. J. Appl. Phys., 60, 1447-1454.
Feyman, R.P. (1960) Lectures on Physics, Vol. 3 section 21.9. Addison Wesley, New York.
Fujmaki, A. Nakajima, K. and Sawada, Y. (1987) Spatiotemporal observation of the Soliton-Antisoliton collision in Josephson Transmission line. Phys. Rev. Lett., 59, 2895-2198.
Fulton, T. A . Magerlein, J. H. and Dynes, R. C. (1976) A Josephson logic design employing current switching junctions. AS 76, 56.
Harris, R. E. (1974) Cosine and other terms in the Josephson tunneling current. Phys. Rev. B. 10, 84-94.
Josephson, B. D. (1962) Possible new effects in superconductor tunneling.Phys. Lett. 1, 251-253.
Josephson, B. D. (1965) Supercurrents through barriers. Adv. Phys. 14, 419-451. see also (1964) Coupled superconductors. Rev Mod Phys, 36, 216-220.

Levring, O. A. Perdersen, N. F and Samulsen, M. R. (1982) Fluxon motion in long overlap and inline Josephson Junctions Appl. Phys. Lett. 40, 846-847.

Likharev,K .K. (1986) Dynamics of Josephson junctions and circuits. Gordon and Breach, New York.

Lomdahl, P. S. (1985) Solitons in Josephson junctions : an overview, J Stat Phys, 39, 5/6, 551-561.

Lomsdahl, P. S. Soerensen, O. H. and Christiansen, P. L. (1982). Soliton excitations in Josephson tunnel junctions . Phys Rev. B, 25, 5737-5748.

Matsuda, A and Kawakami, T. (1983) Fluxon propagation on a Josephson Transmission line. Phys. Rev. Lett., 51, 695-697.

Matsuda, A. (1986) Observation of fluxon-antifluxon collision in a Josephson transmission line. Phys. Rev. B, 34, 3127-3135.

Matsuda, A. and Uheara, S. (1982) Observation of fluxon propagation on Josephson transmission line. Appl. Phys. Lett., 41, 770-772.

Mc Laughlin, D. W . and Scott, A. C. (1978) Perturbation analysis of fluxon motion. Phys. Rev. A , 18 ,1652-1680.

Nitta, J. Matsuda, A and Kawakami, T. (1984) Propagation properties of fluxons in a well damped Josephson transmission line. 55, 2758-2762.

Pagano,S. (1987) Nonlinear dynamics in long Josephson junctions, PhD Thesis, Technical Univ of Denmark, Lingby, Denmark.

Parmentier (1978) .Fluxons in long Josephson Junctions, in Solitons in action, Eds K. Lonngreen and A. C. Scott, Academic Press, New York, pp 173-199.

Pedersen, N. F and Welner, D. (1984) Comparison between experiments and perturbation theory for solitons in Josephson junctions. Phys. Rev.B, 29, 2551.

Pedersen, N. F. (1989) Nonlinear properties of Josephson Junctions. Proceedings of the ASI Summerschool on superconducting electronics, Ciccio, Italy Plenum .

Scott, A. C . Chu, F. Y. F. and Reible, S. A. (1976) Magnetic flux propagation on a Josephson transmission line. J. Appl. Phys. 47, 3272-3286.

Pedersen, N. F. Samuelsen, M. R. and Welner, D . (1984) Soliton annhilation in the perturbed Sine Gordon system, Phys. Rev. B, 30, 4057-4059.

Scott, A. C. (1964) Distributed device application of the superconducting tunnel junction. Solid State Elec 7, 137-146.

Scott, A. C. (1969) A nonlinear Klein-Gordon equation, Am J Phys, 37,52-61.

Swihart, J. C. (1961). Field solution for a thin film superconducting stripline. J. Appl. Phys. 32 461-469.

CHAPTER 3.

Agrawal, G. P. Baldeck, P. L. and Alfano, R. R. (1989) Modulational instability induced by cross-phase modulation in optical fibers. Phys Rev A, 39, 3406-3413.

Beaud,P. Hodel, W. Zysset, B. and Weber,H.P (1987) Ultrashort pulse propagation, pulse breakup and fundamental soliton formation in a single mode optical fiber. IEEE J. Quant. Electron, QE.23,1938-1946.

Blow, J. and Doran, N. J.(1987) Nonlinear effects in optical fibres and fibre devices. IEEE Proc, 134, 138-144.

Blow,K. J. and Doran, N. J. (1985) The asymptotic dispersion of soliton pulses in lossy fibres. Opt Commun, 52, 367-370.

Bourkoff, E. Zhao, W. R.I. Joseph, R. I and Christotoulides, D. N. (1987) Evolution of femtosecond pulses in single mode fibers having higher order nonlinearity and dispersion. Opt. Lett., 12, 272-274.

Cotter, D. (1982) Observation of stimulated Brillouin scattering in low loss silica fiber at 1.3μ m. Electron Lett., 18, 495-496.

Cotter, D. (1982) Transient stimulated Brillouin scattering in long single mode fibers. Electron Lett., 18, 504-506.

Cristodoulides, D. N. and Joseph,R. I. (1985)Femtosecond solitary waves in optical fibers, beyond the slowly varying envelope approximation. Appl Phys Lett, 47, 76-78.

Doran, N. J and Blow, K. J. (1983) . Solitons in optical communications. IEEE J Qant Electron. QE-19, 1883-1888.

Firth, W. Peyghambarian, N. and Tallet, A. (1988) Eds, Proceedings of the Int Conf on optical bistability IV, Aussois, France, Part IV, J Phys, Coll C 2, Suppl 6, 49, C-277-342.

Gloge, D. (1979) The optical fiber as a transmission medium. Rep Prog Phys, 42, 1777-1824.

Hasegawa, A. and Brinkman, W. F.(1980) Tunable coherent IR and FIR sources utilizing modulational instability. IEEE J. Quant Electron, QE-16, 694-697.

Hasegawa, A.(1983) Amplification and reshaping of optical soliton in a glass fiber. 4 Use of stimulated Raman process. Opt. Lett. 8, 650.

Hasegawa, A. (1984) Generation of a train of soliton pulses by induced modulational instability in optical fibers. Opt. Lett, 9, 288290.

Hasegawa,A. and Kodama, Y. (1981) Signal transmission by optical solitons in monomode fiber. Proc. IEEE, 69, 1145-1150.

Hasegawa,A. and Tappert, F. (1973a) Transmission of stationnary nonlinear optical pulses in dispersive dielectric fibers-1. Anomalous dispersion. Appl. Phys. Lett, 23, 142-144.

Hasegawa,A. and Tappert, F. (1973b) Transmission of stationary linear optical pulses in dispersive dielectric fibers-2. Normal dispersion Appl. Phys. Lett. 23, 146-149.

Jain, M. and Tzoar, N. (1978) Propagation of nonlinear optical pulses in inhomogeneous media. J. Appl Phys, 49, 4649-4654. ibid Nonlinear pulse propagation in optical fibers. Opt Lett , 3, 202-204.

Jain,M. and Tzoar, N.(1987) Nonlinear pulse propagation in a monomode dielectric guide. IEEE J Qant Elec, QE-23, 510.

Kaminov, I. P. (1981) Polarization in optical fibers. IEEE Qant Elec, QE-17, 15-22.

Karpman, V. I. (1975), Nonlinear wave in dispersive media. Pergamon Press, New York.

Kodama, Y. (1985) Optical solitons in a monomode fiber. J Phys Stat, 39, 5/6, 597-614.

Kodama, Y. and Ablowitz, M. J. (1980) Perturbations of solitons and solitary waves. Stud. Appl. Math, 64, 225-245.

Kodama, Y. and Hasegawa, A. (1987) Nonlinear pulse propagation in a monomode dielectric guide. IEEE J Quant Elec. QE - 23, 510-524.

Krökel,D. Halas, N. J. Gianlani, G and Grischowsky,D. (1988) dark pulse propagation in optical fibers. Phys Rev Lett, **60**, 29-32.

Menyuk, C.R. (1987a) Nonlinear pulse propagation in birefringent optical fibers. IEEE J. Quant. Electron, QE-23, 174-176.

Menyuk,C. R. (1987b) Stability of solitons in birefringent optical fibers. I. equal propagation amplitudes. Opt Lett, 12, 614-616

Menyuk,C. R. (1988) Stability of solitons in birefringent optical fibers. II Arbitrary amplitudes. Opt Lett., 5, 392-402.

Mitscke, F. M. and Mollenauer, L. F. (1986) Discovery of the soliton self frequency shift. Opt. Lett, 11, 659-661.

Mollenauer and K. Smith. (1988) Demonstration of soliton transmission over more than 4000 km in fiber with loss periodically compensated bh Raman gain. Opt. Lett., 13, 675-677.

Mollenauer,L. F. and Stolen, R. H. (1984) The soliton laser. Opt lett. 9, 13.

Mollenauer,L. F. and Stolen, R.H. (1982) Solitons in optical fibers. Fiberoptic Technology , April, 193-198.

Mollenauer,L. F. Stolen, R. H . and Islam, M. N. (1985) Experimental demonstration of soliton propagation in long films : loss compensated by Raman gain. Opt. Lett. 10, 229-231.

Mollenauer,L. F. Stolen, R.H. and Gordon, J. P . (1980) Experimental observation of picosecond pulse narrowing and solitons in optical fibers,Phys. Rev. Lett, 45, 1095-1098,.

Mollenauer,L. F. Stolen, R.H. and Gordon, J. P. and Tomlinson, W.J.(1983) Extreme picosecond pulse narrowing by means of soliton effect in single mode optical fibers. Optics Letter, 8, 289-291.

Salin, F. Grangier,P. Roger, G and Brun, A. (1986) Observation of high order solitons produced by a picosecond ring laser. Phys. Rev. Lett. 56, 1132-1135.

Satsuma, J. and Yajima, S. (1974) Initial value problems of one dimensional self modulation of nonlinear waves in dispersive media. Prog. Theor. Phys. Suppl., 55, 284-306.

Stegeman, G. I. and Stolen, R. H. (1988) Eds . Nonlinear guided wave phenomena in Optical Physics ,Special issue, J Opt Soc Am B, 5.

Stolen, R. H. Mollenauer, L. F and Tomlinson, W. J. (1983) Observation of pulse restoration at the soliton period in optical fibers Opt. Lett. 8, 186-188 .

Tai,K. Hasegawa, A. and Tomita, A. (1986) Observation of modulation instability in optical fibers. Phys. Rev. Lett., 56, 135-138.

Taniuti, T. (1974) Reductive perturbative method and far fields of wave equations. Phys. Theor. Phys. (Japan), Suppl. 55, 1.
Tratruk, M. V. and Sipe, J. E. (1988) Bound solitary waves in a birefringent optical fiber. Phys. Rev.A 38, 2011-2017.
Wabnitz, C.R. (1988) Modulational polarization instability of light in a nonlinear birefringent dispersive medium. Phys. Rev. A, 38, 2018-2021.
Wai, P. K. Menyuk, C. R. Chen, H and Lee Y.C . (1986) Nonlinear pulse propagation in the neighborhood of the zero-dispersion wavelength of monomode optical fibers. Opt. Lett, 11 , 464-466.
Wai, P. K. Menyuk, C. R. Chen, H and Lee Y.C . (1987) Solitons at the zero dispersion wavelength of a single mode fiber. Opt Lett, 12, 628-630.
Zakharov, V. E. and Shabat, A. B. (1972) Exact theory of two dimensional self focusing and one dimensional self modulation of waves in nonlinear media. Sov. Phys. JETP, 34,62-69.
Zhakarov, V. E. and Shabat, A. B. (1973) Interaction between solitons in a stable medium. Sov. Phys. JETP, 37, 823-828.
Zhao, W and Bourkoff, E. (1988) Femtosecond pulse propagation in optical fibers : higher order effects. IEEE J. Quant Electron., 24, 365-372.

CHAPTER 4.

Balakrishnan, R. and Bishop, A. R. (1985)Nonlinear Excitations on a quantum ferromagnetic chain. Phys Rev Lett, 55, 537-540.
Balucani, U. Lovesey, S. W. . Rasetti, M. G. and Tognetti, V. Eds. (1984) Magnetic excitations and fluctuations . Springer Proceedings in Physics . Springer Verlag_Berlin.
Balucani, U. Lovesey, S. W. . Rasetti, M. G. and Tognetti, V. Eds. (1987) Magnetic excitations and fluctuations II. Springer Proceedings in Physics 23. Springer Verlag.Berlin.
Bhakta, J. C. (1987) A pair of coupled equations for high frequency Langmuir and dispersive ion- acoustic waves with collisional damping. Plasma Phys and Controlled Fusion. 29, 245-255.
Birgeneau, R. J. and Shirane, G (1978). Magnetism in one dimension. Physics Today, 32, Dec.
Borsa, F. Pini, M. G. Rettori, A. and Tognetti, V. (1983) Magnetic specific heat contributions from linear vis à vis nonlinear excitations in the one dimensional antiferromagnet TMMC. Phys Rev B, 28, 5173-5183.
Boucher, J. P. (1980)Solid State Commun, 33, 1025.
Boucher, J. P. (1989). Nonlinear excitations in antiferromagnetic chains. To be published in Proceedings of Nuclear in magnetism. Munich, August 1988.
Boucher, J. P. and Renard, J. P. (1980) Nuclear spin lattice relaxation by solitons in the antiferromagnetic chains $(CH_3)_4$ $NMnCl_3$.Phys Rev Lett, 45, 486-489.
Boucher, J. P. Pynn, R. Remoissenet, M. Regnault, L. P Endoh, Y. and Renard, J. P. (1989). New double-magnons modes in planar

antiferromagnets : a new study by polarised-neutron, inelastic scattering of TMMC in a transverse field, to be published.
Boucher, J. P. Regnault, L. P. and Benner, H. (1987) Soliton dynamics : experiments on magnetic chains, p 24 In Nonlinearity in condensed matter. Eds A.R. Bishop, D.K. Campbell, P. Kumar and S.E. Trullinger. Springer, Berlin.
Boucher, J. P. Regnault, L. P. Rossat Mignot, J. and Henri, Y. (1984) in " Magnetic Excitations and Fluctuations" . Eds F. Lovesey, U. Balucani, F. Borsa and V. Tognetti. Springer, Berlin.
Boucher, J. P. Regnault, L. P. Rossat Mignot, J. Renard, J.P . Bouillot, J and Stirling, W.G. (1981) J Appl Phys, 52, 1956-1960.
Cieplak, M. and Turski, L. A. (1980) Solitons in quantum Heisenberg chain. J Phys C, 13, 5741-5747.
Corones, J . (1977) Solitons as nonlinear magnons. Phys Rev B,16 , 1763-1764.
Cowley, R. A. . Buyers, W. L. Martel, P. andStevenson, R. W. (1969) Two magnon scattering of neutrons. Phys Rev Lett, **23**, 86-89.
De Gronckel, H. A. De Jonge, W. J. Kopinga, K. and Lemmens, L. F. (1988) Thermal conductivity of some soliton-bearing magnetic systems. Phys Rev B, 37, 9915-9918.
De Groot, H. J. M. De Jonge, L. J. Elmassalami, M. Schmitt. H.H. and Thiel, R. C. (1986) Mössbauer relaxation studies of nonlinear dynamics excitations in low-dimensional magnets. Hyperfine Interactions, 27, 93, .
Fogedby, H. C. (1980) Solitons and magnons in the classical Heisenberg chain. J Phys A, 13, 1467-1499.
Gouvea, M. E andPires, A, S. (1986) Nonlinear excitations in the classical one-dimensional antiferromagnet. Phys Rev B, 34, 306-317.
Izyumov, Y. A. (1989) Solitons in quasi one dimensional magnetic materials and their study by neutron scattering. Sov Phys Usp, to be published.
Kopinga, K. and De Jonge, W. J. (1987) Linear and nonlinear excitations in the S=1/2 ferromagnetic chain system [$C_6H_{11}NH_3$] $CuBr_3$ (CHAB), p 167, in Magnetic excitations and fluctuations II. U. Balucani, S.W. Lovesey, M.G. Rasetti and V. Tognetti Eds. Springer Proceedings in Physics 23. Springer Verlag, Berlin.
Kakurai, K. Steiner, M. Pynn, R and Dorner, B. (1986) Study of the linear and nonlinear excitations in $CsNiF_3$, by means of polarized neutron scattering. J Mag and Mag Mat, 54-57, 835-836.
Kumar, P. (1982) Soliton instability in one-dimensional magnet. Phys. Rev. B 25, 483-486.
Laksmanan, M. (1977) Heisenberg continuum system as an exactly solvable dynamical system. Phys Lett, **53**, 53-54.
Lindgârd, P. A. (1984) , p 163, in Condensed Matter research using neutrons. Eds S. W. Lovesey and R. Scherm. Nato ASI series. Plenum, New York.
Magyari, E and Thomas, H. (1982) Kink instability in planar ferromagnets. Phys. Rev. B, 25, 531-533.

Makankov, V. G. and Fedyanin ,V. K. (1984) Nonlinear effects in quasi-one-dimensional models of condensed matter theory. Phys Rep, 104, 1-86.
Maki, K. (1980) Quantum statistics of solitons, p 63 in Physics in one dimension. Eds Bernasconi, J. and Schneider, T. Springer, Berlin.
Maki, K.(1981) Quantum effects in quasi-one dimensional magnetic systems. Phys Rev B, 24, 3991-332.
Mikeska, H. J. (1981) Solitons in one dimensional ferromagnets. J. Appl. Phys., 52, 1950-1955 .
Mikeska, H. J. (1982) Soliton energy in an easy plane quantum spin chain. Phys Rev B, 26, 5213-5222.
Mikeska,H.J. (1978) Solitons in one-dimensional magnet whith an easy plane. J. Phys. C11, L29-L32 .
Mikeska,H.J. (1980) Nonlinear dynamics of classical one dimensional ferromagnets. J. Phys. C, 13, 2913-2923.
Ramirez, A. P. and Wolf, W, P. (1982) Specific of $CsNiF_3$: evidence for spin solitons. Phys Rev Lett , 49, 227-229.
Regnault, L. P. Boucher, J. P. Rossat Mignot, J.Renard, J. P. Bouillot, J. and Stirling, W. G. (1982) A neutron investigation of the soliton regime in the one-dimensional planar antiferromagnet (CD_3) $NMnCl_3$. J Phys C, 15, 1261-1282.
Remoissenet, M. (1986) . Low amplitude breather and envelope solitons in quasi-one-dimensional physical models. Phys Rev B, 33, 2386-2392.
Remoissenet, M. (1989) Real lattices modelled by the nonlinear Schrödinger equation and its generalization, in Proceedings of Workshop "Integrable systems and applications", Oléron, France June 1988. To be published in Lecture Notes in Mathematics (or Physics). Springer, Berlin.
Sahadevan, R. Tamizhmani, K. M. and Lakshmanan, M. (1986) Painlevé analysis and integrability of coumled nonlinear Schrödinger equations. J Phys A: Math Gen, 19, 1783-1791.
Steiner, M. Kakurai, K. and Kjems, J. K. (1983) Experimental studies of the spin dynamics in the 1D ferromagnet with planar anisotropy, $CsNiF_3$, in an external magnetic field J. Magn Mat Mater, 15-18, 1057.
Steiner, M. and Kjems, J. K. (1978) Solitons in $CsNiF_3$: their experimental evidence and their thermodynamics, p 191 in solitons and condensed matter physics. Eds Bishop, A. R. and Schneider J. Springer, Berlin.
Tjon, J. and Wright, J. (1977) Solitons in the continuous Heisenberg chain. Phys Rev B, 15, 3470-3476.
Villain, J. (1975) Physica. Propagative spin relaxation in the Ising-like antiferromagnetic linear chain. 79B,1-12.
Wysin, G and Kumar, P. (1987) Thermomagnetic transport coefficients: solitons in an easy plane magnetic chain. Phys Rev B, **13**, 7063-7070.
Wysin, G. Bishop, A. R and Kumar, P. (1982) Solitons dynamics on a ferromagnetic chain. J. Phys. C, 15, L337-L344.
Wysin, G. Bishop, A. R and Kumar, P. (1984) Soliton dynamics on an easy plane ferromagnetic chain J. PHys. C, 17, 5975-5992.

Wysin, G. Bishop, A. R and Oitmaa, A. R. (1986) Single kink dynamics in an easy plane classical antiferromagnetic chain. J Phys C, 19, 221-235.
Zhakarov, V. E. and Schulman, E.I. (1982)To the integrability of the system of two coupled nonlinear Schrödinger squations. Physica 4D, 270-274.

CHAPTER 5.

Balanovski, E. (1987). The physics of DNA : onset of sloliton-like excitations, chain relative disorder, and basis for a stastical mechanics of the macromolecule. Int J Theor Phys, 26, 49-61.
Balanowski, E. and Beaconsfield, P. (1985). Solitonlike excitations in biological systems. Phys. Rev. A, 32, 3059-3064.
Banerjee, A. and Sobell, H. M. (1983). Presence of nonlinear excitations in DNA structure and their relationnship to DNA permelting and to drug intercalation. J. Biomol struct. Dyn, 1, 253-262.
Barthès, M. (1989) Optical anomalies in acetanalide Davydov solitons, localised modes or Fermi resonance ? To be published in J Mol Liq, special issue.
Baverstock, K. F. andCundall, R. B. (1988). Solitons and energy transfer in DNA. Nature, 322, March 24, 312-313.
Careri, G. Buontempo, U. Galluzi, F. Gratton, E. and Scott, A.C. (1984). Spectroscopic evidence of Davydov like solitons in Acetanilide. Phys. Rev. B, 30, 4689-4703.
Cottingham, J. P. and Scheiwtzer, J. W. (1989) Calculation of the lifetime of a Davydov soliton at finite temperature. Phys Rev Lett, 62, 1792-1795.
Davydov, A. S. (1985). Solitons in Molecular Systems, Reidel, Dordrecht.
Del Giudice, Doglia, S. andMilani, M. (1982) A collective dynamics in metabolically active cells. Phys Script, 26, 232-238.
Dickerson, R. E. (1983). The DNA helix and how it is read. Sci Amer, December, 87-104.
Englander, S.W. Kallenbach, N. R. Heeger, A. J. Krumhansl, J. A. and Litwin, S. (1980) Nature of the open state in long polynucleotide double helixes : possibility of solitons excitations, Proc Nat Acad Sci USA, 777, 7222-7226.
Frank-Kamenitskii, M. D. (1985) Fluctational motility of DNA, p 47 in Structure and motions : menbranes, nucleic acids and proteins. Ed E. Clementi, G; Corongiu , M. H. Sarma & R. H. Sarma. Adenine Press, New York.
Freifelder, D. (1987). Molecular biology. Jones and Bartlett, Publishers, Boston.
Friedland, P. and Kedes, L. H. (1985) Discovering the secrets of DNA. Commun of ACM, 28, 1164-1186.
Fröhlich, H (1969) , Quantum mechanical concepts in biology, p 13 in Theoretical Physics and Biology, Eds Marois, North Holland, Amsterdam.
Fröhlich, H. (1968). Long range coherence and energy storage in biological systems. Int J Quant Chem, II, 641-649.

Heeger, A. J, Kivelson, S, Schrieffer J. R. and Su, W. P. (1988) Solitons in conducting polymers. Rev Mod Phys, 60,781-850.
Homma,S and Takeno, S. (1984). A coupled base-rotator model for structure and dynamics of DNA, Prog Theor Phys, 72, 679-693.
Hyman, J. M. Mc Laughlin,D.W. and Scott, A. C. (1981). On Davydov's α-Helix solitons. Physica D, 3D, 23.
Kim, Y. and Prohofsky, E. W. (1987). Vibrational modes of a DNA polymer at low temperature. Phys. Rev. B, 36, 3449-3451.
Krumhansl, J. A. and Alexander,D. M. (1983). Nonlinear dynamics and conformal excitations in biomolecular materials. In structure and Dynamics : Nuclear acids and proteins. Ed. E. Clemente and R. H Sarma. Adenine Press, New York.
Krumhansl, J. A. Wysin, G.M. Alexander, D. M.Garcia, A. Lomdahl, P.S. Scott P. Layne (1985). Further theoretical studies of (nonlinear) conformational motions in double-helix DNA, 407-415, in Structure and motions : menbranes, nucleic acids and proteins. Ed E. Clementi, G; Corongiu , M. H. Sarma & R. H. Sarma. Adenine Press, New York.
Lewitt, M. (1982). Computer Simulation of DNA Double-helix dynamics Cold Spring Harbor Symp Quant Biol, 46A, 251-262.
Lilley, D. M. (1988). DNA opens up, supercoiling and heavy breathing, TIG, 4, 111-121.
Lomdahl, P.S. Layne, S.P. and Bigio, I. J. Solitons in biology (1985). Los Alamos Sciences, Spring Issue, 4-21.
Lomdahl, P.S. Mac Neil, L. Scott, A. C. Stoneham, M. E. and Webb, S.J. (1982). An assignment to internal soliton vibrations of Laser Raman from living cells. Phys. Lett. 92 A, 207-210.
Mei, W. N. Kohli, M.Prohofsky, E. W. and Van Zandt, L.L. (1981) Acoustic modes and nonbounded interactions of the double helix. Biopolymers, 20, 833-852.
Muto, V. Halding, J. Christiansen P. L. and Scott, A. C. (1988) Solitons in DNA. J Biomol Struct and Dyn, 5, 873-874.
Muto, V. Scott, A. C. and Christiansen, P. L. (1989). Thermally generated solitons in DNA. Preprint
Peyrard, M. and Bishop, A. R. (1989) Statistical mechanics of a nonlinear model for DNA denaturation. Preprint.
Prohofsky, E. W. (1988) Solitons hiding in DNA and their possible significance in RNA transcription. Phys Rev B, 38, 1538-1554.
Saenger, W. (1984). Principles of nucleic acid structure, Springer, New York.
Scott, A. C. (1982a). Dynamics of Davydov solitons. Phys. Rev. B, 26, 578-595.
Scott, A. C. (1982b)The vibrational structure of Davydov soliton. Phys. Script. 35,651-672.
Scott, A. C. (1985a). Solitons in biological molecules. Comments Moll. Cell. Biophys., 3, 15-37.

Scott, A. C. (1985b). Soliton oscillations in DNA, Phys Rev A, 31, 3518-3519.

Scott, A.C.(1985c). Anharmonic analysis of resonant microwave absorption in DNA, Phys Script, 36, 617-638.

Sobell, H. M. (1984). Kink - antikink bound states in DNA structure, p 172 in Structure of biological molecules and assemblies. Vol. II. F. Jurnak and A. Mc Pherson eds, Wiley, New York.

Szent-Giörgyi, A. (1941) The study of energy levels in biochemistry. Nature, 148, 157-159.

Takeno, S. and Homma, S. (1987). Kinks and breathers associated with collective sugar puckering in DNA. Prog. Theor. Phys. 77, 548-562.

Tuszynski, J. A. Paul, R. Chatterjee, R. and Sreenivasan,S.R. (1984). Relationship between Fröhlich an Davydov models of biological order, Phys Rev A, 30, 2666-2675.

Wang, X. Brown, D. W. and Linenberg, K. (1989) Quantum Monte Carlo simulation of the Davydov model. Phys Rev Lett, 1796-1799.

Xiao, J. Lin,J. and Zhang, G. (1987). The influence of longitudinal vibration on soliton excitation in DNA double helices, J Phys A, 20, 2425-2432.

Yomosa, S. (1983). Soliton excitations in deoxyribonucleic acid (DNA) double helices. Phys. Rev. A, 27, 2120-2125.

Yomosa, S. (1984). Solitary excitations in deoxyribonucleic acid (DNA) double helices.Phys. Rev. A, 30, 474-480.

Zhakarov, V.E. (1972). Collapse of Langmuir waves. Sov. Phys. JETP, 72, 908-919.

Zhang, G. (1987) Soliton excitations in deoxyribonucleic acid (DNA) double helices. Phys. Rev. II , 35, 886-891.

Ziman, J. Electrons and phonons : the theory of transport phenomena in Solids. Clarendon, Oxford, 1980.

THE INITIAL-BOUNDARY VALUE PROBLEM FOR THE DAVEY-STEWARTSON 1 EQUATION; HOW TO GENERATE AND DRIVE LOCALIZED COHERENT STRUCTURES IN MULTIDIMENSIONS*

P.M.Santini

Dipartimento di Fisica, Università "La Sapienza", Roma, Italy
I.N.F.N., Sezione di Roma

A.S.Fokas

Department of Mathematics and Computer Science
Clarkson University, Potsdam, N.Y. 13676, USA

ABSTRACT

The inverse scattering transform method is used to linearize the initial - boundary value problem for the Davey- Stewartson 1 equation establishing the genericity, the spectral interpretation and the significantly novel properties of the associated localized coherent structures which we call dromions. In contrast to one-dimensional solitons, generated by the balance of nonlinearity and dispersion, dromions originate from the balance between the focusing effect of the boundary flows and dispersion. Upon interaction dromions not only exhibit a two-dimensional phase shift, but also a change of form and an exchange energy. Furthermore they can be driven everywhere in the plane choosing a suitable motion of the boundaries. However, if this motion is not uniform, energy is lost through radiation.

* This article consists of expanded material of the lectures presented by P. M. Santini.

R. Conte and N. Boccara (eds.), Partially Integrable Evolution Equations in Physics, 223–259.

1. INTRODUCTION

1.1 The Davey-Stewartson Equation as a Universal Model

The Davey-Stewartson (DS) equation [1],[2]

$$i\psi_{1t} + \frac{1}{2}(\psi_{1xx} + \nu\psi_{1yy}) - (\varphi_x - \epsilon|\psi_1|^2)\psi_1 = 0, \quad \varphi_{xx} - \nu\varphi_{yy} - 2\epsilon|\psi_1|^2_x = 0. \quad (1)$$

is a universal model equation; indeed it can be obtained via an asymptotically exact technique of broad applicability from large classes of nonlinear evolution equations (NEE's) in 2+1 dimensions [3],[4], whose linearized versions admit a plane wave solution

$$c\, e^{i(k_1 x' + k_2 y' - \omega(k_1,k_2)t')} \quad (2)$$

(where c, k_1, k_2 are arbitrary constants).

In order to investigate solutions that are modulations over large time and space intervals of the basic plane wave (2), one introduces the following Fourier decomposition

$$u(x,y,t) = \epsilon^{1+r}\psi_0(x,y,t) + (\sum_{n=1} \epsilon^n \psi_n(x,y,t)e^{inz} + c.c.) \quad (3)$$

(which takes account of the appearence of higher harmonics due to the nonlinearity), where

$$x = \epsilon^{p_1}(x' - v_1 t'),\ y = \epsilon^{p_2}(y' - v_2 t'),\ t = \epsilon^q t',\ z = k_1 x' + k_2 y' - \omega(k_1,k_2)t' \quad (4)$$

and (v_1, v_2) is the group velocity of the plane wave, namely $v_1 = \frac{\partial\omega}{\partial k_1}$, $v_2 = \frac{\partial\omega}{\partial k_2}$. Then, focusing the attention on the evolution of the modulation ψ_1 of the leading term of the expansion (3) (the main Fourier mode), one obtains generically (and after a suitable rescaling) equation (1).

Equation (1) for $\nu = -1$ was first derived in fluid dynamics [1] as the shallow water limit of the Benney-Roskes equation [5] and describes the evolution of waves of slowly varying amplitude on a two-dimensional water surface under gravity (ψ_1 is the amplitude of a surface wave packet and φ is the velocity potential which describes the mean motion generated by this surface). If the effect of surface tension dominates gravity, then one obtains equation (1) but now with $\nu = 1$ [2]; for $\nu = 1$ and $\nu = -1$ equation (1) is now called DS1 and DS2 respectively; $\epsilon = 1$ and $\epsilon = -1$ characterize defocusing and focusing regimes for both equations.

The universality of equation (1) explains the large applicability of DS which arises not only in water waves, but also in plasma physics and nonlinear optics. The universality of equation (1) explains also why the integrability of DS is not a coincidence; indeed it is sufficient for it that in the large universe of NEE's in 2+1 dimensions reducing to equation (1) there exists one integrable model, since the exact limit procedure described above preserves integrability [4].

1.2 The Davey-Stewartson 1 Equation and Multidimensional Coherent Structures; the Notion of Dromion

In the following we shall concentrate our attention on DS1. We find it convenient to introduce characteristic coordinates $\xi = x + y$, $\eta = x - y$, $q(\xi,\eta,t) = \psi_1(x,y,t)$, and $U_1 = -\varphi_\eta + \frac{\epsilon}{2}|q|^2$, $U_2 = -\varphi_\xi + \frac{\epsilon}{2}|q|^2$; then DS1 reduces to

$$iq_t + q_{\xi\xi} + q_{\eta\eta} + (U_1 + U_2)q = 0, \tag{5a}$$

$$U_{1\xi} = -\frac{\epsilon}{2}(|q|^2)_\eta, \quad U_{2\eta} = -\frac{\epsilon}{2}(|q|^2)_\xi, \quad \epsilon = \pm 1. \tag{5b}$$

DS1 is the compatibility condition between the following linear equations [6] (Lax pair)

$$\begin{pmatrix} \partial_\xi & q/2 \\ \epsilon q^*/2 & \partial_\eta \end{pmatrix} V = 0, \tag{6a}$$

$$V_t = \hat{P}V, \tag{6.b}$$

$$\hat{P} := i\begin{pmatrix} 1 & 0 \\ 0 & -1 \end{pmatrix}(\partial_\xi - \partial_\eta)^2 - i\begin{pmatrix} 0 & q \\ \epsilon q^* & 0 \end{pmatrix}(\partial_\xi - \partial_\eta) + \begin{pmatrix} iU_1 & -iq_\eta \\ i\epsilon q^*_\xi & -iU_2 \end{pmatrix}, \tag{6b}$$

for the 2x2 matrix eigenfunction V.

Spatially confined solutions of equation (5) were constructed in [7] using a Gel'fand-Levitan-Marchenko (GLM) equation (see also §6). A GLM equation for DS2 and line solitons were constructed in [24]. An initial value problem for the DS1 equation (5) was investigated in [8] using the inverse scattering (or spectral) transform (IST) method (see also [9] and [10]); it has been recently shown in [11] that the case studied in [8] corresponds to $U(\xi,-\infty,t) = U(-\infty,\eta,t) = 0$. In this particular situation arbitrary initial data $q(\xi,\eta,0)$ disperse away. In this particular situation an existence theorem which does not make use of IST has recently been proved [12].

We were motivated to reexamine DS1 and its associated IST formalism because of the important discovery of Boiti et al. [13] that DS1 and its associated class of nonlinear equations admit exponentially localized solitons. The above authors used Bäcklund transformations (BT) to obtain these solitons and investigated the dynamics of a certain two-soliton solution obtained using a nonlinear superposition formula, showing that the only effect of the interaction of the two solitons is a two-dimensional phase-shift.

The spectral interpretation of multidimensional coherent structures, their genericity and several novel properties of these localized coherent structures have been recently established by the authors [14], [15], [16]. In these lecture notes we illustrate the results contained in [14],[15],[16] which can be summarized in the following points.

i) In contrast to integrable equations in 1+1 dimensions like the celebrated Nonlinear Schroedinger (NLS) equation

$$iq_t + q_{xx} - \epsilon|q|^2 q = 0 \tag{7}$$

which is integrated as an initial value problem [17], the integration of DS1 corresponds to the solution of an initial- boundary value problem in which the complex initial condition $q_0(\xi,\eta) := q(\xi,\eta,0)$ and the real boundary conditions $u_2(\xi,t) := U_2(\xi,-\infty,t), u_1(\eta,t) := U_1(-\infty,\eta,t)$ are given, are suitably decaying for large values of ξ,η and are bounded in t [14],[15]. Once the boundary conditions u_2, u_1 are given, then equations (5) are equivalent to

$$iq_t+q_{\xi\xi}+q_{\eta\eta}+[u_1(\eta,t)+u_2(\xi,t)-\frac{\epsilon}{2}\int_{-\infty}^{\xi} d\xi'(|q|^2)_\eta-\frac{\epsilon}{2}\int_{-\infty}^{\eta} d\eta'(|q|^2)_\xi]q = 0, \tag{8a}$$

$$q(\xi,\eta,0) = q_0(\xi,\eta); \qquad q(\xi,\eta,t) \to 0, \ |\xi|,|\eta| \to \infty. \tag{8b}$$

It turns out that the IST formalism introduced in [8] can be generalized to solve this initial-boundary value problem for DS1 [14]; rigorous aspects of the IST formalism for the spectral problem

$$\begin{pmatrix} \partial_\xi & q/2 \\ r/2 & \partial_\eta \end{pmatrix} V = 0$$

in the general case (r and q arbitrary) and in the DS1 reduction $r = \epsilon q^*$ are given in [15]. An IST alternative to [14],[15] has recently been proposed in [18].

ii) The initial condition q_0 and the boundary conditions u_2, u_1 play a substantially different role in this IST formalism and, consequently, in the evolutive process associated with DS1. In contrast to 1+1 dimensional equations like (7), for which an arbitrary initial condition decomposes into a number of solitons (due to a balance between nonlinearity and dispersion), for DS1 arbitrary nonzero boundary flows u_2, u_1 cause a transfering of energy from the mean flow to the surface waves where focusing effects balancing dispersion can be created. The mathematical manifestation of this phenomenon is the fact that DS1 admits localized solutions which decay exponentially in both ξ and η [14],[15]. The basic mechanism for generating these localized solutions of DS1 is due to the boundary conditions u_2, u_1, while the initial condition q_0 plays a secondary role, affecting the amplitudes only. In particular, if $u_2 = u_1 = 0$, dispersion prevails in this two dimensional model and any initial condition (even of solitonic type!) disperses away [14], [15].

iii) From a mathematical point of view these localized solutions are associated with the two - dimensional discrete spectrum of the time-dependent Schroedinger operator

$$L := i\partial_t + \partial_\xi^2 + \partial_\eta^2 + u_2(\xi,t) + u_1(\eta,t). \tag{8c}$$

The notion of (N,M) dromion solution, which corresponds to a purely discrete spectrum of operator (8c), is then introduced [14] and the exponentially localized solitons constructed in [13] using BT are recovered for a special choice of the spectral parameters [14],[15]. Since equation $L\psi = 0$ (with L defined in (8c)) is the linearized version of the DS1 equation (8a), then these localized coherent structures are **nonlinear distorsions of the linear modes** associated with the linearized DS1 equation.

iv) In two space dimensions the presence of nontrivial boundary conditions makes it possible for (8a,b) to exhibit regular localized solutions even in the defocusing case $\epsilon = 1$ [15], [16]; this is not allowed for the one-dimensional analogue (7) of DS1 .

v) In the process of interaction multidimensional localized structures not only exhibit a **two-dimensional phase shift**, but also a **change of form**; such a change of form is associated with an **exchange of energy** [16].Only for a special choice of the spectral parameters this change of form and the associated exchange of energy do not take place. This is one of the novelties of localized coherent structures in multidimensions; while one-dimensional solitons regain their initial amplitude and velocity (and then their identity) after the interaction [19], two-dimensional localized structures exchange energy and do not regain the same form after the interaction. Although they change form, they can still be identified, since their velocity is conserved [16].

It is also interesting to remark that the above coherent structures, although exponentially localized in the plane, **are not solitary**, since they are connected through the mean flow (U_1, U_2), which describes a traveling rectangular network in the (ξ, η) plane, whose nodes coincide with the positions of the localized coherent structures [16].

vi) Localized coherent structures of DS1 can be **driven** everywhere in the (ξ, η) plane choosing a suitable motion of the boundaries u_2, u_1. The price payed for this important freedom (which is absent in one dimension) is that **driven coherent structures radiate energy** if their motion is not uniform [15].

vii) the (N,M) dromion solution is not the only example of explicit, exponentially localized solution of DS1 and, in particular, explicit weak solutions (distributions) are presented. An example of breather-like solution associated with nonlocalized boundaries is also given [16].

We feel that the well-known notion of **soliton** is not adeguate to capture the significantly novel features associated with coherent structures in multidimensions and summarized in i)-vii). We suggest the new name **dromion** to emphasize what are, perhaps, the most significant novel properties of multidimensional coherent structures: their motion on the tracks (in ancient greek **dromos**) described by the mean flow (U_1, U_2) and their ability to be **driven** by the boundaries.

This article is organized as follows.

In §2 we illustrate the IST formalism developed in [14],[15] for the solution of the initial- boundary value problem for the DS1 equation.

In §3 we consider degenerate scattering data. In this case the inverse problem reduces to linear algebraic equations. The additional hypothesis of reflectionless boundaries leads to the (N,M) dromion solution, which generalizes to a multidimensional context the n-soliton solution of 1+1 dimensional systems.

In §4 we investigate the time evolution of the scattering data for stationary and time dependent boundary conditions. We show that dromions are driven by the boundaries and radiate energy if their motion is not uniform.

In §5 we investigate the interaction of dromions studying the longtime behavior of the (N,M) dromion solution. We prove that interacting dromions not only exhibit a two-dimensional phase shift, but also exchange energy.

In §6 we consider several examples of physically relevant localized and non-localized boundaries u_2, u_1 which give rise to explicit, finite energy, generalized solutions (distributions) of DS1.

2. THE IST FORMALISM ASSOCIATED WITH THE INITIAL-BOUNDARY VALUE PROBLEM FOR THE DS1 EQUATION

The IST method to solve the initial- boundary value problem associated with equation (5) is based on the use of the eigenfunctions $\Psi = exp(ik\begin{pmatrix}\eta & 0\\ 0 & -\xi\end{pmatrix})M$ of the spectral problem (6.a) defined by the integral equations [14],[15]

$$M_{11}^{\pm} = 1 - \frac{1}{2}\int_{-\infty}^{\xi} d\xi' q M_{21}^{\pm}; \quad M_{12}^{\pm} = -\frac{1}{2}\int_{\mp\infty}^{\xi} d\xi' q M_{22}^{\pm} e^{-ik(\xi'-\xi)};$$

$$M_{21}^{\pm} = -\frac{\epsilon}{2}\int_{\pm\infty}^{\eta} d\eta' q^* M_{11}^{\pm} e^{ik(\eta'-\eta)}; \quad M_{22}^{\pm} = 1 - \frac{\epsilon}{2}\int_{-\infty}^{\eta} d\eta' q^* M_{12}^{\pm}. \tag{9}$$

Equations (9) (defined for $k \in R$) can be analytically extended to the complex k-plane and imply that M^+ and M^- are analytic in the upper and lower half k-plane respectively. In addition these eigenfunctions satisfy the following scattering equations

$$\mu^+(k) - \mu^-(k) = \int_R dl T(k,l) e^{-i(l\xi + k\eta)} \hat{\mu}^+(l),$$

$$\hat{\mu}^-(k) - \hat{\mu}^+(k) = \int_R dl S(k,l) e^{i(l\eta + k\xi)} \mu^-(l), \tag{10}$$

where $\mu^{\pm}, \hat{\mu}^{\pm}$ denote the vectors $\mu^{\pm} := (M_{11}^{\pm}, M_{21}^{\pm})^T, \hat{\mu}^{\pm} := (M_{12}^{\pm}, M_{22}^{\pm})^T$, and the scattering data S, T are given by

$$S(k,l) := \int_{R^2} \frac{d\xi d\eta}{4\pi} q M_{22}^{-}(k) e^{-i(k\xi+l\eta)}, \quad T(k,l) = \epsilon S^*(l,k). \tag{11}$$

Given $S(k,l)$, equations (10) and the analyticity properties of $M^{\pm}$ define a linear Riemann- Hilbert problem in the complex k- plane. Its solution is given in terms of the linear Fredholm equations:

$$\mu^{\pm}(k) = \begin{pmatrix} 1 \\ 0 \end{pmatrix} + \epsilon \int_{R^2} \frac{dk'dl}{2\pi i(k' - (k \pm i0))} S^*(l,k') e^{-i(l\xi+k'\eta)} \hat{\mu}^{+}(l), \tag{12a}$$

$$\hat{\mu}^{\pm}(k) = \begin{pmatrix} 0 \\ 1 \end{pmatrix} - \int_{R^2} \frac{dk'dl}{2\pi i(k' - (k \pm i0))} S(k',l) e^{i(k'\xi+l\eta)} \mu^{-}(l). \tag{12b}$$

Comparing the asymptotics of equations (12)

$$\mu^{\pm}(k) = \begin{pmatrix} 1 \\ 0 \end{pmatrix} - \frac{\epsilon}{k} \int_{R^2} \frac{dk'dl}{2\pi i} S^*(l,k') e^{-i(l\xi+k'\eta)} \hat{\mu}^{+}(l) + O(k^{-2}), \quad |k| >> 1$$

$$\hat{\mu}^{\pm}(k) = \begin{pmatrix} 0 \\ 1 \end{pmatrix} + \frac{1}{k} \int_{R^2} \frac{dk'dl}{2\pi i} S(k',l) e^{i(k'\xi+l\eta)} \mu^{-}(l) + O(k^{-2}), \quad |k| >> 1$$

with the asymptotics of equations (9)

$$\mu^{\pm}(k) = \begin{pmatrix} 1 \\ 0 \end{pmatrix} + \frac{i\epsilon}{4k} \begin{pmatrix} -\int_{R^2}^{\xi} d\xi' |q|^2 \\ 2q^* \end{pmatrix} + O(k^{-2}),$$

$$\hat{\mu}^{\pm}(k) = \begin{pmatrix} 0 \\ 1 \end{pmatrix} + \frac{i}{4k} \begin{pmatrix} -2q \\ \epsilon \int_{R^2}^{\eta} d\eta' |q|^2 \end{pmatrix} + O(k^{-2}),$$

we finally obtain the reconstruction formulae for the solution q, U_1, U_2 of the DS1 equation:

$$q = \frac{1}{\pi} \int_{R^2} dkdl S(k,l) e^{i(l\eta+k\xi)} M_{11}^{-}(l) = \frac{1}{\pi} \int_{R^2} dkdl S(k,l) e^{i(k\xi+l\eta)} M^{+}{}_{22}^{*}(k), \tag{13}$$

and

$$|q|^2 = -\frac{2}{\pi} \partial_\xi \int_{R^2} dkdl S(k,l) e^{i(k\xi+l\eta)} M_{12}^{+*}(k) =$$

$$-\frac{2\epsilon}{\pi} \partial_\eta \int_{R^2} dkdl S(k,l) e^{i(k\xi+l\eta)} M_{21}^{-}(l), \tag{14a}$$

$$U_2 = u_2 + \partial_\xi \int_{R^2} \frac{dkdl}{\pi} S(k,l) e^{i(k\xi+l\eta)} M_{21}^{-}(l),$$

$$U_1 = u_1 + \epsilon \partial_\eta \int_{R^2} \frac{dkdl}{\pi} S(k,l) e^{i(k\xi+l\eta)} M_{12}^{+*}(k). \tag{14b}$$

The IST scheme is completed noticing that if q evolves according to equation (5), then the Fourier transform $\hat{S}(\xi,\eta)$ of $S(k,l)$ evolves according to

$$i\hat{S}_t + \hat{S}_{\xi\xi} + \hat{S}_{\eta\eta} + (u_1+u_2)\hat{S} = 0; \quad \hat{S}(\xi,\eta) := \int_{R^2} \frac{dkdl}{2\pi} S(k,l)e^{i(k\xi+l\eta)}. \quad (15)$$

Notice that if $|q| << 1$ then $q \sim 2\hat{S}$ and equation (15) is the weak field limit of (5).

To obtain equation (15) we remark that the time evolution of the solution Ψ of (6a) defined via the integral equation (9) is given by equation (6b) only if $u_2 = u_1 = 0$ [11]; for nontrivial boundary conditions one has to add to the RHS of equation (6b) a suitable solution of the spectral problem (6a). For examples the time evolution associated with vector

$$\begin{pmatrix}\Psi_1\\ \Psi_2\end{pmatrix} := \begin{pmatrix}M_{12}^-\\ M_{22}^-\end{pmatrix} e^{-ik\xi}$$

reads [14],[15]

$$\begin{pmatrix}\Psi_1\\ \Psi_2\end{pmatrix}_t = \hat{P}\begin{pmatrix}\Psi_1\\ \Psi_2\end{pmatrix} + \begin{pmatrix}\tilde{\Psi}_1\\ \tilde{\Psi}_2\end{pmatrix}, \quad (16a)$$

where the additional solution of (6a) is given by

$$\begin{pmatrix}\tilde{\Psi}_1\\ \tilde{\Psi}_2\end{pmatrix} = -ik^2\begin{pmatrix}\Psi_1\\ \Psi_2\end{pmatrix} + \int_R dl\gamma(k-l)\begin{pmatrix}\Psi_1\\ \Psi_2\end{pmatrix}(l), \quad (16b)$$

$$\gamma(k) := \frac{i}{2\pi}\int_R d\xi e^{-ik\xi}u_2(\xi,t). \quad (16c)$$

To prove equations (16) we first substitute equation (16a) into the RHS of the identity

$$\begin{pmatrix}\Psi_1\\ \Psi_2\end{pmatrix}_t = \begin{pmatrix} -\int_\xi^\infty d\xi'(\Psi_{1t})_{\xi'} \\ \int_{-\infty}^\eta d\eta'(\Psi_{2t})_{\eta'} \end{pmatrix}$$

(which follows from $\Psi_{1t} \xrightarrow[\xi\to\infty]{} 0$, $\Psi_{2t} \xrightarrow[\eta\to-\infty]{} 0$). We then use the asymptotics of $\Psi_1, \Psi_2, U_1, U_2, q$ and the fact that $(\tilde{\Psi}_1, \tilde{\Psi}_2)^T$ is a solution of (6a) to obtain the following integral equation

$$\begin{pmatrix}\tilde{\Psi}_1\\ \tilde{\Psi}_2\end{pmatrix} = \begin{pmatrix} \frac{1}{2}\int_\xi^\infty d\xi' q\tilde{\Psi}_2 \\ -\frac{\epsilon}{2}\int_{-\infty}^\eta d\eta' q^*\tilde{\Psi}_1 - i(k^2-u_2)e^{-ik^2\xi} \end{pmatrix},$$

which implies equations (16b,c).

Having obtained the correct t-part of the Lax pair, it is now straightforward to obtain the t- evolution of the scattering data. Indeed, if

$$\sigma(k,\eta) := \int_R dl e^{il\eta} S(k,l),$$

then $\sigma(k,\eta) \to e^{-ik\xi} M_{12}^{-}$, as $\xi \to -\infty$ and then

$$\sigma_t = iu_1\sigma + i\sigma_{\eta\eta} - ik^2\sigma + \int_R dl\gamma(k-l)\sigma(l,\eta).$$

Taking the Fourier transform of this equation we finally obtain equation (15).

Equations (9)-(15) provide in principle the linearization of the initial- boundary value problem for the DS1 equation. Given q_0, u_1 and u_2, then equations (9) and (11) yield $S(k,l,0)$, and equation (15) yields $S(k,l,t)$. The solution (12) of the RH problem (10) yields $M^{\pm}(\xi,\eta,t)$ and equations (13) and (14) finally imply $q(\xi,\eta,t)$, $U_2(\xi,\eta,t)$, $U_1(\xi,\eta,t)$.

Rigorous aspects of the IST formalism illustrated in this section are presented in [15]; here we briefly mention that for the direct problem one needs to establish the solvability of equations (9); however these equations are Volterra integral equations, thus if $q \in L_1$ they are always solvable. For the inverse problem one can show that the integral equations (12) cannot have homogeneous solutions; then existence and uniqueness follow from the compactness of the underlying operators. If $\epsilon = -1$ the inverse problem is always solvable without the assumption of small norms. If $\epsilon = 1$ such an assumption for the scattering data is needed; however it is possible to shown that scattering data coming from the direct problem satisfy automatically this condition.

3. DEGENERATE SCATTERING DATA

It turns out that degenerate scattering data play an important role in the determination of the asymptotic behavior in t of solutions of DS1 and in the construction of closed form solutions.

If the scattering data $S(k,l,t)$ are degenerate, then $\hat{S}(\xi,\eta,t)$ is also degenerate, namely

$$\hat{S} = \sum_{n=1}^{N}\sum_{m=1}^{M} \rho_{nm} X_n(\xi,t) Y_m(\eta,t), \quad \rho_{nm} \in C, \tag{17a}$$

$$iX_{jt} + X_{j\xi\xi} + u_2 X_j = 0, \;\; j=1,..,N; \;\; iY_{jt} + Y_{j\eta\eta} + u_1 Y_j = 0, \;\; j=1,..,M; \tag{17b}$$

then the Fredholm equations (12) reduce to the algebraic equation [14],[15]

$$(I - \epsilon A)Z = \rho \tag{18}$$

which defines matrix Z in terms of matrices ρ and A, where

$$A(\xi,\eta,t)=\rho\beta\rho^{+}\alpha^{*},\ (\rho)_{ij}=\rho_{ij},\ (\rho^{+})_{ij}=\rho^{*}_{ji}, \tag{19a}$$

$$\beta_{ij}(\eta,t):=\int_{-\infty}^{\eta}d\eta' Y_i(\eta',t)Y_j^{*}(\eta',t),\quad \alpha_{ij}(\xi,t):=\int_{-\infty}^{\xi}d\xi' X_i(\xi',t)X_j^{*}(\xi',t). \tag{19b}$$

Equations (13), (14) finally yield

$$q=2\sum_{n=1}^{N}\sum_{m=1}^{M}X_nY_mZ_{nm}, \tag{20}$$

$$|q|^2=-4\epsilon\partial_\xi\partial_\eta ln\ det\ (I-\epsilon A), \tag{21a}$$

$$U_2=u_2+2\partial_\xi{}^2 ln\ det\ (I-\epsilon A),\quad U_1=u_1+2\partial_\eta{}^2 ln\ det\ (I-\epsilon A), \tag{21b}$$

3.1 The (N,M) Dromion Solution

The solution (20),(21) of DS1 is then in closed form if the eigenfunctions X_j, Y_j are in closed form; if the boundary conditions u_2, u_1 are reflectionless potentials of the Schroedinger equations (17b), then X_j and Y_j can be found in closed form through the system [14],[15]

$$\phi_n+\sum_{j=1}^{N}C_{nj}\phi_j=c_ne^{-p_n(x-ip_nt)},\quad p_{nR}>0, \tag{22a}$$

$$C_{nj}(x,t,c,p):=\frac{c_nc_j^{*}}{p_n+p_j^{*}}e^{-(p_n+p_j^{*})[x-i(p_n-p_j^{*})t]}, \tag{22b}$$

and

$$u(x,t)=-2\partial_x\sum_{j=1}^{N}c_j^{*}e^{-p_j^{*}(x+ip_j^{*}t)}\phi_j, \tag{23}$$

(where $\phi_i=X_i,\ u=u_2$ if $x\to\xi,\ p_i\to\mu_i$ and $\phi_i=Y_i,\ u=u_1$ if $x\to\eta,\ p_i\to\lambda_i,\ c_i\to\tilde{c}_i$). In this case equations (22),(23),(18)-(21) yield what we call an (N,M) dromion solution of DS1 [14], [15]. In the subcase $p_{j_I}=0$ and $c_j\in R$, then $\phi_j(x,t)=f_j(x)e^{ip_j{}^2t}$, and

$$f_n+\sum_{j=1}^{N}\frac{c_nc_j}{p_n+p_j}e^{-(p_n+p_j)x}f_j=c_ne^{-p_nx},$$

$$u(x)=-2\partial_x\sum_{j=1}^{N}c_je^{-p_jx}f_j;$$

then the solution q of DS1 oscillates in time and we obtain what we call an (N,M) breather solution [14],[15].

We remark that, if X_n, Y_m are the solutions of the linear system (22),then the explicit calculation of the integrals (19b) can be avoided, indeed

$$\beta = (I + C(\eta, t, \tilde{c}, \lambda))^{-1}, \quad \alpha = (I + C(\xi, t, c, \mu))^{-1} \tag{24}$$

(where the matrix function C is defined in (22b)), since the solutions ϕ_i of the linear system (22a) satisfy the following equation [15]

$$\phi_i \phi_j^* = ((I + C(x, t, p, c))^{-1})_{ijx}. \tag{25}$$

Then from the limits $(I+C)^{-1} \xrightarrow[x \to -\infty]{} 0$, $(I+C)^{-1} \xrightarrow[x \to +\infty]{} I$, equation (24) follows together with the orthonormality of the eigenfunctions of the time-dependent Schroedinger equation

$$\int_R dx \phi_i \phi_j^* = \delta_{ij}, \tag{26}$$

where δ_{ij} is the Kroenecker tensor.

It is well-known that the total energy

$$E := \int_{R^2} d\xi d\eta |q|^2 \tag{27}$$

of the wave packet of amplitude q is a constant of the motion for the DS equation. It turns out that the energy associated with the (N,M) dromion solution of DS1 is given by [15],[16]

$$E = -4\epsilon ln\ det(I - \epsilon \rho \rho^+) \tag{28}$$

and depends on the spectral matrix ρ only. Equation (28) can be obtained using equations (21a) -(24) which imply that

$$\int_{R^2} d\xi d\eta |q|^2 = -4\epsilon ln\ det(I - \epsilon A(\infty, \infty)) = -4\epsilon ln\ det(I - \epsilon \rho \rho^+),$$

since equations (22b) and (24) imply that $\beta(-\infty) = \alpha(-\infty) = 0$ and $\beta(\infty) = \alpha(\infty) = I$.

3.2 The (1,1) Dromion Solution of DS1

If N=M=1 equations (22)-(25) yield

$$X = \sqrt{\frac{\mu_R}{2}} \frac{e^{i[-\mu_I(\tilde{\xi}+\xi_1)+|\mu|^2 t + arg \frac{c}{\sqrt{2\mu_R}}]}}{cosh(\mu_R \tilde{\xi})}, \tag{29a}$$

$$u_2 = 2{\mu_R}^2 / cosh^2(\mu_R \tilde{\xi}), \tag{29b}$$

$$Y = \sqrt{\frac{\lambda_R}{2}} \frac{e^{i[-\lambda_I(\tilde{\eta}+\eta_1)+|\lambda|^2 t + arg \frac{\tilde{c}}{\sqrt{2\lambda_R}}]}}{cosh(\lambda_R \tilde{\eta})}, \tag{29c}$$

$$u_1 = 2\lambda_R{}^2/cosh^2(\lambda_R\tilde{\eta}), \tag{29d}$$

$$\tilde{\xi} := (\xi+2\mu_I t-\xi_1),\ \tilde{\eta} := (\eta+2\lambda_I t-\eta_1),\ \ \xi_1 := \frac{1}{\mu_R} ln\frac{|c|}{\sqrt{2\mu_R}},\ \ \eta_1 := \frac{1}{\eta_R} ln\frac{|\tilde{c}|}{\sqrt{2\eta_R}}. \tag{30}$$

The degenerate datum $\hat{S} = \rho XY, \rho \in C$ gives rise, via equations (18)-(20) and (24) to the (1,1) dromion solution

$$q = 2X(\xi,t)Y(\eta,t)Z(\xi,\eta,t), \tag{31}$$

where X, Y are given in (29a), (29c) and

$$Z := \frac{\rho}{1-\epsilon|\rho|^2\tau\tilde{\tau}}, \tag{32}$$

$$\tau := \frac{1}{2}(1+tanh(\mu_R(\xi+2\mu_I t-\xi_1)), \qquad \tilde{\tau} := \frac{1}{2}(1+tanh(\lambda_R(\eta+2\eta_I t-\eta_1)). \tag{33}$$

In equation (31) the factor XY indicates that the motion of the (1,1) dromion solution of DS1 on the (ξ,η) plane is completely described by the motion of the boundaries u_2 and u_1 on the ξ and η axes respectively. The ξ and η components of the velocity of the dromion, given by $-2\mu_I$ and $-2\lambda_I$ respectively, are independent, can be positive or negative, and are not related to the amplitude and width, which depend on μ_R, λ_R. If $\epsilon = -1$, Z is bounded for any choice of the spectral parameter ρ; if $\epsilon = 1$, Z is bounded if $|\rho| < 1$; the bounded function Z determines an asymmetry of q with respect to its center.

The (N,M) dromion solution is considered in detail in §5.

4. THE EVOLUTION OF THE SCATTERING DATA

The time-dependent Schroedinger equation (15) in two space dimensions describes the evolution of the scattering data and then plays a central role in the theory associated with DS1. In the following we investigate the initial-value problem for equation (15) for suitably decaying initial conditions $\hat{S}(\xi,\eta,0)$.

4.1 Time Independent Boundaries

We first consider the case when u_1, u_2 are time independent and suitably decaying. Looking for separable solutions $\hat{S} = T(t)X(\xi)Y(\eta)$ of (15), it follows that

$$T' + i(k^2 + k'^2)T = 0,\ X'' + (u_2 + k^2)X = 0,\ Y'' + (u_1 + k'^2)Y = 0; \tag{34}$$

then the analysis of (15) is intimately related to the stationary Schrodinger equation

$$\psi_{xx} + (u(x) + k^2)\psi = 0 \tag{35}$$

for the potentials u_2, u_1.

We recall that the above equation plays an important role in the integration of the Korteweg-de Vries equation. Let $k_j = ip_j$, $p_j \in R^+$ and ϕ_j, $1 \le j \le N$ be the discrete eigenvalues and eigenfunctions associated with $u(x)$; let $\phi(x,k)$ be the associated continuous eigenfunctions. Then any function $f(x) \in L_2$ can be expanded in terms of this orthonormal set [20],

$$f(x) = \sum_{n=1}^{N} \rho_n \phi_n(x) e^{ip_n^2 t} + \int_R \rho(k)\phi(x,k)e^{-ik^2 t}, \tag{36a}$$

$$\rho_n = \int_R dx f(x)\phi_n^*(x), \quad \rho(k) = \int_R dx f(x)\phi^*(x,k). \tag{36b}$$

Using the above result it is easy to show that the solution of the initial-value problem associated with equation (15) can be written in the form

$$\hat{S}(\xi,\eta,t) = \sum_{n=1}^{N}\sum_{m=1}^{M} \rho_{nm} X_n(\xi) Y_m(\eta) e^{i(\mu_n^2+\lambda_m^2)t} +$$
$$\int_{R^2} dk dl \rho(k,l) X(\xi,k) Y(\eta,l) e^{-i(k^2+l^2)t} +$$
$$\int_R dk [\sum_{n=1}^{N} \rho_n(k) e^{i(\mu_n^2-k^2)t} X_n(\xi) Y(\eta,k) + \sum_{m=1}^{M} \tilde{\rho}_m(k) e^{i(\lambda_m^2-k^2)t} X(\xi,k) Y_m(\eta)], \tag{37}$$

where $\{X(\xi,k), X_j(\xi), \mu_j, 1 \le j \le N\}$ and $\{Y(\eta,k), Y_j(\eta), \lambda_j, 1 \le j \le M\}$ are the orthonormal bases of eigenfunctions and eigenvalues of (35), corresponding to $u_2(\xi)$ and $u_1(\eta)$ respectively. Furthermore ρ_{nm}, $\rho(k,l)$, $\rho_n(k)$ and $\tilde{\rho}_m(k)$ are given in terms of the initial condition $\hat{S}_0$ via

$$\int_{R^2} d\xi d\eta \hat{S}(\xi,\eta,0) F(\xi,\eta,k,l), \tag{38}$$

where $F = X_n(\xi)Y_m(\eta)$, $X(\xi,k)Y(\eta,l)$, $X_n(\xi)Y(\eta,k)$ and $X(\xi,k)Y_m(\eta)$ respectively. The stationary phase method implies the following asymptotics in time

$$\hat{S} \to \sum_{n=1}^{N}\sum_{m=1}^{M} \rho_{nm} X_n(\xi) Y_m(\eta) e^{i({\mu_n}^2+{\lambda_m}^2)t}, \quad t \to \pm\infty, \tag{39}$$

thus the scattering data are asymptotically degenerate and equations (18)-(21) yield a localized solution q of DS1 which oscillates in time (breather)[14],[15].

From the above it follows that if at least one of the two boundaries does not give rise to bound states of the Schroedinger operator, then every initial condition q_0 will disperse away. If bound states do exist, the asymptotic behaviour is essentially determined by these bound states; the initial condition only fixes the constants ρ.

4.2 Time Dependent Boundaries and Driven Coherent Structures

If u_1, u_2 are time-dependent, then separation of variables implies that the solution of equation (15) is intimately related to the time-dependent Schroedinger equation

$$i\psi_t + \psi_{xx} + u(x,t)\psi = 0. \tag{40}$$

We recall that this equation plays an important role in the integration of the Kadomtsev-Petviashvili (KP) equation where t is replaced by y [21]-[22]. However we now demand u to be decreasing in x only as opposed to the case of KP where u is decreasing in both x and y. Equation (40) was also used to construct explicit solutions of several physical models [25]. We expect that a completeness result is also valid for the above equation, so that equation (37) is appropriately generalized, however very interesting results can be obtained considering special cases; in particular it is possible to show that **localized coherent structures of DS1 can be piloted everywhere** in the (ξ, η) plane choosing a suitable motion of the boundaries u_2, u_1 [15]. In §4.3 we show that the price payed for this important freedom (which is absent in one dimension) is that **driven coherent structures radiate energy if their motion is not uniform** [15].

These results, whose applications are potentially very wide, can be illustrated in simple mathematical experiments.

We first notice that, if

$$u(x,t) = \bar{u}(x - vt - x_0), \tag{41}$$

where $\bar{u} \in L_{2+\epsilon}$ is given as well as the real constant v, then the solution of the initial value problem for equation (40) is

$$\psi(x,t) = e^{\frac{iv}{2}(x-\frac{v}{2}t-x_0)}\Big(\sum_{n=1}^{N} \rho_n\phi_n(x-vt-x_0)e^{ip_n{}^2t} + \int_{-\infty}^{\infty} dk\rho(k)\phi(x-vt-x_0,k)e^{-ik^2t}\Big), \tag{42a}$$

$$\rho_n = \int_{-\infty}^{\infty} dx'\phi_n^*(x')\psi_0(x'+vt_0+x_0)e^{-i\frac{v}{2}(x'+\frac{v}{2}t_0)-ip_n^2t_0},$$

$$\rho(k) = \int_{-\infty}^{\infty} dx'\phi^*(x',k)\psi_0(x'+vt_0+x_0)e^{-i\frac{v}{2}(x'+\frac{v}{2}t_0)+ik^2t_0}, \tag{42b}$$

where $\psi_0(x) = \psi(x,0)$ and $\{\phi(x,k), \phi_n(x),\ p_n,\ n = 1,..,N\}$ is the orthonormal basis of eigenfunctions of the stationary Schroedinger operator $\partial_x{}^2 + \bar{u}(x)$. To obtain (42) one has to observe that if $u(x,t)$ is given by (41), then function $f(x',t')$, defined by

$$\psi(x,t) = e^{i\frac{v}{2}(x+\frac{v}{2}t')}f(x',t'), \quad x' = x - vt - x_0, \quad t' = t, \tag{43}$$

evolves according to equation $if_{t'} + f_{x'x'} + \bar{u}(x')f = 0$, and then the results of §4.1 can be used.

The stationary phase method implies that

$$\psi(x,t) \to e^{i\frac{v}{2}(x-\frac{v}{2}t-x_0)} \sum_{n=1}^{N} \rho_n \phi_n(x - vt - x_0) e^{ip_n{}^2 t}, \quad t \to \infty \tag{44}$$

and ψ is asymptotically localized where u is localized.

If now $u(x,t)$ has a more complicated time dependence but asymptotically

$$u(x,t) \to \bar{u}(x - vt - x_0), \quad t \to \infty, \tag{45}$$

then, for large times, the solution $\psi(x,t)$ of equation (40) is also given by equation (44), but now the coefficients $\rho_n' s$ depend in a more complicated way on the "history" of the process, namely on ψ_0 and $u(x,t)$.

Equations (41)-(45) can be used to prove the following results. If

$$u_2(\xi,t) = \bar{u}_2(\xi - vt - \xi_0), \quad u_1(\eta,t) = \bar{u}_1(\eta - wt - \eta_0), \tag{46}$$

then

$$\hat{S}(\xi,\eta,t) = e^{i\frac{v}{2}(\xi-\frac{v}{2}t-\xi_0)+i\frac{w}{2}(\eta-\frac{w}{2}t-\eta_0)} \times$$

$$(\sum_{n=1}^{N}\sum_{m=1}^{M} \rho_{nm} X_n(\xi - vt - \xi_0) Y_m(\eta - wt - \eta_0) e^{i(\mu_n^2+\lambda_m^2)t} +$$

$$\int_{R^2} dk dl \rho(k,l) X(\xi - vt - \xi_0, k) Y(\eta - wt - \eta_0, l) e^{-i(k^2+l^2)t} +$$

$$\int_R dk[\sum_{n=1}^{N} \rho_n(k) e^{i(\mu_n^2-k^2)t} X_n(\xi - vt) Y(\eta - wt, k) +$$

$$\sum_{m=1}^{M} \tilde{\rho}(k) e^{i(\lambda_m^2-k^2)t} X(\xi - vt, k) Y_m(\eta - wt)]), \tag{47}$$

where $\{X(\xi,k), X_j(\xi), \mu_j, 1 \le j \le N\}$ and $\{Y(\eta,k), Y_j(\eta), \lambda_j, 1 \le j \le M\}$ are the orthonormal sets of eigenfunctions and eigenvalues of (35), corresponding to $\bar{u}_2(\xi)$ and $\bar{u}_1(\eta)$ respectively. Furthermore ρ_{nm}, $\rho(k,l)$, $\rho_n(k)$ and $\tilde{\rho}_m(k)$ are given in terms of the initial condition $\hat{S}_0(\xi,\eta,0) = \hat{S}(\xi,\eta,t_0)$ via

$$\int_{R^2} d\xi' d\eta' \hat{S}_0(\xi' + vt_0 + \xi_0, \eta' + wt_0 + \eta_0) F(\xi',\eta',k,l) e^{-i\frac{v}{2}(\xi'+\frac{v}{2}t_0)-i\frac{w}{2}(\eta'+\frac{w}{2}t_0)}, \tag{48}$$

where $F(\xi,\eta,k,l) = X_n^*(\xi) Y_m^*(\eta) e^{-i(\mu_n^2+\lambda_m^2)t_0}$, $X^*(\xi,k) Y^*(\eta,l) e^{i(k^2+l^2)t_0}$, $X_n^*(\xi) Y^*(\eta,k) e^{-i(\mu_n^2-k^2)t_0}$ and $X^*(\xi,k) Y_m^*(\eta) e^{-i(\lambda_m^2-k^2)t_0}$ respectively.

Moreover

$$\hat{S} \to e^{i\frac{v}{2}(\xi-\frac{v}{2}t-\xi_0)+i\frac{w}{2}(\eta-\frac{w}{2}t-\eta_0)}$$

$$\sum_{n=1}^{N}\sum_{m=1}^{M}\rho_{nm}X_n(\xi-vt-\xi_0)Y_m(\eta-wt-\eta_0)e^{i(\mu_n^2+\lambda_m^2)t},\quad t\to\pm\infty. \tag{49}$$

For a more general time dependence of the boundaries, characterized by arbitrary functions $u_2(\xi,t),\ \ u_1(\eta,t)$ satisfying the asymptotics

$$u_2(\xi,t)\to\bar{u}_2(\xi-vt-\xi_0),\ \ u_1(\eta,t)\to\bar{u}_1(\eta-wt-\eta_0),\ \ v,w,\xi_0,\eta_0\in R,\ \ t\to\infty, \tag{50}$$

$\hat{S}(\xi,\eta,t)$ satisfies the asymptotic equation (49), where the coefficients ρ_{nm} depend on the history of the process, namely on $\hat{S}(\xi,\eta,0)$ and on $u_2(\xi,t),\ u_1(\eta,t)$.

Equations (49),(50) show that, asymptotically, the spectral data are degenerate and then equations (18)-(21) give the asymptotics of the corresponding DS1 solution. Such a solution is localized where the mean flows in the ξ and η directions intersect, namely the coherent structure "follows" the motion of the boundaries u_2,u_1 or, equivalently, it is piloted by the boundaries [15].

From equation (28) it follows that the energy associated with this asymptotic coherent structure is

$$E=-4\epsilon ln\ det(1-\epsilon\rho\rho^+),\quad (\rho)_{ij}:=\rho_{ij} \tag{51}$$

and then depends on the history of the process.

This mathematical experiment contains of course the particular case in which the boundaries are asymptotically stationary (v=w=0); furthermore it can be easily generalized to the case in which

$$u_2(\xi,t)\to\sum_j\bar{u}_2^{(j)}(\xi-v_jt),\ \ u_1(\eta,t)\to\sum_j\bar{u}_1^{(j)}(\eta-w_jt),\ \ v_j,w_j\in R,\ \ t\to\infty. \tag{52}$$

4.3 Dromions and Radiation of Energy

In order to show that piloted coherent structures radiate energy if their motion is not uniform, we consider the following simplified version of the mathematical experiment described above, in which the traveling boundaries u_2,u_1 are 1-soliton solutions of (40) and change their velocity at a finite time t_0, namely

$$u_2=\frac{2{\mu_R}^2\theta(t_0-t)}{cosh^2(\mu_R(\xi-v_{in}t-\xi_{in}))}+\frac{2{\mu_R}^2\theta(t-t_0)}{cosh^2(\mu_R(\xi-v_{fin}t-\xi_{fin}))}, \tag{53a}$$

$$u_1=\frac{2{\lambda_R}^2\theta(t_0-t)}{cosh^2(\lambda_R(\eta-w_{in}t-\eta_{in}))}+\frac{2{\lambda_R}^2\theta(t-t_0)}{cosh^2(\lambda_R(\eta-w_{fin}t-\eta_{fin}))}, \tag{54}$$

where θ is the Heaviside function. We also assume the continuity of u_2,u_1 at $t=t_0$ which implies

$$(v_{in}-v_{fin})t_0=\xi_{fin}-\xi_{in},\ (w_{in}-w_{fin})t_0=\eta_{fin}-\eta_{in}, \tag{55}$$

and we choose the corresponding solution of (15) for $t < t_0$ to be given by

$$\hat{S}(\xi,\eta,t) = \rho_0 X(\xi,t) Y(\eta,t), \tag{56}$$

where

$$X = \sqrt{\frac{\mu_R}{2}} \frac{e^{i[\frac{v_{in}}{2}(\xi - \frac{v_{in}}{2}t) + \mu_R{}^2 t]}}{cosh(\mu_R(\xi - v_{in}t - \xi_{in}))}, \tag{57}$$

$$Y = \sqrt{\frac{\lambda_R}{2}} \frac{e^{i[\frac{w_{in}}{2}(\eta - \frac{w_{in}}{2}t) + \lambda_R{}^2 t]}}{cosh(\lambda_R(\eta - w_{in}t - \eta_{in}))}. \tag{58}$$

Then, for $t < t_0$ this solution describes the evolution of the (1,1) dromion solution (29),(30) of DS1, and the energy carried by this coherent structure is (for $\epsilon = -1$)

$$E_0 = 4\, ln(1 + |\rho_0|^2). \tag{59}$$

For $t \geq t_0$, $\hat{S}(\xi,\eta,t)$ is the solution of (15) corresponding to the boundaries

$$u_2 = \frac{2\mu_R{}^2}{cosh^2(\mu_R(\xi - v_{fin}t - \xi_{fin}))},$$

$$u_1 = \frac{2\lambda_R{}^2}{cosh^2(\lambda_R(\eta - w_{fin}t - \eta_{fin}))}$$

and to the initial condition

$$\hat{S}_0(\xi,\eta) = \hat{S}(\xi,\eta,t_0) = \rho_0 X(\xi,t_0) Y(\eta,t_0). \tag{60}$$

This solution is then given by equations (47), (48) and, asymptotically,

$$\hat{S} \to \rho_1 e^{i\frac{v_{fin}}{2}(\xi - \frac{v_{fin}}{2}t - \xi_{fin}) + i\frac{w_{fin}}{2}(\eta - \frac{w_{fin}}{2}t - \eta_{fin})} \times$$

$$\tilde{X}(\xi - v_{fin}t - \xi_{fin}) \tilde{Y}(\eta - w_{fin}t - \eta_{fin}) e^{i(\mu_R^2 + \lambda_R^2)t}, \quad t \to \infty, \tag{61a}$$

$$\rho_1 = \int_{R^2} d\xi' d\eta' \hat{S}_0(\xi' + v_{fin}t_0 + \xi_{fin}, \eta' + w_{fin}t_0 + \eta_{fin}) \times$$

$$e^{-i\frac{v_{fin}}{2}(\xi' + \frac{v_{fin}}{2}t_0) - i\frac{w_{fin}}{2}(\eta' + \frac{w_{fin}}{2}t_0) - i(\mu_R^2 + \lambda_R^2)t_0} \tilde{X}^*(\xi') \tilde{Y}^*(\eta'), \tag{61b}$$

$$\tilde{X}(\xi) = \sqrt{\frac{\mu_R}{2}} \frac{1}{cosh(\mu_R \xi)}, \quad \tilde{Y}(\eta) = \sqrt{\frac{\lambda_R}{2}} \frac{1}{cosh(\lambda_R \eta)}. \tag{62}$$

Equation (57),(58) and (61),(62) imply that, driven by the boundaries u_2, u_1, the dromion of DS1 has a transition from an initial state of energy $E_0 = 4\, ln\, det(1 + |\rho_0|^2)$ to a final state of energy $E_1 = 4\, ln\, det(1 + |\rho_1|^2)$. On the other hand the RHS of equation (61b) can be evaluated explicitely:

$$|\rho_1| = |\rho_0| \frac{V}{sinhV} \frac{W}{sinhW}, \tag{63a}$$

$$V = \frac{\pi}{4} \frac{v_{in} - v_{fin}}{\mu_R}, \quad W = \frac{\pi}{4} \frac{w_{in} - w_{fin}}{\lambda_R}; \tag{63b}$$

then $|\rho_1| < |\rho_0|$ and equation (51) implies that the dromion radiates energy. We observe that radiation does not occur if and only if $v_{in} = v_{fin}$ and $w_{in} = w_{fin}$, namely when the motion is uniform, while positive or negative accelerations give rise to radiation.

We also observe that complicated trajectories in the (ξ, η) plane can be obtained composing the elementary straight motions of this experiment [15]. If, for example, we consider a sequence of transitions of the type described in the previous experiment, then the energy carried by the dromion after n transitions is given by

$$E_n = 4\, ln\ (1 + \gamma_n^2 |\rho_0|^2),$$

where

$$\gamma_n := \prod_{j=1}^{n} \frac{V_j}{sinhV_j} \frac{W_j}{sinhW_j}, \quad \gamma_0 = 1,$$

$$V_j = \frac{\pi}{4} \frac{v_{j-1} - v_j}{\mu_R}, \quad W_j = \frac{\pi}{4} \frac{w_{j-1} - w_j}{\lambda_R},$$

and v_j and w_j are the velocities of the boundaries u_2, u_1 respectively after the jth transition.

If $v_n \to v_\infty$, $w_n \to w_\infty$ as $n \to \infty$, then the motion of the dromion tends asymptotically to a uniform motion; its asymptotic energy is given by E_∞ and the coefficient γ_∞ contains the relevant informations concerning the history of the process.

In conclusion, **localized coherent structures (dromions) of DS1 can be piloted everywhere in the (ξ, η) plane choosing a suitable motion of the boundaries; the price payed for this important freedom is that dromions radiate energy if their motion is not uniform.**

5. INTERACTION OF DROMIONS

In this section we present the asymptotic behavior (as $t \to \pm\infty$) of the (N,M) dromion solution of DS1 defined in §3.1, which describes the interaction of localized coherent structures in multidimensions. We shall show that, in the limit $t \to \pm\infty$, the (N,M) dromion solution consists of N times M widely separated dromions and the set of velocities of these localized coherent structures is the same at $+\infty$ and $-\infty$. Identifying each dromion on the basis of its velocity, the effect of the interaction with all the others results not only in a two-dimensional phase shift, but also in a **change of form**. Such a change of form is associated with an **exchange of energy** [16].

The (N,M) dromion solution introduced in §3 corresponds to the case in which i) the boundary conditions u_2, u_1 are reflectionless potentials of the Shrodinger equation (whose solutions are then given by equations (22)); ii) the scattering data are degenerate, namely the solution of DS1 is obtained solving the algebraic equation (18). Equations (22) and (18) are then the starting point of the asymptotic investigation of this section.

5.1 Asymptotics of the Time Dependent Schroedinger Soliton Solution

The N-soliton solution of the time-dependent Schroedinger equation (40) is obtained solving the linear system (22). This system has a unique and regular solution [14]; indeed matrix C has positive eigenvalues, since the associated quadratic form is positive:

$$< v, Cv >= \int_x^\infty dx' \Big| \sum_{n=1}^{N} v_n c_n^* e^{-p_n^* x' - i p_n^{*2} t} \Big|^2 > 0,$$

where

$$< w, v >:= \sum_i w_i^* v_i. \tag{64}$$

Then I+C has also positive eigenvalues and (22) has a unique and regular solution.

For arbitrary N,M we order the imaginary part of the eigenvalues of the Schroedinger equation (40): $\mu_{1I} < \mu_{2I} < .. < \mu_{NI}$, $\lambda_{1I} < \lambda_{2I} < .. < \lambda_{MI}$, and we transform to a moving reference frame with spatial coordinates

$$\xi + 2\mu_{kI} t, \quad \eta + 2\lambda_{hI} t, \quad 1 \le k \le N, \quad 1 \le h \le M. \tag{65}$$

The asymptotics of system (22) are obtained following the (by now standard) derivation scheme used (for instance) in [19] to analyze the algebraic system associated with the n-soliton solution of the Korteweg- de Vries equation. They read:

$$X_i \xrightarrow[t \to \pm\infty]{} \delta_{ik} X_k^{\pm}, \qquad Y_j \xrightarrow[t \to \pm\infty]{} \delta_{jh} Y_h^{\pm}, \tag{66a}$$

$$u_2 \xrightarrow[t \to \pm\infty]{} \bar{u}(\xi, t, \mu_k, \xi_k^{\pm}), \quad u_1 \xrightarrow[t \to \pm\infty]{} \bar{u}(\eta, t, \lambda_h, \eta_h^{\pm}), \tag{66b}$$

where

$$X_k^{\pm} := \sqrt{\frac{\mu_{kR}}{2}} \frac{e^{i[-\mu_{kI}(\xi + 2\mu_{kI} t) + |\mu_k|^2 t + arg\gamma_k^{\pm}]}}{cosh(\mu_{kR}(\xi + 2\mu_{kI} t - \xi_k^{\pm}))}, \tag{67a}$$

$$Y_h^{\pm} := \sqrt{\frac{\lambda_{hR}}{2}} \frac{e^{i[-\lambda_{hI}(\eta + 2\lambda_{hI} t) + |\lambda_h|^2 t + arg\tilde{\gamma}_h^{\pm}]}}{cosh(\lambda_{hR}(\eta + 2\lambda_{hI} t - \eta_h^{\pm}))}, \tag{67b}$$

$$\bar{u}(x, t, p, \bar{x}) := \frac{2 p_R{}^2}{cosh^2(p_R(x + 2p_I t - \bar{x}))}, \tag{68}$$

$$\gamma_k^{\pm} := \frac{c_k}{\sqrt{2\mu_{kR}}} \prod_{i\pm}^{N,k} \frac{\mu_k - \mu_i}{\mu_k + \mu_i^*}; \quad \tilde{\gamma}_h^{\pm} := \frac{\tilde{c}_h}{\sqrt{2\lambda_{hR}}} \prod_{j\pm}^{M,h} \frac{\lambda_h - \lambda_j}{\lambda_h + \lambda_j^*}, \tag{69a}$$

$$\xi_k^{\pm} := \frac{1}{\mu_{kR}} ln|\gamma_k^{\pm}|; \quad \eta_h^{\pm} := \frac{1}{\lambda_{hR}} ln|\tilde{\gamma}_h^{\pm}|. \tag{69b}$$

$$\prod_{i+}^{N,k} := \prod_{i=1}^{k-1}, \quad \prod_{i-}^{N,k} := \prod_{i=k+1}^{N}, \quad \prod_{i=1}^{0} = \prod_{i=N+1}^{N} = 1. \tag{70}$$

Equations (66b) and (68) indicate that the boundary u_2 describes the interaction of N one-dimensional solitons traveling on the ξ axis. Each soliton (say the ith) preserves its velocity $-2\mu_{iI}$ and shape after interacting with all the others, and exhibits the phase shift

$$\delta_i := \xi_i^+ - \xi_i^- = \frac{1}{\mu_{iR}} \left(\sum_{l=1}^{i-1} - \sum_{l=i+1}^{N}\right) ln\left|\frac{\mu_i - \mu_l}{\mu_i - \mu_l^*}\right| \tag{71}$$

which also indicates the typical pairwise character of the interaction. The asymptotic states at $t \to \pm\infty$ of each X_i are the same and describe the motion of a single localized bump of velocity $-2\mu_{iI}$ given by (67a), displaced from its free motion position of the quantity δ_i. If some of the velocities are positives and others are negative, then in their pairwise interaction two solitons "pass" each other or "collide", depending on whether their velocities have the same or different signs.

Of course the same considerations hold for the boundary u_1, which describes the interaction of M one-dimensional solitons traveling on the η-axis, and for the eigenfunctions $Y_j's$. In particular the phase shift associated with the jth soliton is given by

$$\epsilon_j := \eta_j^+ - \eta_j^- = \frac{1}{\lambda_{jR}} \left(\sum_{l=1}^{j-1} - \sum_{l=j+1}^{M}\right) ln\left|\frac{\lambda_j - \lambda_l}{\lambda_j + \lambda_l^*}\right|. \tag{72}$$

5.2 Asymptotics of the (N,M) Dromion Solution of DS1

We first notice that the hermitian matrices α and $\rho\beta\rho^+$ have positive eigenvalues since the associated quadratic forms are positive, for instance

$$< v, \alpha v > = \int_{-\infty}^{\xi} dp \left|\sum_{n=1}^{L} v_n X_n^*(p)\right|^2 > 0.$$

Since the product of two positive matrices is a matrix with positive eigenvelues, it follows that A has positive eigenvalues. Then, in the focusing case $\epsilon = -1$, matrix $I - \epsilon A$ has positive eigenvalues and equation (18) admits a unique and regular solution for any choice of the spectral parameters [15]; in the defocusing case $\epsilon = 1$ matrix $I - \epsilon A$ has positive eigenvalues if the norm of A is less than 1, condition satisfied if $||\rho\rho^+||_2 :=$ max eigenvalue of $\rho\rho^+ < 1$ [15]. A perhaps more convenient sufficient condition for the regularity of the solutions of equation (18) is given by [23]

$$max \sum_{j=1}^{N} |(\rho\rho^{+})_{ij}| < 1.$$

In order to study the asymptotics of equation (18) we first remark that equations (67) and (68) imply that for $|t| >> 1$ the eigenfunctions X_i and X_j represent two widely separated bumps (unless $i = j$); then

$$X_i X_j^* \underset{t \pm \infty}{\longrightarrow} \delta_{ij} \frac{\mu_{iR}}{2} \frac{1}{cosh^2(\mu_{iR}(\xi + 2\mu_{iI}t - \xi_i^{\pm})}$$

uniformely in x and

$$\alpha_{ij} = ((I + C(\xi, t, \mu, c))^{-1})_{ij} \underset{t \pm \infty}{\longrightarrow} \delta_{ij} \frac{1}{2}(1 + tanh(\mu_{iR}(\xi + 2\mu_{iI}t - \xi_i^{\pm}))). \quad (73)$$

Then, in the reference frame with spatial coordinates (65),

$$\alpha_{ij} \underset{t \to +\infty}{\longrightarrow} \delta_{ij} \begin{cases} 0, & \text{if } j < k, \\ \tau_k^{+}(\xi, t), & \text{if } j = k, \\ 1, & \text{if } j > k, \end{cases} \quad \alpha_{ij} \underset{t \to -\infty}{\longrightarrow} \delta_{ij} \begin{cases} 1, & \text{if } j < k, \\ \tau_k^{-}(\xi, t), & \text{if } j = k, \\ 0, & \text{if } j > k, \end{cases} \quad (74a)$$

and, analogously,

$$\beta_{ij} \underset{t \to +\infty}{\longrightarrow} \delta_{ij} \begin{cases} 0, & \text{if } j < h, \\ \tilde{\tau}_h^{+}(\eta, t), & \text{if } j = h, \\ 1, & \text{if } j > h, \end{cases} \quad \beta_{ij} \underset{t \to -\infty}{\longrightarrow} \delta_{ij} \begin{cases} 1, & \text{if } j < h, \\ \tilde{\tau}_h^{+}(\eta, t), & \text{if } j = h, \\ 1, & \text{if } j > h, \end{cases} \quad (74b)$$

where

$$\tau_k^{\pm} := \frac{1}{2}(1 + tanh(\mu_{kR}(\xi + 2\mu_{kI}t - \xi_k^{\pm}))), \quad \tilde{\tau}_h^{\pm} := \frac{1}{2}(1 + tanh(\lambda_{hR}(\eta + 2\eta_{hI}t - \eta_h^{\pm}))), \quad (45)$$

From equations (18)-(21),(67)-(70), and (74) it follows that

$$q \underset{t \to \pm\infty}{\longrightarrow} q_{kh}^{\pm} := 2X_k^{\pm}(\xi, t) Y_h^{\pm}(\eta, t) Z_{kh}^{\pm}(\xi, \eta, t), \quad (76)$$

where $Z_{rs}^{\pm}$ are the solutions of the asymptotic linear systems

$$Z_{rs}^{\pm} - \epsilon\tau_k^{\pm}(\tilde{\tau}_h^{\pm}\rho_{rh}\rho_{kh}^* - \epsilon \sum_{j\pm}^{M,h} \rho_{rj}\rho_{kj}^*) Z_{ks}^{\pm} + \sum_{i\pm}^{N,k} (\tilde{\tau}_h^{\pm}\rho_{rh}\rho_{ih}^* + \sum_{j\pm}^{M,h} \rho_{rj}\rho_{ij}^*) Z_{is}^{\pm} = \rho_{rh},$$

$$1 \le r \le N, \quad (77)$$

$$\sum_{j+}^{N,k} := \sum_{j=k+1}^{N}, \quad \sum_{j-}^{N,k} := \sum_{j=1}^{k-1}, \quad \sum_{j=1}^{0} = \sum_{j=N+1}^{N} = 0. \quad (78)$$

Equations (77),(76) provide the longtime behavior of the (N,M) dromion solution of DS1. It is rather remarkable that the solutions $Z^{\pm}_{kh}$ of the linear systems (77) can be written in the following invariant form

$$Z^{\pm}_{kh} = \frac{N^{\pm}_{kh}}{D^{\pm}_{kh}(\xi,\eta,t)}, \tag{79a}$$

$$N^{\pm}_{kh} :=< \mathbf{w}^{(k,h)}_{\pm}, \mathbf{v}^{(k,h)}_{\pm} >, \tag{79b}$$

$$D^{\pm}_{kh}(\xi,\eta,t) := \lim_{t\to\pm\infty} det\ (I-\epsilon A) = 1 + \sum_{i\pm}^{N,k}\sum_{j\pm}^{M,h} |\mathbf{v}^{(i,j)}_{\pm}|^2 + |\mathbf{v}^{(k,h)}_{\pm}|^2 \tau^{\pm}_k \tilde{\tau}^{\pm}_h +$$

$$\sum_{j\pm}^{M,h} |\mathbf{v}^{(k,j)}_{\pm}|^2 \tau^{\pm}_k + \sum_{i\pm}^{N,k} |\mathbf{v}^{(i,h)}_{\pm}|^2 \tilde{\tau}^{\pm}_h, \tag{79c}$$

where $|\mathbf{v}|^2 :=< \mathbf{v},\mathbf{v} >$ and the vectors $\mathbf{v}^{(i,j)}_{\pm}$ and $\mathbf{w}^{(i,j)}_{\pm}$ are constructed in terms of the elements $\rho_{ij}'s$ of the NxM spectral matrix ρ in the following way.

Given n,m such that $1 \le n \le N$, $\quad 1 \le m \le M$, we select all possible rxr matrices (with $1 \le r \le min(N-n, M-m)$) obtained from matrix $(-\epsilon)^{\frac{1}{2}}\rho$ intersecting the rows $i_s(n = i_1 < i_2 < .. < i_r \le N)$ with the columns $j_s(m = j_1 < j_2 < ..j_r \le M)$. Then the determinants of these matrices are the components of vector $\mathbf{v}^{(n,m)}_{+}$ and the corresponding n,m-minors of the above matrices (with the prescription that the minor of a matrix of rank 1 is 1) are the components of vector $(-\epsilon)^{-\frac{1}{2}}\mathbf{w}^{(n,m)}_{+}$. $\mathbf{v}^{(n,m)}_{-}, \mathbf{w}^{(n,m)}_{-}$ are defined as above, choosing now $1 \le r \le min(n,m)$, $\ 1 \le i_1 < i_2 < .. < i_r = n$, $\ 1 \le j_1 < j_2 < .. < j_r = m$.

It is also remarkable that, asymptotically, the factorization (76) takes place (compare (76) with the (1,1) dromion formula (31)); the factor $X^{\pm}_k Y^{\pm}_h$ indicates that the dynamics of the (N,M) dromion solution of DS1 in the plane (ξ,η) is completely characterized (and generated) by the one-dimensional dynamics of the boundaries $u_2(\xi,t)$ and $u_1(\eta,t)$ (represented in formulas (66b),(68)), which describe the motion of N, respectively M one-dimensional interacting solitons. Then in the limit $t \to \pm\infty$ the (N,M) dromion solution decomposes into N times M widely separated bumps locally represented by the functions $q^{\pm}_{ij}$ defined in (76), with $1 \le i \le N, 1 \le j \le M$, and the $(i,j)^{th}$- dromion travels with velocity $(-2\mu_{i_I}, -2\lambda_{j_I})$ and exhibits the phase shift $(\delta_i, \varepsilon_j)$ defined in (71) and (72).

The nonlocal nature of the interaction, whose mathematical evidence is in the integrals (19b), implies that $Z^{+}_{ij} \ne Z^{-}_{ij}$ (the asymptotic systems $(77)^{+}$ and $(77)^{-}$ are essentially different); then the $(i,j)^{th}$-dromion **changes form** during the interaction with all the others! This is a novel feature of multidimensional coherent structures and in the next section we will show that such a change of form is associated with an **exchange of energy**.

The following two subcases are of particular interest.

i) If $\rho_{ij} = \rho_i\tilde{\rho}_j$, then all minors of degree greater than one of the matrix ρ are zero and $Z^{\pm}_{kh}$ take the simpler form

$$Z^{\pm}_{kh} = \frac{\rho_k\tilde{\rho}_h}{1-\epsilon(|\rho_k|^2\tau^{\pm}_k + \sum_{i\pm}^{N,k}|\rho_i|^2)(|\tilde{\rho}_h|^2\tilde{\tau}^{\pm}_h + \sum_{j\pm}^{M,h}|\tilde{\rho}_j|^2)}, \tag{80}$$

still exhibiting all the richness of the general case.

ii) If $\rho_{ij} = \delta_{ij}\rho_i$, then the vectors $\mathbf{v}^{(i,j)}_{\pm} = 0$ if $i \neq j$, and

$$Z^{\pm}_{kh} = \frac{\delta_{kh}\rho_k}{1-\epsilon|\rho_k|^2\tau^{\pm}_k\tilde{\tau}^{\pm}_h}. \tag{81}$$

In this last case the number of dromions emerging at $t = \pm\infty$ is the minimum between N and M, the form of each dromion is the same at $t = \pm\infty$ and coincides with the one of the (1,1) dromion solution (31)-(33); then dromions do not exchange energy and the only effect of the interaction is a two-dimensional phase shift. This subcase should correspond to the particular soliton solutions obtained in [13] via BT and superposition formulae.

For completeness we also report the asymptotics of the velocity field (U_1, U_2).

If $\xi + 2\mu_{kI}t = O(1)$, then

$$U_2 \xrightarrow[t\to+\infty]{} \begin{cases} \bar{u}(\xi,t,\mu_k,\xi^+_k), & \text{if } \eta < -2\lambda_{MI}t, \\ \bar{u}(\xi,t,\mu_k,\xi^+_k) + 2\partial_\xi{}^2 lnD^+_{kh}(\xi,\infty), & \text{if } -2\lambda_{hI}t < \eta < -2\lambda_{h-1\,I}t, \\ \bar{u}(\xi,t,\mu_k,\xi^+_k) + 2\partial_\xi{}^2 lnD^+_{k1}(\xi,\infty), & \text{if } -2\lambda_{1I}t < \eta; \end{cases} \tag{82a}$$

$$U_2 \xrightarrow[t\to-\infty]{} \begin{cases} \bar{u}(\xi,t,\mu_k,\xi^-_k), & \text{if } \eta < -2\lambda_{1I}t, \\ \bar{u}(\xi,t,\mu_k,\xi^-_k) + 2\partial_\xi{}^2 lnD^-_{k,h-1}(\xi,\infty), & \text{if } -2\lambda_{h-1\,I}t < \eta < -2\lambda_{hI}t, \\ \bar{u}(\xi,t,\mu_k,\xi^-_k) + 2\partial_\xi{}^2 lnD^-_{kM}(\xi,\infty), & \text{if } -2\lambda_{MI}t < \eta. \end{cases} \tag{82b}$$

If $\eta + 2\lambda_{hI}t = O(1)$, then

$$U_1 \xrightarrow[t\to+\infty]{} \begin{cases} \bar{u}(\eta,t,\lambda_h,\eta^+_h), & \text{if } \xi < -2\mu_{NI}t, \\ \bar{u}(\eta,t,\lambda_h,\eta^+_h) + 2\partial_\eta{}^2 lnD^+_{kh}(\infty,\eta), & \text{if } -2\mu_{kI}t < \xi < -2\mu_{k-1\,I}t, \\ \bar{u}(\eta,t,\lambda_h,\eta^+_h) + 2\partial_\eta{}^2 lnD^+_{1h}(\infty,\eta), & \text{if } -2\mu_{1I}t < \xi; \end{cases} \tag{83a}$$

$$U_1 \xrightarrow[t\to-\infty]{} \begin{cases} \bar{u}(\eta,t,\lambda_h,\eta^-_h), & \text{if } \xi < -2\mu_{1I}t, \\ \bar{u}(\eta,t,\lambda_h,\eta^-_h) + 2\partial_\eta{}^2 lnD^-_{k-1,h}(\infty,\eta), & \text{if } -2\mu_{k-1\,I}t < \xi < -2\mu_{kI}t, \\ \bar{u}(\eta,t,\lambda_h,\eta^-_h) + 2\partial_\eta{}^2 lnD^-_{Nh}(\infty,\eta), & \text{if } -2\mu_{NI}t < \xi; \end{cases} \tag{83b}$$

Equations (82),(83) show that the vector field (U_1, U_2) describes a traveling rectangular network on the (ξ,η) plane whose nodes coincide with the positions of the dromions. The velocity field (U_1, U_2) is constant on each branch between two nodes and it has a discontinuity at the nodes (notice the analogy with an electric network). It is worthwhile to stress again that the **dromions**, although exponentially localized in the (ξ, η) plane, **are not solitary**, being connected through the network described by the mean field (U_1, U_2).

Equations (82), (83) follow directly from the asymptotic limit of equations (21b) and from equation (79c).

We end this section writing down explicitly $Z^{\pm}_{kh}$ for the (2,2) dromion case.

$$Z^{-}_{11} = \frac{\rho_{11}}{1 - \epsilon|\rho_{11}|^2 \tau_1^- \tilde{\tau}_1^-},$$

$$Z^{+}_{11} = \frac{(\rho_{11} - \epsilon\rho^*_{22}(\rho_{11}\rho_{22} - \rho_{12}\rho_{21}))}{1 - \epsilon(|\rho_{22}|^2 + (|\rho_{11}|^2 - \epsilon|\rho_{11}\rho_{22} - \rho_{12}\rho_{21}|^2)\tau_1^+\tilde{\tau}_1^+ + |\rho_{12}|^2\tau_1^+ + |\rho_{21}|^2\tilde{\tau}_1^+)},$$

$$Z^{-}_{12} = \frac{\rho_{12}}{1 - \epsilon(|\rho_{12}|^2\tilde{\tau}_2^- + |\rho_{11}|^2)\tau_1^-}, \quad Z^{+}_{12} = \frac{\rho_{12}}{1 - \epsilon(|\rho_{12}|^2\tau_1^+ + |\rho_{22}|^2)\tilde{\tau}_2^+},$$

$$Z^{-}_{21} = \frac{\rho_{21}}{1 - \epsilon(|\rho_{21}|^2\tau_2^- + |\rho_{11}|^2)\tilde{\tau}_1^-}, \quad Z^{+}_{21} = \frac{\rho_{21}}{1 - \epsilon(|\rho_{21}|^2\tilde{\tau}_1^+ + |\rho_{22}|^2)\tau_2^+},$$

$$Z^{-}_{22} = \frac{(\rho_{22} - \epsilon\rho^*_{11}(\rho_{11}\rho_{22} - \rho_{12}\rho_{21}))}{1 - \epsilon(|\rho_{11}|^2 + (|\rho_{22}|^2 - \epsilon|\rho_{11}\rho_{22} - \rho_{12}\rho_{21}|^2)\tau_2^+\tilde{\tau}_2^+ + |\rho_{21}|^2\tau_2^- + |\rho_{12}|^2\tilde{\tau}_2^-)},$$

$$Z^{+}_{22} = \frac{\rho_{22}}{1 - \epsilon|\rho_{22}|^2\tau_2^+\tilde{\tau}_2^+}.$$

In the sequence illustrated in Figure 1.a-f below the absolute value $|q|$ of the (2,2) dromion solution of DS1 for $\epsilon = -1$ is plotted at various times for generic values of the spectral parameters.

Fig. 1.a

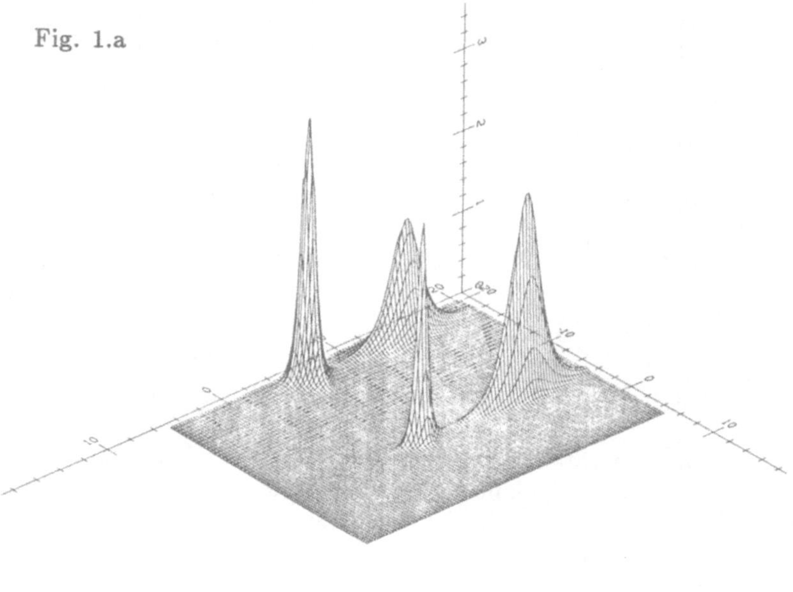

Fig. 1.b

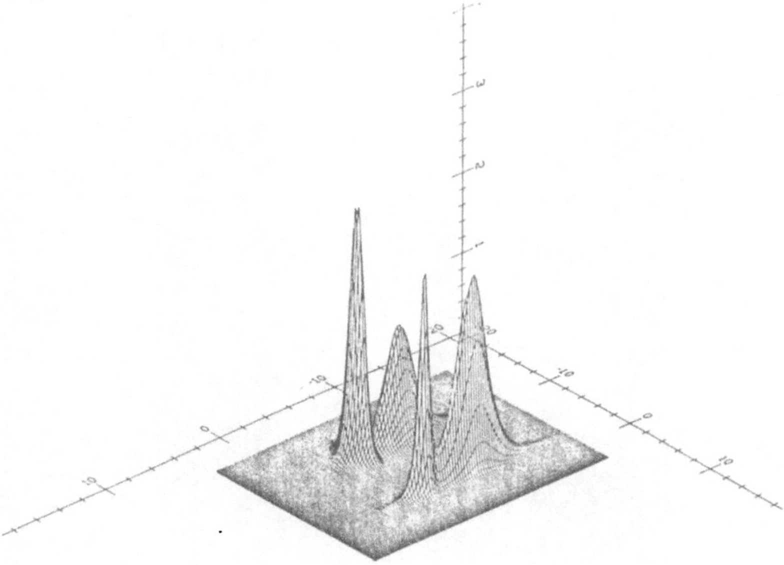

Fig.1.c

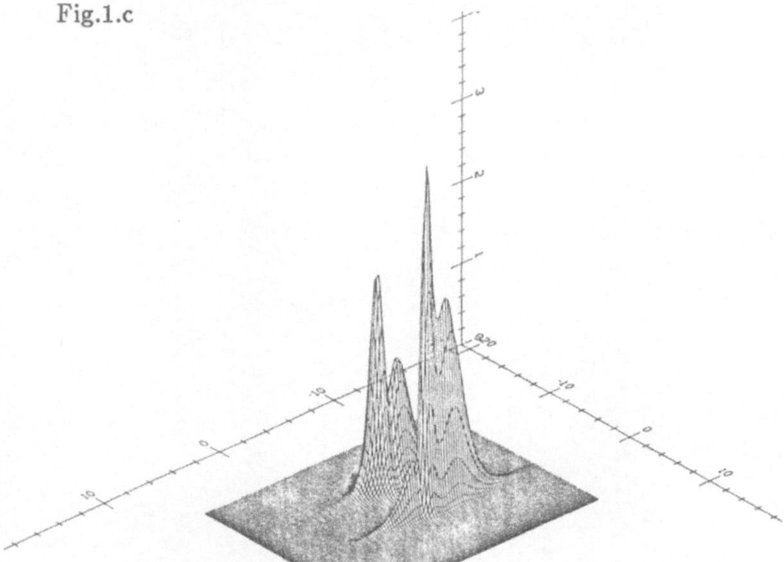

Fig. 1.d

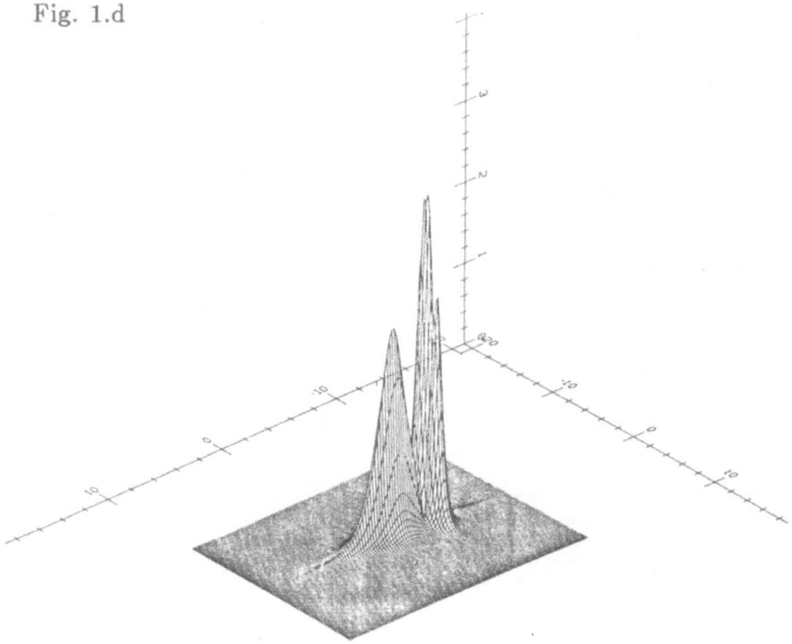

Fig. 1.e

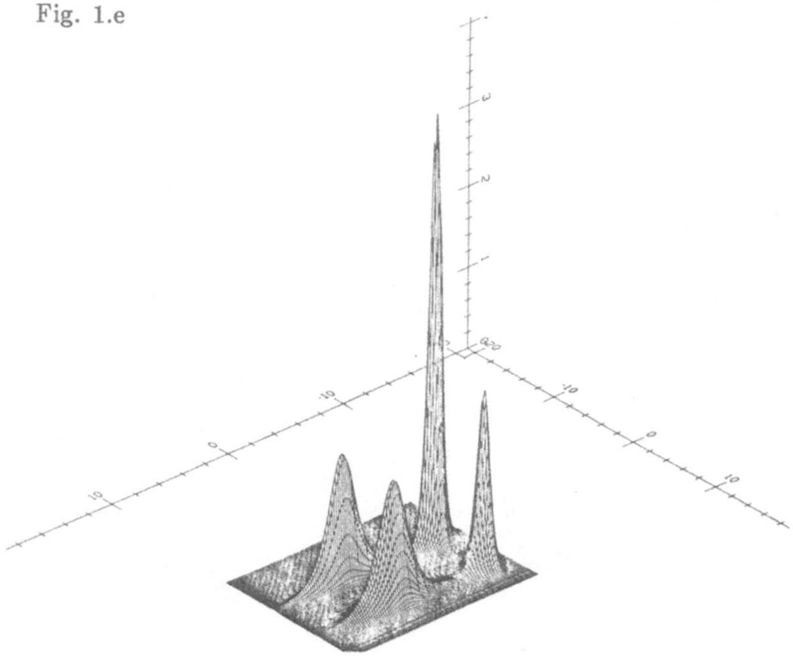

Fig. 1.f

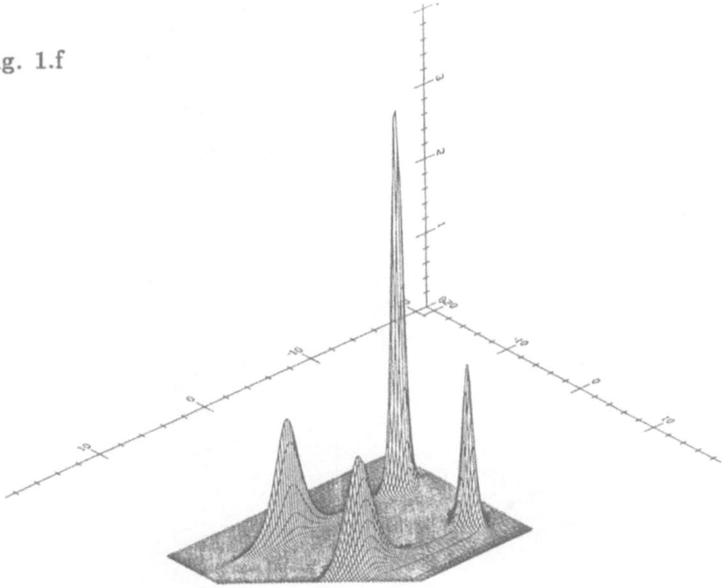

Fig.1; *Plotting of the absolute value of the (2,2) dromion solution of DS1* ($\epsilon = -1$) *for the following generic values of the spectral parameters:* $\mu_1 = 1-1.5i$, $\mu_2 = 3-.3i$, $\lambda_1 = 2-2i$, $\lambda_2 = 4-.5i$, $c_1 = 1+i$, $c_2 = .0001+.001i$, $\tilde{c}_1 = 2+i$, $\tilde{c}_2 = 1+2i$, $\rho_{11} = 1+i$, $\rho_{12} = 3+2i$, $\rho_{21} = 4+i$, $\rho_{22} = 2+3i$. $|q|$ *is plotted at* (*a*) $t = -4.5$, (*b*) $t = -2.5$, (*c*) $t = -1.6$, (*d*) $t = 0$, (*e*) $t = 1.8$, (*f*) $t = 2.8$. *The* ξ *and* η *axes vary in the interval [-21,19].*

Figures 1.a-f show the following.

i) The striking change of form of each dromion is evident comparing the asymptotic states described by Figures 1.a and 1.f.

ii) In Figure 1.c dromions q_{11} and q_{12} overtake dromions q_{21} and q_{22} respectively. This is the moment in which the two one-dimensional solitons described by the boundary u_2 on the ξ axis pass each other.

iii) In Figure 1.d dromions q_{21} and q_{11} overtake dromions q_{22} and q_{12} respectively. This is the moment in which the two one-dimensional solitons described by the boundary u_1 on the η axis pass each other.

In the sequence illustrated in Figures 2.a-f below the absolute value $|q|$ of the (2,2) dromion solution of DS1 for $\epsilon = -1$ is plotted at various times for the same values of the spectral parameters as for Figure 1, except that now $\rho_{12} = \rho_{21} = 0$.

Fig. 2.a

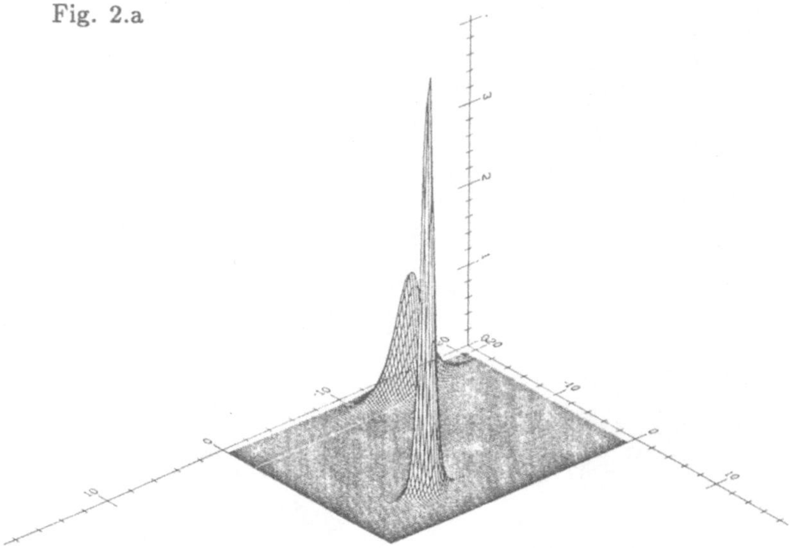

Fig. 2.b

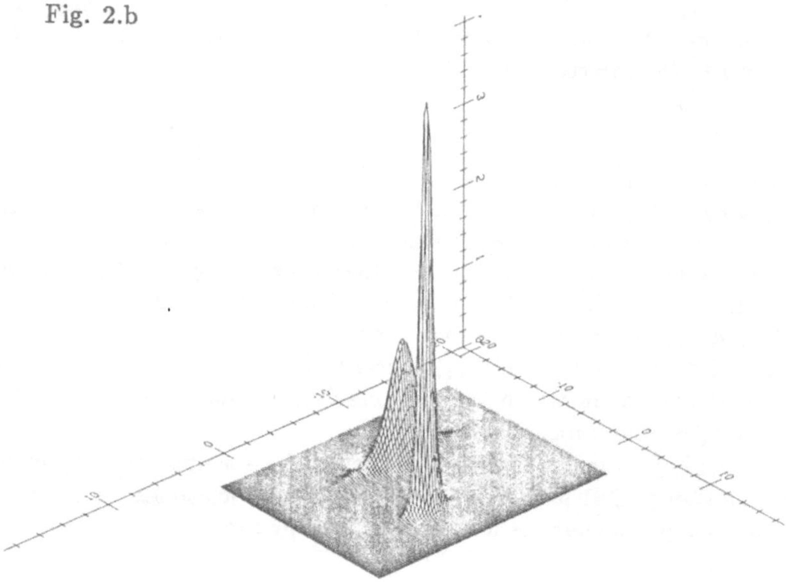

Fig. 2.c

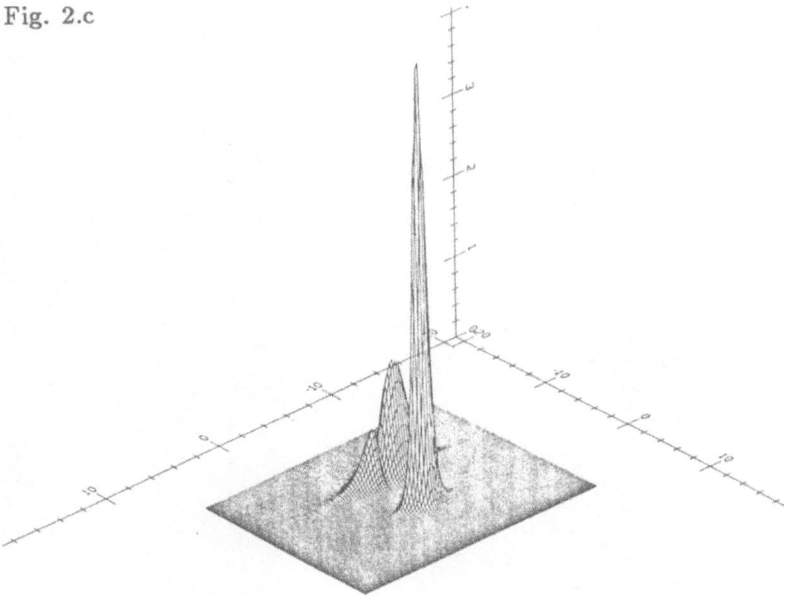

Fig. 2.d

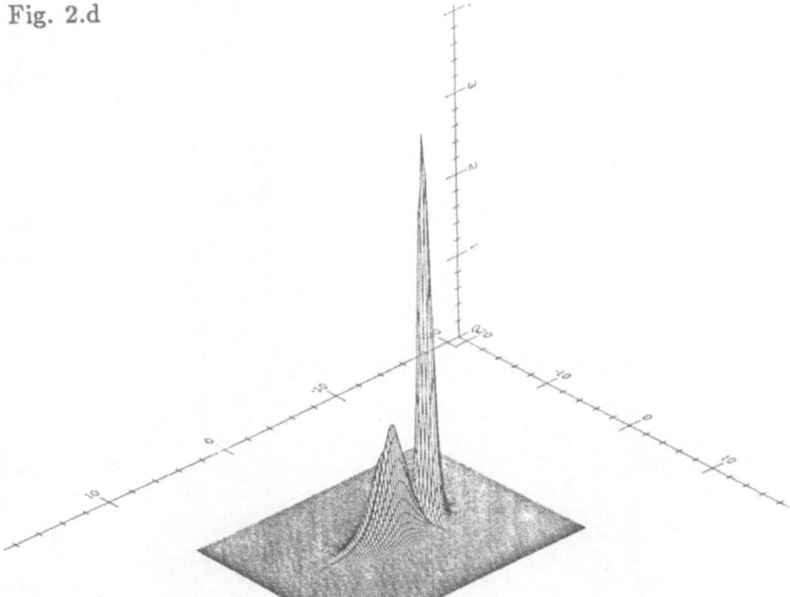

Fig. 2.e

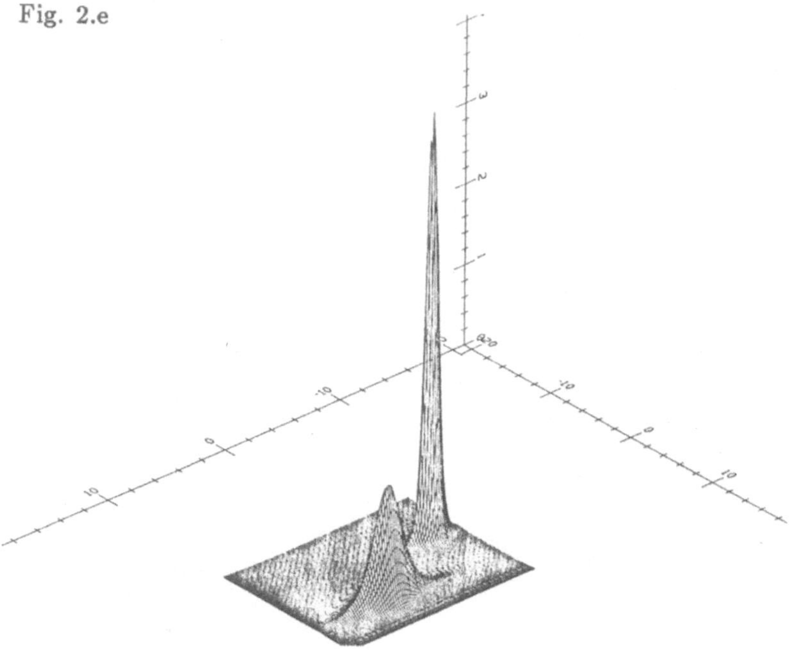

Fig. 2.f

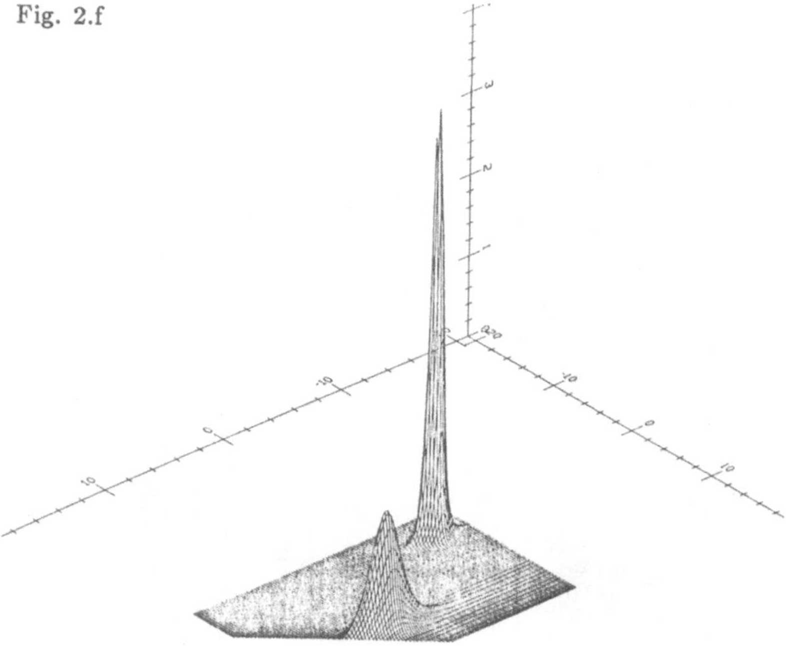

Fig.2; *Plotting of the absolute value of the (2,2) dromion solution of DS1* ($\epsilon = -1$) *for the same values of the spectral parameters as in Fig.1 except that now* $\rho_{12} = \rho_{21} = 0$*.*$|q|$ *is plotted at* (*a*) $t = -4.5$, (*b*) $t = -2.5$, (*c*) $t = -1.6$, (*d*) $t = 0$, (*e*) $t = 1.8$, (*f*) $t = 2.8$. *The* ξ *and* η *axes vary in the interval [-21,19].*

Figures 2.a-f show the following.

i) In agreement with the theory (formula (81)) if $\rho_{12} = \rho_{21} = 0$ the number of dromions emerging asymptotically in Figures 2.a and 2.f is two; furthermore they do not exhibit a change of form.

ii) Figures 2.b,c,d show that during the process of interaction other little bumps appear where the flows U_2, U_1 intersect, namely where, in the generic case $\rho_{12}, \rho_{21} \neq 0$, q_{12} and q_{21} would be located.

5.3 Dromions Exchange Energy

In the previous section we have seen that one of the novelties associated with localized coherent structures in multidimensions is given by the change of form that dromions exhibit during the process of interaction. It turns out that such a change of form is associated with an exchange of energy [16].

In order to measure the increase or loss of energy of each dromion,one has to evaluate the energies

$$E^{\pm}_{k,h} := \int_{R^2} d\xi d\eta |q^{\pm}_{kh}|^2 \tag{84}$$

associated with the $(k,h)^{th}$ dromion emerging at $t = \pm\infty$ from the process of interaction. It turns out that

$$E^{\pm}_{kh} = -4\epsilon ln\ det\{[1 + |\mathbf{v}^{(k,h)}_{\pm}|^2 + \sum_{j\pm}^{M,h} |\mathbf{v}^{(k,j)}_{\pm}|^2 + \sum_{i\pm}^{N,k} |\mathbf{v}^{(i,h)}_{\pm}|^2 + \sum_{i\pm}^{N,k}\sum_{j\pm}^{M,h} |\mathbf{v}^{(i,j)}_{\pm}|^2]$$

$$[1+\sum_{i\pm}^{N,k}\sum_{j\pm}^{M,h} |\mathbf{v}^{(i,j)}_{\pm}|^2]/[1+\sum_{j\pm}^{M,h}(|\mathbf{v}^{(k,j)}_{\pm}|^2+\sum_{i\pm}^{N,k} |\mathbf{v}^{(i,j)}_{\pm}|^2)][1+\sum_{i\pm}^{N,k}(|\mathbf{v}^{(i,h)}_{\pm}|^2+\sum_{j\pm}^{M,h} |\mathbf{v}^{(i,j)}_{\pm}|^2)]\} \tag{85}$$

and, consequently, that

$$\sum_{i=1}^{N}\sum_{j=1}^{M} E^{+}_{ij} = \sum_{i=1}^{N}\sum_{j=1}^{M} E^{-}_{ij} = -4\epsilon ln(1 + \sum_{i=1}^{N}\sum_{j=1}^{M} |\mathbf{v}^{(i,j)}_{\pm}|^2) = -4\epsilon ln\ det(I - \epsilon\rho\rho^{+}). \tag{86}$$

Equations (29),(84)-(85) express the following results.

i) While the energy associated with the $(k,h)^{th}$ dromion is **different** at $t=+\infty$ and $t=-\infty$ $(E^+_{kh}\neq E^-_{kh})$, the sum of the energies of each dromion is the same at $t=+\infty$ and $t=-\infty$. Then **dromions exchange energy.**

ii) The sum of the energies of each single bump coincides with the total energy (29) of the system. This is not obvious a priori, since the fact that q is exponentially small in every reference frame (except the finite number with the dromion velocities) could not exclude the existence of additional energy, because the gaps between dromions is large (see similar considerations in [19]).

To prove equation (85) we consider the asymptotic limit of equation (21)

$$|q|^2 \xrightarrow[t\to\pm\infty]{} |q^\pm_{kh}|^2 = -4\epsilon\partial_\xi\partial_\eta \; ln \; D^\pm_{kh}(\xi,\eta), \tag{87}$$

using also equation (79c). Then

$$\int_{R^2} d\xi d\eta |q^\pm_{kh}|^2 = -4\epsilon ln \frac{D^\pm_{kh}(\infty,\infty)D^\pm_{kh}(-\infty,-\infty)}{D^\pm_{kh}(\infty,-\infty)D^\pm_{kh}(-\infty,\infty)} \tag{88}$$

and equation (85) follows from (88), using the $\xi,\eta\to\pm\infty$ limits of equation (79c). Equation (86) follows from (88) using the identities

$$D^\pm_{kh}(\mp\infty,\eta) \;=\; D^\pm_{k+1,h}(\pm\infty,\eta), \;\; D^\pm_{kh}(\xi,\mp\infty) \;=\; D^\pm_{k,h+1}(\xi,\pm\infty).$$

We end this section writing down explicitly the energies $E^\pm_{ij}$ associated with the (2,2) dromion solution.

$$E^-_{11} = -4\epsilon ln(1-\epsilon|\rho_{11}|^2),$$

$$E^+_{11} = -4\epsilon ln \frac{(1-\epsilon|\rho_{22}|^2)(1-\epsilon(|\rho_{11}|^2+|\rho_{22}|^2+|\rho_{12}|^2+|\rho_{21}|^2-\epsilon|\rho_{11}\rho_{22}-\rho_{12}\rho_{21}|^2))}{(1-\epsilon(|\rho_{22}|^2+|\rho_{12}|^2))(1-\epsilon(|\rho_{22}|^2+|\rho_{21}|^2))}$$

$$E^-_{12} = -4\epsilon ln\frac{1-\epsilon(|\rho_{12}|^2+|\rho_{11}|^2)}{1-\epsilon|\rho_{11}|^2}, \quad E^+_{12} = -4\epsilon ln\frac{1-\epsilon(|\rho_{12}|^2+|\rho_{22}|^2)}{1-\epsilon|\rho_{22}|^2},$$

$$E^-_{21} = -4\epsilon ln\frac{1-\epsilon(|\rho_{21}|^2+|\rho_{11}|^2)}{1-\epsilon|\rho_{11}|^2}, \quad E^+_{21} = -4\epsilon ln\frac{1-\epsilon(|\rho_{21}|^2+|\rho_{22}|^2)}{1-\epsilon|\rho_{22}|^2},$$

$$E^-_{22} = -4\epsilon ln \frac{(1-\epsilon|\rho_{11}|^2)(1-\epsilon(|\rho_{11}|^2+|\rho_{22}|^2+|\rho_{12}|^2+|\rho_{21}|^2-\epsilon|\rho_{11}\rho_{22}-\rho_{12}\rho_{21}|^2))}{(1-\epsilon(|\rho_{11}|^2+|\rho_{21}|^2))(1-\epsilon(|\rho_{11}|^2+|\rho_{12}|^2))}$$

$$E^+_{22} = -4\epsilon ln(1-\epsilon|\rho_{22}|^2),$$

$$\sum_{i=1}^{2}\sum_{j=1}^{2} E^\pm_{ij} = -4\epsilon ln \; det(I-\epsilon\rho\rho^+) = -4\epsilon ln(1-\epsilon(|\rho_{11}|^2+|\rho_{22}|^2+|\rho_{12}|^2+|\rho_{21}|^2-$$

$$\epsilon|\rho_{11}\rho_{22}-\rho_{12}\rho_{21}|^2)).$$

6. OTHER EXPLICIT SOLUTIONS OF DS1

The first examples of explicit spatially confined solutions of DS1 were found in [7] via a GLM equation and correspond to $u_2 = u_1 = 0$ (thus they are not dromions). If, for example, we consider the following explicit solution

$$\hat{S}(\xi,\eta,t) = \rho X(\xi,t)Y(\eta,t), \tag{89}$$

$$X(\xi,t) = (\alpha+it)^{-\frac{1}{2}}e^{-\frac{\xi^2}{4(\alpha+it)}}, \quad Y(\eta,t) = (\beta+it)^{-\frac{1}{2}}e^{-\frac{\eta^2}{4(\beta+it)}}, \tag{90}$$

of equation (15) corresponding to $u_2 = u_1 = 0$, then equations (18)-(22) yield the simplest solution of DS1 obtained in [7]; such a solution describes the evolution of a Gaussian initial condition into a dispersive wave, in agreement with the result that any initial condition disperses away if $u_2 = u_1 = 0$ [14].

The (N,M) dromion and breather solutions are not the only explicit, exponentially decaying solutions of DS1. Let's illustrate this fact in the case of time-independent boundaries u_2 and u_1 [16].

In this case, if $\hat{S}$ is degenerate, namely if

$$\hat{S}(\xi,\eta,t) = \sum_{n=1}^{N}\sum_{m=1}^{M} X_n(\xi)Y_m(\eta)e^{i({\mu_n}^2+{\lambda_m}^2)t} \tag{91}$$

(where X_i, μ_i and Y_i, λ_i are the exponentially decaying eigenfunctions and eigenvalues of the stationary Schroedinger operator (35) corresponding to the potential $u_2(\xi)$ and $u_1(\eta)$ respectively), then equation (18)-(21) imply that the corresponding solution of DS1 is an exponentially localized breather even if u_2 and u_1 give rise to nontrivial reflection coefficients of the Schroedinger operator. This breather solution is in closed form if X_i, Y_j are in closed form. The literature on the Schrodinger equation (35) provides us with several potentials $u(x)$ which give rise to explicit localized eigenfunctions (see for example [20]); in the DS1 context these potentials are special boundaries u_2, u_1 which give rise to explicit exponentially localized solutions of DS1 and in the following we list few of them.

i) δ-boundary.

If $u(x) = a\delta(x-x_0), x_0 \in R, a > 0$, there exists one eigenvalue $k = ip, p = a/2$ and the associated eigenfunction reads

$$\phi(x) = \sqrt{p}(\theta(x-x_0)e^{-p(x-x_0)} + \theta(x_0-x)e^{p(x-x_0)}), \tag{92}$$

where δ and θ are the Dirac and Heaviside distributions respectively.

ii) Rectangular boundary.

If $u(x) = a(\theta(x-x_1) - \theta(x-x_2)), x_1 < x_2, a > 0$, there exist N eigenvalues $k_j = ip_j, p_j > 0$ defined by the equation

$$cotg((x_2-x_1)\nu_j) = (a-2{p_j}^2)/2p_j\nu_j; \quad \nu_j := \sqrt{a-{p_j}^2}$$

and the associated eigenfunctions read

$$\phi_j(x) = \frac{\nu_j}{\sqrt{a}} \frac{2p_j}{\sqrt{2+p_j(x_2-x_1)}}[\theta(x_1-x)e^{p_j(x-x_1)}+$$

$$(\theta(x-x_1)-\theta(x-x_2))(cos(\nu_j(x-x_1))+\frac{p_j}{\nu_j}sin(\nu_j(x-x_1)))+\theta(x-x_2)e^{-p_j(x-x_2)}]. \tag{93}$$

iii) Exponential boundary.

If $u(x) = ae^{-2|x|}$, there exist N eigenvalues $k_j = ip_j$ defined by the equation

$$J_{p_j}(\gamma)J'_{p_j}(\gamma) = 0$$

where $J_\nu(x)$ is the Bessel function, and the corresponding eigenfunctions read

$$\phi_j(x) = c_j(\theta(x)J_{p_j}(\gamma e^{-x}) - \pi\gamma\frac{J_{p_j}(\gamma)J'_{-p_j}(\gamma)}{2sin\pi p_j}\theta(-x)J_{p_j}(\gamma e^{x})). \tag{94}$$

If u_2 and u_1 are any of the three cases listed above, then X_i and Y_j are correspondingly defined by (92)-(94) and the algebraic equations (18)-(21) yield explicit breather-like solutions of DS1.

If, for example, $u_2 = a_2\delta(\xi-\xi_0)$ and $u_1 = a_1\delta(\eta-\eta_0)$, then

$$X(\xi) = \sqrt{\mu}(\theta(\xi-\xi_0)e^{-\mu(\xi-\xi_0)} + \theta(\xi_0-\xi)e^{\mu(\xi-\xi_0)})$$

$$Y(\eta) = \sqrt{\lambda}(\theta(\eta-\eta_0)e^{-\lambda(\eta-\eta_0)} + \theta(\eta_0-\eta)e^{\lambda(\eta-\eta_0)})$$

and equations (18)-(21) yield

$$q = 2XYZe^{i(\mu^2+\lambda^2)t}$$

$$Z(\xi,\eta) := \rho/[1-\epsilon\frac{|\rho|^2}{4}(\theta(\eta-\eta_0)(2-e^{-2\lambda(\eta-\eta_0)})+$$

$$\theta(\eta_0-\eta)e^{2\lambda(\eta-\eta_0)})(\theta(\xi-\xi_0)(2-e^{-2\mu(\xi-\xi_0)})+\theta(\xi_0-\xi)e^{2\mu(\xi-\xi_0)}]$$

This breather is an interesting example of localized weak solution (distribution) of DS1.

We remarke that equations (18)-(21) enable one to construct solutions of DS even if X_j, Y_j are not exponentially localized, provided that integrals (19b) make sense. This is illustrated in the following example.

iv) Step-boundary.

If $u_2 = a_2\theta(\xi-\xi_0)$, $u_1 = a_1\theta(\eta-\eta_0)$, we consider the corresponding solutions of equation (35)

$$X(\xi) = \theta(\xi_0-\xi)e^{\mu\xi} + \theta(\xi-\xi_0)\frac{\mu e^{\mu\xi_0}}{\sqrt{a_1-\mu^2}}[sin\sqrt{a_1-\mu^2}(\xi-\xi_0)+$$

$$\frac{\sqrt{a_1-\mu^2}}{\mu}cos\sqrt{a_1-\mu^2}(\xi-\xi_0)],$$

$$Y(\eta) = \theta(\eta_0 - \eta)e^{\lambda\eta} + \theta(\eta - \eta_0)\frac{\lambda e^{\lambda\eta_0}}{\sqrt{a_2 - \lambda^2}}[sin\sqrt{a_2 - \lambda^2}(\eta - \eta_0) +$$

$$\frac{\sqrt{a_2 - \lambda^2}}{\lambda}cos\sqrt{a_2 - \lambda^2}(\eta - \eta_0)],$$

Then equations (18)-(21) yield

$$q = XYZe^{i(\mu^2+\lambda^2)t}, \tag{95a}$$

$$Z = 2\rho/\{1 - \epsilon|\rho|^2(\theta(\xi_0 - \xi)\frac{e^{2\mu\xi}}{2\mu} + \theta(\xi - \xi_0)\frac{e^{2\mu\xi_0}}{2(a_1 - \mu^2)}[(\xi - \xi_0)a_1 + \frac{a_1}{\mu} +$$

$$\frac{a_1 - 2\mu^2}{2\sqrt{a_1 - \mu^2}}sin2\sqrt{a_1 - \mu^2}(\xi - \xi_0) - \mu cos2\sqrt{a_1 - \mu^2}(\xi - \xi_0)])(\theta(\eta_0 - \eta)\frac{e^{2\lambda\eta}}{2\lambda} +$$

$$\theta(\eta - \eta_0)\frac{e^{2\lambda\eta_0}}{2(a_2 - \lambda^2)}[(\eta - \eta_0)a_2 + \frac{a_2}{\lambda} + \frac{a_2 - \lambda^2}{2\sqrt{a_2 - \lambda^2}}sin2\sqrt{a_2 - \lambda^2}(\eta - \eta_0) -$$

$$\lambda cos2\sqrt{a_2 - \lambda^2}(\eta - \eta_0)])\}. \tag{95b}$$

This breather solution of DS behaves like

$$q = 0(\frac{1}{(\xi - \xi_0)(\eta - \eta_0)})[sin\sqrt{a_1 - \mu^2}(\xi - \xi_0) + \frac{\sqrt{a_1 - \mu^2}}{\mu}cos\sqrt{a_1 - \mu^2}(\xi - \xi_0)]$$

$$[sin\sqrt{a_2 - \lambda^2}(\eta - \eta_0) + \frac{\sqrt{a_2 - \lambda^2}}{\lambda}cos\sqrt{a_2 - \lambda^2}(\eta - \eta_0)]$$

for $\xi - \xi_0, \eta - \eta_0 >> 1$, and decays exponentially in the rest of the plane. It is interesting to notice that the non-localized step-boundaries u_2, u_1 support a finite energy breather!

This breather is plotted at $t = 0$ for $\epsilon = -1$ in Figure 3 below.

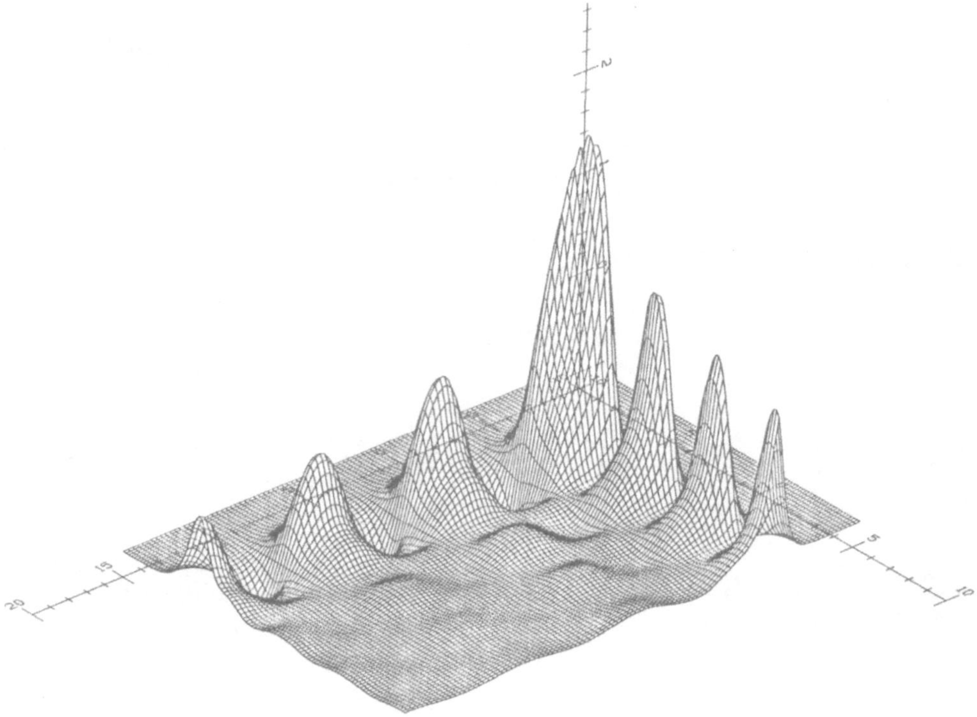

Fig.3; *Plotting of the breather solution (95) of DS1* ($\epsilon = -1$) *for* $a_1 = 10$, $a_2 = 8$, $p_1 = 3$, $p_2 = 2$, $\xi_0 = -.3$, $\eta_0 = -.4$, $\rho = 3$ *at* $t = 0$. ξ *and* η *vary in the intervals [20,-5] and [10,-5] respectively* .

REFERENCES

1. A. Davey and K. Stewartson, Proc. R. Soc. london Ser. A **338**, (1974) 101.
2. M.J. Ablowitz and H. Segur, J. Fluid Mech., **92**, (1979) 691.
3. V.E. Zakharov and E.A. Kuznetsov, Physica **18D**, (1986) 455. V. E. Zakharov and A. M. Rubenchik, Prikl. Mat. Techn. Phys. **5**, (1972) 84. V. E. Manakov, talk presented at the Vth Workshop NEED's 1989, Kolymbari, Crete.
4. F. Calogero and V. Eckhaus, Inverse Problems **3**, (1987) 27; F. Calogero and A. Maccari, on Inverse Problems; an Interdisciplinary Study, ed. P. Sabatier, Academic Press, New York, 1987;
5. D.J. Benney and G.J. Roskes, Stud. Appl. Math. **48**, (1969) 377.

6. M.J. Ablowitz and R. Haberman, Phys. Rev. Lett., **35**, (1975) 691. V. E. Zakharov and A. B. Shabat, Funct. Anal. Appl. **8**, (1974) 43.
7. H. Cornille, J. Math. Phys. **20**, (1979) 199.
8. A.S.Fokas, Phys. Rev. Lett. **51**, (1983) 3.
9. A.S. Fokas and M.J. Ablowitz, Phys. Rev. Lett., **51**, (1983) 7. A.S. Fokas and M.J. Ablowitz, J. Math. Phys., **25**, (1984) 8.
10. D.J. Kaup, Physica **1D**, (1980) 45.
11. C. Shultz and M.J. Ablowitz, Phys. Lett., **135** (1989) 433.
12. J.-M. Ghidaglia and J.-C. Saut, Sur le Problem de Cauchy pour les Equations de Davey- Stewartson, C. R. Acad. Sci. Paris, t. 308, Serie I, (1989) 115.
13. M.Boiti, J.J-P. Leon, L. Martina and F. Pempinelli, Phys. Lett. A, **132**,(1988) 432.
14. A.S Fokas and P.M.Santini, Phys. Rev. Lett., **63** (1989) 1329.
15. A.S. Fokas and P.M. Santini, Coherent Structures and a Boundary-Value Problem for the Davey-Stewartson 1 Equation, Preprint INS 121, Clarkson University, 1989. Physica D (submitted to).
16. P.M. Santini, Energy Exchange of Interacting Coherent Structures in Multidimensions, Preprint 662, Dipartimento di Fisica, Università di Roma I, 1989; Physica D (in press).
17. V.E. Zakharov and P.B. Shabat, Sov. Phys. JETP, **34** (1972) 62.
18. M. Boiti, J.J.-P. Leon and F. Pempinelli, A New Spectral Transform for the Davey- Stewartson Equation, PM /89 -10, Montpellier.
19. C.S. Gardner, J. M. Green, M. D. Kruskal and R.M. Miura, Comm. Pure Appl. Math., **27**, (1974) 97.
20. F. Calogero and A. Degasperis, Spectral Transform and Solitons I, North Holland, 1982.
21. S.V. Manakov, Physica **3D**, (1981) 420.
22. A.S. Fokas and M.J. Ablowitz, Stud. Appl. Math., **69**, (1983) 211.
23. I.S. Gradshtein and I.M. Ryzhik, Tables of Integrals, Series and Products, Academic Press, New York, 1980.
24. D. Anker and N. C. Freeman, Proc. R. Soc. Lond. A. **360**, (1978) 529.
25. B. A. Dubrovin, T. M. Malanyuk, I. M. Krichever and V. G. Makhan'kov, Sov. J. Part. Nucl. **19**, (1988) 252.

Ginzburg-Landau models of non-equilibrium

P. Coullet and **L. Gil**
Laboratoire de Physique Théorique
Parc Valrose. 06034 Nice Cedex.

Abstract. This paper is devoted to study the basic symmetry breaking phenomena that one encounters in non-equilibrium physics. We discuss the formation of patterns which break spatial, temporal and spatio-temporal translations.

1. INTRODUCTION

Transitions in physics are generally associated with symmetry changes. In particular, phase transitions in condensed matter physics provide the most famous examples of spontaneously broken symmetry [1]. When one drives a physical system far from its thermodynamical equilibrium, another kind of transitions takes place, which leads to the spontaneous appearance of patterns. These transitions, induced by instabilities, are also accompanied by symmetry changes. The paradigm of such an instability is the Rayleigh-Bénard convection [2] (see Fig. 1). A horizontal fluid layer of infinite extention is submitted to a vertical temperature gradient. For low heating the fluid is at rest and the heat transfert is provided by the conduction of

R. Conte and N. Boccara (eds.), Partially Integrable Evolution Equations in Physics, 261–275.

the material. Above a critical temperature gradient, this state becomes unstable and the convection sets in. The convective state, characterized by a finite wavevector k_0, breaks the spatial homogeneity of the system. This symmetry breaking transition can be described in the spirit of Landau [3]. A complex order parameter A, which measures the "amount of broken symmetry" is defined as follows: near the convection onset, the vertical velocity W reads

$$W(x,y,z,t) = A(x,y,t)e^{(ik_0x)}\Phi(z) + c.c + ...$$

where Φ takes into account the vertical motion of the fluid and the dots represent corrections which involve the harmonics of the fundamental wavenumber k_0. The next step consists in looking for an equation for the order parameter. In equilibrium physics, such an equation is obtained by minimizing the free energy of the system. In non-equilibrium situations, the existence of such a variational principle is not typical. The equation for the order parameter takes the form of a partial differential equation, which reads

$$\frac{\partial A}{\partial t} = \mu A + \xi(\partial_x - \frac{i}{2k_0}\partial_y^2)^2 A - |A|^2 A \tag{1.1}$$

where μ is related to the distance from the instability threshold and ξ is a positive constant. This equation describes the weakly nonlinear convection. Although first derived in the case of the Raleigh-Bénard convection [4] [5], this equation has a much wider range of application: it describes the generic breaking down of the invariance under space translation [6].

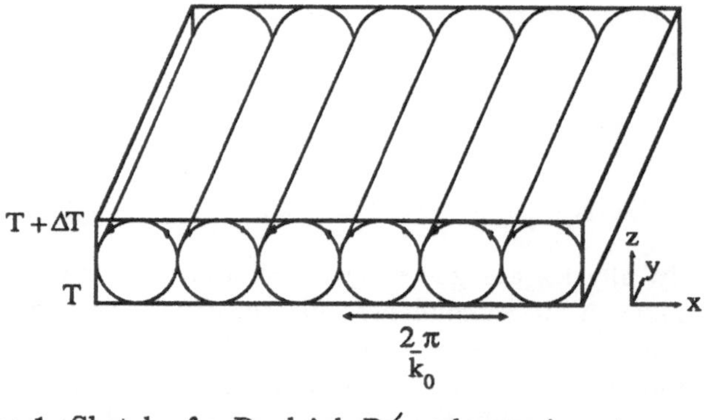

Figure 1: Sketch of a Rayleigh-Bénard experiment.

In this paper we are interested in the basic symmetry breaking phenomena that one encounters in non-equilibrium physics. We successively study the appearance of patterns which break spatial, temporal and spatio-temporal translations.

2. BREAKING SPACE TRANSLATIONS

A. Symmetry considerations.

In order to derive the form of the amplitude equation (1.1) let us consider a physical system invariant with respect to space translation, space rotation, parity and time translation and whose dynamics is governed by a partial differential equation

$$\frac{\partial C}{\partial t} = F(C, \mu) \tag{2.1}$$

where $C = \{C_1, ..C_N\}$ is a set of scalar fields and μ is a given control parameter. Let C_0 be a stationary, homogeneous solution of equation (2.1). Its linear stability is related to the eigenvalues σ of the jacobian matrix $L = \partial F/\partial C|_{C_0}$. The translational invariance enable us to search solutions of the linear part of equation (2.1) of the form $C = c_{\mathbf{k}} \exp(\sigma_{\mathbf{k}} t) \exp(i\mathbf{kr})$. Because of the rotational invariance the eigenvalues $\sigma_{\mathbf{k}}$ are only function of $\mathbf{k}^2$. For μ slightly positive we assume an instability to develop modes with wave-vector such that $\mathbf{k}^2 \simeq \mathbf{k_0}^2$. The associated branch of eigenvalues reads (See Fig. 2)

$$\sigma_{\mathbf{k}} = \mu - \xi(\mathbf{k}^2 - \mathbf{k_0}^2)^2 + O((\mathbf{k}^2 - \mathbf{k_0}^2)^3) \tag{2.2}$$

where ξ is a given positive constant.

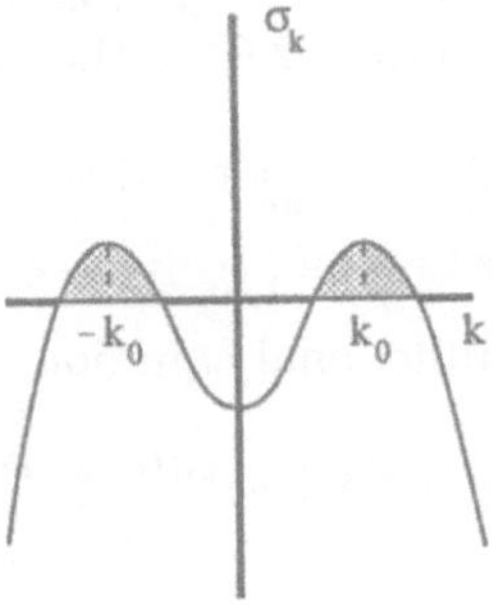

Figure 2: Graph of σ_k.

We next assume that the non-linear evolution selects a single mode solution whose wave-vector can be choosen along the x axis Near $\mu = 0$ we look for a solution of equation (2.1) under the form

$$C = C_0 + A(x, y, t)\chi \exp(ik_0 x) + cc. + H(A, \bar{A}; x, y, t; \mu) \tag{2.3}$$

where $\chi \equiv \chi_{k_0}$ and the amplitude A is supposed to be small and slowly varying in x, y and t and H represents terms of higher order. This expression is a solution of equation (2.1) only if A satisfies some adequate condition. This condition can be found in a systematic way by expanding all quantities in terms of a small parameter which measures the smallness of A, $\partial_t A$ and its gradients and solving linear equations at successive orders. Here we first assume that the solvability condition takes the form of a partial differential equation

$$\frac{\partial A}{\partial t} = f(A, \bar{A}; x, y, t) \tag{2.4}$$

and show that the symmetries of the physical system strongly restrict the form of f. The translational invariance in space and time implies that

$$e^{(-ik_0 a)} f(Ae^{(ik_0 a)}, \bar{A}e^{-(ik_0 a)}; x + a, y + b, t + \theta) = f(A, \bar{A}; x, y, t) \tag{2.5}$$

The parity symmetry implies that

$$f(\bar{A}, A; -x, y, t) = \bar{f}(A, \bar{A}; x, y, t) \tag{2.6.a}$$

$$f(A, \bar{A}; x, -y, t) = f(A, \bar{A}; x, y, t) \tag{2.6.b}$$

One derives these equalities from the fact that if $C(x, y, t)$ is a solution of equation (2.1) then $C(x + a, y + b, t + \theta)$, $C(-x, y, t)$ and $C(x, -y, t)$ are also solutions. We search solutions of equations (2.5) (2.6) of the form

$$f(A, \bar{A}, x, y, t) = \sum_{p,q,n_p,m_p,n_q,m_q} f_{pq}^{n_p,m_p,n_q,m_q}(x, y, t) \partial_x^{n_p} \partial_y^{m_p} A^p \partial_x^{n_q} \partial_y^{m_q} \bar{A}^q \tag{2.7}$$

The f_{pq} have to satisfy the functional equation

$$f_{pq}^{n_p,m_p,n_q,m_q}(x + a, y + b, t + \theta) e^{i((p-q-1)k_0 a)} = f_{pq}^{n_p,m_p,n_q,m_q}(x, y, t)$$

This implies that

$$\frac{\partial f_{pq}}{\partial t} = \frac{\partial f_{pq}}{\partial y} = 0$$

$$\frac{\partial f_{pq}}{\partial x} + ik_0(p - q - 1) f_{pq} = 0$$

From these equations it follows that

$$f_{pq}(x, y, t) = f_{pq} e^{-i((p-q-1)k_0 x)} \tag{2.8}$$

Unless $p = q+1$, the amplitude equation would contain terms which are not slowly varying in x. In order to satisfy the hypothesis of slow variations of A, all the $f_{p,q}$ with $p \neq q+1$ have to be taken to zero. The vanishing of these coefficients is a direct consequence of the averaging involved in the systematic derivation of equation (2.4). Finally when equations (2.6) are also taken into account we are left only with terms of the form

$$(i)^{n_p+n_q} f_{p,q}^{n_p,m_p,n_q,m_q} \partial_x^{n_p} \partial_y^{m_p} A^p \partial_x^{n_q} \partial_y^{m_q} \bar{A}^q \tag{2.9}$$

where $m_p + m_q = 2n$ and $f_{p,q}^{(n_p,m_p,n_q,m_q)}$ real.

Combining the information due to the linear analysis and some scaling arguments the linear part of the amplitude equation can be obtained easily. Let $A(x,y,t) = a_{Q,P} \exp i(Qx+Py) \exp(\sigma_{Q,P})t$. Linearizing equation (2.4) allows one to get $\sigma_{Q,P}$ as an expansion in term of P, Q. A similar expression can be obtained directly from the eigenvalue (2.2) of the original equation. Let $\mathbf{k} = (k_0 + Q)\mathbf{i} + P\mathbf{j}$, where $\mathbf{i}$ and $\mathbf{j}$ are respectively the unit vectors along the x and y axis. Using this expression in equation (2.1) one readily identifies the coefficients of $\sigma_{Q,P}$.

$$\sigma_{\mathbf{k}} = \sigma_{Q,P} = \mu - \xi(2Qk_0 + Q^2 + P^2)^2 + .. \tag{2.10}$$

The optimal balance between the various terms in equation (2.10) implies that Q^2 and P^4 must be of the order μ. Then at the order μ the eigenvalue $\sigma_{Q,P}$ reads

$$\sigma_{Q,P} = \mu - 4k_0^2\xi(Q + \frac{P^2}{2k_0})^2 \tag{2.11}$$

The simplest non-linear term in equation (2.4) which is compatible with the symmetry contraints is $|A|^2A$. Choosing $|A|^2$ of the order μ and introducing appropriate scalings of order unity we finally get the amplitude equation (1.1)

$$\frac{\partial A}{\partial t} = \mu A + (\partial_x - \frac{i}{2k_0}\partial_y^2)^2 A \pm |A|^2 A \tag{2.12}$$

In the following we will restrict ourselves to the case with the minus sign, since it leads to the appearance of stable patterns.

B. Perfect Patterns.

Perfect periodic patterns, solutions of $\mu A - |A|^2 A + \frac{\partial^2 A}{\partial x^2} = 0$ are given by (see Fig.3)

$$A_Q(x) = \sqrt{\mu - Q^2} \exp i(Qx + \phi) \tag{2.13}$$

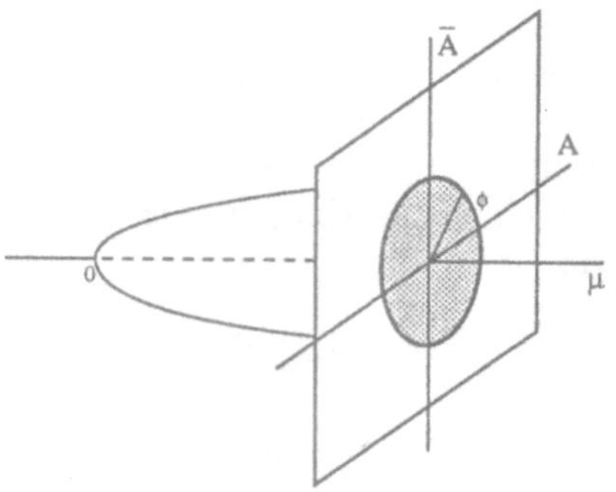

Figure 3: Bifurcation diagram associated with the family of solutions given by equation (2.13)

In order to study their stability, we look for a solution of equation (2.12) under the form $A = \sqrt{\mu - Q^2}(1 + a(x, y, t)) \exp i(Qx + \phi)$. The corresponding linear equation reads

$$\frac{\partial a}{\partial t} = -(\mu - Q^2)(a + \bar{a}) + 2iQ(\partial_x - \frac{i}{2k_0}\partial_y^2)a + (\partial_x - \frac{i}{2k_0}\partial_y^2)^2 a \qquad (2.14)$$

We first consider perturbations depending only on the variable x. One looks for a solution of equation (2.14) under the form

$$a = \tilde{a} \exp i(k_x x) \exp(\sigma t) \qquad (2.15)$$

The corresponding eigenvalues $\sigma_\pm$ are given by

$$\sigma_+ = -\frac{\mu - 3Q^2}{\mu - Q^2} k_x^2 + O(k_x^4) \qquad (2.16.a)$$

$$\sigma_- = -2(\mu - Q^2) - k_x^2 + O(k_x^4) \qquad (2.16.b)$$

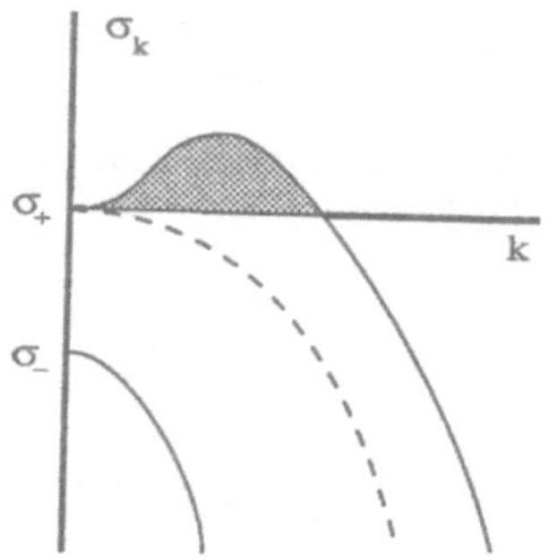

Figure 4: Graph of σ_k. The dashed σ_+-curve corresponds to the stable case, while the solid one is associated with phase instability.

Thus periodic patterns with wavelength too far ($|Q| > \sqrt{\frac{\mu}{3}}$) from the band center ($Q = 0$) are found to be unstable with respect to longwavelength modes. This instability, kwown as the Eckhaus instability [7], has received recently a lot of interest both from theoretical [8] and experimental [9] points of view. Perturbations in the y direction lead to another type of instability. Let $a = \tilde{a} \exp i(k_y y) \exp(\sigma t)$. The two eigenvalues $\sigma_\pm$ are now given by (see Fig. 4)

$$\sigma_+ = -\frac{Q}{2k_0} k_y^2 + O(k_y^4) \tag{2.17.a}$$

$$\sigma_- = -2(\mu - Q^2) - \frac{Q}{2k_0} k_y^2 + O(k_y^4) \tag{2.17.a}$$

For negative Q (subcritical wavenumber) an instability takes place and gives rise to a large scale wavy perturbation transverse to the pattern. These two longwavelength instabilities have a common origin. The marginal mode ($k_x = k_y = 0$) corresponds to a phase translation of the order parameter. Thanks to the translational invariance both $A_Q = \sqrt{\mu - Q^2} \exp i(Qx)$ and $A_Q = \sqrt{\mu - Q^2} \exp i(Qx + \phi)$ are solutions of equation (2.12). For ϕ small enough

$$A_Q = \sqrt{\mu - Q^2} \exp i(Qx + \phi) = \sqrt{\mu - Q^2}(1 + i\phi + ..) \exp i(Qx) \tag{2.18}$$

The instabilities thus appear to be related with the existence of this marginal mode. This remark leads to a very elegant formalism which consists in interpreting the instability as a *phase instability* [10] [11] [12]. It is convenient to let ϕ depend slowly on x and y. From equation (2.12) it is then possible to derive an equation for the phase only using similar arguments which allow one to reduce equation (2.1) to equation (2.4). This *phase equation*, first derived in the context of the Rayleigh-Bénard convection [12] reads

$$\frac{\partial \phi}{\partial t} = \frac{\mu - 3Q^2}{\mu - Q^2} \frac{\partial^2 \phi}{\partial x^2} + \frac{Q}{2k_0} \frac{\partial^2 \phi}{\partial y^2} + ... \tag{2.19}$$

C. Defects.

Besides the regular solutions described above, the amplitude equation possesses more singular solutions, which correspond to defects [13]. The core of a defect is a point where the complex order parameter vanishes (see Fig. 5.a).

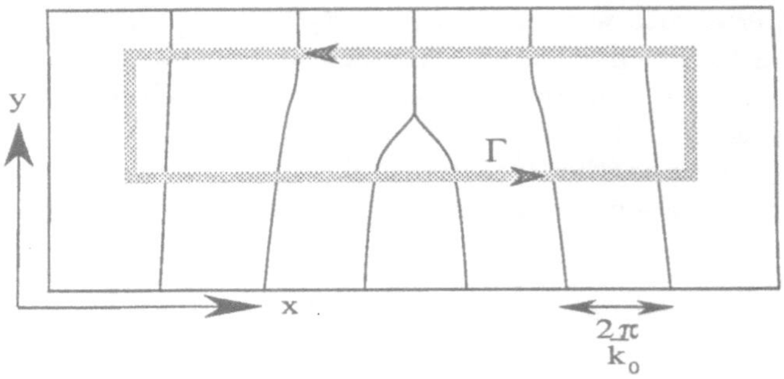

Figure 5.a: Zeroes of a complex field in two dimensions.

The existence and stability of such a solution follow from elementary topological considerations. Far from the core, the defect reaches the homogeneous solution $A_0 = \sqrt{\mu}\exp(i\phi)$. The corresponding solution of equation (2.12) is given by [14]

$$A_D(x,y) = R(x,y)\exp(i\phi(x,y)) \tag{2.20}$$

where $R(x,y)$ goes to $\sqrt{\mu}$ for large x, y and vanishes at the core of the defect. The gradient of its phase has a circulation of $\pm 2\pi$ on any path Γ surrounding its core

$$\oint_\Gamma \nabla\phi \bullet d\mathbf{l} = \pm 2\pi \tag{2.21}$$

The defect solution takes the form of a dislocation. It corresponds to the insertion of an elementary periodic cell.

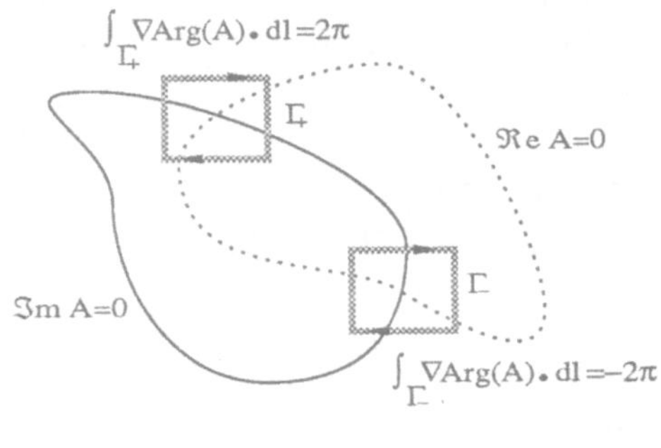

Figure 5.b: A dislocation.

Since a defect is characterized by a zero of the complex order parameter, and since zeroes of a complex field exist generically in two dimensions, one can expect that a typical pattern, for $\mu > 0$, will contain many defects. Spontaneous creation of defects is also likely to occur near the Eckhaus instability [15].

3. BREAKING TIME TRANSLATIONS

A. Temporal order.

Instead of a spatial order, a pattern may induce a temporal order. This is the case for instance in chemical reactors, where beyond a critical value of the flux of reactives, a reference concentration C_i begins oscillating periodically in time [16]. This concentration can be written as:

$$C_i = C_{i,0} + A(x, y, t) \exp i\omega_0 t + c.c. + ..., \tag{3.1}$$

where $\frac{2\pi}{\omega_0}$ is the period of the temporal oscillation. Here again, A is slowly varying in space and time. The symetries of the physical system are assumed to be the same as in the previous case, but since the bifurcated solution does not break any spatial symmetry, the amplitude equation for the order parameter A has to be isotropic and invariant under space translations. The symmetry breaking of time translations implies that this equation has to be invariant under the transformation $A \to A \exp i\phi$. Thus, the amplitude equation reads:

$$\frac{\partial}{\partial t} A = \mu A + (1 + i\alpha)\Delta A - (1 + i\beta)|A|^2 A \tag{3.2}$$

This equation admits a class of homogeneous solutions $A_0 = \sqrt{\mu} \exp i(-\beta\mu t + \phi)$, where ϕ is arbitrary. As before, we study the stability of such solutions with respect to small perturbations. The associated eigenvalues problem possesses two solutions, which read in the limit of small $|k|$:

$$\sigma_\pm = -\mu - ({k_x}^2 + {k_y}^2) \pm \mu[1 - \frac{\alpha\beta}{\mu}({k_x}^2 + {k_y}^2) + ...] \tag{3.3}$$

The eigenvalue σ_- is negative and corresponds to a stable amplitude mode. If $1 + \alpha\beta$ is negative, σ_+ is positive, and the associated perturbation exponentially increases. Since the eigenvalue σ_+ is associated with the phase mode, a phase instability occurs [11].

Figure 6.a: A numerical simulation of equation (3.2) showing a spiral wave solution.

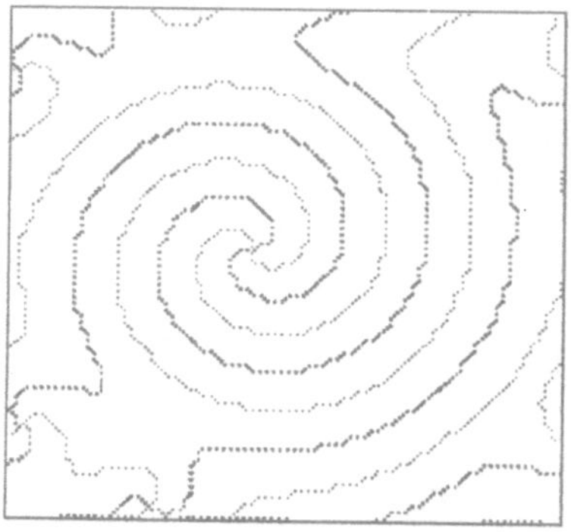

Figure 6.b: Equiphases of the solution presented in Fig. 6.a.

The defects associated with the temporal order take the form of spiral waves, which propagate out of the core [11] (see Fig. 6). They are the analogs of the dislocations in the roll patterns. The order parameter vanishes at the core of the defect, and the lines of the plane (x,y) where $\Re e A = 0$ and $\Im m A = 0$ cross themsleves at this point.

b. Temporal disorder.

A simpler answer to the problem of the creation of defects is that they are induced by some inhomogeneities of the initial conditions. Actually the dynamics of a non-equilibrium system can itself produce defects. In spatially extended system, phase instability acts as a deterministic noise generator which allows the creation of pairs. Some of these pairs quickly dissociate while new pairs appear and other annihilate.

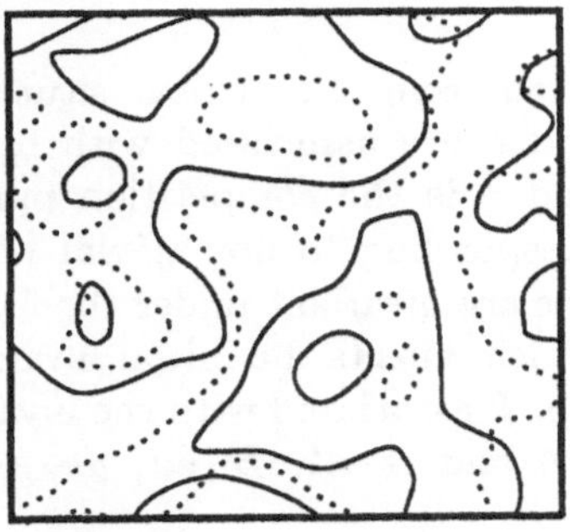

Figure 7: A numerical simulation of equation (3.2) in the unstable phase regime ($1 + \alpha\beta < 0$). The dashed lines represent the locus of points where $\Re e A = 0$ while the solid lines correspond to those where $\Im m A = 0$. The figure displays a complex pattern with many defects.

This spatio-temporally complex state which consists in an assembly of free topological defects moving in a "chaotic" phase field has been termed as "topological turbulence" [17] (see Fig. 7).

4. BREAKING SPACE-TIME TRANSLATIONS

Our last example is associated with the appearance of a wavy structure. This occurs for instance in a Rayleigh-Bénard experiment with a binary mixture [18] [19] [20]. Beyond a critical value of the difference of temperature between the two plates, a system of propagating rolls can be observed. For example, the temperature in a median plane between the two plates can be written as:

$$Q = Q_0 + A \exp i(\omega_0 t + k_0 x) + B \exp i(\omega_0 t - k_0 x) + c.c. + \dots \qquad (4.1)$$

The order parameters A and B obey the following amplitude equations [21] [22] [23]

$$\begin{aligned} \frac{\partial}{\partial t}A = &-c\frac{\partial}{\partial x}A + \mu A + (1+i\alpha)\left(\frac{\partial}{\partial x} - \frac{i}{2k_0}\frac{\partial^2}{\partial y^2}\right)^2 A \\ &+ i\epsilon\left(\frac{\partial^2}{\partial x^2} + \zeta\frac{\partial^2}{\partial y^2}\right)A - (1+i\beta)|A|^2 A - (\gamma + i\delta)|B|^2 A \end{aligned} \qquad (4.2.a)$$

$$\frac{\partial}{\partial t}B = c\frac{\partial}{\partial x}B + \mu B + (1+i\alpha)(\frac{\partial}{\partial x} + \frac{i}{2k_0}\frac{\partial^2}{\partial y^2})^2 B$$
$$+ i\epsilon(\frac{\partial^2}{\partial x^2} + \zeta\frac{\partial^2}{\partial y^2})B - (1+i\beta)|B|^2 B - (\gamma + i\delta)|A|^2 B \tag{4.2.b}$$

where μ measures the deviation from the critical situation, c, α, ϵ and ζ describe dispersive effects, β and δ are associated with nonlinear renormalization of the temporal frequency, and γ is the competition parameter between travelling and standing waves, corresponding to non trivial homogeneous solutions of equation (4.2). These equations are invariant under the following transformations: $A \to A\exp -i\Phi, B \to B\exp i\Phi$ which reflects the initial invariance under space translations, $A \to A\exp i\Psi, B \to B\exp i\Psi$ associated with the invariance under time translations, $x \to -x$, $A \to B$, $B \to A$ and $A \to A$, $B \to B$, $y \to -y$ related to the parity symmetries.

The homogeneous solutions of equation (4.2) are of two kinds: the travelling waves:

$$A = \sqrt{\mu}\exp i(-\beta\mu t + \phi_a), \qquad B = 0 \tag{4.3.a}$$

and

$$A = 0, \qquad B = \sqrt{\mu}\exp i(-\beta\mu t + \phi_b) \tag{4.3.b}$$

and the standing waves:

$$A = \sqrt{\frac{\mu}{1+\gamma}}\exp i(-\frac{\beta+\delta}{1+\gamma}\mu t + \phi_a), B = \sqrt{\frac{\mu}{1+\gamma}}\exp i(-\frac{\beta+\delta}{1+\gamma}\mu t + \phi_b) \tag{4.4}$$

The former are stable with respect to homogeneous perturbations when $\gamma > 1$, the latter when $-1 < \gamma < 1$. The real part of eigenvalues of the linearized problem near the travelling wave solutions reads :

$$\Re e(\sigma)_\pm = -\mu - ({k_x}^2 + \frac{{k_y}^4}{4k_0}) \pm \mu[1 - \frac{\beta\epsilon\zeta}{k_0\mu}{k_y}^2 - \frac{\beta\epsilon}{k_0\mu}{k_x}^2$$
$$- \frac{(\epsilon\zeta)^2}{2\mu^2}(1+\beta^2){k_y}^4 - \frac{\alpha\beta}{\mu}({k_x}^2 + \frac{{k_y}^4}{4{k_0}^2}) + ...] \tag{4.5}$$

The real part of σ_- is always negative when γ is greater than one. The real part of σ_+ becomes positive if $1 + \alpha\beta + \frac{\epsilon\beta}{k_0}$, or $\beta\epsilon\zeta$, or $\frac{1}{{k_0}^2}[1 + \alpha\beta + (1+\beta^2)\frac{2(\epsilon\zeta k_0)^2}{\mu}]$ are negative, and thus a phase instability may occur.

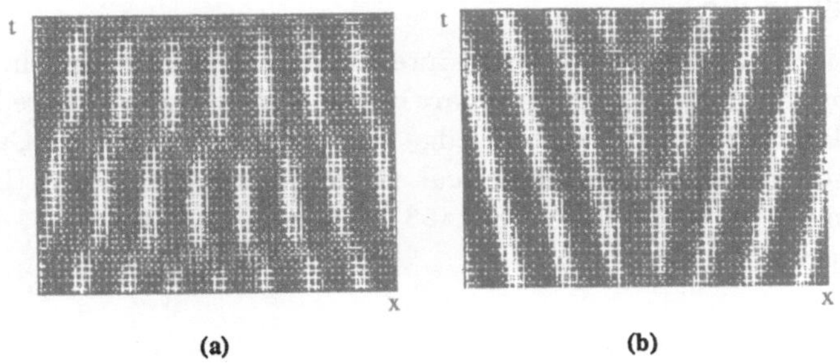

Figure 8.a: $x - t$ diagrams of a solution of equations (4.2) displaying (a) a sink, (b) a source of travelling waves.

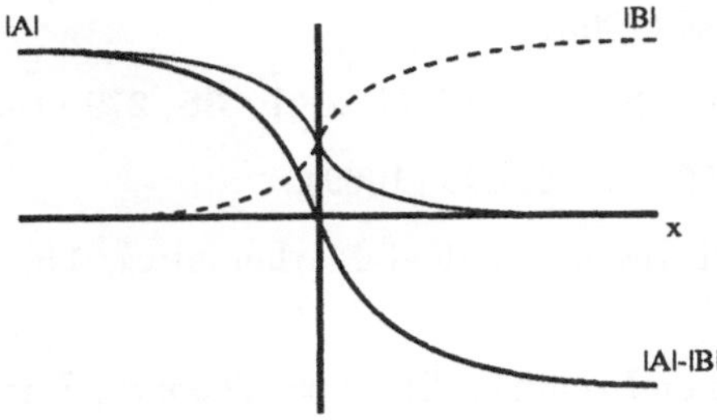

Figure 8.b: Sketch of the kink-like defect corresponding to the figure 8.a.

Defects of travelling waves are of two kinds [23]. The first kind, associated with the spontaneous breaking down of the parity can be seen as domain walls separating left and right propagating waves (see Fig. 8). Theses defects can be either sources or sinks. The other kind of defects is associated with the breaking down of spatio-temporal translations. It corresponds to point defects. At the core of the defect the complex amplitude of the wave vanishes as in the cases of pure static and temporal patterns. Such defects take the appearance of moving dislocations. Phase instability, as in the case of temporal patterns, is likely to generate a turbulent state associated with these topological defects.

Acknowledgements

J. Lega is ackowledged for many interesting discussions and her help in correcting this manuscipt. The original part of the work presented here has been done with her collaboration. We acknowledge the CCVR (Centre de Calcul Vectoriel pour la Recherche) where the numerical simulations presented in this paper have been performed, the DRET (N^0 881453) and the EEC (N^0 SC1-0035-C) for a financial support.

References

1. P.W. Anderson, "Basic Notions of Condensed Matter Physics" in Frontiers in Physics (1984).

2. F.H Busse, Rep. Prog. Phys. **41**, 1929 (1978).

3. L.D. Landau, "On the Theory of Phase Transitions" in "Collected Paper of L.D. Landau" Pergamon Press (1965).

4. A. C. Newell and Whitehead, J. Fluid Mech. **38**, 279 (1969).

5. L.A. Segel, J. Fluid Mech. **38**, 203 (1969).

6. A.C. Newell in Lectures in Applied Mathematics, **15**, Am. Math. Society, Providence (1974).

7. W. Eckhaus , "Studies on Nonlinear Stability Theory", Springer Tracts in Natural Philosophy, **6**, Springer Berlin (1965).

8. L. Kramer and W. Zimmermann, Physica **16D**, 221 (1985).

9. M. Lowe and J.P. Gollub, Phys. Rev. Lett. **55**, 2575 (1985).

10. P. Ortovela and J. Ross, J. Chem. Phys. **58**, 5673 (1973).

11. Y. Kuramoto, Prog. Theor. Phys. **71**, 1182 (1984).

12. Y. Pomeau and P. Manneville, J. Phys. Lettres **40**, 609 (1979).

13. "Physics of Defects", Ed. R.Balian, M.Kleman and J.P.Poirier (North Holland, Amsterdam, 1980)

14. J. Toner and D.R. Nelson, Phys. Rev. **23B**, 316 (1981).

15. "Structure and dynamics of dislocations in anisotropic pattern forming systems", E. Bodenschatz, L. Kramer and W. Pesch, (Preprint 88).

16. Y. Kuramoto, "Chemical Oscillations, Waves and Turbulence", Springer Series in Synergetics **19** (1984).

17. P. Coullet, L. Gil and J. Lega, Phys. Rev. Lett. **62**, 1619 (1989).

18. E.Moses and V.Steinberg, Phys. Rev. **A 34**, 693 (1986).

19. R.W.Walden, P.Kolodner, A.Passner and C.M.Surko, Phys. Rev. Lett. **55**, 496 (1985).

20.I.Rehberg and G.Ahlers, Phys.Rev.Lett. **55**, 500 (1985)

21. P.Chossat and G.Iooss, Japan J. Appl. Maths. **2**, 37 (1985).

22. P.Coullet, S.Fauve and E.Tirapegui, J.Phys.Lett. **46**, 787 (1985).

23. P.Coullet, C.Elphick, L.Gil, and J.Lega, Phys. Rev. Lett. **59**, 884 (1987).

Nonlinear Ginzburg-Landau equation and its applications to fluid-mechanics
An abstract by

Yves Pomeau*,
Laboratoire de Physique Statistique de l'ENS,
24, Rue Lhomond, F-75231 Paris Cedex 05,
France.

The nonlinear Ginzburg-Landau equation is a convenient tool for studying the growth of unstable perturbations in shear flows. In particular, it allows [1] to understand the link between the subcritical character of the instabilities and the formation of localized patches of turbulence. This is consistent with the early observations[6] that those patches grow at constant speed in space with a rather homogeneous state inside. However when applied to real flows, this picture may be changed for two reasons. First there is a kind of very long range interaction through the pressure field in incompressible fluid mechanics and this, together with symmetry considerations explains the remarkable phenomenon of spiral turbulence [2] between counterrotating Taylor-Couette cylinders. Then other complications arise because of the basically non variational structure of the equations of fluid mechanics as well as of Ginzburg-Landau. This has two main consequences, both related to some experimental features:

1) The sign of the velocity of expansion alluded above may depend on the orientation in a basically anisotropic system as a shear flow [3].

2) Then the Ginzburg -Landau equations may have stable solutions in the form of solitary waves [4,5]. This can be understood [5] by pertubation near the two opposite limits: unitary evolution and pure gradient flow and those solitary waves merge continuously with expanding domains by becoming infinitely wide.

References:

[1] Y.Pomeau, Physica 23D,3, (1986);

[2]J.J.Hegseth, C.D.Andereck, F.Hayot, Y.Pomeau, Phys.Rev.Letters,62, 257(1989);

[3]Y.Pomeau, Proceedings of the Conference on "Fluctuations and Instabilities" , held

R. Conte and N. Boccara (eds.), Partially Integrable Evolution Equations in Physics, 277–278.

at Valparaiso (Chile), Dec. 1987, ed. by E.Tirapegui and L.Villaroel, Reidel Pub;
[4] O.Thual and S.Fauve, J.de Phys. 49,1829(1988);
[5] V.Hakim, P.Jakobsen and Y.Pomeau, preprint (July 1989);
[6] I.J. Wygnansky and F.H.Champaigne, J.of Fluid Mech. 135, 27 (1973), and ref. quoted therein.

*Part of this work has been done when the author was visiting jointly the departement of mathematics and the Center for Complex systems at the University of Arizona, Tucson in the Spring of 1989. This is the summary of a talk presented at the Winter school at Les Houches, France , March 21-29, 1989.

DEFECTS AND DISORDER OF NON-LINEAR WAVES IN CONVECTION

R. Ribotta and A. Joets
Laboratoire de Physique des Solides, Bât.510
Université de Paris–Sud
91405 Orsay Cedex
France

ASTRACT . The traveling–wave convection of an anisotropic fluid (a nematic liquid crystal) constitutes a novel model of nonlinear waves. We show experimentally that instabilities of the phase arise in form of localized states inducing shocks, which in turn trigger the nucleation of defects. The conditions for the nucleation, the topology and the role of these defects in the progressive disorganization of the basic wavetrain are studied. The main results can be satisfactorily simulated by using an evolution equation of the complex Newell–Landau–Ginzburg type.

1. INTRODUCTION

The evolution of non-linear dynamical systems in extended space is described by the use of partial differential equations. For instance fluid systems, subjected to an external stress R, undergo transitions to finite amplitude states above a critical value of the stress parameter R_c. These systems usually exhibit either a spatial stationary state or a spatio-temporal one and their stability can be studied by the evolution of the amplitude of the linear basic state.

The Landau-Ginzburg (L.G.) equation is certainly one of the most studied models since it is known to govern the evolution of a large number of non-equilibrium extended systems close to onset [1–4]. It has been the subject of numerous numerical and analytical studies, most often when restricted to one space dimension [5–9]. Particularly interesting solutions to the L.G. equation are time–dependent states in form of propagative space–periodic structures ("traveling waves"). Besides the study of the transitions to spatially homogeneous states of different symmetry, the important problem that concerns the growth of localized initial disturbances has also been recently raised [10]. On the other hand, singular solutions giving rise to defects in extended homogeneous states have been shown to exist stably as topological singularities [11,12].

In this paper, we report on both experimental and numerical studies of the disordering (disintegration) of a one dimensional traveling-wave that is the simplest time–dependent state in convection [13]. It is shown that localized disturbances occur as modulations of the phase and destabilize the wave. Defects of the ordering result from these local instabilities and they are found to mediate a transition to disorder.

R. Conte and N. Boccara (eds.), Partially Integrable Evolution Equations in Physics, 279–297.

2 THE EVOLUTION EQUATION FOR NONLINEAR WAVES

The local value of a physical quantity Q representing the state of the system can be written as :

$$Q(x,z,t) = \epsilon\left\{\mathrm{Re}[A(X,T)\exp i(\omega_0 t - k_0 x) + B(X,T)\exp i(\omega_0 t + k_0 x)].f(z)\right\} + \\ + O(\epsilon^2).$$

in which A and B are assumed to vary slowly in space and in time and f(z) represents the fast variation along z (i.e. across the layer of fluid). The slow scales X,T are such that $X = \sqrt{\epsilon}\,x$, $T = \epsilon t$, and ϵ represents the relative stress parameter $(R-R_c)/R_c$. A and B are associated with respectively a right and a left traveling wave, and satisfy the two coupled Newell-Landau-Ginzburg equations [12,14]:

$$\frac{\partial A}{\partial T} = \mu A - c\frac{\partial A}{\partial X} + (1+i\alpha)\frac{\partial^2 A}{\partial X^2} - (1+i\beta)\,|A|^2A - (\gamma+i\delta)\,|B|^2A$$

$$\frac{\partial B}{\partial T} = \mu B + c\frac{\partial B}{\partial X} + (1+i\alpha)\frac{\partial^2 B}{\partial X^2} - (1+i\beta)\,|B|^2B - (\gamma+i\delta)\,|A|^2B$$

The coefficient α measures the wave dispersion, β couples the amplitude and the frequency of the wave, and $\mu = \epsilon^2$. The dispersion and the nonlinearities can compete and if $1+\alpha\beta < 0$, the basic solution $A = \sqrt{\mu}\exp(-i\beta\mu t)$, $B = 0$, is unstable to sideband long–wavelength perturbations [15], in a way analogous to the Benjamin-Feir instability of water waves [16,17]. Solutions to this equation correspond to :

–single traveling wave (TW): $A \neq 0$, $B=0$ (right–going wave), or $B \neq 0$ with $A = 0$ (left–going wave),

–standing waves (SW): $|A| = |B|$.

Intermediate states $|A| \neq |B| \neq 0$ are often encountered as connecting the two previous solutions and they correspond to modulated traveling waves (MTW) although they would rather appear as solutions of an equation with higher-order terms in the expansion.

Besides the stability analysis of the above mentioned regular solutions against homogeneous perturbations, one may look for the effect of local perturbations of the wavetrains solutions. Bretherton and Spiegel [18] have shown that localized states of phase instability fluctuating in time are also solutions when the initial data is a noise. Also Bekki and Nozaki [10] who studied the natural evolution in the absence of initial noise have proposed that soliton–like solutions as well as shocks may be obtained from a Landau–Ginzburg equation. However it is only recently that the nonlinear phase dynamics of a propagating pattern were studied in a slightly different model by Kuramoto [19] who showed that the main nonlinearity of the phase equation comes from local velocity changes. Coullet et al [12] have studied the sinks and sources as topological defects of waves and have numerically shown the essential feature of these defects using a set of two coupled TDLG equations. Nozaki and Bekki [6,10] have examined the effect of initially localized disturbances on the global stability. More recently, following an approach first established by Whitham [20] and used by Howard and Kopell [21], Bernoff has studied for the Landau-Ginzburg equation, the evolution of slowly varying modulations of the phase of the waves [22]. Some of his results concerned the appearance of shocks following phase modulations, but defects do not appear in this study.

Our concern is to study the stability of uniform wavetrains which are spatio–temporal ordered structures, against slow and localized modulation of the phase $\varphi = (\omega t \pm kx)$. We shall show experimentally and numerically, that slow wavetrain modulations may become locally unstable and trigger shocks. As a consequence of a shock, spatio–temporal defects are created in the wave pattern. One particularly interesting feature of some defects is the reversal of the velocity over a well–defined portion of the space. The progressive disintegration of the initial wave may further be achieved after the random creation of an increasing number of such defects, along with an erratic motion in both directions $\pm \mathbf{x}$. It is then found that this increasing complex behavior may lead in a finite time to a chaotic state [13].

The experiments are made using the state of traveling convective rolls that can be developed in a thin layer of fluid. More precisely, our system consists of a nematic liquid crystal, which is an anisotropic fluid, driven to convection by an electric field.

The experimental results are compared with numerical simulations of the 1–D coupled Newell–Landau–Ginzburg. We use a split–step integration method involving the calculation of Fourier transforms over 256 (or 512 when necessary for a better definition) collocation points.

3. DEFECTS IN TWO–DIMENSIONAL STRUCTURES DEFINED BY A SINGLE WAVEVECTOR

The evolution of the 1–D traveling waves, as well as their defects, will be represented for convenience in the {x,t} space. The corresponding space–time ordered structure is then defined by a single "wavevector". At this point it is useful to recall the elementary defects that exist inside two–dimensional structures (in real x,y space) which are defined by a single wavevector $\mathbf{k} = k\,\mathbf{x}$. This description applies to patterns of parallel rolls as well as to layered systems. We consider, as an example, the case of defects occuring in the ordered structures of *stationary* rolls that are also developed in the same system but under different experimental conditions [23].

Defects correspond to singularities in the ordering and hence are not expressed as analytical solutions to the governing equations for the convective structure. In the layer of stationary convective rolls, the flow is described by the fluid variables Q(r), which vary over distances equal to the roll diameter d, and by the wavevector $\mathbf{k}$ which is, at first, assumed to be constant over the horizontal coordinates. The magnitude of $\mathbf{k}$ is defined as $|\mathbf{k}| = |\nabla\theta(\mathrm{r})|$, where $\theta(\mathrm{r})$ is the phase in space (i.e. the relative position of the roll), associated with the spatially periodic flow of stationary amplitude $Q(\mathrm{r}) \sim A\,(\mathrm{r}).\exp(i\theta)$. In such a 2–D structure defined by a unique wavevector, two classes of point–defects may exist and they correspond either to a singularity in the wavevector direction (orientational dislocation) or to a discontinuity in its magnitude (translational dislocation). The first type is a disclination, while the second one is an "edge–islocation" which corresponds to a phase jump of $\Delta\theta = 2m\pi$. In the ordered rolls structure of a convective isotropic fluid (for instance Rayleigh–Bénard) the two types of point–defects are allowed because of the orientational degeneracy around the vertical axis. On the contrary, in anisotropic systems such as liquid crystals this degeneracy is raised and the direction of the wavevector is unambiguously fixed in the plane. Hence edge–dislocations are the only possible defects.

A dislocation is defined by a non–zero circulation along a positively oriented contour C surrounding its core :

$$\oint_C \mathbf{k}\cdot d\mathbf{s} = 2\pi.m.$$

The circulation is zero everywhere inside a perfect lattice of parallel rolls. A value $m = \pm 1$ corresponds to a sudden jump of 2π in the phase, which means that an extra period has been added on one of the two half–planes (Fig. 1). Indeed it turns out that the observed dislocations are always such that $m = \pm 1$.

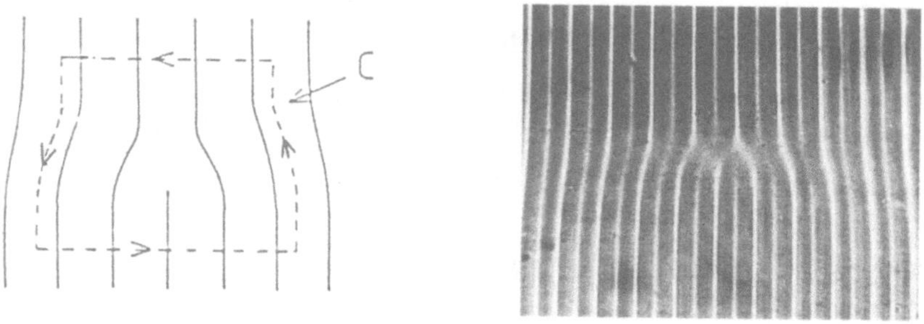

Fig. 1: An edge–dislocation in a 2–D structure of stationary rolls

4. TRAVELING WAVE CONVECTION IN A NEMATIC LIQUID CRYSTAL

A nematic liquid crystal is an anisotropic fluid which can develop convective instabilities when driven by an AC electric field [24]. A layer of nematic becomes unstable to convection above some well defined voltage threshold V_{th} which is frequency dependent and diverges at a cutoff f_c. The range of frequency limited by f_c has been named the "conduction regime". The anisotropy raises the degeneracy of the direction of the wavevector, as opposed to the case of isotropic convection. Therefore, the ordered homogeneous pattern consists of equally spaced rolls which are aligned along the perpendicular (hereafter **y**) direction to the anisotropy axis (**x**). It is found that for values of d typical of usual samples ($5\mu m < d < 100\mu m$), the periodic modulation in space of the molecular alignement associated with the convective pattern is never stationary, but propagates along **x** with a uniform velocity u [25,26]. Typically ,for $d \simeq 50\mu m$, $u \simeq 4.10^{-2}\mu m/s$ in the middle frequency range of the conduction regime, i.e. usually from 20 Hz to 100 Hz, the cut–off being at $f \simeq 120$ Hz. Close to the cut–off and close to DC, a typical value for the velocity is $u \simeq 10\mu m/s$. At high frequencies the pattern is not usually homogeneous in space : the envelope is strongly modulated in space and the convective rolls form isolated domains separated by a steady (non–convective) state. In some cases the domains can be quite elongated in the **x** direction with a rather small extension L_y along the roll axis ($L_y \simeq$ 1 to 2 d). Thus, one has an example of quasi 1-D traveling-wave pattern. Another way to obtain a 1-D pattern is to restrict the size of the electrode along **y** while keeping it very large along **x**. We shall describe hereafter the defects which are spontaneously created in quasi 1-D patterns propagating at relatively high velocity [25b]. We shall compare thes defects with those occuring in the stationary states of the same system where they also mediate a transition to a chaotic state [27–28].

5. ORDERED STRUCTURES OF TRAVELING WAVE CONVECTION

For a uniform or slowly modulated wave, the local wavevector **k** and the frequency ω are defined as $\mathbf{k} = \frac{\partial\varphi}{\partial \mathbf{x}}$, $\omega = \frac{\partial\varphi}{\partial t}$, where φ is the phase of the wave. The phase velocity is $u = \omega/k$. Therefore, the ordering involves both the space through k and the time through ω, and any perturbation either in space or in time is reflected by phase changes. Typical experimental values are $\lambda \simeq$ 60–70 μm (i.e. $k \simeq 10^3 cm^{-1}$) and $u \simeq 10\ \mu m/sec.$ (i.e. $\omega \simeq 1$ rad/sec.or $\tau = 2\pi/\omega \simeq 6.3$ sec.).

It is experimentally quite convenient to adopt the space-time representation for the propagative rolls. A nematic liquid crystal is a strongly birefringent material and the periodic modulation in space of the molecular alignment correlated to the convective flow, focusses periodically a transmitted beam of light onto parallel lines on a plane parallel to the layer (as an example see the photograph of fig.1). The position of these focal lines is related to that of the vertical motion (up and down).

An optical intensity profile recorded along **x** at some position y_0, then consists of a series of peaks separated by the wavelength λ, over a flat baseline. The successive profiles recorded after equal time intervals (here $\delta t \simeq 0.04$ sec.) for the same $y = y_0$, are plotted equally spaced on top of each other along the t axis. One obtains a space-time diagram on which the peaks are aligned on oblique lignes, the slope of which is $dt/dx = u^{-1}$. For a uniform motion (progressive wave), the space–time diagram is a perfectly ordered 2–D structure of oblique lines (Fig.2). Such space–time diagrams can be recorded over times up to some hours and are then particularly convenient to characterize states in which the phase is localized in space and slowly varying in time. The accuracy of the measurements is limited by the spatial resolution of the individual peaks and by the spatial homogeneity of the structure (absence of moving defects). In our experiments we have been able to measure velocities u as low as $1.5\ 10^{-5}\lambda$/sec.($\simeq 10$ Å/sec.) in uniform motion, for times up to 10^4 sec [25b].

6. LOCALIZED MODULATIONS OF THE PHASE AND NUCLEATION OF DEFECTS

An initially uniform progressive wave, i.e. a fully ordered wave with a unique wavevector k, may become unstable against long wavelength disturbances as it is the case for modulational Benjamin-Feir instabilities of water waves [16,17]. Experimentally, the conditions for such perturbations are not yet well understood although they are found to be effective relatively close to the threshold

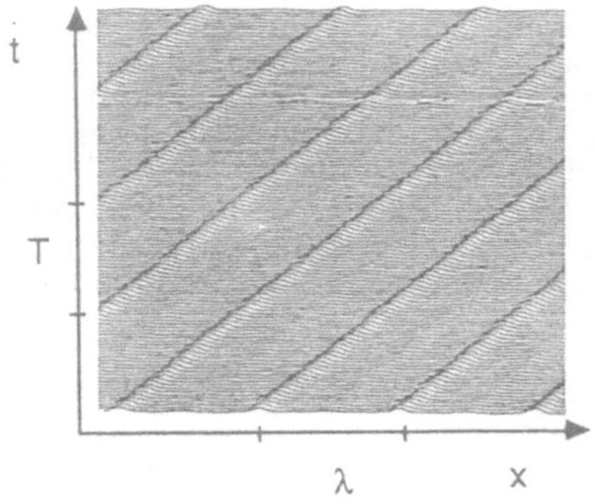

Fig. 2: Space–time representation of a uniform wave propagating to the right

for the basic wave. We usually find that if $\epsilon = (V^2 - V_{th}^2)/V_{th}^2$, where V_{th} is the threshold value for the basic state, is higher than some 2 - 3.10^{-2}, then the basic wave may become unstable. Localized modulations of the phase φ develop, apparently at random in space and in time. They are evidenced by a sudden change Δu in the phase velocity u over some wavelengths λ and might have a large relative amplitude $\Delta u/u \simeq 0.2$. Typically, two types of phase perturbation profiles are found:

a) the velocity amplitude jumps suddenly at some point and its profile along **k** is kink-like (Fig.3a);

b) the profile of the velocity jump is symmetrical and has a lump shape (Fig.3b). The phase variation $\Delta\varphi = \varphi - (\omega_0 t \pm k_0 x)$ has a kink–shape in the first case and a lump–shape in the latter one (Fig. 3).

One can also represent the variation of the local "wavevector" as $\delta k = \partial(\Delta\varphi)/\partial x$. It has a lump-shape in the first case while it is antisymmetrical in the latter one.

6.1 THE NUCLEATION OF SPACE–TIME DISLOCATIONS

In the first case (**a**) hereabove mentioned, the wavetrain is separated into two half trains with a different velocity and a shock occurs at the interface as in usual waves [20] (Fig.3c,d). The effect of the shock is to change the magnitude of the local wavevector k so as to drive it beyond its stability limits. In a compressive shock k increases and when it hits the upper limit of the stability domain k_b, right below the upper branch of the Eckhaus-Benjamin-Feir (E.B.F.) instability domain [17] (Fig. 4a), restabilization is obtained by expelling one spatial period. In an expansive shock a new period is created when the local wavevector k hits the lower limit k_a of the stability domain (above the lower branch of EBF unstable domain). Thus, the local convective period, if pulled into the E.B.F. instability domain, will suddenly either disappear if $k \geq k_b$, or give rise to a new period if $k \leq k_a$, within a time of the order the period of the wave $\tau = 2\pi/\omega$. The creation or annihilation of a wave-period is accompanied by a sudden

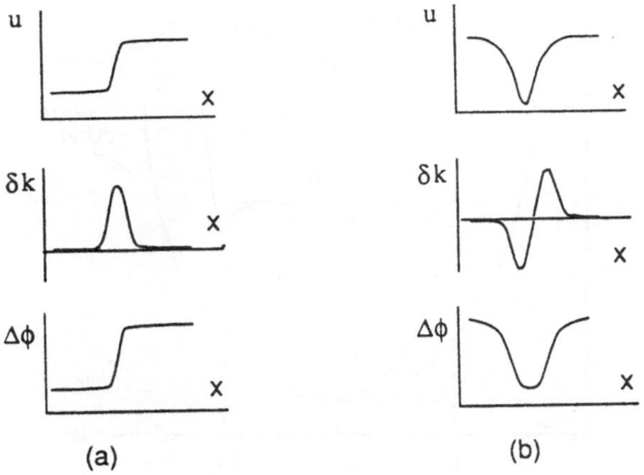

Fig. 3 a,b: Typical localized modulations of the phase. The velocities are u, the local wavevector variation is δk and the phase difference is $\Delta\varphi$.

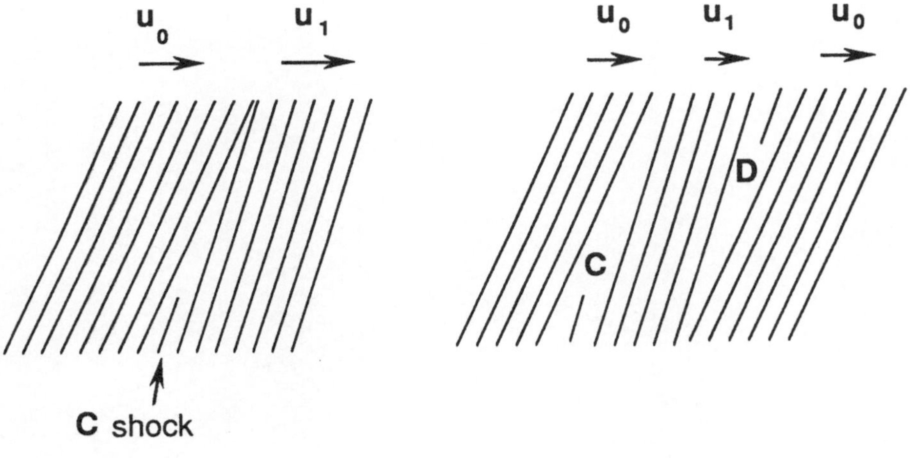

Fig. 3c, d: Sketch of localized perturbations of the phase of an homogeneous traveling wave. The cases c and d correspond to perturbations profiles a and b respectively. The shock is either a compressive one (C) or a dilative one (D).

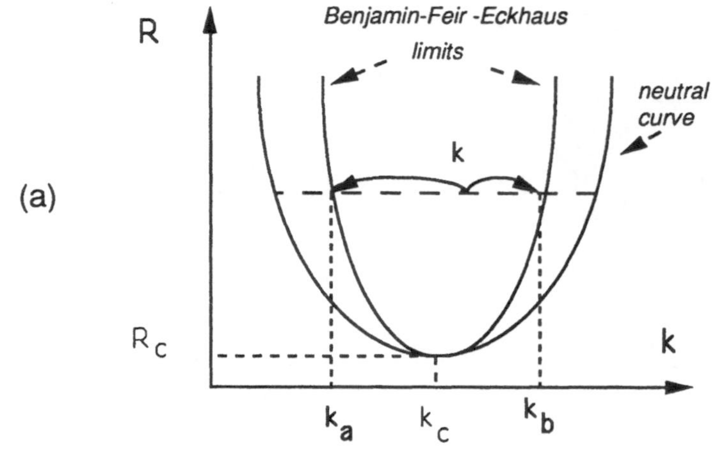

Fig.4a: In a shock the local wavevector amplitude may hit the limits of the Benjamin-Feir-Eckhaus instability of the phase. Restabilization occurs after expelling one period (when $k \rightarrow k_b$), or by creating an extra one ($k \rightarrow k_a$).

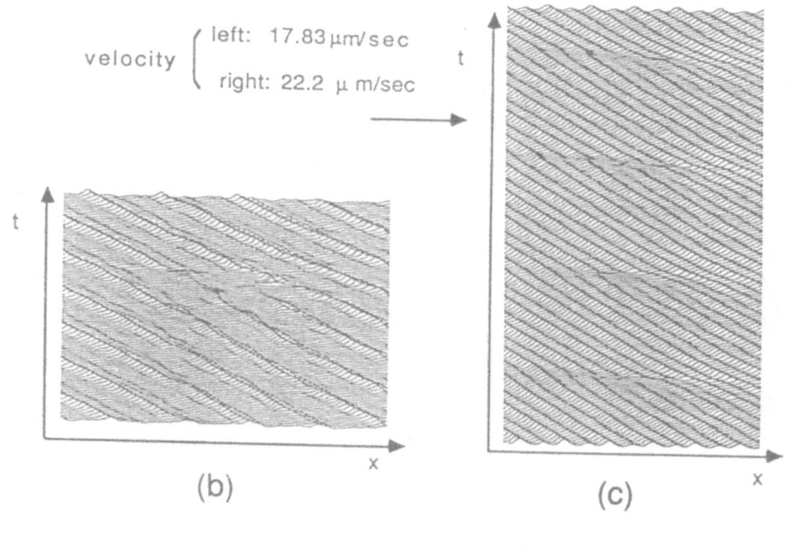

Fig. 4: b) Dislocation in a progressive wave (exper.)
c) A periodic array of dislocations results from a continuous shock between two waves with different velocities (exper.).

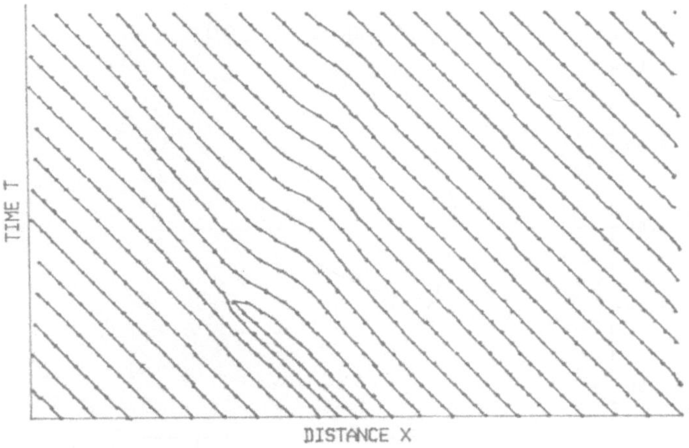

(d)

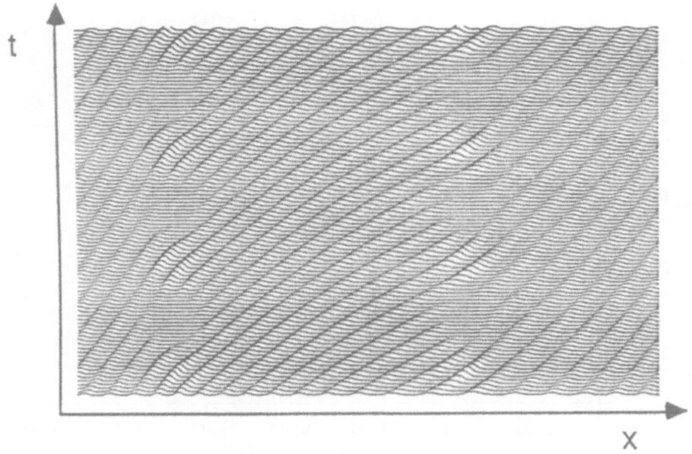

(e)

Fig. 4: d) Numerical simulation of the nucleation of an isolated dislocation by a localized phase winding (compare with Fig. 4a)

e) Time-periodic nucleation of dislocations by the shock between two waves with different velocities (compare with Fig. 4b).

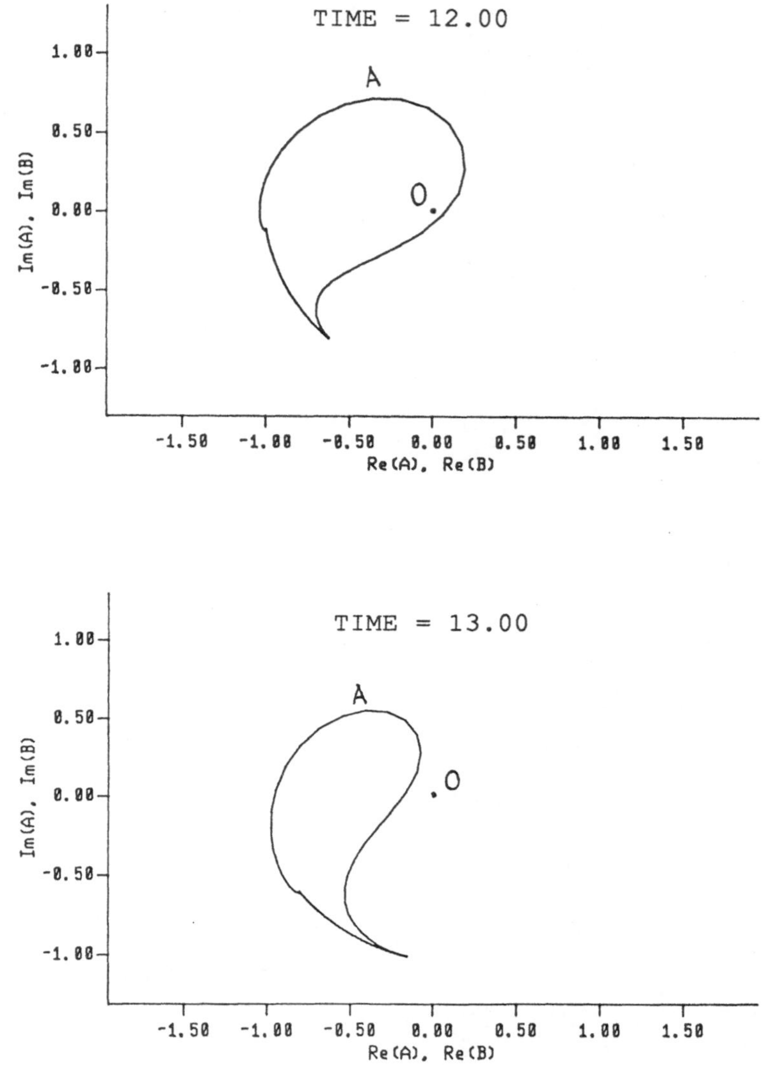

(f)

(g)

Fig. 4f,g : The amplitude A(x) in the complex plane before (f) and after (g) the nucleation of a dislocation. The modulus of A passes through 0,0.

drop of A to zero. This is equivalent to the creation or annihilation respectively of a dislocation, and indeed, it appears so in the {x,t} diagram (Fig.4b). If the two wavetrains keep traveling with a different velocity for a time long compared to τ, the creation or annihilation of dislocations occurs periodically in time (Fig.4c). However, such a case seems not to be a typical one even if it may be observed.

Numerically, periodic arrays of edge–dislocations are produced by imposing as an initial condition the coexistence of two adjacent waves propagating in the same direction with slightly different phase velocity u. The paramater values for the equation are: $c = 0$, $\mu = 1$, $\alpha = 1$, $\beta = .4$, $\gamma = 1.2$, $\delta = 0$ (Fig.4e). The boundary conditions are periodic and the waves are represented with a basic state such that $k_0 = .628$, $\omega_0 = .8$ (the length of the box is $L_0 = 100$). The isolated dislocations are produced with a phase winding in a uniform wave A but localized over a small length $L \simeq 2\lambda$ (fig.4d,f,g).

6.2 SOURCES, SINKS AND SPACE–TIME GRAIN–BOUNDARIES:

The case of a localized perturbation of the phase of the type (b) is a more typical one, since it corresponds to a localized symmetric variation of the velocity (or of the total phase difference $\Delta\varphi$), most likely to occur inside a initially homogeneous wave which is constrained to keep its average wavevector (conservation of the total phase). In other words the wave is unaltered at large distances away from the defect. Then, two shocks, a compression and a dilation ones, occur simultaneously close to each other (Fig. 3d). They break the wave and trigger at their location, both the creation and annihilation of one period. This corresponds to the appearance of two opposite singularities: a sink S_i and a source S_0. The new important effect is that the local wave velocity strongly decreases in the space separating the two defects. A source is a topological singularity [12] where two fronts of counterpropagating waves A and B meet in such a way that A is on the right side and B on the left side (Fig. 5a). The relative position of the two half waves B and A is reversed for a sink. It is found that usually, once they are formed, the two singularities are stable in time, i.e. they remain for a time usually much larger than the natural period τ. They are in fact, the topological boundaries that limit a counterpropagating wavetrain (say left-going B) inside the initial one (right-going A). In the {x,t} diagram this is represented as a domain embedded in the initial one (Fig.7a). In the language of crystallography the two counterpropagating waves are domains of a (cristallographic) phase of different orientation and they are separated by two "grain–boundaries" (or domain walls). This geometry is also found in the spatial 2–D patterns of stationary rolls inside the so–called Oblique Rolls structure [23] (Fig.5b).

A source, or a sink, is composed of a core which extends over about one period λ and of an intermediate region where the phase undergoes a damped oscillation over some periods until the wave is uniform [29]. The core can be simply understood as the point where two half waves B and A with a kink-like profile connect (B and A mutually exclude). It represents the area (the boundary width) over which a topological constraint forces the two waves B and A to be simultaneously present with equal amplitudes. Thus, it is a local standing wave. In fact, it is restricted to the point where the amplitude oscillates around zero and this fact can be confirmed by considering either the structure of the flow or the local variable (the velocity component u_z for instance). The intermediate zone is a damped modulation of the basic progressive wave (say B) wave under the

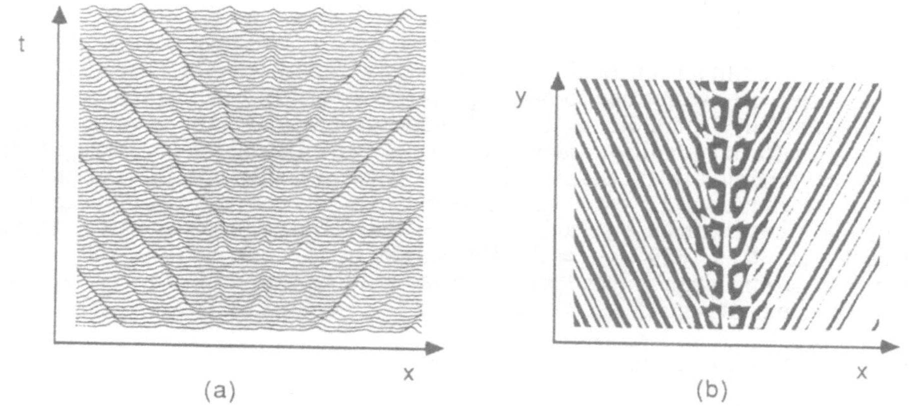

Fig. 5: a) A source separating two counterpropagating waves represents a grain_boundary . b) A real grain_boundary in the (stationary) Oblique Rolls

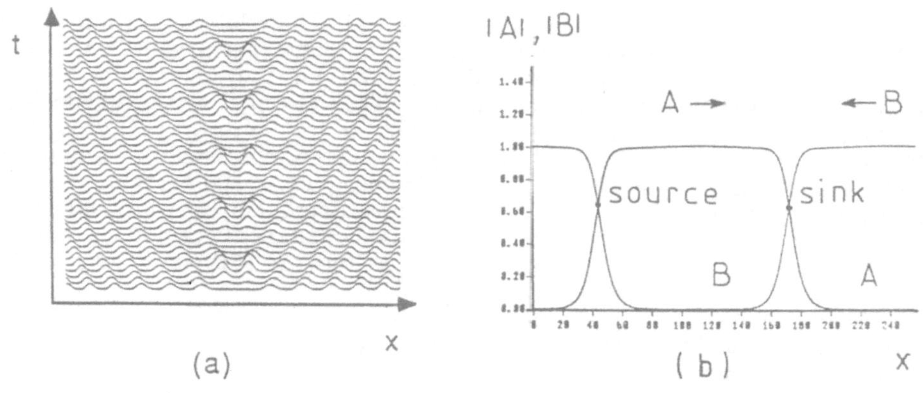

Fig. 6: a) Numerical simulation of a grain_boundary (see text).
b) Modulus of A(x) and of B(x). On each side of the defect the presence of the opposite wave with a smaller amplitude modulates the stable wave

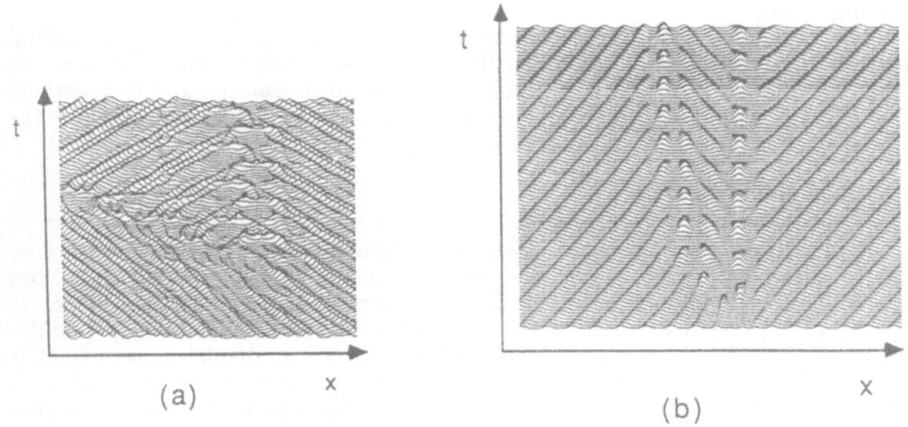

Fig. 7: a) The source and sink enclose a counterpropagating wave "domain".
b) Numerical simulation of nucleation of a pair of defects after a localized phase modulation for A (see text). Two defects (grain boundaries) of opposite topological charge are formed.

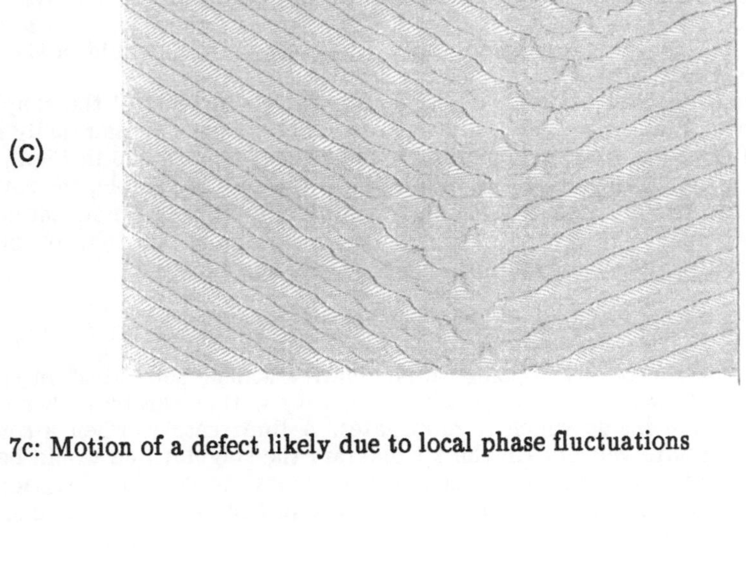

Fig. 7c: Motion of a defect likely due to local phase fluctuations

decreasing coupling with the its opposite (A) since only the wave B is a stable state in this region.

The sink and the source can be simulated by simply taking as initial conditions two amplitudes A and B oscillating in space with a phase difference of $\pi/2$ (Fig.6a). The kink profile is attained within a short time ($\simeq$ 200 time–steps) and one clearly sees that the finite width of the kinks is responsible for the modulated wave region next to the core (Fig.6b).

6.3 STABILITY OF THE DEFECTS:

As the time goes the new "domain" of counterpropagating wave B may develop or disappear. This raises the problem of the motion of the sinks and sources. Preliminary experimental studies indicate that a topological defect is easily unstable in space as soon as the external stress increases slightly ($\epsilon\simeq.05$). The defect can move in either direction by steps that are multiple values of half a period (Fig. 7c). The stepping is simply explained by the fact that the defect is the place where both the phase and the amplitude of the two waves must be equal. This motion is ascribed to long wavelength modulations of at least, one of the waves velocities u_A or u_B, which change locally the amplitude of that wave. Whenever this modulation is continuous in time, it makes the defect move with an average velocity that can be higher than the velocity of anyone of the waves B or A. In that case, the moving defect is similar to the front of a shock wave [30]. It is usually found that one of the two defects is relatively stable in space while the other one moves so as to widen the domain of the new wave (see left side of Fig.7a). However, sometimes, a counterpropagating domain may disappear and the initial state is recovered. Here again, the stability of the new complex state seems to depend on the existence of localized phase modulations i.e. on the dispersivity of the system.

It is easy to numerically generate a spatially growing state inside the initial one by taking as initial conditions a state A with a localized phase winding of 2π and at the same point a localized hump of amplitude for B. The result is shown Fig.7b for $c = 0$, $\alpha = 1$, $\beta = .4$, $\gamma = 1.2$, $\delta = 0$. A value of $\gamma > 1$ is needed to insure stability of the progressive wave, whereas $\gamma < 1$ would make the standing wave state stable.

In order to test the stability of the defects the coefficients of the coupling terms were varied. When $\gamma = 1.02$ and $\delta = .7$, one of the defects splits rapidly into three defects of the same nature. For example in the case represented in Fig.8, the sink splits into a source and two sinks. While the two waves are stable, the core of the defect is made unstable and acts as a germ for the growth of a new ensemble of two counterpropagating waves. This situation is similar to that of phase transition induced by defects in equilibrium structures of solids.

6.4 LARGE–SCALE PHASE MODULATIONS AND QUASI–COHERENT MODES:

A sink or a source is a topological singularity which "separates" in space two waves B and A and it may move with a velocity w that can be higher than that of the wavetrain inside which it propagates. A limit case is when a source propagates with infinite velocity (in the x,t diagram the singular line would be an horizontal). Then at a sudden, the wavetrain reverses its propagation direction as shown in Fig. 9a. If this reversal occurs periodically in time and on a large space

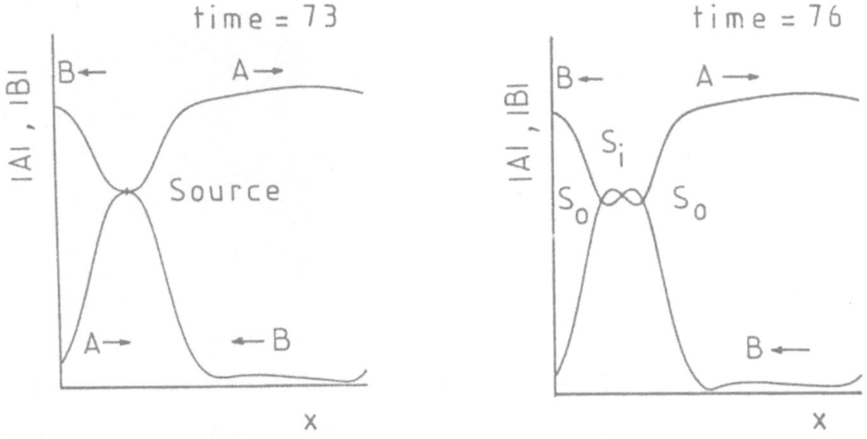

Fig. 8 a: Destabilization of the core of a defect. The source S_0 splits into a sink S_i plus two sources S_0 located at the crossing points.

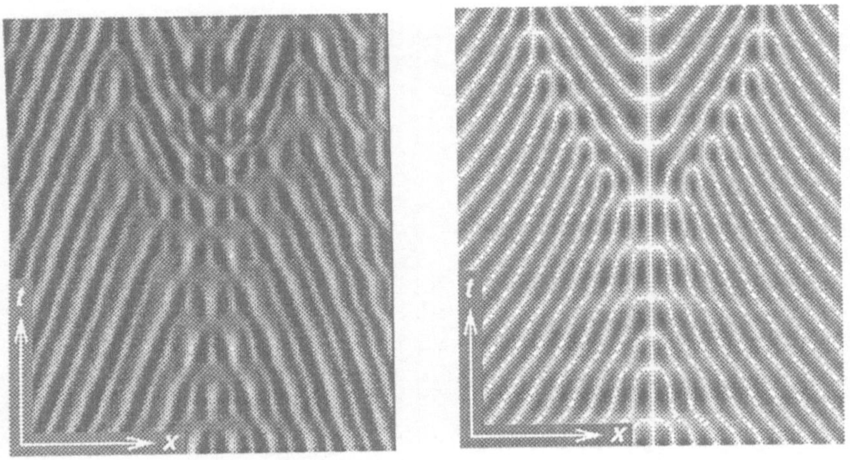

Fig. 8 b: Experimental observation (left) and numerical simulation of the splitting of a defect into three ones.

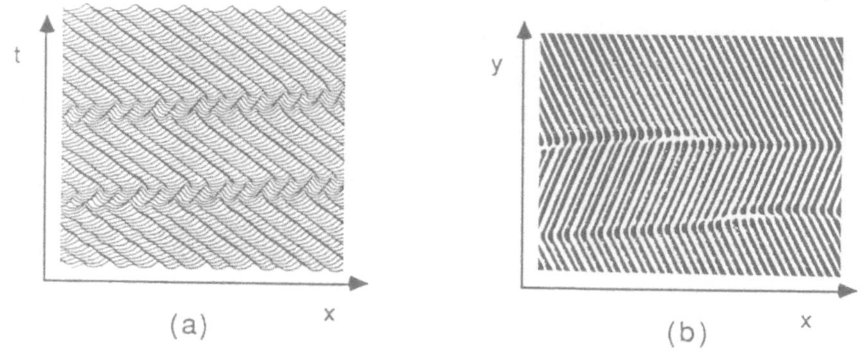

Fig. 9: a) Quasi–collective mode of velocity reversal that can be interpreted as due to a fast moving source (see text).

b) "Twin–boundaries" in the Oblique Rolls

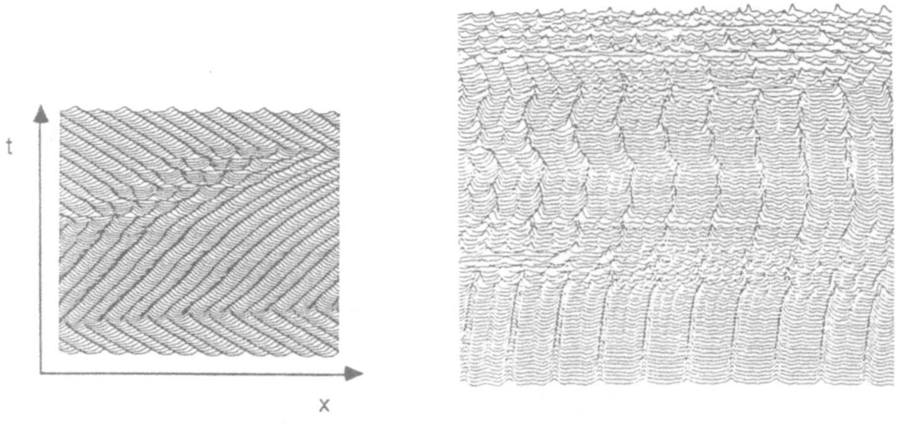

Fig.10: Progressive disorganization of the basic wave by defects and by quasi–collective modes of oscillation

scale, one has a quasi–collective longitudinal mode of oscillation of the structure (bistable state). Therefore, there is an interesting apparent continuity between a pure mode of oscillation in time and the motion of a defect. The space–time representation of this case is quite analogous to that of the twin–boundaries in the Oblique Rolls structure (Fig.9b). There exists also, under some other conditions, another longitudinal collective mode of oscillation of the structure: the periodic compression–dilation (therefore named "time–dependent Eckhaus" mode) with a wavelength that can be as low as that of the basic structure of convection. Reports on its observation are given elsewhere (Joets and Ribotta, preprint 1989).

6.5 SPACE–TIME COMPLEXITY:

The progressive wave is the basic state which can be unstable against either localized or homogeneous perturbative modes. The complexification of the basic state (disordering both in time and space) can be thought as the result of either homogeneous modes or localized states as in the case of quasi–stationary structures. In our experiments as the control parameter is increased just above threshold, so that $\epsilon \simeq 0.05$, defects appear randomly both in time and in space and their density increases with ϵ. The grain boundaries may reverse the direction of the propagation of the basic wave over space scales that can either grow or decrease to zero whenever they mutually annihilate. Mixed states with grain boundaries, dislocations and twin–boundaries are shown in Figure 10. Further increasing ϵ, makes the behaviour become unpredictable and the result is a fully erratic motion with no apparent temporal correlations on small time scales. Usually the disordered state is reached for a value of ϵ around 1.

If we now consider the more general case of 2–D structures, then the progressive wave may become unstable to transverse modes that can be either propagative such as the Busse oscillatory instability [31], or standing. This latter case is most often found and as for the stationary states it seems to be the most dangerous one in the process of a rapid transition to the chaos, because it generates real dislocations of the 2–D spatial structures (we name such a process "Martensitic Transformation" by analogy with the so–called displacive transitions in solids) [27,28].

7 CONCLUSION

The traveling–wave state of convection inside a layer of an anisotropic fluid offers quite rich examples of localized states and of singularities in 1–D non–linear waves. Even right above the threshold for the basic state, localized modulations of the envelope occur on scales comparable to the spatial period of the wave. In this case, the localized phase perturbations give rise to shocks. Because of the finite bandwidth of the unstable state, topological singularities are created. These singularities correspond to defects in the space–time ordering. A fully disordred state may be rapidly reached through the random nucleation of the defects and their incoherent motion inside the homogeneous structure. We have also mentioned the striking similarity of the transition to a fully disordered state both in time–dependent and in quasi–stationary structures, mediated by the defects that are nucleated from localized modulations of the phase. However, a complete equation for the envelope, which could account for heterogeneous time-dependent states has not yet been derived in the case of liquid crystals from the basic microscopic system of equations. Nevertheless, we have found that it is possible to reproduce numerically most of the spatio–temporal effects hereabove described by use of a set of coupled envelope equations of the Landau–Ginzburg type for the two counterpropagative waves. Although up to now, the Landau–Ginzburg model has been the subject of intensive work, it seems that the important question of the effects of strongly localized states resulting from long–wavelength disturbances has not been raised yet in the case of hydrodynamics of convection.

This work was supported by the Direction des Recherches et Etudes Techniques (DRET). We thank P. Coullet, L. Gil and J. Lega (C.G.L.)for making available to us a numerical code and C.G.L. and A.C. Newell for useful discussions.

References

[1] Newell, A.C. and Whitehead, J.A.,(1969) "Finite bandwidth, finite amplitude equation", J. Fluid. Mech., 38, 279.
Segel, L.A. (1969) " Distant side–walls cause slow amplitude modulation of cellular convection", J. Fluid Mech., 38, 203.
Newell, A.C. (1974) "Envelope equations", Lect. Appl. Math. 15, 157.

[2] Kogelman, S. and DiPrima, R.C. (1970) "Stability of spatially periodic supercritical flows in hydrodynamics", Phys. Fluids 13, 1.

[3] Blennerhassett, P.J. (1980) "On the generation of waves by wind", Phil. Trans. Roy. Soc. Lond. A298, 451.

[4] Stewartson, K., and Stuart, J.T. (1971) "Non–linear instability of plane Poiseuille flow", J. Fluid Mech. 48, 529.

[5] Moon, H. T., Huerre, P., Redekopp, L. G. (1983) "Transitions to chaos in the Ginzburg–Landau equation", Physica 7D, 135.

[6] Nozaki, K. and Bekki, N. (1983) "Pattern selection and spatiotemporal transition to chaos in the Ginzburg–Landau equation", Phys. Rev. Lett. 51, 2171; and (1984) "Exact solutions of the generalized Ginzburg–Landau equation", J. Phys. Soc. Jap. 53, 1581.

[7] Sirovich, L., and Newton, P.K. (1986) "Periodic solutions of the Ginzburg–Landau equation, Physica 21D, 115.

[8] Keefe, L.R. (1985) "Dynamics of perturbed wavetrain solutions to the Ginzburg–Landau equation, Stud in Appl. Math. 73, 91;also (1986) "Integrability and structural stability of solutions to the Ginzburg–Landau equation, Phys. Fluids 29, 3135.

[9] Ghidaglia, J.M., and Héron, B. (1987) "Dimension of the attractors associated to the Ginzburg–Landau partial differential equation", Physica 28D, 282.

[10] Bekki, N. and Nozaki, K. (1985) "Formations of spatial patterns and holes in the generalized Ginzburg–Landau equation, Phys. Lett. A110, 133.

[11] Kawasaki, K.and Ohta, T. (1982) "Kink dynamics in one dimensional nonlinear systems, Physica 116A, 573.

[12] Coullet, P., Elphick, C., Gil, L. and Lega, J. (1987) "Topological defects of wave patterns, Phys. Rev. Lett. 59, 884; Coullet, P. and Lega, J.(1988) "Defect–mediated turbulence in wave patterns", Europhys. Lett. 7, 511.

[13] Joets, A. and Ribotta, R. (1989) " Defects and transition to disorder in space–time patterns of nonlinear waves, in P. Coullet and P. Huerre (eds.), New trends in nonlinear dynamics and pattern forming phenomena: the geometry of nonequilibrium, Nato Asi series B, Plenum Press (Cargese Workshop Aug. 1988); also (1989), in the Proceedings of Nonlinear coherent structures in physics, mechanics and biological systems (Paris, June 1988), J. de Phys. C3.
[14] Brand, H.R., Lomdahl, P. and Newell, A. (1986) "Evolution of the order parameter in situations with broken rotational symmetry, Phys. Let. A118, 67.
[15] Lange, C.G. and Newell, A.C. (1974) SIAM J of Appl. Math. 27, 441 (1974)
[16] Benjamin, T. B. and Feir, J. E. (1967) "The disintegration of wave trains on deep water, J. Fluid Mech. 27, 417.
[17] Stuart, J.T. and DiPrima, R.C. (1978) "The Eckhaus and Benjamin–Feir resonance mechanisms", Proc. R. Soc. Lond. A 362, 27.
[18] Bretherton, C. S. and Spiegel, E. A. (1983) "Intermittency through modulational instability", Phys. Lett. 96A, 152.
[19] Kuramoto, Y. (1984) "Phase dynamics of weakly unstable periodic structures", Prog. Theor. Phys. 71, 1182.
[20] Whitham, G.B. (1974) Linear and Nonlinear Waves, John Wiley & Sons, New York.
[21] Howard, L.N. and Kopell, N. (1977) "Slowly varying waves and shock structures in reaction–diffusion equations", Stud. in Appl. Math. 56, 84.
[22] Bernoff, A. (1988) "Slowly Varying Fully Nonlinear Wavetrains in the Ginzburg–Landau Equation", Physica D30, 363.
[23] Joets, A. and Ribotta, R. (1986) "Hydrodynamic transitions to chaos in the convection of an anisotropic fluid", J. Phys. (Paris) 47, 595.
[24] de Gennes, P. G. (1974) The Physics of Liquid Crystals, Clarendon, Oxford.
[25] Joets, A. and Ribotta, R. a)– (1988) "Propagative structures and localization in the convection of a liquid crystal", in , J.E. Wesfreid, H.R. Brand, P. Manneville, G. Albinet and N. Boccara, (eds.), Propagation in Systems far from Equilibrium Springer, Berlin;
and (1988) b)– "Propagative patterns in the convection of a nematic liquid crystal", in the Proceedings of the 12th International Liquid Crystals Conference, Aug. 1988, Freiburg), Liquid Crystals, London.
[26] Joets, A. and Ribotta, R. (1988) "Localized time–dependent state in the convection of a nematic liquid crystal", Phys. Rev. Lett. 60, 2164.
[27] Yang, X. D., Joets, A. and Ribotta, R. (1986) "Singularities in the transition to chaos of a convective anisotropic fluid", Physica 23D, 235.
[28] Ribotta, R. (1988) "Solitons, defects and chaos in dissipative systems", in L. Kubin, G. Martin, (eds.), Non–linear phenomena in materials science, Trans Tech Publications, Switzerland.
[29] Joets, A. and Ribotta, R. (1989) "Structure of defects in nonlinear waves", preprint.
[30] Joets, A. and Ribotta R. (1989) "Stability of defects of nonlinear waves in traveling–wave convection", preprint.
[31] Busse, F. H. (1972) "The oscillatory instability of convection rolls in a low Prandtl number", J. Fluid Mech. 52, 97.

AN INTRODUCTION TO KOWALEVSKI'S EXPONENTS

D. BESSIS
Service de Physique Théorique de Saclay
Laboratoire de l'Institut de Recherche Fondamentale
du Commissariat à l'Energie Atomique
91191 Gif-sur-Yvette Cedex, France

ABSTRACT. We give an introduction to the Kowalevski exponents theory for non linear similarity invariant differential equations. The connection between the arithmetic nature of the exponents and the algebraic integrability is discussed in detail. In a second part, the importance of the Riemann surface associated to the singularities generated by complex Kowalevski exponents is emphasized.

1. INTRODUCTION

The problem of determining the relationship between the structure of the singularities in the complex time plane, and the integrability or non integrability of a given dynamical system is still an open question, although much progress has been made. The structure of the complex singularities of a dynamical system ranges from regular lattices for the simplest integrable systems to complicated fractals for non integrable systems such as for the Hénon-Heiles Hamiltonian.

One would like to be able to claim statements as simple and strong as "non integrable systems are characterized by the fact that they produce natural boundaries on the complex time plane, with singularities sitting on fractals". It is to soon to know if such statements are correct, although they may contain some kind of truth. The possibility to reconstruct to a very good approximation the set of singularities of a non integrable system, using Kowalevski exponents (which are *local* data) lets one think of a possible reconstruction of the solution itself, using complex variable technics.

Historically, the idea that singularities in the complex time plane may directly influence the real time behaviour of Dynamical Systems was illustrated by the work of U. Frisch and R. Morf[1] and of M. Tabor and J. Weiss[2].

Analyzing the Hénon-Heiles Hamiltonian, Y.F. Chang, M. Tabor and J. Weiss[3][4] have singled out the notion of an integrable dynamical system having the Painlevé property, that is, only moving singularities. They have also investigated the case for which the system is non integrable, and they discovered that the singularities for this system show in

R. Conte and N. Boccara (eds.), Partially Integrable Evolution Equations in Physics, 299–320.

the complex time-plane a self similar, and therefore fractal structure. In [5] I.C. Percival and J.M. Green have considered the problem of whether or not the presence of a natural boundary is a generic property of non-integrable system.

The concept of integrability varies through the literature. J. Weiss, M. Tabor and G. Carnevole[6] associate it with the Painlevé property. A Ramani, B. Dorizzi, B. Grammaticos and T. Bountis discuss a notion of integrability under which the Painlevé property can be absent. T. Bountis and H. Segur[8,9] weaken the Painlevé property by admitting logarithmic moving singularities and introducing the concept of weak chaos linked to them.

H. Yoshida[10] investigates a precise but restricted notion of integrability, the algebraic integrability. Yoshida gives the necessary conditions for algebraic integrability : no Kowalevski exponent be irrational or complex. The extension of those results to Hamiltonian systems is not straightforward as shown by M. Kummer, R. Churchill and D. Rod[12]. These authors provide explicit examples, in which the Kowalevski exponents are *not* rational numbers and nevertheless the system is algebraically integrable in the following generalized sens.

An Hamiltonian system with f degrees of freedom is said to be algebraically integrable if one can construct f algebraic first integrals, the other f integrals being given by integration of 1-forms using Liouville's theorem. One would like to be able to recognize from some arithmetic property of the Kowalevski exponents, the non-integrability of a given Hamiltonian. H. Yoshida for Hamiltonians with the kinetic energy in standard form was able to make progresses in this direction[13]. These extensions are built on Ziglin analysis, for which one can refer to the contribution of A. Ramani and B. Grammaticos in "Singularity Analysis and its Relation to Complete, Partial and Non-integrability". A generalization of the concept of Kowalevski exponents, the Painlevé resonances can be found in [14].

This paper is organized in the following way:

- In Section II, the theory of Kowalevski exponents is exposed, with various simple examples to illustrate the content of this approach.
- In Section III, the structure of the complex plane singularities is discussed and the role of the Riemann surface associated to these singularities is emphasized[14].

2. KOWALEVSKI'S EXPONENTS

2.1. Homogeneous systems

We shall consider a system of differential equations:

$$(S)\left\{\begin{array}{ll} \dfrac{dx_1}{dt} & = F_1(x_1, x_2, \cdots, x_N) \\ \dfrac{dx_2}{dt} & = F_2(x_1, x_2, \cdots, x_N) \\ \cdots & \cdots\cdots\cdots\cdots\cdots\cdots \\ \dfrac{dx_N}{dt} & = F_N(x_1, x_2, \cdots, x_N) \end{array}\right. \qquad (II.1.1)$$

where the $F_i(\mathbf{x})$ do *not* depend explicitly on t (time), that is the system is autonomous. The $F_i(\mathbf{x})$ are rational functions, ratios of two polynomials:

$$F_i(x_1, x_2, \cdots, x_N) = \frac{P_i(x_1, x_2, \cdots, x_N)}{Q_i(x_1, x_2, \cdots, x_N)} \qquad (II.1.2)$$

where P_i and Q_i are polynomials in the variables x_k.

Furthermore, we shall suppose that the system (S) is *invariant* under a similarity transformation

$$t \longrightarrow \alpha^{-1}t; \quad x_1 \longrightarrow \alpha^{G_1}x_1; \cdots; \quad x_N \longrightarrow \alpha^{G_N}x_N$$

for any constant α.

For this to be true the F_i should satisfy:

$$F_i\left(\alpha^{G_1}x_1, \alpha^{G_2}x_2, \cdots, \alpha^{G_N}x_N\right) = \alpha^{G_i+1}F_i\left(x_1, x_2, \cdots, x_N\right) \qquad (II.1.3)$$

When they exist, the G_i are clearly specified rational numbers.

Example

Take

$$\begin{cases} F_1\left(x_1, x_2, x_3\right) & = \varepsilon_1 x_2 x_3 \\ F_2\left(x_1, x_2, x_3\right) & = \varepsilon_2 x_3 x_1 \qquad \varepsilon_1\varepsilon_2\varepsilon_3 \neq 0 \\ F_3\left(x_1, x_2, x_3\right) & = \varepsilon_3 x_1 x_2 \end{cases} \qquad (II.1.4)$$

Let us see if it is possible to find G_1, G_2, G_3 such that:

$$\begin{cases} \varepsilon_1\alpha^{G_2}x_1\alpha^{G_3}x_2 & = \alpha^{G_1+1}\varepsilon_1 x_2 x_3 \\ \varepsilon_2\alpha^{G_3}x_3\alpha^{G_1}x_1 & = \alpha^{G_2+1}\varepsilon_2 x_3 x_1 \\ \varepsilon_3\alpha^{G_1}x_1\alpha^{G_2}x_2 & = \alpha^{G_3+1}\varepsilon_3 x_1 x_2 \end{cases} \qquad (II.1.5)$$

We see that

$$\begin{cases} G_2 + G_3 & = 1 + G_1 \\ G_3 + G_1 & = 1 + G_2 \\ G_1 + G_2 & = 1 + G_3 \end{cases} \qquad (II.1.6)$$

The system (II.1.6) has a unique solution $G_1 = G_2 = G_3 = 1$. Each variable x_i has associated to it, a degree of homogeneity G_i. The polynomials P_i and Q_i are built with monomials which must have the same global degree of homogeneity. Finally we obtain for the G_i, equations of the form:

$$\sum_{k=1}^{k=N} p_i^{(k)} G_k = G_i + 1 \qquad (II.1.7)$$

where the $p_i^{(k)}$ are integers (positive, negative or zero).

A system such that exists a set of $\{G_i\}_{i=1}^{N}$ is an homogeneous system. We shall be concern only with such systems.

2.2. Particular solutions

We can look for a *special* type of particular solutions of (S), of the form

$$(S_0)\left\{x_1 = c_1 t^{-G_1} \; ; \; x_2 = c_2 t^{-G_2} \; ; \cdots ; \; x_N = c_N t^{-G_N}\right\} \qquad (II.2.1)$$

with precise $c_1, c_2, \cdots, c_N$ constants. Plugging the solution (S_0) into (S) we get

$$\begin{aligned} -c_i G_i t^{-(1+G_i)} &= F_i\left(c_1 t^{-G_1}, \cdots, c_N t^{-G_N}\right) \\ &= t^{-(1+G_i)} F_i\left(c_1, c_2, \cdots, c_N\right) \end{aligned} \qquad (II.2.2)$$

the last equality is obtained using (II.1.3) for $\alpha = t^{-1}$.

Therefore, provided the system

$$F_i(c_1, c_2, \cdots, c_N) = -c_i G_i \qquad i = 1, 2, \cdots, N \tag{II.2.3}$$

has at least a solution, we can built the scaling solutions (S_0). (II.2.3) is a system of N polynomial equations in the c_i's. It will have at most a finite number of solutions. Each set of solutions $\{c_i\}_{i=1}^{N}$ defines a *scaling* solution.

Example

Let us go back to the previous example

$$F_i = \varepsilon_i x_j x_k, \qquad i, j, k \text{ cyclic permutations of } 1, 2, 3 \tag{II.2.4}$$

here $G_i = 1\ \forall i$ and (II.2.3) becomes

$$\varepsilon_i c_j c_k = -c_i \tag{II.2.5}$$

or

$$\begin{cases} \varepsilon_1 c_2 c_3 & = -c_1 \\ \varepsilon_2 c_3 c_1 & = -c_2 \\ \varepsilon_3 c_1 c_2 & = -c_3 \end{cases} \tag{II.2.6}$$

which gives the solutions:

$$\begin{cases} c_1 = \eta_1 (\varepsilon_2)^{-1/2} (\varepsilon_3)^{-1/2} & \eta_1 = \pm 1 \\ c_2 = \eta_2 (\varepsilon_3)^{-1/2} (\varepsilon_2)^{-1/2} & \eta_2 = \pm 1 \\ c_3 = \eta_3 (\varepsilon_1)^{-1/2} (\varepsilon_3)^{-1/2} & \\ \eta_1 \eta_2 \eta_3 = -1 & \eta_3 = \pm 1 \end{cases} \tag{II.2.7}$$

which means that we get four different scaling solutions, corresponding to the choices $(+,+,-)$, $(+,-,+)$, $(-,+,+)$ and $(-,-,-)$ for the η_i factors.

2.3. Variational solutions near a scaling solution

An interesting question is to know if there exist other solutions of (S) near by the set (S_0). The answer to this question was given by Kowalevski.

Let us write

$$x_i(t) = x_i^{(0)}(t) + \delta x_i(t) \tag{II.3.1}$$

then (S) becomes, at first order in δx_i :

$$\begin{aligned} \frac{\mathrm{d}}{\mathrm{d}t}(\delta x_i) &= \sum_{k=0}^{k=N} \frac{\partial F_i}{\delta x_k} \left(x_1^0, x_2^0, \cdots, x_N^0\right) \delta x_k \\ &= \sum_{k=1}^{k=N} \frac{\partial F_i}{\delta x_k} \left(c_1 t^{-G_1}, c_2 t^{-G_2}, \cdots, c_N t^{-G_N}\right) \delta x_k \qquad i = 1, 2, \ldots N. \end{aligned} \tag{II.3.2}$$

However from the identity:

$$\alpha^{(G_i+1)}F_i(x_1,x_2,...x_N) = F_i(x_1\alpha^{G_1},x_2\alpha^{G_2},...,x_N\alpha^{G_N}). \qquad (II.3.3)$$

we get, differentiating with respect to x_k :

$$\frac{\partial F_i}{\partial x_k}(x_1\alpha^{G_1},x_2\alpha^{G_2},...x_N\alpha^{G_N}) = \alpha^{G_i-G_k+1}\frac{\partial F_i}{\partial x_k}(x_1,x_2,...x_N) \qquad (II.3.4)$$

and therefore (II.3.2) can be rewritten:

$$\frac{d}{dt}(\delta x_i) = \sum_{k=1}^{k=N}\frac{\partial F_i}{\partial x_k}(\mathbf{C})t^{[G_k-G_i-1]}\delta x_k. \qquad (II.3.5)$$

where $\mathbf{C}$ stands for the vector $(c_1,c_2,...,c_N)$. Let us seek for a solution of (II.3.5) of the form:

$$\delta x_i(t) = \xi_i^{(\rho)}t^{\rho-G_i} \qquad (II.3.6)$$

where $\boldsymbol{\xi}^{(\rho)}$ is an N-dimensional vector (constant) and ρ a fixed number. We get:

$$\xi_i^{(\rho)}(\rho - G_i) = \sum_{k=1}^{k=N}\frac{\partial F_i}{\partial x_k}(\mathbf{C})\xi_k^{(\rho)} \qquad (II.3.7)$$

(II.3.7) shows clearly that ρ and $\boldsymbol{\xi}^{(\rho)}$ are respectively the eigenvalue and the eigenvector of the Kowalevski matrix K, whose elements are:

$$K_{ij} = \frac{\partial F_i}{\partial x_j}(\mathbf{C}) + \delta_{ij}G_i \qquad (II.3.8)$$

If K is diagonalizable we get N independent solution of the varied equations (II.3.2).

The N roots of the determinantal equation:

$$\det \mid K_{ij} - \rho\delta_{ij} \mid= 0 \qquad (II.3.9)$$

$\rho_1,\rho_2,...\rho_N$ are called the Kowalevski exponents. In first order we can write:

$$X_i^{(\rho)}(t) = c_i t^{-G_i} + \xi_i^{(\rho)}t^{\rho-G_i} + ... \qquad (II.3.10)$$

Example

Let us compute the Kowalevski exponents for the system

$$\frac{dx_1}{dt} = \varepsilon_1 x_2 x_3 + \text{ cyclic permutations} \qquad (II.3.11)$$

Here

$$K = \begin{bmatrix} 1 & \varepsilon_1 c_3 & \varepsilon_1 c_2 \\ \varepsilon_2 c_3 & 1 & \varepsilon_2 c_1 \\ \varepsilon_3 c_2 & \varepsilon_3 c_1 & 1 \end{bmatrix} \qquad (II.3.12)$$

The eigenvalues are solution of

$$(1-\rho)^3 - (1-\rho)\left[\varepsilon_2\varepsilon_3 c_1^2 + \varepsilon_3\varepsilon_1 c_2^2 + \varepsilon_1\varepsilon_2 c_3^2\right] + 2\varepsilon_1\varepsilon_2\varepsilon_3 c_1 c_2 c_3 = 0 \qquad (II.3.13)$$

which, taking into account (II.2.6) reduces to:

$$(1-\rho)^3 - 3(1-\rho) - 2 \equiv -(1+\rho)(\rho-2)^2 = 0 \qquad (II.3.14)$$

It appears, in this special case, that the exponents are the same for all the various choices of c_i, we get the set:

$$\rho = \{-1, 2, 2\} \qquad (II.3.15)$$

we shall see that -1 is *always* a Kowalevski exponent.

Notice that the set $\{\rho_i\}_{i=1}^N$ is associated with a given scaling solution $(\mathbf{G}, \mathbf{C})$ and in general changes with it.

2.4. Algebraic first integrals

A first integral is a function $\phi(x_1, x_2, ..., x_N)$ such that when $x_1, x_2, ..., x_N$ are replaced by *any* solution of (S), we have:

$$\phi[x_1(t), x_2(t), ..., x_N(t)] = \psi(t) = \text{constant independent of } t \qquad (II.4.1)$$

when $\phi(\mathbf{x})$ is a polynomial in x_k, this is called a polynomial first integral.

Let us contemplate a first integral:

$$\phi(x_1, x_2, ..., x_N) = a \qquad (II.4.2)$$

and suppose that there exists an *algebraic* equation of finite degree k, such as:

$$a^k + \phi_1(x_1, x_2, ..., x_N)a^{k-1} + ... + \phi_k(x_1, x_2, ... x_N) = 0 \qquad (II.4.3)$$

where the $\phi_1(x_1, ... x_N), ..., \phi_k(x_1, ... x_N)$ are rational functions. $\phi(x_1, ... x_k)$ is called an algebraic first integral of order k.

For instance for the system

$$\frac{dx_1}{dt} = \varepsilon_1 x_2 x_3 + \text{cyclic permutations} \qquad (II.4.4)$$

one checks that

$$\phi(x_1, x_2, x_3) = \frac{x_1^2}{\varepsilon_1} - \frac{x_2^2}{\varepsilon_2} + \sqrt{\frac{x_2^2}{\varepsilon_2} - \frac{x_3^2}{\varepsilon_3}} = a \qquad (II.4.5)$$

is an algebraic first integral of order 2. It is in fact a solution of

$$a^2 - 2a\left(\frac{x_1^2}{\varepsilon_1} - \frac{x_2^2}{\varepsilon_2}\right) + \left(\frac{x_1^2}{\varepsilon_1} - \frac{x_2^2}{\varepsilon_2}\right)^2 - \left(\frac{x_2^2}{\varepsilon_2} - \frac{x_3^2}{\varepsilon_3}\right) = 0 \qquad (II.4.6)$$

here

$$\begin{cases} \phi_1 = -2\left(\frac{x_1^2}{\varepsilon_1} - \frac{x_2^2}{\varepsilon_2}\right) & \text{and} \\ \phi_2 = \left(\frac{x_1^2}{\varepsilon_1} - \frac{x_2^2}{\varepsilon_2}\right)^2 - \left(\frac{x_2^2}{\varepsilon_2} - \frac{x_3^2}{\varepsilon_3}\right) \end{cases} \tag{II.4.7}$$

are polynomials in $\mathbf{x}$, therefore the criterion is fulfilled. To show that $\phi(x_1, x_2, x_3)$ is a first integral, notice that (II.4.4) can be rewritten:

$$\frac{x_1}{\varepsilon_1}\frac{dx_1}{dt} = \frac{x_2}{\varepsilon_2}\frac{dx_2}{dt} = \frac{x_3}{\varepsilon_3}\frac{dx_3}{dt} = x_1 x_2 x_3 \tag{II.4.8}$$

which implies

$$\frac{x_1^2}{\varepsilon_1} - \frac{x_2^2}{\varepsilon_2} = C_1 \ ; \quad \frac{x_2^2}{\varepsilon_2} - \frac{x_3^2}{\varepsilon_3} = C_2 \ ; \quad \frac{x_3^2}{\varepsilon_3} - \frac{x_1^2}{\varepsilon_1} = C_3 \ . \tag{II.4.9}$$

The last first integral is not independent of the first two.

We notice that the algebraic first integral, was built up from the rational first integrals. $\frac{x_1^2}{\varepsilon_1} - \frac{x_2^2}{\varepsilon_2}, \ldots$, this is *not* fortuitous, because we have the following.

Lemma 1 BRUNS (1887) The assumption that the set $F_i(x_1, \ldots x_N)$ has all its elements rational functions, implies that each coefficient $\phi_p(x_1, \ldots x_N)$, $p = 1, \ldots, k$ is itself a rational first integral : that is to say each algebraic first integral of order k is built up from k rational first integrals (not necessarily independent).

Therefore we can reduce the study of algebraic first integrals to rational ones.

We shall define an *Algebraically integrable system* (S), a system (S) which has a sufficient amount of algebraic integrals to be reducible to quadratures. The given example is clearly in this class. It is possible to reduce furthermore the search of rational first integrals by invoking Bruns lemma 2.

To start with, let us express every rational first integral as a ratio of polynomial

$$\phi = \sum_m \phi_m(x_1, x_2, \ldots x_N) / \sum_{m'} \phi_{m'}(x_1, x_2, \ldots x_N) \tag{II.4.10}$$

where $\phi_m(x_1, x_2, \ldots, x_N)$ is a polynomial homogeneous of weighted degree m that is:

$$\phi_m(\alpha^{G_1} x_1, \ldots, \alpha^{G_N} x_N) = \alpha^m \phi(x_1, x_2, \ldots, x_N) \tag{II.4.11}$$

(clearly m is necessarily a rational number). For instance using the previous example, the polynomial

$$3x_1^2 x_2 + x_3^3 + x_2 \tag{II.4.12}$$

can be decomposed into the sum of two homogeneous polynomials of weighted degree respectively 3 and 1.

$$3x_1^2 x_2 + x_3^3 \quad \text{and} \quad x_2. \tag{II.4.13}$$

Then BRUNS Lemma 2 states:

Every rational function $\frac{\phi_m(x_1,...x_N)}{\phi_{m'}(x_1,...x_N)}$ (for any choice of m and m') appearing in the previous expression is itself a first integral.

Combining Lemma 1 and 2, we have:

"Every algebraic first integral is compound algebraically of integrals of the forms $\phi_m(\mathbf{x})/\phi_{m'}(\mathbf{x})$ which are weighted homogeneous of degree $m-m'$ (necessarily a rational number).

2.5. Poincaré's Lemma on variational Equations

We shall for convenience consider a slightly more general autonomous system:

$$\frac{dx_i}{dt} = F_i(x_1, ...x_N) \qquad i = 1, ...N. \qquad (II.5.1)$$

with smooth functions F_i. Let $x_i^o(t)$ be a special solution of (II.5.1) and define variations by:

$$x_i(t) = x_i^o(t) + \xi_i(t) \qquad i = 1, ...N. \qquad (II.5.2)$$

We then get, neglecting the terms of order ξ^2, the variational equations:

$$\frac{d\xi_i}{dt} = \sum_{k=1}^{k=N} \frac{\partial F_i}{\partial x_k}(x_1^o(t), ..., x_N^o(t))\xi_k(t) \qquad i = 1, ...N \qquad (II.5.3)$$

we remark incidently that

$$\xi_i = \frac{dx_i^o}{dt} \qquad i = 1, ...N. \qquad (II.5.4)$$

is always a solution of (II.5.3). This remark will be used later on to show that -1 is always a Kowalevski exponent. The existence of a first integral $\phi(x_1, ...x_N) =$ constant for the original system (II.5.1), interconnects the varied solutions ξ_i and ϕ in the following manner:

Lemma 3 (Poincaré)

Let $\phi(x_1, x_2, ...x_N)$ =const be a first integral of the original system (II.5.1). Then for any solution of the variational equations (II.5.3) one has:

$$\sum_{i=1}^{i=N} \frac{\partial\phi}{\partial x_i}(x_1^o(t), ..., x_N^o(t))\xi_i(t) = \text{constant (in } t) \qquad (II.5.5)$$

Proof:

$$\begin{aligned}\frac{d}{dt}\sum_{i=1}^{i=N}\frac{\partial\phi}{\partial x_i}(\mathbf{x}^o(t))\xi_i(t) &= \sum_{i=1}^{i=N}\sum_{J=1}^{J=N}\frac{\partial^2\phi}{\partial x_i\partial_j}(\mathbf{x}^o(t))F_j^{(o)}\xi_i + \frac{\partial\phi}{\partial x_i}(\mathbf{x}^o(t))\frac{\partial F_i}{\partial x_j}(\mathbf{x}^o(t))\xi_j\\ &= \sum_{i=1}^{N}\xi_i\left\{\sum_{i=1}^{N}\frac{\partial^2\phi^{(o)}}{\partial x_i\partial x_j}F_j^{(o)} + \frac{\partial\phi^{(o)}}{\partial x_j}\frac{\partial F_j^{(o)}}{\partial x_i}\right\}\\ &= \sum_{i=1}^{N}\xi_i\frac{\partial}{\partial x_i}\left\{\sum_{J=1}^{N}\frac{\partial\phi}{\partial x_j}F_j\right\}\Bigg|_{x_i=x_i^o(t)}\\ &= \sum_{i=1}^{N}\xi_i\frac{\partial}{\partial x_i}\left\{\frac{d\phi}{dt}\right\}^{(o)} \equiv 0\,.\end{aligned}$$

$$(II.5.6)$$

Before getting to the consequence of Poincaré's lemma, let us point out that a consequence of (II.5.4) is that:

$\rho = -1$ is always a Kowalevski exponent. In fact (II.5.4) tells us that $\{-c_i G_i t^{-(G_i+1)}\}_{i=1}^N$ is a solution of the varied equations, that is $\{-c_i G_i\}_{i=1}^N$ is an eigenvector of the Kowalevski matrix with the eigenvalue -1.

A more fundamental way to understand this result is to remark that the system (S) being *autonomous*, if $\{c_i t^{-G_i}\}_{i=1}^N$ is a scaling solution then $\{c_i(t-t_0)^{-G_i}\}_{i=1}^N$ is for any $t_0$an other solution; for t_0 small, we are back to the varied set $\{-c_i G_i t^{-(G_i+1)}\}_{i=1}^N$ which correspond to the eigenvalue -1. Therefore the eigenvalue -1 is always present, and correspond to the time invariance of the system (S).

We are now going to combine Brun's lemma and Poincaré' Lemma, to get Yoshida's theorems.

2.6. Yoshida's theorems

Theorem I Let $\phi(x_1, x_2, ..., x_N)$ =const be a weighted homogeneous first integral of degree m suppose the components of the vector:

$$\nabla\phi(\mathbf{C}) = \left[\frac{\partial\phi}{\partial x_1}(C_1, C_2, ..., C_N), ... \frac{\partial\phi}{\partial x_N}(C_1, C_2, ..., C_N)\right] \qquad (II.6.1)$$

are finite and not all of them simultaneously zero. Then m is necessarily a Kowalevski exponent.

We shall suppose first that K is diagonalizable. Lemma 3 applied to the Kowalevski variational solutions gives:

$$\sum_{i=1}^N \frac{\partial\phi}{\partial x_i}(C_1 t^{-G_1}, ..., C_N t^{-G_N}).\xi_i^{(\rho)} t^{\rho-G_i} = \text{constant in } t \qquad (II.6.2)$$

But ϕ is weighted of degree m

$$\phi(C_1 t^{-G_1}, ..., C_N t^{-G_N}) = t^{-m}\phi(C_1, C_2, ...C_N) \quad \forall C_1, C_2, ..., C_N \qquad (II.6.3)$$

which implies, deriving with respect to C_i :

$$\frac{\partial\phi_i}{\partial x_i}(C_1 t^{-G_1}, ..., C_N t^{-G_N}) = t^{G_i - m}\frac{\partial\phi}{\partial x_i}(C_1, C_2, ..., C_N) \qquad (II.6.4)$$

Therefore (II.6.2) reads:

$$\left\{\sum_{i=1}^N \frac{\partial\phi}{\partial x_i}(\mathbf{C})\xi_i^{(\rho)}\right\} t^{\rho-m} = \text{constant in } t \qquad (II.6.5)$$

which gives $m = \rho$ provided it exists at least one eigenvector of K : $\xi^{(\rho)}$ such that

$$\xi^{(\rho)}.\nabla\phi(\mathbf{C}) \neq 0 \qquad (II.6.6)$$

But K being diagonalizable, all the N eigenvectors $\{\boldsymbol{\xi}^{\rho_j}\}_{j=1}^{N}$ are linearly independent. If it exist a non zero vector $\nabla\phi(\mathbf{C})$ such that (II.6.6) hold, it would contradict the linear independence of the $\boldsymbol{\xi}^{\rho_j}$. Q.E.D.

In the case when K is not diagonalizable, the N independent solutions of (II.3.5) are given by the expressions (II.3.6) with distinct eigenvalues, and by expressions

$$\xi_i(t) = t^{\rho - G_i} \xi_i^{*(\rho)}(\log t) \qquad (II.6.7)$$

where $\xi_i^{*(\rho)}(u)$ is a polynomial in u whose degree is less than the multiplicity of the eigenvalue ρ. Also in this case it is possible to produce N linearly independent vectors $\left\{\hat{\xi}^{(\rho_i)}\right\}_{i=1}^{N}$ and to repeat a proof similar to the previous one. We shall state now without proof.

Theorem II

Let $\phi_i(x)$ =const and $\phi_2(x)$ =const be two independent weighted homogeneous first rational integrals of the *same* degree m. Suppose the two vectors $\nabla\phi_1(\mathbf{C})$ and $\nabla\phi_2(\mathbf{C})$, are both finite, not identically zero, and furthermore linearly independent. Then $\rho = m$ is a Kowalevski exponent of multiplicity at least 2.

Remark 1

This generalizes to more than two independent first integrals of the same degree.

Remark 2

If ϕ_1 and ϕ_2 are functionally dependent $\nabla\phi_1(\mathbf{C})$ and $\nabla\phi_2(\mathbf{C})$ are *necessarily* linearly dependent. The converse is *not* true, $\nabla\phi_1(\mathbf{C})$ and $\nabla\phi_2(\mathbf{C})$ can be linearly dependent in a special point $\mathbf{C}$, even if $\phi_1(\mathbf{x})$ and $\phi_2(\mathbf{x})$ are functionally independent.

Theorem III

If any Kowalevski exponent is irrational or complex then the system (S) is not algebraically integrable.

That is, it is impossible to find $N-1$, weighted homogeneous rational first integrals.

The proof can be found in ref. 10.

Remark

For an Hamiltonian system with f degrees of freedom, the Kowalevski exponents appear in pairs. More precisely we have the following result[14] :

Let $\phi(\mathbf{q},\mathbf{p})$ =const be a weighted homogeneous first integral of weighted degree m for a similarity invariant Hamiltonian system, where the weighted degree of the Hamiltonian is h. Suppose that for the given choice of C_i $(i = 1, 2, ..., 2f)$ $\nabla\phi(\mathbf{C})$ is finite and not identically zero, then $\hat{m} = h - 1 - m$ is a Kowalevski exponent. (The conjugated Kowalevski exponent).

For instance for $m = h$, $\hat{m} = -1$. One would like to make a natural extension of algebraic integrability to Hamiltonian systems, taking into account Liouville theorem.

Therefore, we shall say that an Hamiltonian system with f degrees of freedom is algebraically integrable, if one can construct f rational first integral of weighted homogeneous degree. The other f first integrals being given by integration of 1-forms using Liouville theorem.

Yoshida thought he could extend Theorem III, to the Hamiltonian case. This appears to be impossible, the reason being clearly connected with the condition $\nabla\phi(\mathbf{C}) \neq 0$. Without this condition, each rational first integral would produce a corresponding Kowalevski rational exponent.

Unfortunately the following examples produce algebraically integrable Hamiltonian, with complex or irrational Kowalevski exponents. The interested reader can refer to the paper of M. Kummer, R. Churchill and D. Rod[12]. Here we shall only discuss in detail one of the examples they provide.

2.7. An example of an integrable system with non rational Kowalevski exponent

Kummer et al, give the following examples:

Let $H(q_1, q_2, p_1, p_2) = -p_1 q_1^2 + \alpha(q_2^2 - p_2^2)q_1 + p_1^3 \quad \alpha \in \mathbb{R}$, be an Hamiltonian, then $G = q_2^2 - p_2^2$ is a polynomial integral of degree 2 for any value of α. However the Kowalevski exponents at $\mathbf{C} = (1, 0, 0, 0)$ are $\{-1, 3, 1 + 2\alpha\}$. If α is choosen irrational, we see that this provides a counter example to a possible extension theorem. Of course one checks that $\nabla G(\mathbf{C}) \equiv 0$.

Let us discuss in more details, the second example. Here we have:

$$H(q_1, q_2, p_1, p_2) = p_1(q_1^2 + p_1^2) + q_1(q_2^2 + p_2^2) \tag{II.7.1}$$

The equations of motion read:

$$\begin{cases} \dfrac{\partial H}{\partial q_i} = -\dfrac{dp_i}{dt} \\ \dfrac{\partial H}{\partial p_i} = \dfrac{dq_i}{dt} \end{cases} \tag{II.7.2}$$

That is :

$$\begin{cases} \dfrac{dq_1}{dt} = q_1^2 + 3p_1^2 \\ \dfrac{dq_2}{dt} = 2q_1 p_2 \\ \dfrac{dp_1}{dt} = -2p_1 q_1 - (q_2^2 + p_2^2) \\ \dfrac{dp_2}{dt} = -2q_1 q_2 \end{cases} \tag{II.7.3}$$

We check immediately that the *unique* scaling solution is:$(q_1^o, q_2^o, p_1^o, p_2^o) = (C_1, C_2, C_3, C_4)t^{-1}$. with:

$$\begin{cases} C_1^2 + 3C_3^2 = -C_1 \\ 2C_1 C_4 = -C_2 \\ -2C_1 C_3 - (C_2^2 + C_4^2) = -C_3 \\ -2C_1 C_2 = -C_4 \end{cases} \tag{II.7.4}$$

The solutions of (II.7.4) are:

$$\begin{cases} C_0 = (-1, 0, 0, 0) \\ C_\pm = (1/2, 0, \pm i/2, 0) \end{cases} \tag{II.7.5}$$

The Kowalevski exponent corresponding to the solution C_0 are:

$$\{-1, 3, 1 \pm 2i\} \tag{II.7.6}$$

and those to the solution $C_\pm$ are:

$$\{-1, 3, 1 \pm i\} \tag{II.7.7}$$

Whatever the choosen solution for $\mathbf{C}$, we see that we always get complex Kowalevski exponent. The exponent 3 correspond to the first integral H itself which is of degree 3. In fact

$$\nabla H(\mathbf{C}) = \left[2C_1C_3, 0, C_1^2 + 3C_3^2, 0\right] \tag{II.7.8}$$

which is never identically zero. Therefore 3 has to be found, among the exponent as well as -1.

It is easy to check multiplying the second equation in (II.7.3) by q_2, the forth by p_2 and adding that

$$G \equiv q_2^2 + p_2^2 \tag{II.7.9}$$

is a polynomial first integral of degree 2 and that, therefore, the system (II.7.3) is *integrable*. We should expect to see the exponent 2 show up. However

$$\nabla G(\mathbf{C}) = (0, C_2, 0, C_4) \tag{II.7.10}$$

but $C_2 = C_4 = 0$ and therefore $\nabla G(\mathbf{C}) = 0$. To conclude it is not clear if there are possibilities to improve this situation, for instance by making some field extension of the rationals.

A different answer to this question has been given by Yoshida himself, in some special cases, in which the kinetic energy in the Hamiltonian has to be in standard form[13].

Finally the extension to non homogeneous system, using Painlevé resonances is discussed in ref. 14.

2.8. Expansions involving Kowalevski exponents

We have the following theorems:

Let ρ be a Kowalevski exponent, then there exists an expansion involving an arbitrary constant I_ρ

$$x_i(t) = t^{-G_i}\left[c_i + P_i(I_\rho t^\rho)\right] \tag{II.8.1}$$

More generally, if $\rho_1, \rho_2, ...\rho_k$ are Kowalevski exponent with *positive* real parts, then:

$$x_i(t) = t^{-G_i}\left[c_i + P_i(I_{\rho_1}t^{\rho_1}, I_{\rho_2}t^{\rho_2}, ..., I_{\rho_k}t^{\rho_k}\right] \tag{II.8.2}$$

where $P_i(z_1, z_2, ..., z_k)$ is a convergent power series and $P_i(0, 0, ...0) = 0$.

This solution converges to $c_i t^{-G_i}$ when $t \longrightarrow 0$. It is clearly the extension of the varied solution:

$$x_i(t) = c_i t^{-G_i} + \delta x_i \tag{II.8.3}$$

where

$$\delta x_i = I_{\rho_1}\xi_i^{(\rho_1)}t^{\rho_1 - G_i} + I_{\rho_2}\xi_i^{(\rho_2)}t^{\rho_2 - G_i} + \dots + I_{\rho_k}\xi_i^{(\rho_k)}t^{\rho_k - G_i} . \quad (II.8.4)$$

We remark that if there is a weighted first integral $\phi(x_1, x_2, \dots x_N)$ of positive degree m and that appears among the positive exponents, let us say

$$m = \rho_m . \quad (II.8.5)$$

Then I_{ρ_m} has to be proportional to the value of ϕ. In particular if we choose the special value $\phi = 0$, then the corresponding I_ρ will also be zero, this remark may simplify some argument later on.

In the same way; we have an analogous statement, for the Kowalevski exponents with negative real part and $t \longrightarrow \infty$.

2.9. Further examples

Let us consider the N-body astronomical problem in dimension 2, for simplicity. Here:

$$\begin{cases} H = \frac{1}{2}\sum_{j=1}^{N} \frac{\vec{p}_j^{\,2}}{m_j} - \sum_{i,j} \frac{m_i m_j}{\mid \vec{q}_i - \vec{q}_j \mid} & \vec{q}_i = \begin{bmatrix} x_i \\ y_i \end{bmatrix} \\ \frac{d\vec{q}_i}{dt} = \frac{\partial H}{\partial \vec{p}_i} \; ; \; \frac{d\vec{p}_i}{dt} = -\frac{\partial H}{\partial \vec{q}_i} \qquad i = 1, 2, \dots n & \vec{p}_i = \begin{bmatrix} p_{x_i} \\ p_{y_i} \end{bmatrix} \end{cases} \quad (II.9.1)$$

Introducing the $\frac{N(N-1)}{2}$ auxilliary variables:

$$S_{ij} = \mid \vec{q}_i - \vec{q}_j \mid^{-1} \quad (II.9.2)$$

this allows to rationalize the system, which becomes:

$$\begin{cases} \frac{d\vec{q}_i}{dt} = \frac{\vec{p}_i}{m_i} \\ \frac{d\vec{p}_i}{dt} = -\sum_j m_i m_j \; S_{ij}^3(\vec{q}_i - \vec{q}_j) \\ \frac{dS_{ij}}{dt} = -S_{ij}^3(\vec{q}_i - \vec{q}_j)(\frac{\vec{p}_i}{m_i} - \frac{\vec{p}_j}{m_j}) \end{cases} \quad (II.9.3)$$

(II.9.3) admits $\frac{N(N-1)}{2}$ first integrals

$$\mid \vec{q}_i - \vec{q}_j \mid - \frac{1}{S_{ij}^2} = \gamma_{ij} = \text{const} \quad (II.9.4)$$

When $\gamma_{ij} = 0$, we have (II.9.1). The system (II.9.3) scales following:

$$\begin{cases} t \longrightarrow \alpha^{-1}t \; ; \; q \longrightarrow \alpha^{-2/3}q \; ; \; p \longrightarrow \alpha^{1/3}p \\ s \longrightarrow \alpha^{2/3}s . \end{cases} \quad (II.9.5)$$

For $N = 2$, the Kowalevski determinant reads:

$$K(\rho) = (\rho+1)(\rho+4/3)(\rho+2/3)^2(\rho+1/3)\rho(\rho-1/3)^2(\rho-2/3) \qquad (II.9.6)$$

All exponents are rational, it is known that the system is integrable (Kepler), among the first integrals one recognizes:

$$\begin{cases} p_{x_1} + p_{x_2} = \text{const} \quad (\text{degree } 1/3) \\ p_{y_1} + p_{y_2} = \text{const} \quad (\text{degree } 1/3) \\ m_1(x_1 p_{y_1} - y_1 p_{x_1}) + m_2(x_2 p_{y_2} - y p_{x_2}) = \text{const} \quad (\text{degree } -1/3) \\ H = \text{constant} \quad (\text{degree } 2/3) \end{cases} \qquad (II.9.7)$$

The other first integrals are given by Liouville's theorem. For the three-body problem $N = 3$, one has:

$$K(\rho) = (\rho+4/3)^3[\rho+2/3]^2[\rho-1/3]^2[\rho+1/3]\rho(\rho+1)(\rho-2/3)(\rho-\rho_\pm^{(1)})(\rho-\rho_\pm^{(2)}) \quad (II.9.8)$$

where

$$\rho_\pm^{(1)} = \frac{1}{6}[-1 \pm \sqrt{13+12\chi'}] \ ; \ \rho_\pm^{(2)} = \frac{1}{6}[-1 \pm \sqrt{13-12\chi'}] \qquad (II.9.9)$$

and

$$\chi' = \frac{1}{m_1+m_2+m_3}\sqrt{1/2[(m_1-m_2)^2+(m_2-m_3)^2+(m_3-m_1)^2} \qquad (II.9.10)$$

we have all together 15 Kowalevski exponent corresponding to $4N + \frac{N(N-1)}{2}$ variables (here $N = 3$).

In general those last exponents are irrational or complex, and the 3-body problem is not algebraically integrable (in the restricted sens).

We come now, after this short introduction to the analytical structure (singularities) in the complex t-plane of the integrable and non-integrable systems. Chaotic system are linked with non-integrability: how chaos is reflected in the complex plane ? There is a general belief that non integrable systems produce natural boundaries in the complex t-plane. Singularities form dense sets sitting on fractals associated to complex Kowalevski exponent.

3. SINGULARITIES IN THE COMPLEX PLANE OF NON-INTEGRABLE SYSTEMS

3.1. Moving singularities

Singularities in the complex plane can be :

- either "fixed"
- or "moving".

The fixed singularities have their position not depending on initial datas, the values of x_i at $t = 0$. The moving ones change in the complex plane with the initial conditions and are, therefore, difficult to catch. The simplest "moving" case is when the only moving singularities are poles (Painlevé type) and the solution be single valued.

3.2. Detection of singularities

The method used is the construction of the "Star of holomorphy" of Weirstrass. This construction has been computerized by Chang[16]. Starting from a regular point in **C** of the solution, the program computes recursively the Taylor series of the solution at this point (truncated at a certain order). Then along a prescribed path, calculate for a further point of this path (inside the first circle of convergence), the necessary values (new Taylor coefficients) which will permit to repeat the procedure. The program allows also to detect the location and the nature of the singularity lying on the circle of convergence of the computed Taylor series.

One has

$$x_i(t) = (t - t_*)^{\alpha} h_i(t) \qquad (III.2.1)$$

where $h_i(t)$ is regular near t_* and if:

$$\alpha = -P \qquad P = 1, 2, 3, \ldots \qquad (III.2.2)$$

then t_* is a pole of order P. If:

$$\alpha = P/Q \qquad P, Q \in \mathbb{Z} \qquad (III.2.3)$$

then t_* is an algebraic singularity of order Q.

When α is an irrational or complex number, then α is called a transcendant singularity (III.2.4)

There can also exist isolated essential singularities. Then:

$$x_i(t) = e^{-\frac{1}{t-t_*}} h_i(t) \qquad (III.2.5)$$

3.3. An example

Frisch and Thual[17] have studied the Kuramoto's model. This model describes the propagation of flame fronts in explosives

$$\begin{cases} u_t + uu_x + u_{xx} + u_{xxxx} = 0 \qquad x \in [0, L] \\ +\text{Boundary conditions and initial conditions} \end{cases} \qquad (III.3.1)$$

This model presents a chaotic behaviour depending on L. The stationary solution satisfies :

$$\left(\frac{u^2}{2} + u_x + u_{xxx}\right)_x = 0 \qquad (III.3.2)$$

and restricting to a constant of integration choosen to be zero, one has:

$$\frac{u^2}{2} + u_x + u_{xxx} = 0 \qquad (III.3.3)$$

Here the Kowalevski exponents are found to be

$$\left\{-1, \frac{13}{2} \pm \frac{i\sqrt{71}}{2}\right\} \tag{III.3.4}$$

The Chang program allows to compute the singularities. They are found to form a natural boundary. Furthermore this natural boundary is well represented as a "Kowalevski fractal", generated simply by the Kowalevski exponents. The next example will illustrate more clearly the concept of Kowalevski fractal.

3.4. A Kowalevski fractal

When the Kowalevski exponents are complex, the singularities generateds by the Chang program are very well approximated[3,4] by a fractal generated simply from the Kowalevsky exponents, and that we shall call a "Kowalevsky fractal". Here is an example.

Consider the Hénon-Heiles Hamiltonian:

$$H = \frac{1}{2}(p_1^2 + p_2^2) + q_1^2 q_2 + \frac{\varepsilon}{3} q_2^3 \tag{III.4.1}$$

For the initial values $t = 0$, $q_1 = 5$, $q_2 = p_1 = p_2 = 0$, Yoshida[18] has worked out the singularities. They sit on the natural boundaries shown in fig.1.

Discussion

Here one finds applying the general theory that $G_i = 2$. However there are *two* sets of C's:

$$(i)\ C_1 = \pm 3\sqrt{2-\varepsilon} \qquad C_2 = 3 \tag{III.4.2}$$

$$(ii)\ C_1 = 0 \qquad C_2 = -6/\varepsilon \tag{III.4.3}$$

The Kowalevski exponents corresponding to the first set are roots of

$$\rho^2 - 5\rho + 6(2-\varepsilon) = 0 \tag{III.4.4}$$

while for the second set, one gets:

$$\rho_2 - 5\rho + 6(1 - 2/\varepsilon) = 0 \tag{III.4.5}$$

For $-48 < \varepsilon < 0$ all Kowalevski exponents are complex and H is a constant of motion of homogeneous degree 6.

In the vicinity of a singularity t_*, we therefore have using the theorem of Section II.8.

$$q_i(t) = (t-t_*)^{-2}\left[C_i + P_i(I_{\rho_1}(t-t_*)^{\rho_1}, I_{\rho_2}(t-t_*)^{\rho_2}, I_6(t-t_*)^6)\right] \tag{III.4.6}$$

And H is easily seen to be proportional to I_6. Choosing $H = 0$ this implies $I_6 = 0$, and setting $\tau = t - t_*$, we get:

$$q_i = \tau^{-2}\left[C_i + P_i(I_\rho \tau^\rho, I_{\rho^*}\tau^{\rho^*})\right] \tag{III.4.7}$$

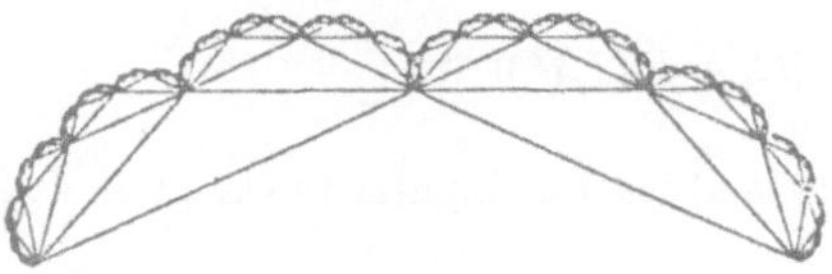

Geometrical construction of natural boundary (ε = −1).

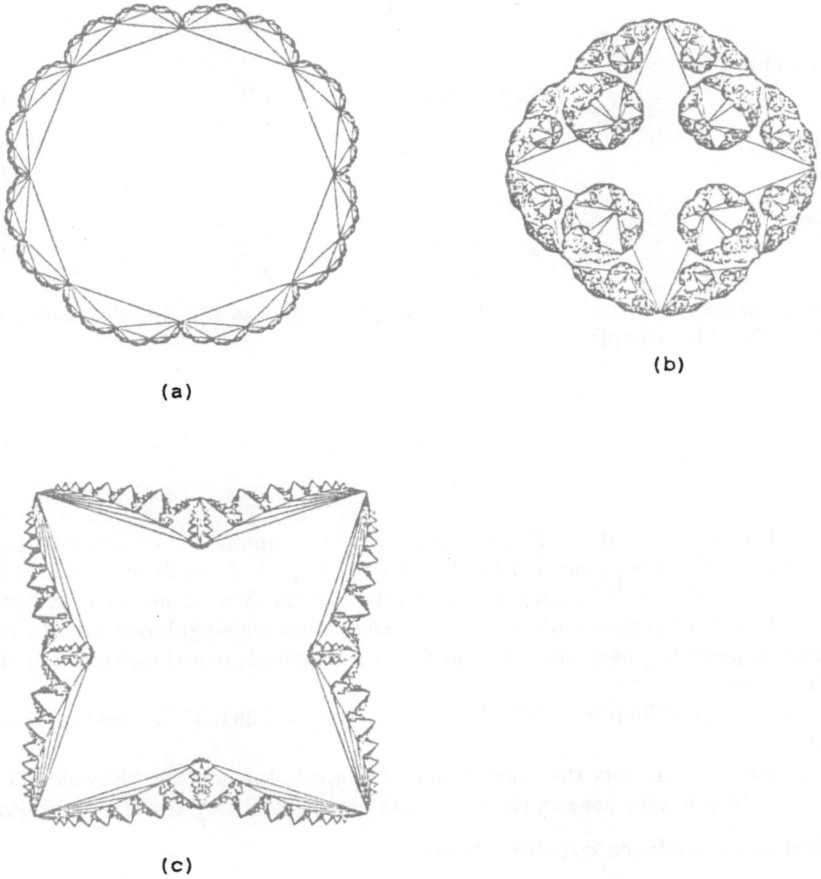

Global natural boundaries. (a) ε = −1, (b) ε = −8, (c) ε = −0.1.

FIGURE 1

where ρ^* is the complex conjugate of ρ.

Suppose $I_{\rho^*} = 0$, to start with, then:

$$q_i = \tau^{-2}\left[C_i + P_i(I_\rho t^\rho)\right] \qquad (III.4.8)$$

On the circle of convergence of $P_i(z)$, there is at least a singularity say at $z = z_0$, therefore when:

$$I_\rho(t - t_*)^\rho = z_0 \qquad (III.4.9)$$

q_i will be singular. The corresponding values of t :

$$t_n = t_* + \left(\frac{z_0}{I_\rho}\right)^{1/\rho} e^{\frac{2i\pi n}{\rho}} \qquad (III.4.10)$$

are on a spiral

$$t_n = t_* + M_0 S^n \qquad S = e^{\frac{2i\pi}{\rho}} \qquad (III.4.11)$$

Setting

$$S = r\, e^{i\theta} \qquad (III.4.12)$$

we have

$$r = \exp\frac{2\pi\rho_I}{|\rho|^2} \qquad \theta = \frac{2\pi\rho_R}{|\rho|^2} \qquad (III.4.13)$$

ρ_R and ρ_I being respectively the real and imaginary part of the Kowalevski exponent ρ.

Here, for this example

$$\begin{cases} r_1 = \exp -\dfrac{\pi\sqrt{23 - 24\varepsilon}}{6(2-\varepsilon)} \\ \theta_1 = \dfrac{5\pi}{6(2-\varepsilon)} \end{cases} \qquad (III.4.14)$$

In fact when $I_{\rho_*} \neq 0$, if $| I_\rho t^\rho | \gg | I_{\rho^*} t^{\rho^*} |$, the complete set of singularities is well approximated by the previous spiral. In $| I_\rho t^\rho | \ll | I_{\rho^*} t^{\rho_*} |$ substitute $\rho \longrightarrow \rho^*$, that is $S = r\, e^{i\theta} \longrightarrow S = r^{-1} e^{i\theta}$. Now we must take the negative values of n to keep $t - t_*$ small. The new spiral goes anti-clockwise. Hence, around a singularity with two complex conjugate exponents, there are *two convergent* semi-spiral, one clockwise, and the other anti-clockwise.

For $\varepsilon = -1$, one finds $\theta_1 = 50°$, $\theta_2 = 25°$, $r_1 = r_2^2 = 0.302$. This gives the construction of fig.2.

By continuing ad infinitum the construction of figure 1, one gets the "Kowalevski fractal" of figure 1 which is very near to the set of singularities given by the Chang program.

3.5. Riemann surfaces considerations

An important question, in trying to characterize those Kowalevski fractal as linked to non integrability is to know if these singularities are sitting on the *same* Riemann sheet or the fractal aspect could just be an illusion comming from the *projection* on C of *isolated*

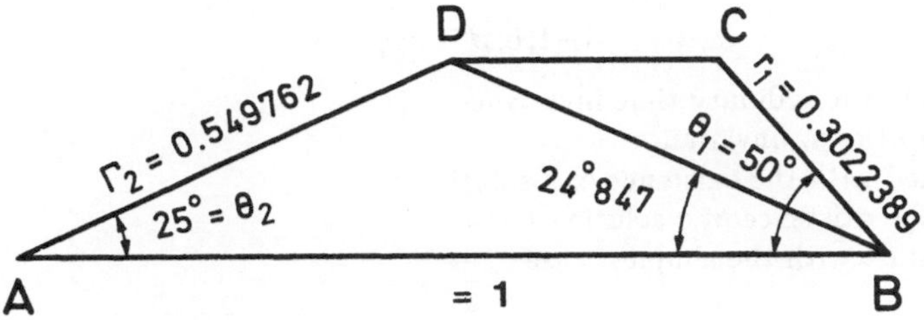

Fig.2 – Construction of the successive singularities $(AB \cdot CD = AD^2)$

singularities on the Riemann surface. To discuss this question, we shall consider a simple example, for more detail see ref. 15.

3.6. An example

Bessis and Chafee have studied the following system

$$\begin{cases} \dfrac{dx_1}{dt} = -\dfrac{x_1^2}{x_2} \\ \dfrac{dx_2}{dt} = -\dfrac{x_2^2}{x_3} \\ \dfrac{dx_3}{dt} = ax_3x_4 + b\dfrac{x_3^2x_4^2}{x_2} + c\dfrac{x_3^2x_4^3}{x_1} \\ \dfrac{dx_4}{dt} = -x_4^2 \end{cases} \qquad (III.6.1)$$

a, b, c being constants. The general solution of this system is:

$$\begin{cases} x_1^{-1} = A(t-t_*)^\alpha + B(t-t_*)^\beta + C(t-t_*)^\gamma \\ x_2^{-1} = A\alpha(t-t_*)^{\alpha-1} + B\beta(t-t_*)^{\beta-1} + C\gamma(t-t_*)^{\gamma-1} \\ x_3^{-1} = A\alpha(\alpha-1)(t-t_*)^{\alpha-2} + B\beta(\beta-1)(t-t_*)^{\beta-2} + C\gamma(\gamma-1)(t-t_*)^{\gamma-2} \\ x_4^{-1} = t-t_* \\ \alpha, \beta, \gamma \text{ roots of } \rho^3 + (a-3)\rho^2 + (2-a+b)\rho + c = 0 \end{cases} \qquad (III.6.2)$$

The Kowalevski exponents associated to the scaling solution:

$$\begin{cases} x_1 = \mu_\alpha t^{-\alpha} \\ x_2 = \dfrac{\mu_\alpha}{\alpha} t^{-\alpha+1} \\ x_3 = \dfrac{\mu_\alpha}{\alpha(\alpha-1)} t^{-\alpha+2} \\ x_4 = t^{-1} \end{cases} \qquad \mu_\alpha \text{ arbitrary} \qquad (III.6.3)$$

are

$$\{-1; 0; \beta - \alpha; \gamma - \alpha\} \tag{III.6.4}$$

-1 is linked to the ordinary time invariance
0 is linked to the μ_α invariance
$\beta - \alpha$ is linked with the existence of a solution $t^{-\beta}$
$\gamma - \alpha$ with the existence of a solution $t^{-\gamma}$.

Let us start with the simplest case

$$(i) \qquad B = C = 0 \tag{III.6.4}$$

The solution is :

$$x_1(t) = A(t - t_*)^{-\alpha} \tag{III.6.5}$$

There are two transcendental singularities: one at t_*, the other at infinity. The Riemann surface of the log uniformizes the solution

$$(ii) \qquad C = 0 \tag{III.6.6}$$

$$x_1(t) = \left[A(t - t^*)^{-\alpha} + B(t - t_*)^{\beta}\right]^{-1} = \frac{B^{-1}(t - t_*)^{-\alpha}}{AB^{-1} + (t - t_*)^{\beta - \alpha}} \tag{III.6.7}$$

Again t_* and infinity are transcendent singularities, however we get also moving *poles* at :

$$(t - t_*)^{\delta} = \lambda \qquad \text{where} \quad \begin{array}{l} \lambda = -AB^{-1} \\ \delta = \beta - \alpha \end{array} \tag{III.6.8}$$

that is at :

$$t_n = t_* + \lambda^{1/\delta}\, e^{\frac{2i\pi n}{\delta}} \tag{III.6.9}$$

$$= t_* + \lambda^{1/\delta}\, e^{-2\pi n \frac{\text{Im}\delta}{|\delta|^2}}\, e^{\frac{2i\pi n \text{Re}\delta}{|\delta|^2}} \tag{III.6.10}$$

If $\text{Im}\delta \neq 0$, the poles sit on a spiral; however due to the transcendent singularity at $t = t_*$, the poles belong to *different* sheets. The number of poles per sheet being integer part of $\frac{|\delta|^2}{\text{Re}\delta}$. If δ is real, then

$$t_n = t_* + \lambda^{1/\delta}\, e^{\frac{2i\pi n}{\delta}} \tag{III.6.11}$$

The poles are on an *helix* on the Riemann surface, and there is still $\left[\frac{2\pi}{\delta}\right]$ poles per sheet. However their projection form a *dense* set if δ is irrational. That is, if any of the Kowalevski exponent is irrational. If one looks only to *projections* of the singularities on $\mathbf{C}$, one would conclude to the existence of a *natural boundary* associated to an irrational Kowalevski exponent.

(iii) We come now to the more general case

$$x_1(t) = \left[A(t - t_*)^{\alpha} + B(t - t_*)^{\beta} + C(t - t_*)^{\gamma}\right]^{-1} \tag{III.6.12}$$

Again we have a transcendent singularity at $t = t_*$. Furthermore we have attached to this singularity an infinite number of poles which move with it, and the initial condition parameters A, B, C. Setting

$$\begin{cases} \alpha - \gamma = \bar{\alpha} \\ \beta - \gamma = \bar{\beta} \end{cases} \qquad (III.6.13)$$

and choosing A, B, C real positive, $\bar{\alpha}, \bar{\beta}$, irrationals, and $0 < \bar{\alpha} < 1$ as well as $0 < \bar{\beta} < 1$, then in the plane cut from t_* to $-\infty$, there are no poles. The reason is that the poles are solution of:

$$D(t) \equiv A(t - t_*)^{\bar{\alpha}} + B(t - t_*)^{\bar{\beta}} + C = 0 \qquad (III.6.14)$$

However

$$\begin{aligned} \mathrm{Im}[D(t)] = A \mid t - t_* \mid^{\bar{\alpha}} \sin(\bar{\alpha} \ \mathrm{Arg}(t - t_*)) \\ + B \mid t - t_* \mid^{\bar{\beta}} \sin(\bar{\beta} \ \mathrm{Arg}(t - t_*)) > 0 \end{aligned} \qquad (III.6.15)$$

It is not difficult, however, to see that the projection of the poles condensate on two closed curves. [see ref. 15].

REFERENCES

[1] U. Frisch and R. Morf, Phys. Rev. **A23**, 5, 1981.

[2] M. Tabor and J. Weiss, Phys. Rev. **A24**, 4, 1981.

[3] Y.F. Chang, M. Tabor and J. Weiss, J. Math. Phys. **23**, 6, 1982.

[4] J. Weiss, Analytic Structure of the Hénon-Heiles System, in Mathematical Methods in Hydrodynamics and Integrability in Related Dynamical systems, ed. by M. Tabor and Y. Treve, AIP Conference Proc., 88, AIP (1982).

[5] I.C. Percival and J.M. Green, Hamiltonians Maps in the complex plane, Princeton Plasma Phys. Lab. Preprint PPL 1744 Jan 1981.

[6] J. Weiss, M. Tabor and G. Carnevole, J. Math. Phys. **24**, 3, 1983.

[7] A. Ramani, B. Dorizzi, B. Grammaticos and T. Bountis, J. Math. Phys. **25**, 4, 1984.

[8] T. Bountis and H. Segur, Logarithmic Singularities and chaotic behaviour, in Hamiltonian systems in Mathematical Methods in Hydrodynamics and Integrability in Related Dynamical Systems, ed. by M. Tabor and Y. Treve AIP Conference Proc. 88 AIP 1982.

[9] T. Bountis, A singularity analysis of integrability and chaos in dynamical systems in singularities and dynamical systems, ed. by S. Pneumatikos North Holland 1984.

[10] H. Yoshida, Necessary conditions for the existence of algebraic first integrals, Celestial Mechanics 31 (1983).

[11] S. Kowalevski, Acta Math. Acad. Sci., Hungaria **14**, 81 (1890).

[12] M. Kummer, R. Churchill and D. Rod, On Kowalevski exponents - Preprint Department of Mathematics and Statistics, University of Calgary, Calgary Alberta T2NIN4 Canada, 1988.

[13] H. Yoshida, Non existence of an Additional Analytic Integral in Hamiltonian Systems with a N-dimensional Homogeneous Potential, Preprint May 1988.

[14] H. Yoshida, B. Grammaticos and A. Ramani, Painlevé Resonances versus Kowalevski exponents: Some Exact Results on Singularity Structure and Integrability of Dynamical Systems, Acta Applicandre Mathematicoe, **8**, 75-103, 1987.

[15] D. Bessis and N. Chafee, On the Existence and Non-existence of Natural Boundaries for Non-integrable Dynamical Systems in Chaotic Dynamics and Fractal, M. Barnsley and S. Demko editors, Academic Press 1986.
[16] Y.F. Chang, Program for ATOMCC, Claremont-McKenna College, Claremont CA, USA.
[17] O. Thual and U. Frisch, Natural Boundaries for the Kuramoto model, in Workshop on Combustion Flames and Fires, Les Houches, France, 1984.
[18] H. Yoshida, Self-Similar natural boundaries of non-integrable dynamical systems in the complex t-plane, Preprint Dept. of Astronomy, Univ. of Tokyo, Tokyo, Japan.

SINGULARITY ANALYSIS AND ITS RELATION TO COMPLETE, PARTIAL AND NON-INTEGRABILITY

M. D. KRUSKAL
Department of Mathematics
Fine Hall, Princeton University
Princeton, NJ 08544
U.S.A.

A. RAMANI
C.P.T. Ecole Polytechnique
91128 Palaiseau
France

B. GRAMMATICOS
L.P.N. Université Paris VII
Tour 24-14, 5° étage
75251 Paris
France

ABSTRACT. The aim of this course is to present and illustrate the connection between integrability and the singularity structure of the solutions of nonlinear dynamical systems. We start by reviewing the various aspects of integrability and then introduce the notions of partial and constrained integrability. Next we present the methodology of the singularity ("Painlevé") analysis and apply it to the study of various systems. A similar approach, due essentially to Yoshida, is presented by the course of D. Bessis which is complementary to ours. Finally we present a detailed review of the recent progress in the domain of nonintegrability. Based on Ziglin's theorem, we prove rigorously the nonexistence of integrals for several systems of physical interest. As a non-linear extension of Ziglin's approach we present the "poly-Painlevé" criterion of (non-)integrability, illustrate it through some example, and propose a practical method for its implementation.

R. Conte and N. Boccara (eds.), Partially Integrable Evolution Equations in Physics, 321–372.

1. Integrability concepts

1.1 TOWARDS A WORKING DEFINITION OF INTEGRABILITY

Integrability is a term widely used in the domain of dynamical systems, and despite this fact (or because of it) the various practitioners do not seem to agree on its definition. The word itself implies some relation to a system of differential equations. The domain of dynamical systems is of course much wider : integral, difference, and functional equations are often used in modelling physical phenomena. However, in the present course we will deal almost entirely with dynamical systems described by differential equations and the question of their integrability. The latter notion should not be confused with solvability which is in fact one of its aspects and means the existence of solution expressed in terms of known functions. Still, this is not a clear-cut definition as there is tremendous arbitrariness in what is considered as a "known" function. Consider as an example of solvability a simplified version of the Rikitake system,

$$\begin{aligned} \dot{x} &= y\,z \\ \dot{y} &= -x\,z \\ \dot{z} &= -x\,y. \end{aligned} \tag{1.1}$$

Its solutions cannot be written in terms of *elementary* functions but instead are given as elliptic functions

$$\begin{aligned} x &= A\,\mathrm{sn}(p(t-t_0)) \\ y &= A\,\mathrm{cn}(p(t-t_0)) \\ z &= p\,\mathrm{dn}(p(t-t_0)) \end{aligned} \tag{1.2}$$

with parameter $m=A^2/p^2$.

While explicit knowledge of the solutions of the system may be very useful it is clear that most times we must content ourselves with less. In fact, one of the main uses of integrability lies in the fact that it allows us to obtain global information on the long-time behavior of the system, usually through the existence of conserved quantities, i.e. quantities the value of which is constant throughout the time-evolution of the system. Thus, integrability is characterized by the existence of "constants of the motion," "integrals," or "invariants." For instance, the Hamiltonian system [1]

$$H = \frac{1}{2}\left(p_x^2 + p_y^2\right) + y^4 + 3\,x^2y^2/4 + x^4/8 \tag{1.3}$$

has two constants of motion, the Hamiltonian, itself and a second invariant :

$$\begin{aligned} I = p_x^4 + (3x^2y^2 + x^4/2)p_x^2 - 2\,x^3\,y\,p_x\,p_y + x^4p_y^2/2 + \\ + (x^4\,y^4 + x^6y^2)/4 + x^8/16. \end{aligned} \tag{1.4}$$

Integrals are used to reduce equations of motion. A particularly simple case is that of one-dimensional Hamiltonian systems :

$$H = p_x^2/2 + V(x) \tag{1.5}$$

The energy H_0 is the first constant of motion and this allows us to obtain the second constant :

$$t - t_0 = \int \frac{dx}{\sqrt{2(H_0 - V(x))}} \tag{1.6}$$

Still t_0 is not as useful as H_0 : in order to obtain it we must solve the problem. Moreover t_0 almost always depends on the integration path and not only on the value of x, while the energy is a local function in phase-space.

In Ref [2], we have attempted a classification of the various aspects of integrability. We have distinguished three different situations.

a) The system can be solved by quadratures. (In [3], we have used the term "explicit integrability" for this case). For instance, the two-dimensional Hamiltonian system

$$H = \frac{1}{2}\left(p_x^2 + p_y^2\right) + F(\rho) + G(\varphi)/\rho^2 \tag{1.7}$$

where $\rho=(x^2+y^2)^{1/2}$ and $\varphi=\arctan(y/x)$, has the second integral

$$I = (xp_y - yp_x)^2 + 2\,G(\varphi). \tag{1.8}$$

This allows the equations of motion to be reduced first to a quadrature for ρ :

$$\dot{\rho}^2 = 2H_0 - 2F(\rho) - I_0/\rho^2 \tag{1.9}$$

where H_0 and I_0 are the conserved values of H and I respectively. Once $\rho(t)$ is obtained from (1.9), the equation for φ can also be reduced to a quadrature :

$$\int \frac{d\varphi}{\sqrt{I_0 - 2\,G(\varphi)}} = \pm \int \frac{dt}{\rho^2(t)} \tag{1.10}$$

b) The equations of motion can be reduced to a system of linear equations which are considered to be integrable because of their nice properties. In fact their solutions can be superimposed linearly and there exist global representations for them in terms of contour integrals. The simplest example is that of the well known Riccati equation :

$$\dot{x} = a(t)\,x^2 + b(t)\,x + c(t) \tag{1.11}$$

which linearizes to

$$\ddot{y} + (\frac{\dot{a}}{a} - b)\dot{y} + acy = 0 \tag{1.12}$$

through the transformation

$$x = -\frac{\dot{y}}{ay}. \tag{1.13}$$

Some PDE's are also integrable through linearization, Burger's equation being the archetype :

$$u_t + u_{xx} + 2uu_x = 0. \tag{1.14}$$

The Cole-Hopf transformation

$$u = v_x / v \tag{1.15}$$

reduces its solution to that of the heat equation :

$$v_t + v_{xx} = 0. \tag{1.16}$$

c) The system can be linearized in terms of integro-differential equations. This is, for example, the case of the Painlevé transcendental equations [4]. Six are known at order two but more surely exist at higher orders. The first four of the Painlevé equations read :

$$
\begin{aligned}
&\text{P I)} \quad && w'' = 6\, w^2 + z \\
&\text{P II)} \quad && w'' = 2\, w^3 + z\, w + a \\
&\text{P III)} \quad && w'' = \frac{w'^2}{w} - \frac{w'}{z} + \frac{aw^2 + b}{z} + cw^3 + \frac{d}{w} \\
&\text{P IV)} \quad && w'' = \frac{w'^2}{2w} - \frac{3w^3}{2} + 4zw^2 + 2(z^2 - a)w + \frac{b}{w}
\end{aligned}
\tag{1.17}
$$

A good candidate for a higher order transcendent is given by the (once integrated) similarity reduction of the fifth order equation from the modified Korteweg-deVries hierarchy :

$$w'''' = 10\, w^2\, w'' + 10\, w\, w'^2 - w^5 - zw. \tag{1.18}$$

The very idea of linearization through integro-differential equations comes from the Inverse Scattering Transform (IST) technique which has been developed in the context of PDE's. In fact, this is the main mechanism of integrability for PDE's and its study has produced a vast and steadily growing amount of results. [5]

1.2 COMPLETE INTEGRABILITY

The notion of integrability, meaning the existence of integrals of motion, is so vague as to be almost useless. Thus, we must refine it further in order to obtain a coherent definition, particularly in view of the relation which we will study between integrability and the singularity structure of the solutions of the equations of motion. As far as the singularity analysis practitioners are concerned, the term integrability implies the existence of *complex analytic* (functionally independent) integrals of motion. Thus the kind of integrability we are interested in could have been dubbed "complex analytic integrability."

Complete integrability means that these integrals exist in sufficient number. For a system of N first order autonomous ordinary differential equations (ODE), sufficient means N-1 time-independent invariants (whereupon the system can be reduced to a single quadrature) or N time-dependent ones (in which case the solutions can be obtained by solving just an algebraic problem). In general terms, it seems reasonable to ask for integrals respecting the invariance of the initial system. For example, for autonomous systems we should look for integrals which are either time-independent or form-invariant under time translation (see 1.31 below). Hamiltonian systems are special because, for the complete integrability of a Hamiltonian system with M degrees of freedom the existence of (M-1) single-valued first integrals I_i ("actions"), in involution (i.e. with vanishing Poisson bracket), in addition to the Hamiltonian itself, allows the construction of (M-1) additional integrals Ω_i ("angles"), following the Hamilton-Jacobi procedure. A system of N first order ODE's may sometimes be similar to a Hamiltonian system [6]. Indeed, complete integrability can be interpreted as the existence of k ($1 \leq k \leq N-1$) first integrals I_i, provided that (N-k-1) more, say Ω_j, can be computed by integration of closed differential forms obtained from the I_i. This includes of course the Hamiltonian case for which k=M. Moreover, if a differential system of order N admits an invariant density measure, then it also falls into that category for k=N-2 since a last invariant can be obtained through the Jacobi Last Multiplier theorem[7].

For instance, the three-dimensional Lotka-Volterra system :

$$\begin{aligned} \dot{x} &= x\,(Cy+z+\lambda) \\ \dot{y} &= y\,(x+Az+\mu) \\ \dot{z} &= z\,(Bx+y+\nu) \end{aligned} \tag{1.19}$$

always admits an invariant density measure 1/xyz [8]. Therefore, in the special case A=B=1, $\lambda=\mu=\nu=0$, where there exists the invariant :

$$I = (x-Cy)(1-z/y) \tag{1.20}$$

we obtain the second invariant

$$\Omega = \frac{xz}{y}\left(1 - \frac{y}{z}\right)^{C+1} \tag{1.21}$$

Of course, analytic integrability is not the only possibility. In [6], in order to relate integrability to the rationality of the Kowalevskaya exponents (KE), Yoshida has introduced the notion of "algebraic integrability". What he meant by that is that the constants of motion should be rational functions. However he imposed rationality on both the "action" and the "angle"-type invariants. Requiring rationality of the latter is a bit awkward, although it is satisfied for some well-known systems. For instance, the Coulomb Hamiltonian in two dimensions :

$$H = \frac{1}{2}(p_x^2 + p_y^2) + (x^2+y^2)^{-\frac{1}{2}} \tag{1.22}$$

has *three* additional single-valued integrals

$$\begin{aligned} I_1 &= xp_y - yp_x \\ I_2 &= p_x(xp_y - yp_x) - y(x^2+y^2)^{-\frac{1}{2}} \\ I_3 &= p_y(xp_y - yp_x) + x(x^2+y^2)^{-\frac{1}{2}} \end{aligned} \tag{1.23}$$

(but no two of them are in involution!).

Still, it is clear that in general the "angle" invariants are multivalued (see eq. (1.21) for irrational C). If Yoshida's hypothesis as far as the "angles" are concerned is relaxed, one can construct counterexamples with irrational (or complex) Kowalevskaya exponents [9,10]. Thus the latest version of Yoshida's theorem [11] based on Ziglin's approach abandons this notion of algebraic integrability for the sake of the analytic integrability introduced previously.

Algebraic integrability has also been the object of a series of deep studies by Adler and van Moerbecke [12]. Their notion of "algebraic complete integrability" is related not only to the existence of a sufficient number of invariants in involution but to a further demand that the solutions be expressible in terms of Abelian integrals. The advantage of this restriction was that they were able to prove some fundamental theorems relating the complex structure of the solutions with integrability and explain the relation between Painlevé (see section 2) analysis and the dynamics on complex algebraic tori.

The two types of algebraic integrability presented above (Yoshida and Adler-van Moerbecke) clearly put more restrictions on the system than analytic integrability but, in return, they insure that the solutions don't behave too badly. Going in the other direction, we can introduce less restrictive types of integrability.

The simplest generalization one can think of is to ask for integrals that are analytic only within a given domain rather than globally analytic integrals. Such integrals are not uncommon. For instance, the Lotka-Volterra system (1.19) has many subcases where such integrals exist. For example, when $ABC+1=0$ and $\lambda=\mu=\nu$, the conserved quantity

$$I = x^{AB}\, y^{-B}\, z\, (x-Cy+ACz)^{-AB+B-1} \qquad (1.24)$$

is defined in any domain not containing the plane $x-Cy+ACz=0$ and the coordinate planes $x=0$, $y=0$. In fact, for a strictly real-time evolution of the system, integrals which are analytic in some such domains may well suffice.

"Integrable" systems of another kind are those which describe a particle in a potential decreasing sufficiently fast for the particle to be asymptotically free [13]. Thus, the asymptotic momenta are constants of the motion (here "asymptotic" means as time goes to infinity). However though this is true for real trajectories, it ceases to be so when one considers complex time.Then the asymptotic momenta are not single-valued functions of the initial point and asymptotically free systems are not "analytically" integrable. In section 3 we give two examples (3.30 and 3.44) of asymptotically free potentials which are generically nonintegrable as proven by the application of Ziglin's theorem.

Going one step further we can say that every differential equation is integrable in a trivial sense. In fact, let us consider the system of ODE's :

$$\frac{dx_i}{dt} = F_i(t, x_1,\dots, x_n) \qquad i = 1, \dots, n \qquad (1.25)$$

with initial conditions $x_i(t_0)=c_i$. These initial conditions can be taken as the constants of motion of the system. The general solution of the system is :

$$x_i = \Phi_i(t, c_1,\dots, c_n) \qquad (1.26)$$

and by inverting (1.26), which is equivalent to integrating back in time from t to t_0, we can write the n constants as :

$$c_i = I_i(t, x_1,\dots, x_n) \qquad (1.27)$$

However the inversion is not at all guaranteed to be single valued. Indeed this will be the case if the integration path wanders in the complex plane around "bad" singularities.

Whether "real-time" information on the system is sufficient is essentially a philosophical question. While it is clear that the "physical" time is real, it is equally clear that the structure of the solution in complex time heavily influences the real-time behaviour [9]. Singularity analysis techniques are based exclusively on complex time and it is at this price only that one can prove exact theorems on analytic (or algebraic) integrability. The enormous difference between real and complex time as far as integrability is concerned can be easily grasped in the following example.

Consider the following simple ODE in the complex domain :

$$\dot{x} = \frac{\alpha}{t-a} + \frac{\beta}{t-b} + \frac{\gamma}{t-c} . \tag{1.28}$$

Its integration (by quadratures) is straightforward :

$$I = x - \alpha \log (t-a) - \beta \log (t-b) - \gamma \log (t-c) . \tag{1.29}$$

However, since the logarithm of a complex number is defined only up to an integer multiple of $2i\pi$, the right-hand-side of (1.29) (and thus the value of I also) is determined only up to an additive term $2i\pi (k\alpha+m\beta+n\gamma)$, with k, m, n, arbitrary integers. Now if one (resp. two) of the α, β, γ are zero, one can construct a two- (resp. one-) dimensional lattice, and define I in a unique way, in the interior of an elementary lattice cell. But, if $\alpha\beta\gamma \neq 0$ and α, β, γ are linearly independent over the integers the indeterminacy in (1.29) is very high : for given x and t, the possible values of I *densely* fill the plane. Conversely, the knowledge of t and I does not suffice to determine x in any useful way. Thus, the integral (1.29) is useful only in the presence of "discrete" multivaluedness while *"dense"* multivaluedness will be viewed as incompatible with integrability.

The relation between integrability and complex dynamics has received further attention recently in the work of Flashka [14]. He has shown that even when a Hamiltonian system possesses a full set of constants of motion in involution the motion of the system (in phase space) does not necessarily take place on a torus or even on a manifold that is of the type "circles×lines". This was materialized by the following example of two-dimensional Hamiltonian :

$$H = p_x^2 + 3\, x\, p_y^2 - x^3 - y^2 \tag{1.30}$$

which admits the invariant :

$$I = p_y^3 + 2\, y\, p_x - 3\, x^2\, p_y. \tag{1.31}$$

In this case the orbits of the system for fixed values of H and I lie on a torus with one tube extending to infinity. This happens because the solutions of the equations of motion do not exist for all time, i.e. they blow up in finite time. Thus one of the basic assumptions of the Arnol'd-Liouville theorem [15] is not satisfied.

This is not the only case where something untoward happens. In fact the existence of "integrable" systems in which the motion takes place on surfaces more complicated than simple tori has been known for some years [16]. Examples have been given of such systems, known as pseudo-integrable, where the orbits lie on surfaces of genus higher than two, i.e. "multiply-holed tori." Perhaps the simplest example one can present is two-dimensional polygonal billiards with angles equal to rational multiples of π when reflections are not uniquely defined everywhere. This is the case of reflections in an angle larger than π [17]. Whether some "chaotic behaviour" is possible in the case of pseudo-integrable systems (or in Flashka's systems) is not known, but it is clear that these systems are not fully integrable.

1.3 PARTIAL AND CONSTRAINED INTEGRABILITY

In the previous section, we have examined the various forms of complete integrability. While in some cases the conditions on the integrals were weak to the point of allowing the question of random behaviour of the system to be raised, it was always assumed that the system possessed a complete set of integrals. Relaxing this assumption introduces naturally the notion of partial integrability. Thus one possible form of partial integrability is to have an insufficient number of integrals of motion. In the case of Hamiltonian systems with M degrees of freedom, insufficient means less than M-1 "action"-like integrals in involution (in addition to the Hamiltonian). Consider, for instance, the famous Hamiltonian (M=3) for an asymmetric spinning top with gravity :

$$H= (A\Omega_1^2 +B\Omega_2^2 +C\Omega_3^2)/2+ x_0\,\alpha+ y_0\,\beta+ z_0\,\gamma \tag{1.32}$$

where A, B, C are the principal moments of inertia, Ω_i the angular velocities, α, β, γ the direction cosines and x_0, y_0, z_0 are related to the position of the center of mass. For general values of the moments of inertia only one additional invariant exists, namely :

$$I = A\Omega_1\alpha +B\Omega_2\beta+C\Omega_3\,\gamma \tag{1.33}$$

which is the component of the angular momentum in the direction of gravity.

Partial integrability can also be associated with the existence of integrals of inadequate form. For a system of N first order differential equations the existence of N-1 *time-dependent* first integrals is not sufficient for complete integrability : the system can be reduced to a first order nonautonomous differential equation $\dot{x}$ =f(x,t) which is in general nonintegrable. Still, time-dependent invariants offer some long-term global information about the system. In the case of the Lorenz system

$$\begin{aligned} \dot{x} &= \sigma\,(y-x) \\ \dot{y} &= -y + \rho\,x - x\,z \\ \dot{z} &= -\,b\,z + x\,y \end{aligned} \tag{1.34}$$

for b=2σ, one time-dependent integral exists :

$$x^2 - 2\,\sigma\,z = C\,e^{-2\sigma t}. \tag{1.35}$$

We readily see that, for σ>0, as time goes to (positive) infinity the motion is attracted, to the paraboloidal surface x^2-2σz=0. The integral (1.35) is precisely of the type "form-invariant under time translation" we alluded to previously. In fact a shift in time $t \to t+t'$ conserves the form of the integral, the new value of the constant being just $C'=C\,e^{-2\sigma t'}$.

Hamiltonian systems are, as always, special. Time-dependent Hamiltonians do not conserve energy (of course) so even one-dimensional ones are generally not integrable.

However, the existence of just one conserved quantity, even an explicitely time-dependent one, suffices for integrability. Indeed, for any Hamiltonian $H(x, p_x, t)$ one can introduce a new variable z, and define the (time-independent, two-degrees-of-freedom) Hamiltonian $H'=H(x, p_x, p_z) - z$ whereupon the time derivative of p_z is just unity. If H has a conserved quantity $I(x, p_x, t)$, then H' has a second invariant $I(x, p_x, p_z)$ and so is integrable.

Another type of incomplete integrability is "constrained" integrability. One well-known example is the fixed-energy integrability of Hamiltonian systems. In Ref[18], Hietarinta has investigated the question in detail and presented several examples of two-dimensional Hamiltonian systems which are integrable at zero energy only. For instance, the Hamiltonian :

$$H = \frac{1}{2}\left(p_x^2 + p_y^2\right) + 4\,[\,a(x^6+y^6) + (4b-a)(x^2-y^2)x^2y^2] \tag{1.36}$$

possesses, when H=0, the second invariant

$$I = \frac{x^2 - y^2}{x^2 + y^2}\,(\,p_x^2 - p_y^2\,) - \frac{4xy}{x^2 + y^2}\,p_x p_y + 8\,a(x^2-y^2)^2 - 32bx^2y^2 \tag{1.37}$$

Another famous example belonging to the same class is furnished by the equations of the Kadomtsev-Petviashvili hierarchy of Jimbo and Miwa [19]. When Jimbo and Miwa presented their hierarchy of equations, it was generally assumed that *all* these equations were integrable. This, however, is not true, and the question of existence of *unconditionally* integrable equations in higher dimensions remains open. In fact, what happens is that the *entire* Kadomtsev-Petviashvili hierarchy, taken together, is integrable [20], while individual equations taken in isolation, apart from the first one, fail the usual integrability tests. So the concept of conditional integrability can be summarized as follows : Any given equation in the hierarchy must be considered together with the lower equations of the hierarchy ; the common solutions of *all* these equations have the usual integrability properties.

In order to illustrate this point we will consider the second equation in the hierarchy :

$$u_{xxxy} + 3\,u_{xy}\,u_x + 3\,u_y\,u_{xx} + 2\,u_{yt} - 3\,u_{zx} = 0, \tag{1.38}$$

in relation with the first one which is precisely the Kadomtsev-Petviashvili (KP) equation itself, written in potential form :

$$u_{xxxx} + 6\,u_{xx}\,u_x + 3\,u_{yy} - 4\,u_{xt} = 0. \tag{1.39}$$

In particular, we consider solitary wave solutions of equation (1.38) and ask whether they behave like solitons with respect to mutual interactions. Indeed, solitary waves exist for numerous nonlinear PDE's and two-solitary wave solutions exist for any equation that can be cast into Hirota's bilinear form [21]. The existence of N-soliton solutions for $N \geq 3$ is a quite nontrivial phenomenon that can be considered to be an indication of integrability[22]. We show that solitary wave solutions of (1.38) can be composed into 3-soliton solutions only in the case when the obtained solution also solves the (potential) KP equation (1.39). Following Hirota [21] we look for the "N soliton" solutions in the following way. We define the quantities

$$\eta_i = k_i x + l_i y + m_i t + n_i z, \quad (i = 1, 2, \ldots)$$

where the sets of numbers (k_i, l_i, m_i, n_i) satisfy the following dispersion relations

$$P_{KP}(k_i, l_i, m_i) = k_i^4 + 3l_i^3 - 4l_i m_i = 0 \tag{1.40}$$

for the KP equation (1.39) and

$$P_{JM}(k_i, l_i, m_i, n_i) \equiv (k_i^3 + 2m_i) l_i - 3k_i n_i = 0 \tag{1.41}$$

for the "Jimbo-Miwa" (JM) equation (1.38). Using the variables η_i we define the following solutions :

"One-soliton" solutions :

$$\tau = 1 + e^{\eta_1}, \text{ where } u = 2\frac{\partial}{\partial x}\log \tau$$

This provides a solution of the KP equations (1.39) if (1.40) is satisfied (z is then a parameter that can be absorbed into the soliton phase). The function τ is a solution of equation (1.38) if (1.41) is satisfied and a common solution of both equations if (1.40) and (1.41) hold simultaneously.

"Two-soliton" solutions :

$$\tau = 1 + e^{\eta_1} + e^{\eta_2} + A_{12}\, e^{\eta_1+\eta_2}$$

with

$$A_{12} = \frac{P(k_1 - k_2, l_1 - l_2, m_1 - m_2, n_1 - n_2)}{P(k_1 + k_2, l_1 + l_2, m_1 + m_2, n_1 + n_2)}. \tag{1.42}$$

If P in (1.42) is taken to be P_{KP} and (1.40) is satisfied for both sets (k_i, l_i, m_i) $(i = 1, 2)$, then τ is a solution of the KP equation. Similarly, if P is P_{JM} and (1.41) is satisfied, then τ is a solution of the JM equation (whether (1.40) is satisfied or not). If both (1.40) and (1.41) are satisfied then the two expressions for A_{12} coincide and τ is a common solution of the KP equation *and* the JM equation. The common value of A_{12} in this case is

$$A_{12} = \frac{k_1^2\, k_2^2 (k_1 - k_2)^2 - (k_1 l_2 - k_2 l_1)^2}{k_1^2\, k_2^2 (k_1 + k_2)^2 - (k_1 l_2 - k_2 l_1)^2}. \tag{1.43}$$

"Three-soliton" solutions :

If a three soliton solution exists in the Hirota formalism then its form must be

$$\tau = 1 + e^{\eta_1} + e^{\eta_2} + e^{\eta_3} + A_{12}\, e^{\eta_1+\eta_2} + A_{13}\, e^{\eta_1+\eta_3} + A_{23}\, e^{\eta_2+\eta_3} + A_{12} A_{13} A_{23}\, e^{\eta_1+\eta_2+\eta_3}. \tag{1.44}$$

Moreover, (1.44) is a solution only if the quantity

$$Q = P(k_1 + k_2 + k_3, \ldots)\, A_{12} A_{13} A_{23} + P(k_1 + k_2 - k_3, \ldots)\, A_{12} + P(k_1 - k_2 + k_3, \ldots)\, A_{13} + P(-k_1 + k_2 + k_3, \ldots)\, A_{23} \tag{1.45}$$

vanishes identically.

If P is P_{KP} in (1.45) and (1.40) is satisfied for all three sets (k_i, l_i, m_i) $(i = 1, 2, 3)$ then the quantity Q does indeed vanish. This is of course well known : the KP equation does have three soliton solutions (and N soliton solutions for any N). If, on the other hand, P is P_{JM} and we request *only* that (1.41) be satisfied for the three sets (k_i, m_i, n_i, l_i) $(i = 1, 2, 3)$

then Q is *not* identically zero. We do not reproduce its value here since it is ridiculously long. Thus, in general τ is *not* a solution of the JM equation.

Let us now require that (1.40) be satisfied, in addition to (1.41), and let P be P_{JM}. For A_{12} we obtain the expression (1.43) (for both P_{KP} and P_{JM}). In this case we find that the quantity Q does vanish and hence τ is a solution of the JM equation. This was to be expected : we simply reobtain the N soliton solutions (for N = 3) obtained more generally by Jimbo and Miwa [19] for the entire KP hierarchy.

The crucial point that we are making is that the variables η_i for these N-soliton solutions must satisfy both (1.40) and (1.41). Solitary waves of the JM equation do not, in general, interact as solitons in collisions of three or more at a time. Thus, the criterion of the existence of multisoliton solutions leads to the "conditional integrability" of the JM equation (1.38). Integrability in the usual (unconditional) sense would imply that all solitary wave solutions of the equation should interact like solitons in interactions of arbitrary multiplicity.

2. Singularity structure and the Painlevé property [23]

2.1 FIXED AND MOVABLE SINGULARITIES

Although historically the initial approach to integrability of differential equations was to obtain *global* solutions in terms of elementary functions, very soon, following the work of Cauchy, the interest shifted towards considering the equations *locally* in the complex domain. Thus one could start by looking for solutions in some domain of the complex plane of the independent variable z and then obtain a more global result by analytic continuation [4]. To do all this sucessfully, however, one ought to know where the singularities of the problem are located in the z-plane.

There are two types of singularities : the ones termed *fixed,* because their location is determined by the equation itself, and those called *movable,* the location of which depends on the initial conditions. Linear equations can only have fixed singularities. Consider, for exemple, the ODE ($w' \equiv dw/dz$)

$$(z-c)\, w' = \lambda w \quad , \tag{2.1}$$

the solution of which reads $w=K(z-c)^{\lambda}$. When λ is a positive integer the solution is holomorphic. For negative integers λ, $z = c$ is a fixed pole of the solution. For rational λ, it is an algebraic branch point, while for all other λ it is a transcendental branch point. "Worse" singularities can also arise, as in the example

$$w' = -w(z-c)^{-2} \tag{2.2}$$

the solution $w = K \exp[1/(z-c)]$ of which has an essential singularity at $z = c$.

Nonlinear equations can have both fixed and movable singularities. Consider the ODE

$$w' - \lambda\, w^{1-1/\lambda} = 0 \qquad (2.3)$$

with solution :

$$w = (z - z_0)^{\lambda}\,. \qquad (2.4)$$

Here z_0 cannot be read off the equation (2.3) itself and is arbitrary. Depending on the value of λ, $z = z_0$ is an analytic point, a movable pole, or a (movable) branch point. More complicated singularities, such as logarithmic branch points, or essential singularities are also seen to arise in simple examples of nonlinear ODE's.

It was Painlevé who first sought to determine all first order ODE's

$$w' = f(z,w)\ , \qquad (2.5)$$

with f rational in w and analytic in z, the only movable singularities of which are *poles*. The idea was that equations having this so called *Painlevé property* might be easier to integrate or solve analytically. Consider, for example, the case of an f polynomial in w, i.e.

$$w' = \sum_{n=0}^{m} P_n(z) w^n\ , \qquad (2.6)$$

where the $P_n(z)$'s are analytic in z . Its singularities are, in general, branch points near which the solution can be written as a series expansion

$$w = a_0 (z - z_0)^{-\frac{1}{m-1}} \left[1 + \sum_{n=1}^{\infty} a_n (z - z_0)^{\frac{n}{m-1}} \right] \qquad (2.7)$$

The case of particular interest here is m=2, for which the singularities at $z = z_0$ are simple poles. This is the well-known case of the Riccati equation

$$w' = f_0(z) + f_1(z)\ w + f_2(z)\, w^2\ , \qquad (2.8)$$

(with $f_i(z)$ analytic in z) , which is proved to be the *only first order ODE* (2.5) with the Painlevé property. In fact, with the substitution $w = -y'/[f_2(z)y]$, equation (2.8) can be transformed into a *linear,* second order ODE for y. This is, in a precise sense, one form of global integrability, since the *fixed* nature of the singularities of linear ODE's permits global integral representations of their solutions.

At second order, in a remarkable series of papers (see Ref. [4] for more details) Painlevé and his co-workers performed an exhaustive singularity analysis of *all* ODE's of the form

$$w'' = F(z, w, w')\ , \qquad (2.9)$$

with F rational in w' , algebraic in w and analytic in z , the *critical points* (branch points and essential singularities) of which are fixed. In other words, they were able to identify all such equations, the only movable singularities of which are poles. Forty-four of these equations were shown to be integrable in terms of elementary functions, by quadratures, or by linearization. For the solution of the remaining six equations, new transcendental functions (the so-called Painlevé transcendents) had to be introduced. The first four of these transcendental equations are given in Eq.16. We have seen in Section 1.1 how these equations can be linearized, which exemplifies their integrable character.

2.2 THE ABLOWITZ-RAMANI-SEGUR CONJECTURE

The discovery of the first integrable PDE, the Korteweg-de Vries (KdV) equation, by Kruskal and coworkers, has led to the development of special techniques for the treatment of nonlinear PDE's which soon produced scores (and even infinite families) of new integrable equations.

This is where the work of Ablowitz and Segur came in. While studying similarity reductions of known integrable PDE's, they observed that they all led to ODE's of the Painlevé type [24] : sometimes, these ODE's were solvable directly by known functions. In the more interesting cases, however, they were among P I - P VI of eq. (1.16), for which Painlevé transcendents are needed. Thus using the methods of IST, they obtained the *"linearization"* of one of Painlevé's irreducible equations, P II, with the aid of the Gel'fand - Levitan - Marchenko linear integral equation. They noted that the Painlevé property was present in *all known* reductions of integrable PDE's. Let us consider some examples to illustrate this point. We start with the Korteweg-de Vries equation

$$u_t + 6\, u\, u_x + u_{xxx} = 0 \tag{2.10}$$

and consider a time-independent reduction, i.e. look for u=u(x). The equation becomes

$$6\, u\, u_x + u_{xxx} = 0 \tag{2.11}$$

which can be integrated twice to yield (in terms of u'=du/dx)

$$u'^2 = C_1 + C_2\, u - 2\, u^3 \tag{2.12}$$

The last equation defines an elliptic function, the singularities of which are just poles. Next, we consider the Boussinesq equation

$$u_{tt} = u_{xx} + (u^2/2)_{xx} + u_{xxxx} \tag{2.13}$$

and obtain the reduction to travelling-wave solutions u=u(x-ct)=u(ξ). Equation (2.13) becomes, with u'=du/dξ :

$$(1 - c^2)\, u'' + (u^2/2)'' + u''''/4 = 0 \tag{2.14}$$

which can be integrated twice to give (after rescaling and translation) :

$$u'' + 2\, u^2 + \xi = 0 \tag{2.15}$$

Equation (2.15) is just the first of the six Painlevé transcendental equations. Finally let us introduce a different type of reduction. We start with the modified Korteweg-de Vries equation :

$$u_t + 6\, u^2 u_x + u_{xxx} = 0 \tag{2.16}$$

and look for a self-similar solution : $u(x,t)=(3t)^{-1/3}u(\xi)$ with $\xi=x/(3t)^{1/3}$.We find thus :

$$-(\xi u)' + 6\, u^2 u' + u''' = 0 \tag{2.17}$$

which can be integrated once to the second Painlevé transcendental equation

$$u'' + 2\, u^3 - \xi\, u + C = 0 \tag{2.18}$$

The above observation and scores of others led to the formulation of the Ablowitz-Ramani-Segur (ARS) conjecture :

Every ordinary differential equation obtained by an exact reduction of a nonlinear partial differential equation solvable by the IST method, has the Painlevé property.

Of course, transformations of variables are allowed and a given equation may pass the test only after some transformations. Integrability through a more trivial linearization, as is the case of Burger's equation should be considered as also included in the ARS conjecture. This conjecture, therefore, would provide a necessary condition for the integrability of a given partial differential equation.

Still, for the conjecture to be applicable, two different ingredients are needed : first a method to obtain *all* the reductions of a given PDE and second a way to test a given ODE for the Painlevé property. Ablowitz, Ramani and Segur proposed an algorithm to get at least a necessary condition.

2.3 THE ABLOWITZ-RAMANI-SEGUR ALGORITHM

The ARS algorithm was originally developed in order to determine whether a nonlinear ODE (or system of ODE's) admits movable branch points, either algebraic or logarithmic [25, 26]. It is important to keep in mind that this algorithm provides a necessary condition for the absence of such movable branch points. Thus, the (somewhat atypical) occurence of movable essential singularities cannot be detected by this procedure.

Let us consider a system of ODE's of the form

$$w_i' = F_i(w_1, w_2, \ldots, w_n; z) \qquad i = 1,\ldots,n \qquad (2.19)$$

The main assumption on which the ARS algorithm rests is that the dominant behaviour of the solutions in the neighbourhood of a movable singularity is of the form

$$w_i \sim \alpha_i (z-z_0)^{p_i} \qquad z \to z_0 \qquad (2.20)$$

Dominant logarithmic branches can also exist, of course, and while apparently excluded by the ansatz (2.20) they are also taken into consideration in our study.

Most often, the system at hand is not written in the form (2.19) of first order ODE's but includes higher derivatives as well. In that case, the functions w_i may well not tend to infinity as $z \to z_0$ (for $p_i > 0$ in (2.20)), but some of the higher derivatives can become singular. This seemingly innocent technical point can be of capital importance when it comes to finding *all* possible singular behaviours of the system.

The ARS algorithm proceeds in three steps, dealing with the dominant behaviours, the resonances and the conditions at the resonances, respectively.

2.3.1 Step 1 : Dominant behaviours.

Let us look for a solution of eq. (2.19) of the form :

$$w_i = \alpha_i (z - z_0)^{p_i} \quad , \qquad (2.21)$$

where for simplicity we can assume that $\mathrm{Re}(p_i) < 0$ for some i and z_0 is arbitrary. Substituting (2.21) in (2.19) one finds all possible p_i's , for which two or more terms in each equation *balance,* while the rest can be ignored, as arising at higher orders in powers of $(z-z_0)$. For each such choice of the p_i's, the balance of these so called leading terms also determines the corresponding values of the α_i's .

This first step is also the most delicate of the ARS algorithm. In order to arrive at the correct conclusions, one must find and examine separately *all possible* dominant behaviours. Extra care is needed here, as omissions can easily lead to erroneous results.

First of all, note that several choices of p_i's are possible. Let us illustrate this with an example :

$$x' = x\,(-x+ay+\lambda)\ ,$$
$$y' = y\,(bx-y+\mu)\ . \qquad (2.22)$$

Setting $p \equiv p_1$ and $q \equiv p_2$ and starting with

$$x = A\tau^p\ ,\quad y = B\tau^q \ \text{ with } \tau \equiv z-z_0\ , \qquad (2.23)$$

we obtain from (2.22) the following three possibilities : $q > p = -1$, $p > q = -1$, and $p = q = -1$, leading to three distinct leading behaviour :

(i) The leading terms in the equations read

$$x' = -x^2\ ,\quad y' = bxy \qquad (2.24)$$

thus we find A=1 and q=b while B is free i.e.

$$x = \tau^{-1}\ ,\quad y = B\tau^b.$$

The requirement for the Painlevé property to hold is that b be an integer.

(ii) The leading terms in the equations now read

$$x' = axy\ ,\ y' = -y^2 \qquad (2.25)$$

and we find B=1, p=a and A is free thus :

$$x = A\tau^a\ ,\quad y = \tau^{-1}.$$

Similarly, the requirement for the Painlevé property to hold is that a also be an integer.

(iii) The leading terms are now

$$x' = x\,(-x+ay),\quad y' = y\,(bx-y)$$

$x = A\tau^{-1}$, $y = B\tau^{-1}$ where A and B can be readily computed :

$$A = \frac{1+a}{1-ab}\ ,\ B = \frac{1+b}{1-ab}\ .$$

The requirement that both a and b be integers guarantees the nonvanishing of the denominator unless a=b=1 or a=b=-1, which must be examined separately. In the first case the system writes

$$x' = x\,(-x+y+\lambda)\ ,$$
$$y' = y\,(x-y+\mu)\ .$$

Dividing the first equation by x and the second by y and upon adding we immediately see that the logarithmic derivative of xy is constant, thus :

$$xy = C\,e^{(\lambda+\mu)\,z},$$

where C is some constant. Substituting the last quantity back into the equation we find that the system reduces to two Riccati equations and thus indeed possesses the Painlevé property.

The second case, a=b=-1, is quite similar. Here we find :

$$x/y = C\, e^{(\lambda-\mu)\, z}$$

and thus this case also reduces to Riccati equations and possesses the Painlevé property.

If p and/or q had turned out *not* to be integers then z_o would have been an algebraic branch point at leading order, and that would appear to discourage any further application of the algorithm. It may turn out, however, that a simple change of variables suffices to turn the system into one with no movable branch points. Even if this is not the case, if the p (or q) in question is a rational of a special type, the algorithm is again applicable and may be related to the so-called "weak Painlevé" concept [1].

If all possible p_i's are integers, then for each of them the leading behaviour can be viewed as the first term of a Laurent series around a movable pole, i.e. our ansatz now becomes:

$$w_i = (z - z_0)^{p_i} \sum_{m=0}^{\infty} a_i^{(m)} (z - z_0)^m \qquad (a_i^{(0)} \equiv \alpha_i)\ , \tag{2.26}$$

the location z_0 of the singularity being the first free (integration) constant of the system (2.19). For an nth order system there are still n–1 such arbitrary constants to be sought among the $a_i^{(m)}$'s in (2.26). If they are all found to be present there, the expansion (2.26) will be referred to as *generic*. The powers m at which these constants arise are termed *"resonances"* of the series (2.26), and it is to their determination that we now turn.

2.3.2 Step 2 : Resonances.

At this step, we start by keeping only the leading terms in the original equations and substitute into them for every w_i the simplified expression:

$$w_i = \alpha_i\, \tau^{p_i} (1+\gamma_i\, \tau^r), \qquad r > 0\ , \qquad i = 1,\ldots,n\ . \tag{2.27}$$

We then retain in (2.19) only the terms *linear* in γ_i , which we write as :

$$\mathbf{Q}\,(r).\gamma = 0 \quad , \quad \gamma \equiv (\gamma_1, \ldots, \gamma_n) \quad , \tag{2.28}$$

where $\mathbf{Q}$ (r) is an $n \times n$ matrix, with r entering only in its diagonal elements, at most linearly. Clearly then, some of the γ_i 's will be arbitrary (and hence free constants will enter in (2.26)) at the r roots of the algebraic equation :

$$\det\ \mathbf{Q}\,(r) = 0\ , \tag{2.29}$$

where $r = -1$ is always a root, related to the one free constant we have ab initio, namely the location z_0 of the singularity.

Let us illustrate all this, again on the previous example (2.22), with $\gamma \equiv \gamma_1$, $\delta \equiv \gamma_2$. The expression (2.27) for a leading behaviour of case (i), i.e.

$$x = \tau^{-1}(1+\gamma\,\tau^r)\ , \quad y = B\tau^b(1+\delta\,\tau^r)\ , \tag{2.30}$$

yields

$$\begin{pmatrix} r+1 & 0 \\ 1 & -r \end{pmatrix} \begin{pmatrix} \gamma \\ \delta \end{pmatrix} = \begin{pmatrix} 0 \\ 0 \end{pmatrix} \tag{2.31}$$

cf. (2.28), with the resonances being $r = -1$ and $r = 0$ cf. (2.29). Similarly in case (ii), we again find that the resonances are -1 and 0.

For the case (iii), on the other hand, we put :

$$x = A\tau^{-1}(1+\gamma\,\tau^r) \quad , \quad y = B\tau^{-1}(1+\delta\,\tau^r) \quad ,$$

and find

$$\det \mathbf{Q}\,(r) = \begin{vmatrix} A(r+1) - aB & -aB \\ -bA & b(r+1) - bA \end{vmatrix}$$

The two resonances can again be readily obtained : $r=-1$ and $r=r'\equiv(1+a)(1+b)/(1-ab)$. For the equation to be of Painlevé type, we must ask that r' be an integer. This results to the following couples of values for a and b (up to a simultaneous permutation of (a,b) and (x,y))

(a,b) :	r' :
(0,n)	-n-1
(n,-2-n)	1
(1,2)	6
(1,3)	4
(1,5)	3
(2,2)	3
(2,5)	2
(3,3)	2

Some general remarks are in order here :

(a) The resonance $r = 0$ corresponds to the coefficient of one of the leading terms being arbitrary, cf. (2.30), (2.31).

(b) As we have already seen, $r = -1$ is always a root of (2.29), as can be seen by perturbing $z-z_0$ to $z-z_0+\varepsilon$ at leading order in (2.21), expanding in powers of ε, and observing that the first contribution enters at order $(z-z_0)^{p_i-1}$.

(c) Any resonance with $\mathrm{Re}(r) < 0$ (except $r =-1$), must be ignored, since they violate the hypothesis that the p_i's are the powers of the leading terms in the series (2.26). Such resonances imply that the corresponding singular expansions are not generic, in the sense defined above.

(d) Any resonance with $\mathrm{Re}(r) > 0$, but r not an integer, indicates that $z=z_0$ is a movable branch point. The algorithm terminates at this stage. If r is real and rational, it must still be checked, whether this algebraic branch point can be removed by a simple coordinate transformation, as was already mentioned in Step 1.

(e) In the case where p_i is itself rational, the appearance of a rational r, with the *same denominator* as p_i indicates a finite branching with multiplicity determined by the leading singularity and is directly related to the so-called "weak–Painlevé" concept [1].

Thus a singular expansion (2.26) will be called generic if it is associated with (n–1) non negative (real) integer resonances. If, for every leading behaviour one finds less than n–1 such resonances, then all the solutions found are non-generic. This usually indicates that the ansatz (2.21) misses an essential part of the solution, most probably, a leading logarithmic singularity. If for every leading singular behaviour of step 1, all the resonances with non-negative real part are integer and provided *at least one* leading behaviour is generic, (i.e. involves (n–1) non-negative resonances) we may then proceed to :

2.3.3 Step 3 : The conditions at the resonances.

In this step, we shall check for the occurrence of non-dominant logarithmic branch points. To do this, we substitute into the full equation (2.19), for every different leading behaviour (2.20) the truncated expansion

$$w_i = \alpha_i \tau^{p_i} + \sum_{m=0}^{r_s} a_i^{(m)} \tau^{p_i+m} \quad , \tag{2.37}$$

(we recall that $\tau=z-z_0$) where r_s is the largest positive root of (2.29). We then identify the terms order by order in powers of τ. We obtain equations reminiscent of (2.28) but with, in general, terms of lower order appearing on the right-hand-side to get:

$$\mathbf{Q}\,(m)\,\mathbf{a}^{(m)} = \mathbf{R}^{(m)}\,(z_0\,;\,\mathbf{a}^{(j)}) \quad , \; j=1,\ldots,m-1 \; , \tag{2.38}$$

with $m=1, \ldots, r_s$,$\mathbf{R} \equiv (R_1, \ldots, R_n)^T$.

Then :

(i) for $m<r_1$, (r_1 being the smallest positive resonance), (2.20) determines $\mathbf{a}^{(m)}$.

(ii) At $m=r_1$, for (2.38) to have a solution (i.e $\mathbf{a}^{(r_1)}$ to have one arbitrary component, assuming r_1 is a simple root of (2.29)) the following compatibility conditions must be satisfied :

$$\det\, \mathbf{Q}^{(k)}(r_1) = 0 \quad , \quad k=1, 2, \ldots, n \tag{2.39}$$

where $\mathbf{Q}^{(k)}(r_1)$ is the matrix $\mathbf{Q}(r_1)$ with its kth column replaced by $\mathbf{R}^{(r_1)}$.

(iii) If (2.39) is satisfied, then for $r_1< m <r_2$, the next smallest positive resonance, (2.38) again determines $\mathbf{a}^{(m)}$.

(iv) The same procedure must be repeated successively at each higher resonance up to the largest one. (The case of multiple roots does not present any particular difficulty : one must ensure that the number of arbitrary components of $\mathbf{a}^{(r)}$ be equal to the multiplicity of the resonance r).

However it may turn out that for some resonance r condition (2.39) is not satisfied. Then one or more of the expansions (2.26) will have to be altered in the following way :

$$w_i = \sum_{m=0}^{r-1} a_i^{(m)} \tau^{p_i+m} + (\, a_i^{(r)} + b_i^{(r)} \ln \tau)\, \tau^{p_i+r} + \ldots , \tag{2.40}$$

with $\ln \tau$, $(\ln\tau)^2$, etc possibly entering at higher orders. The logarithms introduce new terms in the expansion : we determine the coefficients $b_i^{(r)}$ by demanding that the coefficient of the appropriate power of τ vanishes while the $a_i^{(r)}$ are free.

Let us illustrate all of this on our simple 2-dimensional example (2.22). Recall that for its two first leading behaviours (i) and (ii), the resonances were $r=0$, $r=-1$, and the two integration constants z_0 and B (resp. A). The associated Laurent series, therefore, contains only integer powers of τ throughout and no compatibility conditions need to be checked. For the third leading behaviour (iii), however, there exists one nontrivial resonance r' and thus, to find out whether logarithmic terms are needed, we must go to order r' computing all the coefficients of the series :

$$x=\frac{A}{\tau}\sum_{n=0}^{r'}\alpha_n\tau^n, \quad y=\frac{B}{\tau}\sum_{n=0}^{r'}\beta_n\tau^n \tag{2.41}$$

along the way.

The case (a,b)=(0,n) need not be examined in further detail. It suffices to remark that when one of the (a,b) vanishes, the corresponding equation decouples from the other one and is just a Riccati. Once it is solved, its solution can be substituted into the other equation which becomes a non-autonomous Riccati.

Let us do explicitely one of the simplest, namely the second case a=n,b=-n-2, which has a resonance r=1. We expand x and y up to first order :

$$x=\frac{1}{(n+1)\tau}+\alpha_1$$
$$y=-\frac{1}{(n+1)\tau}+\beta_1$$

and substitute into the full equations (2.22). We find that at the resonance the two equations for α_1 and β_1 are :

$$(2+n)\,\alpha_1 - n\,\beta_1 = \lambda$$
$$(2+n)\,\alpha_1 - n\,\beta_1 = \mu$$

The resonance condition is thus $\mu=\lambda$.

The remaining six cases can be examined in a similar way. These tedious calculations can be executed on a computer with the help of a language for algebraic manipulations. For the sake of completeness we give here the results :

(1,2)	$\mu=\lambda,-3\lambda/2,-4\lambda,-8\lambda/7,-13\lambda/7$
(1,3)	$\mu=\lambda,-2\lambda,-5\lambda$
(1,5)	$\mu=\lambda,-3\lambda,-7\lambda$
(2,2)	$\mu=\lambda,-\lambda$
(2,5)	$\mu=\lambda,-2\lambda$
(3,3)	$\mu=\lambda,-\lambda$

In summary, we shall say that a system of ODE's (2.19) satisfies the necessary conditions for the *Painlevé property* (i.e. for having no movable critical points other than poles), if its solutions can be expanded in pure Laurent series (2.36), near every one of their movable singularities at $z=z_0$. In other words, following the ARS algorithm as outlined above we must come across *no* algebraic branch points and *no* logarithmic singularities. This turns out be a rare occurence as we discovered already in the simple example (2.22), where it is true only for some special values of the parameters.

We are not going to confirm the integrability of the Painlevé cases of the above example by explicit construction of the integrals. The interested reader can find their integration in [2], where a more general system has been treated.

2.4 COMPLETE AND PARTIAL INTEGRABILITY OF ODE'S

As we have just seen, the ARS conjecture was originally formulated for PDE's. Still, it was clear from the onset that an analogous conjecture could be proposed for ODE's and in fact Segur [27] was the first to apply this tacit extension of the ARS conjecture to the integrability of ODE's. "First", in this context, means of course first in modern times, as Kowalevskaya has applied the method of singularity analysis to the investigation of the integrability of the anisotropic spinning top already a full century ago [28]. The equivalent of the ARS conjecture for ODE's would read :

Every system of ODE's which is completely integrable (in the sense of analytic integrability introduced in 1.2) has the Painlevé property.

We are not going to discuss here whether this conjecture can be turned into a rigorous theorem. This question has been addressed in detail in [9]. The conclusion of these studies is that in order to have a chance to find a rigorous proof of the conjecture, a drastic modification of its form would be needed, either on the type of integrability [12] or the acceptable singular behaviours. Indeed, there exist systems which, although analytically integrable do not pass the Painlevé test. We do not know, however, of any system which possesses the full Painlevé property and is not integrable. Concerning the "weak-Painlevé" property, introduced in [1], we believe that it is just an indicator of (possible) integrability, and thus no strong conjecture should be formulated around it. As for the integrable weak-Painlevé systems encountered, it has often turned out that they could be turned into full Painlevé ones through a transformation of both dependent and independent variables [29]. Ercolani and Siggia [30] propose another mechanism for the weak-Painlevé property : it may be due to the (integrable) reduction of a higher dimensional (integrable) full Painlevé system to lower dimensionality. Still, as has been argued in [23] the weak Painlevé criterion can be of considerable heuristic value. On the other hand, there are more restricted versions of the ARS conjecture which can (and have) been rigorously proved. In section 1 we have discussed Yoshida's attempt at a proof of the connection between singularity structure and algebraic integrability of weight-homogeneous systems. A more useful approach relating singularity structure and analytic integrability, also due to Yoshida and based on Ziglin's theorem, will be presented in section 3.

As we have seen in section 2.2 for a system to possess the Painlevé property the resonances must be integers and the compatibility conditions at each of the positive ones must be satisfied. However for a Nth order system it may turn out that this constraint is only partially satisfied : some resonances may turn out to be irrational or complex or some compatibility conditions may fail to check in general. However, when the free coefficients corresponding to the "bad" resonances are put to zero, while some of the others take special values in order to satisfy the compatibility conditions, the singular expansions of the solutions are of Laurent type and may still contain some free parameters, i.e. they may be not completely trivial. A priori, this does not guarantee anything beyond the existence of these particular solutions. Still, in some cases, this partial fulfillment of the Painlevé condition can be associated to the existence of some integrals of motion.
The situation is the same as for complete integrability, only worse : partial integrability may well exist without any relation to the Painlevé property whatsoever and vice versa. Still, as we will see in the examples that follow, such relations exist for many systems of physical interest which again stresses the great heuristic value of the singularity analysis approach.

2.4.1 The Lorenz system : One of the very first applications of the Painlevé analysis on systems of ODE's was the study of the Lorenz equations :

$$\begin{aligned} \dot{x} &= \sigma(y - x) \\ \dot{y} &= \rho x - x z - y \\ \dot{z} &= x y - b z \end{aligned} \tag{2.42}$$

which arises in simple models of hydrodynamic turbulence. For general values of b, ρ, σ, the solutions present a chaotic behaviour. Still, there exist values for which the behaviour of the system becomes regular. This can be most easily seen when one rewrites the system through a rescaling of x, y, z and t.

$$\begin{aligned} \dot{x} &= y - \sigma \varepsilon x \\ \dot{y} &= x - x z - \varepsilon y \\ \dot{z} &= x y - \varepsilon b z \end{aligned} \tag{2.43}$$

where $\varepsilon=(\sigma\rho)^{-1/2}$. At the limit of infinite Reynolds number ρ, i.e. $\varepsilon\to 0$, one obtains a system which can be integrated in terms of elliptic functions. Thus one may wonder for which other values of the parameters such a regular behaviour can occur. Segur [27] has studied the system from the point of view of Painlevé analysis. Starting from the leading singular behaviours of the type $x \sim 1/\tau$, $y \sim 1/\tau^2$, $z \sim 1/\tau^2$ he found that the resonances were r=-1, 2 and 4. Two compatibility conditions resulted. Apart from the (expected) solution $\varepsilon=0$, the following parameter values also ensured the Painlevé property for the system.

i) $\sigma = 0$

ii) $\sigma = 1/2, b = 1, \rho = 0 \quad (\varepsilon = \infty)$

iii) $\sigma = 1, b = 2, \rho = 1/9$

iv) $\sigma = 1/3, b = 0, \rho$ free

All of them are integrable. In case i), the equations are linear. In case ii), two time-dependent integrals are found :

$$y^2 + z^2 = C_1 e^{-2t}$$
$$x^2 - z = C_2 e^{-t}$$

and thus the system can be reduced to a quadrature, and the solutions can be expressed in terms of elliptic functions. In case iii), one integral exists

$$x^2 - 2z = C e^{-2t}$$

and after a change of variables one obtains the second Painlevé transcendent. Finally in case iv), the equations can be combined to a single third order equation for x which can be integrated once to give

$$x \ddot{x} - \dot{x}^2 + x^4/4 = C e^{-4t/3}.$$

A simple change of variables, $X \equiv x e^{t/3}$, $T \equiv e^{-t/3}$ suffices to transform this equation into the third Painlevé transcendent. We remark that complete integrability in the last two cases is of the third type introduced in section 1.1 : linearization of the system through integrodifferential equations.

In their analysis of the Lorenz system, Tabor and Weiss[31] made the following remark. When $b=2\sigma$, the first resonance condition is satisfied but not the second one. Interestingly enough, one integral exists in this case :

$$x^2 - 2\sigma z = C e^{-2\sigma t}$$

but no other integral is known. Thus in this case, partial integrability is related to the partial fulfillment of the Painlevé conditions. Still, the same condition is satisfied also for the value $b=1-3\sigma$, but no integral is known in this case.

On the other hand, there exists integrals which were initially discovered independently of any relation to the singularity structure [32]

a) for $b=1, \rho=0, \sigma$ free one has $y^2 + z^2 = C e^{-2t}$

b) for $b=4, \sigma=1, \rho$ free : $4(1-\rho)z + \rho x^2 + y^2 - 2xy + x^2 z - x^4/4 = C e^{-4t}$

c) for $b=1, \sigma=1, \rho$ free : $-\rho x^2 + y^2 + z^2 = C e^{-2t}$

d) for $b=6\sigma-2, \rho=2\sigma-1$: $(2\sigma-1)^2 x^2/\sigma + y^2 - (4\sigma-2)xy + x^2 z - x^4/4\sigma = C e^{-4\sigma t}$

Following this discovery Levine and Tabor have shown, in [33], how these integrals could be obtained from the singular expansions of the solutions and went on to formulate the hypothesis that they are the only ones possible for the Lorenz system.

Thus, for the system under study, it appears clearrly that complete integrability is related to the Painlevé property, while partial integrability bears some relation to the latter, without being strongly conditioned by it. At best, the partial fulfillment of the Painlevé condition serves to identify some values of the parameters for which partial integrability may exist. On the other hand, the pattern of singularities in the complex plane is closely related to the behaviour of the system. Indeed, this has been shown by Tabor and Weiss [31], who have studied in conjunction the behaviour of the system in real time (for both regular and chaotic regimes) and the location of the singularities in the complex time plane.

2.4.2 The three-wave interaction : This model corresponds to the interaction of three quasi-synchronous waves in a plasma with quadratic non-linearities [34]. The equations write :

$$\dot{x} = -2y^2 + \gamma x + z - \frac{\delta^2}{2}$$
$$\dot{y} = 2xy + \gamma y + \frac{\gamma\delta}{2} \qquad (2.44)$$
$$\dot{z} = -2zx - 2z$$

Two leading behaviours exist :
$x \sim 1/\tau$, $y \sim 1/\tau$, $z \sim \tau$ and $x \sim 1/\tau$, $y \sim \tau$, $z \sim 1/\tau^2$. The first type of singularities has resonances -1, 0 and 1, and the resonance conditions are automatically satisfied. The second type has resonances at -1, 1 and 2. The compatibility condition at resonance 1 writes :

$$\gamma\delta = 0 \qquad (2.45)$$

Two cases can be distinguished :

a) $\gamma = 0$. The second compatibility at r=2 is then automatically satisfied and one integral exists :

$$zy = C\,e^{-2t}$$

Using this integral, one can reduce the system to :

$$\ddot{y} = \frac{\dot{y}}{4} - 4\,y^3 + 2\,C\,e^{-2t} - 2\,\delta\,y^2$$

The transformation $Y=e^t y$, $T=e^{-t}$ allows us to reduce this equation to the third Painlevé transcendent.

b) $\delta=0$ ($\gamma\neq 0$). The compatibility at resonance r=2 reads :

$$\gamma(\gamma+1) = 0.$$

For $\gamma=-1$, we find one integral :

$$x^2 + (y+\delta/2)^2 + z = D\,e^{-2t}$$

valid for all δ, while for $\delta=0$ we have in addition

$$zy = C\,e^{-3t}.$$

Using those two integrals *and* the change of variables introduced in a) above, we reduce the system to a single quadrature :

$$\left(\frac{dY}{dT}\right)^2 = 4DY^2 - 4\,Y^4 - 4\,C\,Y$$

and the solutions can be expressed in terms of elliptic functions.

If $\delta=0$ but γ is free, the Painlevé conditions are partially fulfilled and one integral does exist :

$$zy = C\, e^{(\gamma-2)t},$$

but no further integration appears possible.

If $\gamma=-1$ and $\delta\neq 0$, as we have just seen, one integral also exists. However, the violation of the compatibility at r=1 has already introduced logarithmic terms in the singular expansion of the variables.

2.4.3 The Rikitake system : The Rikitake two disk dynamo model, proposed for the description of the time variation of the earth's magnetic field [34] is

$$\begin{aligned} \dot{x} &= -\mu x + \beta y + yz, \\ \dot{y} &= -\mu y + \beta x + xz, \\ \dot{z} &= -xy + \alpha. \end{aligned} \tag{2.46}$$

The analysis of (2.46) is particularly simple since it has only one type of singularity, near which the leading order behavior of its solutions is

$$x \sim i/\tau, \;\; y \sim -i/\tau, \;\; z \sim 1/\tau \;\; (\tau \equiv t - t_0). \tag{2.47}$$

Developing the asymptotic series to higher orders one easily find that two more free constants are expected simultaneously at the second higher order. The compatibility conditions for these free constants to enter with only integer powers of τ yield

a) $\alpha = 0$ and

b) either $\beta = 0$, or $\mu = 0$. (2.48)

The two cases possessing the Painlevé property can be easily integrated : with $\alpha = 0$ and $\beta = 0$ there is one integral,

$$x^2 - y^2 = C^2 e^{-2\mu t}, \tag{2.49}$$

and after the variable transformation

$$x + y \equiv Cu\, e^{-\mu t}, \;\; T \equiv e^{-\mu t} \tag{2.50}$$

the Rikitake system leads to a single ODE of second order,

$$\frac{d^2u}{dT^2} = \frac{1}{u}\left(\frac{du}{dT}\right)^2 - \frac{1}{T}\frac{du}{dT} - \frac{C^2}{4\mu^2}\left(u^3 - \frac{1}{u}\right), \tag{2.51}$$

which is again a special case of the third Painlevé transcendental equation. On the other hand, the case $\alpha = 0$ and $\mu = 0$ is even simpler. Multiplying the first and second equations (2.46) by $\dot{x}$ and $\dot{y}$ respectively, adding and substracting yields the two integrals

$$x^2 - y^2 + 2z^2 = C, \tag{2.52a}$$

$$x^2 - y^2 + 4\beta z = D. \tag{2.52b}$$

Using (2.52) the complete integration of the system can be performed in terms of elliptic functions.

Its is interesting now to ask whether the partial fulfillment of conditions (2.48) yields partially integrable models. The answer is that inded it does. Take, for example, the case $\beta = 0$ (with $\alpha\mu \neq 0$). The integral (2.49) still exists and the motion as $t \to \pm\infty$ (for $\mu \gtrless 0$) is attracted by any of the four limit cycles

$$x_1 = \varepsilon_1 y,$$

$$(z - \varepsilon_1\mu)^2 + y^2 = 2\varepsilon_2\alpha \ln|y| \quad (\varepsilon_i = \pm 1).$$

On the other hand, if $\mu = 0$, one integral also exists for all α and β :

$$x^2 - y^2 + 4\beta\,(z - \alpha t) = D, \tag{2.53}$$

but no further integration appears possible.

But the pole-like behavior at leading order accompanied by only $\ln\tau$ terms at higher orders may yield more than partial integrability. For example if both μ and β vanish, the Rikitake model possesses two integrals and hence is completely integrable : The first integral is (2.53) (with $\beta = 0$). A second integral can be derived by noting first that the operations for deriving the integral (2.52a) here give

$$\frac{d}{dt}(x^2 + y^2 + 2z^2) = 4z\alpha. \tag{2.54}$$

Now the $\dot{x}$ and $\dot{y}$ equations in (3.14) with $\beta = 0$ and $\mu = 0$ give $z = (\dot{x}+\dot{y}) / (x+y)$, which, when substituted in (2.54) yields the second integral

$$x^2 + y^2 + 2z^2 - 4\alpha \ln|x + y| = E. \tag{2.55}$$

We observe, therefore, at least for these models that it is often possible to identify partially integrable cases, with only one integral by partially satisfying the conditions of the Painlevé property. This does not seem to hold always. For example, for $\alpha = 0$ and any β, μ no simple integral appears to exist.

2.4.4 The Rabinovich system : This system models an interaction of three waves :

$$\begin{aligned} \dot{x} &= -\omega y - \nu_1 x - y z \\ \dot{y} &= \omega x - \nu_2 y - x z \\ \dot{z} &= -\nu_3 z - x y \end{aligned} \tag{2.56}$$

where ν_1, ν_2, ν_3 are the damping rates and ω is proportional to the amplitude of the "feeder" wave. The leading singularities are [35]

$$x=\varepsilon_1/t,\ y=\varepsilon_2/\tau,\ z=\varepsilon_3/\tau,$$

where $\varepsilon_i = \pm 1$ and the three ε_i are related through $\varepsilon_1\,\varepsilon_2\,\varepsilon_3 = 1$. The resonances are -1 and a double resonance at r=2. A first resonance condition reads :

$$(\nu_1 + \nu_2 - \nu_3)\,(\nu_1 - \nu_2 + 2\varepsilon_3\,\omega) = 0. \tag{2.57}$$

If the first factor does not vanish this means that both $\nu_1 - \nu_2$ and ω must be zero as equation (2.57) must hold for both values of ε_3. Thus we have

a) $\nu_1+\nu_2=\nu_3$. In this case, the second condition at r=2 is just

$$2\nu_1\nu_2=\varepsilon_3(\nu_2-\nu_1)\,\omega$$

which leads, because of the two choices for ε_3 to either :

$$\omega=0 \text{ and } \nu_1\nu_2=0 \text{ which gives } \nu_1=0, \nu_2=\nu_3 \text{ or } \nu_2=0, \nu_1=\nu_3$$

$$\text{or } \omega\neq 0,\ \nu_1=\nu_2=\nu_3=0.$$

b) $\nu_1=\nu_2$, $\omega=0$. In this case we find that the second condition is either $\nu_3=0$ or $\nu_1+\nu_2=2\nu_3$. This gives either :

$$\omega=0,\ \nu_1=\nu_2,\ \nu_3=0$$

$$\text{or } \omega=0,\ \nu_1=\nu_2=\nu_3.$$

In summary there are three essentially distinct cases :

i) $\nu_1=\nu_2=\nu_3=0$

for which we have two integrals

$$x^2+y^2-2z^2=C, \quad x^2-y^2+4z=D$$

and the system can be reduced to a quadrature. The solutions can be expressed in terms of elliptic functions.

ii) $\omega=0$ and $\nu_1=\nu_2=\nu_3=\nu$

where we can eliminate the ν depending terms through the change of variables $W=w\,e^{\nu t}$ for w=x,y,z and $T=e^{-\nu t}$ whereupon we fall back to the previous case.

iii) $\omega=0$, $\nu_1=\nu_2$ and $\nu_3=0$ and its two circular permutations.

To integrate this case we first obtain an integral

$$x^2-y^2=C\,e^{-2\nu t} \tag{2.58}$$

The use of this integral together with a change of variables similar to the one introduced in ii) above reduces the equations of motion to the equation P III for the variable X+Y.

Partially integrable cases exist here also. For instance, in the case $\nu_1=\nu_2=\nu_3=\nu$ but $\omega\neq 0$, a single integral exists :

$$x^2+y^2-2z^2=C\,e^{-2\nu t}.$$

Again, the integral (2.58) exists even for $\nu_3\neq 0$ but the resulting non-autonomous second order equation does not have the Painlevé property and cannot be further integrated.

In the above two cases, the Painlevé compatibility conditions are partially satisfied. However there are other cases where this occurs but no integral is known :

a') $\nu_1+\nu_2=\nu_3$. Here, no integral is found in general, but if in addition, $\nu_1=\nu_2=\nu$, one integral does exist

$$x^2-y^2-4\omega z=C\,e^{-2\nu t}.$$

b') $\nu_1=\nu_2$, $\omega\neq 0$ but $\nu_3=0$

b") $\omega=0$, $\nu_3(2\nu_3-\nu_1-\nu_2)=(\nu_1-\nu_2)^2$.

2.4.5 The Lotka-Volterra system : This model of the population dynamics of competing species is one of the best-known models of nonlinear dynamics. Integrable cases of various multidimensional generalizations are known [36], while Bogoyavlensky has presented its relation to the KdV equation itself [37]. Here, we will limit ourselves to the study of the three-dimensional case and we will draw heavily on Ref [8] where we have presented an exhaustive study of the subject. The three-dimensional Lotka-Volterra system writes, after a suitable scaling of the variables :

$$\begin{aligned} \dot{x} &= x\,(Cy + z + \lambda) \\ \dot{y} &= y\,(x + Az + \mu) \\ \dot{z} &= z\,(Bx + y + \nu) \end{aligned} \tag{2.59}$$

where A, B, C, λ, μ, ν are real or complex parameters.

It is clear that system (2.59) has two kinds of singularities, one where x, y, and z all diverge as τ^{-1} and one where two of the x, y, z have a simple pole while the third diverges less (or not at all). The latter singularity induces a first condition on the A, B, C. In fact, for this leading behaviour to be of Painlevé type, the exponent of the less diverging variable must be integer. Taking into account the fact that this can be the case for any of the x, y, z we get that all three numbers

$$\alpha = -C - \frac{1}{A},\ \beta = -A - \frac{1}{B},\ \gamma = -B - \frac{1}{C} \tag{2.60}$$

must be non-negative integers. We distinguish two cases, depending on whether ABC+1 vanishes or not.

When ABC+1 = 0 there are only four possible Painlevé cases :

A	B	C	α	β	γ
-3/2	-2	-1/3	1	2	5
-3	-1/2	-2/3	1	5	2
-2	-1	-1/2	1	3	3
-1	-1	-1	2	2	2

The resonances are equal to -1, 0 and 1. The compatibility condition can be easily computed and it leads to $\lambda=\mu=\nu$.

When ABC+1 $\neq$ 0, the resonances for the first type of singularities are -1 and the two solutions of :

$$\frac{1}{r(r-1)} = \frac{1}{\alpha+1} + \frac{1}{\beta+1} + \frac{1}{\gamma+1} - 1$$

where α, β, γ are the integers introduced above. The requirement that these resonances be integer leads to the following solutions :

α	β	γ	r
1	1	1	-1,2
1	2	2	-2,3
1	2	3	-3,4
1	3	2	-3,4

1	2	4	-5,6
1	4	2	-5,6

up to cyclic permutation. In the first case, A is free ($\neq 0,-1$) and $B=-1/(A+1)$, $C=-(A+1)/A$. In the five remaining cases, A, B and C are complex numbers uniquely determined, up to complex conjugation, by solving back equation (2.60).

For the second type of singularities, the resonances are again -1, 0, 1 and the compatibility conditions again $\lambda=\mu=\nu$. Once this constraint is imposed, we readily see that the system can be tranformed into one where the linear terms are absent through the now familiar transformation $W = w\, e^{-\lambda t}$ ($w = x, y, z$), $T = e^{\lambda t}$. Once λ, μ and ν are set to zero, it is straightforward to convince oneself that the resonance condition will automatically be satisfied at any r.

All systems identified by the Painlevé test are indeed integrable. The integrals are homogeneous polynomials (in X, Y and Z) of degree given by the positive resonance r, in perfect agreement with a theorem by Yoshida which relates the order of the invariant to the KE (which here coincides with the positive resonance).

When $ABC+1=0$, we can readily find a linear integral :

$$I = ABX + Y - A\,Z.$$

When $ABC+1\neq 0$, in the first subcase, r=2 and one quadratic integral is easily found :

$$I = (AX + (A+1)Y + A(A+1)Z)^2 - 4\,A(A+1)^2YZ. \qquad (2.61)$$

The expressions of the remaining five cases with r=3, 4 and 6 can be found in Ref.[8].

Still, the Painlevé cases are not the only ones for which integrals exist. When A=1, $\lambda=\mu=\nu=0$, we get :

$$I = \frac{Z^C\,(CY-X)^{BC+1}}{X\,Y^{BC}} \qquad (2.62)$$

unless C=0 in which case the integral tends (singularly) to :

$$I = \frac{Y}{X} + B\log\frac{Y}{X} - \log Z. \qquad (2.63)$$

The question may arise as to how this case was found. In Ref.[8], we have used the Frobenius compatibility approach introduced in [38]. The essence of this method is the following. We investigate the values of the parameters for which the quadratic differential system (2.59) has one integral in common with some linear system. Representing the right-hand-side of equation (2.59) by the quadratic vector field Q(x,y,z) and writing L(x,y,z) for a general linear vector field we have from the Frobenius compatibility theorem that Q, L and their commutator [Q,L] must be linearly dependent at every point. Thus :

$$\det\{\,Q, L, [Q,L]\,\} = 0. \qquad (2.64)$$

The compatibility condition (2.64) leads to a system of overdetermined nonlinear equations for the coefficients of Q and L which, however, does have several solutions. For instance, if A=1, $\lambda=\mu=\nu=0$, a compatible linear system is $L(x,y,z) = (x,y,0)$. The integrals of the quadratic Lotka-Volterra system must be expressed solely in terms of the integrals $u=x/y$ and $v=z$ of the above linear system. Using (2.59) and taking for simplicity C=0, we find :

$$\dot{u} = -\,x\,u,\ \dot{v} = (Bx+y)\,v$$

or, equivalently

$$\frac{\dot{v}}{v} + B\frac{\dot{u}}{u} + \frac{\dot{u}}{u^2} = 0$$

which can be readily integrated once to the integral (2.63) given above.

Moreover, the Lotka-Volterra system is particular inasmuch as it possesses an invariant measure density i.e. there exists a function M(x,y,z) satisfying :

$$\frac{\partial}{\partial x}(M\dot{x}) + \frac{\partial}{\partial y}(M\dot{y}) + \frac{\partial}{\partial z}(M\dot{z}) = 0$$

where $\dot{x}$, $\dot{y}$ and $\dot{z}$ are expressed through equation (2.59) in terms of x, y and z. In fact almost by inspection one can find M=1/xyz. When such an invariant measure density exists the Jacobi Last Multiplier Theorem insures that, starting from the initial integral I, one can find a second integral Ω. When ABC+1=0 we find

$$\Omega = X^{AB}\,Y^{-B}\,Z.$$

In the case B= -1/(A+1), C= -(A+1)/A, we introduce p = AX + (A+1)Y + A(A+1)Z and q= p - 2A(A+1)Z, to find

$$\Omega = \log\frac{p+\sqrt{I}}{p-\sqrt{I}} + A\log\frac{q+\sqrt{I}}{q-\sqrt{I}}$$

where I is given by equation (2.61).

For A=1, $\lambda=\mu=\nu=0$, we get for $C\neq 0$

$$\Omega = x - C\,y + I^{1/C}\int^{\frac{x}{y}} w^{-1+1/C}\,(C-w)^{-B-1/C}\,dw$$

while for C=0 we obtain

$$\Omega = x - e^{-I}\int^{\frac{y}{x}} w^{B-1}\,e^{w}\,dw,$$

where I is, in each case, the corresponding first integral given by (2.62), (2.63). Note that the last two integrals can be expressed in terms of incomplete Beta and Gamma functions.

2.5 INTEGRABILITY OF PARTIAL DIFFERENTIAL EQUATIONS

For the original A.R.S. conjecture to be applicable, one must find *all* the reductions of a given PDE. Sometimes, all the reductions *one can find* are just too trivial to yield an interesting information.Weiss and collaborators made a major progress in this domain [39] by doing away with reductions and introducing the Painlevé property for the PDE's themselves. In fact, according to Weiss, a PDE will possess the Painlevé property if its solutions are single-valued about any singular manifold $\phi(z_1,z_2,\ldots,z_n)$ (z_i being the independent variables) which is non-characteristic [40].

Two different attitudes can be adopted at this point. The first is to use the Weiss algorithm just to test for the Painlevé property. In this direction, one of us (MDK) proposed an improvement which led to considerable simplification. Instead of defining the singularity manifold by $\phi=0$, one could solve for one of the variables, say z_1 (chosen in practice as the one in which the order of the equation is highest) and defining the singularity manifold by $z_1+\psi(z_2,\ldots z_n)$. This made the practical implementation of the algorithm not much more difficult than inthe case of an ODE. A different orientation was chosen by Weiss himself. Starting from a Painlevé expansion of the form :

$$u = \phi^{-m} \sum_{n=0}^{\infty} u_n \phi^n \tag{2.64}$$

he truncated the expansion at "constant term" level i.e. :

$$u = u_o\, \phi^{-m} + u_1\, \phi^{-m+1} + \ldots + u_{m-1}\, \phi^{-1} + u_m. \tag{2.65}$$

Substituting back into the PDE, one obtains an overdetermined system of equations for ϕ and the u_i ($i=0,1,\ldots,m$). The equation for u_m is exactly the same as for u and thus the truncated expression (2.65) is just an autoBäcklund transformation between u and u_m (provided the overdetermined system has nontrivial solutions). Lax pairs can subsequently be obtained provided the autoBäcklund transformation includes a free constant, although the procedure is not quite straightforward. We will not go into any further detail as this topic will be covered in sufficient detail by J. Weiss himself [5].

Another direction which we will not examine here is partial integrability for PDE's. This subject will be treated, from the point of view of the existence of "soliton-like " solutions, in the course of J. Hietarinta [41]. What we will concentrate on, in the following paragraphs, is integrability under constraints. The Kadomtsev-Petviashvili hierarchy of Jimbo and Miwa will be examined again from the point of view of singularity analysis.

In order to apply the Painlevé test to the equation (1.34) of the Jimbo-Miwa hierarchy we write the solution w(x,y,z,t) in the form (2.64) with

$$\phi = x + \psi\,(y,t,z), \quad u_k = u_k\,(y,t,z). \tag{2.66}$$

Substituting (2.66) into (1.34) and following the usual procedure [39], we find

$$m = 1,\ u_0 = 2.$$

The "resonances", i.e. the values at which the recursion relations obtained from (1.34) do not determine the functions u_k, are found to be k = -1, 1, 4, and 6. The final step is to compute the coefficients u_2 , u_3 and u_5 from the recursion relation and to verify that the resonance conditions, i.e. the compatibility conditions for the existence of the free functions u_1 , u_4 and u_6, are satisfied. The result is

$$u_1\,(y,\,t,\,z) = \text{free}$$

$$u_2 = (-2\phi_t\phi_y + 3\phi_z - 6u_{1,y})\,/12\phi_y \tag{2.67}$$

$$u_3 = (-2u_{1,yy}\phi_y\phi_y - 2u_{1,y}\phi_{yy} + 2\phi_{ty}\phi_{yy} - \phi_{yz}\phi_y + \phi_{yy}\phi_z)/16\phi_y^3$$
$$u_4 = \text{free},$$

while the expression for u_5 is too long to reproduce here. At the resonance $k = 6$ we obtain a condition which is not satisfied identically. Indeed the resonance condition here is

$$R = -2u_{1,yz}\phi_{yy} + 2u_{1,yy}\phi_{yz} + 2\phi_{ty}\phi_{yz} - 2\phi_{tz}\phi_{yy}\phi_y - 2\phi_{tz}\phi_{yy}\phi_y - \phi_{yz}^2 + \phi_{yy}\phi_{zz} = 0.$$

We see that the condition $R = 0$ is an equation relating the functions $u_1(y,t,z)$ and $\phi(y,t,z)$, rather than an identity. The conclusion is that the JM equation (1.34), taken on its own, does not satisfy the Painlevé criterion and is therefore presumably not integrable.

Let us now consider the JM equation together with the preceding equations in the hierarchy, which in this case is simply the (potential) KP equation (1.35). The KP equation does not involve the variable z, so we fix $z = z_0$ and write a singular expansion for the solutions of the KP equation (which is well known to satisfy the Painlevé requirement) :

$$u(x,y,t) = \tilde{\phi}^{-m} \sum_{n=0}^{\infty} \tilde{u}_n \tilde{\phi}^n,$$

where $\tilde{\phi} = x + \tilde{\psi}(y,t,z_0)$, $\tilde{u}_n = \tilde{u}_n(y,t,z_0)$. As in the case of the JM equation, we find :

$$m = 1, \quad \tilde{u}_0 = 2$$

and resonances at $K = -1, 1, 4$, and 6. We have

$$\tilde{u}_1 = \text{free}, \quad \tilde{u}_2 = \frac{1}{3}\tilde{\phi}_t - \frac{1}{4}\tilde{\phi}_y^2 .$$

In the spirit of "conditional integrability" of a nonlinear PDE, introduced in section 1, we now require that $u(x,y,z,t)$ be a solution of the JM equation and simultaneously, for a fixed value z_0 of z, a solution of the KP equation. This means that we must have :

$$\tilde{\psi}(y,t,z_0) = \psi(y,t,z_0) \text{ and } \tilde{u}_k(y,t,z_0) = u_k(y,t,z_0)$$

for all values of k. In particular $\tilde{u}_2 = u_2$ implies

$$\phi_z = 2u_{1,y} + 2\phi_t\phi_y - \phi_y^3 . \tag{2.68}$$

Using condition (2.68) we can show that $\tilde{u}_3 = u_3$; we then choose $\tilde{u}_4 = u_4$ (since both are free) and obtain $\tilde{u}_5 = u_5$. At order $k = 6$ the compatibility condition for the KP equation is satisfied automatically (as the KP is integrable). Moreover, the compatibility condition for the JM equation at $k = 6$ is now also satisfied, i.e. $R = 0$ is a consequence of (2.68).

We see that "conditional integrability" in the case at hand means that the JM equation satisfies the necessary conditions for the Painlevé property only for a subclass of solutions. These are solutions for which the evolution of the singularity manifold ϕ in the z direction is determined by equation (2.68) for initial data given at some $z = z_0$ by an arbitrary function of y and t. For integrability in the usual sense ϕ should be an arbitrary function of all three variables y, t and z.

3. Nonintegrability results : Ziglin's and poly-Painlevé approaches

In the previous sections, we have investigated the question of integrability (complete and partial) and its relation to the singularity structure. It is clear from our presentation that integrable systems are quite rare: most systems are not integrable. However, this is not a rigorous stetement. In fact, most systems are just *presumably* nonintegrable. For some of them, an observed chaotic behaviour was interpreted as an undeniable indication of non-integrability. But what was lacking till recently was rigorous proofs. This has been remedied recently thanks to the pioneering work of Ziglin [42]. (Before proceeding any further, we should mention, in passing, another non-integrability theorem based on the self-intersection of separatrices and which is known as the Melnikov method [43]). Ziglin's theorem has established the necessary condition for an analytic Hamiltonian system to possess single-valued first integrals. The main idea of Ziglin was to relate the existence of analytic constants of motion for the complete Hamiltonian to that of a reduced system (obtained by variation around a given trajectory) and ultimately to the monodromy group of the latter system.

In what follows, we are going to present a sketch of the proof of Ziglin's theorem together with a review of results of non-integrability [44]. One interesting recent result concerns an alternate proof of Yoshida's theorem, based on the Ziglin approach. In fact, as explained in section 1, the use of complex-analytic integrability has made it possible to relate in a simple manner integrability to the rational character of the Kowalevski exponents for homogeneous Hamiltonian systems.

3.1 ZIGLIN'S THEOREM

In the original formulation, Ziglin's theorem furnishes necessary conditions for the existence of additional analytic first integrals for a given N-dimensional Hamiltonian system H and in particular for the existence of N integrals (including H) i.e. the Liouville integrability of the system.

However, the presentation can be greatly simplified if we restrict ourselves to the case of N=2 dimensions, where the existence of only one additional integral is investigated. In what follows, we will illustrate the various stages of the algorithm using the Hénon-Heiles cubic Hamiltonian

$$H = (p_x^2 + p_y^2) /2 + C\, x^2 y + y^3/3 \tag{3.1}$$

We start from the equations of motion:

$$\begin{aligned} \ddot{x} &= -2Cxy \\ \ddot{y} &= -y^2 - Cx^2 \end{aligned} \tag{3.2}$$

and we look for particular solutions, and more specifically, straight-line ones. Solutions of this kind always exist for homogeneous potentials. Their advantage will become clear later. In the present case, we obtain two kind of straight-line solutions :

i) $\quad x = 0 \;$ (and $p_x = 0$)

ii) $\quad y + \mu x = 0 \quad$ with $\quad \mu^2 = C/(2C - 1)$ (3.3)

(remark that for $C = 1/2$ only one straight-line solution exists)

On the type i) straight-line solution, the equation of motion for y reduces to :

$$\ddot{y}_0 = - y_0^2$$

which can be integrated to

$$\frac{1}{2}\dot{y}_0^2 + \frac{1}{3} y_0^3 + K = 0 \tag{3.4}$$

The latter equation defines a Jacobi elliptic function $y_0(t; K, t_0)$, with two generically complex periods and one double pole per unit-period parallellogram. This situation is typical for the reduced equations of motion in homogeneous potentials: one obtains multiply periodic functions (generally defined on some complicated Riemann surface).

In the case ii) we get

$$\ddot{y}_0 = - 2\, C/\mu\, y_0^2$$

which leads to an equation similar to (3.4).

Next, we linearize around the solution x=0, $y=y_0$ for case i) (and analogously for case ii)). We put $y=y_0+\eta$, $x=0+\xi$ and obtain:

$$\ddot{\eta} = - 2\, y_0\, \eta \tag{3.5a}$$

$$\ddot{\xi} = - 2\, C\, y_0\, \xi \tag{3.5b}$$

Here, the equations for η and ξ separate. This is the advantage of the straight-line solutions. The variational equations of motion always separate into the tangential (here (3.5a)) and the normal (here (3.5b)) parts through a fixed (i.e. time independent) rotation. The solutions of the tangential variational equation are $\eta_1=\partial y_0/\partial K$, and $\eta_2=\partial y_0/\partial t_0= -\partial y_0/\partial t$.

The main ingredient in the proof of Ziglin's theorem is that an integral of the initial Hamiltonian (other than H itself) leads, when expanded around the straight-line solution, to an integral for the normal variational equations of motion (NVE). Thus a necessary condition for the integrability of the initial Hamiltonian is the integrability of the NVE.

In the case at hand, a second integral exists for instance for C=1, given by

$$I = p_x\, p_y + \frac{1}{3} x^3 + x\, y^2 \tag{3.6}$$

Linearizing around type i) solutions, we find that this constant leads to

$$J = \dot{y}_0\, \dot{\xi} + y_0^2\, \xi$$

while for type ii) (here $\mu=1$)

$$J = \dot{y}_0\, \dot{\xi} + 2\, y_0^2\, \xi$$

where y_0 is given by the appropriate equation in each case. In the present example the integral J is linear in $\xi, \dot{\xi}$ but this need not always be the case. If, however I is analytic and independent of the Hamiltonian, there exists some finite higher order term in its expansion which does not vanish and is an integral of the NVE.

Looking at equation (3.5b) we remark that it is a Hill's equation, with periodic coefficient y_0. Its evolution along a path Γ in t-plane connecting point t_0 to any other point (maybe t_0 itself) where both y_0 and $\dot{y}_0$ take the same value, defines an area preserving mapping :

$$\begin{pmatrix} \xi \\ \dot{\xi} \end{pmatrix}_{final} = M_\Gamma \begin{pmatrix} \xi \\ \dot{\xi} \end{pmatrix}_{initial} \tag{3.7}$$

The mappings defined above form a group, induced by the fundamental group of the Riemann surface defined by y_0, called the monodromy group of the NVE, and they are represented by matrices M of determinant unity, which are called monodromy matrices. Note that because the Riemann surface defined by y_0 is generally not simply connected a closed path in t plane need not be represented by the unit matrix.

One definition is needed at this point. Let us assume that M is diagonalizable (the non-diagonalizable case will be treated apart) with eigenvalues ρ and ρ^{-1}. We call M non-resonant if ρ is not a root of unity. (The extension of the definition of non-resonance for N dimensions is straightforward : the eigenvalues $\rho_1, \rho_1^{-1}, \rho_2, \rho_2^{-1}, \ldots, \rho_{N-1}, \rho_{N-1}^{-1}$, must not satisfy any relation $\rho_1^{k_1}\rho_2^{k_2} \ldots \rho_{N-1}^{k_{N-1}} = 1$ with integer k's unless all the k's vanish).

We can now formulate Ziglin's theorem in its original version.

Theorem 1 : Suppose that a two-dimensional Hamiltonian system has an additional integral which is analytic (at least in a neighborhood of a given straight line solution of the equations of motion). Furthermore suppose that there exists a non-resonant monodromy matrix M_1. Then any other monodromy matrix M_2 must

- either commute with M_1 i.e. $M_1M_2M_1^{-1}M_2^{-1} = \mathbb{1}$
- or be such that $M_1M_2M_1^{-1}M_2^{-1} = M_1^2$
  (which means that M_2 permutes the eigenvectors of M_1
  in which case Tr M_2=0)

A generalization of the theorem, given in [45], concerns the extension to non-diagonalizable matrices. The extended theorem can be formulated as follows :

Theorem 2 : With the same assumptions as Theorem1 above, suppose that there exists a non-diagonalizable monodromy matrix M_1. Any other monodromy matrix M_2 must be resonant and moreover, in a basis where M_1 is upper (resp. lower) triangular, it must also be upper (resp. lower) triangular.

Below, we give a sketch of the proof of Ziglin's theorem. We have seen earlier that an analytic integral I of the original Hamiltonian induces an integral for the NVE. Thus :

Lemma 1 : If the Hamiltonian system under consideration has an additional analytic integral then there exists a nontrivial homogeneous polynomial

$$J(\xi,\dot{\xi},t) = \sum_{j=0}^{n} \psi_j(t)\, \xi^j\, \dot{\xi}^{n-j} \tag{3.8}$$

which is an integral of the NVE, where the $\psi_j(t)$'s depend on t *only* through y_0 and $\dot{y}_0$ and therefore are single valued on Γ.

This polynomial $J(\xi,\dot{\xi},t)$ is invariant under the action of the monodromy group. Indeed, suppose we now fix a point $(y_0, \dot{y}_0)$ on Γ and integrate the equations of motion over a loop in Γ. The quantities ξ and $\dot{\xi}$ are thus modified by the action of the monodromy group while the ψ_k's recover their initial values, and J has kept its value constant throughout. Thus, we have

Lemma 2 : With the same assumptions as above, there exists a homogeneous polynomial

$$\Phi(\xi, \dot{\xi}) = \sum_{j=0}^{n} \varphi_k\, \xi^j\, \dot{\xi}^{n-j} \tag{3.9}$$

which is invariant under the action of the monodromy group of the NVE. Using the quantity Φ we can prove Ziglin's theorem.

Suppose that there exists some matrix M_1 which is diagonalizable and non-resonant. Changing to the basis of the eigenvectors of M_1, with new coordinates ζ and η, we get :

$$\Phi = \sum_{j=0}^{n} \omega_j\, \zeta^j\, \eta^{n-j} \tag{3.10}$$

The action of M_1 changes Φ to

$$\Phi = \sum_{j=0}^{n} \omega_j\, \rho^{2j-n}\, \zeta^j\, \eta^{n-j} \tag{3.11}$$

But ρ^{2j-n} cannot be unity unless n=2j, as M_1 is non-resonant. Thus for Φ to be invariant under M_1 we must have :

$$\Phi = (\zeta\eta)^j \tag{3.12}$$

Now, we act on Φ with some other member M_2 of the monodromy group, which, in the basis which diagonalizes M_1, can be written :

$$M_2 = \begin{pmatrix} a & b \\ c & d \end{pmatrix} \tag{3.13}$$

(recall that $\det(M_2)=ad-bc=1$). Since M_2 must also leave Φ invariant we find :

$$[(a\zeta + b\eta)(c\zeta + d\eta)]^j = (\zeta\eta)^j \tag{3.14}$$

which entails :

$$ac = bd = 0 \quad \text{and} \quad (ad + bc)^j = 1$$

The only solutions are :

i) b=c=0 and ad = 1 i.e. M_2 commutes with M_1

thus $M_1 M_2 M_1^{-1} M_2^{-1} = 1$

ii) a=d=0 and bc = 1 i.e. M_2 permutes the eigenvectors of M_1

thus the "commutator" $M_1M_2M_1^{-1}M_2^{-1}$ is equal to M_1^2

In the latter case, Tr M_2=0, which implies that the eigenvalues of M_2 are i and -i, and M_2 is thus resonant. Therefore, if two non-resonant matrices exist, they must commute for an additional integral to exist.

We now turn to non-diagonalizable matrices. In an appropriate basis, M_1 can be written :

$$M_1 = \pm\begin{pmatrix} 1 & 0 \\ \alpha & 1 \end{pmatrix} \tag{3.15}$$

with $\alpha \neq 0$. As a consequence, in this frame, Φ can only assume the form $\Phi = \zeta^n$. For the matrix M_2 to leave ζ^n invariant, we must have $a^n = 1$, b = 0 and ad=1. If a is neither 1 nor -1, the two eigenvalues a and d are distinct and M_2 is therefore diagonalizable, and moreover resonant. If however a=d=±1, then M_2 is not diagonalizable (unless c=0, in which case it is just ±1) and assumes a lower triangular form *in the same basis* as M_1. The "commutator" $M_1M_2M_1^{-1}M_2^{-1}$ is unity only if a=d=±1, but its trace is always 2.

Ziglin's original theorem was also formulated in the N-dimensional case [42].

Theorem 3 : Suppose that a given Hamiltonian H has N-1 analytic integrals, which are functionally independent together with H, and moreover that there exists a non-resonant matrix M. Then it is necessary that any other monodromy matrix M' either : a) commute with M or b) permute the eigenspaces of M.

In particular whenever M' is also non-resonant it must also necessarily commute with M.

The problem which arises at this point is that of the practical applicability of Ziglin's theorem. The knowledge of the monodromy matrices is necessary. Of course, one can always compute them numerically with some precision. For N>2, however, the numerically obtained eingenvalues are of little help if one wants to investigate the resonance condition. Still several possibilities exist. One of the simplest ones can be implemented whenever the reduced equation of motion for the straight-line solution leads to a doubly periodic trajectory with only one singularity per unit period parallellogram. In this case, the path corresponding to the "commutator" $M_s=M_1M_2M_1^{-1}M_2^{-1}$ is homotopic to a closed loop around the singularity. Now, as was explained in [9] and [46], the eigenvalues of M_s can be obtained *explicitely* from singularity analysis considerations. (This statement will become clearer in the next section dealing with the theorem of Ref [6]). The explicit knowledge of the 'commutator' can be of great help. Of course, if $M_s=1$ the M_1,M_2 matrices are not constrained in any way. On the other hand, if $M_s \neq 1$ (even if $M_s= -1$, a case that often arises with polynomial potentials), then we can prove non-integrability by numerically computing the traces of M_1 and M_2. If any of them has a trace depending on the energy, then by continuity it cannot be always resonant, nor can the (energy independent) commutator always be its square [46]. Similarly, if at least one of the two matrices has a trace larger than 2 in absolute value it is certainly not resonant, and one can check numerically whether the commutator can be its square.

3.2 EXPLICIT COMPUTATION OF THE MONODROMY MATRICES

As we have seen in the previous section, one must compute the monodromy matrices, so as to be able to conclude on the resonance condition. The ideal situation is when we can get the monodromy matrices explicitely. It turns out that this can be done for a large class of Hamiltonians, in particular for Hamiltonians involving homogeneous potentials [47]. This result is obtained by transforming the NVE into a Gauss hypergeometric equation, the monodromy properties of which are exactly known. Let us illustrate this calculation through a specific example [48].

We start with the Hamiltonian :

$$H = (p_x^2 + p_y^2)/2 + \rho + \kappa y \tag{3.16}$$

where $\rho = \sqrt{x^2 + y^2}$. The equations of motion are :

$$\ddot{x} = - x / \rho$$
$$\ddot{y} = - y / \rho - \kappa \tag{3.17}$$

The straight-line solutions are $x=0$, $\ddot{y}= -(\kappa\pm1)$, the $\pm$ sign being due to the square root in the definition of ρ). We rescale the time so as to obtain $\ddot{y}+1=0$, integrated to $y=h-t^2/2$ (with the appropriate initial conditions and $h\neq0$). The NVE is then written as :

$$\ddot{\xi} + \lambda \xi / y = 0 \tag{3.18}$$

where $\lambda = (1 \pm \kappa)^{-1}$. We introduce a new independent variable :

$$z =(1+t/\sqrt{2h})/2 \tag{3.19}$$

and obtain for the NVE :

$$z (1-z) d^2\xi/dz^2 + 2 \lambda \xi = 0 \tag{3.20}$$

This is a special case of Gauss hypergeometric equation :

$$z (1-z) d^2\xi/dz^2 + [c-(a+b+1)z] d\xi/dz - ab \xi = 0 \tag{3.21}$$

From the monodromy properties of the Gauss hypergeometric equation we can conclude that, for integrability, λ must be of the form $n(n-1)/2$. Moreover, we can use the fact that λ is given by $1/(1+\kappa)$ in one case and $1/(1-\kappa)$ in the other. Thus the only possible cases are $1/(1+\kappa)=n(n-1)/2$ and $1/(1-\kappa)=m(m-1)/2$, with n and m integer. This is clearly impossible unless $\kappa=0$ (or infinite). Thus, the system is never integrable apart from these trivial cases.

In fact, the transformation to an equation of the type (3.21) can be performed for any homogeneous potential $V(x_1,x_2)$ of degree k. Indeed, we start by looking for straight-line solutions of the form $x_1=c_1\phi(t)$, $x_2=c_2\phi(t)$. We find a solution where c_1, c_2 and $\phi(t)$ satisfy

$$(c_1,c_2) = \text{grad } V(c_1,c_2) \tag{3.22a}$$

and

$$\ddot{\phi} + \phi^{k-1} = 0 \tag{3.22b}$$

which can be integrated to (by rescaling the integration constant, namely the energy, to 1/k)

$$t=\sqrt{\frac{k}{2}}\int \frac{d\phi}{\sqrt{1-\phi^k}} \tag{3.23}$$

The denominator of the integral defines a Riemann surface, the branch points of which are $w_n=e^{2i\pi n/k}$ (n=0, ..., k-1). The NVE takes the form :

$$\ddot{\xi} + \lambda\, \phi^{k-2}\, \xi = 0 \tag{3.24}$$

with λ given by :

$$\lambda = \nabla^2\, V(c_1,c_2) - (\, k-1) \tag{3.25}$$

Introducing the change of independent variable $z=\phi^k$ we transform (3.24) into a Gauss hypergeometric equation (3.21) with

$$a+b = 1/2 - 1/k, \qquad ab = -\lambda/2k, \qquad c = 1 - 1/k \tag{3.26}$$

The basic monodromy matrices of the Gauss hypergeometric equation correspond to anticlockwise loops around the points z=0 and z=1. Their expressions are [49]:

$$M_0 = \begin{pmatrix} 1 & e^{-2i\pi b} - e^{-2i\pi c} \\ 0 & e^{-2i\pi c} \end{pmatrix} \text{ and } M_1 = \begin{pmatrix} e^{2i\pi(c-a-b)} & 0 \\ 1 - e^{2i\pi(c-a)} & 1 \end{pmatrix} \tag{3.27}$$

We now consider two closed circuits on the Riemann surface, C_1 and C_2 enclosing respectively (counterclockwise) 1 and $e^{2i\pi/k}$, and $e^{-2i\pi/k}$ and 1, with a common base point w on the real axis, 0<w<1. These circuits are mapped in the complex z plane through $z=\phi^k$ and we get the corresponding monodromy matrices:

$$M(C_1) = M_1M_0M_1M_0^{-1}$$
$$M(C_2) = M_0^{-1}M_1M_0M_1$$

From the above expressions it is fairly easy to implement Ziglin's theorem. We have

$$E(\lambda)= \mathrm{Tr}\, M(C_1)= \mathrm{Tr}\, M(C_2)= 2 \cos \left(\frac{2\pi}{k}\right) + 4 \cos^2\left(\frac{\pi}{2k}\sqrt{(k-2)^2 + 8\, k\, \lambda}\right) \tag{3.28}$$

and moreover $M(C_1)$ commutes with $M(C_2)$ only when $k=\pm 2$ or $E(\lambda) = 2$ or $2\cos(2\pi/k)$. So if λ is such that both $M(C_1)$ and $M(C_2)$ are non-resonant then we can conclude on the non-integrability of H. This occurs in particular whenever $|E(\lambda)|>2$. Using this inequality, we can define regions for λ in which no additional analytic integral exists [47]. We obtain thus non-integrability in the regions :

- for $k\geq 3$ $\quad \lambda<0,\ 1<\lambda<k-1,\ k+2<\lambda<3k-2, \ldots$
- for $k=1$ $\quad \mathbb{R} - \{ 0, 1, 3, 6, 10, \ldots \}$
- for $k=-1$ $\quad \mathbb{R} - \{ 1, 0, -2, -5, -9, \ldots \}$
- for $k\leq -3$ $\quad \lambda>1,\ 0>\lambda>k+2,\ k-1>\lambda>3k+3, \ldots$

(The case $k=-2$ is integrable for all values of λ).

We remark that for $|k|>3$ we obtain whole regions in which non-integrability *cannot* be proved. However, as we have seen for Hamiltonian (16), by using information from several straight-line solutions we can further limit the possible regions for integrability and in some cases constrain the problem to the point where an exact answer is reached. For the potential

$$V = x^4 + y^4 + 2\,\varepsilon\, x^2 y^2 \tag{3.29}$$

we find that the non-integrability region is :

$$\lambda<0,\ 1<\lambda<3,\ 6<\lambda<10,\ \ldots$$

where λ is defined for the two different straight-line solutions as

$$\lambda=\varepsilon \ \text{ and } \ \lambda = \frac{3-\varepsilon}{\varepsilon+1}.$$

It is easy to check that only the well known integrable cases ε=0,1,3 survive. For all other values of ε the above quartic potential is not integrable, (in particular for ε infinite which corresponds to a well-known ergodic potential, see 3.38).

Another case where the monodromy matrices can be explicitely computed is a generalized Toda lattice of the form [45]:

$$H = (p_x^2 + p_y^2)/2 + e^{\alpha x - y} + e^{-\alpha x - y} \tag{3.30}$$

straight-line solutions exist for this type of Hamiltonians. In particular taking $x=p_x=0$, we get the NVE :

$$\ddot{\xi} + 2\,\alpha^2 e^{-y} \xi = 0 \tag{3.31}$$

with y given by the equation

$$\ddot{y} + 2\ e^{-y}\, y = 0 \tag{3.32}$$

By introducing a new independent variable $z=2e^{-y}/h$ (where h is the energy) we can transform the NVE into a Gauss hypergeometric equation with

$$a+b=1/2, \quad ab=-\alpha^2/2 \quad \text{and } c=1.$$

The monodromy matrices obtained (for a particular path) can be simultaneously cast into upper triangular form provided

$$\alpha^2 = n\,(n-1)/2$$

with integer n. The integrable cases can only be found among these values of α, which include the known ones α^2 =0,1,3.

For the Hénon-Heiles case we introduced at the beginning of this paper, however, we cannot reach a clear-cut conclusion. Implementing the non-integrability condition on both straight-line solutions, we find :

$$C^{-1}< 1,\ 1< C^{-1}< 2,\ 2< C^{-1}< 3,\ 6 < C^{-1}< 8,\ 13 < C^{-1}< 16,\ \ldots$$

Thus, apart from the discrete values C=1, C=1/2 we have whole regions on which non-integrability cannot be deduced although possible integrable cases have zero measure even in these regions. This is indeed a typical situation in Ziglin's approach.

3.3 NON-INTEGRABILITY IN TERMS OF KOWALEVSKI EXPONENTS

The main difficulty with Ziglin's theorem lies in the fact that the resonance condition cannot be practically exploited : one must check whether some expressions are the cosines of angles rationally related. Even in the best case where these expressions are explicitely known, this is a transcendental problem.

This difficulty has been overcome by a new version, given in Ref.[11], of Yoshida's theorem [6]. The first step is to obtain a new condition for the non-integrability regions through a different choice of paths C_1 and C_2 on the Riemann surface which will lead to simpler expressions for the traces of the monodromy matrices. See [44] for a detailed description. The following monodromy matrices are thus obtained :

$$M(C_1) = (M_0 M_1)^{4k}$$

$$M(C_2) = (M_1 M_0)^{4k}$$

The advantage of this choice lies in the fact that the common value of the traces of these matrices is :

$$2\cos\left(2\pi\sqrt{(k-2)^2 + 8k\lambda}\right) \tag{3.33}$$

and moreover $M(C_1)$ and $M(C_2)$ commute only when their trace is ± 2, or when $k=\pm 2$. The non-resonance condition is now easier to check. What is more interesting is that this theorem can be expressed in a particularly simple way in terms of the KE's, and thus furnishes a further proof of Yoshida's theorem originally proposed in [6], (although here in a slightly less general context)

In particular we consider a Hamiltonian of the form

$$H=\Sigma p_i^2/2 + V(x_1,x_2,\dots,x_N) \tag{3.34}$$

with V(x) a homogeneous, algebraic potential of degree k. The equations of motion admit singular solutions of the form

$$x_j=c_j\phi(t) \tag{3.35}$$

where the c_j can be chosen as a solution of

$$(c_1, c_2, \dots, c_N) = \text{grad}\, V(c_1, c_2, \dots, c_N) \tag{3.36}$$

and $\phi(t)$ is a singular solution αt^{-g} of equation (3.22a) for ϕ where $g=2/(k-2)$ and $\alpha=(kg^2/2)^{g/2}$. The KE's are introduced when one looks for a variational solution around (3.35) through

$$x_j= (c_j+\chi_j)\,\alpha t^{-g}$$

Linearizing one obtains for χ solutions of the form t^{ρ_i} where the ρ_i's are the KE's. Due to the Hamiltonian character of the equations of motion the KE's come by pairs ρ_i and ρ_{i+N}. To obtain them we compute the eigenvalues λ_i of the Hessian matrix $\partial^2 V(x_1,x_2,\dots,x_N)/\partial x_j \partial x_m$ evaluated at $x_j=c_j$. The set of the KE's ρ_i and ρ_{i+N} are given by the roots of the equation

$$\rho^2-(2g+1)\rho+g(g+1)(1-\lambda_i)=0$$

We denote by $\Delta\rho_i$ the difference of the two roots, namely $\Delta\rho_i=\sqrt{1+8k\lambda_i/(k-2)^2}$.
Note that in equation (3.33), the expression that appears is just

$$2\cos(2\pi(k-2)\Delta\rho). \qquad (3.37)$$

This is by no means a coincidence. Indeed, the paths C_1 and C_2 correspond to going aroud the singularity in t-plane 2(k-2) times. Along any of these paths, the eigensolution (for a KE ρ_i) acquires a phase factor $e^{4i\pi(k-2)\rho_i}$ which is equal to $e^{2i\pi(k-2)\Delta\rho_i}$. Equation (3.37) expresses just that fact.

We define the rational independence of the $\Delta\rho_i$'s if no relation of the form $\Sigma m_i\Delta\rho_i=0$ exists for integer m_i's unless all $m_i=0$. We thus have the following theorem :

Theorem 4 : If the $\Delta\rho_i$ are rationally independent the N-dimensional Hamiltonian system cannot possess any additional analytic integral besides the Hamiltonian itself.

In the two-dimensional case, there are only two $\Delta\rho_i$ one of which is always rational. Thus rational independence means just that the other $\Delta\rho_i$ is not a rational number. Moreover the sum $\rho_i+\rho_{i+N}=2g+1$ is always rational. This allows the following corollary to the above theorem .

If there exists an irrational or complex KE the two-dimensional Hamiltonian cannot have any additional analytic integral.

The main advantage of this theorem is that the singularity analysis leads directly to the result. No transcendental non-resonance condition need be checked.

Consider, as an example, the following Hamiltonian :

$$H=(p_x^2+p_y^2)/2+x^2y^2/2 \qquad (3.38)$$

The solution of (3.36) is $c_1^2=c_2^2=1$. The eigenvalues of the Hessian matrix can be readily computed and we find $\lambda_1=3$, $\lambda_2=-1$. Thus , $\Delta\rho_1=5$, $\Delta\rho_2=\sqrt{-7}$. Because of the latter, it is clear that the Hamiltonian (3.38) does not have any additional analytic integral.

On the other hand Hamiltonians which are not of the form (3.34) are not covered by this theorem. In particular, Rod et al. [10] have presented the example of :

$$H=p_1(p_1^2+\mu_1x_1^2)+\varepsilon x_1(p_2^2+\mu_2x_2^2) \qquad (3.39)$$

which has irrational KE's : $1\pm2i\omega$, $\omega^2=\varepsilon^2\mu_2/\mu_1^2$, while still being integrable. Still, no contradiction arises as the monodromy group is a commutative one [11]. It might well turn out that this is always the case whenever irrational or imaginary KE's appear for integrable Hamiltonians. On the other hand the Hamiltonian

$$H=(p_x^2+p_y^2)/2+p_xxy+p_y(ax^2+by^2) \qquad (3.40)$$

studied by Roekaerts and Yoshida [50] possesses a non-commutative group. However no explicit analytic expression for the monodromy matrices of the associated monodromy group exists yet. Still, the known integrable cases indeed correspond to values of the parameters a and b for which the KE's are rational.

3.4 EXTENSION TO NON-HOMOGENEOUS HAMILTONIANS

In the case of homogeneous Hamiltonians analytic expressions for the monodromy matrices can often be obtained, thus allowing the application of Ziglin's theorem. In the non-homogeneous case, the situation becomes more difficult as no closed-form results exist. Still, the results on the homogeneous potentials can be of great help [51,52]. By taking the (zero- or) infinite-energy limit of the equations of motion (through appropriate scalings) one can reduce the system to one where only the highest degree homogeneous part contributes. (Similarly, at the zero-energy limit we recover the lowest-degree part). Thus, the two limits can be treated exactly and if either one turns out to be non-integrable, then the same holds for the whole potential. Moreover, numerical computations can be of some help as an energy dependence of the trace of a monodromy matrix indicates that it is non-resonant for a dense set of values of the energy. One situation, typical for polynomial potentials, is when the lowest order term is quadratic, of the form $V_2 = a x^2 + b y^2$. In this limit, the trace of one monodromy matrix is just

$$\mathrm{Tr}\, M = 2 \cos 2 \pi \sqrt{a/b}.$$

If one can compute also the high energy limit and if the trace of the monodromy matrix for the 'same' path [52] (which is not always that easy to ascertain) is different in this limit from the one computed at zero energy, then this shows that the trace depends on the energy and that the matrix is non-resonant over a dense set of energy values h.

Apart from numerics, there exist cases where this energy dependence is established through a perturbation expansion of the solutions of the NVE around an equilibrium point [46] or by very general considerations [53]. Thus Ito [46] and Rod [53] examined the Hénon-Heiles potential with isotropic quadratic part of the form :

$$V = (x^2 + y^2)/2 + C x^2 y + y^3/3 \tag{3.41}$$

They have shown that this potential is not integrable for *all* values of C except 0, 1/6, 1/2 and 1. Indeed, the integrability of V is established for C=0, 1/6 and 1 as an additional invariant is known in these cases. The case C=1/2 is presumably *not* integrable but Ziglin's analysis cannot deal with it.In fact, the NVE is the same as the one resulting from the (integrable) potential

$$V' = (x^2 + y^2)/2 +(x^2 + y^2)^{3/2}/3 \tag{3.42}$$

and no linear analysis can distinguish between the two.

Various results have been obtained using a combination of the above methods. A typical example is the truncated Toda three particle system. By reducing to the center of mass frame, we get :

$$H = (p_x^2 + p_y^2)/2 + e^{\sqrt{3}x-y} + e^{-\sqrt{3}x-y} + e^{2y} \tag{3.43}$$

Suppose we expand the exponential potential in Taylor series and truncate it at some order, leading to a polynomial potential [54]. The integrability of this truncated system has been investigated from Ziglin's point of view in [52], where it was proven that by examining the higher order term we can prove the non-integrability at any order of truncation except quadratic. Only the fourth-order truncation necessitates a special study (performed in [55]) as the fourth order term *alone* is integrable and only its combination with the lower order terms leads to a non-integrable potential. Thus, the Toda Hamiltonian does not possess any integrable truncation apart from the quadratic one.

As a further example, we can consider the potential (k>0)

$$V = 1/(x^2+\lambda y^2+\mu)^{k/2} \tag{3.44}$$

Taking the low-energy limit we can put μ to zero and thus we are left with an homogeneous potential of degree -k. Two straight line solutions exist which give constraints on the possible integrability domain. In particular we can prove that the potential is *not* integrable when $\lambda<1$ and also when $\lambda>1$. This leaves $\lambda=1$ as the only possible integrable value, in which case integrability is trivial.

Another example of non-homogeneous potential which can be treated exactly within Ziglin's theorem framework has been given in Ref.[56]. The potential in question writes :

$$V = W(x,y) + \frac{1}{y^2} \tag{3.45}$$

where W is a homogeneous polynomial potential of degree n, of the form

$$W = - y^n - \lambda\, x^2\, y^{n-2} \ldots$$

The importance of integrable potentials of the form (3.45) lies in the fact that they can be extended to higher dimensions (through a radial extension $y^2 \rightarrow \Sigma y_i^{\,2}$) without losing their integrable character. The NVE writes

$$\ddot{\xi} = 2\,\lambda\, y^{n-2}\,\xi$$

where y is given by

$$\dot{y}^2 = 2\,(\,y^n - \frac{1}{y^2} + E\,)$$

At *zero energy*, it is possible to transform the NVE into a hypergeometric equation and thus obtain the monodromy matrices analytically. It results that the integrable potentials of the form (3.45) obtained in [56] are the only ones of this type, namely :

a) a potential of the form $V= \{(y+\rho)^{n+1} + (y-\rho)^{n+1}\}/\rho + \mu/x^2$ is always integrable

b) the potential $V= \{(y+\rho)^{n+1} + (y-\rho)^{n+1}\}/\rho + \nu/y^2$ is integrable for $n=\pm2,\pm4$

c) the potential $V= (x+y)^n + (x-y)^n + \mu/x^2 + \nu/y^2$ is integrable only for $n=\pm2$ and 4.

3.5 THE POLY-PAINLEVE METHOD.

As we have seen in the previous subsections, Ziglin's method is particularly useful in constructing proofs of nonintegrability. Still, one of its main advantages, the fact that it deals with first order expansions, also becomes a limitation. As the NVE is linear one cannot investigate departures from integrability which appear at higher orders. A typical example of such a case is the Hénon-Heiles Hamiltonian (3.41) for C=1/2. No Ziglin-based method would ever distinguish it from a genuinely integrable one. Thus the need for a non-linear extension of Ziglin arises.

As a matter of fact, an attempt toward such a nonlinear approach already exists. Indeed one of us (MDK) has introduced the poly-Painlevé method which is based on the singularity analysis (Painlevé method) but has non-local features built in it as it takes care of the "interaction" between singularities. As in the case of Ziglin, multivaluedness is incompatible with integrability *unless* the "interaction" between singularities satisfies some commutation properties.

No precise theorem has been formulated to date, related to this method (although this can undoubtedly be taken care of) , moreover neither do we have a general algorithm for its implementation. Still, some results already exist [57] and in any case it constitutes a possible way of extending the present non-integrability approaches to the fully non-linear domain.

The "poly-Painlevé" criterion, consists in checking whether the *"interaction"* between two of the system's singularities leads to a multivaluedness of the "dense" type. The method is asymptotic and a parameter ε will be introduced, which may be taken arbitrarily small. The main idea is the following : if one can show that the some trajectory can be characterized by *two* values of an integration constant, which differ by an additive quantity proportional to ε^n (for some n), this will mean that the integration constant, for this trajectory, is hopelessly (i.e. "densely") multivalued.

Let us illustrate the "poly-Painlevé" criterion with an example of Abel's equation :

$$\frac{dy}{dx} = y^3 + x \tag{3.46}$$

This equation has a leading order singularity of the type

$$y \sim (x-x_0)^{-1/2} \quad , \qquad \text{as } x \to x_0 \ ,$$

where x_0 is an arbitrary, movable branch point. Note that it is not a Painlevé, but rather a "weak-Painlevé" case, since the only resonance is at $r=-1$ (i.e. only x_0 is free) and only powers of $(x-x_0)^{-1/2}$ appear. Thus, the "weak-Painlevé" method cannot disprove integrability for (3.46) (or, for any other first order ODE of that type).

By contrast, according to the "poly-Painlevé" criterion eq. (3.46) will be classified as *non-integrable*. To see why this is so, let us introduce a small parameter in the equation, in such a way that the variation of x from x_0, will be small compared to the variation of y near its singularity. This can be done by a coordinate transformation

$$x = \varepsilon^{-3/5}(1+\varepsilon X) \ , \ y = \varepsilon^{-1/5}\, Y \ ,$$

in terms of which (3.46) becomes

$$\frac{dY}{dX} = Y^3 + 1 + \varepsilon X \tag{3.47}$$

We now suppose that Y has a singularity at $X = X_*$ finite, and proceed to study the multivaluedness of X.

If $\varepsilon = 0$, eq. (3.47) can be integrated

$$X - \bar{X} = \int_{Y_0}^{Y} \frac{dY'}{Y'^3 + 1} , \tag{3.48}$$

$\bar{X}$ being an integration constant. Expanding the integrand in (3.48)

$$\frac{1}{Y'^3 + 1} = \frac{1}{3}\left[\frac{1}{Y'+1} + \frac{j}{Y'+j} + \frac{j^2}{Y'+j^2}\right] , \qquad j = e^{2i\pi/3} ,$$

we see that, given an initial point defining one trajectory, the possible values for the integration constant $\bar{X}$ corresponding to this trajectory differ by

$$2i\pi\,(k + mj + n\,j^2)/3 \quad ; \ k, m, n \ \text{any integer.} \tag{3.49}$$

However, since $1 + j + j^2 = 0$, (3.49) defines a two-dimensional lattice in the complex domain, infinite but discrete. According to the "poly-Painlevé"criterion, therefore, the equation (3.47), for $\varepsilon = 0$, is an integrable one.

The next step in this program is to study the multivaluedness of X, at each order in ε. Integrability will be preserved if and only if no extra multivaluedness is introduced at higher orders. For if ε is taken arbitrary small, any additive term proportional to ε^n would lead to "dense" multivaluedness. Of course, it is impossible, in the absence of an explicit recursion relation, to carry out this program to arbitrarily high orders. In many cases, however, one can find such an additive term already at low orders, and thus can disprove integrability.

This is indeed what happens with eq. (3.47) . Expanding $X = X_0 + \varepsilon X_1 + \varepsilon^2 X_2 + \ldots$ we integrate (3.47) order by order in ε , starting with :

$$X_0 - \bar{X} = \int_{Y_0}^{Y} \frac{dY'}{Y'^3 + 1} , \tag{3.50}$$

while for higher than zeroth order we write :

$$dX = \frac{dY}{Y^3+1+\varepsilon X} = \frac{dY}{Y^3+1}\left[1 - \frac{\varepsilon X}{Y^3+1} + \frac{\varepsilon^2 X^2}{(Y^3+1)^2} + \ldots\right]$$

This leads to the relations :

$$dX_1 = -\frac{X_0 dY}{(Y^3+1)^2}\ , \quad dX_2 = -\frac{X_1 dY}{(Y^3+1)^2} + \frac{X_0^2 dY}{(Y^3+1)^3}\ , \ \ldots \tag{3.51}$$

As was said earlier, X_0 is multivalued on a two-dimensional lattice. What must be checked now is that, when we integrate over some complicated path in Y space, which brings us back to the same value of X_0, all the X_i's $(i \geq 1)$ vanish. Now, since the integrand is small at infinity, cf. (3.50), going around one of its poles in Y-space is equivalent to going around any of the other two. Thus, it is sufficient to consider only the poles at $Y=-1$ and $Y=-j$.

Consider a path going once counterclockwise around $Y=-1$ (subpath Γ_1), then counterclockwise around $Y=-j$ (subpath Γ_2) once clockwise around $Y=-1$ (subpath Γ_3), and finally once clockwise around $Y=-j$ (subpath Γ_4). Clearly, over this whole integration path, X_0 does indeed return to its initial value. The variation of X_1, of course, is the contribution from all four paths ; but since X_0 takes the same value (up to an additive constant) on subpaths Γ_1 and Γ_3 (resp. Γ_2 and Γ_4) we can write

$$\Delta X_1 = \oint_{\Gamma_1} \frac{X_0(1) - X_0(3)}{(Y^3+1)^2} dY + \oint_{\Gamma_2} \frac{X_0(2) - X_0(4)}{(Y^3+1)^2} dY\ , \tag{3.52}$$

with

$$X_0(1) - X_0(3) = -\oint_{\Gamma_2} \frac{dY}{Y^3+1}\ , \quad X_0(2) - X_0(4) = \oint_{\Gamma_1} \frac{dY}{Y^3+1}\ , \tag{3.53}$$

$X_0(i)$ being the value of X_0 on subpath Γ_i .

The integrals in (3.53) are easy to compute using residues. The result is :

$$\Delta X_1 = -\frac{2i\pi}{3}\, j \oint_{\Gamma_1} \frac{dY}{\left(Y^3+1\right)^2} + \frac{2i\pi}{3} \oint_{\Gamma_2} \frac{dY}{\left(Y^3+1\right)^2}\ , \tag{3.54}$$

upon substitution into (3.52). The two new integrals in (3.54) are also easily evaluated by residues and yield

$$\Delta X_1 = -\frac{2i\pi}{3}\, j\left(\frac{1}{9}\right) + \frac{2i\pi}{3}\left(\frac{j}{9}\right) = 0\ ,$$

which shows that no modification of the multivaluedness at $X=X_*$ occurs at order ε, when following the path described above. Of course, this does not prove that ΔX_1 vanishes along *any* path. Rather than trying on a more complicated path, however, we prefer here to carry out this procedure to order ε^2.

We thus find, in a similar way,

$$\Delta X_2 = \oint_{\Gamma_1} \frac{X_1(3) - X_1(1)}{(Y^3+1)^2} dY + \oint_{\Gamma_2} \frac{X_1(4) - X_1(2)}{(Y^3+1)^2} dY$$

$$+ \oint_{\Gamma_1} \frac{X_0^2(1) - X_0^2(3)}{(Y^3+1)^3} dY + \oint_{\Gamma_2} \frac{X_0^2(2) - X_0^2(4)}{(Y^3+1)^3} dY \qquad (3.55)$$

cf. (3.51). Using $X_0(3) = X_0(1) + 2i\pi j / 3$, $X_0(4) = X_0(2) - 2i\pi / 3$, we find that (3.55) becomes

$$\Delta X_2 = \frac{4i\pi}{9} [X_1(3) - X_1(1)] + \frac{4i\pi j}{9} [X_1(4) - X_1(2)] +$$

$$+ \oint_{\Gamma_1} \frac{dY}{\left(Y^3+1\right)^3} \left[-\frac{4i\pi j}{3} X_0(1) + \frac{4\pi^2 j^2}{9} \right] + \oint_{\Gamma_2} \frac{dY}{\left(Y^3+1\right)^3} \left[\frac{4i\pi}{3} X_0(2) + \frac{4\pi^2}{9} \right] \qquad (3.56)$$

Note here, however, that the terms involving $X_0(1)$ and $X_0(2)$ are difficult to compute, since $X_0(1)$ and $X_0(2)$ are themselves functions of Y , and do not appear in combinations which yield constants along a given path.

The second one of these terms (along Γ_2) can be computed in terms of the first (along Γ_1) arguing as follows : let us choose the origin as a common point for both paths Γ_1 and Γ_2; Γ_2 can then be obtained from Γ_1 by a simple rotation of $2\pi/3$. This shows that (up to an additive constant $2i\pi / 3$) the value of X_0 , at a point of Γ_2 , is j times its value at the corresponding point of Γ_1 , since $1 / (Y^3+1)$ is invariant under this rotation, while $dY \to j\, dY$. Accordingly (up to the integral of the appropriate constant times the rest of the integrand) the integral involving X_0 on Γ_2 , in (3.56), is j^2 times its value along Γ_1 , since both X_0 and dY are multiplied by j.

It therefore remains to compute the first integrand with X_0 in (3.56) along Γ_1. Although this can be done analytically for the problem at hand, we prefer to carry out this integration numerically for two reasons. First, because the analytical expressions become too long and inconvenient to handle and second in order to show how this criterion may be used in more complicated systems.

The simplest approach is to go back to the definitions of X_0 , X_1 and X_2 and integrate them numerically around their singularities, for example, as follows. Starting from $Y = 0$, we go around $Y=-1$, counterclockwise, on a circle of unit radius, then similarly around $Y=-j$, and then clockwise around the first one and clockwise around the second. In this way, the integration path stays clear from all singularities and numerical errors are minimized. We thus find that X_0 and X_1 indeed go to zero (within the precision of the calculation), while the *real* part of X_2 also turns out to be compatible with zero. Within the same precision, however, the imaginary part of X_2 does *not* vanish, but gives:

$$\Delta X_2 = 20\, i\pi / 81\sqrt{3} \neq 0 \, .$$

We therefore conclude that for a given integration constant $\bar{X}$ and a given Y, X takes its values on a set which is dense in $\mathbb{C}$. Hence, according to the "poly-Painlevé" criterion, $\bar{X}$ is a "useless" constant of integration, and equations (3.47) and (3.46) are termed non-integrable in that sense.

We wish to end this section with two remarks : first of all, the reader must have noticed that the (movable) singularity of the original problem $x=x_0$ did not actually show up in the above analysis. What really mattered was the behavior around the zeroes of Y^3+1. Moreover, throughout the discussion, the working independent variable was Y rather than X. This shows that the "poly-Painlevé" approach does away with such, possibly artificial, distinctions between dependent and independent variables, fixed and movable singularities, or singularities in the finite domain versus those at infinity. What is important, as in Ziglin's approach, is certain commutation properties, according to which multivaluedness, in and of itself, need not be an obstacle to integrability.

As a final remark, we must point out that the "poly-Painlevé" criterion, although intuitively appealing in its association of "dense" multivaluedness to non-integrability, does not yet constitute a powerful algorithm. Its dependence on quadratures and contour integrals suggests that it will be difficult to implement analytically in higher dimensional systems (unless the integrals of the $\varepsilon=0$ case can be used to reduce the system to quadratures). Still, a combination of analytical and numerical techniques as we have proposed in this section is expected to help significantly in making the "poly-Painlevé" criterion more easily applicable. It would be interesting, in this context, to undertake the systematical study of first order ODE's using the poly-Painlevé method, just as P. Painlevé himself did a century ago while developing the methods which were to lead to the modern sigularity analysis approach.

4. Conclusion

In this course we have presented a panoramic view of the concepts of integrability. Several definitions of integrability exist ranging from the mathematically rigorous to the intuitively plausible. However, it is clear that if one wishes to use the singularity analysis tools for its detection, then the most convenient definition of integrability is the one we dubbed "complex analytic integrability". The most precise theorems existing to date, due to Ziglin and Yoshida, are related to just this kind of integrability. In fact, they establish an exact relation between this property and the structure of the singularities of the solutions of the equations of motion. Thus we can conclude epigrammatically that the singularity analysis approach, initially intoduced as a heuristic integrability detector, has at last attained the status of a rigorous mathematical tool, useful in an expanding domain of applications.

REFERENCES

1 A. Ramani, B. Dorizzi and B. Grammaticos, Phys. Rev. Lett., 49 (1982) 1539.

2 A. Ramani, B. Grammaticos, B. Dorizzi and T. Bountis, J. Math. Phys. 25 (1984) 878

3 B. Dorizzi, B. Grammaticos and A. Ramani, J. Math.Phys. 25 (1984) 481.

4 E.L. Ince, "Ordinary Differential Equations" (Dover, London, 1956).

5 J. Weiss, Lecture Notes in present volume.

6 H. Yoshida, Celest. Mech. 31 (1983) 363 and 381.

7 E. T. Whittaker, Analytical Dynamics of Particles, (Cambridge U.P.,Cambridge, 1959)

8 B. Grammaticos, J. Moulin-Ollagnier, A. Ramani, J.M Strelcyn and S. Wojciechowski 'Integrals of Quadratic Ordinary Differential Equations in $\mathbb{R}^3$: the Lotka-Volterra system', submitted for publication.

9 H. Yoshida, B. Grammaticos and A. Ramani, Acta Appl. Math. 8 (1987) 75.

10 M. Kummer, R.C. Churchill and D.L. Rod, 'On Kowalevski exponents', preprint.

11 H. Yoshida, 'Non-existence of an additional analytic integral in Hamiltonian systems with an N-dimensional homogeneous potential' (1988), preprint.

12 M. Adler and P.van Moerbecke, Algebraic completely integrable systems: a systematic approach, (Perspective in Mathematics, Academic Press, New York, 1988).

13 E. Gutkin, Physica 16D (1985) 235.

14 H. Flashka, Phys. Lett. A 131 (1988), 505.

15 R. Abraham and J. E. Marsden, Foundation of Mechanics, Benjamin-Cumming, Reading (1978)

16 P. J. Richens and M. V. Berry, Physica 2D (1981) 495.

17 A. Ramani, A. Kalliterakis, B. Grammaticos and B. Dorizzi, Phys. Lett. A 115 (1986) 25

18 J. Hietarinta, Phys. Rep. 147 (1987), 87.

19 M. Jimbo and T. Miwa, Publ. Res. Inst. Math. Sci., Kyoto Univ. 19 (1983), 943.

20 B. Dorizzi, B. Grammaticos, A. Ramani and P. Winternitz, J. Math. Phys. 27 (1986) 2848.

21 R. Hirota, Direct Methods in Soliton Theory in "Solitons", R.K. Bullough and P.J. Caudrey, eds. Topics in Modern Physics, Springer Verlag, New York (1980), and references therein.

22 M. Ito, J. Phys. Soc. Japan, 49, (1980) 771.

23 A. Ramani, B. Grammaticos and T. Bountis, 'The Painlevé property and singularity analysis of integrable and non-integrable systems', Physics Reports, to appear (1989).

24 M.J. Ablowitz and H. Segur, Phys. Rev. Lett. 38 (1977), 1103.

25 M.J. Ablowitz, A. Ramani and H. Segur, Lett. al Nuovo Cimento 23 (1978) 333.

26 M.J. Ablowitz, A. Ramani and H. Segur, J. Math. Phys. 21 (1980) 715 and 1006.

27 H. Segur, Lectures at International School "Enrico Fermi", Varenna, Italy (1980).

28 V.V. Golubev, Lectures on the Integration of the Equation of a Rigid Body About a Fixed Point, State Publishing House, Moscow (1953).

29 J. Hietarinta, B. Grammaticos, B. Dorizzi and A. Ramani, Phys. Rev. Lett. 53 (1984) 1707.

30 N. Ercolani and E. D. Siggia, Physica 34D (1989) 303.

31 M. Tabor and J. Weiss, Phys. Rev. A 24 (1981), 2157.

32 M. Kùs, J. Phys. A. Math. Gen. 16 (1983) L689.

33 G. Levine and M. Tabor, Physica 33D (1988) 189.

34 T. Bountis, A. Ramani, B. Grammaticos and B. Dorizzi, Physica 128A (1984) 268.

35 The analysis presented for this case in [34] was incomplete and stands here corrected.

36 S.V. Manakov, Sov. ZETPh. 67 (1974), 543; M. Kac and P. van Moerbecke, Adv. Math. 16 (1975), 160.

37 O.I. Bogoyavlensky, Phys. Lett. A 134 (1989) 34.

38 J.-M. Strelcyn and S. Wojciechowski, Phys. Lett. A 133 (1988), 207.

39 J. Weiss, J. Math. Phys. 24 (1983), 1405 ; 25 (1984), 13 ; 25 (1984), 2226 ; 26 (1985), 258 ; 26 (1985) 2174 ; 27 (1986) 1293 ; 27 (1986), 2467 ; 28 (1987), 2025.

40 R.S. Ward, Phys. Lett. 102A (1984), 279.

41 J. Hietarinta, Lecture Notes in present volume.

42 S.L. Ziglin, Funct. Anal. Appl. 16 (1983), 181 and 17 (1983), 6.

43 S.L. Ziglin, Trans. Moscow Math. Soc. 1 (1982), 283; V.K. Mel'nikov, Trans. Moscow Math. Soc. 12 (1963), 1.

44 A. Ramani, B. Grammaticos and H. Yoshida, 'Rigorous non-integrability results related to singularity analysis', Proceedings of thr International Workshop on Nonlinear Evolution Equations, Como, July 1988, to appear.

45 H. Yoshida, A. Ramani, B. Grammaticos and J. Hietarinta, Physica 144A (1987), 310.

46 H. Ito, Kodai Math. J. 8 (1985), 120.

47 H. Yoshida, Physica 21D (1986),163 and 29D (1987) ,128.

48 B. Grammaticos, A. Ramani and H. Yoshida, Phys. Lett. 124A (1987), 65

49 J. Plemelj, Problems in the sense of Riemann and Klein, Interscience, New York (1962), chap. 6.

50 D. Roekaerts and H. Yoshida, J. Phys. A. 21 (1988), 3547.

51 H. Yoshida, 'Ziglin analysis for proving non-integrability of Hamiltonian systems', in Finite Dimensional Integrable Nonlinear Dynamical Systems, eds. P.G.L. Leach and W.H. Steeb, World Scientific, Singapore, 74-93 (1988).

52 H. Yoshida, Comm. Math. Phys. 116 (1988), 529.

53 D.L. Rod, 'On a theorem of Ziglin in Hamiltonian dynamics', preprint.

54 H. Yoshida, Phys. Lett. 120A (1987), 388.

55 H. Yoshida, A. Ramani and B. Grammaticos, Physica 30D (1988), 151.

56 B. Grammaticos and A. Ramani, 'Why some integrable (2-D) Hamiltonians cannot be extended to higher dimensions', to appear in Phys. Lett. A (1989).

57 M. D. Kruskal and P. A. Clarkson, in preparation.

A CONCEPT OF INTEGRABILITY BASED ON THE SYMMETRY APPROACH

Alexandre V. Mikhailov
Landau Institute for
Theoretical Physics
Academy of Sciences
MOSCOW 117940

ABSTRACT. We present some motivations of the definition of integrability that come from the symmetry approach. The concept of formal symmetry will be discussed. The formal symmetry approach claims to answer the question whether a given PDE possesses higher symmetries and/or conservation laws. We can formulate necessary conditions only, but these conditions are so strong that they provide a possibility to classify integrable equations. All computations are algorithmic and for special cases a symbolic computer code for IBM-PC is available.

R. Conte and N. Boccara (eds.), Partially Integrable Evolution Equations in Physics, 373.

Bäcklund transformations and the Painlevé property

John Weiss
6 Lockeland Avenue
Arlington, Mass 02174

Abstract. For systems with the Painlevé Property, Bäcklund transformations can be defined. These appear as specializations (truncations) of certain expansions of the solution about its *singular manifold.* With reference to the Lax pair for a system, the Bäcklund transformations are equivalent to transformations of linear systems developed by Darboux (Bäcklund-Darboux transformations).

For specific systems the Bäcklund-Darboux transformations lead to a reformulation of these systems in terms of the Schwarzian derivative. We find the Bäcklund transformations of these systems and study their periodic fixed points.

The periodic fixed points of the Bäcklund transformations determine a finite dimensional invariant manifold for the flow of the system. The resulting (ordinary) differential equations have a hamiltonian structure and the flow of the (partial) differential system is represented by commuting flows on the finite dimensional manifold.

Contents.

1. Introduction: The Painlevé Property.

We are interested in the behavior of solutions of nonlinear ordinary and partial differential equations. Among the simpler properties that a solution can have is whether the solution is single or multiple valued as a function of its (complex) independent variables. We ask, at least locally, where the solution lives. Historically, this question was found to be interesting.

1.1 SURVEY AND DEFINITIONS.

Kovalevsky [1] found that when the solutions of the spinning top equations are single valued the equations are completely integrable. Painlevé [2] considered a wide

R. Conte and N. Boccara (eds.), Partially Integrable Evolution Equations in Physics, 375–411.

class of second order equations and classified these according to the nature of their singularities. Since the coefficients of the equations are allowed to depend on the independent variable, *fixed* singularities can arise at the fixed locations of singularities of the coefficients. Painlevé and his coworkers found essentially six different equations within the class considered whose solutions are single valued as functions of the complex independent variable, except possibly at the fixed singularities of the coefficients. These are known as the Painlevé transcendents and have a great variety of interesting properties and applications. Singularities that are not fixed are said to be *movable*.

An ordinary differential equation is said to have the Painlevé Property when every solution is single valued, except at the fixed singularities of the coefficients. That is, the Painlevé Property requires that the movable singularities are no worse than poles.

Ablowitz et al [3] found that when certain integrable partial differential equations have reductions to ordinary differential equations, the ordinary differential equations have the Painlevé Property. They conjecture that when a system is *integrable* its reductions will have the Painlevé Property. This conjecture is supported by the results of McLeod and Olver [4]. Integrable here means there exists a nontrivial Lax pair for the system.

The major difference between analytic functions of one complex variable and several complex variables is that, in general, the singularities of a function of several complex variables cannot be isolated [5]. If $f = f(z_1, \cdots, z_n)$ is a meromorphic function of n complex variables ($2n$ real variables), the singularities of f occur along analytic manifolds of (real) dimension $2n - 2$. These manifolds are determined by conditions of the form

$$\phi(z_1, \cdots, z_n) = 0, \tag{1.1}$$

where ϕ is an analytic function of $(z_1, \cdots, z_n)$ in a neighborhood of the manifold.

With reference to the above, we say that [6] *a partial differential equation has the Painlevé Property when the solutions of the PDE are single valued about the movable, singularity manifolds.* For partial differential equations we require that the solution be a *single-valued functional* of the data, i.e. *arbitrary functions.* This is a formal property and not a restriction on the data itself.

To verify if a PDE has the Painlevé Property we [6] introduce a method for expanding a solution of a nonlinear PDE about a movable, singular manifold (1.1). Let $u = u(z_1, \cdots, z_n)$ be a solution of the PDE and assume that

$$u = \phi^\alpha \sum_{j=0}^{\infty} u_j \phi^j \tag{1.2}$$

where ϕ and

$$u_j = u_j(z_1, \cdots, z_n)$$

are analytic functions of $(z_1, \cdots, z_n)$ in a neighborhood of the manifold (1.1). Substitution of (1.2) into the PDE determines the possible values of α and defines the recursion relations for $u_j, j = 0, 1, 2, \cdots$. When α is an integer and (1.2) is a valid and general expansion about the manifold (1.1), then the solution has a single valued representation about (1.1). If this representation is valid for all allowed movable singularity manifolds then the PDE has the Painlevé Property. For a specific PDE it is necessary to identify all possible values for α and then find what the form of the resulting *Psi* series [7] is.

A point that we will emphasize is that the Psi series for nonlinear PDE contain a lot of information about the solutions of the PDE. For equations which have the Painlevé property we have developed a method for finding the Lax pairs and Bäcklund transformations [8,9,10]. An outline of the *singular manifold method* is presented in section 2. For equations that do not have the Painlevé Property it is still possible to obtain single valued expansions by specializing the arbitrary functions that appear in the Psi series expansions. This specialization leads to a system of partial differential equations for the formally arbitrary data. For specific systems, and we conjecture in general, these equations are integrable. The form of the resulting reduction enables the identification of integrable reductions of the original system [11]. This is examined in section 3.

1.2 EXAMPLES AND COUNTER-EXAMPLES.

To illustrate the nature of the Painlevé Property it is worthwhile to examine a few examples of equations with and without the Painlevé Property.

A simple case of an equation with the Painlevé Property is **Burgers Equation.**

$$u_t + uu_x = u_{xx} \tag{1.3}$$

It is not difficult to find the Psi series

$$u = \phi^{-1} \sum_{j=0}^{\infty} u_j \phi^j \tag{1.4}$$

is valid for (1.3). Examination of the recursion relations for the u_j obtains a system of the form

$$(j-2)(j+1)\phi_x^2 u_j = F_j(u_{j-1}, \cdots, u_0, \phi_t, \phi_x, \cdots).$$

For (1.4) to be valid F_2 must vanish identically. Evaluation of the recursions obtains

$$j = 0, \; u_0 = -2\phi_x$$

$$j = 1, \; \phi_t + u_1\phi_x = \phi_{xx}$$

$$j = 2, \; \partial_x(\phi_t + u_1\phi_x - \phi_{xx}) = 0.$$

The relation (compatibility condition) at $j = 2$ is satisfied identically and the expansion (1.4) is valid with *arbitrary* functions ϕ and u_2.

The Korteweg-de Vries equation

$$u_t + u_{xxx} + 3uu_x = 0 \tag{1.5}$$

has singularities of the form

$$u = \phi^{-2} \sum_{j=0}^{\infty} u_j \phi^j \tag{1.6}$$

with arbitrary functions ϕ, u_4 and u_6. The KdV equation has the Painlevé Property about singularities of the form (1.6).

The Schwarzian KdV equation [8]

$$\frac{\psi_t}{\psi_x} + \{\psi; x\} = \lambda \tag{1.7}$$

where

$$\{\psi; x\} = \frac{\psi_{xxx}}{\psi_x} - \frac{3}{2}\left(\frac{\psi_{xx}}{\psi_x}\right)^2$$

is the Schwarzian derivative , has singularities of the form

$$\psi = \phi^{-1} \sum_{j=0}^{\infty} \psi_j \phi^j \tag{1.8}$$

with arbitrary ϕ, u_0, u_1 if the non-characteristic condition $\psi_x \simeq \phi_x \neq 0$ is verified. If $\psi_x = 0$ the expansion about the *characteristic manifold* has the form

$$\psi = f(t) + \phi^3 \sum_{j=0}^{\infty} \psi_j \phi^j \tag{1.9}$$

where $\phi = x - x_0(t)$ and $\psi_j = \psi_j(t)$.

The Schwarzian KdV equation has single valued expansions about both characteristic and noncharacteristic manifolds. *The Painlevé Property requires all movable singularity manifolds to be single valued, whether characteristic or not.* The above result runs counter to the observation of Ward [12,13] that direct consideration of expansions about characteristic manifolds cannot be allowed in the definition of the Painlevé Property since, for linear systems, arbitrarily bad singularities propagate along characteristics. For general systems, expansions about characteristics, when they exist, introduce certain arbitrary data [14]. If the data is *bad*, the expansion is still required to be a single valued functional of that data. In this sense, expansion (1.9) is a single valued functional of the data $f(t)$, $x_0(t)$, however multiple valued that data as a function of t may be. Of course, the same observation applies to the non-characteristic expansion (1.8). *The Painlevé Property is a statement of how the solutions behave as functionals of the data in a neighborhood of a singularity manifold and not a statement about the data itself.* The following example will illustrate this point.

A derivative Schwarzian equation

$$\frac{\psi_t}{\psi_x} + \frac{\partial}{\partial x}\{\psi; x\} = 0 \tag{1.10}$$

has non-characteristic singularities of the form

$$\psi = \phi^{-1} \sum_{j=0}^{\infty} \psi_j \phi^j \tag{1.11}$$

where $\phi, \psi_0, \psi_1, \psi_2$ are arbitrary. Therefore, (1.10) has a single valued expansion depending on the maximum number of arbitrary functions allowed for by the order of the equation. However, about the characteristic manifold where $\psi_x = 0$

$$\psi = f(t) + \phi^4 \sum_{j=0}^{\infty} \sum_{k=0}^{\infty} \psi_{jk} \phi^j \phi^{k\alpha} \tag{1.12}$$

where $\alpha = \frac{7}{2} + i\sqrt{11}/2$ and $\phi = x - x_0(t)$. The expansions (1.12) are highly multilple valued *as functionals of* ϕ.

In general, to verify that an equation has the Painlevé Property it is necessary to show that all the allowed singularities are single valued (as functionals of the data). This requirement is often overlooked and has lead to some wrong conclusions.

Doktorov and Sakovich [15] claim that

$$\triangle \psi = P_n(\psi) \tag{1.13}$$

has the Painlevé property for any polynomial of degree n since if $\psi = u^{-1}$ then the equation for u has single valued poles depending on two arbitrary functions and therefore has the Painlevé property. In effect, they are only claiming that (1.13) has a single valued expansion about the simple zeros of ψ as is locally valid by the Cauchy-Kovalevsky Theorem. It is not difficult to see that (1.13) has singularities that are not single valued. For instance, the reduction to an ODE is within the class originally studied by Painlevé.

It is also thought that the Clarkson equation

$$u_t^2 = 2uu_x^2 - (1+u^2)u_{xx} \tag{1.14}$$

has only meromorphic psi-series and has Painlevé Property [16,17]. However, this is not the case. Consider the points where

$$u^2 + 1 = 0. \tag{1.15}$$

Rewriting the above we have

$$u_t^2 - 2iu_x^2 = (u-i)(2u_x^2 - (u+i)u_{xx})$$

and letting $G = u + i$

$$G_t^2 - 2iG_x^2 = (G-2i)(2G_x^2 - GG_{xx}).$$

To leading order

$$G = 2i + G_0\phi^\alpha + \cdots$$

where $\Re\alpha \geq 0$.

By substitution in the above

$$\alpha^2 G_0^2(\phi_t^2 - 2i\phi_x^2)\phi^{2\alpha-2} =$$
$$\alpha(\alpha+1)G_0^3\phi_x^2\phi^{3\alpha-2} - 2i\alpha(\alpha-1)G_0^2\phi_x^2\phi^{2\alpha-2}.$$

Since $\Re\alpha \geq 0$ we have

$$\alpha = 2i\phi_x^2/\phi_t^2.$$

For this leading order the balance equations are

$$G_t^2 + 2i(GG_{xx} - G_x^2) = 0$$

and the resonances

$$G = 2i + G_0\phi^\alpha + G_1\phi^{\alpha+r}$$

obtains

$$\alpha(\alpha+r)\phi_t^2 + i\{\alpha(\alpha-1) + (\alpha+r)(r-\alpha-1)\}\phi_x^2 = 0.$$

Using the leading order

$$r = -1, 0.$$

Thus G_0 is arbitrary. From the above, the Clarkson equation has a movable essential singularity and does not have the Painlevé Property.

A variant of the preceding, a **modified Clarkson Equation** [18], also has non-meromorphic singularities. The equation is

$$u_t = (1+u^2)u_{xx} + (1-2u)u_x^2. \tag{1.16}$$

Again, expand about the points where $1+u^2=0$. To see this, let $G = u + i$ and find

$$G_t = (G-2i)(GG_{xx} - 2G_x^2) + (1-2i)G_x^2.$$

Let $G = 2i + G_0\phi^\alpha + \cdots$ and get to leading order

$$\alpha = 2i$$

and a resonance at $r=0$.

The equation

$$(1+u^2)u_{xx} = (2u-1)u_x^2 \tag{1.17}$$

is the traveling wave of (1.14) and the steady state of (1.16). This equation was shown by Painlevé to have the general solution

$$u = \tan(\log(ax+b)).$$

The movable essential singularities are shown by the previous examples.

The steady state of (1.14) is,

$$(1-u^2)u_{xx} + 2uu_x^2 = 0. \tag{1.18}$$

The paraphrase of the analysis for the first example detects no non-meromorphic singularity. The general solution is found to be

$$u = \tan(ax+b).$$

An example of an equation with a non-constant resonance is the **Rand equation** [18]

$$u^2 u_{xxx} = 3u_t^3 \tag{1.19}$$

It has the leading order $u = u_0\phi^\alpha + \cdots$ where it can be shown that

$$(\alpha-1)(\alpha-2) = 3(\phi_t^3/\phi_x^3)\alpha^2.$$

This quadratic equation for α determines the leading order. Of course α is a non-constant functional of ϕ_t and ϕ_x.

The resonance condition

$$u = u_0\phi^\alpha + u_1\phi^{\alpha+r} + \cdots$$

easily determines the resonances

$$r = -1, 0, 4 - 3\alpha.$$

It is the case here that one resonance, $4 - 3\alpha$, is a functional of the singular manifold, ϕ.

Finally, we consider the **inviscid Burgers equation**

$$u_t + uu_x = 0 \tag{1.20}$$

has a leading order of the form

$$u = u_0 + u_1\phi^\alpha + u_2\phi^{2\alpha} + \cdots .$$

By substitution into the above a solution is

$$\alpha = 1/2$$

and

$$\phi_t + u_0\phi_x = 0.$$

The next term in the expansion is

$$u_{0,t} + u_0u_{0,x} + 1/2\phi_x u_1^2 = 0.$$

If $u_{0,t} + u_0u_{0,x} = 0$ then $u_1 = u_2 = \cdots = 0$ and from the above

$$\phi_x^2\phi_{tt} - 2\phi_x\phi_t\phi_{xt} + \phi_t^2\phi_{xx} = 0. \tag{1.21}$$

This equation can be linearized by a Legendre transformation, as we shall show in section 3.

In the preceding paragraphs our intent is to illustrate both the definition of the Painlevé Property and the variety of singularities revealed by the functional Psi series. This approach is capable of substantial generalization. In this paper we will, for the most part, describe the applications to integrable systems.

2. The Singular Manifold Method.

If an equation has the Painlevé Property we propose to calculate the Băcklund transformations, Lax pair, Modified equations and Miura transformations through the expansions of the solutions about the singularity manifold. The *Singular Manifold Method* consists in truncating the Laurent Psi series after the *constant level* term. By construction, this forms a possible Bäcklund transformation. Depending on the distribution of *resonances*, a generally overdetermined system of equations are defined by the recursion relations for the coefficients of the Laurent expansion. Reduction of this system to consistent form defines the Bäcklund-Darboux transformation, the Schwarzian form of the modified equation and the related Miura transformation to the original system. The Lax pair can be found by linearizing the Miura transformation and modified equation, using the invariance of the Schwarzian derivative under the Möbius group to motivate the substitution for *linear* variables. The invariance under the Möbius group and the *discrete* symmetries of the modified equations are found to constitute a nontrivial Bäcklund transformation for these systems. We will examine the use of these transformations in finding explicit solutions ,i.e. rational, finite-zone. We will also present the two cases where the method finds partial results. In one case, the Bullough-Dodd equation does not have a Bäcklund transformation, and in the other case, the Bäcklund transformation for the *modified Nonlinear Schrödinger* equations is a reduction of the system. Plausible extensions of the method, including applications to ODEs, are examined.

Since this is a method for the discovery of the structures associated with a given equation, we present the analysis for the KdV equation in detail.

2.1 THE KORTEWEG-DE VRIES EQUATION.

The Korteweg-de Vries equation

$$u_t + u_{xxx} + 3uu_x = 0 \tag{2.1}$$

has meromorphic singularities of the form

$$u = \phi^{-2} \sum_{j=0}^{\infty} u_j \phi^j \tag{2.2}$$

about the singularity manifold $\phi(x,t) = 0$. In the above we find that ϕ , u_4 and u_6 are arbitrary functions, and it is required that, for (2.2) to be well defined, ϕ be *non-characteristic*. That is, $\phi_x \neq 0$ when $\phi = 0$. For locally analytic data we will show that (2.2) converges in a neighborhood of $\phi = 0$.

It is also of interest to consider the slight generalization of (2.1)

$$(\partial/\partial x)(u_t + u_{xxx} + 3uu_x) = 0 \tag{2.3}$$

which has the expansion (2.2) with arbitrary functions ϕ, u_4, u_5 and u_6.

Now, we truncate (2.2) after the constant term to obtain

$$u = u_0/\phi^2 + u_1/\phi + u_2. \tag{2.4}$$

Substitution into (2.1), or (2.3), obtains a system of four equations in the four functions ϕ, u_0, u_1, u_2. This is most readily seen for (2.3) since setting the *arbitrary* functions $u_4 = u_5 = u_6 = 0$ and requiring $u_3 = 0$ obtains (2.4) and four equations. From these we find

$$u = -4\phi_x^2/\phi^2 + 4\phi_{xx}/\phi + u_2 \tag{2.5}$$

$$u_2 + \lambda = -(\phi_{xxx}/\phi_x) + \frac{1}{2}(\phi_{xx}/\phi_x)^2 \tag{2.6}$$

$$\phi_t/\phi_x + \{\phi; x\} = \lambda \tag{2.7}$$

where u and u_2 satisfy the KdV equation. Again,

$$\{\phi; x\} = \phi_{xxx}/\phi_x - \frac{3}{2}(\phi_{xx}/\phi_x)^2 \tag{2.8}$$

i.e. the Schwarzian derivative [7,8], is the unique differential invariant [19] of the Möbius group

$$\phi = (a\psi + b)/(c\psi + d). \tag{2.9}$$

The Möbius group is the unique group of conformal (monodromny preserving) automorphisms of the complex ϕ (Riemann) sphere.

The relations (2.6) and (2.7) imply that u_2 satisfy (2.1) since

$$u_{2,t} = -(\frac{\partial}{\partial x} + V)\frac{\partial}{\partial x}(\frac{\partial}{\partial x} - V)\frac{\phi_t}{\phi_x}$$

where

$$V = \frac{\phi_{xx}}{\phi_x}. \tag{2.10}$$

This definition of V obtains the modified KdV equation

$$V_t + V_{xxx} - \frac{3}{2}V^2V_x = \lambda V \tag{2.11}$$

from (2.7).

The Bäcklund-Darboux equation (2.5) may be written in the form

$$u = 4\frac{\partial^2}{\partial x^2}\ln\phi + u_2 \tag{2.12}$$

where

$$u_2 + \lambda = -\frac{\partial}{\partial x}(\phi_{xx}/\phi_x) - \frac{1}{2}(\phi_{xx}/\phi_x)^2 \tag{2.13}$$

and ϕ satisfies the Schwarzian-KdV equation (2.7).

We regard (2.13) as a *Miura transformation* from (2.7) to (2.1). It also has the form of a *Riccati equation* in the variable $W = \phi_{xx}/\phi_x$ and can be linearized by the substitution $W = -2v_x/v$. This obtains the linear equation for v

$$2v_{xx} = (u_2 + \lambda)v \tag{2.14}$$

and the identification $\phi_x = v^{-2}$. From (2.7) the additional linear equation

$$v_t = (2\lambda - u_2)v_x + \frac{1}{2}u_{2,x}v \tag{2.15}$$

is found. By construction (2.14) and (2.15) are the Lax pair for the KdV equation and imply u_2 is a KdV solution.

The linearizing substitution for ϕ has the form

$$\phi = v_1/v_2$$

where v_1 and v_2 are solutions of (2.14).

In terms of the linear equation (2.14) the Bäcklund transformation is the classical Darboux transformation for adding elements to the spectra [20,21].

Now, consider the Bäcklund transformations for the Schwarzian KdV equation (2.7). From the invariance of the Schwarzian derivative we have the invariance of (2.7) under the Möbius group (2.9). This invariance is also found by examining the singularities of (2.7), (1.8), and finding that the truncated expansion

$$\phi = \psi^{-1}\phi_0 + \phi_1 \tag{2.16}$$

requires constant ϕ_0, ϕ_1 and ψ satisfies (2.7). In other words, we find the Möbius invariance. An additional Bäcklund transformation is found from the *discrete symmetries* of the modified equation (2.11). That is, $V \Rightarrow -V$ implies the invariant transformation

$$\phi_x = 1/\psi_x \tag{2.17}$$

for (2.7). The time dependent form of this transformation is found from (2.7) and (2.17). That is,

$$\phi_t/\phi_x + \psi_t/\psi_x + (\phi_{xx}/\phi_x)(\psi_{xx}/\psi_x) = 2\lambda \tag{2.18}$$

and (2.17) imply by $\phi_{xt} = \phi_{tx}$ and $\psi_{xt} = \psi_{tx}$ that ϕ and ψ both satisfy (2.7). Therefore, we have a strong Bäcklund transformation for (2.7). In section 4 we will show how the periodic fixed points of the Bäcklund transformations for (2.7) define finite dimensional invariant manifolds and as commuting hamiltonian flows *factor* the KdV flow on this invariant manifold.

Using the Bäcklund transformations, it is simple to show that the Laurent expansion

$$\phi = \xi^{-1}\phi_0 + \phi_1 + \cdots$$

converges in a neighborhood of $\xi(x,t) = 0$. The symmetry $\psi = \phi^{-1}$ maps the *pole* into a simple *zero* and the data satisfies the conditions of the *Cauchy-Kovalevsky Theorem*, being non-characteristic and locally analytic. Therefore, both expansions converge in a (punctured) neighborhood. Furthermore, the symmetry (2.17) maps the characteristic, non-conformal singularity

$$\phi = f(t) + \xi^3\phi_3 + \cdots$$

into a simple pole

$$\psi = \xi^{-1}\psi_0 + \cdots$$

and by the previous argument , this also converges. Using this result and the relation between the modified and KdV systems, the pole singularities for the KdV equation also converge. The characteristic singularities for the KdV are trivial.

The convergence of the Laurent series for the KdV equation

$$u = \phi^{-2}\sum_{j=0}^{\infty} u_j\phi^j$$

implies, by the recursion relations, the existence of the conservation laws

$$\frac{\partial}{\partial t}A_j = \frac{\partial}{\partial x}B_j$$

where

$$A_j = \sum_{k=0}^{\infty} \frac{(j+k)!}{j!k!} u_{j+2+k}\phi^k.$$

Therefore the Painlevé property directly implies the existence of formal integrals.

The KdV (2.1) and modified KdV (2.11) have a hamiltonian structure [9,10]

$$u_t + \frac{\partial}{\partial x}(u_{xx} + \frac{3}{2}u^2) = 0 \tag{2.19}$$

$$V_t + \frac{\partial}{\partial x}(V_{xx} - \frac{1}{2}V^3) = 0 \tag{2.20}$$

and are connected by the Miura transformation

$$u = \pm V_x - \frac{1}{2}V^2. \tag{2.21}$$

Note

$$u_{xx} + \frac{3}{2}u^2 = \delta_u \int (-u_x^2 + \frac{1}{2}u^3)dx$$

$$V_{xx} - \frac{1}{2}V^3 = \delta_V \int (-V_x^2 - \frac{1}{8}V^4)dx.$$

Using the Miura transformation finds the *second hamiltonian* structure for the KdV equation from the first hamiltonian structure of (2.20). By the change of variable formula $\Omega = (\delta_V u)\partial_x(\delta_V u)^t$. It is

$$\Omega = \partial_x^3 + 2u\partial_x + u_x = (\partial_x - V)(\partial_x)(\partial_x + V) \tag{2.22}$$

where $u = V_x - \frac{1}{2}V^2$ and the KdV equation is

$$u_t + \Omega\delta_u H_2 = 0$$

where $H_1 = \int \frac{1}{2}u^2$.

The gradient of the integrals for the KdV equation satisfy the Lenard recursion formula

$$\partial_x \delta_u H_{j+1} = \Omega\delta_u H_j \tag{2.23}$$

and the higher order equations are

$$u_t + \partial_x \delta_u H_j = 0 \tag{2.24}$$

for $j = 1, 2, \cdots$. Now, putting together the above we have the following result [9].

The sequence of *higher-order KdV equations*

$$u_t + \partial_x \delta_u H_{j+1} = 0 \tag{2.25}$$

for $j = 1, 2, 3, \cdots$ has the Bäcklund-Darboux transformation

$$u = 4\frac{\partial^2}{\partial x^2}\ln\phi + u_2 \tag{2.26}$$

where

$$u_2 = -\frac{\partial}{\partial x}(\phi_{xx}/\phi_x) - \frac{1}{2}(\phi_{xx}/\phi_x)^2 \tag{2.27}$$

and

$$\frac{\phi_t}{\phi_x} + \delta_u H_j(\{\phi; x\}) = 0. \tag{2.28}$$

Furthermore,

$$u_3 = \{\phi; x\}$$

and u_2 satisfy (2.25) and the sequence (2.28) is invariant under the Möbius group and the symmetry (2.17).

Note that the first few gradients are:

$$\delta_u H_1 = u$$

$$\delta_u H_2 = u_{xx} + \frac{3}{2}u^2$$

$$\delta_u H_3 = u_{xxxx} + 5uu_{xx} + \frac{5}{2}u_x^2 + \frac{5}{2}u^3.$$

Now, using a simple leading order argument and the Bäcklund transformations to raise and lower the *weight* of the Laurent expansions it is not difficult to see that the higher-order systems must have the Painlevé Property and the formal Laurent expansions converge in a punctured neighborhood of the singularity manifold when the data is locally analytic in this neighborhood [9]. That is, the highest weight singularity for the Nth Schwarzian equation is of the form

$$\phi = f(t) + \xi^{2N+1}\phi_{2N+1} + \cdots .$$

The odd order poles are, for $k < N$,

$$\phi = \xi^{-2k-1}\phi_0 + \cdots$$

and the odd order zeroes are, for $k < N$,

$$\phi = \xi^{2k+1}\phi_0 + \cdots .$$

The symmetry $\phi = 1/\psi$ maps poles $\Leftrightarrow$ zeroes. The symmetry $\phi_x = 1/\psi_x$ maps a zero of order $2k+1 \Leftrightarrow$ a pole of order $2k-1$. Combining the two transformations any singularity can be mapped into a simple zero $\phi = \xi\phi_0 + \cdots$ and by the *Cauchy-Kovalevsky Theorem* this is single-valued and convergent. The map from this form into the complete set of singularities can be shown to introduce no multiple-valued $\ln \xi$ terms and to depend on the maximum number of allowed arbitrary functions.

Therefore, *the KdV sequence has the Painlevé Property with convergent Laurent series.*

To summarize, the Painlevé expansion truncated after the constant level term defines a form of modified equation that is expressed in terms of the Schwarzian derivative. By linearizing the Miura transformation the Lax pair is found. The discrete symmetries of the modified equations and the Möbius group are Bäcklund transformations for the Schwarzian equations. The Miura transformation allows the calculation of second hamiltonian structures and tha associated recursion operators for the gradients of conserved densities. The action of Bäcklund transformations on the singularity structure of equation sequences allows the conclusion that the sequence is Painlevé and the formal Laurent expansions converge.

2.2 THE BOUSSINESQ EQUATION.

The analysis for the Boussinesq equation [10]

$$u_{tt} + \frac{\partial^2}{\partial x^2}(\frac{1}{3}u_{xx} + u^2) = 0 \tag{2.29}$$

finds the Bäcklund - Darboux transformation

$$u = 2\frac{\partial^2}{\partial x^2}\ln\phi + u_2 \tag{2.30}$$

where $\phi = v_1/v_2$ and v_1, v_2 satisfy

$$4v_{xxx} + 6uv_x + 3(u_x + h)v = 0$$

$$v_t = v_{xx} + (u + \lambda)v$$

$$h_x = u_t.$$

The Schwarzian modified equation is

$$\frac{\partial}{\partial t}(\phi_t/\phi_x) + \frac{1}{3}\frac{\partial}{\partial x}(\{\phi; x\} + \frac{3}{2}(\phi_t/\phi_x)^2) = 0. \tag{2.31}$$

The discrete symmetry for this system has the form

$$\frac{\phi_{xx}}{\phi_x} = -\frac{1}{2}\frac{\psi_{xx}}{\psi_x} \mp \frac{3}{2}\frac{\psi_t}{\psi_x}$$

$$\frac{\phi_t}{\phi_x} = \pm\frac{1}{2}\frac{\psi_{xx}}{\psi_x} - \frac{1}{2}\frac{\psi_t}{\psi_x}. \tag{2.32}$$

2.3 THE SINE-GORDON EQUATION.

The Sine-Gordon equation [11]

$$u_{xt} = \sin(u) \tag{2.33}$$

has the Painlevé property in the variable $V = e^{iu}$ where

$$VV_{xt} - V_xV_t = \frac{1}{2}(V^3 - V). \tag{2.34}$$

The expansion

$$V = \phi^{-2}\sum_{j=0}^{\infty} V_j\phi^j$$

with resonances $j = -1, 2$ and $V_0 = 4\phi_x\phi_t$ and $V_1 = \phi_{xt}$. The Painlevé transformation is

$$V = -4\frac{\partial^2}{\partial x t}\ln\phi + V_2 \tag{2.35}$$

where $V_2 = \phi_{xt}^2/(\phi_x\phi_t)$.

The Schwarzian modified equations are

$$\Omega_1 = \{\phi; t\} + 2Z_{tt}/Z = \alpha \tag{2.36}$$

$$\Omega_2 = \{\phi; x\} + 2W_{xx}/W = \beta \tag{2.37}$$

where $Z^2 = \phi_x/\phi_t$, $W^2 = \phi_t/\phi_x$, and $\alpha\beta = \frac{1}{4}$. The identity $\phi_x(\partial/\partial x)\Omega_1 + \phi_t(\partial/\partial t)\Omega_2 = 0$ demonstrates the equivalence of the above two equations.

To find the *discrete symmetries* let $\Theta = -\phi_{xt}/\phi_t$ and $\Phi = -\phi_{xt}/\phi_x$ and get the system

$$\Theta_t + \frac{1}{2}\Theta\Phi + \frac{\lambda}{2}\Theta/\Phi = 0 \tag{2.38}$$

$$\Phi_x + \frac{1}{2}\Theta\Phi + \frac{1}{2\lambda}\Phi/\Theta = 0. \tag{2.39}$$

The discrete symmetries of these equations imply the strong Bäcklund transformations for the Schwarzian equations.

$$\frac{\phi_{xt}}{\phi_t}\frac{\psi_{xt}}{\psi_t} = \frac{1}{\lambda}, \quad \phi_x\psi_x = 1 \tag{2.40}$$

$$\frac{\phi_{xt}}{\phi_x}\frac{\psi_{xt}}{\psi_x} = \lambda, \quad \phi_t\psi_t = 1 \tag{2.41}$$

$$\frac{\phi_{xt}}{\phi_t}\frac{\psi_{xt}}{\psi_t} = -\frac{1}{\lambda}, \quad \frac{\phi_{xt}}{\phi_x}\frac{\psi_{xt}}{\psi_x} = -\lambda \tag{2.42}$$

For instance, $\psi_t = -\phi_t^{-1}$ and $\psi_x = -(1/\lambda)\phi_{xt}^2/(\phi_t^2\phi_x)$ imply by the condition $\psi_{xt} = \psi_{tx}$ that ϕ satisfies the Schwarzian equation (2.36).

The Möbius group is a point symmetry and composition with the BT above generates the solutions

$$\phi_0 = e^{\sigma t + x/\sigma}$$

$$\phi_1 = \tanh(\sigma t/2 + x/(2\sigma))$$

$$\phi_2 = \sinh(\sigma t + x/\sigma) + \sigma t - x/\sigma.$$

The Sine-Gordon equation is also the *minus one* equation of the KdV sequence. In (2.34) let $V = \phi_x$ and find

$$\frac{\partial}{\partial t}\{\phi; x\} = -\frac{\partial}{\partial x}\left(\frac{1}{\phi_x}\right). \tag{2.43}$$

The recursion operator of the KdV sequence has the form

$$\frac{\partial}{\partial x}\delta_u H_{j+1} = \Omega\delta_u H_j$$

where $\Omega = (\partial_x - W)(\partial_x)(\partial_x + W)$ and $W = \phi_{xx}/\phi_x$. The minus one functional satisfies the condition, $\delta_u H_0 = 1$,

$$\partial_x\delta_u H_0 = \Omega\delta_u H_{-1} = 0$$

and the general solution is

$$\delta_u H_{-1} = (a\phi^2 + b\phi + c)/\phi_x \tag{2.44}$$

where $u = \{\phi; x\}$. Therefore, the Sine-Gordon equation has the form

$$u_t = \partial_x\delta_u H_{-1} \tag{2.45}$$

where $u = \{\phi; x\}$ and $\delta_u H_{-1} = -1/\phi_x$.

2.4 HENON-HEILES SYSTEM.

The singular manifold method can also be applied to ordinary differential equations [22,23]. For instance, the Bãcklund transformation for the Hénon-Heiles system

$$x_{tt} = -Ax - 2dxy$$
$$y_{tt} = -By + cy^2 - dx^2 \tag{2.46}$$

with $d/c = -\frac{1}{6}$ has the form

$$x = \phi^{-1}x_0 + x_1$$
$$y = \phi^{-2}y_0 + \phi^{-1}y_1 + y_2 \tag{2.47}$$

where $y_0 = -\phi_t^2$, $y_1 = \phi_{tt}$, $y_2 = \frac{1}{12}(4\lambda - B - 3V - 3(\phi_{tt}/\phi_t)^2)$ and $x_0^2 = \phi_t^2 V$, $x_1 = -\frac{1}{2}(V_t/V + \phi_{tt}/\phi_t)V^{1/2}$. The variable V is

$$V = \{\phi; t\} + \lambda$$

where

$$\frac{1}{2}V_t^2 + \frac{1}{2}V^3 + (\frac{1}{3}B - 2A + \frac{2}{3}\lambda)V^2 + (\frac{1}{6}B^2 - \frac{2}{3}\lambda^2)V = 0. \tag{2.48}$$

This defines V as a Weierstrass elliptic function and $\phi = u_1/u_2$, where u_1, u_2 are solutions of the linear equation

$$u_{tt} = -\frac{1}{2}(V + \lambda)u.$$

2.5 THE HARRY DYM EQUATION.

The Schwarzian KdV equation [24] is

$$\frac{\phi_t}{\phi_x} + \{\phi; x\} = \lambda. \tag{2.49}$$

Under the change of variable

$$x \to \phi \quad t \to t \quad \phi \to x$$

$$\phi_x = 1/x_\phi \quad x_t = -\phi_t/\phi_x$$
$$\{\phi; x\} = -\phi_x^2\{x; \phi\}$$

and find

$$x_t + \lambda + x_\phi^{-2}\{x; \phi\} = 0. \tag{2.50}$$

The Harry Dym equation is found by substituting $W = x_\phi^2$, setting $\lambda = 0$,

$$W_t = 2\frac{\partial^3}{\partial \phi^3}W^{-1/2}.$$

The Schwarzian KdV sequence of equations has a Hamiltonian structure

$$\phi_t + M_x^n \circ \phi_x = 0 \tag{2.51}$$

where

$$M_x = \Omega_1 J_1$$

and

$$\Omega_1 = \phi_x \partial_x^{-1} \phi_x$$

$$J_1 = \partial_x \phi_x^{-1} \partial_x \phi_x^{-1} \partial_x.$$

The change of variable, with $\partial_\phi = \phi_x^{-1}\partial_x$ induces a Hamiltonian structure for the Harry Dym sequence [24]

$$x_t + L_\phi^n \circ 1 = 0 \tag{2.52}$$

where

$$L_\phi = \Omega_2 J_2$$

and

$$\Omega_2 = \partial_\phi^{-1}$$

$$J_2 = x_\phi^{-1} \partial_\phi^3 x_\phi^{-1}.$$

The rational solutions for the KdV equation can be found by the BT

$$\phi_{j+1,x} = \phi_j^2 / \phi_{j,x} \tag{2.53}$$

$$\phi_{j+1,t}/\phi_{j+1,x} + \phi_{j,t}/\phi_{j,x} = (\phi_{j,xx}/\phi_{j,x})^2 - 4\frac{\partial^2}{\partial x^2} \ln \phi_j$$

where

$$\phi_0 = x$$

$$\phi_1 = x^3 + 12t$$

$$\phi_2 = (x^6 + 60tx^3 - 720t^2)/x$$

etc. These also implicitly define algebraic solutions of the Harry Dym equation, ie. $x^3 = \phi - 12t$, etc. The N-Soliton solutions of the KdV correspond to the *N-Cusp* solutions of the Harry Dym equation.

2.6 SOME OBSERVATIONS.

In general and as illustrated above the Singular Manifold Method finds connections between equations, modified equations, equations in a sequence and the related cusp soliton equations. A primary application of the method would be to classify integrable systems and their relationships to each other.

The procedure as described has been applied to many systems. Among these are the Boussinesq sequence, the Caudrey-Dodd-Gibbon sequence and Hirota-Satsuma sequences. As described above the method does not directly apply to systems that do not have Bäcklund transformations. A system of this type is the Bullough-Dodd equation

$$U_{xt} = e^{U} - e^{-2U}$$

Surprisingly, the Bullough-Dodd equation is a specialization of an equation in the Caudrey-Dodd-Gibbon sequence. The Bäcklund transformation for this sequence does not preserve the form of the specialization. Hence, there is no Bäcklund transformation in the usual form for this equation [24,25].

A different sort of problem occurs for the nonlinear Schrödinger equation [10]. Here, the method works in so far as finding the Lax pair and Schwarzian modified equations are concerned. However, the modified equations themselves have a Bäcklund transformation that acts as a reduction on the system to a pair of Burgers equations.

The singular manifold method also applies to ordinary differential equations [23]. We have found for the Hénon-Heiles system a Bäcklund transformation and a three parameter class of associated solutions. The general solution depending on four parameters is not directly found by the singular manifold method. However, by direct formulation in terms of the Schwarzian derivative we have also found an identification of the integrable instances of the Hénon-Heiles system with the KdV and Caudrey-Dodd-Gibbon equations. This allows an immediate identification of the Lax pairs for the Hénon-Heiles system.

When the singular manifold method does not find a Lax pair with a parameter, it is possible to effect a resummation of terms in the truncation so as to allow an explicit dependence on the singular manifold itself. This procedure was introduced by Weiss in reference [9] and applied to the Caudrey-Dodd-Gibbon equation. Later, B. Gaffet in references [26,27] found a Bäcklund transformation for the Bullough-Dodd equation that is in the form of an expansion about the singular manifold. Whether this transformation is an involution or not, the form of the transformation involves a resummation of terms involving an explicit dependence on the singular manifold. Finally, Newell et al [28] and Gibbon et al [29] apply this same method to introduce a spectral parameter in the Lax pair for the Hénon-Heiles system. Therefore, a resummation of terms involved in the singular manifold truncation seems to be required for certain problems. In this sense the method requires further development to be truly algorithmic.

A recent paper by R. Conte [30] examines the question of why the *Painlevé-Bäcklund equations* that arise on application of the singular manifold method are invariant under the Mõbius group. This would explain why the Schwarzian derivative naturally occurs in the formulation of integrable systems, since, as explained earlier, the Schwarzian derivative is the unique differential invariant of the Mõbius group. This question is also examined in reference [9] where it is conjectured that the appearence of the Schwarzian derivative is related to the conformal geometry of the *complex Riemann sphere* and not as might be supposed, from an associated second order Lax pair operator. This question is also considered by Wilson [31].

3. Conditional or Partial Integrability.

For systems without single-valued expansions (Painlevé Property) , it is possible to constrain the *arbitrary* functions in the expansion so as to restore the single-valued behavior. Depending on the number of constraints the resulting expansion will represent a solution of reduced dimensions.

The systems of constraints are expressed as a system of partial differential equations for the previously arbitrary functions (data) in the expansion. For this system of partial differential equations we make the following conjecture.

Conjecture: *The constraint equations are completely integrable.*

To illustrate this point we will present several examples from reference [11].

The first example is the **Double Sine-Gordon** equation

$$u_{xt} = 4a\sin(u/2) + 4\sin u. \tag{3.1}$$

To apply the Painlevé analysis we set

$$V = e^{iu/2}$$

and find

$$VV_{xt} - V_xV_t = a(V^3 - V) + V^4 - 1. \tag{3.2}$$

The expansion about the singular manifold takes the form

$$V = \phi^{-1}\sum_{j=0}^{\infty}\sum_{k=0}^{\infty} V_{jk}\phi^j(\phi^2\ln\phi)^k. \tag{3.3}$$

From the recursion relations

$$V_{00}^2 = \phi_x\phi_t$$

$$V_{10} = -\frac{1}{2}\frac{\phi_{xt}}{\phi_x\phi_t}V_0 - \frac{a}{2}.$$

The arbitrary functions are ϕ and V_{20}. At the *resonance* $j = 2$ and $k = 0$ we can cancel the $\phi^2\ln\phi$ terms by requiring that ϕ satisfy the constraint equation

$$\phi_x^2\phi_{tt} - 2\phi_x\phi_t\phi_{xt} + \phi_t^2\phi_{xx} = 0. \tag{3.4}$$

This equation is identical to (1.21) and is integrable by a Legendre transformation [11]

$$\xi = \phi_x,\ \eta = \phi_t$$

$$x = W_\xi,\ t = W_\eta$$

$$\phi(x,y) + W(\xi,\eta) = x\xi + y\eta$$

and has the result that

$$\xi^2 W_{\xi\xi} + 2\xi\eta W_{\xi\eta} + \eta^2 W_{\eta\eta} = 0. \tag{3.5}$$

Let

$$\frac{d}{ds} = \xi\frac{\partial}{\partial\xi} + \eta\frac{\partial}{\partial\eta} \tag{3.6}$$

and (3.4) becomes

$$\frac{d^2}{ds^2}W = \frac{d}{ds}W. \tag{3.7}$$

The complete solution of (3.7) is

$$W = W_0 + W_1$$

where W_0 is homogenous of degree zero in (ξ,η) and W_1 is homogenous of degree one. This implies, using the definition of the Legendre transformation, that

$$\phi(x,t) = -W_0(\xi,\eta)$$

and

$$\xi x + \eta t = W_1(\xi,\eta).$$

The Legendre transformation is invertible when $\omega = \phi_{xx}\phi_{tt} - \phi_{xt}^2 \neq 0$. When $\omega = 0$, then $\phi = f(x+t)$ and we have a traveling wave form that is integrable for (3.1). Some simple closed form solutions when $\omega \neq 0$ are

$$\phi = f(x/t)$$

for arbitrary f.

In the cases where ϕ satisfies the constraint equation the expansion (3.3) becomes single-valued

$$V = \phi^{-2}\sum_{j=0}^{\infty} V_j\phi^j \tag{3.8}$$

where V_2 is arbitrary.

Next, we consider the N dimensional elliptic Sine-Gordon equation [11]

$$-\triangle u = \sin u \tag{3.9}$$

where

$$\triangle = \sum \partial_{x_j}^2 = \nabla^t \nabla.$$

Using $V = e^{iu}$

$$-V \triangle V + \nabla V \cdot \nabla V = \frac{1}{2}(V^3 - V) \tag{3.10}$$

The Painlevé expansion

$$V = \phi^{-2} \sum_{j=0}^{\infty} V_j \phi^j \tag{3.11}$$

is valid with arbitrary V_2 iff

$$\nabla\phi \cdot D \nabla\phi = 0 \tag{3.12}$$

where

$$D_{ii} = \frac{1}{2} \sum_{l=1, l\neq i}^{N} \sum_{m=1, m\neq i}^{N} (\phi_{lm}^2 - \phi_{ll}\phi_{mm}) \qquad (3.13)$$

$$D_{ij} = \sum_{m=1}^{N} (\phi_{ij}\phi_{mm} - \phi_{im}\phi_{jm}). \qquad (3.13)$$

The matrix D is symmetric and equation (3.12) is invariant under arbitrary scalings and translations in the independent variables, and orthogonal changes of independent variables

$$\nabla = B \nabla'$$

where

$$B^t = B^{-1}.$$

Using these properties it can be shown that the hypersurface M defined by the level sets

$$M = \{\hat{x}; \phi(\hat{x}) = \phi_0\}$$

has the property that principal curvatures of M as a manifold in R^N , $K_j; j = 1, \cdots, N-1$ verify the condition

$$K_1 K_2 + K_1 K_3 + \cdots + K_{N-2} K_{N-1} = 0. \tag{3.14}$$

When $N = 2$ the condition is trivial and (3.9) is integrable. When $N = 3$ equation (3.12) is

$$\phi_t^2(\phi_{xx}\phi_{yy} - \phi_{xy}^2) + \phi_x^2(\phi_{tt}\phi_{yy} - \phi_{yt}^2) + \phi_y^2(\phi_{tt}\phi_{xx} - \phi_{xt}^2)$$

$$+2\phi_x\phi_t(\phi_{ty}\phi_{yx} - \phi_{xt}\phi_{yy})$$
$$+2\phi_y\phi_t(\phi_{tx}\phi_{xy} - \phi_{yt}\phi_{xx})$$
$$+2\phi_x\phi_y(\phi_{xt}\phi_{yt} - \phi_{xy}\phi_{tt}) = 0. \tag{3.15}$$

Equation (3.15) may be integrated by a Legendre transformation, $\xi_1 = \phi_t$, $\xi_2 = \phi_x$, $\xi_3 = \phi_y$ and $t = W_{\xi_1}$, $x = W_{\xi_2}$, $y = W_{\xi_3}$ where

$$\phi(t, x, y) + W(\xi_1, \xi_2, \xi_3) = t\xi_1 + x\xi_2 + y\xi_3$$

and

$$W = W_0 + W_1$$

where W_0 and W_1 are homogenous of degree zero and one, respectively. Again, the form of ϕ might be used to find integrable reductions.

When $N \geq 4$ it is not known if (3.12) is integrable. *Our conjecture states that it is integrable for all N.*

4. Periodic fixed points of Bäcklund transformations.

A Bäcklund transformation (BT) maps solutions of a nonlinear system into solutions of the same nonlinear system. The BT is applied iteratively to define a sequence of solutions beginning from a known *seed* solution. Rather than study solutions defined in this manner we consider the periodic fixed points of the Bäcklund transformations. These are integrable systems of ordinary differential equations that define finite dimensional invariant manifolds of the flow associated with the infinite dimensional partial differential system. We find this flow is expressed as commuting flows on the invariant manifold [32].

4.1 THE KdV FIXED POINTS. In section 2 we have seen that the Schwarzian KdV equation

$$\phi_t/\phi_x + \{\phi; x\} = \lambda \tag{4.1}$$

has the Bäcklund transformations

$$\phi = (a\psi + b)/(c\psi + d) \tag{4.2}$$

and

$$\phi_x = 1/\psi_x. \tag{4.3}$$

$$\phi_t/\phi_x + \psi_t/\psi_x + (\frac{\phi_{xx}}{\phi_x})(\frac{\psi_{xx}}{\psi_x}) = 2\lambda. \tag{4.4}$$

We compose $\phi = -1/\psi$ and (4.3), (4.4) and get

$$\phi_{j+1,x} = \phi_j^2/\phi_{j,x} \tag{4.5}$$

$$\frac{\phi_{j+1,t}}{\phi_{j+1,x}}+\frac{\phi_{j,t}}{\phi_{j,x}}=(\frac{\phi_{j,xx}}{\phi_{j,x}})^2-4\frac{\partial^2}{\partial x^2}\ln\phi_j+2\lambda \tag{4.6}$$

where $j=1,2,3,\cdots,\mod N$. The periodic fixed points continue to define a strong Bäcklund transformation since

$$\phi_{j+1,xt}=\phi_{j+1,tx}$$

continues to imply that the set $\{\phi_j, j=1,2,3,\cdots,\mod N\}$ are solutions of (3.1).

Now, define the variables $\xi_j=\phi_{j,x}/\phi_j$ and find that (4.5) is $\xi_{j+1}\xi_j=\phi_j/\phi_{j+1}$. By logarithmic differentiation

$$\xi_{j+1,x}/\xi_{j+1}+\xi_{j,x}/\xi_j=\xi_j-\xi_{j+1} \tag{4.7}$$

where $j=1,2,3,\cdots\mod N$.

We claim that (4.7) is a completely integrable hamiltonian system.

First, we introduce the *circulant matrices* [33] A and B where

$$A=\mathbf{circ}[1,1,0,\cdots,0]$$

$$B=\mathbf{circ}[1,-1,0,\cdots,0]. \tag{4.8}$$

We note that $\det B=0$ for any N and the null vector of B is $(1,1,\cdots,1)^t$. We also define the variable $\hat{\beta}=(\beta_1,\beta_2,\cdots,\beta_N)^t$ where $\beta_j=\ln\xi_j$. Then, system (4.7) is

$$A\hat{\beta}_x=B\hat{\xi}. \tag{4.9}$$

Applying the null vector of B to (4.9) we find the *Casimir* integral

$$H_N=\prod_{j=1}^{N}\xi_j. \tag{4.10}$$

The complete set of independent integrals in involution for the system (4.9) are

$$H_{N-2m}=L^m\circ H_N \tag{4.11}$$

where

$$L=\sum_{j=1}^{N}\frac{\partial^2}{\partial\xi_j\partial\xi_{j+1}}. \tag{4.12}$$

To see this we need the following identities, for each j

$$\xi_j\partial_jL^m\circ H_N=L^m\circ H_N-m\partial_j(\partial_{j-1}+\partial_{j+1})L^{m-1}\circ H_N \tag{4.13}$$

$$\xi_j\partial_j(\partial_{j-1}-\partial_{j+1})L^m\circ H_N=(\partial_{j-1}-\partial_{j+1})L^m\circ H_N. \tag{4.14}$$

Then

$$\partial/\partial_x(L^m \circ H_N) = \sum \xi_{j,x}(\partial_j \circ H_N)$$
$$= \sum \frac{\xi_{j,x}}{\xi_j} \circ \xi_j \partial_j L^m \circ H_N$$
$$= -m \sum \frac{\xi_{j,x}}{\xi_j} \circ \partial_j(\partial_{j-1} + \partial_{j+1}) L^{m-1} \circ H_N$$
$$= -m \sum (\frac{\xi_{j,x}}{\xi_j} + \frac{\xi_{j+1,x}}{\xi_{j+1}}) \partial_j \partial_{j+1} L^{m-1} \circ H_N$$
$$= -m \sum (\xi_j - \xi_{j+1}) \partial_j \partial_{j+1} L^{m-1} \circ H_N$$
$$= -m \sum \xi_j \partial_j (\partial_{j+1} - \partial_{j-1}) L^{m-1} \circ H_N$$
$$= -m \sum (\partial_{j+1} - \partial_{j-1}) L^{m-1} \circ H_N$$
$$= 0.$$

The higher order flows

$$A\hat{\beta}_x = B diag[\xi_1, \cdots, \xi_N] \nabla H_k \tag{4.15}$$

where $k = N, N-2, N-4, \cdots$, $\beta_j = \ln \xi_j$ can be shown to commute by an argument similar to that above.

If N is odd then A is invertible and $\Omega = A^{-1}B = [0, -1, 1, -1, 1, \cdots, -1, 1]$ is anti-symmetric. The anti-symmetric matrix

$$M_{\xi} = diag[\xi_1, \cdots, \xi_N] \Omega diag[\xi_1, \cdots, \xi_N] \tag{4.16}$$

defines the bracket

$$\{G, H\} = (\nabla G)^t M_{\xi} \nabla H \tag{4.17}$$

and the systems (4.15) are

$$\hat{\xi}_x = M_{\xi} \nabla_{\xi} H_k. \tag{4.18}$$

4.2 CONSTRAINED FLOWS. If N is even the flow is *constrained* by the contraction with the null vector of A, $a_0 = [1, -1, \cdots, 1, -1]$, with (4.7). The constraint is

$$C_1 = \sum_{j=1}^{N} (-1)^j \xi_j = 0. \tag{4.19}$$

It can be shown that the flow for even N is equivalent to two copies of the periodic Toda lattice of period $N/2$ [34].

The constrained flow can be shown to have the form

$$\hat{\xi}_x = J_\xi \nabla_\xi H_1 - \frac{1}{H_1}\{C_1; H_1\} diag[\xi_1, \cdots, \xi_N] \nabla_\xi C_1$$

where

$$J_\xi = diag[\xi_1, \cdots, \xi_N] \Omega_{k+1} diag[\xi_1, \cdots, \xi_N]$$

$$\Omega_{k+1} = BG$$

$$G = \frac{1}{2n} \mathbf{circ}[0, 2, -4, 6, -8, \cdots, -2 + 2n]$$

$$B = \mathbf{circ}[1, -1, 0, \cdots, 0].$$

The constrained higher order flows have the form

$$\hat{\xi}_x = \Omega \nabla_\xi L^{m-1} \circ H_N$$

where

$$\Omega = BA^t = \mathbf{circ}[0, -1, 0, \cdots, 0, 1]$$

$$\Omega \nabla_\xi H_j = J_\xi \nabla_\xi L \circ H_j - \frac{1}{H_1}\{C_1; L \circ H_j\} diag[\xi_1, \cdots, \xi_N] \nabla_\xi C_1.$$

This identity verifies that the above form of the equations automatically preserves the constraint.

4.3 THE KAC-VAN MOERBEKE SYSTEM. In general, the periodic fixed points for the KdV equation can be transformed into the *Kac-Van Moerbeke* system. That is, let

$$A \begin{pmatrix} \xi_{1,x}/\xi_1 \\ \vdots \\ \xi_{N,x}/\xi_N \end{pmatrix} = B \begin{pmatrix} \xi_1 & & \\ & \ddots & \\ & & \xi_n \end{pmatrix} \nabla_\xi H$$

where

$$H = \sum \frac{1}{\xi_j \xi_{j+1}} = L \circ H_N / H_N.$$

Let $\theta_j = \ln(\xi_{j+1}/\xi_j)$ and get the Kac-Van Moerbeke system

$$\hat{\theta}_x = \Omega \nabla_\theta H$$

where $\Omega = BA^t$ and

$$H = \sum e^{\theta_j}.$$

For even N the constraint for equivalence with the KdV equation is

$$\sum (-1)^j \theta_j = 0.$$

4.4 FACTORING THE KdV FLOW. The KdV flow *factors* on the finite dimensional invariant manifold

$$A\hat{\beta}_x = Bdiag[\xi_1, \cdots, \xi_N]\nabla H_1 \tag{4.19}$$

$$A\hat{\beta}_t = Bdiag[\xi_1, \cdots, \xi_N]\nabla H_3. \tag{4.20}$$

The flows of the higher order flows factor according to (4.19) and

$$A\hat{\beta}_t = Bdiag[\xi_1, \cdots, \xi_N]\nabla H_k. \tag{4.21}$$

Again, for odd N the KdV flow is

$$\hat{\xi}_x = M_{\hat{\xi}}\nabla H_1 \tag{4.21}$$

$$\hat{\xi}_t = M_{\hat{\xi}}\nabla H_3 \tag{4.22}$$

where $\{H_1, H_3\} = 0$.

4.5 TRANSFORMATION TO CANONICAL FORM. The hamiltonian systems defined above may be transformed to the canonical form for hamiltonian systems as follows. To be specific, let N be odd,

$$\hat{\xi}_x = M_{\hat{\xi}}\nabla H,$$

and $\xi_j = e^{\theta_j}$. Find

$$\hat{\theta}_x = \Omega\nabla_\theta H$$

where $\Omega = A^{-1}B$ and $H = H(e^\theta)$. The eigenvalues of A, B, Ω are

$$\hat{\beta}_k = (1, r^k, \cdots, r^{k(N-1)})^t$$

where $r = e^{(2\pi i)/N}$. That is,

$$\Omega\hat{\beta}_k = \lambda_k\hat{\beta}_k$$

$$\lambda_k = (1 - r^k)/(1 + r^k) = -i\sigma_k$$

where

$$\hat{\beta}_k = \hat{s}_k + i\hat{t}_k$$

$$\hat{s}_k = (1, c_k, c_{2k}, \cdots, c_{k(n-1)})$$

$$\hat{t}_k = (0, s_k, s_{2k}, \cdots, s_{k(n-1)})$$

$$c_k = \cos(2\pi k/N)$$

$$s_k = \sin(2\pi k/N)$$

Then, with $n = 2m+1$

$$\Phi = \begin{pmatrix} 1 & 1 & \cdots & 1 & 0 & \cdots & 0 \\ 1 & c_1 & \cdots & c_m & s_1 & \cdots & s_m \\ \cdots & \cdots & \cdots & \cdots & \cdots & \cdots & \cdots \\ 1 & c_{2m} & \cdots & c_{2m^2} & s_{2m} & \cdots & s_{2m^2} \end{pmatrix}$$

find

$$\Phi^t \Omega \Phi = N \begin{pmatrix} 0 & & & & & & \\ & -\sigma_1 & & & & & \\ & & \ddots & & & & \\ & & & -\sigma_m & & & \\ & & & & \sigma_1 & & \\ & & & & & \ddots & \\ & & & & & & \sigma_m \end{pmatrix} \begin{pmatrix} 0 & I \\ -I & 0 \end{pmatrix}$$

Let

$$J_m = \begin{pmatrix} \sigma_m^{1/2} & & \\ & \ddots & \\ & & \sigma_m^{1/2} \end{pmatrix}$$

$$F = \frac{1}{n^{1/2}} \begin{pmatrix} 0 & \cdots & 0 \\ \vdots & -J_m & \\ 0 & & J_m \end{pmatrix}$$

$$\begin{pmatrix} 0 \\ \hat{p} \\ \hat{q} \end{pmatrix} = F\hat{\theta}$$

Then we find the *canonical equations*

$$\begin{pmatrix} \hat{p} \\ \hat{q} \end{pmatrix}_x = \begin{pmatrix} 0 & -I \\ I & 0 \end{pmatrix} \nabla H(\hat{p}\hat{q})$$

where

$$p_j = \sigma_j^{1/2} \sum_{k=0}^{2m} c_{jk}\theta_k$$

$$q_j = \sigma_j^{1/2} \sum_{k=0}^{2m} s_{jk}\theta_k$$

$$\theta_{j+1} = \sum_{k=1}^{m} \sigma_k^{-1/2}(s_{jk}q_k - c_{jk}p_k)$$

Therefore, our Bäcklund variables are related to the canonical variables by the finite Fourier transform.

5. Caustic surfaces, and Factoring the Laplace-Darboux transformation.

We are going to show, as a natural extension of the previous section, that the periodic fixed points of Bäcklund transformations factor the flow of the two dimensional periodic Toda lattice. The Toda lattice arises in the study of ***invariants*** of caustic surfaces and the geometry of the *Laplace-Darboux* transformation.

5.1 BÄCKLUND TRANSFORMATIONS. The KdV and Boussinesq systems are instances of the general system in component form [34]

$$\frac{\xi_{j,x}}{\xi_j} + \frac{\xi_{j+1,x}}{\xi_{j+1}} + \cdots + \frac{\xi_{j+p,x}}{\xi_{j+p}} = \xi_j - \xi_{j+p} \tag{5.1}$$

where $j = 1, 2, \ldots \mod N$. The KdV systems correspond to $p = 1$ and the Boussinesq to $p = 2$. Let the circulant forward shift matrix [33] be

$$C = \mathbf{circ}[0, 1, 0, 0, \ldots, 0].$$

In the N-vector form equations (5.1) are

$$A \begin{pmatrix} \frac{\xi_{1,x}}{\xi_1} \\ \vdots \\ \frac{\xi_{N,x}}{\xi_N} \end{pmatrix} = B\hat{\xi} \tag{5.2}$$

with

$$\begin{aligned} A &= I + C + \cdots + C^p \\ B &= I - C^p. \end{aligned}$$

The Casimir integrals of (5.2) correspond to the null vectors of B. The null vectors of A produce the constraints.

Associated with the principal Casimir, for any N

$$H_N = \prod_{j=1}^{N} \xi_j$$

we find the principal integrals of (5.2)

$$H_{N-pm-m} = L^m \circ H_N \tag{5.3}$$

where $m = 0, 1, 2, \ldots$ and

$$L = \sum_{j=1}^{N} \partial_{\xi_j} \partial_{\xi_{j+1}} \cdots \partial_{\xi_{j+p}}.$$

The systems (5.2) have a Hamiltonian structure

$$A\begin{pmatrix}\frac{\xi_{1,x}}{\xi_1}\\ \vdots \\ \frac{\xi_{N,x}}{\xi_N}\end{pmatrix} = B\begin{pmatrix}\xi_1 & & \\ & \ddots & \\ & & \xi_N\end{pmatrix}\nabla_{\hat{\xi}} H_1 \tag{5.4}$$

where $H_1 = \sum_{j=1}^{N} \xi_j$.

The higher-order equations associated with the integrals (5.3) are

$$A\begin{pmatrix}\frac{\xi_{1,x}}{\xi_1}\\ \vdots \\ \frac{\xi_{N,x}}{\xi_N}\end{pmatrix} = B\begin{pmatrix}\xi_1 & & \\ & \ddots & \\ & & \xi_N\end{pmatrix}\nabla_{\hat{\xi}} H_{N-pm-m}. \tag{5.5}$$

When A is invertible, then

$$\Omega = A^{-1}B$$

is an antisymmetric circulant matrix.

We have the systems

$$\hat{\xi}_{,x} = M_{\hat{\xi}}\nabla_{\hat{\xi}} H_1 \tag{5.6}$$

and

$$\hat{\xi}_{,x} = M_{\hat{\xi}}\nabla_{\hat{\xi}} H_{N-pm-m} \tag{5.7}$$

where

$$M_{\hat{\xi}} = \begin{pmatrix}\xi_1 & & \\ & \ddots & \\ & & \xi_N\end{pmatrix}\Omega\begin{pmatrix}\xi_1 & & \\ & \ddots & \\ & & \xi_N\end{pmatrix}$$

is the co-symplectic form.

5.2 THE LAPLACE-DARBOUX TRANSFORMATION. Darboux [20] has shown that the parameters (x, y) for surfaces in three dimensions can be defined so the coordinates (z_j) of the surface satisfy a partial differential equation of the form:

$$\partial^2 z/\partial x\partial y + a\partial z/\partial x + b\partial z/\partial y + cz = 0 \tag{5.8}$$

where (a, b, c) are functionals of the first fundamental form in the (x, y) parameters.

Under the gauge transformation $z \to \lambda z$, the form of (5.8) is preserved and:

$$h = \partial a/\partial x + ab - c$$

$$k = \partial b/\partial y + ab - c$$

are invariant.

The *Laplace transformation* of a surface is a partial factorization of (5.8) in terms of the *invariants*[42].

$$z_1 = \partial z/\partial y + az \tag{5.9}$$

$$\partial z_1/\partial x + bz_1 = hz$$

Equations (5.9) imply that z satisfies (5.8) and z_1 satisfies the system

$$\partial^2 z_1/\partial x\partial y + a_1 \partial z_1/\partial x + b_1 \partial z_1/\partial y + c_1 z_1 = 0$$

where

$$a_1 = a - \partial \ln h/\partial y \tag{5.10}$$

$$b_1 = b$$

$$c_1 = c - \partial a/\partial x + \partial b/\partial y - b\partial \ln h/\partial y.$$

From (5.10) the Laplace transformation of the invariants is

$$h_1 = 2h - k - \partial^2 \ln h/\partial x\partial y \tag{5.11}$$

$$k_1 = h.$$

Darboux [20] studied the periodic fixed points of the Laplace transformation and found that these surfaces are related as a sequence of *focal surfaces*. From (5.11), the periodic fixed points are

$$\partial^2 \ln h_j/\partial x\partial y = -h_{j+1} + 2h_j - h_{j-1} \tag{5.12}$$

where $j = 1, 2, 3, \cdots \pmod n$ and n is the order of the fixed point. The substitution

$$h_j = e^{\theta_{j+1} - \theta_j}$$

obtains the *two dimensional periodic Toda lattice*

$$\theta_{j,xy} = -e^{\theta_{j+1} - \theta_j} + e^{\theta_j - \theta_{j-1}} \tag{5.13}$$

We next consider some results of Laplace, Darboux and Moutard [20] that attempt to solve equation (5.8) through factorization of systems related by the Laplace transformation. In the process one finds solutions of various fixed and free end Toda lattices. The work of Laplace, as described in [20], dates from a paper published in 1773. The extensions of this work by Darboux and Moutard appeared from 1870 to 1880. The connections with surface theory and singularities of ray systems (caustics) are described in [20] and proposed as a method for the classification of surfaces. The simplest surfaces in this classification are surfaces whose *caustics* reduce to space curves. These are the *Cyclides of Dupin*. Pictures of these can be found in Hilbert and Cohn-Vossen [35]. These are conformally equivalent to tori. On the other hand, the period two fixed point of the Laplace transformation describes surfaces of constant negative curvature. The extension of this classification to the higher order fixed points was a goal of Darboux's program. In so far as I am aware this goal remains unrealized.

An example of the transformation of *invariants* is provided by the *Euler* equation

$$z_{xy} + \frac{\alpha}{x+y} z_x + \frac{\beta}{x+y} z_y + \frac{\gamma}{(x+y)^2} z = 0.$$

Under the Laplace transformation the sequence of invariants are

$$h_j = \frac{(\alpha + j)(\beta - j - 1) - \gamma}{(x+y)^2}$$

$$k_j = h_{j-1}$$

When h_j vanishes identically the associated equation for z_j factors and can be integrated by quadrature.

The period one fixed point of the Laplace transformation, $h_1 = h$ obtains for z

$$z_{xy} = z$$

and with, $h = e^w$, the *Liouville equation*, for the invariant,

$$w_{xy} = e^w.$$

The period two fixed point equation for the invariant $h_2 = h$, with $h = e^w$, obtains the *Sinh-Gordon* equation.

$$w_{xy} = \sinh(w).$$

When an invariant vanishes it is possible to integrate the entire sequence through the equation with null invariant by quadrature. Let $h_n = 0$ be the invariant for

$$z_{xy} + a_n z_x + b_n z_y + c_n z = 0.$$

Then, using equation (5.12)

$$h_{j-1} = 2h_j - h_{j+1} - \frac{\partial^2}{\partial xy} \ln h_j$$

and, if $h_{n-1} = \theta$, $h_{n-2} = 2\theta - \frac{\partial^2}{\partial xy} \ln \theta$, etc. Let a, b, c be the coefficients of the original system

$$z_{xy} + az_x + bz_y + cz = 0.$$

Then $b = b_n$, $a = a_n + \frac{\partial}{\partial y}\ln(hh_1\cdots h_n)$ and $c = a_x + ab - h$. Let $\gamma = \int(b_n dx - a_n dy)$, then the general integral of the original equation is

$$z = e^{-\int b_n dx}\frac{\partial}{h\partial x}\frac{\partial}{h_1\partial x}\cdots\frac{\partial}{h_{n-1}\partial x}\left(e^{\gamma}(X + \int Ye^{-\gamma}dy\right)$$

where X and Y are arbitrary functions of x and y, resp.

In general it is said that an equation with invariant $h = h_0$ has *rank* n when $h_n = 0$. An equation has *doubly finite rank* if $h_n = 0$ and $h_{-m} = 0$ for some n, m. The *characteristic* for equations of doubly finite rank is $\mu = m + n$. The characteristic is an invariant of the sequence of equations for $z_{-m}, \cdots, z_n$. It can be shown [20] that z_{-m} and z_n satisfy determinate ordinary differential equations in y and x, resp. The solution of the equation with invariant h can be expressed in a form (without partial quadrature)

$$z = AX + A_1X^{(1)} + \cdots + A_nX^{(n)} + BY + B_1Y^{(1)} + \cdots + B_mY^{(m)}$$

where X and Y are arbitrary functions of x and y, resp. and the A_j and B_j are determinate functionals of X, Y.

The above results of Laplace were reformulated by Darboux. For an equation of finite rank, $h_n = 0$ the forms

$$H_0 = \alpha$$

$$H_1 = \begin{vmatrix} \alpha & \alpha_x \\ \alpha_y & \alpha_{xy} \end{vmatrix}$$

$$H_2 = \begin{vmatrix} \alpha & \alpha_x & \alpha_{xx} \\ \alpha_y & \alpha_{yx} & \alpha_{yxx} \\ \alpha_{yy} & \alpha_{yyx} & \alpha_{yyxx} \end{vmatrix}$$

etc., where α is an arbitrary function of x, y, define the invariants, for $j = 0, 1, 2, \cdots$,

$$h_n = 0$$

$$h_{n-1} = -\frac{\partial^2}{\partial x\partial y}\ln(H_0)$$

$$h_{n-j} = -\frac{\partial^2}{\partial x\partial y}\ln(H_{j-1}).$$

These invariants satisfy the Toda lattice

$$\frac{\partial^2}{\partial x\partial y}\ln h_j = h_{j+1} - 2h_j + h_{j-1}$$

with fixed end $h_n = 0$.

If the lattice is of doubly finite rank with characteristic $\mu = \nu - 1 = m + n$ then $h_{-m} = 0$ implies $H_\nu = 0$. Or, the arbitrary function α satisfies the equation

$$\begin{vmatrix} \alpha & \alpha_x & \cdots & \frac{\partial^\nu \alpha}{\partial x^\nu} \\ \alpha_y & \frac{\partial^2 \alpha}{\partial x y} & \cdots & \cdots \\ \cdots & \cdots & \cdots & \cdots \\ \frac{\partial^\nu \alpha}{\partial y^\nu} & \cdots & \cdots & \frac{\partial^{2\nu} \alpha}{\partial x^\nu \partial y^\nu} \end{vmatrix} = 0$$

From the functional form of this equation, the general solution is

$$\alpha = \eta_1 \epsilon_1 + \cdots + \eta_\nu \epsilon_\nu$$

where η_j and ϵ_j are linearly independent functions of x and y, resp. In effect, this solves the fixed end Toda lattice.

If the invariants are equal, $h = k$, the equation is self adjoint and of the form

$$z_{xy} = \lambda z.$$

If this has finite rank in one direction, it has finite rank in both directions, Say, $h_n = h_{-n} = 0$. The equal invariant condition $h = k = h_{-1}$ corresponds to a free end Toda lattice

$$\frac{\partial^2}{\partial x \partial y} \ln h_j = h_{j+1} - 2h_j + h_{j-1}$$

with free end at $j = 0$. The characteristic of this system is $\mu = 2n$. Moutard found the general solution for systems of the above type by raising the characteristic. This is content of

Moutard's Theorem

Let ς be the general integral and let w be an integral of

$$\Delta(z) = \frac{1}{z} \frac{\partial^2 z}{\partial x \partial y} = \lambda.$$

Then, the general integral of

$$\Delta(z) = \frac{1}{z} \frac{\partial^2 z}{\partial x \partial y} = \Delta(\frac{1}{w}) = \lambda_1$$

is

$$z = \frac{1}{w} \int \{(w\varsigma_x - \varsigma w_x)dx - (w\varsigma_y - \varsigma w_y)dy\}.$$

If the rank of the first equation is n, the rank of the second equation is $n+1$. That is, $h_n = h_{-n} = 0$ for the first, implies $h_{n+1} = h_{-n-1} = 0$ for the second.

Furthermore, the general solution of any equal invariant system of finite rank can be found by this procedure.

The work of Moutard is described in [20]. According to Darboux Moutard's major work on this topic was lost in the fires set during the suppression of the Paris Commune in 1871. The work presented in [20] is a reconstruction based on earlier published notes.

5.3 FACTORING THE TODA LATTICE. We now find Bäcklund transformations for the *Darboux equations* (5.12) and the Toda lattice equations (5.13). With reference to systems (5.6) and (5.7), without loss of generality normalize the Casimir, $H_N = 1$, and set

$$\hat{\xi}_{,x} = M_{\hat{\xi}} \nabla_{\hat{\xi}} H_1 \tag{5.14}$$

$$\hat{\xi}_{,y} = M_{\hat{\xi}} \nabla_{\hat{\xi}} (H_{N-p-1}/H_N) \tag{5.15}$$

where $H_{N-p-1} = L \circ H_N$. Then, let $\xi_j = e^{\psi_j - \psi_{j+1}}$ and find that (5.14), (5.15) imply

$$\psi_{j,xy} = e^{\psi_{j+p} - \psi_j} - e^{\psi_j - \psi_{j-p}} \tag{5.16}$$

where $j = 1, 2, 3, \cdots \pmod N$.

To see this let $\xi_j = e^{\theta_j}$ and find

$$\hat{\theta}_{,x} = \Omega \nabla_\theta H_1$$

$$\hat{\theta}_{,y} = \Omega \nabla_\theta G$$

where

$$G = \sum e^{-\theta_j - \theta_{j+1} - \cdots - \theta_{j+p}}$$

$$= \sum 1/\xi_j \xi_{j+1} \cdots \xi_{j+p} = H_{N-p-1}/H_N.$$

It can be shown that

$$\hat{\theta}_{,xy} = C^{-p}(I - C^p)(I - C) \begin{pmatrix} e^{-\theta_1 - \theta_2 - \cdots - \theta_p} \\ \vdots \\ e^{-\theta_N - \theta_1 - \cdots - \theta_{p-1}} \end{pmatrix}.$$

Let $\hat{\theta} = (I - C)\hat{\psi}$ and find (5.16).

When $p = 1$ (5.16) are the Toda lattice of period N. If N and p are relatively prime (5.16) is again a Toda lattice of length N. If $N = mp$ (5.16) is p distinct lattices of length m. When N and p have common factors (5.16) there is one lattice for each distinct orbit of translation by p (mod N). In all cases the set of fields ξ_j are directly related to the set of invariants H_j. When A is not invertible we find for equations (5.4) and (5.5) a similar connection with the Toda lattice. In this case one must take into account the *constraints* that apply to these systems to obtain a valid correspondence. A more comprehensive analysis of (5.4) and (5.5) as completely integrable hamiltonian systems and the properties of their related surfaces is currently in progress [36,37].

Consideration of the form of (5.6), (5.7) and the possible relations between p and N determine that for a lattice of fixed length m there will exist an infinite sequence of distinct Bäcklund transformations. For instance, we have a Bäcklund transformation for a lattice of length m when $N = pm$ for $p = 1, 2, 3, \cdots$.

Finally, the Bäcklund transformation for the Toda lattice that was reported in ref.[38] corresponds in our formulation to the system (5.7) with $p = 1$

$$\hat{\xi}_{,x} = M_{\hat{\xi}} \nabla_{\hat{\xi}} H_{N-2}$$

$$\hat{\xi}_{,y} = -M_{\hat{\xi}} \nabla_{\hat{\xi}} H_{N-2}.$$

6. Conclusions and Comments.

The *Painlevé test*, as described in section 1, is proposed as a sufficient condition for integrability. The Painlevé Property is a statement about how the solutions behave as functionals of the data in the neighborhood of a singularity manifold and not a statement about the data itself. Examples of this phenomenon are examined. Expansions about *characteristic manifolds* are required to be single-valued. Essential singularities are found to be determined by certain *Psi series* involving non-constant leading orders and resonances.

The *singular manifold method* finds Bäcklund transformations by truncating the functional Laurent series after the constant level term. This results in the formulation of *modified* equations in terms of the *Schwarzian derivative*. The Miura transformation between the modified and given system can be used to determine the *Lax pair* and recursion operators for the gradients of conserved densities. The symmetries of the modified equation and the invariance under the Möbius group are a form of Bäcklund transformation for the modified equation. The periodic fixed points of these Bäcklund transformations are finite dimensional invariant manifolds for the flow of the system. The dynamics occurs as commuting hamiltonian flows on this finite dimensional manifold.

Constrained *Psi* series expansions are applied to non-Painlevé systems. The constraints are expressed as nonlinear partial differential equations. We conjecture that these are integrable and provide the integrable reductions of the original system.

In this volume an extensive review of singularity analysis and its relation to complete and partial integrability is presented by A. Ramani, B. Grammaticos and M. Kruskal. A review of the work of Yoshida on Kovalevsky exponents is presented by D. Bessis.

Acknowledgments.

This work was supported in part by the Institute for Mathematics and its Applications with funds provided by the National Science Foundation, and by the National Science Foundation Research Grant No. DMS-8607576.

Support was also provided by the NATO Advanced Study Institute during the author's stay at Les Houches, March 1989.

REFERENCES

1. S. Kowalevskaya, Acta Mathematica **14** (1890), 81.
2. E.L. Ince, "Ordinary Differential Equations," Dover, New York, 1956.
3. M.J Ablowitz, A. Ramani and H. Segur,, J. Math. Phys. **21** (1981), 715-721.
4. J.B. McLeod and P.J. Olver, SIAM J. Math. Anal. **14** (1983), 488-506.
5. W.F. Osgood, "Topics in the Theory of Functions of Several Complex Variables," Dover, New York, 1966.
6. J. Weiss, M. Tabor and G. Carnevale, J. Math. Phys. **24** (1983), 522-526.
7. E. Hille, "Ordinary Differential Equations in the Complex Domain," John Wiley, New York, 1976.
8. J. Weiss, J. Math. Phys. **24** (1983), 1405-1413.
9. J. Weiss, J. Math. Phys. **25** (1984), 13-24.
10. J. Weiss, J. Math. Phys. **26** (1985), 258-269.
11. J. Weiss, J. Math. Phys. **25** (1984), 2226-2235.
12. R.S. Ward, Phys. Lett. **102A** (1984), 279-282.
13. R.S. Ward, Phil. Trans. R. Soc. Lond. A **315** (1985), 451-457.
14. R. Courant and D. Hilbert, "Methods of Mathematical Physics," Interscience, New York, 1962.
15. E.V. Doktorov and S. Yu. Sakovich, J. Phys. A **18** (1985), 3327-3334.
16. P.A. Clarkson, Phys. Lett. **109A** (1985), 205-208.
17. P.A. Clarkson, Physica **18D** (1986), 209-210.
18. Willy Hereman, private communication.
19. M. Lavie, Canadian J. Math. **21** (1969), 235.
20. G. Darboux, "Théorie Générale des Surfaces, II," Chelsea, New York, 1972.
21. P. Deift and E. Trubowitz, Comm. Pure Appl. Math. **32** (1979), 121-151.
22. J. Weiss, Phys. Lett. **102A** (1984), 329-331.
23. J. Weiss, Phys. Lett. **105A** (1984), 387-389.
24. J. Weiss, J. Math. Phys. **27** (1986), 1293-1305.
25. M.J. Ablowitz and H. Segur, "Solitons and the Inverse Scattering Transformation," SIAM, Philadelphia, 1981.
26. B. Gaffet, Physica **26D** (1987), 123-139.
27. B. Gaffet, J. Phys. A **21** (1988), 2491-2531.
28. A.C. Newell, M. Tabor and Y.B. Zeng, Physica **29D** (1987), 1-68.
29. J.D. Gibbon, A.C. Newell, M. Tabor and Y.B. Zeng, Nonlinearity **1** (1988), 481-490.
30. R. Conte, Phys. Lett. **134A** (1988), 100-104; Phys. Lett. **140A** (1989), 383-390.
31. George Wilson, Phys. Lett. **132A** (1988), 445-450.
32. J. Weiss, J. Math. Phys. **27** (1986), 2647-2656.
33. P. Davis, "Circulant Matrices," Wiley, New York, 1983.
34. J. Weiss, J. Math. Phys. **28** (1987), 2025-2039.
35. D. Hilbert and S. Cohn-Vossen, "Geometry and the Imagination," Chelsea, New York, 1952.
36. J. Weiss, Phys. Lett. **137A** (1989), 365-368.
37. J. Weiss, work in progress.
38. A.P. Fordy and J. Gibbons, Commun. Math. Phys. **77** (1980), 21-30.

DIFFERENTIAL GEOMETRY TECHNIQUES FOR SETS OF NONLINEAR PARTIAL DIFFERENTIAL EQUATIONS

Frank B. Estabrook
169-327 Jet Propulsion Laboratory
California Institute of Technology
4800 Oak Grove Drive
Pasadena, California 91109, U.S.A.

I. Introduction

In Section II of these notes, we will discuss how exterior differential systems geometrically express sets of first order partial differential equations. We emphasize the Cartan integer characters of a system, calculated from the ranks of successive derived sets of homogeneous linear algebraic equations, which show how local integration of generic solutions can proceed. Criteria emerge for the differential system to be well-posed, and for separating coordinates into "independent" and "dependent" variables. Invariant - or coordinate-independent - formulations of associated ordinary differential systems ("Cauchy characteristic" vector fields), internal symmetry groups ("isovectors"), variational principles, etc., can all also be systematically considered.

Although Cartan theory at first appears abstract, the major intent of these notes is to try and show how it is in fact a useful technique of applied mathematics. It leads into a heuristic series of manipulations, that can expose the algebraic structures underlying a set of partial differential equations, structures which can be quite obscured in many, if not all, coordinate systems.

Given an exterior differential ideal, I, generated by a closed set of forms, one of the most important manipulations is the consistent adjoining of additional generators, and perhaps also of additional coordinates or basis forms. In Section III we define the resulting general concepts of specialization and prolongation. These are techniques for finding

R. Conte and N. Boccara (eds.), Partially Integrable Evolution Equations in Physics, 413–434.

consistent sub-families of solutions that still are themselves generically rich and well-posed and for introducing potentials or other usefully consistent auxiliary fields.

In Section IV we will show as an extended example how the calculation goes, starting at the very beginning, for the (of course very well known) Korteweg-de Vries equation. This may be unduly tedious, but only by working through such differential-form calculations can one truly acquire an appreciation of the depth of Cartan's formulation. The KdV is of course quite simple algebraically, as only two independent variables arise, a fact that leads to 1-form prolongation and to Lie algebraic relations. Moreover, the Lie algebra is exceptionally simple, viz., sl(2,R). In the last fifteen years quite a number of approaches to the KdV have been fruitful, and it seems that no one of them serves well to expose all the properties of this remarkable equation. The Cartan approach does, however, lead rather naturally to the sl(2,R) pseudopotential satisfying the related mKdV equation, to the Lax equations, to the "singularity-manifold" equation, to the full sl(2,R)-based Kaĉ-Moody Lie algebra underneath it all, and to the Bäcklund transformation. Yet even this case, as we will see, is by no means completely algorithmic, still requiring insight as one proceeds, like any good technique of applied mathematics.

II. Cartan Theory of Partial Differential Equations [1-4]

Consider a closed differential ideal I in an n-dimensional manifold, generated by sets of 1-forms α, 2-forms β, 3-forms γ, and so on, all of which are explicitly given in a local basis of 1-forms, say ω^i, i = 1...n. This frame need not be "holonomic" - derived from a coordinate patch (in which case we would have $\omega^i = dx^i$). If the ideal is <u>invariantly</u> given, the expansion of this set of generators of I, $\{\alpha,\beta,\gamma \ldots\}$, on the basis ω^i has constant numerical coefficients.

Integral manifolds of I are subspaces such that I, restricted to them, vanishes; that is $\alpha_* = 0$, $\beta_* = 0$, etc.

The simplest integral submanifolds of I are one-dimensional; denote a generic one of these by $\mathcal{V}_1$. They can each in principle be found by integration, beginning at an arbitrary initial point with a vector, say $\vec{V}_1$, chosen so that at successive points along the resultant trajectory $\mathcal{V}_1$

$$\vec{V}_1 \rfloor \alpha = 0 \qquad \text{rank } s_0 \tag{1}$$

This is a set of linear homogenous equations for the components of $\vec{V}_1$ - in any independent basis - of rank, say s_0 (just the number of independent 1-forms α). So a choice of $\ell_1 = n - s_0$ functions of an autonomous variable y^1 will thus have to be made. We could express the result as an equivalent set of autonomous ordinary differential equations for the trajectory in terms of holonomic components of $\vec{V}_1$; i.e., if $V_1^i = \vec{V}_1 \rfloor dx^i$

$$\frac{dx^i}{dy^1} = V_1^i \qquad \vec{V}_1 \rfloor dy^1 = 1 \tag{2}$$

The result is an explicit map of $\mathcal{V}_1 \to X$, the manifold of x^i. We could normalize y^1 as shown, and finally substitute back to find the induced map $\vec{V}_1 \to \frac{\partial}{\partial y^1}$.

Now use $\mathcal{V}_1$ as an initial boundary for a second set of integrations, to find a one parameter family of trajectories that will form a two-dimensional integral manifold $\mathcal{V}_2$. Start at each point of $\mathcal{V}_1$, labeled with y^1, and there and at successive points along each resultant trajectory, choose a vector $\vec{V}_2$ such that

$$\begin{aligned} &\vec{V}_2 \rfloor \alpha = 0 \\ &\vec{V}_2 \rfloor \vec{V}_1 \rfloor \beta = 0 \end{aligned} \qquad \text{rank } s_0 + s_1 \tag{3}$$

Given $\vec{V}_1$, this is a set of linear homogeneous equations for the components of $\vec{V}_2$. The rank cannot be less than s_0, so with Cartan we denote it $s_0 + s_1$. Now one solution of (3) can be just $\vec{V}_1$, so for the construction we must have $\ell_2 = n - s_0 - s_1 > 1$, and then that many functions of a second autonomous variable, y^2, can be chosen during the integrations along the trajectories of $\vec{V}_2$. This results in a two-dimensional integral manifold $\mathcal{V}_2$, as clearly from (1) and (3) all of I vanishes when restricted to $\mathcal{V}_2$.

This second set of integrations along $\vec{V}_2$ may not appear to have been completely specified, however, as we only had previously determined $\vec{V}_1$ at a boundary. We also need the components of $\vec{V}_1$, as parameters, at each point of the integrations along the trajectories of $\vec{V}_2$. Cartan's deep insight was that since we are working with a <u>closed</u> ideal I, it is consistent to require $[\vec{V}_1, \vec{V}_2] = 0$ as we go. This determines $\vec{V}_1$ as being "dragged along" the $\vec{V}_2$ congruence. Now since

$$\pounds_{V_2}(\vec{V}_1 \rfloor \alpha) = [\vec{V}_2, \vec{V}_1] \rfloor \alpha + V_1 \rfloor (d(\vec{V}_2 \rfloor \alpha) + \vec{V}_2 \rfloor d\alpha)) \tag{4}$$

we see that all three right-hand terms vanish by our construction, and so the condition $\vec{V}_1 \rfloor \alpha = 0$, true initially, is itself dragged along $\vec{V}_2$ and preserved. $\vec{V}_1$ thus is simultaneously constructed throughout the 2-dimensional manifold as everywhere belonging to 1-dimensional integral manifolds, as indeed we initially took it. The result is that one has intersecting families of 1-dimensional integral manifolds $\mathcal{V}_1$ from both $\vec{V}_1$ and $\vec{V}_2$, that these are 2-forming, and that the 2-manifolds $\mathcal{V}_2$ are also integral manifolds. Normalized autonomous variables y^1 and y^2 can be introduced by writing the holonomic components of the vectors as $V^i_1 = \frac{\partial x^i}{\partial y^1}$, $V^i_2 = \frac{\partial x^i}{\partial y^2}$ and the construction guarantees that these are consistent. They map $\vec{V}_1$ and $\vec{V}_2$ to $\frac{\partial}{\partial y^1}$ and $\frac{\partial}{\partial y^2}$.

The construction of 3-dimensional integral manifolds proceeds entirely analogously. This time we begin with a bounding 2-manifold $\mathcal{V}_2$, everywhere containing $\vec{V}_1$ and $\vec{V}_2$, and search for a $\vec{V}_3$ at each point of $\mathcal{V}_2$ such that

$$\begin{aligned} &\vec{V}_3 \rfloor \alpha = 0 \\ &\vec{V}_3 \rfloor \vec{V}_1 \rfloor \beta = 0 \\ &\vec{V}_3 \rfloor \vec{V}_2 \rfloor \beta = 0 \\ &\vec{V}_3 \rfloor \vec{V}_2 \rfloor \vec{V}_1 \rfloor \gamma = 0 \end{aligned} \tag{5}$$

The rank is denoted $s_0 + s_1 + s_2$, so s_2 must be ≥ 0, and we can proceed if $\ell_3 > 2$. As we integrate trajectories of $\vec{V}_3$, we drag the integral 2-manifold $\mathcal{V}_2$ along by $[\vec{V}_3, \vec{V}_1] = 0$, $[\vec{V}_3, \vec{V}_2] = 0$, which preserves $[\vec{V}_1, \vec{V}_2] = 0$ and satisfies equations analogous to (4) for its propagation along $\vec{V}_3$.

Integral manifolds constructed in this way, from nested integral manifolds of lower dimensionality, are called regular. Not all integral manifolds are regular - various "singular" manifolds can also exist. But in any event, we recognize with Cartan that the positive integers s_0, s_1, s_2, are numerical concomitants of the closed ideal I and demonstrate in principle the integration of generic solutions.

Now at each integration we add more linear equations and further construction can only become more constrained, $\ell_p \leq \ell_{p-1}$. But if we have a p-1-dimensional integral manifold, we of course have p-1 trivial solutions of the linear homogeneous equations for $\vec{V}_p$ - thus we need $\ell_p > p-1$ to proceed to construct $\mathcal{V}_p$. The process must terminate, so the regular integral manifolds of I must have a maximum dimension: this is Cartan's genus, g. If $\ell_g > g - 1$ but $\ell_{g+1} \leq g$ we cannot proceed past g dimensions. This says in particular that if

$$\ell_{g+1} = \ell_g = n - (s_0 + \ldots s_{g-1}) = g, \quad \text{i.e., } s_g = 0, \tag{6}$$

there is no freedom left - no arbitrary function other than normalization of $\vec{V}_g$ or the last autonomous variable y^g - in the final construction of maximum dimensional integral manifolds, $\mathcal{V}_g$.

Although there is no unique ideal I to represent a given set of partial differential equations, limiting the ideals considered to those that satisfy the criterion of Eq. (6) makes the choice of I as a practical matter quite limited. We denote such ideals as being "well-posed". The associated sets of partial differential equations will include all integrability conditions, and will be neither over nor under determined.

It is also important to explicitly calculate which basis 1-forms remain independent when restricted to the maximal dimension integral submanifolds $\mathcal{V}_g$. A set of g of these - neither they nor any linear combinations of them vanishing - are then a suitable 1-form basis to span $\mathcal{V}_g$. Cartan denotes these as involutory, and when holonomic they belong to a possible set of independent variables for a set of partial differential equations locally equivalent to I.

III. Prolongation and Specialization

Hugo Wahlquist and I originally introduced prolongation [5-6] for sets of partial differential equations with $g = 2$ (two independent variables), in which auxiliary 1-forms, say α', are adjoined to the ideal I, to form an ideal $I' = \{I, \alpha'\}$ that is still closed and well-posed. These also involve new variables, which become essentially potential fields (or, more generally, pseudopotentials or "Miura" transforms) in the integral manifolds of I'. This prolongation process can now be formulated in the general case. We seek to adjoin additional forms, say α', β', γ'..., of degrees 1, 2, 3, etc., involving n' additional coordinates $x' = \{x^{n+1}, \cdots x^{n+n'}\}$ and basis 1-forms, to construct an ideal $I' = \{I, \alpha', \beta', \ldots\}$ with the following properties: (1) $dI' \subset I'$, (2) $\ell_1' \geq \ell_1$, $\ell_2' \geq \ell_2$, $\ell_3' \geq \ell_3$, etc., (3) $g' = g$ and (4) I' and I involutory with respect to the same g bases. In a solution manifold, the adjoined prolongation variables are seen as auxiliary fields, functions of the same independent variables. In recent work on fields in 3 and 4 dimensional Riemannian backgrounds [7] this more general approach is important and used, but it is not as algorithmic as when $g = 2$.

If the second of these conditions is not fulfilled, but the others are, the enlarged exterior differential system I' is itself still well-posed and of interest in its own right. Since the freedom of the Cartan construction is at some point limited, however, solutions of such an I' must be <u>specialized</u>, that is, must be a consistent and well formulated <u>subset</u> of solutions of the ideal I, while at the same time an interesting set of auxiliary or pseudopotential variables is now also solved for. Examples of this are similarity solutions, and, in Riemannian geometry, the Ricci-flat and Einstein-Maxwell solutions.

IV. Manipulations of the KdV Exterior Differential System

1. <u>The KdV Exterior Differential System.</u>

$I = \{\beta\}$ is generated by three 2-forms:

$$\beta_1 \equiv du \wedge dt - z\,dx \wedge dt$$

$$\beta_2 \equiv dz \wedge dt - p\,dx \wedge dt \tag{7}$$

$$\beta_3 \equiv - du \wedge dx + dp \wedge dt + 12uz\, dx \wedge dt$$

n=5, s_0=0, s_1=3, g=2; x and t involutory. These are the generating forms introduced in Reference [5], and in the following we will stay as close as possible to the formulas, symbols for pseudopotentials, etc. of that paper.

We can search algorithmically for a set of 1-forms ω^i, involving an equal number of new auxilliary variables y^i (the range of i is unspecified) that together with I generates a closed prolonged ideal $I'=\{\beta,\omega^i\}$, by setting

$$\omega^i \equiv dy^i + F^i(u,z,p,y)\,dx + G^i(u,z,p,y)\,dt. \tag{8}$$

From $d\omega^i \subset I'$ we find, after considerable algebra and integration over the initial variables, that F^i and G^i must be of the form

$$\begin{aligned} F^i &= 2X^i_1 + 2uX^i_2 + 3u^2X^i_3 \\ G^i &= -2(p+6u^2)X^i_2 + 3(z^2 - 8u^3 - 2up)X^i_3 \\ &\quad + 8X^i_4 + 8uX^i_5 + 4u^2X^i_6 + 4zX^i_7 \end{aligned} \tag{9}$$

where the X^i_A are functions only of the prolongation variables y^i. Moreover, regarding X^i_A as components of a vector X_A in the space of the y^i, so $X^i_A = X_A \rfloor dy^i$ the remaining pde's that they must satisfy are writeable in terms of Lie brackets:

$$\begin{aligned} &[X_1, X_3] = [X_2, X_3] = [X_1, X_4] = [X_2, X_6] = 0 \\ &[X_1, X_2] = -X_7; \;\; [X_1, X_7] = X_5; \;\; [X_2, X_7] = X_6 \\ &[X_1, X_5] + [X_2, X_4] = 0; \;\; [X_3, X_4] + [X_1, X_6] + X_7 = 0 \end{aligned} \tag{10}$$

This is an <u>incomplete</u> Lie algebra, as not all commutators are given, though more can be sought by introducing higher order - or nested - commutators, as we will see below.

The factored form that is formed by substitution back into Eq. (8) is the same as that which arises in the theory of connections on principal fiber bundles [18], except here the algebra is not completely known. Consequently, the requirement that $d\omega^i \subset I'$ gives not only the curvature 2-forms Ω involving the $d\eta^A$ but also "constraint" 2-forms Σ algebraic in the η^A. The entire set of 2-forms is closed modulo itself and constitutes an invariant differential system [12] for the KdV equation:

$$\omega^i \equiv dy^i + X_A \rfloor dy^i \, \eta^A$$

$$d\omega^i \equiv (d\eta^A - \tfrac{1}{2}\Gamma^A_{BC}\,\eta^B \wedge \eta^C) X_A \rfloor dy^i \mod \omega^i \tag{11}$$

$$\rightarrow \Omega \text{ (structure 2-forms)} + \Sigma \text{ (algebraic 2-forms)}$$

where

$$\begin{aligned} \eta^1 &= 2\,dx \\ \eta^2 &= 2u\,dx - 2(p+6u^2)\,dt \\ \eta^3 &= 3u^2\,dx + 3(z^2 - 8u^3 - 2up)\,dt \\ \eta^4 &= 8\,dt \\ \eta^5 &= 8u\,dt \\ \eta^6 &= 4u^2\,dt \\ \eta^7 &= 4z\,dt \end{aligned} \tag{12}$$

2. The Infinite Lie Algebra (KdV-12+).

Already in [5] the additional Lie algebra generator $X_8 \equiv -[X_3, X_4]$ was usefully introduced and further Lie commutators derived. Shortly thereafter W. Kinnersley pointed out to us that a Heisenberg algebra was generated, and independently W. Shadwick found a consistent set of 11 generators [20]. At each extension, more commutators were explicitly found. Peter Gragert [13] at the Twente Institute of Technology developed a number of symbolic computing programs for handling differential forms and Lie algebraic systems, and pushed on to 14 and eventually 17 dimensional sets for the KdV system. Using this printout, H. van Eck [21] and I independently found the entire prolongation algebra to be the sum of a 5-dimensional Heisenberg radical and the infinite Kaĉ-Moody algebra denoted $A_1^{(1)+}$. To be explicit, define

$$
\begin{aligned}
X_8 &= -[X_3, X_4] = [X_1, X_6] + X_7 \\
X_9 &= -[X_2, X_4] = [X_1, X_5] \\
X_{10} &= [X_5, X_7] = -[X_2, X_9] - X_5 \\
X_{11} &= -[X_1, X_9] \\
X_{12} &= [X_5, X_9]
\end{aligned} \tag{13}
$$

and derive:

$$
\begin{aligned}
&[X_3, X_A] = 0, \ A \neq 4 \\
&[X_2, X_5] - [X_1, X_6] = 0 \\
&[X_6, X_7] = X_6 \\
&[X_4, X_6] + [X_2, X_4] = 0 \\
&[X_1, X_6] + [X_5, X_6] = 0
\end{aligned} \tag{14}
$$

plus many more. The radical is

$$
\begin{aligned}
\overset{1}{H} &= X_1 - X_{10} \\
\overset{2}{H} &= X_2 - X_6 \\
\overset{3}{H} &= X_3 \\
\overset{4}{H} &= X_4 + X_{12} \\
\overset{5}{H} &= X_8
\end{aligned} \tag{15}
$$

$[\overset{1}{H}, \overset{2}{H}] = [\overset{3}{H}, \overset{4}{H}] = -H^5$ and all other commutators zero

and the seven needed (or minimal) set of Kaĉ-Moody generators are

$$
\begin{aligned}
\overset{0}{V} &= X_6 \\
\overset{0}{W} &= X_7 - X_8
\end{aligned}
$$

$$\overset{0}{U} = (X_5 - X_{10})/2$$
$$\overset{1}{V} = X_5 + X_{10} \qquad (16)$$
$$\overset{1}{W} = -X_9$$
$$\overset{1}{U} = (X_{11} + X_{12})/2$$
$$\overset{2}{V} = X_{11} - X_{12}$$

The entire Kac-Moody algebra is graded:

$$[\overset{i}{V}, \overset{j}{W}] = \overset{i+j}{V} \qquad (17)$$
$$[\overset{i}{U}, \overset{j}{V}] = \overset{i+j}{W}$$
$$[\overset{i}{W}, \overset{j}{U}] = \overset{i+j}{U}$$

and has a projection on $s\ell(2,R)$:

$$\overset{i}{W} \to (-\lambda)^i \overset{o}{W}$$
$$\overset{i}{U} \to (-\lambda)^i \overset{o}{U} \qquad (18)$$
$$\overset{i}{V} \to (-\lambda)^i \overset{o}{V}$$

3. Explicit prolongation forms.

Using the original coordinates of [5], viz.

$$y_1 = \xi,\ y_2 = s,\ y_3 = v,\ y_4 = \tau,\ y_5 = r,\ y_6 = q,\ y_7 = w,\ y_8 = y \qquad (19)$$

a realization of the KdV generators is

$$\overset{1}{H} = \frac{1}{2}\frac{\partial}{\partial \xi} + \frac{1}{2} rw \frac{\partial}{\partial r}$$
$$\overset{2}{H} = \frac{1}{2}\frac{\partial}{\partial w}$$

$$\overset{3}{H} = \frac{1}{3}\frac{\partial}{\partial q}$$
$$\overset{4}{H} = \frac{1}{8}\frac{\partial}{\partial \tau} - \frac{3}{4}rq\frac{\partial}{\partial r}$$
$$\overset{5}{H} = \frac{1}{4}r\frac{\partial}{\partial r} \tag{20}$$
$$\overset{0}{V} = \frac{\partial}{\partial y}$$
$$\overset{0}{W} = y\frac{\partial}{\partial y} + \frac{1}{2}\frac{\partial}{\partial v}$$
$$\overset{0}{U} = -\frac{1}{2}y^2\frac{\partial}{\partial y} - \frac{1}{2}y\frac{\partial}{\partial v} - \frac{1}{2}e^{2v}\frac{\partial}{\partial s}$$

with the rest given by Eq. (18). Then from

$$\begin{aligned} X_A \rfloor dy^i\eta^A = {} & (\overset{1}{H} + \tfrac{1}{2}\overset{1}{V} - \overset{o}{U})\rfloor dy^i\eta^1 + (\overset{2}{H} + \overset{o}{V})\rfloor dy^i\eta^2 \\ & + \overset{3}{H}\rfloor dy^i\eta^3 + (\overset{4}{H} - \overset{1}{U} + \tfrac{1}{2}\overset{2}{V})\rfloor dy^i\eta^4 \\ & + (\overset{o}{U} + \tfrac{1}{2}\overset{1}{V})\rfloor dy^i\eta^6 + (\overset{o}{W} + \overset{5}{H})\rfloor dy^i\eta^7 \end{aligned} \tag{21}$$

the first eight (most useful) prolongation forms [5] are explicitly found:

potential prolongations or conservation laws:

$$\begin{aligned} & d\xi + \tfrac{1}{2}\eta^1 \\ & d\tau + \tfrac{1}{8}\eta^4 \\ & dr + \tfrac{1}{2}rw\eta^1 - \tfrac{3}{4}rq\,\eta^4 + \tfrac{1}{4}r\,\eta^7 \\ & dq + \tfrac{1}{3}\eta^3 \\ & dw + \tfrac{1}{2}\eta^2 \end{aligned} \tag{22}$$

pseudopotential prolongations:

$$\begin{aligned} & dy + \tfrac{1}{2}(y^2 - \lambda)\eta^1 + \eta^2 - \tfrac{1}{2}\lambda(y^2 - \lambda)\eta^4 \\ & \qquad - \tfrac{1}{2}(y^2 + \lambda)\eta^5 + \eta^6 + y\,\eta^7 \\ & dv + \tfrac{1}{2}y\,\eta^1 - \tfrac{1}{2}\lambda y\,\eta^4 - \tfrac{1}{2}y\,\eta^5 + \tfrac{1}{2}\eta^7 \\ & ds + \tfrac{1}{2}e^{2v}\eta^1 - \tfrac{1}{2}\lambda e^{2v}\eta^4 - \tfrac{1}{2}e^{2v}\eta^5 \end{aligned} \tag{23}$$

4. Representations of sl(2,R), the Lax and Painlèvé Equations and the Hyperbolic Plane.

The Cartan-Maurer structure equations for $s\ell\,(2,R)$ dual to (the zero grade of) Eq. (17) are:

$$\begin{aligned} d\underset{0}{w} + \underset{0}{v} \wedge \underset{0}{u} &= 0 \\ d\underset{0}{u} + \underset{0}{u} \wedge \underset{0}{w} &= 0 \\ d\underset{0}{v} + \underset{0}{w} \wedge \underset{0}{v} &= 0 \end{aligned} \tag{24}$$

and so we have found the KdV specialization to consist of setting

$$\begin{aligned} \underset{0}{u} &= -2\,dx + 8(u+\lambda)\,dt \\ \underset{0}{v} &= (2u-\lambda)\,dx - 2\left(p + 4u^2 + 2\lambda u - 2\lambda^2\right)dt \\ \underset{0}{w} &= 4z\,dt \end{aligned} \tag{25}$$

The three pseudopotentials y, v and s entering the forms Eq. (23) are a faithful representation of $s\ell\,(2,R)$:

$$\begin{aligned} & dy + \underset{0}{v} + y\,\underset{0}{w} - \tfrac{1}{2}y^2\underset{0}{u} \qquad \text{(the "simple" pseudopotential)} \\ & dv + \tfrac{1}{2}\underset{0}{w} - \tfrac{1}{2}y\,\underset{0}{u} \\ & ds - \tfrac{1}{2}e^{2v}\underset{0}{u} \end{aligned} \tag{26}$$

The first involves only y, so taken by itself is a 1-dimensional representation. Only $s\ell(2,R)$ has such a representation - or projection - on a line, so only $s\ell(2,R)$-related evolution equations will have simple pseudopotentials.

$s\ell\,(2,R)$ has a two-dimensional linear (Lax) pair of forms: set $\psi = e^{-v}$, $\phi = ye^{-v}$ and find

$$\begin{aligned} & d\phi + \psi\,\underset{0}{v} + \tfrac{1}{2}\phi\,\underset{0}{w} \\ & d\psi + \tfrac{1}{2}\phi\,\underset{0}{u} - \tfrac{1}{2}\psi\,\underset{0}{w} \end{aligned} \tag{27}$$

This can also be seen as resulting from the existence of a multiply transitive 2x2 matrix (or linear) representation of the algebra sℓ (2,R):

$$\begin{aligned} \overset{0}{U} &\rightarrow -\frac{1}{2}\begin{pmatrix} 0 & 0 \\ 1 & 0 \end{pmatrix} \\ \overset{0}{V} &\rightarrow -\frac{1}{2}\begin{pmatrix} 0 & 2 \\ 0 & 0 \end{pmatrix} \\ \overset{0}{W} &\rightarrow -\frac{1}{2}\begin{pmatrix} 1 & 0 \\ 0 & -1 \end{pmatrix} \end{aligned} \tag{28}$$

The usefulness of a complete coordinatization of sℓ(2,R) should not however be overlooked: the third coordinate s in Eq. (26) satisfies the "singularity-manifold" equation that arises in Painlèvé analysis [17] [23], and apparently captures features quite concealed in the linear Lax representation.

Another useful representation uses, instead of v and s, coordinates say μ and ν adapted to a general right invariant vector field in SL(2,R). Consider such a field

$$R = (A - Bs + \tfrac{1}{2}Cs^2)\frac{\partial}{\partial s} - \tfrac{1}{2}(B - Cs)\frac{\partial}{\partial v} + \tfrac{1}{2}Ce^{2v}\frac{\partial}{\partial y} \tag{29}$$

where $[\overset{o}{W}, R] = [\overset{o}{U}, R] = [\overset{o}{V}, R] = 0$. Impose $R\rfloor d\mu = R\rfloor d\nu = 0$ and find that we can set

$$\begin{aligned} \mu &= y + \frac{1}{2}\frac{B - Cs}{A - Bs + \frac{1}{2}Cs^2}e^{2v} \\ \nu &= \frac{1}{2}\frac{\left(2AC - B^2\right)^{1/2}}{A - Bs + \frac{1}{2}Cs^2}e^{2v} \\ \bar{y} &= y \end{aligned} \tag{30}$$

In these coordinates (dropping the bar on y) we have

$$\begin{aligned} \overset{o}{W} &= \mu\frac{\partial}{\partial\mu} + \nu\frac{\partial}{\partial\nu} + y\frac{\partial}{\partial y} \\ \overset{o}{U} &= \frac{1}{2}(\nu^2 - \mu^2)\frac{\partial}{\partial\mu} - \nu\mu\frac{\partial}{\partial\nu} - \frac{1}{2}y^2\frac{\partial}{\partial y} \end{aligned} \tag{31}$$

$$\overset{\circ}{V} = \frac{\partial}{\partial\mu} + \frac{\partial}{\partial y}$$

Projection onto μ, ν just drops the last terms. Then the hyperbolic 2-space metric $(d\mu^2 + d\nu^2)/\nu^2$ is invariant under these three, as Killing vectors. The (2-dimensional!) prolongation is a complex version of the simple pseudopotential: set $\zeta = \mu + i\nu$ and find

$$d\zeta + \underset{o}{v} + \zeta \underset{o}{w} - \tfrac{1}{2}\zeta^2 \underset{o}{u} \tag{32}$$

The distinction between these pseudopotentials is ultimately global: in fact y should originally have been taken to be a periodic coordinate expressing the projection of SL(2,R) onto S^1. Exponentiating, SL(2, R) acts on the hyperbolic plane by fractional linear transformation $\zeta \to (a\zeta+b)/(c\zeta+d)$. The cross ratio of two points ζ and κ, $c \equiv (\zeta-\kappa)(\zeta^*-\kappa^*)/(\zeta^*-\kappa)(\zeta-\kappa^*)$ is invariant, and geodesic distance is given by $\frac{1}{2}\ln\{(1+c)/(1-c)\}$.

The higher order partial differential equations satisfied by some of these $s\ell$ (2,R) pseudopotentials can finally be tabulated:

$$\psi_{xx} + (2u - \lambda)\psi = 0 \qquad \text{Schrödinger}$$

$$y_t + y_{xxx} - 6y^2 y_x + 6\lambda y_x = 0 \qquad \text{mKdV} \tag{33}$$

$$v_t + v_{xxx} - 2v_x^3 + 6\lambda v_x = 0$$

$$s_t + s_{xxx} - \tfrac{3}{2}s_{xx}^2/s_x + 6\lambda s_x = 0 \qquad \text{"singularity manifold" or "Ur"}$$

5. <u>The General Case </u>(not projected on $s\ell$ (2,R)) <u>and the Bäcklund Transform.</u>

Let us use a dual 1-form basis $\underset{i}{u}, \underset{i}{v}, \underset{i}{w}$, i = 0, 1, .. to write the Cartan-Maurer equations for the Kaĉ-Moody algebra (non-negative grades only) over $s\ell$ (2,R):

$$d\underset{0}{v} + \underset{0}{w} \wedge \underset{0}{v} = 0$$

$$d\underset{o}{w} + \underset{o}{v} \wedge \underset{o}{u} = 0$$

$$d\underset{0}{u} + \underset{0}{u} \wedge \underset{0}{w} = 0$$

$$A_1^{(1)+}: \qquad d\underset{1}{v} + \underset{1}{w} \wedge \underset{0}{v} + \underset{0}{w} \wedge \underset{1}{v} = 0$$

$$d w_1 + v_1 \wedge u_0 + v_0 \wedge u_1 = 0 \tag{34}$$

$$d u_1 + u_1 \wedge w_0 + u_0 \wedge w_1 = 0$$

$$d v_2 + w_2 \wedge v_0 + w_1 \wedge v_1 + w_0 \wedge v_2 = 0$$

etc.

If we stop with these seven, that is if we consider the ideal I generated by the 1-forms $\{ w_2, u_2, v_3, w_3, \ldots \}$, we get an invariant differential system with 7 Ω's and 4 Σ's for the mKdV equation [12]. For the KdV we have five additional variables and the additional Cartan-Maurer structure relations

$$d h_1 = d h_2 = d h_3 = d h_4 = 0 \tag{35}$$

$$d h_5 + h_1 \wedge h_2 + h_3 \wedge h_4 = 0$$

and the KdV-12 ideal will include the five additional 1-forms

$$\begin{aligned} & u_1 + 2 v_2 \\ & h_5 - w_0 \\ & u_1 + h_4 \\ & 2 h_1 + u_o - 2 v_1 \\ & w_1 \end{aligned} \tag{36}$$

There are now 5 Σ forms in I:

$$\begin{aligned} & h_4 \wedge w_o \\ & h_1 \wedge h_2 + h_3 \wedge h_4 - v_o \wedge u_o \\ & (h_1 + u_o) \wedge w_o \\ & v_o \wedge h_4 - h_1 \wedge u_o \\ & (h_1 + u_o) \wedge h_4 \end{aligned} \tag{37}$$

so $n = 12$, the Cartan characters are $s_o = 5$, $s_1 = 4$, and $g = 2$. The formalism of such invariant differential systems is very close to that of the method of moving frames in differential geometry [7]. It is perhaps not surprising that the KdV equation can be cast as a zero-curvature field equation in two dimensions.

In the deepest algebraic understanding [9][10][16] of the KdV hierarchy, as arising from reductions of the full algebra $A_1^{(1)}$, the above radical of the (third order) KdV equation is seen to come from the absent negative grades. The full infinite algebra $A_1^{(1)}$ has discrete invariances known already to relativity theorists as Geroch transforms, Ehlers transforms, and so on. These persist in the reduction, and give rise to Bäcklund transformations (of the still-infinite algebra $A_1^{(1)+}$) whose existence is not at all clear from inspection of the five or eight (finite) dimensional ideals.

The usual Bäcklund transform of the KdV equation [22] is shown in Figure 1 as a linear transformation matrix M for generators of the above algebra $\aleph_5 \otimes A_1^{(1)+}$. In verifying that the transformed generators satisfy the same algebra - that is, in calculating their commutators - one needs to use $\overset{o}{U} \lrcorner\, dy = -\frac{1}{2}y^2$, $\overset{o}{V} \lrcorner\, dy = 1$, $\overset{0}{W} \lrcorner\, dy = y$, with all other contractions vanishing.

The 12 by 12 transformation matrix M shown extends indefinitely to the right and down: lines 9-11 repeatedly translate down and over to become lines 12-14, and so on. (If these same lines were to be similarly repeatedly translated up and to the left to operate on both + and - grades of $A_1^{(1)}$, <u>and if y were then taken as a constant</u>, corresponding discrete invariance of the entire algebra $A_1^{(1)}$ is found. Introduction of the vectors H_i, and allowing y to vary at the zero grade, can be seen as achieving reduction by left truncation of the entire algebra.)

The matrix M is <u>involutory</u>: MM = I. This property can be preserved when the Galilean transformation A of Figure 2 is used to introduce an "eigenvalue" parameter by setting $\bar{M} = A^{-1} MA$. So $\bar{M}$ is the general 1-soliton Bäcklund transformation of the Bäcklund equation. Since it is involutory, M transposed gives the Bäcklund transformation of the dual basis forms. We find, reading down the columns of Fig. 1,

$$\bar{\underset{o}{v}} = \underset{o}{v} + 2y\,\underset{o}{w} - \tfrac{3}{2}y^2\,\underset{o}{u} - y^2\,\underset{1}{v} - y^3\,\underset{1}{w} + \tfrac{1}{2}y^4\,\underset{1}{u}$$

$$\bar{\underset{o}{w}} = -\underset{o}{w} + y\,\underset{o}{u} + 2y\,\underset{1}{v} + 2y^2\,\underset{1}{w} - y^3\,\underset{1}{u}$$

$$
\begin{aligned}
\bar{u}_0 &= 2v_1 + 2y\,w_1 - y^2 u_1 \\
\bar{v}_1 &= \tfrac{1}{2}u_0 + y\,w_1 - y^2 u_1 - y^2 v_2 \\
\bar{w}_1 &= -\,w_1 + y\,u_1 + 2y\,v_2 \\
\bar{u}_1 &= 2v_2 \\
\bar{v}_2 &= \tfrac{1}{2}u_1
\end{aligned}
\tag{38}
$$

From this, using Eq. (36), we can write the KdV-7 BT in symmetrized form

$$
\begin{aligned}
&\bar{h}_4 = -\,h_4 \\
&\bar{h}_1 = -\,h_1 \\
&\bar{h}_2 - \bar{v}_0 - y\,\bar{w}_0 + \tfrac{1}{2}y^2\bar{u}_0 = -\,h_2 + v_0 + y\,w_0 - \tfrac{1}{2}y^2 u_0 \\
&\bar{h}_3 + \tfrac{3}{2}y^2\bar{v}_0 + \tfrac{3}{2}y^3\bar{w}_0 - \tfrac{3}{4}y^4\bar{u}_0 = -\,h_3 - \tfrac{3}{2}y^2 v_0 - \tfrac{3}{2}y^3 w_0 + \tfrac{3}{4}y^4 u_0 \\
&\bar{w}_0 - y\,\bar{u}_0 = -\,w_0 + y\,u_0 \\
&\bar{u}_0 + \bar{h}_1 + \tfrac{1}{2}y^2\bar{h}_4 = u_0 + h_1 + \tfrac{1}{2}y^2 h_4 \\
&\bar{v}_0 + y\,\bar{w}_0 - \tfrac{1}{2}y^2\bar{u}_0 = v_0 + y\,w_0 - \tfrac{1}{2}y^2 u_0
\end{aligned}
\tag{39}
$$

So we can set

$$
\begin{aligned}
\bar{y} &= y \\
\bar{v} &= -\,v \\
\bar{w} - \tfrac{1}{2}\,\bar{y} &= -\,w + \tfrac{1}{2}\,y \\
\bar{q} + \tfrac{1}{6}\,y^3 &= -\,q - \tfrac{1}{6}\,y^3
\end{aligned}
\tag{40}
$$

and also find the BT for the original coordinates

$$
\begin{aligned}
d\bar{x} &= -\,dx \\
d\bar{t} &= -\,dt \\
\bar{z} - 2\bar{y}\bar{u} &= z - 2yu \\
\bar{u} + \tfrac{1}{2}\,\bar{y}^2 &= -\,u - \tfrac{1}{2}\,y^2 \\
\bar{p} + 4\bar{u}^2 + 2\bar{y}^2\bar{u} - 2\bar{y}\bar{z} &= -\,p - 4u^2 - 2y^2u + 2yz
\end{aligned}
\tag{41}
$$

Figure 1. The "Single soliton" Bäcklund Transformation for KdV

	$\overset{1}{H}$	$\overset{2}{H}$	$\overset{3}{H}$	$\overset{4}{H}$	$\overset{5}{H}$	$\overset{0}{V}$	$\overset{0}{W}$	$\overset{0}{U}$	$\overset{1}{V}$	$\overset{1}{W}$	$\overset{1}{U}$	$\overset{2}{V}$
$\overset{1}{H}$	-1	.	.	.	$2y$							
$\overset{2}{H}$	.	-1	.	.	.							
$\overset{3}{H}$	.	.	-1	.	.							
$\overset{4}{H}$	.	.	.	-1	y^3							
$\overset{5}{H}$	.	.	.	.	1							
$\overset{0}{V}$	.	2	$-3y^2$	.	.	1	.	.	.	.	.	.
$\overset{0}{W}$	.	$2y$	$-3y^3$	.	-2	$2y$	-1	.	.	.	.	.
$\overset{0}{U}$	.	$-y^2$	$\frac{3}{2}y^4$	.	$2y$	$-\frac{3}{2}y^2$	y	.	$\frac{1}{2}$	.	.	.
$\overset{1}{V}$						$-y^2$	$2y$	2	.	.	.	.
$\overset{1}{W}$						$-y^3$	$2y^2$	$2y$	y	-1	.	.
$\overset{1}{U}$						$\frac{1}{2}y^4$	$-y^3$	$-y^2$	$-y^2$	y	.	$\frac{1}{2}$
$\overset{2}{V}$						.	.	.	$-y^2$	$2y$	2	.
$\overset{2}{W}$						.	.	.	$-y^3$	etc.		
$\overset{2}{U}$						.	.	.	$\frac{1}{2}y^4$		(repeats)	

Figure 2. Galilei Invariance

	^1H	^2H	^3H	^4H	^5H	^0V	^0W	^0U	^1V	^1W	^1U	^2V
^1H	1	λ	$\frac{3}{2}\lambda^2$	.	.							
^2H	.	1	3λ	.	.							
^3H	.	.	1	.	.							
^4H	3λ	$\frac{3}{2}\lambda^2$	$\frac{3}{2}\lambda^3$	1	.							
^5H	.	.	.	.	1							
^0V						1	.	.	.	.	.	.
^0W						.	1	.	.	.	.	.
^0U						.	.	1	.	.	.	.
^1V						2λ	.	.	1	.	.	.
^1W						.	2λ	.	.	1	.	.
^1U						.	.	2λ	.	.	1	.
^2V						$4\lambda^2$	.	.	2λ	.	.	1

References

1. Cartan, E. (1945) Les Systèmes Différentiels Extérieurs et Leurs Applications Géométriques, Hermann, Paris.

2. Estabrook, F. B. (1976) 'Some old and new techniques for the practical use of exterior differential forms', in R. Miura (ed), The Inverse Scattering Method, Solitons and Their Application, Lecture Notes in Mathematics No. 515, Springer-Verlag, Berlin, New York, pp. 135-161.

3. Burke, W. L. (1985) Applied Differential Geometry, Cambridge University Press, Cambridge.

4. Harrison, B. K. and Estabrook, F. B. (1971) 'Geometric approach to invariance groups and solution of partial differential systems', J. Math. Phys. 12, 653-666.

5. Wahlquist, H. D. and Estabrook, F. B. (1975) 'Prolongation structures of nonlinear evolution equations', J. Math. Phys. 16, 1-7.

6. Estabrook, F. B. and Wahlquist, H. D. (1976) 'Prolongation structures of nonlinear evolution equations, II' J. Math. Phys. 17, 1293-1297.

7. Estabrook, F. B. and Wahlquist, H. D. (1989) 'Classical geometries defined by exterior differential systems on higher frame bundles', Classical and Quantum Gravity 6, 263-274. Sections II and III of the present lecture notes are edited versions of introductory paragraphs in this paper.

8. Ablowitz, M., Kaup, D., Newell, A. and Segur, H. (1974) 'The inverse scattering transform - Fourier analysis for nonlinear problems', Studies in Appl. Math. L111, MIT, 249-315.

9. Date, E., Jimbo, M., Kashiwara, M. and Miwa, T. (1982) 'Transformation groups for soliton equations - Euclidean Lie algebras and reduction of the KP hierarchy', Publ. Res. Inst. for Math. Sciences, Kyoto University, 18, 1077-1110.

10. Dodd, R. (1988) 'A note on the Lie algebra g(m∞)', Lett. Math. Phys., in press.

11. Estabrook, F. B. (1982) 'Moving frames and prolongation algebras', J. Math. Phys. 23, 2071-2076.

12. Estabrook, F. B. (1983) Invariant Differential Systems and Intrinsic Coordinates, MSRI 065-83, Mathematical Sciences Research Institute, Berkeley.

13. Gragert, P., Kersten, P. and Martini, R. (1983) 'Symbolic Computation in Applied Differential Geometry', Acta Applicandae Mathematicae, 1, 43-77.

14. Hoenselaers, C. (1986) 'More Prolongation Structures', Prog. Theoret. Phys. 75, 1014-1029.

15. Manin, Yu. (1979) 'Algebraic Aspects of Nonlinear Differential Equations', J. Sov. Math. 11, 1-196.

16. Morris, H. and Dodd, R. (1983) 'Infinite dimensional Lie algebras and the direct determination of deformation problems for equations of Painlevé type', Proc. Royal Irish Academy, 83A, 127-143.

17. Nucci, M. (1988) Painlevé Property and Pseudopotentials for Nonlinear Evolution Equations, technical report No. 1988-8, Dipart. di Math., Universitá di Perugia, Perugia.

18. Pirani, F., Robinson, D. and Shadwick, W. (1979) Local Jet Bundle Formulation of Bäcklund Transformations, Reidel, Holland/Boston/London.

19. Reyman, A. and Semenov-Tian-Shansky, M. (1988) 'Compatible Poisson Structures for Lax Equations: an r-Matrix Approach', Physics Lett. A, 130, 456-60.

20. Shadwick, W. (1980), 'The KdV prolongation algebra', J. Math. Phys. 21, 454-461.

21. van Eck, H. (1982) The Explicit Form of the Lie Algebra of Wahlquist and Estabrook, Memorandum Nr. 394, Dept. of Appl. Mathematics, Twente University of Technology, Enschede, Netherlands.

22. Wahlquist, H. and Estabrook, F. (1973) 'Bäcklund transformation for solutions of the Korteweg-de Vries equation', Phys. Rev. Lett. 31, 1386-90.

23. Weiss, J., Tabor, M. and Carnevale, G. (1983) 'The Painlevé property for partial differential equations', J. Math. Phys. 24, 522-526.

INERTIAL MANIFOLDS AND ATTRACTORS OF PARTIAL DIFFERENTIAL EQUATIONS

J.M. Ghidaglia
Laboratoire d'Analyse Numérique
C.N.R.S. et Université Paris-Sud,
91405 Orsay Cédex, France.

Lectures given at the NATO Advanced Studies Institute "Partially integrable nonlinear evolution equations and their physical applications" held in Les Houches, France, March 21-30, 1989.

Résumé. Ces leçons ont pour objet de présenter quelques résultats récents de la théorie des systèmes dynamiques de dimension infinie engendrés par des équations aux dérivées partielles non linéaires dissipatives. On insistera sur les variétés inertielles.

Abstract. Our aim in these lectures is to review on some recent results pertaining to the theory of infinite dimensional dynamical systems generated by nonlinear dissipative partial differential equations. Emphasis will be on inertial manifold theory.

Table of Contents.

R. Conte and N. Boccara (eds.), Partially Integrable Evolution Equations in Physics, 435–458.

1. Introduction.

Partial differential equations (p.d.e.'s), as models of real life phenomena, are obtained as distinguished limits of huge discrete systems. The equations of continuum mechanics represent a typical example of such a procedure. Hence a finite dimensional system is replaced by an infinite dimensional one, for which it is expected that the relevant informations are more tractable.

In fact, it turns out that the question whether or not the mathematical model, i.e. the p.d.e., does actually depend in the long run on a finite number of parameters (or degrees of freedom) is a major issue in theoretical physics for a long time. Concerning the problem of turbulence in fluids, Landau and Hopf have independently proposed in the fourties that the complicated and nonstationary patterns displayed by turbulent flows were caused by the fact that the solution to the underlying p.d.e. (namely the Navier-Stokes equations) was a quasi-periodic function depending on a very large number of rationally independent frequencies. In the same time, Kolmogorov discussed the fact that the presence of dissipation (viscous effects) actually impose a finite dimensional character to the flow. Kolmogorov also proposed a bound on the relevant number of degrees of freedom in term of a nondimensional number related to the physical characteristics of the flow.

More recently these questions have drawn very much attention in relation with numerical simulations of nonstationnary p.d.e.'s. Indeed, since computations are definitely finite dimensional, it is of fundamental importance to estimate the actual number of degrees of freedom on which the long time behavior of the solutions depends.

We have motivated the problem of reducing an infinite dimensional system to a finite dimensional one in the context of fluid dynamics. There are, of course, numerous and various e.g. physical contexts, in which the same question arises.

In order to give a more mathematical formulation to the questions we wish to address here, we write the evolutionary p.d.e. we consider as a differential equation :

$$\frac{du}{dt} = N(u), \; u(0) = u_0. \tag{1.1}$$

The function $t \to u(t)$ assumes its values in an infinite Banach space E and the vector field N is unbounded an nonlinear. We suppose that Problem (1.1) is well posed in the sense that there exists a semi-group $\{S(t), t \in \mathbb{R}_+\}$ that solves this equation. That is, we know that $u(t) = S(t)u_0$ is the solution to (1.1) and

$$S(0) = I \text{ (the identity map)}, \tag{1.2}$$

$$S(t_1 + t_2) = S(t_1) o S(t_2), \tag{1.3}$$

$$\text{the mappings } S(t) \text{ are continuous on } E. \tag{1.4}$$

Our aim is to discuss some features of the long time behavior of the autonomous equation (1.1), i.e. as $t \to +\infty$. For example we can ask whether or not all the trajectories of this equation are attracted (in a suitable sense) by some subset $\mathcal{A}$ of E, which is invariant by the flow :

$$S(t)\mathcal{A} = \mathcal{A}, \ \forall\, t \geq 0. \tag{1.5}$$

If so, such a set will represent the permanent regimes allowed by the dynamics (however see Remarks 2.7). The question on the number of degrees of freedom in the long run can be related, in that context, to the dimension of $\mathcal{A}$. At this point we wish to recall that the capacity (or the fractal dimension) of a subset $\mathcal{E} \subset E$ is the limit

$$d_F(\mathcal{E}) = \limsup_{\epsilon \to 0} \frac{\log N_\epsilon(\mathcal{E})}{\log (1/\epsilon)} \tag{1.6}$$

where $N_\epsilon(\mathcal{E})$ denotes the minimal number of balls of radius $\epsilon > 0$ which are needed to cover $\mathcal{E}$.

Although it is known that if the capacity of $\mathcal{A}$ is finite, then it is homeomorphic to a subset of $\mathbb{R}^k$ for some $k \in \mathbb{N}$; this set could be very complicated and in particular it is unlikely that one could replace the p.d.e. (1.1) by k o.d.e.'s. Moreover the knowledge of the dimension of $\mathcal{A}$ does not give at all a clue towards the determination of the k degrees of freedom which are relevant.

Inertial Manifolds are designed in particular in order to overcome this difficulty. Given a splitting $E = E_1 \oplus E_2$, where E_1 is finite dimensional, the graph $\mathcal{M} = \mathcal{M}(\phi) =$

$\{(p, \phi(p)), p \in E_1\}$ of a Lipschitz function ϕ from E_1 into E_2 is an Inertial Manifold provided we have

$$\begin{aligned} &\text{if } u_0 \in \mathcal{M} \text{ then the solution of (1.1) belongs to } \mathcal{M} \ : \\ &\mathcal{M} \ : \ u(t) \in \mathcal{M}, \ \forall\, t \geq 0, \end{aligned} \tag{1.7}$$

$$\text{for every } u_0 \in E, \ u(t) \text{ converges } \textit{exponentially} \text{ towards } \mathcal{M} \ :$$

$$\limsup_{t \to +\infty} \ (\log d(u(t), \mathcal{M}))/t < 0. \tag{1.8}$$

First we notice that if $u_0 = (p_0, \phi(p_0)) \in \mathcal{M}$, then according to (1.7), $u(t) = (p(t), \phi(p(t))$ and $t \to p(t)$ solves the system of o.d.e.'s on E_1 :

$$\frac{dp}{dt} = P_1 N(p + \phi(p)), \ p(0) = p_0, \tag{1.9}$$

where P_1 denotes the projection on E_1 associated to the splitting $E_1 \oplus E_2 = E$. We note that thanks to (1.8), it is now legitimate to replace the p.d.e. (1.1) by the finite dimensional dynamical system (1.9). It is the goal of the fourth section to discuss with more details the previous construction. Applicability is discussed in the fifth section in relation with the question of robustness. Finally the sixth section is devoted to the illustration of the previous analysis in the context of dissipative perturbations of Schrödinger equations. At the end of these notes we comment slightly the bibliography and give some indications on further readings.

2. Uniform bounds and attractors.

In this section we mainly introduce some definitions and give a usefull criterion for the existence of a global attractor. We recall that the setting is very general and that we are given a semi-group $\{S(t), t \geq 0\}$ on a Banach space E that enjoys (1.2) to (1.4).

Definition 2.1. *A bounded subset $\mathcal{B}_a \subset E$ is an absorbing set for $S(\cdot)$ on E, if for every bounded set $B \subset E$, there exists $T = T(B) \geq 0$ such that for $t = T(B)$, $S(t)\mathcal{B} \subset \mathcal{B}_a$.*

The existence of an absorbing set for the semi-group is stronger than boundedness of trajectories as we see now.

Example 2.2. Let us consider the linear complex o.d.e. $\dot{z} = -\lambda z$, $\lambda \in \mathbf{C}$. If $Re\ \lambda > 0$ every bounded set in $\mathbf{C}$ which contains 0 in its interior is absorbing. On the contrary, if $Re\ \lambda = 0$, although each trajectory is bounded (forward and backward), there is no bounded absorbing set for the flow.

If we assume that $S(\cdot)$ do possess a bounded absorbing set, for every bounded set $X \subset E$, we know that the omega-limit set $\omega(X)$ (see (2.1) below) is included in the closure (in E) of $\mathcal{B}_a$: $\omega(X) \subset cl(\mathcal{B}_a)$. All the trajectories starting from X are attracted by $\omega(X)$ and therefore all the trajectories are attracted by $\omega(\mathcal{B}_a)$. In the case where E is infinite dimensional (this is the case to be considered when $S(\cdot)$ is generated by a p.d.e.) it is not clear that for X bounded in E, its omega-limit set

$$\omega(X) = \bigcap_{s>0} cl\left(\bigcup_{t\geq s} S(t)X\right), \tag{2.1}$$

although bounded, is not empty. This is due to the well known fact that boundedness does not imply compactness in infinite dimensional Banach spaces.

However, as discussed above, the set $\omega(\mathcal{B}_a)$ is a good candidate for being a global attractor in the sense of

Definition 2.3. *A subset $\mathcal{A}$ is the global attractor for $S(\cdot)$ on E if it enjoys the three following properties*

$$\textit{compactness} : \mathcal{A} \textit{ is compact and non empty in } E, \tag{2.2}$$

$$\textit{invariance} : S(t)\mathcal{A} = \mathcal{A},\ \forall t \geq 0, \tag{2.3}$$

attraction : *for every bounded set B in E,*

$$\lim_{t \to +\infty} d(S(t)B, \mathcal{A}) = 0. \tag{2.4}$$

Remark 2.4. In (2.4), we have set

$$d(X,Y) = \text{Sup}\ \{d(x,Y),\ x \in X\}, \tag{2.5}$$

where as usual

$$d(x, Y) = \text{Inf } \{d(x, y),\ y \in Y\}. \tag{2.6}$$

In particular, $d(X, Y)$ is not symmetric and $d(X, Y) = 0$ amounts to say that $X \subset cl(Y)$. It is then clear that there exists at most one global attractor. Indeed, if $\mathcal{A}_1$ and $\mathcal{A}_2$ both satisfy (2.2) to (2.4), we deduce from (2.4) that

$$d(S(t)\mathcal{A}_1, \mathcal{A}_2) = (by\ (2.3)) = d(\mathcal{A}_1, \mathcal{A}_2) \to 0.$$

Hence $d(\mathcal{A}_1, \mathcal{A}_2) = 0$, i.e. $\mathcal{A}_1 \subset cl(\mathcal{A}_2) =$ (by (2.2)) $= \mathcal{A}_2$: $\mathcal{A}_1 \subset \mathcal{A}_2$ and therefore $\mathcal{A}_1 = \mathcal{A}_2$.

It is then standard to derive the following criterion for the existence of a global attractor.

Proposition 2.5. *We are given a semi-group $S(\cdot)$ which enjoys the properties* (1.2) *to* (1.4). *The two following properties are equivalent :*

(i) *$S(\cdot)$ possesses a global attractor,*
(ii) *there exists a compact set $K \subset E$ such that for every bounded set $B \subset E$,*

$$\lim_{t \to +\infty} d(S(t)B, K) = 0.$$

Moreover when (ii) *holds true, the global attractor $\mathcal{A}$ is the omega-limit set of K : $\mathcal{A} = \omega(K)$.*

Comments 2.6. (i) The proof, which is elementary, is left as an exercice.
(ii) The property (ii) turns out to be a classical question in the theory of nonlinear p.d.e.'s i.e. finding a priori estimates on the solutions. See Section 6 for examples.

We would like to end this section with the following

Remark 2.7. A naive approach could lead to consider instead of $\mathcal{A} = \omega(\mathcal{B}_a)$ the set $\mathcal{B} = \bigcup_{u_0 \in E} \omega(\{u_0\})$. Although $\mathcal{B} \subset \mathcal{A}$, this inclusion is strict in general and the set $\mathcal{A}$ is topologically much more involved that $\mathcal{B}$. In particular $\mathcal{A}$ is connected while in general it is not the case for $\mathcal{B}$. The major reason for which $\mathcal{B}$ is not concerned by the long time behavior is related to the question of stability with respect to perturbations. More precisely, assuming that the vector field N in (1.1) depends smoothly on a parameter μ (even analytically on μ), and assuming that one is able to construct in the neighbourhood of $\mu = \mu_0$ the sets $\mathcal{A}_\mu$ and $\mathcal{B}_\mu$; we can ask whether or not the correspondances $\mu \to \mathcal{A}_\mu$ or $\mathcal{B}_\mu$ are continuous (in a natural topology). It turns out that the dependence of $\mathcal{B}_\mu$ can "explode" when μ crosses the value μ_0 (see e.g. the introduction of J.K. Hale [28]). Concerning the dependence of $\mathcal{A}_\mu$ on μ, the situation is less dramatic and the mapping $\mu \to \mathcal{A}_\mu$ is upper-semicontinuous (J.K. Hale and G. Raugel [29]). However, in some cases, the set $\mathcal{A}_\mu$ can "shrink" as μ crosses μ_0 (see e.g. [28]).

3. Lyapunov exponents and fractal dimension.

In what follows we assume that the evolution equation we consider is posed on a Hilbert space H (see also Remark 3 below).

Hence we are given a semi-group $\{S(t)\}_{t \geq 0}$ acting on H and we supplement assumptions (1.2) to (1.4) with

$$u_0 \to S(t)u_0 \text{ is Fréchet differentiable in } H, \tag{3.1}$$

i.e. there exists $L(t, u_0) \in \mathcal{L}(H)$ such that for every $v_0 \in H$,

$$\lim_{\epsilon \to 0} \frac{S(t)(u_0 + \epsilon v_0) - S(t)u_0}{\epsilon} = L(t, u_0)v_0. \tag{3.2}$$

Let us take two initial data u_0^1 and u_0^2 which are close : $| u_0^1 - u_0^2 | << 1$ and look for the evolution of the difference of the two corresponding solutions. Then as long as $| S(t)u_0^1 - S(t)u_0^2 |$ is small (e.g. for small t), we see that the distance between these two trajectories is of the order of $L(t, u_0^1)(u_0^2 - u_0^1)$. Therefore the first characteristic value of the linear operator $(L(t, u_0)^* L(t, u_0))^{1/2}$ (which is equal to the supremum of $\| L(t, u_0)\xi \|$ for $\| \xi \| \leq 1$) will indicate whether the semi-group $S(t)$ shrinks or on the contrary blows perturbations of the initial data u_0 (for small $t \geq 0$) according to whether this number is less or greater than one. More generally, considering $m+1$ intial data $u_0^1, ..., u_0^m$ which form a true $m-$parallelepiped but are close one to each other ; we can study the parallelepiped built on $S(t)u_0^1, ..., S(t)u_0^m$ and ask whether or not it has a greater or a smaller volume.

This question is therefore related to the mth exterior product of $L(t, u_0)$ and motivate the introduction of the following numbers :

$$\bar{\omega}_m(t, u_0) = \sup_{v \in O_+(u_0)} \omega_m(L(t, v)) \tag{3.3}$$

where $O_+(u_0) = \{S(t)u_0, t \geq 0\}$ is a positive orbit starting from u_0 and for $L \in \mathcal{L}(H)$,

$$\omega_m(L) = \| \Lambda^m L \|_{\mathcal{L}(\Lambda^m H)} \equiv \sup_{\text{Gram}(\xi_1, ..., \xi_m) \leq 1} \text{Gram} \quad (L\xi_1, ..., L\xi_m)^{1/2}. \tag{3.4}$$

Here Gram $(\eta_1, ..., \eta_m)$ denotes the Gram determinant $\det_{1 \leq i,j \leq n} (\eta_i, \eta_j)_H$ (recall that it is the square of the volume of the $m-$dimensional parallelepiped defined by the vectors $\eta_1, ..., \eta_m$).

Geometrically speaking, the fact that for some m, say $m = m_0$, we have $\bar{\omega}_{m_0} < 1$, shows that, in the neighborhood of $O_+(u_0)$, $S(t)$ shrinks $m-$dimensional volumes. This will produce a result on the dimension of $\omega(u_0)$ as we argue now.

Let us consider a compact subset X in H which is negatively invariant by $S(\cdot)$ i.e.

$$X \subset S(t)X, \ \forall t \geq 0. \tag{3.5}$$

It turns out that for dynamical systems of interest, the strong differentiability assumption (3.2) is too strong. For what follows it will be sufficient to insure a weaker form of differentiability ; in particular it will be enough to consider trajectories on X as far as the study of the dimension of X is concerned. More precisely we assume that for every $u \in X$, there exists $L(t, u) \in \mathcal{L}(H)$ such that

$$\lim_{\epsilon \to 0} \sup_{\substack{u, v \in X \\ \| u - v \| \leq \epsilon}} \frac{\| S(t)v - S(t)u - L(t, u)(v - u) \|}{\| v - u \|} = 0. \tag{3.6}$$

Assuming moreover that, for every $t \geq 0$,

$$\sup_{u \in X} \| L(t, u) \| < \infty; \tag{3.7}$$

we notice that as a consequence of the chain rule and of (3.3),

$$\bar{\omega}_m(t_1+t_2,u_0) \le \bar{\omega}_m(t_1,u_0)\bar{\omega}_m(t_2,u_0) < \infty$$

which yields the existence of the limit

$$\pi_m(u_0) = \lim_{t\to+\infty} \bar{\omega}_m(t,u_0)^{1/t} = \operatorname{Inf}_{t>0} \bar{\omega}_m(t,u_0)^{1/t}. \tag{3.8}$$

By extension, we set

$$\pi_m(X) = \operatorname{Sup}_{u_0\in X} \pi_m(u_0) \tag{3.9}$$

which lead to the uniform Lyapunov exponents on X :

$$\begin{cases} \mu_1(X) = \operatorname{Log} \pi_1(X), \\ \mu_{m+2}(X) = \operatorname{Log} \pi_{m+2}(X) - \operatorname{Log} \pi_{m+1}(X), \; m \in \mathbb{N}. \end{cases} \tag{3.10}$$

With these notations we can state

Theorem 3.1. *Under the previous hypotheses on $S(\cdot)$ and X (namely* (1.2), (1.3), (1.4), (3.5), (3.6), (3.7)), *if there exists $m_0 \in \mathbb{N}^*$ such that* ($\mu_i = \mu_i(X)$)

$$\mu_1 + \ldots + \mu_{m_0+1} < 0, \tag{3.11}$$

then

$$\{\text{The Hausdorff dimension of } X\} \le m_0 + 1 \tag{3.12}$$

and

$$d_F(X) \le (m_0+1)\left(1 + \operatorname{Max}_{1\le \ell\le m_0} \frac{|\,\mu_1+\ldots+\mu_\ell\,|}{|\,\mu_1+\ldots+\mu_{m_0+1}|}\right). \tag{3.13}$$

Comments 3.2. (i) Concerning a proof of this result we refer to R. Temam [38, p. 287]. The result on the Hausdorff dimension in the case where the mappings $S(t)$ are compact was proved by Douady and Oesterlé [16]. The estimate on the fractal dimension is due to Constantin, Foias and Temam [9]. The general result i.e. the noncompact case was addressed by Ghidaglia and Temam [27].

(ii) Concerning applications of Theorem 3.1 to semi-groups generated by evolution equations, we would like to make the following comments. The linear operator $L(t, u_0)$ is in fact obtained by the linearized flow (see (1.1))

$$\frac{dv}{dt} = DN(u(t))v, \; v(0) = v_0, \tag{3.14}$$

i.e. $L(t, u_0)v_0 = v(t)$ is the solution of the nonautonomous linear differential equation (3.14) where $u(t) = S(t)u_0$. Hence the verification of the crucial hypothesis (3.11) is related to the study of the Gram determinant

$$G_m(t) = \text{Gram}\ (v_1(t), ..., v_m(t)) \tag{3.15}$$

where $\{v_1^0, ..., v_m^0\}$ are fixed independent vectors in H and $v_i(t)$ = the solution of (3.14) with v_0^i instead of v_0. It is in fact possible to deduce from (3.14) an evolution equation on $G_m(t)$. Although this leads in various situations to the desired condition (i.e. (3.11)) for sufficiently large m, see R. Temam [38], it is not always the case for weakly damped equation. We refer to Ghidaglia [23] for a description of an alternate method and to Biler [3], Constantin, Foias and Gibbon [6], Ghidaglia [21,22] and Maâche [32] for applications to specific equations.

4. Inertial Manifolds.

In this Section we first introduce a method of construction of Inertial Manifolds by the so-called Lyapunov-Perron method and then make some comments on other methods which are now available.

4.1. The Lyapunov-Perron method

This is actually the first method and it was introduced by Foias, Sell and Temam [20]. Let us begin by giving the functional framework adapted to this construction.

We let A be a closed unbounded positive self-adjoint linear operator with domain $D(A) \subset H$. We assume that $v \to | Av |$ is a norm on $D(A)$ equivalent to the graph norm, A^{-1} being compact on H. Hence there exists a complete orthonormal family $\{w_j\}_{j=1}^{\infty}$ in

H made of eigenfunctions of A :

$$Aw_j = \lambda_j w_j,\ j = 1, ...$$
$$0 < \lambda_1 \leq \lambda_2 \leq ... \leq \lambda_j \to \infty \text{ as } j \to \infty,$$

where the $\lambda_j's$ are the associated eigenvalues repeated according to their multiplicity. We denote by $\sigma(A) = \{\Lambda_k\}_{k=1}^{\infty}$, $\Lambda_1 < \Lambda_2 < ...$ the set of distinct eigenvalues and by $m_k \in \mathbb{N}^*$ the finite multiplicity of Λ_k. The spectral projections R_Λ and P_Λ are defined as usual :

$$R_\Lambda v = \sum_{j:\lambda_j=\Lambda} (v, w_j)w_j,\quad P_\Lambda = \sum_{\lambda \leq \Lambda} R_\lambda,$$

where $|\cdot|$ and $(,)$ denote the norm and scalar product on H.

We are going to consider the evolution equation on H :

$$\frac{du}{dt} + Au + Cu + F(u) = 0,\quad u(0) = u_0, \tag{4.1}$$

where C is a bounded linear operator from $D(A^{s_0})$ into H and F is a nonlinear mapping from $D(A^\alpha)$ into $D(A^{\alpha-\gamma})$. Here s_0, α and γ are given and $s_0 \in \mathbb{R}$, $\alpha \in \mathbb{R}$ while $\gamma \in [0, 1/2]$. We assume that C is squew-symmetric and commutes with A : $AC = CA$. Concerning F we suppose that it is globally Lipschitzian (see however Remark 6.1):

$$| A^{-\gamma}(F(v) - F(w)) |_\alpha \ \leq L_F \mid v - w \mid_\alpha,\ \forall v, w \in D(A^\alpha), \tag{4.2}$$
$$| A^{-\gamma}F(v) |_\alpha \ \leq L_F(1 + \mid v \mid_\alpha),\ \forall v \in D(A^\alpha), \tag{4.3}$$

where we have denoted

$$(v, w)_\alpha = (A^\alpha v, A^\alpha w),\ \mid v \mid_\alpha = \mid A^\alpha v \mid,$$

the scalar product and norm on $D(A^\alpha)$.

Under these hypotheses, it is a classical matter to show that (4.1) is well-posed on $D(A^\alpha)$ and $D(A^{\alpha+1/2})$ i.e. the mapping $S(t)\ :\ u_0 \to u(t)$ enjoys (1.2) to (1.4) for $E = D(A^\alpha)$ and $E = D(A^{\alpha+1/2})$.

As we have motivated in the Introduction, we are wondering whether or not the infinite dimensional evolution equation (4.1) can be replaced by a finite dimensional dynamical system or even by a system of o.d.e.'s. To this aim, we introduce the following definition.

Definition 4.1. *An Inertial Manifold $\mathcal{M} \subset D(A^\alpha)$ for* (4.1) *is a finite dimensional Lipschitz manifold which is invariant by $S(t)$:*

$$S(t)\mathcal{M} \subset \mathcal{M}, \ \forall t \geq 0, \tag{4.4}$$

and attracts exponentially all its solutions :

$$\begin{aligned} &\forall R > 0, \ \exists \sigma > 0, \ C > 0 \text{ st } , \forall u_0 \in D(A^\alpha), |\, u_0 \,|_\alpha \leq R \\ &d_\alpha(S(t)u_0, \mathcal{M}) \leq Ce^{-\sigma t}, \ \forall t \geq 0. \end{aligned} \tag{4.5}$$

Concerning existence of such sets we have the following result.

Theorem 4.2. *If N is such that*

$$\begin{aligned} \Lambda_{N+1} &\geq 3L_F^2 \Lambda_1^{2\gamma-1}/2, \\ \Lambda_{N+1} - \Lambda_N &\geq 30 L_F(\Lambda_N^\gamma + \Lambda_{N+1}^\gamma), \end{aligned} \tag{4.6}$$

then there exists $\phi \in C(P_{\Lambda_N}H, \ (I - P_{\Lambda_N})D(A^\alpha))$, whose graph is an inertial manifold for (4.1).

Before going into some details of the proof of this result we would like to make some comments.

Comments 4.3. (i) The meaning of condition (4.6) is that the spectrum of A contains large gaps. This condition is refered to in the litterature as the spectral gap condition. It is a very strong condition and actually it limits the range of applications. See the comments in Section 4.2 and Section 6.

(ii) This result is essentially due to Foias, Sell and Temam [20]. In fact these authors have considered the case where $C = 0$ and F has compact support. However their proof can be extended to the case where $C \neq 0$ and F is globally Lipschitzian (see [12]).

Indications on the proof of Theorem 4.2. In a first step we derive necessary conditions on mapping ϕ from PH into $QD(A^\alpha)$ ($P = P_{\Lambda_N}$, $Q = I - P_{\Lambda_N}$, N to be specified) so that its graph

$$\mathcal{M}(\phi) = \{(p, \phi(p)),\ p \in PM\}$$

is an Inertial Manifold for (4.1).

We take $u_0 \in D(A^\alpha)$ and denote by $p = Pu$ and $q = Qu$. According to (4.1) we have

$$\frac{dp}{dt} + Ap + Cp + PF(p+q) = 0, \tag{4.7}$$

$$\frac{dq}{dt} + Aq + Cq + QF(p+q) = 0. \tag{4.8}$$

Now since $\mathcal{M}(\phi)$ is invariant, if we take $u_0 = (p_0, \phi(p_0))$, we must have

$$q = \phi(p). \tag{4.9}$$

And therefore (4.7) reads

$$\frac{dp}{dt} + Ap + Cp + PG(p) = 0, \tag{4.10}$$

where

$$G(p) \equiv F(p + \phi(p)). \tag{4.11}$$

As said previously, ϕ is a Lipschitz function on PH :

$$|\,\phi(p)\,|_\alpha \leq \ell(1 + |\,p\,|_\alpha). \tag{4.12}$$

Therefore using (4.2) we see that

$$|\,G(p_1) - G(p_2)\,|_\alpha \leq \Lambda_N^\gamma L_F(1+\ell)\,|\,p_1 - p_2\,|_\alpha\,. \tag{4.13}$$

It follows that the system of o.d.e.'s (4.10) can be solved forward and backward in time and the study of the evolution of $|\,p\,|_\alpha^2$ leads to the estimate

$$1 + |\,p(t)\,|_\alpha \leq (1 + |\,p(0)\,|)e^{-(\Lambda_N + \Lambda_N^\gamma L_F(1+\ell))t},\ \forall t \leq 0. \tag{4.14}$$

On the other hand, we can write (4.8) as

$$\frac{d}{dt}\{e^{(A+C)t}q(t)\} = -e^{(A+C)t}QG(p(t)),$$

and this leads after integration to

$$e^{(A+C)t}q(t) = q(0) - \int_0^t e^{(A+C)s}QG(p(s))ds. \tag{4.15}$$

We are going now to make $t \to -\infty$ in (4.15). Indeed, since $q(t) = \phi(p(t))$ we see that thanks to (4.14) we have

$$| q(t) |_\alpha \leq Ce^{-(\Lambda_N + \Lambda_N^\gamma L_F(1+\ell))t}.$$

Now for $t \leq 0$,

$$| e^{(A+C)t}q |_\alpha \leq e^{\Lambda_{N+1}t} | q |_\alpha,$$

and therefore, under condition (4.6), the left hand side of (4.15) goes to 0 as $t \to -\infty$. It follows that

$$q(0) = \phi(p(0)) = \int_0^{-\infty} e^{(A+C)\sigma}QG(p(\sigma))d\sigma$$

and since $p(0) \in PH$ was arbitrary we conclude that for every $p_0 \in PH$,

$$\phi(p_0) = -\int_{-\infty}^0 e^{(A+C)\sigma}QF(p(\sigma) + \phi(p(\sigma)))d\sigma. \tag{4.16}$$

This is an integral equation for ϕ (recall that $p(\cdot)$ is the solution to (4.10) starting from p_0). And then this equation is solved by the classical contraction fixed point theorem. We will not go further into the details which are somewhat technical but let us mention that the condition $(4.6)_2$ is crucial when one aims to show the contracting property.

It is straightforward to check that if ϕ satisfy (4.16) then its graph $\mathcal{M}(\phi)$ is indeed invariant, and in fact

$$S(t)\mathcal{M}(\phi) = \mathcal{M}(\phi), \ \forall t \geq 0.$$

The exponential attraction contained in the definition of an Inertial Manifold is not a by-product and must be proved directly. It relies on a special property of equation (4.1) under the conditions (4.6) : the so-called Cone property, see. e.g. R. Temam [38, p. 409].

4.2. <u>A few comments about other methods</u>.

At the present time there are four other methods which lead to the construction of Inertial Manifold for equations of the type (4.1) :

- a variant of the previous method due to Chow and Lu [4],
- the method of integral manifolds due to Constantin, Foias, Nicolaenko and Temam [7],
- the method using the Principle of Spatial Averaging due to Mallet-Paret and Sell [33],
- Sacker's method due independently to Fabes, Luskin and Sell [17] and Debussche [13].

It is worthwhile to mention that all these methods do not exactly apply to the same type of equations and have their own range of applications. However it is remarkable that all these methods require a condition similar to (4.6) i.e. that the spectrum of the dissipative operator A contains large gaps. It must also be pointed out that in some particular cases, the method of Mallet-Paret and Sell [33] is superior to the others since these authors are able to construct inertial manifolds in cases where the spectrum of A does not contain arbitrarilyy gaps.

5. Stability of Inertial Manifolds.

We have pointed out in Section 2 that when the vector field N in (1.1) depends smoothly on a parameter μ, it cannot be expected that the global attractor $\mathcal{A}_\mu$ will depend continuously on μ. As it is well known in the theory of dynamical systems this is due to a lack of hyperbolicity (see e.g. Palis and De Melo [36]). On the contrary Inertial Manifolds are stable with respect to perturbations. More precisely three types of perturbations will be considered below, with emphasis on that generated by time discretization (see the end of this Section for a singularly perturbed equation).

The first kind of perturbation belongs to the class of regular ones, i.e. we consider instead of (4.1) the following equation

$$\frac{du}{dt} + Au + Cu + F(u,\mu) = 0, \tag{5.1}$$

where μ varies in a certain compact topological space Λ. We suppose that $F(\cdot,\mu)$ depends smoothly on μ and that (4.2) is satisfied for $\mu \in \Lambda$ with F replaced by $F(\cdot,\mu)$. Assuming that (4.6) holds true, we are able to construct using Theorem 4.2 a family $\{\Phi_\mu\}_{\mu\in\Lambda}$ such that the graph of Φ_μ is an Inertial Manifold for (5.1). Since Φ_μ is obtained via the contraction principle, it is straightforward to see that for every $\mu_0 \in \Lambda$,

$$\lim_{\mu \to \mu_0} \quad \operatorname{Sup}_{p \in P_{\Lambda_N} H} \frac{| \phi_\mu(p) - \phi_{\mu_0}(p) |_\alpha}{1 + | p |_\alpha} = 0.$$

that is the Inertial Manifolds depend continuously on μ.

The second type of perturbation corresponds to a Galerkin approximation of (4.1), associated with the eigenfunctions $\{w_j\}$ of A. Hence for any $M \geq 1$, the perturbed equation reads

$$\frac{du}{dt} + Au + P^M F(u) = 0, \tag{5.2}$$

where $P^M = P_{\Lambda_M}$.

Let us assume that (4.6) holds true for some $N \geq 1$. Then for $M \geq N$, according again to Theorem 4.2, (5.2) possesses an Inertial Manifold $\mathcal{M}_M$ which is the graph of a function ϕ_M from $P_{\Lambda_N} H$ into $(I - P_{\Lambda_N}) D(A^\alpha)$. In this setting too, ϕ_M converges to ϕ as $M \to \infty$ and more precisely

$$\lim_{M \to \infty} \quad \operatorname{Sup}_{p \in P_{\Lambda_N} H} \frac{| \phi_M(p) - \phi(p) |_\alpha}{1 + | p |_\alpha} = 0, \tag{5.3}$$

(see Temam [38, p. 443]).

Finally, the third type of perturbation is related to time discretization. We consider the following numerical scheme

$$\begin{cases} \dfrac{u^{n+1/2} - u^n}{\tau} + Au^{n+1/2} + F(u^n) = 0, \\[2ex] \dfrac{u^{n+1/2} - u^{n+1/2}}{\tau} + C\dfrac{u^{n+1} + u^{n+1/2}}{2} = 0, \end{cases} \tag{5.4}$$

where $\tau > 0$ denotes the time step. This produces a discrete dynamical system $u^n \to u^{n+1} = S^\tau u^n$ and we expect that u^n will accurately represent $u(n\tau)$, the solution at time $n\tau$ of (4.1), provided $u(0) = u^0$. It is straightforwad to extend Definition 4.1 to a discrete dynamical system :

Definition 5.1. *An Inertial Manifold $\mathcal{M} \subset D(A^\alpha)$ for (5.4) is a finite dimensional Lipschitz manifold which is invariant by S^τ :*

$$S^\tau \mathcal{M} \subset \mathcal{M} \text{ (here } \tau \text{ is fixed),}$$

and attracts exponentially the solutions to (5.4) :

$$\forall R > 0,\ \exists \sigma > 0,\ K \geq 0, \text{ such that } \forall u^0 \in D(A^\alpha), | u^0 |_\alpha \leq R$$

$$d_\alpha((S^\tau)^n u^0, \mathcal{M}) = d_\alpha(u^n, \mathcal{M}) \leq K e^{-\sigma\tau n},\ \forall n \geq 0.$$

Concerning existence of Inertial Manifolds for (5.4) we have the following result.

Theorem 5.1. ([10]). *We assume that N is such that (4.6) holds true. For every τ satisfying $\tau\Lambda_{N+1} \leq 1$, the discrete dynamical system (5.4) possesses an Inertial Manifold $\mathcal{M}_\tau$ which is the graph of a function from $P_{\Lambda_N} H$ into $(I - P_{\Lambda_N})(D(A^\alpha)$.*

As τ goes to 0, the Inertial Manifolds $\mathcal{M}_\tau$ converge to that obtained in Theorem 2.1 :

Theorem 5.2. ([10]). *Under the assumptions of Theorem 5.1, there exists a constant K such that for every τ satisfying $0 < \tau\Lambda_{N+1} \leq 1$, we have*

$$\underset{p \in P_{\Lambda_N} H}{\text{Sup}} \frac{| \phi(p) - \phi_\tau(p) |_\alpha}{1 + | p |_\alpha} \leq K\tau^\varsigma \,|\, \text{Log}\, \tau \,| \tag{5.5}$$

where $\varsigma = 1 - \gamma$ ς for $s_0 \leq 1$ and $\varsigma = (1-\gamma)/(2s_0 - 1)$ for $s_0 \geq 1$.

Concerning the proof of these two results, we refer to Demengel and Ghidaglia [10]. Let us notice that (5.5) is an error estimate between the exact Inertial Manifold $\mathcal{M}$ and the approximate one $\mathcal{M}_\tau$. Since (5.4) is a first order scheme, one could expect that (5.5) will be of order one too (which is indeed the case for $\gamma = 0$, [10]). It is then natural to ask whether the error estimate could be of higher order provided we replace (5.4) by an appropriate higher order scheme. In general the answer is negative. Indeed, even for the continuous case, i.e. equation (4.1), if we assume that (4.6) holds true and F is C^k $(k \geq 2)$; one cannot expect that ϕ will be C^k too (see Demengel and Ghidaglia [11]). Now the evolution equation (4.1) restricted to the Inertial Manifold is given by a system of o.d.e.'s (see (4.10)):

$$\frac{dp}{dt} + Ap + Cp + P_{\Lambda_N} F(p + \phi(p)) = 0.$$

Since ϕ is not necessarily C^k, it is not possible to obtain error estimate of order k for the discretization of this equation.

Another type of perturbation was considered by Mora and Sorà-Morales [35]. With the notations of Section 4, they consider the second order in time equation

$$\epsilon \frac{d^2u}{dt^2} + \frac{du}{dt} + Au + F(u) = 0,$$

where $\epsilon > 0$ is bounded to tend to zero. The Authors show the existence of a family of Inertial Manifolds $\{\mathcal{M}_\epsilon\}$ for small ϵ and study its limit as $\epsilon \to 0$ (see also [4] and [14]).

6. A short illustration via perturbed Schrödinger equations.

In this Section we shall consider two types of perturbations of the cubic Schrödinger equation :

$$iu_t + u_{xx} + \mid u \mid^2 u = 0, \tag{6.1}$$

associated with periodic boundary conditions

$$u(x+L,t) = u(x,t), \ x \in \mathbb{R}, \ t \in \mathbb{R}. \tag{6.2}$$

6.1. A damped and driven cubic Schrödinger equation.

Here we study the following equation

$$iu_t + u_{xx} + \mid u \mid^2 u + i\gamma u = f \tag{6.3}$$

where $\gamma > 0$ is a damping factor and $f = f(x)$ is an external driving force. We restrict the exposition to the case of an autonomous equation and refer to Ghidaglia [21,24] for the case of time periodic driving terms and more general nonlinearities.

Inspired by the two first invariants of (6.1) i.e. the mass $\int_0^L \mid u \mid^2 dx$ and the energy $\int_0^L (2 \mid u_x \mid^2 - \mid u \mid^4) dx$, one deduce easily the existence of a bounded absorbing set (see Definition 2.1) for (6.2) and (6.3) in $H^1_L = \{v, \int_0^L (\mid v \mid^2 + \mid v_x \mid^2) dx < \infty,$

$v(x+L) = v(x)\}$. In order to show that the long time behavior of this dynamical system is described by a global attractor, we have to show, according to Proposition 2.4, that trajectories are uniformly attracted by a compact set in H^1_L. But since (6.3) is time-reversible, the evolution is described by a group and this property is not obvious (and not known yet) using the strong topology. Instead, we use the weak topology for which boundedness implies compactness and are able to construct a global attractor $\mathcal{A}_1$ in H^1_L endowed with its weak topology and $\mathcal{A}_2$ in $H^2_L = \{v \in H^1_L, \int_0^L \mid v_{xx} \mid^2 dx < \infty\}$ endowed with its weak topology too. We notice that $\mathcal{A}_2 \subset \mathcal{A}_1$, while it is not known, but otherwise expected, that equality holds true.

The set $\mathcal{A}_2$ is compact in H^1_L and invariant hence we can ask whether it is finite dimensional. According to Section 3, we have to consider the linearized flow which reads here as follows :

$$iv_t + v_{xx} + 2 \mid u \mid^2 v + u^2 \bar{v} + i\gamma v = 0.$$

Estimates on the Lyapunov exponents are delicate since the dissipation effects are weak. Here again it seems necessary to use the invariants (mass and energy) of the unperturbed equation as a guide in the study of the Gram determinant (3.15). We refer to [21] and [24] for details. As a matter of fact it can be shown that $\mathcal{A}_2$ has finite dimension and explicit bounds on this dimension in terms of the data can be given. For example a particular case was considered in [24] where $\gamma = O(\epsilon^{-2})$, $f = O(\epsilon^{-3})$, $\epsilon > 0$ being a small parameter. It is found that there exists two constants κ_1 and κ_2 such that

$$\kappa_1 \epsilon^{-1} \leq d_H(\mathcal{A}_2) \leq d_F(\mathcal{A}_2) \leq \kappa_2 \epsilon^{-6}, \; 0 < \epsilon \leq 1.$$

Existence of Inertial Manifolds is an open problem here. For example if we write (6.3) as (4.1), we will set $Av = \gamma u$ and then $\Lambda_N = \gamma$, $\forall N$. Hence the spectral gap condition (4.6) is far from being achieved!

6.2. The Ginzburg-Landau equation.

In this paragraph we consider the following equation

$$u_t - (1 + i\alpha)u_{xx} + (1 + i\beta) \mid u \mid^2 u = \gamma u. \tag{6.4}$$

together with periodic B.C. (6.2). After a change of dependent and independent variables, this equation can indeed be seen as a particular perturbation of (6.1). By contrast with (6.3), (6.4) contains a diffusion term and this will allow for the construction of an Inertial Manifold for (6.4).

It is easy to show that the semi-group $S(t)$ generated by (6.4)-(6.2) possesses a bounded absorbing set ([25, 15] in H^1_L :

$$\mathcal{B}_a = \{v \in H^1_L, \int_0^L (|v|^2 + |v_x|^2)dx \leq \rho\}.$$

Thanks to a classical smoothing property of parabolic equations, for every $t > 0$, $S(t)$ is a compact mapping and since $\mathcal{B}_a$ is a bounded absorbing set, $K = cl(S(t_0)\mathcal{B}_a)$ is also absorbing and compact in H^1_L. Therefore the point (ii) in Proposition 2.4 is fullfiled and $\mathcal{A} = \omega(K) = \omega(\mathcal{B}_a)$ is the global attractor. It can also be shown that this set is finite dimensional ([25, 15]).

Since after a transient time the dynamics lies into $\mathcal{B}_a$, it is legitimate to modify equation (6.4) outside of $\mathcal{B}_a$. More precisely using a cutoff function $\theta \in C^\infty([0,\infty])$ such that $\theta(s) = 1$ for $s \in [0,2]$ and $\theta(s) = 0$ for $s \geq 3$; we replace (6.4) by

$$u_t - (1+i\alpha)u_{xx} + (1+i\beta)\theta\left(\frac{\int_0^L(|u|^2 + |u_x|^2)dx}{\rho}\right)|u|^2 u = \gamma u. \tag{6.5}$$

Now (6.5) enters the setting of Section 4. To that aim, we set

$$H = L^2_L = \left\{v, \int_0^L (|v|^2)dx < \infty, v(x+L) = v(x) \ for \ a.e. x\right\},$$

$$D(A) = H^2_L,\ Av = v - v_{xx},\ Cv = -iv_{xx},$$

$$F(v) = (1+i\beta)\theta\left(\frac{\int_0^L(|v|^2 + |v_x|^2)dx}{\rho}\right)|v|^2 v - \gamma v - v.$$

It is then easy to see that (4.3) and (4.3) hold true with $\alpha = 1/2$, $\gamma = 0$; and that $s_0 = 1$. Here the eigenvalues of A are explicitely known :

$$\Lambda_k = 1 + 4(k-1)^2\pi^2 L^{-2},\ k = 1,2,...$$

It is clear that (4.6) holds true for sufficiently large N showing the existence of an Inertial Manifold for (6.5).

Remark 6.1. In the previous construction we have modified the original equation by localizing in a neighbourhood of an absorbing set. This procedure is classical in the theory of invariant manifolds. Now the hypotheses (4.2)-(4.3) which could look very restrictive at first sight become much clear. Indeed in each case after finding a suitable bounded absorbing set in $D(A^{\alpha})$, and assuming that F is locally Lipschitzian in $D(A^{\alpha})$, we can always applies the localization device to construct a new F which satisfy (4.2)-(4.3). This procedure does not affect the equation inside the absorbing set in which the dynamics occur after a transient time.

7. Indications on further readings and references.

These notes were by no means exhaustive, and the rapid examples of Section 6 were inspired by some of the works of the author. In particular we have not mentionned regular attractors (see e.g. Babin and Vishik [2]), lower bounds on the dimension of attractors (see e.g. Babin and Vishik [1] and Ghidaglia and Héron [25]), Approximate Inertial Manifolds (see e.g. Foias, Manley and Temam [18], Titi [39]).

There is also a few monographs on the subject : Henry [30], Ghidaglia and Saut [26], Temam [38], Hale [28], Constantin, Foias, Nicolaenko and Temam [7] to which we refer for complements, other aspects, proofs and references. Recently Sell [37] has published a very usefull bibliography with a classification according to the topic.

As far as nonlinear p.d.e.'s are concerned, these studies began in relation with the Navier-Stokes equations. Then other parabolic equations like reaction-diffusion equations and Kuramoto-Sivashinski equations were investigated (see the references in Temam [38]). Later existence of a global attractor for damped nonlinear wave equations was proved in Babin-Vishik [1,2]. These attractors were shown to be finite dimensional in Ghidaglia-Temam [27]. In these equations the difficulties arise from a lack of compactness of the group describing the evolution. Nonlinear damped and driven dispersive equations like the Korteweg-de Vries equation or nonlinear Schrödinger equations were addressed in

Ghidaglia [22,21,23]. Extensions to nonlocal operators are made in Maâche [32].

The span of applications of the theory of Inertial Manifolds is much more narrow, it does not include at the present time the Navier-Stokes equations, wave equations, nonlinear damped and driven dispersive equations. Let us mention that Mora and Solà-Morales [34] have given a counterexample in the setting of semilinear damped wave equations. However existence of Inertial Manifolds have been proved for reaction-diffusion equations (e.g. in [20], [33], [7]), for the complex Ginzburg-Landau equation (e.g. in [8], [10], [5]), for the Kuramoto-Sivashinski equation ([19]), and for the pseudo-Korteweg-de Vries-Burgers equation ([10]).

References

[1] A.V. Babin and M.I. Vishik, Attractors of partial differential equations and estimate of their dimension, *Russian Math. Surveys 38,* 151-213, 1983.

[2] A.V. Babin and M.I. Vishik, Regular semigroups and evolution equations, *J. Math. Pures Appl. 62,* 441-491, 1983.

[3] P. Biler, Attractors for the system of Schrödinger and Klein-Gordon equations with Yukawa coupling, *Preprint Orsay,* 1988.

[4] S.N. Chow and K. Lu, Invariant manifolds for flows in Banach spaces, *J. Diff. Equ. 74,* 285-317, 1988.

[5] Constantin, A construction of inertial manifolds, Preprint, 1989.

[6] P. Constantin, C. Foias and J.D. Gibbon, Finite dimensional attractors for the laser equations, *Nonlinearity 2,* 241-269, 1989.

[7] P. Constantin, C. Foias, B. Nicolaenko and R. Temam, *Integral manifolds and inertial manifolds for dissipative partial differential equations,* Springer, New York, 1989.

[8] P. Constantin, C. Foias, B. Nicolaenko and R. Temam, Spectral barriers and inertial manifolds for dissipative partial differential equations, *J. Dynamics and Differential Equations, 1,* to appear.

[9] P. Constantin, C. Foias and R. Temam, Attractors representing turbulent flows, *Memoirs of A.M.S. 53, n° 314,* 1985.

[10] F. Demengel and J.M. Ghidaglia, Inertial manifolds for partial differential evolution

equations under time-discretization : existence, convergence and applications, *Preprint Orsay* 1989 ; see also *C.R. Acad. Sci. Paris, Série I, 307,* 1988.

[11] F. Demengel and J.M. Ghidaglia, Some remarks on the smoothness of inertial manifolds, *Preprint* 1989.

[12] F. Demengel and J.M. Ghidaglia, Construction of inertial manifolds via the Lyapunov-Perron method, to appear.

[13] A. Debussche, Inertial manifolds and Sacker's equation, *Differential and Integral Equations,* to appear.

[14] A. Debussche, Inertial manifolds for flows in Banach spaces using backward integration, to appear.

[15] C.R. Doering, J.N. Elgin, J.D. Gibbon and D.D. Holm, Low dimensional behavior in the complex Ginzburg-Landau equation, *Nonlinearity,* 1988.

[16] A. Douady and J. Oesterlé, Dimension de Hausdorff des attracteurs, *C.R. Acad. Sci. Paris, Série A, 290,* 1135-1138, 1981.

[17] E. Fabes, M. Luskin and G.R. Sell, Construction of inertial manifolds by elliptic regularization, *IMA Preprint 459*, Minneapolis, 1988.

[18] C. Foias, O. Manley and R. Temam, Modelization of the interaction of small and large eddies in turbulent flows, *Math. Mod. and Numer. Anal. (M2AN),* 93-114, 1988.

[19] C. Foias, B. Nicolaenko, G.R. Sell and R. Temam, Inertial manifolds for the Kuramoto-Sivashinsky equation and an estimate of their lowest dimension, *J. Math. Pures Appl. 67*, 197-226, 1988.

[20] C. Foias, G.R. Sell and R. Temam, Inertial manifolds for nonlinear evolutionary equations, *J. Diff. Equ. 73*, 309-353, 1988.

[21] J.M. Ghidaglia, Finite dimensional behavior for weakly damped driven Schrödinger equations, *Ann. Inst. Henri Poincaré, Analyse Non Linéaire 5,* 365-405, 1988.

[22] J.M. Ghidaglia, Weakly damped driven Korteveg-de Vries equations behave asymptotically as a finite dimensional system in the long time, *J. Diff. Equ. 74,* 369-390, 1988.

[23] J.M. Ghidaglia, Upper bounds on the Lyapunov exponents for dissipative perturbations of infinite dimensional Hamiltonian systems, *Lect. Notes in Physics,* P. Lochak Ed., to appear in 1989.

[24] J.M. Ghidaglia, Explicit upper and lower bounds on the number of degrees of freedom for damped and drivent cubic Schrödinger equations, *Math. Mod. and Numer. Anal.*

(M2AN) 23, 433-443, 1989.

[25] J.M. Ghidaglia and B. Héron, Dimension of the attractors associated to the Ginsburg-Landau partial differential equation, *Physica 28D*, 282-304, 1987.

[26] J.M. Ghidaglia and J.C. Saut, *Equations aux dérivées partielles non linéaires dissipatives et systèmes dynamiques*, Hermann, Paris, 1988.

[27] J.M. Ghidaglia and R. Temam, Attractors for damped nonlinear hyperbolic equations, *J. Math. Pures Appl. 66*, 273-319, 1987.

[28] J.K. Hale, *Asymptotic behavior of dissipative systems*, A.M.S. Providence, 1988.

[29] J.K. Hale, X.B. Lin and G. Raugel, Upper semicontinuity of attractors for approximations of semigroups and partial differential equations, *J. Math. Comp.*, *50*, 89-123, 1988.

[30] D.B. Henry, *Geometric theory of semilinear parabolic equations*, Lect. Notes in Math. 840, Springer Verlag, New York, 1981.

[31] M. Luskin and G.R. Sell, Approximation theories for inertial manifolds, *Math. Mod. and Numer. Anal. (M2AN) 23*, 445-462, 1989.

[32] S. Maâche, In preparation.

[33] J. Mallet-Paret and G.R. Sell, Inertial manifolds for reaction-diffusion equations in higher space dimensions, *J. of the A.M.S. 1*, 805-866, 1988.

[34] X. Mora and J. Solà-Morales, Existence and non-existence of finite-dimensional globally attracting invariant manifolds in semilinear damped wave equations, in *Dynamics of infinite dimensional systems*, S.N. Chow and J.K. Hale eds., Springer-Verlag, New York, 187-210, 1987.

[35] X. Mora and J. Solà-Morales, Diffusion equations as singular limits of damped wave equations, *J. Diff. Equ. 78*, 262-307, 1989.

[36] J. Palis and W. De Mels, *Geometric theory of dynamical systems : an introduction*, Springer-Verlag, New York, 1982.

[37] G.R. Sell, References on dynamical systems, *IMA Preprint*, Minneapolis, 1989.

[38] R. Temam, *Infinite dimensional dynamical systems in mechanics and physics*, Springer-Verlag, New York, 1988.

[39] I. Titi, On approximate inertial manifolds to the 2D Navier-Stokes equation, MSI Preprint 88-119, 1988 and *J. Math. Anal. Appl.*, to appear.

HIROTA'S BILINEAR METHOD AND PARTIAL INTEGRABILITY

J. HIETARINTA
Department of Physical Sciences
University of Turku
20500 Turku
Finland

ABSTRACT. We discuss Hirota's bilinear method from the point of view of partial integrability. Many different levels of integrability are shown to exist.

1. Introduction

In these lectures we show that Hirota's bilinear method, which is usually applied to completely integrable systems, is well suited for partially integrable equations as well. We will elaborate on the many levels of partial integrability that appear for bilinear equations.

The bilinear method was introduced by Hirota [1] in 1971, and has been actively studied since (for reviews see e.g. [2,3]). The fundamental idea in this formalism is to use some *dependent variable transformation* to put the nonlinear evolution equation (NEE) in a form where the unknown function appears *bilinearly*. In the process one usually has to extract one or two derivatives from the equation, sometimes even add new independent variables. This transformation is discussed in Sec. 2 with several examples.

A bilinear form has been constructed for most, if not all, completely integrable systems by Hirota and others. One can indeed make the conjecture that *all* completely integrable NEE's can be put into a bilinear form. The converse, however, is not true: a bilinear form can also be constructed for many equations that are not integrable. In particular it is not true that any NEE that can be written in bilinear form automatically has N-soliton solutions (NSS) for any N. In fact even the existence of two-soliton solutions (2SS) is nontrivial for a completely general bilinear equation, although there are generic types where 2SS always exist (see Sec. 5-7).

In these lectures we will give examples of bilinear equation that are partially integrable in the sense that for them one can construct some soliton solutions but not all. We present also some new models that have 3SS and 4SS, but whose complete integrability has not (yet) been established [4-8].

R. Conte and N. Boccara (eds.), Partially Integrable Evolution Equations in Physics, 459–478.

2. Deriving the bilinear form

2.1. General

The first step that we must do when an equation is given is to write it in the bilinear form. This is also the most difficult step, because there is no systematic method for it. In this section we will give some examples and general comments on what to do with a new equation.

To introduce notation let us consider the standard (integrable) example, the Korteweg–de Vries (KdV) equation

$$u_{xxx} - 6uu_x + u_t = 0. \tag{2.1}$$

It is often more convenient to write it in the once integrated potential form using $u = v_x$:

$$v_{xxx} - 3{v_x}^2 + v_t = 0. \tag{2.2}$$

The Gel'fand–Levitan–Marchenko equation gives the multisoliton solution of (2.1) in the form $u = -2\partial_x^2 \log\det M$, where M is a matrix in which the x, t dependency enters through e^{ax+bt}. Hirota made [1] the logical step of introducing a new dependent variable F by

$$u = -2\partial_x^2 \log F, \text{ or } v = -2\partial_x \log F, \tag{2.3}$$

and obtained for F the equation

$$F_{xxxx}F - 4F_{xxx}F_x + 3{F_{xx}}^2 + FF_{xt} - F_xF_t = 0. \tag{2.4}$$

This can be written in a more compact form if we define new bilinear operators $D_x, D_t, \ldots$ (collectively denoted by $D_{\vec{x}}$) by [2,3]

$$P(D_{\vec{x}})F{\cdot}G = P(\partial_{\vec{x}} - \partial_{\vec{x}'})F(\vec{x})G(\vec{x}')\big|_{\vec{x}'=\vec{x}}, \tag{2.5}$$

(For a list of properties of the D-operator, see Appendix I of Ref. [9].) Eq. (2.4) can then be written as

$$(D_x^4 + D_xD_t)F{\cdot}F = 0. \tag{2.6}$$

We formalize this procedure as follows:

Definition: We say that a NEE can be written in the Hirota bilinear form if it is equivalent to

$$\sum_{\alpha,\beta=1}^{s} P^{\gamma}_{\alpha\,\beta}(D_{\vec{x}})f_\alpha{\cdot}f_\beta = 0, \quad \gamma = 1,\ldots,r. \tag{2.7}$$

for some s, r and functions $P^{\gamma}_{\alpha\,\beta}$.

Several comments should be made about the above definition:

a) Some results have also been obtained when the functions $P^{\gamma}_{\alpha\beta}$ depend on the independent variables $\vec{x}$ or ordinary derivatives, but these non-Hirota bilinear forms are not discussed at length here.
b) The above set of equations may contain more independent and dependent variables than there are in the original NEE.
c) If P is not a polynomial then the original NEE was nonlocal. An example of this is the Toda lattice, which has e^{D_x} terms.
d) The correspondence from NEE to Hirota form is *not* 1-1, but rather many to few.

Since the derivation of the bilinear form is a difficult step where considerable ingenuity is needed, we will look at a few more examples.

2.2. EXAMPLES

2.2.1. The modified KdV equation. The mKdV is given by

$$v_{xxx} + 24v^2 v_x + v_t = 0. \tag{2.8}$$

We will show that this system has not one but two bilinear forms. Let us introduce two new dependent variables G and F by $v = G/F$ [2]. When this is substituted into (2.8) we get first

$$\begin{aligned}&[G_{xxx}F - 3G_{xx}F_x + 3G_xF_{xx} - GF_{xxx} + G_tF - GF_t]F^2 \\ &+ 6[-F^2G_xF_{xx} + FGF_xF_{xx} + F_x{}^2G_xF - GF_x{}^3 + 4G^2FG_x - 4G^3F_x] = 0,\end{aligned} \tag{2.9}$$

but this can be written in terms of the D-operator alone as

$$[(D_x^3 + D_t)G\cdot F]F^2 - 3[D_xG\cdot F][D_x^2F\cdot F - 8G^2] = 0. \tag{2.10}$$

Since we have an extra dependent variable we may define one of them in terms of the other as we please, let us define G by

$$D_x^2F\cdot F = 8G^2, \tag{2.11a}$$

then from (2.10) we find the other equation

$$(D_x^3 + D_t)G\cdot F = 0, \tag{2.11b}$$

so that (2.8) is equivalent to (2.11).

For the other bilinear form we first introduce a new dependent variable w by $v = w_x$. Eq. (2.8) can then be written as

$$\partial_x[w_{xxx} + 8w_x{}^3 + w_t] = 0. \tag{2.12}$$

We will require that the integrated form of (2.12) is valid, i.e. the part in square brackets vanishes by itself. The substitution [10]

$$w = \arctan(g/f) \tag{2.13}$$

leads to

$$(f^2+g^2)[(D_x{}^3+D_t)g\cdot f]-3(D_xg\cdot f)[D_x{}^2(f\cdot f+g\cdot g)]=0 \tag{2.14}$$

which can be separated into two bilinear equations as

$$\begin{cases} (D_x^3-D_t)g\cdot f=0 \\ D_x^2(f\cdot f+g\cdot g)=0, \end{cases} \tag{2.15}$$

or with a complex rotation $F=f+ig$, $G=f-ig$,

$$\begin{cases} (D_x^3-D_t)G\cdot F=0 \\ D_x^2G\cdot F=0, \end{cases} \tag{2.15'}$$

2.2.2. The nonlinear Schrödinger equation. The nlS is given by

$$y_{xx}+i\,y_t+|y|^2y=0 \tag{2.16}$$

where y is complex. One uses $y=G/F$, where F is real, and obtains [11] first

$$F[(D_x{}^2+D_t)G\cdot F]-G[D_x{}^2F\cdot F-G^*G]=0, \tag{2.17}$$

which separates as

$$\begin{cases} (D_x^2+iD_t-\lambda)G\cdot F=0, \\ (D_x^2-\lambda)F\cdot F=|G|^2. \end{cases} \tag{2.18}$$

We have here included the separation constant (it could even be a function), which is needed for the "envelope-hole" solutions.

Eq. (2.18) is the standard form, but it is perhaps not so well known that (2.16) has also another bilinear form. If we introduce two new complex functions f and g by [12]

$$y=\frac{\sqrt{2}D_xg\cdot f}{|f|^2+|g|^2} \tag{2.19}$$

then (2.16) is equivalent to

$$\begin{cases} D_x(f^*\cdot f+g^*\cdot g)=0, \\ (D_x{}^2+iD_t)f\cdot g^*=0, \\ (D_x{}^2-iD_t)(f^*\cdot f-g^*\cdot g)=0. \end{cases} \tag{2.20}$$

The derivation of this last result is not that easy if you do not know the answer.

2.2.3. Shallow water wave equation. The above examples illustrate how dependent variables can be introduced, even several of them if needed. Sometimes one has to introduce new *independent* variables. Consider the following equation:

$$v_{xxxt}+\alpha v_xv_{xt}+\beta v_tv_{xx}-v_{xx}-v_{xt}=0. \tag{2.21}$$

If $\alpha = \beta = 3$ (2.21) can be integrated once and the substitution $v = 2\partial_x \log f$ leads to

$$(D_x^3 D_t - D_x^2 - D_x D_t) f \cdot f = 0. \tag{2.22}$$

If $\alpha = 4$, $\beta = 2$ the above substitution yields first [13]

$$3D_x[D_x^3 D_t - D_x^2 - D_x D_t) f \cdot f] \cdot f^2 - D_t(D_x^4 f \cdot f) \cdot f^2 + D_x(D_x^3 D_t f \cdot f) \cdot f^2 = 0. \tag{2.23}$$

To proceed further we introduce a new independent variable τ by

$$(D_x^4 + D_x D_\tau) f \cdot f = 0. \tag{2.24}$$

Next we use the identity

$$D_t(D_x D_\tau f \cdot f) \cdot f^2 = D_x(D_t D_\tau f \cdot f) \cdot f^2 \tag{2.25}$$

and find that (2.21) is in this case equivalent to (2.24) and

$$(2D_x^3 D_t - 3D_x^2 - 3D_x D_t - D_t D_\tau) f \cdot f = 0. \tag{2.26}$$

In this case we had originally 1 equation for 1 function of 2 variables, but the bilinear form is given by 2 equations for 1 function of 3 variables. This method of introducing new independent variables is typical for higher members of a hierarchy. (In the generic case it is not possible to have even 2SS for such a pair of equation for one dependent variable, but for the above case it does work.)

2.2.4. The Kaup equation. There are two NEE's with dispersion relation $p^5 + \Omega$, the Sawada-Kotera equation and Kaup's equation:

$$u_{xxxxx} + 20(u u_{xxx} + 2u_x u_{xx}) + 120u^2 u_x + u_t = 0. \tag{2.27}$$

The substitution $u = \partial_x{}^2 \log f$ leads to [14]

$$3D_x[D_x{}^6 + D_x D_t) f \cdot f] \cdot f^2 - 5D_x(D_x{}^4 f \cdot f) \cdot (D_x{}^2 f \cdot f) = 0. \tag{2.28}$$

As above one introduces a new independent variable τ and the constraint (2.24). Using (2.25) and the further identity

$$3D_x(D_x D_\tau f \cdot f) \cdot (D_x{}^2 f \cdot f) = D_\tau(D_x{}^4 f \cdot f) \cdot f^2 - D_x(D_x{}^3 D_t f \cdot f) \cdot f^2 \tag{2.29}$$

one finds that (2.27) is equivalent to (2.24) and

$$[3D_x{}^6 - 5D_x{}^3 D_\tau + 3D_x D_t - 5D_\tau{}^2] f \cdot f = 0. \qquad (2.30).$$

It is interesting to note that if (2.24) is taken as the basic or first equation of a hierarchy then it looks like *two* hierarchies can be built on to of it. In the Jimbo-Miwa classification [15] (2.24) is the degree 4 member of the KdV hierarchy, while (2.26) and (2.30) are the degree 5 and 6 members, respectively. Note that (2.26) and (2.30) are also integrable alone.

2.2.5. Dissipative systems. It is in no way guaranteed that one can convert a given equation into the Hirota bilinear form. For dissipative systems one usually gets extra ordinary derivatives. As an example take the dissipative KdV equation

$$v_{xxx} - 2{v_x}^2 + v_t = \epsilon v. \tag{2.31}$$

The standard substitution (2.3) yields

$$(D_x^4 + D_x D_t) F \cdot F = \epsilon F_x F. \tag{2.32}$$

For the generalization of sine–Gordon equation

$$u_{xt} + \alpha u_x + \beta u_t = \sin(u) + 2\lambda \sin(u/2) \tag{2.33}$$

one uses $u = 4\arctan(g/f)$, which yields [16-18]

$$\begin{aligned} &[D_x D_t g \cdot f](f^2 - g^2) - gf[D_x D_t (f \cdot f - g \cdot g)] - gf(f^2 - g^2) \\ &+ [\alpha(D_x g \cdot f) + \beta(D_t g \cdot f) - \lambda g f](f^2 + g^2) = 0, \end{aligned} \tag{2.34}$$

Here we have bilinear expressions with three different kinds of coefficients, which would lead to three equations for only two dependent variables. However, we can use the indentity

$$(D_x g \cdot f)(f^2 + g^2) = \partial_x (gf)(f^2 - g^2) - \partial_x (f^2 - g^2) fg \tag{2.35}$$

and then we get from the coefficients of $f^2 - g^2$ and gf the following two equations

$$\begin{cases} (D_x D_t + \mu + \alpha \partial_x + \beta \partial_t - 1) g \cdot f = 0 \\ (D_x D_t + \mu + \alpha \partial_x + \beta \partial_t)(f \cdot f - g \cdot g) = -\lambda (f^2 + g^2). \end{cases} \tag{2.36}$$

This is not in the Hirota bilinear form due to the ordinary derivatives. If the parameters α and β vanish we get the double sine–Gordon equation, and if also $\lambda = 0$ we get the standard sine–Gordon equation (sG). Only this last one has 2SS [18].

2.2.6. Linearizable equations. A suitable transformation of the type used here may sometimes even linearize the equation. Examples of this include Burger's equation

$$u_{xx} + u u_x - u_t = 0, \tag{2.37}$$

which linearizes with $u = \partial_x \log F$, and Thomas equation

$$u_{xy} + \alpha u_x + \beta u_t + u_x u_y = 0, \tag{2.38}$$

which linearizes with $u = \log F$. For the Sharma-Tasso-Olver equation

$$u_t + 3{u_x}^2 + 3u^2 u_x + 3u u_{xx} + u_{xxx} = 0, \tag{2.39}$$

one obtains first with $u = \partial_x \log F$

$$F_{xt}F - F_tF_x + F_{xxxx}F - F_{xxx}F_x = 0, \tag{2.40}$$

but this is quickly integrated to

$$F_{xxx} + F_t = a(t)F. \tag{2.41}$$

2.3. Signals of the proper transformation

The above examples give some hints as to which kinds of transformations may succeed. In a new situation the following steps should be useful:

2.3.1. Isolate e^{x+vt}. First one should find a one-soliton solution for the equation. This can be done by assuming the ansatz $u = u(x + vt)$, substituting it into the equation and solving the remaining ODE. If this step cannot be completed the system is probably not even partially integrable.

After a 1SS has been found it should be written in a form where the $x + vt$ and v dependence enters only through e^η and its derivatives [$\eta = \alpha(v)(x - vt)$, $\alpha(v)$ is some function of the only free parameter v]. Terms that are often met include $f_\pm = 1 \pm e^\eta$, and $g = e^\eta$. There is considerable amount of ambiguity in this step. When the solution is written in terms of $f_\pm$, g and their derivatives one can then try to use this as a bilinearizing substitution.

If one wants to go further along this direction the next step would be to check if the equations allows 2SS of Hirota type [19]

2.3.2. Painlevé analysis. In the Hirota form the soliton solutions are expressed in terms of finite sums of exponentials (see Sec. 5-7) and the solution is therefore as regular as one can hope. This suggests the idea that the Hirota substitution for a NEE could be found by looking for a transformation that eliminates *all* singularities [20].

As an illustration consider P_{II} given by

$$u'' = 2u^3 + zu + \alpha. \tag{2.42}$$

This is known to have two kinds on singularities,

$$u \sim \frac{\pm 1}{z - z_o} + O(z - z_o). \tag{2.43}$$

Let us split u by $u = P + Q$ where P has only poles of type $\frac{+1}{z-z_o}$ and Q of type $\frac{-1}{z-z_o}$ (the splitting is otherwise arbitrary). Then $\exp(\int P\,dz)$ is an entire function, which will denoted by p, similarly $q = \exp(-\int Q\,dz)$ is entire. In terms of the entire functions p and q

$$u = \frac{p'}{p} - \frac{q'}{q}, \tag{2.44}$$

and then a substitution into (2.42) yields

$$(D_z^3 - zD_z - \alpha)q\cdot p + u[D_z^2 q\cdot p] = 0. \tag{2.45}$$

Here the singularities enter through u so we must require that each part in the expression vanishes separately. Thus the bilinear form of (2.42) is

$$\begin{cases} (D_z^3 - zD_z - \alpha)q\cdot p = 0, \\ \qquad\qquad D_z^2 q\cdot p = 0. \end{cases} \tag{2.46}$$

The main point in the above exercise is that we have *derived* the necessary substitution (2.44) from the singularity structure. The above example was for an ODE, but this idea should work for NEE's as well. In fact, along similar lines it has been noted that for many integrable systems the truncating Weiss-Tabor-Carnevale expansion leads to the correct substitution [21]

3. Nonlinearization of bilinear equations

Above we discussed the problem of finding a bilinear form for a given NEE, but sometimes the bilinear form is the primary one. In that case one wants to find a transformation to a NEE where the new dependent variables have the correct properties. For example, the 1SS to a bilinear equation is usually of type $F = 1+e^{\eta}$, or $F = 1+e^{\eta}$, $G = 1-e^{\eta}$, or $F = 1$, $G = e^{\eta}$ etc., but these functions do not behave like solitons, but rather grow exponentially in some directions. To get a genuine soliton interpretation we must express F and G in terms of new variables, which approach zero (or at least a finite constant) at infinities. There is no algorithmic way to do this, but there is a more or less canonical first step:
For KdV type equations one defines a new dependent variable u by

$$F = e^u, \tag{3.1}$$

for mKdV and SG one uses

$$F = e^u \cos w, \quad G = e^u \sin w, \tag{3.2}$$

or

$$F = e^{u+v}, \quad G = e^{u-v}, \tag{3.3}$$

depending on the choice of vacuum, and for nlS

$$F = e^u, \quad G = we^u. \tag{3.4}$$

(Definitions (3.2) and (3.4) are chosen so that the 0SS $F = 1$, $G = 0$ corresponds to $u = w = 0$, while in (3.3) $u = v = 0$ gives the $F = G = 1$ vacuum.)

The substitution above is often just the first step. Since e.g. u grows linearly as x, y or t goes to infinity, one needs to define a new variable $\bar{u}$ obtained by taking one or two derivatives, e.g. as in (2.3). Or if w is the primary dependent variable, one may define v by $v = w_x$ as was done for the mKdV equation. After such changes one can often extract derivatives of the equations (c.f. (2.12)). Since these steps are purely ad hoc we will not pursue them further here and leave open the problem of finding the 'best' NEEs corresponding to the bilinear equations.

4. Integrability vs. partial integrability

In the following sections we will discuss the existence of various multi-soliton solutions to bilinear equations. Partial integrability will in this context mean that the system allows only some soliton-solutions but not all kinds.

To construct solutions we must start by considering the vacuum solution or the 0SS, on top of which the multisoliton solutions are built. Since the nonlinearizing substitution often involves expressions like $\log F$, the 0SS is usually a nonzero constant solution. Depending on the equation there might even be a continuously parametrized collection of vacuum solutions.

Next we fix the vacuum and consider the possible 1SS's. These are usually linear in e^η ($\eta = \vec{p}\cdot\vec{x}+\eta^0$, and the parameters $\vec{p}$ are restricted by a dispersion relation). The existence of 1SS imposes only minimal conditions on the bilinear equations, in fact equations in Hirota bilinear form generically have 1SS's, sometimes even several different types of 1SS's.

For completely integrable systems these 1SS's can be combined to make NSS's for any N, so we formalize the combination idea as follows:

> DEFINITION: A set of bilinear equations is *Hirota integrable* if one can combine N one-soliton solutions of any type to make an NSS, and the combination in question is a finite polynomial in the e^η's involved.

The finite polynomial mentioned in the definition is determined from the equations recursively, so the key question is whether the recursive definition yields a truncated or infinite expression. (An infinite expansion might be acceptable if it converges, but in the cases that have been studied nontruncating expressions has been found to diverge.) This introduces several kinds of partial integrability, for the expression might truncate only for particular number or kinds of starting 1SS's, or parameter values.

Even the existence of 2SS is not automatic for all equations of Hirota form, however there are several classes of bilinear equations which are *generic* in the sense that for them 2SS's always exists for any dispersion relation that satisfies certain symmetry properties. In the rest of these lectures we will restrict our discussion to the following basic types:

$$P(D_{\vec{x}})F\cdot F = 0, \qquad \text{KdV}$$

$$\begin{cases} B(D_{\vec{x}})G\cdot F = 0, \\ A(D_{\vec{x}})(F\cdot F + G\cdot G) = 0, \end{cases} \qquad \text{mKdV and SG}$$

$$\begin{cases} B_c(D_{\vec{x}})G\cdot F = 0, \\ A(D_{\vec{x}})F\cdot F = |G|^2, \end{cases} \qquad \text{nlS}$$

In the above B is either odd (mKdV) or even (SG) function of its variables, P and A are even, and B_c is complex with the property $[B_c(\vec{X})]^* = B_c(-\vec{X}^*)$. For nlS-type the function G is complex. The mKdV type equation can also be written as the pair $B(D_{\vec{x}})G{\cdot}F = 0$, $A(D_{\vec{x}})G{\cdot}F = 0$, after a rotation in the (F, G) space. The above equations are generic in the sense that they have 1SS and 2SS for any P, B, or A (in this context a complex 1SS for nlS counts as a 2SS).

5. KdV-type equations

5.1. The vacuum solution and the one-soliton solution

In Sec.1 we showed how the KdV-equation can be written as a bilinear equation (2.6), which is of the generic type

$$P(D_{\vec{x}})F{\cdot}F = 0. \tag{5.1}$$

We will now construct the 0SS and 1SS for (5.1).

From KdV equation we know that it has constant solutions, even if we take the integrated form. This corresponds to the solutions $F = e^{\vec{q}\cdot\vec{x}+const.}$. However, this $\vec{x}$-dependency in not significant in the construction of multisoliton solutions, for bilinear equations are invariant under an overall change of all dependent variables:

$$P(D_{\vec{x}})(e^{\eta}F){\cdot}(e^{\eta}G) = e^{2\eta}P(D_{\vec{x}})F{\cdot}G, \quad \eta = \vec{q}\cdot\vec{x} + \text{ const.} \tag{5.2}$$

Thus we can take $F = 1$ as the vacuum or 0SS. This imposes for P in (5.1) the condition

$$P(0) = 0. \tag{5.3}$$

The 1SS is built on top of the 0SS, i.e. we assume that

$$F = 1 + e^{\eta}, \quad \eta = \vec{p}\cdot\vec{x} + \text{ const.} \tag{5.4}$$

When (5.4) is substituted into (5.1) we find that it holds if

$$P(\vec{p}) = 0. \tag{5.5}$$

This is a condition on the parameters $\vec{p}$ of η, and is called the *dispersion relation.* If the parameter space is n-dimensional then (5.5) defines an $n-1$-dimensional submanifold, which we call the *dispersion manifold.* Many subsequent equations hold only if the parameters are in the dispersion manifold, such equalities are denoted by $\doteq$ and the dispersion manifold in question should be evident from context.

In Sec. 2 we found that equation (2.21) with $\alpha = 4$, $\beta = 2$ has a bilinear representation using not one but *two* equations, (2.24) and (2.26). For the generic case let us therefore suppose that we have a set of bilinear equations

$$P^k(D_{\vec{x}})F{\cdot}F = 0, \quad k = 1, \ldots, r, \tag{5.6}$$

for the one dependent variable F. For the 0SS $F = 1$ we get the natural generalization of (5.3) for each k. For the 1SS (5.4) we similarly find the conditions

$$P^k(\vec{p}) = 0, \quad k = 1, \ldots, r. \tag{5.7}$$

Thus in the generic case the dispersion manifold is an $n - r$ dimensional manifold. If some of the polynomials P^k have common factors the dispersion manifold may have higher dimensional components.

If in the example 2.2.3 we associate p with x and q with t, then the dispersion relation of the original equation is $p^3 q - p^2 - pq = 0$. For (2.24) and (2.26) the parameter space is 3-dimensional (r associated with τ), but we also have two conditions, $p^4 + pr = 0$ and $2p^3 q - 3p^2 - 3pq - qr = 0$. Upon eliminating r we get the dispersion relation of the original equation.

Finally we would like to note that some bilinear equations that are not of Hirota type may still have 1SS's of the above type. Assume the bilinear equation is of the more general type

$$P(\partial_{\vec{x}} - \partial_{\vec{x}'}, \partial_{\vec{x}} + \partial_{\vec{x}'})\, F(\vec{x})\, F(\vec{x}')\big|_{\vec{x}'=\vec{x}} = 0. \tag{5.8}$$

If the ansatz (5.4) is substituted into (5.8) we get the conditions:

$$P(0,0) = 0, \tag{5.9a}$$
$$P(\vec{p},\vec{p}) + P(-\vec{p},\vec{p}) = 0, \tag{5.9b}$$
$$P(0,2\vec{p}) = 0. \tag{5.9c}$$

In the proper Hirota bilinear form there is no dependence on the second variable and therefore (5.9c) is satisfied by virtue of (5.9a) and (5.9b) gives dispersion relation. In the present case (5.9b) and (5.9c) may both give nontrivial conditions on the parameter space, but for some cases interesting solutions may nevertheless be constructed this way [22].

5.2. The two-soliton solution

Hirota's ansatz for a 2SS is the truncated series

$$F = 1 + e^{\eta_1} + e^{\eta_2} + K_{12} e^{\eta_1 + \eta_2}, \tag{5.10}$$

where the η's are defined as before in (5.4), and K may be a function of $\vec{p}_1$ and $\vec{p}_2$. There are now two different parameter vectors $\vec{p}_i$, each one giving the coordinates of some point in the dispersion manifold. Eq. (5.10) is a natural generalization of (5.4) and allows some interaction through the K-term.

When (5.10) is substituted to (5.1) and we use (5.3) and (5.5) there will be one more condition,

$$K_{12}\, P(\vec{p}_1 + \vec{p}_2) + P(\vec{p}_1 - \vec{p}_2) \doteq 0. \tag{5.11}$$

In the generic case we can solve for K and thus construct the 2SS.

Special analysis is necessary if either or both of the polynomials $P(\vec{p}_1 \pm \vec{p}_2)$ vanish on the manifold, i.e. the solitons *resonate* [23]. If both vanish we get no condition on K, and if $P(\vec{p}_1 - \vec{p}_2) \doteq 0$ we must take $K = 0$, i.e. $F = 1 + e^{\eta_1} + e^{\eta_2}$. If $P(\vec{p}_1 + \vec{p}_2) \doteq 0$ but $P(\vec{p}_1 - \vec{p}_2) \not\doteq 0$ it seems that we should take $K = \infty$. However, we could have introduced the free constant K equally well as the coefficient of the e^{η_2} term, say, by redefining the constant additive term in η_2. In (5.11) K would then have appeared as the coefficient of $P(\vec{p}_1 - \vec{p}_2)$ and should be set to 0 in this special case, thus $F = 1 + e^{\eta_1} + e^{\eta_1 + \eta_2}$. This solution is of the same type as above for we can write it also as $F = e^{\eta_1}[1 + e^{-\eta_1} + e^{\eta_2}]$ and the overall factor is irrelevant, as was shown in (5.2).

If we have r bilinear equations for the one function F as in (5.6) we also get r equations for the constant K. Such sets of bilinear equation do not therefore generically have 2SS, but only if the various P^k's are compatible in the sense that the K's calculated from them agree on the dispersion manifold. (For the example above (2.24, 2.26 and 2.30) one can indeed verify that all of the bilinear equations yield $K = [(p_1 - p_2)/(p_1 + p_2)]^2$.) Even if the polynomials fail this condition they may be partially integrable if there are some particular parameter pairs, for which compatibility is obtained.

5.3. The three-soliton condition

We showed above that a 2SS can always be constructed for a single bilinear equation of type (5.1). However, the existence of higher soliton solutions is not automatic. A natural extension of the above procedure is to use the expansion [2]

$$F = 1 + \epsilon \sum_{i=1}^{N} e^{\eta_i} + O(\epsilon^2), \tag{5.12}$$

where ϵ is a formal expansion parameter. When (5.12) is substituted into (5.1) the terms with higher ϵ-powers are obtained recursively. For known integrable systems this yields for F an expansion that truncates with ϵ^N being the highest order term [1]:

$$F = \sum_{\mu_i \in \{0,1\}} \exp\left(\textstyle\sum_{0<i<j\le N} A_{ij}\mu_i\mu_j + \sum_{i=1}^{N} \eta_i\mu_i\right), \tag{5.13}$$

where $\exp A_{ij} = K_{ij}$ given by (5.11). If one attempts to use (5.13) for the general equation (5.1) one finds that the expansion truncates only if [1]

$$\sum_{\sigma_i = \pm 1} P\left(\textstyle\sum_{i=1}^{k} \sigma_i \vec{p}_i\right) \prod_{0<i<j\le k} \left[P(\sigma_i \vec{p}_i - \sigma_j \vec{p}_j)\sigma_i\sigma_j\right] \doteq 0, \tag{5.14}$$

for $k = 3, \ldots, N$.

5.4. Some results

Since all equations of type (5.1) have 2SS we must next consider the possibilities of satisfying condition (5.14) for the easiest nontrivial case $k = 3$. Normally (5.14) should be taken as a condition for P, so that if P satisfies (5.14) for any set $\vec{p}_i$ on the dispersion manifold, then the system has 3SS of any type. The condition (5.14) can also be taken for the parameters. In that case we get another restriction on the parameter space, but it will be possible to get 3SS's at least for some parameter values. Examples of this include higher dimensional models, which then have 3SS's if the plane waves are at fixed angles to each other.

The existence of 3SS is quite demanding so it is interesting to find out which special equations satisfy the three-soliton condition. In [4] we have studied this question and searched for all polynomials satisfying the three soliton condition (3SC). The necessary computations are very cumbersome and impractical without computer algebra. We used REDUCE [24] for our search. This search allowed the possibility of the polynomial P having multiple factors, which needs special care when the dispersion relation is implemented for a computer algebra system. In the search we first looked for the possible leading monomials and then for the remaining terms. Surprisingly few polynomials passed the test, the results are as follows: We found four different truly nonlinear models

a) Kadomtsev-Petviashivili equation [25]:

$$(D_x^4 - 4D_xD_t + 3D_y^2)F\cdot F = 0 \tag{5.15}$$

b) Ito equation [26]:

$$(D_x^3D_t + aD_x^2 + D_tD_y)F\cdot F = 0 \tag{5.16}$$

c) A new 1+1 dimensional combination of a) and b):

$$(D_x^4 - D_xD_t^3 + aD_x^2 + bD_xD_t + cD_t^2)F\cdot F = 0 \tag{5.17}$$

d) Ramani equation [27]:

$$(D_x^6 + 5D_x^3D_t - 5D_t^2 + D_xD_y)F\cdot F = 0. \tag{5.18}$$

It should be noted that all *projections* of the above results pass the test as well, where by projection we mean any linear transformation of the variables which does not increase the number of variables. For example, the KdV-equation is obtained from a) by $(x,t,y) \to (x,t,0)$, Boussinesq by $(x,t,y) \to (x, \frac{1}{2}it, (x+it)/\sqrt{3})$, Sawada-Kotera from d) by $(x,t,y) \to (x,0,t)$ and so on.

In addition to the above we obtained also several equations with dispersion manifolds consisting of linear subspaces:

e) One-dimensional models, including

$$D_x^2[D_x^2(D_x^2+a)^2+1]F\cdot F = 0. \tag{5.19}$$

f) $$(D_x^2-1)^N D_x^M D_t^P F\cdot F = 0, \text{ for } M \text{ and } P \text{ odd.} \tag{5.20}$$

g) $$D_x^m D_t^n D_y^p F\cdot F = 0, \text{ when two of the powers are odd.} \tag{5.21}$$

h) $$D_xD_tD_yD_zF\cdot F = 0. \tag{5.22}$$

In [4] we checked only for the existence of 3SS, but we have now verified that cases a) to g) also have 4SS [cases f) and g) were checked up to degree 8]. However, case

h) and its projections have 4SS only when the projection is to one of the leading terms of cases a) to c) or g). Thus the existence of 4SS and the existence of 3SS with arbitrary quadratic terms in the polynomial impose in some way similar additional conditions on the leading part.

If (5.14) is not satisfied in general it may hold on a component of the manifold. This approach was taken by Newell and Yungbo [28]. They fixed the dispersion manifold by $p = (k, -k^3, k^5, \ldots)$ and K in (5.11) to be given by the KdV hierarchy, and then searched for polynomials having NSS under these circumstances.

6. mKdV- and SG-type equations

In this section we consider the pair

$$\begin{cases} B(D_{\vec{x}})G\cdot F = 0, \\ A(D_{\vec{x}})(F\cdot F + G\cdot G) = 0. \end{cases} \tag{6.1}$$

Here the polynomial B is either even or odd, and A is even.

6.1. The vacuum solution and the one-soliton solution

For the vacuum solution we assume $F = 1$, $G = 0$. The ansatz for a 1SS is therefore

$$F = 1 + \alpha e^{\eta}, \quad G = \beta e^{\eta}, \quad \eta = \vec{p}\cdot\vec{x} + \eta^0, \tag{6.2}$$

which, when substituted to (6.1), yields the conditions

$$\beta B(\vec{p}) = 0, \quad \alpha A(\vec{p}) = 0. \tag{6.3}$$

There are now two obvious (normalized) ways to satisfy this equation:

α) $\alpha = 1$, $\beta = 0$ with the dispersion relation $A(\vec{p}) = 0$, and

β) $\alpha = 0$, $\beta = 1$ with the dispersion relation $B(\vec{p}) = 0$.

It might sometimes happen that $A(\vec{p})$ and $B(\vec{p})$ vanish at the same time, in which case the constants α and β are left undetermined. This case will be discussed elsewhere [29].

If the A equation is auxiliary in the above pair of bilinear equations then it might not be meaningful to include the α possibility, but if A is truly nonlinear the α-solitons may be physical and should be included.

6.2. The two-soliton solutions

We can now make 2SS from α- and/or β-type 1SS. The part that is linear in e^{η} is taken directly from above, in addition we will have an interaction term $e^{\eta_1+\eta_2}$. There are three different cases:

$\alpha+\alpha$) The ansatz is

$$\begin{cases} F = 1 + e^{\eta_1} + e^{\eta_2} + K\, e^{\eta_1+\eta_2}, \\ G = L\, e^{\eta_1+\eta_2}, \end{cases} \tag{6.4}$$

with $A(\vec{p}_i)=0$ and after substituting to (6.1) we find

$$K=-\frac{A(\vec{p}_1-\vec{p}_2)}{A(\vec{p}_1+\vec{p}_2)},\quad L=0. \tag{6.5}$$

Note that in this case the system reduces to KdV-type.

$\alpha+\beta$) Now we start with

$$\begin{cases} F=1+e^{\eta_1}+K\,e^{\eta_1+\eta_2}, \\ G=e^{\eta_2}+L\,e^{\eta_1+\eta_2}, \end{cases} \tag{6.6}$$

with $A(\vec{p}_1)=0$ and $B(\vec{p}_2)=0$ and obtain

$$K=0,\quad L=-\frac{B(\vec{p}_1-\vec{p}_2)}{B(\vec{p}_1+\vec{p}_2)}. \tag{6.7}$$

$\beta+\beta$) Finally we can have

$$\begin{cases} F=1+K\,e^{\eta_1+\eta_2}, \\ G=e^{\eta_1}+e^{\eta_2}+L\,e^{\eta_1+\eta_2}, \end{cases} \tag{6.8}$$

with $B(\vec{p}_i)=0$, and as in the $\alpha+\alpha$ case we find

$$K=-\frac{A(\vec{p}_1-\vec{p}_2)}{A(\vec{p}_1+\vec{p}_2)},\quad L=0. \tag{6.9}$$

The above constructions work for any polynomials A and B with correct symmetry properties. In this way (6.1) is a generic bilinear equation that is at least partially integrable since it always has 2SS of several kinds.

6.3. The three-soliton condition

The existence of 3SS and in general NSS is again much more restrictive. As before there is a recursively defined NSS ansatz, If the solitons $1,\ldots,K$ are of type β and $K+1,\ldots,N$ are of type α the ansatz is [8]

$$F=\sum_{\mu\in S^0}\exp\left(\textstyle\sum_{0<i<j\le N}A_{ij}\mu_i\mu_j+\sum_{i=1}^N\eta_i\mu_i\right), \tag{6.10a}$$

$$G=\sum_{\mu\in S^1}\exp\left(\textstyle\sum_{0<i<j\le N}A_{ij}\mu_i\mu_j+\sum_{i=1}^N\eta_i\mu_i\right), \tag{6.10b}$$

where

$$S^a=\{(\mu_1,\ldots,\mu_N)|\mu_i\in\{0,1\},\ \textstyle\sum_{i=1}^K\mu_i=a\ (\mathrm{mod}\ 2)\}. \tag{6.11}$$

The coefficients $e^{A_{ij}}$ were obtained in the construction of the 2SS above,

$$e^{A_{ij}}=\begin{cases} -A(\vec{p}_i-\vec{p}_j)/A(\vec{p}_i+\vec{p}_j), & \forall K<i<j,\ \text{or}\ i<j\le K \\ -B(\vec{p}_i-\vec{p}_j)/B(\vec{p}_i+\vec{p}_j), & \forall i\le K<j. \end{cases} \tag{6.12}$$

As usual these ansatze do not give NSS for arbitrary polynomials A and B but only for those that satisfy the additional conditions [8]

$$\sum_{\sigma} P^*((\prod_{i=1}^{K}\sigma_i)(\sum_{i=1}^{N}\sigma_i\vec{p}_i)) \prod_{0<i<j\leq N} [P_{ij}(\vec{p}_i - \sigma_i\sigma_j\vec{p}_j)\sigma_i\sigma_j] \doteq 0, \qquad (6.13a)$$

where

$$P^* = \begin{cases} B, & \text{when } K \text{ is odd,} \\ A, & \text{when } K \text{ is even,} \end{cases} \qquad (6.13b)$$

$$P_{ij} = \begin{cases} A, & \text{for } 0 < i,j \leq K \text{ or } K < i,j \leq N, \\ B, & \text{for } 0 < i \leq K < j \leq N, \end{cases} \qquad (6.13c)$$

and the dispersion relations are given by

$$\begin{cases} B(\vec{p}_i) = 0, & \text{for } 0 < i \leq K, \\ A(\vec{p}_i) = 0, & \text{for } K < i \leq N. \end{cases} \qquad (6.13d)$$

6.4. Some results

In [5,6,8] we have searched for the polynomials A and B which pass the 3SC and 4SC. In [5,6] the polynomial A was quadratic and treated as auxiliary, therefore the α type solitons were ignored. In [8] we included the possibility of α-type solitons; this means that for the existence of 3α and 4α solitons the A polynomial must be quadratic or in the list a)-d) (5.15-18) or its projections. We did not consider the degenerate cases e)-g). The following list contains all pairs of polynomials (up to projections) that have at least two variables and 3SS and 4SS of all combinations:

a) $$A = D_x^2, \quad B = aD_x^7 + bD_x^5 + D_x^2 D_t + D_y, \qquad (6.14)$$

b) $$A = D_x D_t, \quad B = aD_x^3 + bD_t^3 + D_y, \qquad (6.15)$$

c) $$A = D_x D_t, \quad B = D_x D_t D_y + aD_x + bD_t, \qquad (6.16)$$

d) $$A = D_x^3 D_t + aD_x^2 + D_t D_y, \quad B = D_x^3 + D_y, \qquad (6.17)$$

e) $$A = D_x^6 + 5D_x^3 D_t - 5D_t^2 + D_x D_y, \quad B = D_x^3 + D_t. \qquad (6.18)$$

f) $$A = D_x D_t, \quad B = aD_x^3 D_t + D_t D_y + b, \qquad (6.19)$$

g) $$A = D_x^3 D_t + 3bD_x^2 + D_t D_y, \quad B = D_x D_t + b, \qquad (6.20)$$

All of these are previously unknown generalizations of the well known mKdV, and SG equations: a)-e) are mKdV type while f) and g) are SG-type. Note the nice symmetry properties between the x and t variables in b) and c). d) and g) contain an Ito-type A-polynomial (5.16) within the mKdV- and SG-equation, respectively. e) is a combination of Ramani-equation (5.18) with mKdV.

We found also a number of equations that allowed certain 3-soliton combinations but not others. For example

$$A = D_x D_t, \quad B = D_x^2 - D_t^2 + c \qquad (6.21)$$

and

$$A = D_x^4 + 2D_xD_t, \quad B = D_x^3 - D_t \tag{6.22}$$

have all 3SS and 4SS combinations *except* $\alpha + 2\beta$ and $2\alpha + 2\beta$. Similarly

$$A = D_x^3D_t + 2D_t^2, \quad B = D_x^3 - D_t \tag{6.23}$$

and

$$A = D_x^3D_t + D_xD_t - 2D_t^2, \quad B = D_x^3 + D_x + D_t \tag{6.24}$$

have 3β, 4β, and $\alpha + 3\beta$ but not $\alpha + 2\beta$, $2\alpha + \beta$, $2\alpha + 2\beta$, nor $3\alpha + \beta$.

As usual the 3SC can be taken as a condition on parameters, if it is not satisfied on the whole dispersion manifold. For example the multidimensional SG-equation has been treated in [30]. Multisoliton solutions are usually allowed only if the solitons are parallel, which effectively reduces the problem to a standard 1+1 dimensional one.

7. nlS-type equations

Finally we present our results for nlS-type equations (for further details see [8,31]. We consider equations of the following general form

$$\begin{cases} B(D_x, \ldots)G \cdot F = 0, \\ A(D_x, \ldots)F \cdot F = |G|^2. \end{cases} \tag{7.1}$$

The vacuum is obviously given by $F = 1$, $G = 0$ and on top of it we can again have two kinds of solitons:

α) KdV-type solitons

$$G = 0, \ F = 1 + e^{\eta_A}, \tag{7.2}$$

with dispersion relation given by $A(\vec{p}) = 0$

β) The standard nlS-soliton

$$G = e^{\eta_B}, \ F = 1 + Ke^{\eta_B + \eta_B^*}, \tag{7.3}$$

with dispersion relation from $B(\vec{p}) = 0$, and $K = 1/[2A(\vec{p} + \vec{p}^*)]$. Note that this looks like a 2SS.

Multisoliton solutions combining the above are constructed along the lines given in Sec. 6. There is the added complication due to the fact that η and η^* both enter. The proper ansatz for M solitons of type α and N solitons of type β is [8]

$$\begin{aligned} F &= \sum_{\mu \in S^0} \exp\left(\textstyle\sum_{i<j} A_{ij}\mu_i\mu_j + \sum_i \eta_i\mu_i\right), \\ G &= \sum_{\mu \in S^1} \exp\left(\textstyle\sum_{i<j} A_{ij}\mu_i\mu_j + \sum_i \eta_i\mu_i\right). \end{aligned} \tag{7.4}$$

where the indices run from 1 to $2N+M$, and in this case

$$S^a = \{(\mu_1, \ldots, \mu_{2N+M}) | \mu_i \in \{0,1\}, \textstyle\sum_{i=1}^{N} \mu_i - \sum_{i=N+1}^{2N} \mu_i = a.\} \tag{7.5}$$

Here η_i is associated with a nlS-type soliton β for $i = 1, \ldots, N$, $\eta_{i+N} = \eta_i{}^*$, and for $i = 2N+1, \ldots, 2N+M$ η_i is associated with a KdV-type soliton α, thus the dispersion relations are

$$\begin{cases} B(\vec{p}_i) = 0, & \text{for } i = 1, \ldots, N \\ B(-\vec{p}_i) = 0, & \text{for } i = N+1, \ldots, 2N \\ A(\vec{p}_i) = 0, & \text{for } i = 2N+1, \ldots, 2N+M. \end{cases} \tag{7.6}$$

The middle equation follows from the first, because of the assumed property of the B polynomial $[B(\vec{p}_i)]^* = \pm B(-\vec{p}_i{}^*)$. For the coefficients A_{ij} we have

$$e^{A_{ij}} = \begin{cases} 2A(\vec{p}_i - \vec{p}_j), & \text{for } 1 \le i < j \le N, \\ 1/[2A(\vec{p}_i + \vec{p}_j)], & \text{for } 1 \le i \le N < j \le 2N, \\ 2A(\vec{p}_i - \vec{p}_j), & \text{for } N < i < j \le 2N, \\ -B(\vec{p}_i - \vec{p}_j)/B(\vec{p}_i + \vec{p}_j), & \text{for } 1 \le i \le N,\ 2N+1 \le j \le 2N+M. \end{cases} \tag{7.7}$$

Using the above ansatz one can derive conditions for the existence of multi-soliton solutions [8]. We have searched for polynomials A and B having 3SS and 4SS. For A we had a quadratic polynomial in [7] and one of the KdV-type results in [8]. The *complete* results are contained in the following three equations (various transformations and projections are allowed, as usual):

a) Melnikov's equation:

$$B = D_x^2 + iD_y + c, \quad A = a(D_x^4 - 3D_y^2) + D_x D_t. \tag{7.8}$$

The special case with $a = 0$ was obtained in [7], it goes to nlS if $(x,t,y) \to (x,x,t)$. For $G = 0$ the second equation reduces to the KP-equation so the result is a combination of nlS and KP.

b) A combination of KP and DS:

$$B = i\alpha D_x^3 + 3D_xD_y - 2iD_t + c,\ A = a(\alpha^2 D_x^4 - 3D_y^2 + 4\alpha D_xD_t) + bD_x^2. \tag{7.9}$$

In contrast to (7.8) both equations in (7.9) have 3 variables. Thus this equation is a true 2+1 combination of KP and DS equations.

c) Generalization of Ito's equation:

$$B = i\alpha D_x^3 + 3cD_x^2 + i(bD_x - 2dD_t) + g,\ A = \alpha D_x^3 D_t + aD_x^2 + (b+3c^2)D_xD_t + dD_t^2. \tag{7.10}$$

This is a combination of Ito's equation (5.16) and Hirota's generalization of nlS [11].

8. Conclusions

In these lectures we have considered partial integrability from the point of view of Hirota's bilinear method. Partial integrability enters naturally with the levels of integrability characterized by the number of solitons that can be combined to an N-soliton solution:

1) First there are equations that cannot be put into the Hirota bilinear form at all. These are the least integrable systems in this classification.
2) All bilinear equations have 1SS's for some dispersion manifold, which is in fact defined by this statement. These equations may or may not have 2SS.
3) There are generic classes of equations the have 2SS irrespective of the actual form of the dispersion relation. Other types of equations may have 2SS for specific polynomials.
4) The existence of 3SS is very rare and equations that have 3SS combinations of any type are probably integrable. Some equations can only have 3SS of certain types.
5) Those equations that have 3SS of any type also have 4SS. [The only known counterexamples have dispersion manifolds that are composed of linear subspaces.] These systems are conjectured to be completely integrable.

At each of the above levels it is possible to have higher soliton solutions by restricting the dispersion manifold further.

As we have seen in these lectures, Hirota's bilinear method is very useful in the study of partial integrability. We have given here only the basic results along with some examples; other applications will no doubt be found.

Acknowledgements
I would like to thank B. Grammaticos, M. Kruskal and A. Ramani for discussions.

REFERENCES

[1] R. Hirota, Phys. Rev. Lett. **27**, 1192 (1971).
[2] R. Hirota, in: *Bäcklund Transformations, the Inverse Scattering Method, and Their Applications*, ed R.M. Miura, p. 40. Springer-Verlag, Berlin, (1976).
[3] R. Hirota, in: *Solitons*, eds R.K. Bullough and P.J. Caudrey, p. 157. Springer-Verlag, Berlin (1980).
[4] J. Hietarinta, J. Math. Phys. **28**, 1732 (1987).
[5] J. Hietarinta, J. Math. Phys. **28**, 2094 (1987).
[6] J. Hietarinta, J. Math. Phys. **28**, 2586 (1987).
[7] J. Hietarinta, J. Math. Phys. **29**, 628 (1988).
[8] J. Hietarinta, work in progress (1989).
[9] R. Hirota and J. Satsuma, Suppl. Prog. Theor. Phys. **59**, 64 (1976).

[10] R. Hirota, J. Phys. Soc. Jpn. **33**, 1456 (1972).

[11] R. Hirota, J. Math. Phys. **14**, 805 (1973).

[12] R. Hirota, J. Phys. Soc. Jpn. **51**, 323 (1982).

[13] R. Hirota and J. Satsuma, J. Phys. Soc. Jpn. **40**, 611 (1976).

[14] J. Satsuma and D.J. Kaup, J. Phys. Soc. Jpn. **43**, 692 (1977).

[15] M. Jimbo, and T. Miwa, Publ. RIMS, Kyoto Univ. **19**, 943 (1983).

[16] R. Hirota, J. Phys. Soc. Jpn. **33**, 1459 (1972).

[17] R. Hirota, Prog. Theor. Phys. **52**, 1498 (1974).

[18] P.J. Caudrey, in *Nonlinear Equations in Physics and Mathematics*, ed. A.O. Barut, p. 177. Reidel, Dordrecht (1978).

[19] F. Lambert and R. Wilcox, these proceedings.

[20] M. Kruskal and J. Hietarinta, work in progress

[21] J.D. Gibbon, P. Radmore, M. Tabor and D. Wood, Stud. Appl. Math. **72**, 39 (1985); A.C. Newell, M. Tabor and Y.B. Zeng, Physica **29D**, 1 (1987).

[22] K. Nozaki and N. Bekki, J. Phys. Soc. Jpn. **53**, 1581 (1984); R. Hirota, in *Dynamical Problems in Soliton Systems*, ed. S. Takeno, p. 42. Springer-Verlag (1985).

[23] R. Hirota and M. Ito, J. Phys. Soc. Jpn. **52**, 744 (1983).

[24] A.C. Hearn, *REDUCE User's Manual Version 3.2*, Publ. CP78, Rev 4/85. Rand, Santa Monica (1985).

[25] J. Satsuma, J. Phys. Soc. Jpn. **40**, 286 (1976).

[26] M. Ito, J. Phys. Soc. Jpn. **49**, 771 (1980).

[27] A. Ramani, in *Fourth International Conference on Collective Phenomena*, ed. J. Lebowitz, p. 54. New York Academy of Sciences, New York (1981).

[28] A.C. Newell and Z. Yungbo, J. Math. Phys. **27**, 2016 (1986).

[29] J. Hietarinta, B. Grammaticos and A. Ramani, work in progress.

[30] R. Hirota, J. Phys. Soc. Jpn. **35**, 1566 (1973); K. Kobayashi and M. Izutzu, J. Phys. Soc. Jpn. **41**, 1091 (1976).

[31] J. Hietarinta, in *Nonlinear Evolution Equations: Integrability and Spectral Methods*, eds. A. Degasperis and A.P. Fordy, Manchester University Press (1989).

GENERALIZED SYMMETRIES, RECURSION OPERATORS AND BIHAMILTONIAN SYSTEMS

Y. KOSMANN-SCHWARZBACH
U.F.R. de Mathématiques
Université de Lille I
F-59655 Villeneuve d'Ascq
France

ABSTRACT. We outline the symmetry approach to the theory of integrable systems, and its geometrical interpretation. We indicate the role of the generalized symmetries and recursion operators, in particular of the Nijenhuis recursion operator of a bihamiltonian system.

It is generally agreed that the "amount" of symmetry that an evolution equation possesses is an indication of "how integrable it is". In the lectures that we briefly summarize below, we outlined the main definitions and methods that figure in the theory of generalized symmetries. In principle, a knowledge of the generalized symmetries of an equation permits a determination of some of its exact solutions by the method of reduction. In addition, the introduction of the generalized symmetries is indispensable in order to give a simple and general formulation of Nœther's theorem, which establishes a one-to-one correspondence between classes of generalized symmetries and classes of conservation laws [14] [1] [29]. Thus the tools provided by the theory of symmetries are employed in the definition, study and solution, to the extent that it may be possible, of the "partially integrable evolution equations". Whatever the precise definition of these equations may be, the Lie algebra of their symmetries already figures in their study and will doubtless play a still more important rôle in the further investigations of this field. Since it is impossible to furnish a complete bibliography on so vast a subject we have only provided references to basic works where further references may be found, and to some recent articles whose results are quoted here.

R. Conte and N. Boccara (eds.), Partially Integrable Evolution Equations in Physics, 479–489.

1. Classical and generalized infinitesimal symmetries

We consider a system of p' partial differential equations in p unknowns in n independent variables

$$Du = 0.$$

Such an equation has symmetries of various types, and we shall consider them in order of increasing generality: 1. *classical* (or *Lie-point*) symmetries to which correspond the classical infinitesimal symmetries, 2. in the case of a scalar equation ($p = p' = 1$), *contact* symmetries to which correspond the infinitesimal contact symmetries, 3. *generalized infinitesimal symmetries* (commonly called Lie-Bäcklund transformations, although they are by no means special types of Bäcklund transformations), 4. *nonlocal infinitesimal symmetries.* The last two types of infinitesimal symmetries do not in fact correspond to groups of global transformations.

We shall consider u to be a section of the trivial vector bundle $F = \mathbb{R}^n \times \mathbb{R}^p \longrightarrow \mathbb{R}^n$. We use the notations $x = (x^1,\dots,x^n) = (x^i)$ for the independent variables, and $u = (u^1,\dots,u^p) = (u^\alpha)$ for the dependent variables.

1.1. CLASSICAL INFINITESIMAL SYMMETRIES

The infinitesimal generator of a one-parameter group of symmetries can be viewed either as a vector field,

$$X = X^i(x,y) \frac{\partial}{\partial x^i} + Y^\alpha(x,y) \frac{\partial}{\partial y^\alpha} ,$$

or as a differential operator,

$$(Xu)^\alpha(x) = Y^\alpha(x,u(x)) - X^i(x,u(x)) \frac{\partial u^\alpha}{\partial x^i}(x). \tag{1}$$

The fundamental criterion for the determination of the infinitesimal symmetries of the equation $Du = 0$ is whether the differential operator X has the property that

$$Du = 0 \quad \text{implies} \quad VD(u,Xu) = 0, \tag{2}$$

where

$$VD(u,v) = \frac{d}{dt}D(u(t))\Big|_{t=0} \quad , \text{ with } u(0) = u, \ \frac{du}{dt}(t)\Big|_{t=0} = v,$$

is the Fréchet derivative of D. In coordinates, x^i, u^α, u^α_i, ..., $u^\alpha_{I(k)}$,

$$VD(u,v) = \frac{\partial D}{\partial u^\alpha} v^\alpha + \frac{\partial D}{\partial u^\alpha_i} v^\alpha_i + \dots + \frac{\partial D}{\partial u^\alpha_{I(k)}} v^\alpha_{I(k)} .$$

When X has that property, then it is called a *classical infinitesimal symmetry* of the equation $Du = 0$. This criterion is equivalent to the one using prolongations of vector fields [1]. Many results on the determination of Lie algebras of classical infinitesimal symmetries for ordinary differential equations go back to Sophus Lie. For evolution equations, see the references given in the bibliography, where most of the examples are in the case of the dimension 1+1. For the 1+2 case, see Winternitz's survey in this volume [11].

1.2. GENERALIZED INFINITESIMAL SYMMETRIES

More generally, we may consider a nonlinear differential operator X on the sections of the trivial bundle F, with values in F. Formula (1) becomes a special case of this more general situation, corresponding to a quasilinear first-order differential operator X. Criterion (2) is formally valid and yields a necessary and sufficient condition for a nonlinear differential operator X to be a *generalized infinitesimal symmetry* of the equation $Du = 0$.

Geometrically, nonlinear differential operators of order k on the sections of F are interpreted as *evolution fields* of order k, *i. e.*, vector-bundle mappings from the k-th jet bundle of F to the vertical bundle VF. They are vector fields on the infinite jet bundle of F which are vertical (with respect to the standard projection of the infinite jet bundle onto the base manifold $\mathbb{R}^n$) and which commute with the n vector fields of total derivation

$$\frac{\partial}{\partial x^i} + u_i^\alpha \frac{\partial}{\partial u^\alpha} + \dots + u^\alpha_{I(k)+1_i} \frac{\partial}{\partial u^\alpha_{I(k)}} + \dots , \qquad i = 1, 2, \dots, n.$$

(Here $I(k) + 1_i$ indicates the multiindex of length k+1 obtained from I(k) by adding 1 in the i-th place.) The Lie bracket of two evolution vector fields (called the *vertical bracket* in [24] - [28]) is the Lie bracket of vector fields on the infinite jet bundle of F. In terms of Fréchet derivatives, the Lie bracket [X,Y] of evolution fields X and Y is

$$[X,Y] = VY \circ X - VX \circ Y , \tag{3}$$

the opposite of the commutator of the nonlinear operators X and Y. (We have denoted the differential operator $u \longrightarrow VX(u,Yu)$ by $VX \circ Y$.) In the case of a single independent variable, it coïncides with the Gel'fand-Dikii bracket.

On a bundle, $F = \mathbb{R}^n \times \mathbb{R} \longrightarrow \mathbb{R}^n$, that has a one-dimensional fiber, the first prolongation of an evolution field of order 1 is a *contact vector field* on the first jet bundle of F. To determine the *infinitesimal contact symmetries* of a scalar partial differential equation, one must find its generalized infinitesimal symmetries of order 1. See [9] for the determination of the Lie algebra of the infinitesimal contact symmetries for evolution equations of order at least 2.

1.3. FLOWS. SYMMETRIES OF EVOLUTION EQUATIONS

If the autonomous evolution equation

$$u_t = Xu \tag{4}$$

has a unique solution, $U(t,x)$, with initial condition $u(x)$, we say that μ_t defined by

$$U(t,x) = (\mu_t u)(x)$$

is the *flow* of the evolution vector field X. The symmetries of (4) are the evolution fields Y such that

$$\frac{\partial Y}{\partial t} + [X,Y] = 0.$$

In particular, the *time-independent generalized infinitesimal symmetries* of $u_t = Xu$ are the evolution fields Y that commute with X.

1.4 ALGORITHM

The computation of the Lie algebra of the generalized infinitesimal symmetries of a given differential equation is simplified by the use of *gradings*. See section VIII.2 of [7] where results, previously announced by Krasil'shchik and Vinogradov, are explained in detail. For Burgers's equation, the Lie algebra of generalized infinitesimal symmetries is isomorphic to the Lie algebra of polynomial divergence-free vector fields on the plane. For the Korteweg-de Vries equation, the Lie algebra of generalized infinitesimal symmetries is the semi-direct product of a 2-dimensional solvable Lie algebra and the infinite-dimensional Abelian ideal of the higher-order KdV equations. The determination of the generalized infinitesimal symmetries of a given order of an equation is algorithmic, and the computations can be performed by computer [1] [18].

1.5. NONLOCAL INFINITESIMAL SYMMETRIES

In dimension 1+1, one can formally introduce *integro-differential* operators. See [16] [27] [4]. The theory of coverings recently developped by Vinogradov and Krasil'shchik [17] yields a theoretical framework for the previous computations. It is obviously an apt framework for the study of recursion operators (see below), which are often of a nonlocal nature. This is a rich theory from which one can derive the inverse scattering formulation and the theory of Bäcklund transformations. Several equations have been shown to be coverings of equations that had appeared to be unrelated. (See the forthcoming publications of M. Marvan.)

1.6. INVARIANT SOLUTIONS

As in the case of the classical infinitesimal symmetries (see [11]), one can search for solutions of an equation that are invariant under a given generalized symmetry, or under a Lie algebra of such symmetries. See *e. g.*, [7]. Thus, the determination of the generalized infinitesimal symmetries of an equation will, to some extent, permit its integration, *i. e.*, the determination of exact solutions of the equation.

2. Recursion operators

2.1. RECURSION OPERATORS. THE CASE OF EVOLUTION EQUATIONS

A *recursion operator* (or *strong symmetry* in the terminology of [22]) for the equation Du = 0 is an operator $\mathcal{R}$ on the evolution fields, which preserves the Lie algebra of generalized infinitesimal symmetries of the equation, *i. e.*, which maps a generalized infinitesimal symmetry, X, into another, $\mathcal{R}X$. (The case where $\mathcal{R}$ is a multiple of the identity is trivial and will be henceforth excluded.) See [1][6][12].

Let $u_t = Xu$ be an evolution equation. Let $\mathcal{R}$ be invariant under X, *i. e.*, $\mathcal{L}_X\mathcal{R} = 0$, where $(\mathcal{L}_X\mathcal{R})(Y) = [X,\mathcal{R}Y] - \mathcal{R}([X,Y])$. Then R is a recursion operator for $u_t = Xu$. In particular, the sequence of evolution fields X, $\mathcal{R}X$, $\mathcal{R}^2X$, ..., $\mathcal{R}^kX$, ... is a sequence of symmetries of $u_t = Xu$, that are, in general, non commuting.

One can also define time-dependent recursion operators by the condition

$$\frac{\partial\mathcal{R}}{\partial t} + \mathcal{L}_X\mathcal{R} = 0.$$

The existence of a recursion operator is strongly related to the integrability properties of the equation, since an equation which admits a recursion operator admits "many" symmetries. However, the existence of a commuting family of symmetries, familiar in the theory of completely integrable Hamiltonian systems, does not follow from the existence of a recursion operator alone but requires the additional property that it be a Nijenhuis operator (see below), whence the importance of the bihamiltonian systems in the theory of integrability: such systems necessarily possess recursion operators which are Nijenhuis operators.

2.2. MASTERSYMMETRIES. FORMAL SYMMETRIES

Mastersymmetries are recursion operators of adjoint type, *i. e.*, defined by

$$\mathcal{R}Y = [\tau, Y],$$

where τ is a time-dependent symmetry of $u_t = Xu$. See [12] and references therein.

The notion of *formal symmetry* is closely related to that of recursion operators. It permits the determination of all generalized inifinitesimal symmetries of an equation. See [10].

2.3. NIJENHUIS OPERATORS

An operator $\mathcal{R}$ on the evolution fields is called a *Nijenhuis operator* if its Nijenhuis torsion, $T(\mathcal{R})$, defined by

$$T(\mathcal{R})(X,Y) = [\mathcal{R}X, \mathcal{R}Y] - \mathcal{R}([\mathcal{R}X, Y] + [X, \mathcal{R}Y]) + \mathcal{R}^2([X,Y]),$$

vanishes. The fundamental observation is that when a recursion operator, $\mathcal{R}$, is a Nijenhuis operator, then all $\mathcal{R}^k X$, $k = 0, 1, 2, \ldots$ commute in pairs.

2.4. HAMILTONIAN AND BIHAMILTONIAN STRUCTURES

The Hamiltonian structures defined on finite-dimensional manifolds have analogues in the infininite-dimensional case of the sections of the trivial vector bundle, $F = \mathbb{R}^n \times \mathbb{R}^p \longrightarrow \mathbb{R}^n$. We outline the approach of [26], which is equivalent to others found in the references [21] [22] [1] and to the formal variational approach of Gel'fand and Dorfman. The functions are replaced by the *functionals*, equivalence classes modulo divergence of differential operators on the sections of F with values in the n-forms on $\mathbb{R}^n$. The differential 1-forms are replaced by the *simple 1-forms*, differential operators on the sections of F with values in the dual of F, tensored with the n-forms on $\mathbb{R}^n$. (These objects are forms on the infinite jet bundle of F.) There is a pairing between forms and vectors: to a simple 1-form, λ, and an evolution field, X, there corresponds a functional $\langle\lambda, X\rangle$. The variational derivative $\frac{\delta H}{\delta u}$ of a functional H is a simple 1-form and, by definition,

$$X.H = \langle\frac{\delta H}{\delta u}, X\rangle.$$

A *Hamiltonian* (or *Poisson*) *structure* on F is a linear, antisymmetric map, P, from the simple 1-forms to the evolution fields such that the Jacobi identity is satisfied for the Poisson bracket of functionals $\{\ ,\ \}$ defined by

$$\{H,K\} = -P(\frac{\delta H}{\delta u}).K .$$

Two Hamiltonian structures, P and P', are called *compatible* if the sum of the corresponding Poisson brackets satisfies the Jacobi identity. If P and P' are compatible, and if P is invertible, then P and P' are said to define a *bihamiltonian structure* on F. Such pairs of compatible Hamiltonian structures play an important role in the theory of integrability (see [21] [22] [19]), because they define a Nijenhuis operator which is a recursion operator for all evolution equations that are Hamiltonian systems with respect to both Hamiltonian structures.

2.5. THE NIJENHUIS OPERATOR OF A BIHAMILTONIAN STRUCTURE

When P and P' define a bihamiltonian structure, the operator $\mathcal{R} = P' \circ P^{-1}$ is a *Nijenhuis operator* acting on the evolution fields. Moreover, P, $P' = \mathcal{R}P$, $\mathcal{R}^2P$, ..., $\mathcal{R}^kP$, ... is a sequence of Hamiltonian structures which are compatible in pairs. See [19] [20].

2.6. BIHAMILTONIAN SYSTEMS AND THEIR RECURSION OPERATOR

An evolution field X is called Hamiltonian with respect to a Hamiltonian structure P if X is the image under P of the variational derivative of a functional H. Let X be Hamiltonian with respect to Hamiltonian structures P and P', where P is invertible. Set $\mathcal{R} = P' \circ P^{-1}$. Then, since both $\mathcal{L}_X P$ and $\mathcal{L}_X P'$ vanish, $\mathcal{L}_X \mathcal{R} = 0$. Thus $\mathcal{R}$ is a recursion operator for $u_t = Xu$.

An evolution field X is called a *bihamiltonian system* if it is Hamiltonian with respect to Hamiltonian structures P and P' which define a bihamiltonian structure. In this case, $\mathcal{R} = P' \circ P^{-1}$ is a Nijenhuis recursion operator for the evolution equation $u_t = Xu$. In particular, there is a hierarchy of evolution fields, $X_0 = X$, $X_1 = \mathcal{R}X$, $X_2 = \mathcal{R}^2X$, ..., $X_k = \mathcal{R}^kX$, ... , which in fact commute in pairs. Moreover, each evolution field X_k is locally bihamiltonian, *i. e.*, locally there exist functionals H_k, k = 0, 1, 2, ..., such that

$$X_k = P'(\frac{\delta H_k}{\delta u}) = P(\frac{\delta H_{k+1}}{\delta u}).$$

These functionals are integrals of the motion (conserved densities) for each evolution equation $u_t = X_k u$, and these integrals of the motion are in involution with respect to P and P'. See [21] [22] [19], and also [1] and [30].

2.7. THE R-MATRIX APPROACH

The solutions of the classical Yang-Baxter equation give rise to bihamiltonian structures on Lie groups. Let G be a Lie group with Lie algebra $\mathfrak{g}$. Any antisymmetric, invertible solution r of the *classical Yang-Baxter equation* on $\mathfrak{g}$ (see [32] [33] and the references therein) yields a bihamiltonian structure on the Lie group G. In fact, if r in $\mathfrak{g} \otimes \mathfrak{g}$ is an antisymmetric solution of the classical Yang-Baxter equation, the left- and right-invariant bivectors that it generates are both Poisson bivectors which are necessarily compatible. If, moreover, r is invertible, we obtain a bihamiltonian structure on G.

The R-matrix approach to the Kostant-Symes theorem is the following. Consider an operator R on $\mathfrak{g}$ that satisfies the *modified Yang-Baxter equation*

$$[RX,RY] - R([RX,Y] + [X,RY]) = - [X,Y].$$

(The modified Yang-Baxter equation is closely related to the classical Yang-Baxter equation but its solutions R are linear operators on $\mathfrak{g}$, as opposed to r wich is an element in $\mathfrak{g} \otimes \mathfrak{g}$.) Then $[X,Y]_R = [RX,Y] + [X,RY]$ is a Lie bracket on $\mathfrak{g}$ yielding two Lie-Poisson structures on $\mathfrak{g}^*$, $\{\ ,\ \}$ and $\{\ ,\ \}_R$. Two Casimir functions for $\{\ ,\ \}$ are in involution with respect to $\{\ ,\ \}_R$. See [31].

Magri has shown (see [33]) that from a solution of the modified Yang-Baxter equation on an associative Lie algebra $\mathfrak{g}$, one can obtain a quadratic Poisson structure compatible with the Lie-Poisson structure on $\mathfrak{g}^*$.

Thus, solutions of the modified Yang-Baxter equation will yield bihamiltonian structures on duals of Lie algebras. Such constructions are also applicable in the case of the infinite-dimensional Lie algebras of matrix-valued functions of a spectral parameter. They should permit the study of the integrability properties of some of the evolution equations of interest in physics.

Bibliography

The best reference for most of the material in these lectures is:

[1] P. J. Olver, *Applications of Lie Groups to Differential Equations*, Springer-Verlag (1986). [Contains an extensive bibliography.]

Other *books* on the subject of symmetries are:

[2] R. L. Anderson and N. H. Ibragimov, *Lie-Bäcklund Transformations in Applications*, SIAM (1979).

[3] G. W. Bluman and J. D. Cole, *Similarity Methods for Differential Equations*, Springer-Verlag (1974).

[4] W. I. Fushchich and A. G. Nikitin, *Symmetries of Maxwell's Equations*, Reidel (1987).

[5] N. H. Ibragimov, *Transformation Groups Applied to Mathematical Physics*, Reidel (1985).

[6] B. G. Konopelchenko, *Nonlinear Integrable Equations (Recursion Operators, Group-Theoretical and Hamiltonian Structures of Soliton Equations)*, *Lecture Notes in Physics* **270**, Springer-Verlag (1987).

[7] I. S. Krasil'shchik, V. V. Lychagin and A. M. Vinogradov, *Geometry of Jet Spaces and Nonlinear Partial Differential Equations*, Gordon and Breach (1986).

[8] L. V. Ovsiannikov, *Group Analysis of Differential Equations*, Academic Press (1982).

For a thorough study of the classical and generalized symmetries of scalar evolution equations in dimension 1+1, see:

[9] V. V. Sokolov, On the symmetries of evolution equations, *Russ. Math. Surveys* **43** (1988), 165-204.

The following article contains the definition and study of the *formal symmetries* of systems of evolution equations in dimension 1+1:

[10] A. V. Mikhailov, A. B. Shabat and R. I. Yamilov, The symmetry approach to the classification of non-linear equations. Complete lists of integrable systems, *Russ. Math. Surveys* **42** (1987), 1-63.

Regarding the classical symmetries and invariant solutions of differential equations, including evolution equations in dimension 1+2, see:

[11] P. Winternitz, *this volume*.

For a recent *survey* including the theory and applications of *mastersymmetries*, see:

[12] A. S. Fokas, Symmetries and integrability, *Studies in Applied Math.* **77** (1987), 253-299. [Contains an extensive bibliography.]

The following articles give an overview of the *geometric approach* to symmetries and its applications:

[13] T. Tsujishita, On variation bicomplexes associated to differential equations, *Osaka J. Math.* **19** (1982), 311-363. [A very clear but mathematically sophisticated article.]

[14] A. M. Vinogradov, Local symmetries and conservation laws, *Acta Appl. Math.* **3** (1984), 21-78.

[15] A. M. Vinogradov, Symmetries and conservation laws of partial differential equations: Basic notations and results, *Acta Appl. Math.* **15** (1989), 3-21. [See also the other articles in the same issue, which contain many applications.]

For the theory of nonlocal symmetries, see:

[16] I. S. Krasil'shchik and A. M. Vinogradov, Nonlocal symmetries and the theory of coverings, *Acta Appl. Math.* 2 (1984), 79–96.

[17] I. S. Krasil'shchik and A. M. Vinogradov, Nonlocal trends in the geometry of differential equations: symmetries, conservation laws and Bäcklund transformations, *Acta Appl. Math.* 15 (1989), 161–209.

For the algorithmic determination of symmetries, see the following review article which contains a useful but incomplete bibliography. (There are also published or unpublished programs by Rosenau and Schwarzmeier, Steinberg, Rosencrans, Champagne and Winternitz, D. Gutkin, and others. See [1] and volume 16 (1989) of *Acta Appl. Math.*, edited by A. M. Vinogradov.)

[18] F. Schwarz, Symmetries of differential equations: From Sophus Lie to computer algebra, *SIAM Review* 30 (1988), 450–481.

On the hierarchies of Poisson structures and their applications, see:

[19] F. Magri and C. Morosi, A geometrical characterization of integrable Hamiltonian systems through the theory of Poisson-Nijenhuis manifolds, *Quaderno* S 19 (1984), University of Milan.

[20] Y. Kosmann-Schwarzbach and F. Magri, Poisson-Nijenhuis structures, *Ann. Inst. Henri Poincaré*, to appear.

On the topic of *bihamiltonian systems*, the basic references are (see also [1] and [30]):

[21] F. Magri, A simple model of the integrable Hamiltonian equation, *J. Math. Physics* 19 (1978), 1156–1162.

[22] B. Fuchssteiner and A. S. Fokas, Symplectic structures, their Bäcklund transformations and hereditary symmetries, *Physica* 4D (1981), 47–66.

These lectures have been based in part on the following articles by the author:

[23] Y. Kosmann-Schwarzbach, Sur les transformations de similitude des équations aux dérivées partielles, *Comptes rendus Acad. Sci. Paris* 287A (1978), 953–956.

[24] —— , Generalized symmetries of nonlinear partial differential equations, *Lett. Math. Physics* 3 (1979), 395–404.

[25] —— , Vector fields and generalized vector fields on fibered manifolds, *Lecture Notes in Math.* 792 (1980), 307–355.

[26] ——, Hamiltonian systems on fibered manifolds (Poisson and vertical brackets in field theory), *Lett. Math. Physics* 5 (1981), 229–237.

[27] ——, Lie algebras of symmetries of partial differential equations, in *Differential Geometric Methods in Mathematical Physics*, S. Sternberg ed., Reidel (1984), 241-277.

[28] ——, On the momentum mapping in field theory, *Lecture Notes in Math.* **1139** (1985), 25-73.

[29] ——, Sur les théorèmes de Noether, in *Géométrie et Physique*, Y. Choquet-Bruhat *et al.* eds., *Travaux en cours* 21, Hermann (1987), 147-160.

[30] ——, Géométrie des systèmes bihamiltoniens, in *Systèmes dynamiques non linéaires*, P. Winternitz ed., *Séminaire de Mathématiques Supérieures* 102, Presses de l'Université de Montréal (1986), 185-216.

Regarding the R-matrix approach to integrable systems, see the following articles and the references cited there:

[31] M. A. Semenov-Tian-Shansky, What is a classical r-matrix? *Funct. Anal. Appl.* 17 (1983), 259-272.

[32] Y. Kosmann-Schwarzbach and F. Magri, Poisson-Lie groups and complete integrbility, Part I. Drinfeld bigebras, dual extensions and their canonical representations, *Ann. Inst. Henri Poincaré* **49** A (1988), 433-460.

[33] Y. Kosmann-Schwarzbach, The modified Yang-Baxter equation and bi-hamiltonian structures, *Differential Geometric Methods in Theoretical Physics*, A. I. Solomon ed., World Scientific (1989), 12-25.

NONLINEAR DISPERSIVE EQUATIONS WITHOUT INVERSE SCATTERING

Jean-Claude SAUT

Université Paris XII and Laboratoire d'Analyse Numérique,
CNRS et Université Paris-Sud, Bâtiment 425, 91405 Orsay

1. INTRODUCTION. Nonlinear dispersive equations occur in various physical contexts, e.g. as envelope equations (see for instance Ablowitz and Segur [5], Newell [70]). Some of them can be viewed as completely integrable infinite dimensional hamiltonian systems, and this leads to deep insights in their structure via the powerful techniques of the inverse-scattering transform. However, many nonlinear dispersive equations which share the same physical interest are not integrable. Other techniques are thus needed to study their properties. These two approaches are not exclusive, even for integrable equations. The aim of the lecture is to review some recent results which do not use inverse scattering with the hope of giving a flavor of the wealth of the subject. This review is the object of Section 2. We do not attempt to describe every single work but rather choose some striking results for each theme. Also, some subjects are not touched here, for instance, scattering theory and scattering at low energy, an analysis of which can be found in the recent monograph of Strauss [79]. The third section will emphasize and develop the case of the Davey-Stewartson systems ; we describe here in some details a recent work of Ghidaglia and Saut [39] [40]. To conclude this introduction we present two examples which illustrate the fact that, for a given physical situation, the completely integrable case occurs for some special or limit values of the parameters involved.

1.1. The general intermediate long wave equation (Kubota, Ko, Dobbs [58]).

Interest is focused here in the motion of a nonlinear internal gravity wave within a

R. Conte and N. Boccara (eds.), Partially Integrable Evolution Equations in Physics, 491–513.

horizontally stratified medium having a representative density distribution as shown below.

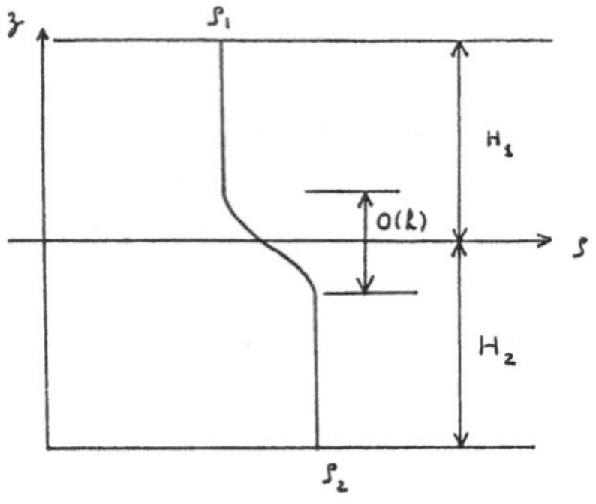

Figure 1

In the undisturbed fluid, the density variation is confined to a thin layer of thickness h located at a distance H_1 below the upper surface, with the total depth defined by $H = H_1 + H_2$. Only weakly nonlinear internal waves with wavelength λ much larger than the characteristic scale for the density profile λ will be considered.

Using the multi-scale method, Kubota, Ko and Dobbs derived the following equation for the amplitude of the perturbation

$$A_t + c_0 A_x + \alpha A A_x - C_0 \frac{\partial^2}{\partial x^2} \int_{-\infty}^{\infty} G(x-\xi) A(\xi, t) d\xi = 0 \qquad (1.1)_1$$

$$G(x) = \frac{\beta_1}{2H_1}\left[\mathrm{Coth}\frac{\pi x}{2H_1} - \mathrm{sgn}(x)\right] + \frac{\beta_2}{2H_2}\left[\mathrm{Coth}\frac{\pi x}{2H_2} - \mathrm{sgn}(x)\right] \qquad (1.1)_2$$

$\alpha, c_0, \beta_1, \beta_2$ are constants which depend on the parameters of the problem.

The equation $(1.1)_1$, $(1.1)_2$ is not completely integrable. On the other hand, in the limit $H_2 \to 0$ (pycnocline located near a boundary) it reduces to

$$A_t + c_0 A_x + \alpha A A_x = \frac{\beta_1 c_0}{2H_1} \frac{\partial^2}{\partial x^2} \int_{-\infty}^{\infty} \left[\mathrm{Coth}\frac{\pi(x-\xi)}{2H_1} - \mathrm{sgn}(x-\xi)\right] A(\xi, \tau) d\xi \qquad (1.2)$$

which is the usual (integrable) intermediate long wave (I.L.W.) equation (see [5]).

Equations (and systems) such as (1.1) are studied in [1].

1.2. <u>The general Davey-Stewartson system</u>.

The Davey-Stewartson system was introduced in [35] as a model for the evolution of weakly nonlinear packets of water waves that travel predominantly in one direction but in which the wave amplitudes are modulated slowly in both horizontal directions. Hence it is a two dimensional generalization of the cubic monodimensional Schrödinger equation. Djordjevic and Redekopp [36] (see also Ablowitz and Segur [4]) have extended the analysis of [35] by including full gravity, surface tension and depth effects. The model is derived from the Euler equations with free boundary under suitable assumptions (small amplitudes, slowly varying modulations, nearly one dimensional waves and a balance of all three effects). The resulting system for the (complex) amplitude $A(\xi,\eta,\tau)$ and the (real) mean flow velocity potential $\phi(\xi,\eta,\tau)$ (which is defined up to an additive constant) writes in dimensionless form

$$iA_\tau + \lambda A_{\xi\xi} + \mu A_{\eta\eta} = \chi \mid A \mid^2 A + \chi_1 A\phi_\xi \tag{1.3}$$

$$\alpha\, \phi_{\xi\xi} + \phi_{\eta\eta} = -\beta(\mid A \mid^2)_\xi \tag{1.4}$$

The (real) coefficients occuring in (1.3), (1.4) depend on the physical parameters of the problem (fluid depth, wave numbers, group velocity,...). μ, β, χ_1, are positive but χ, λ, α can achieve both signs. In particular $\alpha = \frac{gh-C_g^2}{gh}$ (h = fluid depth, C_g =linear group velocity, g = the constant of gravity) is negative if the effects of surface tension are strong enough (see [36]).

While $\mid A \mid\to 0$ as $\xi^2+\eta^2 \to +\infty$ is a natural boundary condition for (1.3) in all cases, the appropriate boundary condition for (1.4) will depend dramatically on the sign of α. It will be $\phi \to 0$ as $\xi^2+\eta^2 \to +\infty$ when $\alpha > 0$, and roughly speaking that ϕ vanishes ahead of the support of A, (if A is compactly supported), when $\alpha < 0$. This will of course lead to different formulations of the mathematical problem to be solved.

It is noteworthy to mention that in the shallow water limit of (1.3) (1.4), i.e. when $kh \to 0$ with $a \mid \kappa \mid << (kh)^2$, where $\kappa = (k,\ell)$ is the wave vector and a the characteristic amplitude of the disturbance (and in this case only), the Davey-Stewartson system is of inverse scattering type. In this case, (1.3), (1.4) can be written after rescaling

$$iA_t - \sigma A_{xx} + A_{yy} = \sigma \mid A \mid^2 A + A\phi_x \tag{1.5}$$

$$\sigma\phi_{xx} + \phi_{yy} = -2(\mid A \mid^2)_x \tag{1.6}$$

with $\sigma = \pm 1$.

On the other hand, (1.3)(1.4) reduces in the deep water limit ($kh \to \infty$ where k is the wave number in the x-direction) to the nonlinear Schrödinger equation in 2 dimensions

$$iA_\tau + \lambda_\infty A_{\xi\xi} + \mu_\infty A_{\eta\eta} = \chi_\infty \mid A \mid^2 A,$$

where λ_∞ is usually negative, leading to an "hyperbolic" Schrödinger equation.

The system (1.5),(1.6) has be derived independently by Ablowitz-Haberman [3], Morris [69] and Cornille [34]. The motivation was in these works to systematically look for two-dimensional generalizations of the cubic onedimensional Schrödinger equation which were of IST type. The study of (1.5)(1.6) under inverse scattering methods has led recently to very interesting issues : existence of special soliton or lump solutions [37], [38], [73], solution of the Cauchy problem ([2][8] and the bibliography of these papers). However, no results on the Cauchy problem for the general system (1.3), (1.4) seemed to be known until those of Ghidaglia and Saut [39][40]) which will be reported in Section 3.

2. DISPERSIVE EQUATIONS : A DIRECT APPROACH.

2.1. <u>The Cauchy problem</u>. The inverse scattering method provides a fashion of constructing (sometimes formally) a solution to the Cauchy problem. Many of the results so obtained concern very smooth (e.g. belonging to the Schwartz classe S) solutions. The question of specifying the data in a given (not necessarily very smooth) class, and to precise the class the solution belongs to is not a very tractable one by the inverse scattering method (some nice results in this direction can be found in Cohen ([28] [29] [30]). On the other hand, specific PDE's methods, taking carefully into account the linear dispersive term, can be very powerful.

.1.1. <u>KdV like equations</u>. We consider here equations of the type

$$\begin{aligned} &u_t + f(u)_x + Lu_x = 0, \quad u = u(x,t), \quad t \geq 0, \quad x \in \mathbb{R} \\ &u(x,0) = u_0(x) \end{aligned} \tag{2.1}$$

where $f(u)$ is a given smooth function (usually $f(u) = \frac{u^{p+1}}{p+1}$) and L is a, local or nonlocal, operator given in Fourier variable by

$$\widehat{Lu}(k) = q(k)\hat{u}(k) \tag{2.2}$$

The symbol $q(k)$ of L is real and is related to the dispersion relation. A paradigm of equations such as (2.1) is the KdV equation ($p = 1$, $q(k) = k^2$).

$$u_t + uu_x + u_{xxx} = 0 \tag{2.3}$$

Note that the generalized KdV equations, GKdV, (p integer ≥ 2, $q(k) = k^2$) are not integrable for $p \geq 3$. Other popular examples are the Benjamin-Ono equation ($p = 1$, $q(k) = \mid k \mid$), the general intermediate long wave equation ($p = 1$, $q(k) = \beta_1(k\ coth\delta_1 k - \frac{k}{\delta_1})$ $+\beta_2(k\ coth\ \delta_2 k - \frac{k}{\delta_2})$,...). Many results have been obtained recently for the Cauchy problem (2.1). Abdelouhab, Bona, Felland and Saut [1] give a rather complete pricture of the Cauchy problem for nonlocal equations (including Benjamin-Ono, I.L.W.) in Sobolev spaces. It is proven in particular that the solutions of the I.L.W. equation converge strongly to solutions of the Korteweg-de Vries equation, or to solutions of the Benjamin-Ono equation, in appropriate asymptotic limits. Other results concerning the Benjamin-Ono equation can be found in Iorio [53] [54]. More precise results are available for the KdV (or generalized KdV) equation. Kato [55], Krushkov-Faminskii [57], using a remarkable smoothing property of the KdV equation (see 2.3 below) have shown the existence of a solution to the Cauchy problem for (2.3) when the initial data is merely in L^2. Tsutsumi [85] constructed solutions having a positive Radon measure as initial data. Very sharp results concerning the uniqueness of (possibly weighted) L^2 or H^1 solutions of the GKdV have been derived by Ginibre and Tsutsumi [44] (see Ginibre, Tsutsumi and Velo [45] for issues concerning the existence of solutions in the uniqueness class), extending previous ones of Krushkov-Faminskii [57]. Typically, for the usual KdV equation (2.3), existence and uniqueness hold provided

$$(1 + x_+)^{\beta/2} u_0 \in L^2, \quad (1 + x_+)^{\gamma/2} \frac{du_0}{dx} \in L^2,$$

where $x_+ = \mathrm{Sup}(x, 0)$, $\beta, \gamma \geq 0$, $\beta + \gamma \geq 1/2$, $\beta \geq \frac{3\gamma}{5}$. Another striking result is that for the GKdV, the H^1 solutions are unique provided $p > 3/2$.

Precise estimates of the propagator of the free evolution group (the Airy function) are crucial in these works.

.1.2. Schrödinger equations

We consider here the Cauchy problem

$$\begin{aligned} &iu_t + \Delta u + f(\mid u \mid^2)u = 0 \\ &u(x, 0) = u_0(x) \end{aligned} \tag{2.4}$$

where $u = u(x,t)$, $x \in \mathbb{R}^n$, $t \geq 0$ is a complex-valued function, and f a real-valued function, typically $f(z) = \alpha z^{\lambda/2}$, $\alpha \in \mathbb{R}$.

After the pionnering papers of Ginibre and Velo ([42], [43]), recent progresses concern the existence and uniqueness of solutions to (2.4) under weaker conditions on f and/or

u_0. They all use the $L^p - L^q$ estimates on the linear group due to Strichartz [80] (Kato [56], Y. Tsutsumi [84], Cazenave and Weissler [22] [23] [24] [25], and the monograph of Cazenave and Haraux [26]). Here is a typical result, dealing with the L^2–critical case $f(|u|^2)u = |u|^{4/n}$ [22] : for any $u_0 \in L^2(\mathbb{R}^n)$, (2.4) possesses a unique maximal L^2 solution, defined on some interval $[0, T^*)$. If $\| u_0 \|_{L^2}$ is small enough, then $T^* = +\infty$.

2.2. Blowing-up of solutions

.2.1. Blow-up of the solutions of Schrödinger equations.

It has been established (Zakharov [93], Glassey [48], M. Tsutsumi [83]) that H^1 solutions of the N.L.S. equation (2.7) with $f(|u|^2) = |u|^\lambda$, $\lambda \geq 4/n$ may blow-up in finite time. Recent issues concern theoretical and numerical analysis of the structure of the blow-up. By determining the best constant in an interpolation inequality, M. Weinstein [88][92] has shown, in the critical case $\lambda = \frac{4}{n}$, that the condition $u_0 \in H^1$, $\| u_0 \|_{L^2} < \| \psi \|_{L^2}$, where ψ is the ground state (see Section 2.4 below) is necessary and sufficient to prevent the blow up of the associated solution.

On the other hand [92] he proved that for a class of initial data in H^1 such that $\| u_0 \|_{L^2} = \| \psi \|_{L^2}$, if the associated solution u satisfies

$$\lim_{t \to T} \int |\nabla u(x,t)|^2 \, dx = +\infty \quad \text{for } T < \infty,$$

then there exist functions $y(t) \in \mathbb{R}^n$ and $\gamma(t) \in \mathbb{R}$ such that as $t \to T$,

$$S_{\lambda(t)} u(\cdot + y(t), t) e^{i\gamma(t)} \to \psi(\cdot) \qquad \text{strongly in } H^1,$$

where $\lambda(t) = \| \nabla \psi \|_{L^2} \, / \, \| \nabla u(t) \|_{L^2}$ and S_λ is the spatial dilation operator $(S_\lambda u)(x,t) = \lambda^{n/4} u(\lambda x, t)$. Thus the formation of the singularity is self-similar with a profile given by the ground state solitary wave. The proof of this result uses in particular the principle of concentration-compactness of P.L. Lions [63] (classification of the ways in which compactness can be lost in variational problems characterizing the ground state).

Other qualitative results on the blow-up have been obtained by F. Merle [65] [66] [67], Merle and Tsutsumi [68] and Papanicolaou et al [60] [61] [62]. For instance, for critical nonlinearities $f(|u|^2)u \sim |u|^{4/n} u$ as $u \to +\infty$, Merle and Tsutsumi show that a solution whose H^1 norm blows up at $t = T$ has not limit in L^2 as $t \to T$. In the critical power case $f(|u|^2)u = |u|^{4/n} u)$, Merle [67] constructs a solution which blows up in a finite time at exactly k points in $\mathbb{R}^n$. Lemesurier, Papanicolaou, C. Sulem and P.L. Sulem ([61]) use a method of dynamic rescaling of variables to investigate numerically the nature of the focusing singularities of the cubic and quintic Schrödinger equations in two and three dimensions. The papers [60][62] are devoted to a construction of classes of singular solutions to the cubic Schrödinger equation in two dimensions.

.2.2. Dispersive blow-up.

We consider here the generalized KdV (GKdV) equation

$$\begin{aligned} u_t + u^p u_x + u_{xxx} &= 0 \\ u(x,0) &= u_0(x) \end{aligned} \tag{2.5}$$

It can be reasonably conjectured that H^1-solutions of (2.5) blow up in finite time in the supercritical case $p \geq 4$. This conjecture is supported by numerical computations [18] [19] but a proof is still missing (an analysis of the expected blow-up can be found in Weinstein [92]).

The issue we want to discuss here is totally different ; it will apply to any value of p, including the KdV case, $p = 1$. It initiated in a remark of Benjamin, Bona, Mahony [10] who noticed that for the linear equation ($p = 0$), a specific initial data can be prescribed in a way that many, widely spaced, short-wave components will coalesce at a single point of some given time and thereby create some loss of spatial smoothness in the solution at that time. This remark is sharpened and generalized to the nonlinear equation (2.5) in [16] [17]. A typical result is the following. Let $T > 0$ and (t_*, x_*), $0 < t_* < T$ be given. For $k = 0$ and $p = 1$, or $k \geq 1$ and p arbitrary, there exists $u_0 \in H^k(\mathbb{R}) \cap C^\infty(\mathbb{R})$ and a solution u of (2.5) which satisfies $u \in L^\infty(0;T;H^k(\mathbb{R})) \cap L^2(0,T;H^{k+1}_{loc}(\mathbb{R}))$ and

$$\partial_x^k u \text{ is continuous on } (0,T) \times \mathbb{R}\backslash\{t_*, x_*\}, \qquad \lim_{x \to x_*, t \to t_*} \mid \partial_x^k u(t,x) \mid = +\infty$$

The idea of the proof is quite simple. Write u as

$$u(t,\cdot) = S(t)u_0 - \int_0^t S(t-s)u^p u(s,\cdot)ds, \tag{2.6}$$

where $S(t)$ is the unitary group associated to the linear equation. A precise study of the Airy function shows that $S(t)u_0$ satisfies the properties of u above. On the other hand, it can be shown (under a mild decay assumption on u_0) that the nonlinear term in (2.6) is "smooth". This justify the word "dispersive blow-up" to describe this phenomena. It is worth noticing that while Cohen [28] has shown by a careful use of inverse scattering techniques, that the solutions of the KdV equation may undergo a loss of smoothness, her analysis does not lead to the sort of specific conclusions we draw here.

2.3. Smoothing properties.

The problem we want to address here is concerned with local smoothing properties of evolution partial differential equations which are reversible and conservative. Such a property is excluded for the wave equation for instance. On the other hand, Kato [55] has shown a local smoothing property of the KdV equation : the soluton of the initial value problem is, locally, and for almost every time, one derivative smoother than the initial datum.

This is quite a general property of dispersive equations and systems. Constantin and Saut [31][32] have studied the general linear Cauchy problem

$$\begin{aligned} &u_t + iP(D)u = 0 \\ &u(0,x) = u_0(x) \end{aligned} \tag{2.7}$$

where $u(t,x)$, $t \in \mathbb{R}$, $x \in \mathbb{R}^n$, $D = \frac{1}{i}\left(\frac{\partial}{\partial x_1}, ..., \frac{\partial}{\partial x_n}\right)$ and $P(D)$ is defined via a real symbol $p(k)$,

$$P(D)u = \int_{\mathbb{R}^n} e^{2i\pi x.k} p(k)(\mathcal{F}_2 u)(k)dk$$

where $\mathcal{F}_2$ is the Fourier transform with respect to the $x-$variables.

The assumptions on $p(k)$, reflecting the strong dispersive nature of (2.7), are that, roughly speaking, $p(k)$ behaves like $\mid k \mid^m$for $\mid k \mid \to +\infty$, with $m > 1$. A typical result is : if u_0 belongs to the Sobolev space $H^s(\mathbb{R}^n)$, then, for almost every $t \neq 0$, the solution $u(t,\cdot)$ belongs to $H^{s+d}_{loc}(\mathbb{R}^n)$, where $d = (m-1)/2$. This property has been independently discovered, for the Schrödinger equation, $p(k) = \mid k \mid^2$, by Sjölin [76] and Vega [86] (in this case, $\frac{m-1}{2} = \frac{1}{2}$). The proofs are harmonic analysis flavoured (restriction of the Fourier transform). Constantin and Saut have recently extended this result to Schrödinger equations with short range potentials [33].

The analysis made in [32] covers most of the (linearized) strongly dispersive equations and systems which occur in physical contexts (KdV, Boussinesq, Schrödinger equations and their generalizations, local or nonlocal,...). The nonlinear situation is slightly less satisfactory. Some (partial) local smoothing results for the nonlinear Schrödinger equations are proved in [32] [74] [77]. For the Benjamin-Ono and I.L.W. equations, the linear theory predicts a local smoothing by 1/2. This is actually the case, as shown recently by Ginibre and Velo [46] [47], extending some previous results of Ponce [71] and M. Tom [72].

We conclude this section by noticing that while global smoothing cannot occur in L^2–Sobolev spaces for equations which are time reversible, there exists the possibility of global smoothing in different spaces. For instance, global smoothing holds provided the initial data decays sufficiently fast at infinity (see e.g. Cohen [28] for the KdV equation by inverse scattering techniques, Kato [55] also for the KdV equation, and Hayashi, Nakamizu and Tsutsumi [51] [52] for a class of nonlinear Schrödinger equations).

2.4. Stability of solitary waves;

A solitary wave is a localized, finite energy solution of a nonlinear evolution equation. It results from a balance of dispersion and a focusing nonlinearity. The nonlinear Schrödinger equation and the KdV equation (together with the related equations : Benjamin-Ono, I.L.W., ...) possess such solutions.

For instance, the work of W. Strauss [78] and H. Berestycki and P.L. Lions [14] show that the equation

$$\Delta u - Eu + f(|u|^2)u = 0 \quad E > 0 \tag{2.8}$$

has a positive, radial, smooth and exponentially decaying solution R, called a ground state. Therefore,

$$\psi(x,t) = R(x;E)e^{iEt}$$

is a solitary wave solution of N.L.S.

Note that in the case of the polynomial nonlinearity, $f(|u|^2)u = |u|^\lambda u$, the ground state was recently shown to be unique (Kwong [59], extending previous results of Coffman [27] and Mc-Leod and Serrin [64]).

On the other hand, a general existence theory of solitary waves for "generalized KdV" equations such as (2.1) is developed in Benjamin, Bona, Bose [11]. Explicit solitary waves are explicitely known for the KdV, GKdV (2.5), Benjamin-Ono, I.L.W. equations.

A pretty large number of papers have been recently devoted to the orbital stability (or instability) of solitary waves (Albert et al [7], Albert and Bona [6], Benjamin [9], Bennett et al [12], Berestycki and Cazenave [9], Bona [15], Bona and Sachs [21], Bona, Souganidis and

Strauss [20], Grillakis [49]; Grillakis, Shatah and Strauss [50], Stubbe [81][82], Weinstein [87] [88] [90] [91],...).

Here are a few typical results. For the N.L.S. with $f(|u|^2)u = |u|^{p-1}u$, the ground state is stable if and only if $1 < p < 1 + \frac{4}{n}$. If $p > 1 + \frac{4}{n}$, the ground state is unstable (nearly initial data lead to blow up in finite time). For the generalized KdV equation (2.5), the solitary wave is stable if and only if $p < 4$ (see Bona, Souganidis, Strauss [20]).

Interesting enough is the I.L.W. equation. Recall (see the Introduction) that it reads

$$u_t + u_x + uu_x - (Mu)_x = 0$$

where M is defined in Fourier variables by

$$\widehat{Mu}(k) = (k \coth kH - \frac{1}{H})\hat{u}(k)$$

An explicit solitary-wave solution $u(x,t) = \varphi(x - Ct)$ was found for any $C > 1$ and $H > 0$ by R.I. Joseph. It is given by

$$\varphi(y) = \frac{b}{\cosh^2(ay) + (b^2/16a^2)\sinh^2(ay)}$$

where $a \in [0, \pi/2H)$ and $b \in (0, \infty)$ are determined uniquely in terms of C and H.

Albert and Bona [6] have recently established the stability of Joseph's solitary waves for all values of $C > 1$ and $H > 0$.

3. THE INITIAL VALUE PROBLEM FOR THE GENERAL DAVEY-STEWARTSON SYSTEMS.

We discuss here in some details the results obtained in [39][40] by J.M. Ghidaglia and J.C. Saut for the general Davey-Stewartson systems (1.3)(1.4).

We shall study a scaled version of (1.3), (1.4), namely

$$i\frac{\partial A}{\partial t} + \delta\frac{\partial^2 A}{\partial x^2} + \frac{\partial^2 A}{\partial y^2} = \chi |A|^2 A + bA\frac{\partial \phi}{\partial x} \tag{3.1}$$

$$\frac{\partial^2 \phi}{\partial x^2} + m\frac{\partial^2 \phi}{\partial y^2} = \frac{\partial}{\partial x}|A|^2 \tag{3.2}$$

$$A(x,y,0) = A_0(x,y) \tag{3.3}$$

The parameters δ, χ, b, m are real, can assume both signs and δ, χ have been normalized in such a way that $|\delta| = |\chi| = 1$.

For what follows, it will be useful to classify (3.1)(3.2) into elliptic-elliptic, hyperbolic-elliptic, elliptic-hyperbolic, hyperbolic-hyperbolic acording to the sign of (δ, m) : $(+,+)$, $(-,+)$, $(+,-)$ and $(-,-)$.

In all these cases, assuming that A and ϕ are smooth enough and decay suitably at infinity, (3.1) (3.2) admits three interesting integrals [4].

$$M(t) \equiv \int |A(t)|^2 \, dxdy = M(0), \ \forall t \tag{3.4}$$

$$E(t) \equiv \int \left[\delta \left|\frac{\partial A}{\partial x}\right|^2 + \left|\frac{\partial A}{\partial y}\right|^2 + \frac{1}{2}\left(\chi |A|^4 + b\left(\left|\frac{\partial \phi}{\partial x}\right|^2 + m\left|\frac{\partial \phi}{\partial y}\right|^2\right)\right)\right] dxdy \tag{3.5}$$
$$= E(0), \ \forall\, t$$

$$\frac{d^2}{dt^2}\int (\delta x^2 + y^2)\,|A|^2\, dxdy = 8E(0), \ \forall\, t \tag{3.6}$$

The two conservation laws (3.4)(3.5) are respectively that of mass and energy while (3.6) express the evolution of the moment of inertia.

3.1. The Cauchy problem in the elliptic-elliptic and hyperbolic-elliptic case.
The first results do not use the conservation of energy.

Theorem 1 *(Existence and uniqueness)*

(i) *Let* $A_0 \in L^2(\mathbb{R}^2)$. *Then there exists a unique maximal solution* (A,ϕ) *of* (3.1) (3.2) (3.3) *on* $[0,T^*)$, $T^* > 0$ *which is such that*
$A \in C([0,T^*)\ ;\ L^2(\mathbb{R}^2)) \cap L^4((0,t) \times \mathbb{R}^2)$,
$\nabla\phi \in L^2((0,t) \times \mathbb{R}^2)$, $\| A(t) \|_{L^2} = \| A_0 \|_{L^2}$, $0 \le t < T^*$.

(ii) *If* A_0 *is sufficiently small in* $L^2(\mathbb{R}^2)$, $T^* = +\infty$: *the solution is global.*

Theorem 2 *(smoothness)*

(i) *Let* $A_0 \in H^1(\mathbb{R}^2)$, *then the solution* (A,ϕ) *satisfies moreover*
$A \in C([0,T^*); H^1(\mathbb{R}^2)) \cap C^1([0,t]; H^{-1}(\mathbb{R}^2))$

$\nabla A \in L^4((0,t) \times \mathbb{R}^2)$, $\nabla\phi \in C([0,t]; L^p(\mathbb{R}^2))$ *and*
$\nabla^2\phi \in L^4(0,t; L^q(\mathbb{R}^2))$, *for every* $t \in [0,T^*)$, $p \in [2,+\infty)$, $q \in [2,4)$.

(ii) *If* $A_0 \in H^2(\mathbb{R}^2)$, *then*
$A \in C([0,T^*); H^2(\mathbb{R}^2)) \cap C^1([0,t]; L^2(\mathbb{R}^2))$
$\nabla\phi \in C([0,T^*); H^2(\mathbb{R}^2))$
$A \in L^2(0,t; H^{5/2}_{\text{loc}}(\mathbb{R}^2))$, $\nabla^2\phi \in L^1(0,t; H^{3/2}_{\text{loc}}(\mathbb{R}^2))$ *for every* $\mathbf{t} \in [\mathbf{0}, \mathbf{T}^*)$.

Theorem 3 (*Continuous dependence).*

Let $I = [0,T]$, *for some* $T > 0$. *The map* $A_0 \to (A, \nabla\phi)$ *is continuous from* $H^1(\mathbb{R}^2)$ *into* $C(I; H^1(\mathbb{R}^2)) \times C(I; L^p(\mathbb{R}^2))$, $2 < p < +\infty$. *More precisely, let* $A \in C(I; H^1(\mathbb{R}^2)$, $\nabla\phi \in C(I; L^p(\mathbb{R}^2))$ *be a solution of* (3.1) (3.2) (3.3) *with* $A(0) = A_0$. *Let* $A_{0n} \to A_0$ *in* $H^1(\mathbb{R}^2)$ *as* $n \to +\infty$. *Then the solution* (A_n, ϕ_n) *with* $A_n(0) = A_{0n}$ *exists on the interval* I *if* n *is sufficiently large, and* $(A_n, \nabla\phi_n) \to (A, \nabla\phi)$ *in* $C(I; H^1(\mathbb{R}^2)) \times C(I; L^p(\mathbb{R}^2))$, $2 < p < +\infty$.

Remarks 1.

1. When $b = 0$ and $\delta = -1$, (1.5) reduces to an "hyperbolic" Schrödinger equation

$$i\frac{\partial A}{\partial t} - \frac{\partial^2 A}{\partial x^2} + \frac{\partial^2 A}{\partial y^2} = \chi \mid A \mid^2 A, \tag{3.7}$$

to which our results apply. Further properties of Schrödinger equations like (3.7) are investigated in [41].

2. In the context of Theorem 1, the energy (3.5) is in general not defined and one cannot expect its conservation. On the other hand, when $A_0 \in H^1(\mathbb{R}^2)$ (Theorem 2), the energy is well defined and conserved. Furthermore, when the mass M and E bound a priori the H^1 norm of A (i.e. $\delta > 0$, $m > 0$, $\chi \geq \max(-b, 0)$, the solution constructed in Theorem 2 is global (see Theorem 5 below).

3. Using inverse scattering techniques, Beals and Coifman have recently proved [8] global existence in $S(\mathbb{R}^2)$ for the Cauchy problem corresponding to (1.5), (1.6) with $\sigma = +1$.

Indications on the proofs. We will just give some comments on the proofs (see [40] for details).

(i) We use (3.2)) to express $\nabla\phi$ in terms of $|A|^2$, by performing successively two Riesz transforms. Hence

$$\|\nabla\phi\|_{L^p(\mathbb{R}^2)} \leq Cp \|A\|^2_{L^{2p}(\mathbb{R}^2)},\ 1 < p < +\infty \tag{3.8}$$

This allows to transform (3.1) (3.2) into a nonlocal nonlinear Schrödinger equation with will be studied on its integral formulation.

$$A(t) = S(t)A_0 - i\int_0^t S(t-s)[\chi\,|A|^2 A + AE(A)](s)ds \tag{3.9}$$

where $S(t)$ denotes the group associated to the linear Schrödinger equation and $E(A) = \phi_x$.

The crux of the proof of Theorems 1 and 2 is to use contraction type arguments in spaces such as $L^4(0,T;L^4(\mathbb{R}^2))$, similar to those used by Cazenave-Weissler [22], Kato [56] in the context of the nonlinear "elliptic" Schrödinger equation. This method needs $L^p - L^q$ estimates for the propagator of the linear Schrödinger equation. They are classical for the usual "elliptic" Schrödinger equation, but are also valid in the more general case of the propagator associated to the linear equation.

$$i\frac{\partial u}{\partial t} + \sum_{i,j=1}^{n} a_{ij}\frac{\partial^2 u}{\partial x_i \partial x_j} = 0 \text{ in } \mathbb{R}^n \tag{3.10}$$

where the real constants a_{ij} are such that the matrix (a_{ij}) is invertible.

(iii) The proof of the smoothness properties follows the same lines. The second part of (ii) in Theorem 2 results from a local smoothing effect for the linear Schrödinger equation (see Section 2), which is still valid for the more general equation (3.10).

So far we have not used the conserved quantities (3.5) (3.6) (except in Remarks 1,2). We will now make use of (3.5) (3.6) to obtain blow-up results in the elliptic-elliptic case, for large initial data.

To begin with we want to justify (1.10), i.e. to prove that the Cauchy problem for (3.1) (3.2) (3.3) is well posed in a weighted Sobolev space.

Theorem 4.
Let $A_0 \in \Sigma = \{v \in H^1(\mathbb{R}^2),\ (x^2+y^2)\,|v|^2 \in L^1(\mathbb{R}^2)\}$. The solution (A,ϕ) of (3.1), (3.2), (3.3) satisfies (3.6) and

$$(x^2+y^2)\,|A|^2 \in C([0,T^*);L^1(\mathbb{R}^2)) \cap L^2((0,T)\times\mathbb{R}^2),\ \forall\, t \in [0,T^*)$$

Idea of the proof. If $A_0 \in H^1(\mathbb{R}^2)$ we know by Theorem 2 that the maximal solution (A, ϕ) on the inverval $[0, T^*)$ given by Theorem 1 is such that $A(t)$ belongs to $H^1(\mathbb{R}^2)$ for $0 \leq t < T^*$. Theorem 4 will follow from a continuation argument which says that if $A_0 \in \Sigma$ then $A(t)$ belongs to Σ as long as $A(t) \in H^1(\mathbb{R}^2)$. Use will be made of a commutation result for the linear Schrödinger equation which we state in the context of the general equation (3.10).

Lemma 1. *Let $A = (a_{ij})$ be a nonsingular $n \times n$ symmetric matrix and let $B = A^{-1}$. The linear operator $J = J_1, ..., J_n)$ given by*

$$J_k v = e^{i\psi(x)/4t}(2it)\frac{\partial}{\partial x_k}(e^{-i\psi(x)/4t}v)$$

where

$$\psi(x) = \sum_{i,j=1}^{n} b_{ij}x_i x_j, \quad B = (b_{ij})$$

commutes with the general linear Schrödinger operator

$$L = i\frac{\partial}{\partial t} + a_{ij}\frac{\partial^2}{\partial x_i \partial x_j}$$

(For the classical Schrödinger operator, $a_{ij} = \delta_{ij}$, this is just the conformal invariance property. See Ginibre and Velo [42]).

We can now state the blow-up result in the elliptic-elliptic case.

Theorem 5. *Let $\Sigma_- = \{v \in \Sigma, E(v) < 0\}$ (see Theorem 4).*
(i) *The set Σ_- is not empty if and only if $\chi < \mathrm{Max}(-b, 0)$*
(ii) *For $\chi \geq \mathrm{Max}(-b, 0)$, the solutions obtained in Theorem are global : $T^* = \infty$.*
(iii) *For $\chi < \mathrm{Max}(-b, 0)$ and for every $A_0 \in \Sigma_-$, the maximal solution obtained in Theorem 1 satisfies $T^* < +\infty$.*

3.2. The Cauchy problem in the elliptic-hyperbolic case.

It will assumed in this section that $\delta = +1$, $m < 0$. We set $c = \sqrt{-m}$. In the previous section, since the equation for φ was elliptic, it was natural to impose that $\nabla\varphi$ vanishes at infinity. In the present case the general boundary condition for the mean flow potential is to prescribe φ in the characteristic directions $\xi = cx - y$ and $\eta = cx + y$. More precisely, given two functions $\varphi_1(t,s)$ and $\varphi_2(t,s)$, $s \in \mathbb{R}$, $t \in \mathbb{R}_+$, these b.c. are

$$\begin{aligned} \lim_{\xi \to +\infty} \varphi(t,x,y) &= \varphi_1(t,\eta) \\ \lim_{\eta \to +\infty} \varphi(t,x,y) &= \varphi_2(t,\xi) \end{aligned} \tag{3.11}$$

We will assume that

$$\varphi_i \in L^\infty(\mathbb{R}_+; C_b(\mathbb{R}), \qquad \lim_{s \to +\infty} \varphi_i(t,s) = 0, \quad i = 1,2$$

The main result is

Theorem 6. Assume that $A_0 \in H^1(\mathbb{R}^2)$ satisfies

$$\left(\frac{2 \mid b \mid}{C} + \mathrm{Min}(-\chi, 0)\right) \int_{\mathbb{R}^2} \mid A_0 \mid^2 dxdy < 1$$

Then there exists a solution (A,φ) of (3.1),(3.2),(3.3),(3.11) in the sense of distributions such that

$$A \in L^\infty(\mathbb{R}_+; H^1(\mathbb{R}^2)) \cap C(\mathbb{R}_+; H^1_w(\mathbb{R}^2))$$
$$\varphi \in L^\infty(\mathbb{R}_+; C_b(\mathbb{R}^2)), \nabla\varphi \in L^\infty(\mathbb{R}_+; L^q_{mpc}(\mathbb{R}^2)),\ 1 \le q < 2.$$

<u>Indications on the proof</u>

(i) For the sake of simplicity, we will assume $\varphi_1 = \varphi_2 = 0$ in (3.1). The general case is harmless, actually (cf [40]). In order to express φ in function of A, we study the following problem. Given $f \in L^1(\mathbb{R}^2)$, find φ solution of the wave equation

$$\frac{\partial^2 \varphi}{\partial x^2} - c^2 \frac{\partial^2 \varphi}{\partial y^2} = f \tag{3.12}$$

$$\lim_{cx+y \to +\infty} \varphi(x,y) = \lim_{cx-y \to +\infty} \varphi(x,y) = 0 \tag{3.13}$$

We state

Proposition 1. *Let $f \in L^1(\mathbb{R}^2)$ and K be the kernel*

$$K(x,y;x_1,y_1) = \frac{1}{2} H(c(x_1 - x) + y - y_1) H(c(x_1 - x) + y_1 - y)$$

where H is the Heaviside function. Then, the function

$$\phi(x,y) = \mathcal{K}f(x,y) = \int_{\mathbb{R}^2} K(x,y,x_1,y_1)f(x_1,y_1)dx_1dy_1 \tag{3.14}$$

is continuous and is a solution of (3.12) (3.13). *Moreover*

$$\operatorname*{Sup}_{(x,y)\in\mathbb{R}^2} \mid \phi(x,y) \mid \leq \frac{1}{2c}\int_{\mathbb{R}^2} \mid f(x,y) \mid dxdy \tag{3.15}$$

$$\int_{\mathbb{R}^2} \left| \left(\frac{\partial\phi}{\partial x}\right)^2 - c^2\left(\frac{\partial\phi}{\partial y}\right)^2 \right| dxdy \leq \frac{1}{2c}\left(\int_{\mathbb{R}^2} \mid f(x,y) \mid dxdy\right)^2 \tag{3.16}$$

Remarks 2.

1. In general $\nabla\phi$ does not belong to $L^2(\mathbb{R}^2)$, even if $f \in \mathcal{D}(\mathbb{R}^2)$.
2. The estimate (3.16) will give a sense to the energy E.

(ii) Construction of approximate solutions and a priori estimates. We consider the regularized problem

$$\begin{cases} i\dfrac{\partial A^\epsilon}{\partial t} + \Delta A^\epsilon + i\epsilon\Delta^2\dfrac{\partial A^\epsilon}{\partial t} = \chi \mid A^\epsilon \mid^2 A^\epsilon + bA^\epsilon\dfrac{\partial\phi^\epsilon}{\partial x}, \\ \dfrac{\partial^2\phi^\epsilon}{\partial x^2} - m\dfrac{\partial^2\phi^\epsilon}{\partial y^2} = +\dfrac{\partial}{\partial x}(\mid A^\epsilon \mid^2) \\ A^\epsilon(\cdot,0) = A_0^\epsilon \in H^2(\mathbb{R}^2) \text{ with } A_0^\epsilon \to A_0 \text{ in } H^1(\mathbb{R}^2). \end{cases} \tag{3.17}$$

The existence of a unique global solution of (3.17) is proved with Segal's Theorem [75]. Moreover, one obtains the following a priori estimates, provided $\| A_0 \|_{L^2}$ is sufficiently small,

$$\begin{aligned} \| A^\epsilon \|_{L^\infty(\mathbb{R}_+;H^1(\mathbb{R}^2))} &\leq C \\ \| \phi^\epsilon \|_{L^\infty(\mathbb{R}_+\times\mathbb{R}^2)} &\leq C \end{aligned} \tag{3.18}$$

where C is independent of ϵ.

(iii) Passing to the limit. Thanks to (3.18) one can extract a subsequence still denotes $(A^\epsilon, \phi^\epsilon)$ such that

$$\begin{aligned} A^\epsilon &\to A \quad \text{in } L^\infty(0,\infty;H^1(\mathbb{R}^2)) \text{ weak}^* \\ A^\epsilon &\to A \quad \text{in } L^p_{loc}((0,\infty)\times\mathbb{R}^2) \text{ strongly } p \geq 2 \\ \phi^\epsilon &\to \psi \quad \text{in } L^\infty((0,\infty)\times\mathbb{R}^2)) \text{ weak}^* \end{aligned}$$

It can be proved that (A, ψ) satisfy the equation (in the sense of distributions)

$$\begin{aligned} iA_t + \Delta A &= \chi \mid A \mid^2 A + bA\psi_x \\ A(0) &= A_0. \end{aligned} \tag{3.19}$$

The fact that $\psi = \mathcal{H}(\mid A \mid_x^2)$ is more subtle. The main ingredient, obtained by the representaiton (3.14) is that ϕ_x^ϵ is bounded in $L^\infty(\mathbb{R}_+; L^q_{loc}(\mathbb{R}^2))$, for $q \in [1, 2)$.

Remarks 3.

1. The hyperbolic-hyperbolic case is totally open. However it does not seem to be relevant to the theory of surface water waves.

2. Some special solutions (solitons, lumps) have been recently found analytically, in the special cases of system (1.5) (1.6) ([2] [38] [75]). Use of inverse scattering is here a crucial point.

References

1 L. Abdelouhab, J.L. Bona, M. Felland, J.C. Saut, Nonlocal models for nonlinear, dispersive waves, *Physica D*, to appear.

2 M.J. Ablowitz, A.S. Fokas, On the inverse scattering transform of multidimensional nonlinear equations related to first-order systems in the plane, *J. Math. Phys., 25(8)*, (1984), 2494-2505.

3 M.J. Ablowitz, R. Haberman, Nonlinear evolution equations, in two and three dimensions, *Phys. Rev. Lett. 35*, (1975), 1185-1188.

4 M.J. Ablowitz, H. Segur, On the evolution of packets of water waves, *J. Fluid Mech. 92(4)*, 1979, 691-715.

5 M.J. Ablowitz, H. Segur, *Solitons and the Inverse Scattering Transform*, S.I.A.M., Philadelphia, 1981.

6 J.P. Albert, J.L. Bona, Total positivity and the stability of internal waves in stratified fluids of finite depth, Preprint, 1989.

7 J.P. Albert, J.L. Bona, D. Henry, Sufficient conditions for stability of solitary-wave solutions of model equations for long waves, *Physica D 24*, 1987, 343-366.

8 R. Beals, R.R. Coifman, The spectral problem for the Davey-Stewartson and Ishimori hierarchies, preprint, 1988.

9 T.B. Benjamin, The stability of solitary waves, *Proc. Royal Soc. London A 328*, 1972, 153-183.

10 T.B. Benjamin, J.L. Bona, J.J. Mahony, Model equations for long waves in nonlinear, dispersive media, *Phil. Trans. Roy. Soc. London A 272*, 1972, 47-78.

11 T.B. Benjamin, J.L. Bona, D.K. Bose, Solitary-wave solutions of nonlinear problems, *Phil. Trans. Roy. Soc. London A*, to appear.

12 D. Bennett, J.L. Bona, R. Brown, S. Stansfield, J. Stoughair, The stability of internal waves, *Math. Soc. Cambridge Phil. Soc. 94*, 1983, 351-379.

13 H. Berestycki, T. Cazenave, Instabilité des états stationnaires dans les équations de Schrödinger et de Klein-Gordon non linéaires, *C. R. Acad. Sci. Paris 293*, 1981, 489-492.

14 H. Berestycki, P.L. Lions, Nonlinear scalar field equations, *Arch. Rational Mech. Anal. 82, 1983, 313-346 and 347-376.*

15 J.L. Bona, On the stability of solitary waves, *Proc. Royal Soc. London A 344*, 1975, 363-374.

16 J.L. Bona, J.C. Saut, Singularités dispersives de solutions d'équations de type Korteweg-de Vries, *C. R. Acad. Sci. Paris Série I, 303*, 1986, 101-103.

17 J.L. Bona, J.C. Saut, Dispersive blow-up of solutions of generalized Korteweg-de Vries equations, *J. Diff. Eq.*, to appear.

18 J.L. Bona, V.A. Dougalis, O.A. Karakashian, Fully discrete Galerkin methods for the Korteweg-de Vries equations, *Comp. Math. with Appl. 12A;* 1986, 859-884.

19 J.L. Bona, V.A. Dougalis, O.A. Karakashian, Conservative high order numerical schemes for the generalized Korteweg-de Vries equation, Preprint, 1988.

20 J.L. Bona, P.E. Souganidis, W.A. Strauss, Stability and instability of solitary waves of Korteweg-de Vries type, *Proc. Royal Soc. London A, 411*, 1987, 395-412.

21 J.L. Bona, R.L. Sachs, Global existence of smooth solutions and stability of solitary waves for a generalized Boussinesq equation *Comm. Math. Phys. 118*, 1988, 15-29.

22 T. Cazenave, F.B. Weissler, Some remarks on the nonlinear Schrödinger equation in the critical case, Proceedings of the Second Howard University Symposium on Nonlinear Semigroups, Partial Differential Equations and Attractors, Washington DC, August 1987, Springer, to appear.

23 T. Cazenave, F. Weissler, The Cauchy problem for the critical nonlinear Schrödinger equation in H^s, *Nonlinear Anal. T.M.A.*, to appear.

24 T. Cazenave, F. Weissler, The structure of solutions to the pseudo-conformally invariant nonlinear Schrödinger equation, preprint 1989.

25 T. Cazenave, F.B. Weissler, The Cauchy problem for the nonlinear Schrödinger equation in H^1, *Manuscripta Math. 61*, 1988, 477-494.

26 T. Cazenave, A. Haraux, Introduction aux problèmes d'évolution semi-linéaires, Publications Mathématiques de la SMAI, Ellipses, Paris, 1990.

27 C.V. Coffman, Uniqueness of the ground state solution for $\Delta u - u + u^3 = 0$ and a variational characterization of other solutions, *Arch. Rational Mech. Anal. 46*, 1972, 81-95.

28 A. Cohen, Solutions of the Korteweg-de Vries equations for irregular data, *Duke Math. J. 45*, 1978, 149-181.

29 A. Cohen, Existence and regularity for solutions of the Korteweg-de Vries equations, *Arch. Rational Mech. Anal. 71*, 1979, 143-175.

30 A. Cohen, Decay and regularity in the inverse scattering problem, *J. Math. Anal. Appl. 87*, 1982, 395-426.

31 P. Constantin, J.C. Saut, Effets régularisants locaux pour des équations dispersives générales, *C. R. Acad. Sci. Paris Série I, 304*, 1987, 407-410.

32 P. Constantin, J.C. Saut, Local smoothing properties of dispersive equations, *Journal of A.M.S. 1,2*, 1988, 413-439.

33 P. Constantin, J.C. Saut, Local smoothing properties of Schrödinger equations, *Indiana Math. J., 38(3)*, 1989, 791-810.

34 H. Cornille, Solutions of the generalized nonlinear Schrödinger equation in two spatial dimensions, *J. Math. Phys. 20(1)*, 1979, 199-209.

35 A. Davey, K. Stewartson, On three-dimensional packets of surface waves, *Proc. R. Soc. London A 338*, 1974, 101-110.

36 V.D. Djordjevic, L.G. Redekopp, On two dimensional packets of capillary-gravity waves, *J. Fluid Mech. 79(4)*, 1979, 703-714.

37 A.S. Fokas, P.M. Santini, Recursion operators and bi-hamiltonian structures in multidimensions I and II, *Comm. Math. Phys. 115*, 1988, 375-419, and *116*, 1988, 449-474.

38 A.S. Fokas, P.M. Santini, Solitons in multidimensions, preprint, Clarkson University, INS # 106, 1988.

39 J.M. Ghidaglia, J.C. Saut, Sur le problème de Cauchy pour les équations de Davey-Stewartson, *C.R. Acad. Sci. Paris 308*, 1989, 115-120.

40 J.M. Ghidaglia, J.C. Saut, On the initial value problem for the Davey-Stewartson systems, *Nonlinearity*, to appear.

41 J.M. Ghidaglia, J.C. Saut, Nonelliptic Schrödinger equations, in preparation.

42 J. Ginibre, G. Velo, On a class of nonlinear Schrödinger equations, Part I and II : *J. Funct. Anal., 32*, 1979, 1-32 and 1-71 ; Part III : *Ann. Inst. Henri Poincaré, Section A 28*, 1978, 287-316.

43 J. Ginibre, G. Velo, The global Cauchy problem for the nonlinear Schrödinger equations revisited, *Ann. Inst. Henri Poincaré Anal. non Linéaire 2*, 1985, 309-327.

44 J. Ginibre, Y. Tsutsumi, Uniqueness of solutions for the generalized Korteweg-de Vries equation, *SIAM J. Math. Anal. 20(6)*, 1989, 1388-1425.

45 J. Ginibre, Y. Tsutsumi, G. Velo, Existence and uniqueness of solutions for the generalized Korteweg-de Vries equation, Preprint, 1988.

46 J. Ginibre, G. Velo, Propriétés de lissage et existence de solutions pour l'équation de Benjamin-Ono, *C. R. Acad. Sci. Paris Série I*, 1989.

47 J. Ginibre, G. Velo, Commutator expansions and smoothing properties of generalized Benjamin-Ono equations, Preprint, 1989.

48 R.T. Glassey, On the blowing up of solutions to the Cauchy problem for the nonlinear Schrödinger equation, *J. Math. Phys. 18*, 1977, 1794-1797.

49 M. Grillakis, Linearized instability for nonlinear Schrödinger and Klein-Gordon equations, *Comm. Pure Appl. Math. 41*, 1988, 747-774.

50 M. Grillakis, J. Shatah, W.A. Strauss, Stability theory of solitary waves in the presence of symmetry I, *J. Funct. Anal. 74*, 1987, 160-197.

51 N. Hayashi, K. Nakamitsu, M. Tsutsumi, On solutions of the initial value problem for the nonlinear Schrödinger equation in one space dimension, *Math. Z. 192*, 1986, 637-650.

52 N. Hayashi, K. Nakamitsu, M. Tsutsumi, On solutions of the initial value problem for the nonlinear Schrödinger equation, *J. Funct. Anal. 71*, 1987, 218-245.

53 R. Iorio Jr, On the Cauchy problem for the Benjamin-Ono equation, *Comm. Partial Diff. Eq. 11*, 1986, 1031-1081.

54 R. Iorio Jr, The Benjamin-Ono equation in weighted Sobolev spaces, Preprint, 1989.

55 T. Kato, On the Cauchy problem for the (generalized) Korteweg-de Vries equations, *Stud. Appl. Math. Adv. in Math., Supplementary Study 18*, 1983, 93-128.

56 T. Kato, On nonlinear Schrödinger equations, *Ann. Inst. Henri Poincaré, Physique Théorique, 46*, 1987, 113-129.

57 S.N. Krushkov, A.V. Faminskii, Generalized solutions to the Cauchy problem for the Korteweg-de Vries equation, *Math. U.S.S.R. Sbornik 48*, 1984, 93-138.

58 T. Kubota, D.R.S. Ko, L. Dobbs, Propagation of weakly nonlinear internal waves in a stratified fluid of finite depth, *J. Hydronautics 12*, 1978, 157-165.

59 M.K. Kwong, Uniqueness of positive solutions of $\Delta u + u + u^p = 0$ in $\mathbb{R}^n$, *Arch. Rational Mech. Anal 105, 3*, 1989, 243-266.

60 M.J. Landman, G.C. Papanicolaou, C. Sulem, P.L. Sulem, Rate of blow-up for solutions of the nonlinear Schrödinger equation at critical dimension, *Physical Review A, 38, 8*, 3837-3843.

61 B.J. Lemesurier, G. Papanicolaou, C. Sulem, P.L. Sulem, Focusing and multi-focusing solutions of the nonlinear Schrödinger equation, *Physica D31*, 1988, 78-102.

62 B.J. Lemesurier, G. Papanicolaou, C. Sulem, P.L. Sulem, Local structure of the self-focusing singularity of the nonlinear Schrödinger equation, *Physica D32*, 1988, 210-226.

63 P.L. Lions, The concentration-compactness principle in the calculus of variations : the locally compact case, Parts 1,2, *Ann. I.H.P. Analyse Non Linéaire 1*, 1984, 109-145 and 223-283.

64 K. Mc Leod, J. Serrin, Uniqueness of positive radial solutions of $\Delta u + f(u) = 0$ in $\mathbb{R}^n$, *Arch. Rational Mech. Anal. 99*, 1987, 115-145.

65 F. Merle, Limit of the solution of the nonlinear Schrödinger equation at the blow-up time, *J. Funct. Anal.*, to appear.

66 F. Merle, Sur la dépendance continue de la solution de l'équation de Schrödinger non linéaire près du temps d'explosion, *C. R. Acad. Sci. Paris 16*, 1987, 479-482.

67 F. Merle, Construction d'une solution de l'équation de Schrödinger non linéaire avec exactement k points d'explosion dans le cas de la puissance critique, Preprint, 1988.

68 F. Merle, Y. Tsutsumi, L^2–concentration of blow-up solutions for the nonlinear Schrödinger equation with the critical power nonlinearity, Preprint, 1988.

69 H.C. Morris, Prolongation structures and nonlinear evolution equations in two spatial dimensions. II A generalized nonlinear Schrödinger equations, *J. Math. Phys., 18(2)*, 1977, 285-288.

70 A. Newell, Solitons in Mathematics and Physics, CBMS, 48, SIAM, 1985.

71 G. Ponce, Smoothing properties of solutions to the Benjamin-Ono equation, Preprint, 1988.

72 M. Tom, Ph.D. Thesis, Penn. State University.

73 P.M. Santini, A.S. Fokas, The initial-boundary value problem for the Davey-Stewartson 1 equation ; how to generate and drive localized coherent structures in multidimensions, These Proceedings.

74 J.C. Saut, Properties of dispersive equations, in Proc. Conf. on Integrable Systems and Applications, M. Balabane, P. Lochak Eds, Springer Lecture Notes in Physics, n° 342, 1989, 295-311.

75 I. Segal, Nonlinear Semigroups, *Annals of Math., 78*, 1963, 339-364.

76 P. Sjölin, Regularity of solutions to the Schrödinger equations, *Duke Math. J. 55*, 1987, 699-715.

77 P. Sjölin, Local regularity of solutions to nonlinear Schrödinger equations, Preprint Uppsala University, 1989.

78 W.A. Strauss, Existence and solitary waves in higher dimensions, *Comm. Math. Phys. 55*, 1977, 149-162.

79 W. Strauss, Nonlinear Wave Equations, Proceedings of the CBMS Lectures, 1989.

80 R. Strichartz, Restriction of Fourier transforms to quadratic surfaces and decay of solutions of wave equations, *Duke Math. J. 44*, 1977, 705-714.

81 J. Stubbe, Stability of ground states in nonlinear classical field theories, Preprint, 1988.

82 J. Stubbe, Existence and stability of solitary waves of Boussinesq-type equations, Preprint, 1988.

83 M. Tsutsumi, Nonexistence of global solutions to nonlinear Schrödinger equations, *SIAM J. Math. Anal. 15*, 1984, 357-366.

84 Y. Tsutsumi, L^2–solutions for nonlinear Schrödinger equations and nonlinear groups, *Funk. Ekva. 30*, 1987, 115-125.

85 Y. Tsutsumi, The Cauchy problem for the Korteweg-de Vries equation with measures as initial data, *SIAM J. Math. Anal.* *20,3*, 1989, 582-588.

86 L. Vega, Schrödinger equations : pointwise convergence to the initial data, *Proc. A.M.S.* *102,4*, 1988, 874-878.

87 M.I. Weinstein, Modulational stability of ground states of nonlinear Schrödinger equations, *SIAM J. Math. Anal.* *16,3*, 1985, 472-490.

88 M.I. Weinstein, Solitary waves of nonlinear dispersive evolution equations with critical power nonlinearities, *J. Diff. Eq.* *69*, 1987, 192-203.

89 M.I. Weinstein, Nonlinear Schrödinger equations and sharp interpolation estimates, *Comm. Math. Phys.* *87*, 1983, 567-576.

90 M.I. Weinstein, Lyapunov stability of ground states of nonlinear dispersive evolution equations, *Comm. Pure Appl. Math.* *39*, 1986, 51-67.

91 M.I. Weinstein, Existence and dynamic stability of solitary-wave solutions of equations arising in long wave propagation, *Comm. Partial Diff. Eq.* *12*, 1987, 1133-1173.

92 M.I. Weinstein, On the structure and formation of singularities in solutions to nonliner dispersive evolution equations, *Comm. Partial Diff. Eq.* *11,5*, 1986, 545-565.

93 V.E. Zakharov, A.B. Shabat, Exact theory of two-dimensional self-focusing and one dimensional self-modulation of waves in nonlinear media, *Sov. Phys. JETP* *34*, 1972, 67-69.

GROUP THEORY AND EXACT SOLUTIONS OF PARTIALLY INTEGRABLE DIFFERENTIAL SYSTEMS

P. WINTERNITZ
Centre de recherches mathématiques
Université de Montréal
C.P. 6128-A, Montréal, Québec, Canada H3C 3J7

ABSTRACT. The application of Lie group theory to obtain exact analytic solutions of nonlinear differential equations is reviewed. The emphasis is on recent developments such as the use of infinite dimensional symmetry groups, on algorithms for classifying finite dimensional subgroups of both finite and infinite dimensional Lie groups and on the combination of Lie group theory with singularity analysis (Painlevé analysis). At each stage the use of computer algebra plays an important role: for finding the symmetry group, for identifying its Lie algebra, classifying its subgroups and performing Painlevé analysis.

TABLE OF CONTENTS

R. Conte and N. Boccara (eds.), Partially Integrable Evolution Equations in Physics, 515–567.

1. Introduction

The purpose of this lecture series is to show how the theory of Lie groups, or more specifically, the theory of local Lie point transformations, can be used to obtain large classes of exact analytical solutions of differential equations. The emphasis, in keeping with the topic of this Advanced Study Institute, will be on nonlinear partial differential equations (PDE). The methods and results do not depend on the integrability, in any sense of the word, of the considered PDE's. In these lectures the words "partially integrable nonlinear evolution equations" will be used in a specific sense. They will signify systems that cannot be integrated by essentially linear techniques, such as inverse scattering, or the dressing method. On the other hand, when certain specifically chosen boundary or initial conditions are imposed, these systems will reduce to integrable subsystems. These can then be solved explicitly and one obtains particular solutions of the original system.

The application of Lie groups to the study of differential equations has a long history. As a matter of part, it is as long as the history of Lie groups themselves, since S. Lie introduced continuous groups of transformations precisely in this context. Several recent textbooks on the general topic of Lie groups and differential equations exist [1–7] as well as conference or school proceedings (see e.g. ref. 8).

In these lectures we restrict our attention to local Lie point transformation groups that either leave the considered differential equations invariant, or transform different equations amongst each other. Such transformations have the form

$$x' = \Lambda_g(x,u), \quad u' = \Omega_g(x,u) \tag{1.1}$$

where $x = (x_1, \dots x_p)$ and $u = (u_1, \dots, u_q)$, are the independent and dependent variables and g denotes the group parameters. The vector valued functions Λ_g and Ω_g thus depend on the variables x and u, but not on the derivatives u_{i,x_k}, nor on higher order derivatives. Moreover, these functions need only be defined locally, for

g close to the identity element and for x and u close to some chosen point (x, u) on the manifold $X \times U \sim M \sim \mathbb{R}^p \times \mathbb{R}^q$. Generalized transformations in which the new variables x' and u' do depend on derivatives of u, are treated in other lectures in this series (e.g. by Y. Kosmann–Schwarzbach [**9**], and I.I. Mikhaîlov) [**10**].

We shall call "the symmetry group" G of a system of differential equations the maximal group of transformations of the form (1.1), leaving the considered equations invariant. The symmetry group G will leave the set of all solutions invariant, i.e. transform solutions amongst each other.

As motivation, let us just mention three of the many uses of Lie groups in the context of differential equations.

1. New solutions from old ones. Applying the symmetry group to a known solution, we obtain a family of new solutions. Quite often one can obtain interesting solutions from trivial ones.

2. Symmetry reduction. Symmetries of ordinary differential equations (ODE) can be used to reduce the order of the equation. In some cases the reduction can go all the way down to an algebraic equation. When applicable, symmetry reduction for an ODE leads to the general solution.

Symmetries of partial differential equations (PDE's) can be used to reduce the number of independent variables in the equation. In particular the reduction can go all the way to an ordinary differential equation, or to an algebraic one. For PDE's symmetry reduction in general provides particular solutions, rather than the general one.

3. Classification of differential equations. Point transformations that do not leave a considered differential equation invariant can be used to classify equations into equivalence classes and to choose simple representatives of such classes. In particular, nonlinear equations may be transformable into linear ones [**11**].

In short, the use of Lie group theory makes it possible to elucidate the structure of differential equations and their solution sets and to construct general, or at least particular, explicit solutions.

Finally, let us sum up some of the recent developments that have contributed to an outburst of new activity in this field.

1. Developments in physics and other natural and applied sciences. Nonlinear phenomena have finally caught up with us. All the fundamental equations of physics are nonlinear (Einstein equations, Yang–Mills equations, Navier–Stokes equations, etc.), as are most of the equations governing the propagation of waves in various media. There is hence an increased interest in obtaining exact analytic solutions of nonlinear systems. Group theoretical method provide such solutions for large classes of equations and do not rely on the "integrability" of these equations.

2. Developments in mathematics

(a) The symmetry groups of quite a few interesting differential equations turn out to be infinite–dimensional. Considerable progress has been made recently in the theory of infinite–dimensional Lie algebras and Lie groups [**12**,13].

(b) In order to make efficient use of the symmetry group of an equation one needs a classification of its subgroups. Methods and algorithms for classifying subgroups of Lie groups and subalgebras of Lie algebras have recently been developed [**14–17**]. The methods were developed for finite–dimensional Lie algebras but have been extended to the classification of finite–dimensional subalgebras of infinite–dimensional Lie algebras [**18**,19,20].

(c) Once group theory is used to reduce a differential equation to a lower dimensional or lower order one, it is still necessary to solve the reduced equation. Even if the original equation is non integrable, the reduced one may be in the class of integrable equations. In particular, a combination of group theory and singularity analysis (Painlevé analysis) turns out to be very fruitful. For a review of Painlevé analysis for ODE's and PDE's see lectures by M. Kruskal, B. Grammaticos, A. Ramani [**21**], D. Bessis [**22**], J. Weiss [**23**] in this volume. For a combination of symmetry reduction with Painlevé analysis, see e.g. refs. 24–26.

3. Developments in computer science

(a) One of the major obstacles to applying group theory to solve differential equations is that the task of finding the symmetry group of a system of equations is a cumbersome one. This task has been made much easier by the development of computer packages in various symbolic languages, such as REDUCE [**27**], or MACSYMA [**28**].

(b) Symbolic manipulations on computers can be used at all stages of the work. Once the Lie algebra of the symmetry group is found, it should be identified (as being decomposable into a direct sum, or indecomposable; as being simple, solvable, or having a nontrivial Levi decomposition, etc). Algorithms [**29**] and computer packages [**30**,31] exist for doing this. The classification of subgroups, discussed above, can be done in a computer assisted manner.

The result of all of these developments is that large classes of solutions of nonlinear differential equations can be obtained in a quite simple manner.

2. Algorithm for Finding the Symmetry Group of a System of Differential Equations

We shall present the algorithm in a very summary and simplified manner. All proofs and details can be found in the literature, in particular in Olver's book [1].

Consider a system of m differential equations of order n

$$\begin{aligned} &\Delta^i(x, u, u_{(1)}, u_{(2)}, \ldots, u_{(n)}) = 0 \\ &x = (x_1, \ldots, x_p) \qquad\qquad i = 1, \ldots, m \\ &u = (u_1, \ldots, u_q) \end{aligned} \tag{2.1}$$

where x denotes the independent variables, u the dependent ones and say $u_{(k)}$ denotes the set of all derivatives of order k. We wish to find the Lie group of local point transformations (1.1) transforming solutions of (2.1) amongst each other. Thus, if $u(x) = f(x)$ is a solution then

$$u'(x') = g \circ f(x) \tag{2.2}$$

should also be a solution.

The group G that we are looking for will act on a space M

$$M \subset X \times U, \quad X \sim \mathbb{R}^p, \quad U \sim \mathbb{R}^q \tag{2.3}$$

where X and U denote the spaces of independent and dependent variables, respectively.

We shall consider functions

$$\begin{aligned} &f : X \to U, \\ &u = f(x), \quad \text{i.e.} \quad u_i = f_i(x_1, \ldots, x_p), \quad i = 1, \ldots q \end{aligned}$$

and prolongations of functions

$$\mathrm{pr}^{(n)} f(x) \ : \{f, f_{x_i}, \ldots, f_{x_{j_1}, \ldots, x_{j_n}}\},$$

i.e. functions together with their derivatives upto order n.

A transformation $g \in G$ transforms points $(x, u) \in M$ and functions $f(x)$ into

$$\begin{aligned} &(x, u) \longrightarrow (\tilde{x}(x, u), \tilde{u}(x, u)) \in M \\ &f(x) \ \longrightarrow \tilde{f}(\tilde{x}) = g \circ f(x). \end{aligned}$$

The n–th prolongation of the group action will also transform all derivatives of the function upto order n:

$$\mathrm{pr}^{(n)}G \ : \{x, u = f(x), f_{x_i}, f_{x_i\,x_k}, \dots, f_{nx}\} \longrightarrow$$
$$\{\tilde{x}, \tilde{f}(\tilde{x}), \tilde{f}_{\tilde{x}_i}, \tilde{f}_{\tilde{x}_i\,\tilde{x}_k}, \dots, \tilde{f}_{n\tilde{x}}\}.$$

The action on the prolonged space $M^{(n)}$ is induced by the action on M. Thus if we know G, then $\mathrm{pr}^{(n)}G$ is completely determined.

Following S. Lie's approach we shall not look directly for the transformations g of (1.1) that leave the system (2.1) invariant and that are, in general, nonlinear. Instead, we shall first linearize the problem by restricting to infinitesimal transformations. Our first aim is thus to find the Lie algebra L of the symmetry group G (the "symmetry algebra" for short). It will be realized by vector fields of the form

$$v = \sum_{i=1}^{p} \xi^i(x,u)\partial_{x_i} + \sum_{\alpha=1}^{q} \phi_\alpha(x,u)\partial_{u_\alpha} \tag{2.4}$$

where x, u are some local coordinates on M. The corresponding local Lie group of 1-parameter local transformations is obtained by integrating the vector field (2.4), i.e. solving the system of first order ODE's:

$$\frac{dx_i'}{d\lambda} = \xi_i(x',u'), \qquad x_i'\Big|_{\lambda=0} = x_i, \quad i = 1,\dots,p, \quad \alpha = 1,\dots,q$$
$$\frac{du_\alpha'}{d\lambda} = \phi_a(x',u'), \qquad u_\alpha'\Big|_{\lambda=0} = u_\alpha. \tag{2.5}$$

This procedure will provide us with the connected component G_0 of the symmetry group G. Further discrete elements may exist and provide further components of G.

The fact that we are restricting ourselves to point transformations is reflected in the fact that the coefficients ξ^i and ϕ_α in (2.4) do not depend on derivatives $u_{x_i}, \dots$.

The vector field v of (2.4) acts on functions of x and u. The n–th prolongation $\mathrm{pr}^{(n)}v$ of v acts on functions of x, u and all derivatives $u_{x_i}, u_{x_i\,x_k}, \dots$, upto order n. An explicit expression for $\mathrm{pr}^{(n)}v$ can be obtained by differentiating the expression for $\mathrm{pr}^{(n)}G$ with respect to λ and setting $\lambda = 0$. Conversely, integrating $\mathrm{pr}^{(n)}v$ as in (2.5) one obtains $\mathrm{pr}^{(n)}G$. Hence the prolongation of the vector field v is completely determined by the field v itself. We shall give an expression for this prolongation in local coordinates, referring to Olver [**1**] for its derivation.

The prolongation formula is

$$\mathrm{pr}^{(n)}v = v + \sum_{\alpha=1}^{q} \sum_{k=1}^{n} \sum_{J} \phi_\alpha^J \frac{\partial}{\partial u_J^\alpha}, \tag{2.6a}$$

where

$$J \equiv J(k) = (j_1, \dots, j_k), \quad 1 \le j_k \le p, \quad k = j_1 + j_2 + \dots + j_k \tag{2.6b}$$

is a multi–index, telling us which derivatives of u^α of order k are involved in a particular term. The coefficients ϕ_α^J in (2.6a) are given in terms of ξ^i, ϕ_α, and their derivatives upto order k. An explicit formula for ϕ_a^J exists [1], but we shall give a recursive formula instead. We define the total derivative D_i with respect to x_i as

$$D_{x_i} = \frac{\partial}{\partial x_i} + \sum_\alpha \frac{\partial u_\alpha}{\partial x_i}\frac{\partial}{\partial u_\alpha} + \sum_{\alpha,j} \frac{\partial u_{\alpha x_j}}{\partial x_i}\frac{\partial}{\partial u_{\alpha x_j}} + \ldots . \tag{2.6c}$$

The coefficient ϕ_α^i, corresponding to first derivatives u_{x_i} in (2.6a) then is

$$\phi_\alpha^i \equiv \phi_\alpha^{x_i}(x, u, u_{x_1}, \ldots, u_{x_p}) = D_{x_i}\,\phi_\alpha - \sum_{k=1}^{p}(D_{x_i}\xi^k)u_{x_k}^\alpha . \tag{2.6d}$$

For the further terms we use a recursive formula, namely

$$\phi_\alpha^{J,k}(x, u, u^{(1)}, \ldots, u^{(k)}) = D_k\,\phi_\alpha^J - \sum_{i=1}^{p}(D_k\,\xi^a)u_{J,a}^\alpha . \tag{2.6e}$$

If we have a Lie algebra L of vector fields of the form v (2.4) on M, then the prolonged vector fields $\mathrm{pr}^{(n)}v$ form the same Lie algebra, since we have

$$\begin{aligned} \mathrm{pr}^{(n)}[v, w] &= [\mathrm{pr}^{(n)}v, \mathrm{pr}^{(n)}w] \\ \mathrm{pr}^{(n)}(av + bw) &= a\,\mathrm{pr}^{(n)}v + b\,\mathrm{pr}^{(n)}w. \end{aligned} \tag{2.7}$$

We see that the expression for $\mathrm{pr}^{(n)}v$ is long and cumbersome. On the other hand it is entirely algorithmic and explicit; moreover, it is the basic tool in the calculation of symmetry groups. Calculating $\mathrm{pr}^{(n)}v$ is thus a prototype of a problem for which computer algebra is ideally suited.

The algorithm for finding the symmetry group of a differential system can now be stated quite simply as follows. The vector field (2.4) is an element of the Lie algebra L of the symmetry group G of the system (2.1), if its n–th prolongation $\mathrm{pr}^{(n)}v$ annihilates the system on its solution surface:

$$\mathrm{pr}^{(n)} \cdot \Delta^i\big|_{\Delta^\ell=0} = 0, \quad i, \ell = 1, \ldots, m. \tag{2.8}$$

In practice this means that the system (2.1) is viewed as a system of algebraic equations for m of the highest order derivatives of u_α, figuring in (2.1). They must be appropriately chosen, so that they can be solved for from (2.1) and substituted into (2.8). The system (2.8) is a system of differential equations for the coefficients ξ^i and ϕ_α in (2.4). Since these depend on x and u only, the coefficient of each linearly independent expression in the derivatives $u^{(1)}, \ldots, u^{(n)}$ that remain in (2.8) after the substitution, must separately vanish. This provides an overdetermined system of linear partial differential equations for the coefficients $\xi_i(x, u)$ and $\phi_\alpha(x, u)$, that is usually quite easy to solve. It is called the system of "determining equations" for the symmetries of the considered differential system.

Three different possibilities can occur:

1. The determining system allows only the trivial solution $\eta_i = 0$, $\phi_\alpha = 0$ $(i = 1, \ldots, p,\ \alpha = 1, \ldots, q)$. We then have $v = 0$, $g = e$ (the identity transformation only).

2. The general solution of the determining equations depends on r integration constants $(r < \infty)$. The dimension of the symmetry algebra L and symmetry group G will then be

$$\dim L = \dim G = r. \tag{2.9}$$

3. The general solution depends on arbitrary functions of x_i and u_α. The symmetry algebra and group are then infinite dimensional.

The MACSYMA program [**28**] derives the determining equations, solves the simplest ones among them, uses the solution to simplify further and then prints out the remaining ones. The REDUCE program [**27**] goes further in the solving and in many cases actually obtains the vector fields in final form.

3. A Pedagogical Example: Linear Heat Equation with Variable Conductivity

As an example, let us find the symmetry group of the equation

$$\Delta(x, t, u, u_x, u_t, u_{xx}, u_{xt}, u_{tt}) \equiv u_t - c(x,t)\, u_{xx} = 0 \tag{3.1}$$

where $c(x,t)$ is the heat conductivity, assumed to be some sufficiently smooth function of x and t. Simplifying notations, we write the vector field (2.4) as

$$v = \xi(x,t,u)\, \partial_x + \tau(x,t,u)\, \partial_t + \phi(x,t,u)\, \partial_u. \tag{3.2}$$

The second prolongation of v has the form

$$\mathrm{pr}^{(2)} v = v + \phi^t\, \partial_{u_t} + \phi^x\, \partial_{u_x} + \phi^{tt}\, \partial_{u_{tt}} + \phi^{tx}\, \partial_{u_{tx}} + \phi^{xx}\, \partial_{u_{xx}}. \tag{3.3}$$

The basic equation (2.8) reduces to

$$\mathrm{pr}^{(2)} v \cdot \Delta \,\big|_{\Delta=0} = \left[\phi^t - c(x,t)\phi^{xx} - \left(\xi\, \tfrac{\partial c}{\partial x} + \tau\, \tfrac{\partial c}{\partial t}\right) u_{xx}\right]\Big|_{u_{xx}=\frac{1}{c}u_t} = 0. \tag{3.4a}$$

Using the prolongation formula (2.6) we find:

$$\phi^t = \phi_t - \xi_t\, u_x + (\phi_u - \tau_t) u_t - \xi_u\, u_x u_t - \tau_u u_t^2 \tag{3.4b}$$

$$\begin{aligned} \phi^{xx} &= \phi_{xx} + (2\,\phi_{xu} - \xi_{xx})\, u_x - \tau_{xx}\, u_t + (\phi_{uu} - 2\,\xi_{xu})\, u_x^2 \\ &\quad - 2\,\tau_{xu}\, u_x u_t - \xi_{uu}\, u_x^3 - \tau_{uu} u_x^2\, u_t \\ &\quad + (\phi_u - 2\,\xi_x)\, u_{xx} - 2\,\tau_x\, u_{xt} - 3\,\xi_u\, u_x u_{xx} - 2\tau_u\, u_x u_{xt} - \tau_u\, u_t u_{xx} \end{aligned} \tag{3.4c}$$

where the subscripts denote partial derivatives.

Substituting for u_{xx} from (3.1) and equating to zero the coefficients of $u_x u_{xt}$, u_{xt}, u_x^3, $u_x u_t$ and u_x^2, we obtain

$$\tau_u = 0, \quad \tau_x = 0, \quad \xi_{uu} = 0, \quad \xi_u = 0, \quad \phi_{uu} = 0. \tag{3.5}$$

Thus, for any $c(x,t)$ we have

$$\tau = \tau(t), \quad \xi = \xi(x,t), \quad \phi = f(x,t)u + h(x,t). \tag{3.6}$$

Using (3.6) and equating to zero the coefficients of u_x, u_t and 1, we obtain a system of equations for $\tau(t)$, $\xi(x,t)$, $f(x,t)$ and $h(x,t)$:

$$\xi_t - c\,\xi_{xx} + 2c\,f_x = 0 \tag{3.7a}$$

$$\dot{\tau} - 2\,\xi_x + \tfrac{c_x}{c}\,\xi + \tfrac{c_t}{c}\,\tau = 0 \tag{3.7b}$$

$$f_t - c\,f_{xx} = 0 \tag{3.7c}$$

$$h_t - c\,h_{xx} = 0 \tag{3.8}$$

Equation (3.8) decouples from the system so that $h(x,t)$ is a solution of the considered equation (3.1). The corresponding element of the Lie algebra is

$$v_s = h(x,t)\,\partial_u \tag{3.9}$$

and the corresponding group transformation, according to (2.5) is

$$t' = t, \quad x' = x, \quad u'(x',t') = u(x,t) + \lambda h(x,t). \tag{3.10}$$

Thus, the presence of (3.9) in the Lie algebra is simply a reflection of the linear superposition principle.

More generally, the symmetry algebra of any linear PDE is infinite–dimensional and contains an infinity of linearly independent elements of the form

$$v_s = \sum_{\alpha=1}^{q} h_\alpha(x_1, \ldots, x_p)\frac{\partial}{\partial u_\alpha} \tag{3.11}$$

where $\{h_\alpha(x)\}$ is any solution of the studied system. The corresponding invariance property is the linear superposition formula.

It remains to solve the system (3.7). For $c(x,t)$ generic the only solution is $\xi = 0$, $\tau = 0$, $f(x,t) = \text{const}$. Integrating the corresponding vector field we obtain the transformation

$$x' = x, \quad t' = t, \quad u'(x',t') = e^\lambda\,u(x,t), \tag{3.12}$$

which is also simply a consequence of the linearity of the equation.

To continue further we shall consider a special case of physical interest, coming from a study of Brownian motion, namely

$$c(x,t) = (1 + \tfrac{b^2}{2}x^2)^2, \quad b = \text{const}, \tag{3.13}$$

For $c(x,t)$ as in (3.13) the general solution of (3.7) depends on 6 real constants. Factoring out the infinite–dimensional ideal (3.9) we obtain a six–dimensional real Lie algebra L_0. A convenient basis is given by the following vector fields:

$$\begin{aligned}
T &= \partial_t \\
D &= 2t\,\partial_t + (1 + \tfrac{b^2}{2}x^2)\tfrac{\sqrt{2}}{b} \arctan \tfrac{bx}{\sqrt{2}}\,\partial_x \\
&\qquad + \left[-\tfrac{1}{2} + b^2 t + \tfrac{\sqrt{2}}{2}\, bx \arctan \tfrac{bx}{\sqrt{2}}\right] u\,\partial_u \\
C &= t^2\,\partial_t + (1 + \tfrac{b^2}{2}x^2)\tfrac{\sqrt{2}}{b}\, t \arctan \tfrac{bx}{\sqrt{2}}\,\partial_x \\
&\qquad + \left\{-\tfrac{1}{2}t + \tfrac{1}{2}b^2t^2 + b\tfrac{\sqrt{2}}{2}\, xt \arctan \tfrac{bx}{\sqrt{2}} - \tfrac{1}{2b^2}\,(\arctan \tfrac{b}{\sqrt{2}}x)^2\right\} u\,\partial_u \\
P &= (1 + \tfrac{b^2}{2}x^2)\,\partial_x + \tfrac{1}{2}b^2 xu\,\partial_u \\
B &= (1 + \tfrac{b^2}{2}x^2)t\,\partial_x + \left[b^2xt - \tfrac{\sqrt{2}}{b} \arctan \tfrac{bx}{\sqrt{2}}\right] \tfrac{u}{2}\partial_u \\
L &= u\,\partial_u
\end{aligned} \tag{3.14}$$

The commutation relations for this Lie algebra are summed up in Table 1.

Table 1. Commutation relations for the algebra (3.14).

	D	T	C	P	B	M
D	0	$-2T$	$2C$	$-P$	B	0
T	$2T$	0	D	0	P	0
C	$-2C$	$-D$	0	$-B$	0	0
P	P	0	B	0	$-\frac{1}{2}M$	0
B	$-B$	$-P$	0	$\frac{1}{2}$M	0	0
M	0	0	0	0	0	0

We see that $\{D,T,C\}$ is the Lie algebra $sl(2,\mathbb{R})$, $\{P,B,M\}$ the Heisenberg algebra and that $\{P,B,M\}$ is the radical (maximal solvable ideal) of the algebra L_0. Moreover, comparing with the well–known symmetry algebra of the constant coefficient heat equation [**1**]

$$u_t - u_{xx} = 0, \tag{3.15}$$

we see that the two algebras are isomorphic.

The existence of such an isomorphism is a necessary (but not sufficient) condition for a local point transformation to exist, transforming the equations into each other. Such a transformation must transform the two Lie algebras into each other. In the case under consideration, the Lie algebra (3.14) must be transformed into the algebra of the heat equation, obtained by taking the limit $b \to 0$ in (3.14). Since the elements T and P in (3.14) commute and are linearly independent (everywhere) an invertible transformation

$$x \longrightarrow \xi(x,t,u), \quad t \longrightarrow \tau(x,t,u), \quad u \longrightarrow w(x,t,u) \tag{3.16}$$

exists, taking $\{T, P\}$ into

$$T = \partial_\tau, \quad P = \partial_\xi. \tag{3.17}$$

Any such transformation will take the considered variable coefficient heat equation into an equation with constant coefficients. The transformation (3.16) is not unique and the freedom in its choice should be used to appropriately transform the remaining basis elements of L_0. The transformation is easy to find and we just present the final inverse transformation. The result is: if $w(\xi,\tau)$ satisfies the heat equation $w_\tau - w_{\xi\xi} = 0$, then

$$\begin{aligned} u(x,t) &= \left(\cos \tfrac{b\xi}{\sqrt{2}}\right)^{-1} \left(\exp \tfrac{b^2\tau}{2}\right) w(\xi,\tau) \\ x &= \tfrac{\sqrt{2}}{b} \tan \tfrac{b\xi}{\sqrt{2}}, \quad t = \tau \end{aligned} \tag{3.18}$$

will satisfy the equation

$$u_t - \left(1 + \tfrac{b^2}{2}x^2\right)^2 u_{xx} = 0. \tag{3.19}$$

We have thus illustrated the method for finding the symmetry algebra of a given PDE. We have also demonstrated one of the important applications of the Lie algebra, namely the identification of classes of equivalent equations. Since the general solution of the heat equation (3.15) is known, the above result provides the general solution of equation (3.19).

4. The Method of Symmetry Reduction for Partial Differential Equations

The main application of the symmetry group G of a PDE is to construct group invariant solutions, or in other words, to perform symmetry reduction. The basic idea is to take a subgroup $G_0 \subset G$ and to require that a solution of the differential system be invariant under the group G_0 (rather than that it be transformed into other solutions). This invariance imposes constraints on the solution that are expressed by first order linear PDE's. The group invariant solution will thus satisfy the original equation, plus further ones, and this leads to the reduction. Different subgroups G_0 lead to different reductions, but two subgroups that are conjugate under the symmetry group G lead to equivalent reductions. It is hence imperative to perform a subgroup classification. Usually we do not need all subgroups of G

but only those with an appropriate action on the space $M \sim X \times U$. Indeed, any subgroup $G_0 \subset G$ will act on M and sweep out certain orbits on M. Let us denote the dimension of a generic orbit d and its codimension $K = p + q - d$. This implies that K functionally independent G_0–invariants exist: let us denote a basis for these invariants $I_1(x,u), \ldots, I_k(x,u)$. The group $G_0 \subset G$ will provide explicit group invariant solutions if the invariants satisfy two conditions:

(i) k invariants exist depending only on the independent variables x_i. We denote them

$$\xi_1(x), \ldots, \xi_k(x), \quad k = K - q, \quad 0 \leq k \leq p-1 \tag{4.1}$$

(ii) q further invariants exist, providing an invertible mapping to the dependent variables u_α. We denote these invariants F_μ and the condition is that the corresponding Jacobian J be nonsingular

$$\det J = \det \left(\frac{\partial F_\mu}{\partial u_\alpha}\right) \neq 0. \tag{4.2}$$

If these two conditions are satisfied, we consider the invariants F_μ to be functions of ξ_i and then solve for u_α $(\alpha = 1, \ldots, q)$ from the expressions:

$$F_\mu(\xi_1, \ldots, \xi_k) = F_\mu(x, u). \tag{4.3}$$

The result is written in the form

$$u_\alpha(x) = U_\alpha(F_1, \ldots, F_q, x_1, \ldots, x_p), \quad F_\mu = F_\mu(\xi_1, ..., \xi_k). \tag{4.4}$$

Substituting $u_\alpha(x)$ back into the original equations we obtain a system of equations for F_μ, involving only the invariants F_μ and ξ_j. The variables x_i that cannot be expressed in terms of the invariants ξ_j will drop out of the reduced equations. The reason for this is that $\{\xi_i, F_\mu\}$ form a complete set of invariants for the transformation group G and hence $\{\xi_i, F_\mu, \frac{\partial F_\mu}{\partial \xi_i}, \frac{\partial^2 F_\mu}{\partial \xi_i \partial \xi_j}, \ldots\}$ form a complete set of invariants of the prolongation $\mathrm{pr}^{(n)}G$. The reduced equations have the form

$$\tilde{\Delta}_j \left(\xi_i, F_\mu, \frac{\partial F_\mu}{\partial \xi_i}, \frac{\partial^2 F_\mu}{\partial \xi_i \partial \xi_j}, \ldots\right) = 0. \tag{4.5}$$

The number of independent variables in (4.5) is k, with $0 \leq k \leq p-1$. Thus if $k = 0$ (4.5) is a system of algebraic equations. If $k = 1$, we have a system of ODE's, etc. If we can solve the system (4.5) we then substitute $F_\mu(\xi_i)$ back into (4.4) and obtain particular solutions of the original system.

The method of symmetry reduction for PDE's consists of several steps.

1. Find the symmetry group G of the system.

2. Classify all subgroups $G_0 \subset G$ having properties (4.1) and (4.2) into conjugacy classes under the action of the symmetry group G. Order the subgroups according

to the codimension k of the projection of the generic orbits onto the space X of independent variables and choose a representative G_0 of each conjugacy class.

3. For each representative subgroup G_0, find a set of $k+q$–functionally independent invariants, and express u_α in terms of invariants, as in (4.4).

4. Substitute u_α as in (4.4) into the original equations and obtain the reduced system.

5. Solve the reduced system, substitute back into (4.4) and apply a general symmetry group transformation to the result. This will produce a family of particular solutions.

We have elaborated on Step 1 of the method in Section 2 above. Step 2 will be treated below in Section 5. Step 3, the calculation of invariants of a given Lie group G_0, acting on a space M, is a standard one. Indeed, let L_0 be the Lie algebra of G_0 and let $\{v_1, \ldots, v_r\}$ be a basis of L_0, where v_i are vector fields in the tangent space of M, i.e. known differential operators of the form (2.4). A function $F(x,u)$ will be invariant under G_0 if it satisfies

$$v_a F(x,u) = 0, \quad a = 1, \ldots, r. \tag{4.6}$$

This is a system of r first order linear PDE's. They can be solved by the method of characteristics.

Step 4 of the method is elementary. Step 5 relies on our ability to solve the reduced lower–dimensional system. If the codimension k satisfies $k = 0$ we already have an integral of the original system. For $k = 1$ we must solve an ODE (or system of ODE's). This can again be done using group theory, or by performing a singularity analysis (see below), or using the entire "bag of tricks" developed for solving ODE's. For $k \geq 2$ we still have PDE's to deal with. They may be integrable, even if the original equations are not. Otherwise, a further reduction to lower dimensions is in order.

5. Classification of Subalgebras of Finite Dimensional Lie Algebras

Let us consider a Lie algebra L and a group G of automorphisms of $L : GLG^{-1} = L$. We shall be interested in the case when L is the Lie algebra of G. The group G is thus either $G = \exp L$, or it may contain further discrete components. Our aim is to obtain a representative list of subalgebras of L, such that every subalgebra of L is conjugate to precisely one algebra in the list (and no two algebras in the list are conjugate). Two algebras, L and L', are conjugate under G if we have

$$GL'G^{-1} = L, \tag{5.1}$$

or elementwise, for all $x' \in L'$ there exists an element $x \in L$ and $g \in G$ such that $gx'g^{-1} = x$. Three different situations occur, each calling for a different classification method:

1. The algebra L is a simple Lie algebra.

2. The algebra L is a direct sum $L = L_1 \oplus L_2$ (for instance if L is semisimple, but not simple).

3. The algebra L is indecomposable and has a nontrivial ideal N:

$$L = F \ni N, \quad [F,F] \subseteq F, \quad [N,N] \subseteq N, \quad [F,N] \subseteq N,$$
$$F \neq \emptyset, \quad R \neq \emptyset \tag{5.2}$$

If L is simple, we must first find its maximal subalgebras, using a method explained in ref. 15. If a maximal subalgebra $\tilde{L} \subset L$ is again simple, we must again find its maximal subalgebras. If $\tilde{L}$ is not simple, we proceed as in case 2 or 3 above.

If L is a direct sum of two (or more) Lie algebras, we follow a method originally proposed by Goursat for direct products of discrete groups [**32**,**33**]. Its adaptation to Lie algebras is presented in ref. 15.

Here we shall concentrate on the third case, when L is a semidirect sum of two Lie algebras [**14**], as in (5.2). In particular (5.2) may be a Levi decomposition [**34**], when F is semisimple and N is the radical (maximal solvable ideal) of L.

Classification algorithm for semidirect sum algebras:

Step. 1. Classify all **subalgebras of the factor algebra F** into conjugacy classes under the action of the Lie group $G_F = \exp F$. Produce a representative list

$$S(F) = \{F_1, F_2, \ldots, F_N\}, \quad F_1 = \{0\}, \ldots, F_N = F \tag{5.3}$$

of G_F conjugacy classes of subalgebras of F. The representative F_i may depend on continuous parameters, reflecting the fact that there may exist infinitely many classes of subalgebras. For each subalgebra F_i in the list $S(F)$ construct its normalizer in G_F:

$$\mathrm{Nor}\,(F_i, G_F) = \{g \in G_F | gF_ig^{-1} \subseteq F_i\}. \tag{5.4}$$

The list $S(F)$ should be a "normalized" list, that is the normalizer in F of each subalgebra F_i in $S(F)$ should also be in $S(F)$; we have

$$\mathrm{nor}\,_F F_i = \{x \in F | [x, F_i] \subseteq F_i\}. \tag{5.5}$$

Step. 2. Classify all **"splitting" subalgebras of L**. A subalgebra $\tilde{L}_0 \subset L$ is called a splitting subalgebra if it is conjugate to a subalgebra L_0 that is a semidirect sum of a subalgebra of F and a subalgebra of N:

$$L_0 = F_0 \oplus N_0, \quad F_0 \subseteq F, \quad N_0 \subseteq N. \tag{5.6}$$

The procedure for finding all splitting subalgebras of L consists of the following steps:

(i) For each subalgebra F_i in the list $S(F)$ find all invariant subspaces $\widetilde{N}_{i,\alpha} \subset N$:

$$[F_i, \widetilde{N}_{i,\alpha}] \subseteq \widetilde{N}_{i,\alpha}, \quad \widetilde{N}_{i,\alpha} \subseteq N \tag{5.7}$$

If the algebra N is abelian, than each subspace $\widetilde{N}_{i,\alpha} \subset N$ is also a subalgebra. If N is not abelian, the subspaces $\widetilde{N}_{i,\alpha}$ that are not subalgebras should be weeded out.

(ii) For each F_i classify the invariant subalgebras $\widetilde{N}_{i,\alpha}$ into conjugacy classes under the action of the normalizer group $\text{Nor}\,(F_i, G_F)$. Choose a representative $N_{i,\alpha}$ of each conjugacy class.

(iii) Form a representative list of all splitting subalgebras of L:

$$S_1(L) = \{L_{i,\alpha} \subset L | L_{i,\alpha} = F_i \oplus N_{i,\alpha}\} \tag{5.8}$$

(iv) For each subalgebra $L_{i,\alpha}$ in the list $S_1(L)$ find its normalizer $\text{Nor}\,(L_{i,\alpha}, G)$ in the group G. The list $S_1(L)$ should again be a normalized one, that is, for each $L_{i,\alpha}$ its normalizer in L

$$\text{nor}\,_L L_{i,\alpha} = \{x \in L | [x, L_{i,\alpha}] \subseteq L_{i,\alpha}\} \tag{5.9}$$

should also be in the list.

For the list $S_1(L)$ to be a representative list of all splitting subalgebras of L we must: 1. Include the trivial subalgebras $F_1 = \{0\}$ and $F_N = F$ in the list $S(F)$; 2. For each F_i we include $N_{i,1} = \{0\}$ and $N_{i,r} = N$ in the list of invariant subalgebras. The list $S_1(L)$ includes the list $S(F)$.

The outlined procedure provides a basis for each splitting subalgebra $L_{i,\alpha} \subseteq L$ such that

$$\begin{aligned} L_{i,\alpha} = \ & \{B_a, X_j\}, \quad B_a \in F, \quad X_j \in N, \\ & 1 \le a \le \dim F_i, \quad 1 \le j \le \dim N_{i,\alpha} \equiv r \end{aligned} \tag{5.10}$$

Step. 3. Classify all **"nonsplitting" subalgebras** of L. A subalgebra $L_0 \subset L$ is nonsplitting if it is not conjugate to any splitting one. Any basis for a nonsplitting subalgebra will contain at least one element with a nonzero projection onto both F and N, i.e. an element of the type

$$B_i + aX_j, \quad a \neq 0. \tag{5.11}$$

Procedure for finding all nonsplitting subalgebras of L.

(i) Run through the list $S_1(L)$ of all splitting subalgebras. For each member of the list, say $L_{i,\alpha}$, choose a basis as in (5.10). Complement the basis $\{X_j\}$ of $N_{i,\alpha}$ to a basis of N

$$N = \{X_1, \dots, X_r; Y_1, \dots, Y_s\}, \quad r = \dim N_{i,\alpha}, \quad r + s = \dim N \tag{5.12}$$

(ii) For a given splitting subalgebra $N_{i,\alpha}$ form the vector space

$$V = \{B_a + \sum_{\mu=1}^{s} c_{a\mu} Y_\mu, X_j\}, \quad 1 \le a \le \dim F_i, \quad 1 \le j \le s, \tag{5.13}$$

where the constants c_{ak} are constrained by the requirement that V be a Lie algebra

$$[V, V] \subseteq V. \tag{5.14}$$

Let us write the commutation relations in the considered basis as

$$\begin{aligned} [B_a, B_b] &= f_{abc} B_c, \quad [B_a, X_k] = \alpha_{ak\ell} X_\ell \\ [B_a, Y_\mu] &= \rho_{a\mu\nu} Y_\nu + \sigma_{a\mu m} X_m, \quad [X_i, X_j] = \omega_{ijm} X_m \\ [Y_\mu, Y_\nu] &= \beta_{\mu\nu\sigma} Y_\sigma + \gamma_{\mu\nu m} X_m \\ [X_i, Y_\mu] &= \lambda_{i\mu\nu} Y_\nu + \tau_{i\mu m} X_m \end{aligned} \tag{5.15}$$

Condition (5.14) then reduces to a system of algebraic equations for the constants $c_{a\mu}$.

$$c_{b\nu}\, \rho_{a\nu\alpha} - c_{a\mu}\, \rho_{b\mu\alpha} - c_{c\alpha}\, f_{abc} = -c_{a\mu}\, c_{b\nu}\, \beta_{\mu\nu\alpha} \tag{5.16}$$

$$c_{a\mu}\, \lambda_{j\mu\alpha} = 0 \tag{5.17}$$

In general these equations are nonlinear and must be solved by inspection. If the ideal N is abelian we have

$$\omega_{ijm} = \beta_{\mu\nu\sigma} = \gamma_{\mu\nu m} = \lambda_{i\mu\nu} = \tau_{i\mu m} = 0 \tag{5.18}$$

and the problem greatly simplifies. Indeed condition (5.17) is satisfied trivially, (5.16) becomes linear and simply expresses the fact that the quantities $c_{a\mu}$ form cocycles. Even if N is not abelian, equations (5.16,17) may be linear: it suffices to have $\beta_{\mu\nu\alpha} = 0$ in (5.15).

(iii) Once the constants $c_{a\mu}$ are chosen to satisfy (5.17) the vector space (5.13) becomes a Lie algebra. We then use the group $\widetilde{G} = \mathrm{Nor}\,(L_{i,\alpha}, G) \ltimes \mathrm{Nor}\,(N_{i,\alpha}, G_N)$, where G_N is the Lie group corresponding to the Lie algebra N, to classify the obtained subalgebras into conjugacy classes. This task also greatly simplifies if N is abelian. Indeed, then the group G_N is used to get rid of all cocycles that are actually coboundaries. At this stage we produce a representative list $S_2(L)$ of all conjugacy classes of nonsplitting subalgebras of L. We annul as many of the coefficients $c_{a\mu}$ as possible. Recall that a subalgebra is nonsplitting only if not all of the coefficients $c_{a\mu}$ can be annuled simultaneously by an action of the group $\tilde{G}$.

Step. 4. Form the **final representative list** $S(L) = S_1(L) \cup S_2(L)$ of all G–conjugacy classes of subalgebras of L by joining together the two disjoint lists $S_1(L)$ and $S_2(L)$ of splitting and nonsplitting subalgebras. It is convenient at this stage to relabel the subalgebras and denote them say $S_{i,j}$ where the first index is the dimension of the subalgebras, whereas the second one labels different subalgebras of the same dimension. Within the same dimension subalgebras should be ordered according to isomorphism class [**35**]. The representatives of conjugacy classes can always be so chosen that the normalizer of each algebra in the representative list $S(L)$ is also contained in $S(L)$.

Several comments on the classification procedure are in order.

1. The procedure is iterative: the subalgebras of F are assumed to be known. Their classification is a lower dimensional problem of the same type.

2. The decomposition $L = F \ni N$ is usually not unique. Whenever possible, N should be chosen to be abelian.

3. Most of the work in the classification procedure concerns nonsplitting subalgebras. Which subalgebras are splitting, and which are not, depends on the original decomposition of L into a direct sum.

4. When considering conjugacy under some Lie group it is helpful to make use of the Baker–Campbell–Haussdorf formula

$$e^{\lambda X} Y e^{-\lambda X} = Y + \lambda[X,Y] + \frac{\lambda^2}{2!}[X[X,Y]] + \frac{\lambda^3}{3!}[X[X[Y,Y]]] + \ldots, \quad (5.19)$$

where $X, Y \in L$ and λ is a parameter. If X is a nilpotent element, the series on the right-hand-side of (5.19) terminates. If X generates a rotation or a dilation, the series is easily summed up explicitly.

5. The classification methods described above have been generalized to the case of finite–dimensional subalgebras of infinite dimensional Lie algebras [**20**] (see also below).

6. Example of the Heat Equation

Let us return to the symmetry algebra of the heat equation (3.15). Its Lie algebra can be written as $\tilde{L} = \{F \ni N\} \ni \{S(h)\}$, where $S(h)$ corresponds to the linear superposition formula. We shall factor out this infinite dimensional ideal and concentrate on the six–dimensional algebra

$$L = \{F \ni N\}, \quad F = \{D, T, C\}, \quad N = \{P, B, M\}. \quad (6.1)$$

The commutation relations are those of Table 1; the corresponding vector fields are

$$D = 2t\,\partial_t + x\,\partial_x - \tfrac{1}{2}u\,\partial_u, \quad T = \partial_t, \quad C = t^2\,\partial_t + xt\,\partial_x - \tfrac{1}{4}(x^2+2t)\,u\,\partial_u,$$
$$P = \partial_x, \quad B = t\,\partial_x - \tfrac{1}{2}xu\,\partial_u, \quad M = u\,\partial_u. \tag{6.2}$$

The element D corresponds to dilations, T to time translations, C to projective transformations, P to space translations, B to Galileian boosts, and M to the multiplication of the solution u by a constant. For purposes of symmetry reduction to an ODE we only need one–dimensional subalgebras, since we are interested in subgroups of the symmetry group G, having generic orbits of codimensional 2 in the space $M \sim X \times U = \{x,t\} \times \{u\}$. To illustrate the classification algorithm, and also in preparation for other applications, we shall classify all subalgebras of L.

Step 1. Subalgebras of $F \sim s\ell(2,\mathbb{R}) \sim o(2,1)$. The subalgebras of $s\ell(2,\mathbb{R})$ are known. A representative list of $SL(2,\mathbb{R})$ conjugacy classes of subalgebras of $s\ell(2,\mathbb{R})$ is given by:

$$F_1 = \{D,T,C\}, \quad F_2 = \{D,T\}, \quad F_3\{D\},$$
$$F_4 = \{C+T\}, \quad F_5 = \{T\}, \quad F_6 = \{\emptyset\}. \tag{6.3}$$

The normalizers of F_i in $G \sim SL(2,\mathbb{R})$ are:

$$\mathrm{Nor}\,(F_1,G) = SL(2,\mathbb{R}) = G, \quad \mathrm{Nor}\,(F_2,G) = \exp F_2,$$
$$\mathrm{Nor}\,(F_3,G) = \exp F_3 \otimes \exp \tfrac{\pi}{2}(C+T), \quad \mathrm{Nor}\,(F_4,G) = \exp F_4,$$
$$\mathrm{Nor}\,(F_5,G) = \exp F_2, \quad \mathrm{Nor}\,(F_6,G) = SL(2,\mathbb{R}). \tag{6.4}$$

Note that the normalizer of F_3 in the group $SL(2,\mathbb{R})$ contains a discrete element: a rotation through $\pi/2$.

Step 2. Splitting subalgebras of L. We write out the representatives of the invariant subalgebra classes for each F_i.

F_1: $N_{1,1} = \{P,B,M\}, \quad N_{1,2} = \{M\}, \quad N_{1,3} = \{\emptyset\}$.

We mention that $\{P,B\}$ is an invariant subspace

$([D,P] = -P, \quad [D,B] = B, \quad [T,P] = 0, \quad [T,B] = P,$

$[C,P] = -B, \quad [C,B] = 0)$

but not a subalgebra, since we have $[P,B] = -\frac{1}{2}M$.

F_2: $N_{2,1} = \{P,B,M\}, \quad N_{2,2} = \{P,M\}, \quad N_{2,3} = \{P\}, \quad N_{2,4} = \{M\},$

$N_{2,5} = \{\emptyset\},$

F_3: $N_{3,1} = \{P,B,M\}, \quad N_{3,2} = \{P,M\}, \quad N_{3,3} = \{P\}, \quad N_{3,4} = \{M\},$

$N_{3,5} = \{\emptyset\}$.

Note that the adjoint action of $\exp\alpha(C+T)$ is

$P' = P\cos\alpha + B\sin\alpha, \quad B' = -P\sin\alpha + B\cos\alpha, \quad M' = M,$

$D' = D\cos 2\alpha + (C-T)\sin 2\alpha, \quad C' - T' = -D\sin 2\alpha + (C-T)\cos 2\alpha,$

$C' + T' = C + T.$

Hence, for $\alpha = \pi/2$ we have $D' = -D$, $P' = B$, $B' = -P$ and the algebra $\{D, B\}$ is conjugate to $\{D, P\}$, $\{D, B, M\}$ to $\{D, P, M\}$.

F_4: $N_{4,1} = \{P, B, M\}, \quad N_{4,2} = \{M\}, \quad N_{4,3} = \{\emptyset\}.$

F_5: $N_{5,1} = \{P, B, M\}, \quad N_{5,2} = \{P, M\}, \quad N_{5,3} = \{P + \epsilon M\}, \quad N_{5,4} = \{M\},$ $N_{5,5} = \{\emptyset\}.$

In $N_{5,3}$ we have $\epsilon = \pm 1$ if the classification group G is $G = \exp L = SL(2, \mathbb{R})$. If we include the discrete operation of parity: $\Pi : x \to -x, t \to t, u \to u$ in G (Π also leaves to the heat equation invariant), then we have $\epsilon = +1$ (since the algebra with $\epsilon = -1$ is then conjugate to the one with $\epsilon = +1$). Note also that $\{T, P + bM\}$ is a subalgebra for any $b \in \mathbb{R}$, but we can scale b by an element of $\mathrm{Nor}\,(F_5, G)$:

$e^{\lambda D}\, T\, e^{-\lambda D} = e^{-2\lambda} T, \quad e^{\lambda D}(P + bM)e^{-\lambda D} = e^{-\lambda}\, P + bM.$

We put $e^{-\lambda} = |b|$, then factor out $|b| \neq 0$, to obtain $P + \epsilon M$.

F_6: $N_{6,1} = \{P, B, M\}, \quad N_{6,2} = \{P, M\}, \quad N_{6,3} = \{P\}, \quad N_{6,4} = \{M\},$ $N_{6,5} = \{\emptyset\}.$

Taking

$$L_{i,\mu} = F_i \dotplus N_{i,\mu}, \quad \forall i, \mu \tag{6.5}$$

we obtain the list $S_1(L)$ of representatives of all conjugacy classes of splitting subalgebras of L.

Step 3. Nonsplitting subalgebras. The ideal $\{P, B, M\}$ is nilpotent, rather than abelian, but the results are still quite simple. Most of the splitting subalgebras (6.5) do not give rise to any nonsplitting ones. Let us just look at those that do.

Consider $L_{2,5} = F_2$ and form

$$\widetilde{D} = D + a_1 P + b_1 B + c_1 M, \quad \widetilde{T} = T + a_2 P + b_2 B + c_2 M. \tag{6.6a}$$

Since $N_{2,5} = \{\emptyset\}$ the normalizer is $\mathrm{Nor}\,(L_{2,5}, G) \rtimes \mathrm{Nor}\,(N_{i,\alpha}, G_N) = \exp F_2 \rtimes \exp N$. Conjugating (6.6a) by $\exp\,(a_1 P - b_1 B)$ we obtain

$$\{\widetilde{D}, \widetilde{T}\} \longrightarrow \{D + cM, T + pP + bB + mM\} = \{D_0, T_0\}. \tag{6.6b}$$

The commutation relation $[D_0, T_0] = -2T_0$ (condition (5.16)) implies $p = b = m = 0$. We obtain the representative nonsplitting subalgebra

$$L_{2,6} = \{D + aM, T\}, \quad a \in \mathbb{R}, \quad a \neq 0. \tag{6.7}$$

Similarly $L_{2,3}$ yields

$$L_{2,7} = \{D + aM, T, P\}, \quad a \in \mathbb{R}, \quad a \neq 0. \tag{6.8}$$

The only other nonsplitting subalgebras are obtained from $L_{3,5}$, $L_{3,3}$, $L_{4,3}$, and $L_{5,5}$. They are

$$\begin{aligned} &L_{3,8} = \{D + aM\}, \quad L_{3,9} = \{D + aM, P\}, \quad L_{4,4} = \{C + T + aM\}, \\ &L_{5,6} = \{T + \epsilon_0 B\}, \quad L_{5,7} = \{T + \epsilon_1 M\}. \end{aligned} \tag{6.9}$$

We have $a \neq 0$, $a \in \mathbb{R}$, in all cases $\epsilon_1 = \pm 1$, $\epsilon_0 = \pm 1$ under $\exp L$, $\epsilon_0 = +1$ if parity Π is included.

Adding the subalgebras (6.7),..., (6.9) to the list (6.5) we obtain the final list $S(L)$ of all subalgebras of the heat equation symmetry algebra. Changing notation, we represent the one–dimensional subalgebras by

$$\begin{aligned} &S_{1,1} = \{D\}, \quad S_{1,2} = \{T\}, \quad S_{1,3} = \{C + T\}, \quad S_{1,4} = \{D + aM\}, \\ &S_{1,5} = \{T + B\}, \quad S_{1,6} = \{T + M\}, \quad S_{1,7} = \{T - \epsilon M\}, \\ &S_{1,8} = \{C + T + aM\}, \quad S_{1,9} = \{P\}, S_{1,10} = \{M\}, \quad a \in \mathbb{R}, \quad a \neq 0. \end{aligned} \tag{6.10}$$

As examples of the use of subalgebras, let us consider several reductions of the heat equation. The algebra $S_{1,10}$ does not satisfy condition (4.2), since the invariants are (x, t). All the other subalgebras $S_{1,1}, \ldots, S_{1,9}$ do provide group invariant solutions. Let us consider two cases.

<u>$S_{1,2}$. $T = \partial_t$.</u>

The equation $TF(x, t, u) = 0$ implies that x and u are invariants:

$$I_1 \equiv \xi = x, \quad I_2 = F = u. \tag{6.11}$$

The reduction formulas are

$$u(x, t) = F(x), \quad F_{xx} = 0, \quad F = ax + b \tag{6.12}$$

and we obtain a static solution

$$u(x, t) = ax + b. \tag{6.13}$$

Applying a general transformation of the symmetry group G we obtain a seven parameter family of solutions

$$u(x,t) = (1+4ct)^{-1/2} \exp\left[\lambda - \frac{2vx+4cx^2-v^2t}{4(1+4ct)}\right] \left\{a\left(\frac{e^{-d}(x+vt)}{1+4ct} - x_0\right) + b\right\} \tag{6.14}$$

where the parameters a and b come from the solution (6.12) and λ, v, c, d and x_0 come from the symmetry group.

$S_{1,4}$. $D + aM = 2t\,\partial_t + x\,\partial_x + (a - \frac{1}{2})u\,\partial_u$.

The invariants are:

$$I_1 = \xi = x\,t^{-1/2}, \quad I_2 = F = t^{1/2-a}u. \tag{6.15}$$

Putting

$$u(x,t) = t^{a-\frac{1}{2}}\,F(\xi), \tag{6.16}$$

we obtain

$$\ddot{F} + \tfrac{1}{2}\xi\,\dot{F} + (\tfrac{1}{2} - a)F = 0. \tag{6.17}$$

The solution of (6.17) can be expressed in terms of two different confluent hypergeometric functions, or in terms of parabolic cylinder functions.

It is quite easy to run through the subalgebras (6.10). Each representative subalgebra provides a different reduction to a different ODE. This can be solved; applying the group G to the solution we obtain a multiparameter family of solution, corresponding to the most general subalgebra $L_0 \subset L$, conjugate to the representative one.

As a final comment, let us mention that the symmetry group of the time dependent Schrödinger equation

$$iu_t + u_{xx} = 0, \quad u \in \mathbb{C} \tag{6.18}$$

is closely related to that of the heat equation. Its subgroups have been classified and used to study symmetry breaking due to an added linear or nonlinear interaction term in the equation [**36**]:

$$iu_t + u_{xx} = F(x,t,u,u^*).$$

7. Kac–Moody–Virasoro Symmetries of Integrable Nonlinear Systems in Three–Dimensions. Example of the Three Wave Resonant Interaction Equations

7.1 INTRODUCTORY COMMENTS

In this section we shall discuss the Lie point symmetries of nonlinear PDE's that involve three indepent variables, and are integrable by linear techniques, such as the inverse scattering transform [**37**], or the dressing method [**38**]. Prime examples of such integrable multidimensional equations are the Kadomtsev–Petviashvili equation [39], the Davey–Stewartson equation [40,41], the three–dimensional three–wave resonant interaction equations [42—46] (3WRI), and several others.

In keeping with the topic of this Advanced Study Institute we shall make no use of the integrability of the studied equations. Instead, we shall concentrate on their Lie point symmetry groups, which in all cases turn out to be infinite–dimensional. The corresponding Lie algebras have a very specific structure, namely that of Kac–Moody–Virasoro algebras. We shall use the example of the 3WRI equations to illustrate the situation and to show how the infinite–dimensional symmetry group can be used to obtain large classes of exact analytic solutions. In general, these group invariant solutions are not the same ones that are obtained by inverse scattering: they do not necessarily decay rapidly at infinity, nor do they have to be periodic. The results of this Section were obtained in collaboration with L. Martina [20].

7.2 THE 3WRI EQUATIONS AND THEIR SYMMETRY ALGEBRA

We write the 3WRI equations in the form

$$u_{j,x_j} = i\epsilon_j u_k^* u_\ell^*, \quad \epsilon_j = \pm 1, \quad (i,j,k) = (1,2,3). \tag{7.1}$$

The functions $u_j(x_1, x_2, x_3)$ are complex wave amplitudes, ϵ_j are normalized coupling constants, the stars denote complex conjugation and (x_1, x_2, x_3) are characteristic coordinates defined so as to give

$$\frac{\partial}{\partial_{x_j}} = \frac{\partial}{\partial t} + (\vec{v}_j, \nabla), \quad \nabla = \left(\frac{\partial}{\partial x}, \frac{\partial}{\partial y}, \frac{\partial}{\partial z}\right), \quad j = 1,2,3 \tag{7.2}$$

where $\vec{v}_1$, $\vec{v}_2$, and $\vec{v}_3$ are three group velocities, that in general are linearly independent.

Equations (7.1) have their origin in several branches of physics (nonlinear optics, fluid dynamics, plasma physics, ...) and are important because they describe the interaction of waves in weakly nonlinear and dispersive media.

Applying the algorithm described in Section 2 and the MACSYMA program [28], we find that the general element of the symmetry algebra can be written as

$$V = \sum_{i=1}^{3} [V_i(f_i(x_i)) + W_i(s_i(x_i))] \tag{7.3a}$$

where

$$V_1(f_1) = f_1(x_1)\partial_{x_1} - \tfrac{1}{2}\dot{f}_1(x_1)(u_2\,\partial_{u_2} + u_3\,\partial_{u_3} + u_2^*\,\partial_{u_2^*} + u_3^*\,\partial_{u_3^*}) \tag{7.3b}$$

$$W_1(s_1) = i\,s_1(x_1)(u_3\,\partial_{u_3} - u_2\,\partial_{u_2} - u_3^*\,\partial_{u_3^*} + u_2^*\,\partial_{u_2^*}). \tag{7.3c}$$

The dot above f_1 denotes differentiation with respect to the argument. The vector fields $V_2(f_2)$, $V_3(f_3)$, $W_2(s_2)$ and $W_3(s_3)$ are obtained by cyclic permutations. The functions $f_i(x_i)$ and $s_i(x_i)$ are arbitrary real functions of one variable each and they belong to the $C(\Omega)$ class, where $\Omega \in \mathbb{R}$ is some open set in $\mathbb{R}$. The symmetry algebra is hence infinite–dimensional. The commutation relations are

$$\begin{aligned} [V_i(f_i), V_j(f_j)] &= \delta_{ij}\, V_i(f_i\dot{f}_j - \dot{f}_i f_j) \\ [V_i(f_i), W_j(s_j)] &= \delta_{ij}\, W_j(f_i\dot{s}_j) \\ [W_i(s_i), W_j(s_j)] &= 0. \end{aligned} \tag{7.4}$$

We see that the Lie algebra (7.1) is the direct sum of three Lie algebras, each of which is a semidirect sum:

$$L = L_1 \oplus L_2 \oplus L_3, \quad L_i = \{V_i(f_i), W_i(s_i)\} \tag{7.5a}$$

$$L_i = \{V_i(f_i)\} \ni \{W_i(s_i)\}, \quad i = 1, 2, 3. \tag{7.5b}$$

Each of the components L_i has an infinite–dimensional abelian ideal $\{W_i(s_i)\}$ and an infinite–dimensional simple factor algebra $\{V_i(f_i)\}$. Thus (7.5b) is a Levi decomposition [**34**] in which $\{W_i(s_i)\}$ is the radical. The fact that $\{V_i(f_i)\}$ is simple follows from the local isomorphism

$$V_i(f_i) \sim J(\mathbb{R}) = \{f(\xi)\partial_\xi | f \in C^\infty(\mathbb{R})\}$$

where $J(\mathbb{R})$ is the algebra of real smooth vector fields on $\mathbb{R}$, one of Cartan's infinite–dimensional simple Lie algebras [47]. The isomorphism is given by the mapping

$$\psi : f(\xi)\partial_\xi \longrightarrow V_i(f(x_i)).$$

A physically important (and obvsious) 12–dimensional subalgebra is obtained by restricting all the arbitrary functions in (7.3) to be first order polynomials. We obtain

$$\begin{aligned} V_i(1) &= P_i = \partial_{x_i} \\ V_i(x_i) &= D_i = x_i\,\partial_{x_i} - \tfrac{1}{2}(u_j\,\partial u_j + u_k\,\partial_{u_k} + c.c.) \\ W_i(1) &= i(u_k\,\partial_{u_k} - u_j\,\partial_{u_j} - c.c.) \\ W_i(x_i) &= i\,x_i(u_k\,\partial_{u_k} - u_j\,\partial_{u_j} - c.c.) \end{aligned} \tag{7.6}$$

where $c.c.$ denotes complex conjugation and the indices (i, j, k) are cyclic permutations of $1, 2, 3$. We see that P_i and D_i correspond to translations and dilations. The vector fields $W_i(s_i)$ correspond to gauge transformations that change the phases of the three waves in a coordinate dependent manner.

If the 3WRI system is modified, for instance by introducing dissipation

$$u_{jx_j} + b_j u_j = i\,\epsilon_j u_k^* u_\ell^*, \quad b_j \in \mathbb{R}, \quad b_j \neq 0 \tag{7.7}$$

the system becomes nonintegrable. The symmetry also gets greatly reduced. The symmetry algebra can still be written in the form (7.3a), however instead of the arbitrary functions $f_i(x_i)$ we obtain f_i = constant, so that (7.3b) reduces to translations $V_i = \partial_{x_i}$. The gauge symmetry (7.3c) remains intact and the symmetry group is hence still infinite–dimensional. It can be used to obtain some exact solutions [**48**].

7.3 THE SYMMETRY ALGEBRA AS A KAC–MOODY–VIRASORO ALGEBRA

A Kac–Moody–Virasoro algebra is an infinite–dimensional vector space with a basis

$$\{L_m, T_m^a, C, K\}, \quad 1 \le a \le N, \quad m \in \mathbb{Z} \tag{7.8a}$$

satisfying the commutation relations

$$\begin{aligned}
[L_m, L_n] &= (m-n)L_{m+n} + \tfrac{1}{12}m(m^2-1)\delta_{m,-n}C \\
[T_m^a, T_m^b] &= f^{abc}T_{m+n}^c + m\,\delta_{ab}\delta_{m,-n}K \\
[L_m, T_n^a] &= -n\,T_{m+n}^a \\
[C, L_m] &= [C, T_m^a] = [K, L_m] = [K, T_m^a] = [K, C] = 0
\end{aligned} \tag{7.8b}$$

where f^{abc} are the structure constants of some finite dimensional real or complex Lie algebra A. In most applications and most mathematical investigations A is assumed to be a simple Lie algebra.

Kac–Moody-Virasoro algebras presently seem to be ubiquitous in theoretical physics and in the theory of integrable systems [**12**,13,49,50]. Their physical applications include dual resonance models and current algebra in particle physics, string theory and string field theory, integrable statistical models nad the integration of soliton equations.

A simple realization of the algebra (7.8) is obtained as follows. We introduce a scalar parameter λ, a finite–dimensional Lie algebra A with basis $\{X^1, \ldots, X^N\}$ and commutation relations

$$[X^a, X^b] = f^{abc}X^c, \quad A = \{X^1, \ldots, X^N\}. \tag{7.9}$$

We put

$$L_m = -\lambda^{m+1}\partial_\lambda, \quad T_m^a = X^a\lambda^m, \quad C = 0, \quad K = 0. \tag{7.10}$$

The commutation relations (7.8) are then satisfied.

Let us now return to the 3WRI algebra (7.3) and expand the $C^\infty(\Omega)$ functions $f_i(x_i)$ and $s_i(x_i)$ into Laurent series. We obtain a basis for the 3WRI symmetry algebra in the form

$$V_{1n-1} = x_1^n - \tfrac{n}{2} X_1^{n-1}(u_2\,\partial_{u_2} + u_3\,\partial_{u_3} + c.c.)$$
$$W_{1n} = i\,x_1^n(u_3\,\partial_{u_3} - u_2\,\partial_{u_2} - c.c.) \tag{7.11}$$

(and cyclically for V_2, V_3, W_2 and W_3). The commutation relations for the algebra L_1 now are

$$[V_{1n}, V_{1m}] = (m-n)V_{1m+n}, [V_{1n}, \quad W_{1m}] = mW_{1m+n},$$
$$[W_{1m}, W_{1n}] = 0. \tag{7.12}$$

The algebra (7.3) is thus identified as the direct sum of three centerless Kac–Moody–Virasoro $\hat{u}(1)$ algebras. By this we mean that the central elements C and K in (7.8) are represented trivially ($C = K = 0$) and that the algebra A of (7.9) is $u(1) = \{X^1\}$, so that $f^{abc} = 0$.

In passing we note that for other integrable systems (Kadomtsev–Petviashvili [**18**], modified Kadomtsev–Petviashvili [**28**], Davey-Stewartson [**19**]) the symmetry algebras are also centerless Kac–Moody–Virasoro algebras. Their Kac–Moody parts T_m^a are however not based on $u(1)$, but on some solvable algebra A (that can be imbedded into $s\ell(n,\mathbb{R})$ for some n; $n = 4$ for the KP equation, $n = 5$ for the Davey–Stewartson equation, ...).

7.4 THE SYMMETRY GROUP OF THE 3WRI EQUATIONS

The Lie group of local point transformations leaving the 3WRI equations invariant, is obtained by integrating the vector fields (7.3). In view of the structure (7.5), this Lie group G will be the direct product of three locally isomorphic groups, each of which is itself a semidirect product:

$$G = G^1 \otimes G^2 \otimes G^3 \tag{7.13a}$$

$$G^i = N^i \rtimes S^i. \tag{7.13b}$$

To obtain a one-parameter subgroup in N^1 we integrate $W_1(s_1)$:

$$\frac{dx_i'}{d\lambda} = 0, \quad \frac{du_1'}{d\lambda} = 0, \quad \frac{du_2'}{d\lambda} = -i\,s_1(x_1')u_2', \quad \frac{du_3'}{d\lambda} = i\,s_1(x_1')u_3'$$
$$x_i'(\lambda = 0) = x_i, \quad u_i'(\lambda = 0) = u_i \tag{7.14}$$

to obtain

$$x_i' = x_i, \quad u_1'(x_i') = u_1(x_1), \quad u_2'(x_i') = e^{-i\lambda s_1(x_1)}u_2(x_i),$$
$$u_3'(x_i') = e^{i\lambda s_1(x_1)}u_3(x_i). \tag{7.15}$$

This is a gauge transformation, changing the phases of the waves u_2 and u_3 in a x_1–dependent manner.

To obtain a one-parameter subgroup in S^1 we integrate $V_1(f_1(x_1))$:

$$\frac{dx_1'}{d\lambda} = f_1(x_1'), \quad \frac{dx_2'}{d\lambda} = 0, \quad \frac{dx_3'}{d\lambda} = 0, \quad \frac{du_1'}{d\lambda} = 0,$$
$$\frac{du_2'}{d\lambda} = -\tfrac{1}{2}\dot{f}_1(x_1')u_2', \quad \frac{du_3'}{d\lambda} = -\tfrac{1}{2}\dot{f}_1(x_1')u_3'. \tag{7.16}$$

We obtain

$$x_1' = F_1^{-1}(\lambda + F_1(x_1)), \quad x_2' = x_2, \quad x_3' = x_3, \quad u_1' = u_1,$$
$$u_2' = [f_1(x_1)/f_1(x_1')]^{1/2}u_2, \quad u_3' = [f_1(x_1)/f_1(x_1')]^{1/2}u_3,$$
$$F_1(x_1) = \int_c^{x_1} \frac{ds}{f(s)}, \quad c = \text{const.} \tag{7.17}$$

This is a quite general transformation (reparametrization) of the variable x_1, accompanied by an appropriate transformation of the waves u_2, and u_3. In particular for $f(x_1) = 1$ we have a translation of x_1, for $f(x_1) = x_1$ a dilation of x_1, u_2 and u_3. For $f(x_1) = x_1^2$ we obtain a projective transformation

$$\exp \lambda V_1(x_1^2) : (x, u) \longrightarrow \left(\frac{x_1}{1-\lambda x_1}, x_2, x_3, u_1, u_2(1-\lambda x_1), u_3(1-\lambda x_1)\right). \tag{7.18}$$

We mention that the 3WRI equations are also invariant under a group of discrete transformations, generated by

$$\begin{aligned} P \ &: x_i \to -x_i, \quad u_i \to -u_i \\ C \ &: x_i \to -x_i, \quad u_i \to u_i^* \\ U_1 \ &: x_i \to x_i, \quad u_1 \to u_1, \quad u_2 \to -u_2, \quad u_3 \to -u_3 \\ U_2 \ &: x_i \to x_i, \quad u_1 \to u_1, \quad u_2 \to u_2, \quad u_3 \to -u_3 \end{aligned} \tag{7.19}$$

and also under simultaneous permutations of (x_i, u_i, ϵ_i).

7.5 LOW DIMENSIONAL SUBALGEBRAS OF THE SYMMETRY ALGEBRA

A. *The Goursat Method for Direct Sums*

In order to make use of the symmetry group G of the 3WRI equations to perform symmetry reduction, we need to classify its low dimensional subgroups into conjugacy classes. More specifically, we need to find all subgroups of G having generic orbits of real codimension $k + 6$ in the space $X \times U$ of independent and dependent variables. The values $k = 1$ and $k = 2$ will yield reductions to ODE's and to PDE's in two variables, respectively. Moreover, group invariant solutions are obtained only if the subgroup invariants satisfy the invertibility condition (4.2).

As usual, we shall classify subalgebras of the symmetry algebra, rather than subgroups of the symmetry group. The above considerations imply that we only need to classify subalgebras $L_0 \subset L$ satisfying:

1. dim $L_0 = 1$, or 2

2. All elements of L_0 are vector fields of the form (7.3) with

$$(f_1, f_2, f_3) \neq (0,0,0). \tag{7.20}$$

Since the considered algebra is a direct sum, $L = L_1 \oplus L_2 \oplus L_3$, we shall apply the Goursat method [**32**,33], adapted to Lie algebras [**15**].

Let us consider the direct sum of two Lie algebras, $L = A \oplus B$, and classify its subalgebras under the corresponding Lie group $G_A \otimes G_B$. Two types of subalgebras of direct sum algebras exist:

(i) Nontwisted subalgebras. These are conjugate to direct sums of subalgebras in each component and can be represented by direct sums

$$L_0 = A_0 \oplus B_0, \quad A_0 \subseteq A, \quad B_0 \subseteq B. \tag{7.21}$$

(ii) Twisted subalgebras. These are subalgebras that are not conjugate to direct sums of subalgebras of A and B. In any basis, a twisted subalgebra will contain elements with nonzero projections onto A and B.

The procedure for finding all subalgebras of a direct sum consists of four steps.

Step 1. Find representatives of all G_A conjugacy classes of subalgebras of A, and G_B conjugacy classes of subalgebras of B. Denote them

$$\begin{aligned} &A_{ja} \subseteq A, \quad j = 0,1\ldots,n_a, \quad a = 1,2,\ldots \\ &B_{kb} \subseteq B, \quad k = 0,1\ldots,n_b, \quad b = 1,2,\ldots \end{aligned} \tag{7.22a}$$

where the dimensions are

$$\dim A_{ja} = j, \quad \dim B_{kb} = k, \quad \dim A = n_a, \quad \dim B = n_b. \tag{7.22b}$$

Put

$$A_{01} = B_{01} = \{\emptyset\}, \quad A_{n_a 1} = A, \quad B_{n_b 1} = B. \tag{7.23}$$

For each representative subalgebra, find its normalizer in G_A or G_B, respectively:

$$\mathrm{Nor}\,(A_{ja}, G_A), \quad \mathrm{Nor}\,(B_{kb}, G_B).$$

Step 2. Form a representative list S_1 of all nontwisted subalgebras

$$\begin{aligned} &S_1 : A_{ja} \oplus B_{kb} \\ &j = 0,1,\ldots,n_a, \ a = 1,\ldots,a_m, \ k = 0,1,\ldots,n_b, \ b = 1,\ldots,b_m. \end{aligned} \tag{7.24}$$

Step 3. Form a representative list S_2 of all twisted subalgebras of L. Two subalgebras $A_{ja} \subseteq A$ and $B_{kb} \subseteq B$ can be twisted together if a homomorphism exists from one to the other:

$$\tau(A_{ja}) = B_{kb}, \quad \text{or} \quad \sigma(B_{kb}) = A_{ja}. \tag{7.25}$$

If a homomorphism τ exists, then construct the most general mapping

$$\tau : a_i \mapsto \tau(a_i) \in B_{kb}. \tag{7.26}$$

A twisted subalgebras is then obtained by taking

$$L_\tau = \{a_i + \tau(a_i)\}, \quad i = 1, \ldots, j. \tag{7.27}$$

The mapping τ will in general involve free parameters. The further classification and simplification of (7.27) is performed using the combined normalizer

$$\mathrm{Nor}\, A_{ja} \otimes \mathrm{Nor}\, B_{kb}. \tag{7.28}$$

Step 4. Form a final list of representatives of all conjugacy classes of subalgebras of $L = A \oplus B$ by joining together the lists S_1 and S_2. The final list should be a normalized one, that is the normalizer algebra of each algebra in the list, should also be in the list. The list should be ordered by dimension and isomorphism class.

As a simple example of the Goursat twist method consider two algebras

$$A_{11} = \{a\}, \quad B_{21} = \{b_1, b_2\}, \quad [b_1, b_2] = b_1. \tag{7.29}$$

The most general homomorphism $\sigma(B_{21}) = A_{21}$ is given by

$$\sigma(b_1) = 0, \quad \sigma(b_2) = \lambda a, \quad \lambda \in \mathbb{R} \tag{7.30}$$

we obtain the twisted subalgebra

$$\{b_1, b_2 + \lambda a\}. \tag{7.31}$$

The range of the parameter λ depends on what are the normalizers of A_{11} and B_{21} in G_A and G_B, respectively.

B. *Subalgebras of the Kac–Moody–Virasoro Algebra $\hat{u}(1)$*

Returning to the problem at hand, namely the subalgebras of the 3WRI equations, we perform the first step of the above algorithm by classifying the subalgebras of one component of the direct sum, namely the Kac–Moody–Virasoro algebra $\hat{u}(1)$.

We have

$$L_1 \approx \hat{u}(1) \approx \{V_1(f_1)\} \ni W_1(s_1)\} \tag{7.32}$$

with commutation relations as in (7.4). Since $\hat{u}(1)$ has a semidirect sum structure, we proceed as in the algorithm described in Section 5.

B1. *Subalgebras of the Centerless Virasoro Algebra $\{V\}$*

Let us consider a *one-dimensional subalgebra* of $\{V\}$ generated by an element $V(f)$

of the form (7.3b). Let us apply a general group transformation $\exp\lambda[\mathrm{ad}\,V(\phi)]$ of the form (7.17) to $V(f)$. We obtain

$$\begin{aligned}\exp\lambda[\mathrm{ad}\,V(\phi)]\cdot V(f) &= f(x_1(x_1'))\tfrac{\phi(x_1')}{\phi(x_1(x_1'))}\partial_{x_1'} \\ &- \tfrac{1}{2}\left\{\tfrac{-\dot\phi(x_1(x_1'))+\dot\phi(x_1')}{\phi(x_1(x_1'))}f(x_1(x_1')) + \dot f(x_1(x_1'))\right\} \\ &\times (u_2'\partial_{u_2'} + u_3'\partial_{u_3'} + c.c.).\end{aligned} \tag{7.33}$$

In (7.33), the function $f(x_1)$ is given, the function $\phi(x_1)$ can be freely chosen, as can the parameter λ. Indeed, their choice amounts to a choice of the equivalence transformation that we are performing. We choose $\phi(x_1)$ to satisfy the functional equation

$$f(x_1(x_1'))\frac{F(x_1')}{F(x_1(x_1'))} = 1, \tag{7.34}$$

and we find that the element $V(f)$ for any $f \in C^\infty(\Omega)$, $\Omega \subset \mathbb{R}$, is conjugate under the group $\exp\{V\}$ to a translation and we obtain a single class of one–dimensional subalgebras, represented by

$$V_{1,1} = \{V(1)\} = \{\partial_x\} \tag{7.35}$$

For a proof that eq. (7.34) does have a solution for any $f(x)$ sufficiently smooth function $f(x)$, see ref. [**51**].

Let us now consider a *two–dimensional subalgebra* $\{V(\tilde f_1),\ V(\tilde f_2)\} \in \{V\}$. We can always transform one element, say $V(\tilde f_1)$ into $V(1)$ and obtain

$$\{V(1),\ V(f)\},\quad [V(1),\ V(f)] = V(\dot f). \tag{7.36}$$

The algebra would be abelian if we had $\dot f = 0$; this is forbidden, since then $V(f)$ and $V(1)$ would not be linearly independent. Hence (7.36) is nonabelian and with no loss of generality we can put $\dot f = 1$. We thus obtain a single class of two–dimensional subalgebras of the Virasoro algebra $\{V\}$, represented by

$$V_{2,1} = \{V(1),\ V(x)\},\quad [V(1),\ V(x)] = V(1). \tag{7.37}$$

All finite–dimensional Lie algebras of dimension $d \geq 3$ except $su(2)$ and $s\ell(2,\mathbb{R})$ contain abelian subalgebras of dimension two or more. Hence they cannot be contained in the Virasoro algebra V. A simple calculation shows that $su(2)$ cannot be realized by real vector fields in one variable (nor by the vector fields (7.3b)). The algebra $s\ell(2,\mathbb{R})$ can, upto equivalence, be realized in a single manner, namely

$$V_{3,1} = \{V(1),\ V(x),\ V(x^2)\}. \tag{7.38}$$

The final result can be summed up as a simple theorem.

THEOREM. *The centerless Virasoro algebra $V(1)$ has precisely three conjugacy classes of finite–dimensional subalgebras. Their dimensions are 1, 2 and 3 and they can be represented by $V_{1,1}$, $V_{2,1}$ and $V_{3,1}$ of (7.35), (7.37) and (7.38), respectively. Their normalizers are* $\operatorname{nor} V_{1,1} = V_{2,1}$, $\operatorname{nor} V_{2,1} = V_{2,1}$, *and* $\operatorname{nor} V_{3,1} = V_{3,1}$.

B2. *Splitting Subalgebras of the $\hat{u}(1)$ Algebra $\{V\} \ni \{W\}$*

Following the algorithm of Section 5, we now classify the invariant subspaces in $\{W\}$ of each representative subalgebra of $\{V\}$.

Invariant subspaces of $V_{0,1} = \{0\}$. The normalizer of $V_{0,1}$ in $\{V\}$ is $\operatorname{nor} V_{01} = \{V\}$. We hence classify subalgebras of $\{W\}$ under the action of $\exp V$. We have

$$\exp[\lambda \operatorname{ad} V(f)]\, W(s) = W(s'), \quad s' = s(x_1(x_1')) \tag{7.39}$$

(see (7.17)). If we have $s = \text{const}$, then s' is also constant. Otherwise, we can choose [51] $f(x)$ so that

$$\frac{ds}{dx}\frac{f(x)}{f(x'(x))} = 1. \tag{7.40}$$

We thus obtain two classes of one–dimensional subalgebras of W, represented by

$$N_{1,1} = \{W(1)\}, \quad N_{1,2} = \{W(x_1)\}. \tag{7.41}$$

For $n \geq 2$ every subalgebra of W is conjugate to precisely one subalgebra of the form

$$\{W(1)\,, W(x),\ W(f_3), \ldots, W(f_n)\}$$

or

$$\{W(x), W(f_2), \ldots, W(f_n)\}$$

where $\{1, x, f_3, \ldots, f_n\}$ and $\{1, x, f_2, \ldots, f_n\}$ are, respectively, linearly independent sets of functions of x.

Invariant subspaces of $V_{1,1}$, $V_{2,1}$ and $V_{3,1}$. Since we only need 1 and 2–dimensional subalgebras of the symmetry algebra of the 3WRI equations, we can restrict ourselves to one–dimensional invariant subspaces of $V_{1,1}$ alone. We have

$$[V(1),\ W(s)] = W(\dot{s}). \tag{7.42}$$

Depending on whether we have $\dot{s} = 0$, or $\dot{s} = s$, we obtain two classes of splitting subalgebras represented by

$$\{V(1),\ W(1)\}, \quad [V(1),\ W(1)] = 0 \tag{7.43}$$

and

$$\{V(1), W(e^x)\}, \quad [V(1),\ W(e^x)] = W(e^x). \tag{7.44}$$

We note that the commutation relations for a two–dimensional invariant subspace of $V_{1,1}$ and hence a three–dimensional splitting subalgebra, can be written as

$$\begin{pmatrix} [V(1),\ W(s_1)] \\ [V(1),\ W(s_2)] \end{pmatrix} = A \begin{pmatrix} W(s_1) \\ W(s_2) \end{pmatrix}, \quad A \in \mathbb{R}^{2\times 2}. \tag{7.45}$$

The matrix A can be chosen in its Jordan canonical form and moreover can be multiplied by some chosen nonzero constant. For each standard form, except $A = 0$, we obtain one conjugacy class of subalgebras [**20**].

For invariant subspaces of $V_{2,1}$ and $V_{3,1}$ in $\{W\}$, see ref. [20].

B3. *Nonsplitting Subalgebras of* $\hat{u}(1)$

A nonsplitting subalgebra of $\hat{u}(1)$ obtained from $V_{1,1}$ would involve an element of the form

$$\{V(1) + W(s)\}. \tag{7.46}$$

We have

$$\exp\left[\operatorname{ad} \lambda W(S)\right] \cdot (V(1) + W(s)) = V(1) + W(s) - \lambda W(\dot{S}).$$

Choosing $\lambda \dot{S} = s$, we see that (7.46) is conjugate to $V(1)$ and the algebra is a splitting one.

On the other hand, starting from $V_{2,1}$, we have $\{V(1) + W(s_1),\ V(x_1) + W(s_2)\}$. As above, we can transform $s_1 \to 0,\ s_2 \to s$; the requirement that we have a Lie algebra then implies $s_2 = a =$ const. We thus obtain a class of two–dimensional Lie algebras, represented by

$$\{V(1),\ V(x) + aW(1)\}, \quad a \in \mathbb{R}, \quad a \neq 0. \tag{7.47}$$

Similarly, we obtain a nonsplitting three–dimensional Lie algebra

$$\{V(1),\ V(x) + aW(x),\ W(1)\}, \quad a \in \mathbb{R}, \quad a \neq 0. \tag{7.48}$$

All subalgebras of the Kac–Moody–Virasoro algebra $\hat{u}(1)$ of dimension d for $1 \leq d \leq 3$ are summed up in Table 2.

Table 2. Representatives of conjugacy classes of subalgebras of dimension d of the Kac-Moody-Virasoro algebra $\hat{u}(1)$ for $1 \leq d \leq 3$.

We denote $\hat{u}(1) = F \supset N,\ F = \{V(f)\},\ N = \{W(s)\}$ and $\dot{+}$ signifies a direct sum of vector spaces. Asterisks refer to comments in the text.

d	No	Isomorphism Class	Basis	$\mathrm{nor}_u S_{i,k}$
1	$S_{1,1}$	A_1	$V(1)$	$S_{3,2}$
	$S_{1,2}$	A_1	$W(1)$	$\hat{u}(1)$
	$S_{1,3}$	A_1	$W(x)$	$V(x)\dot{+}N$
2	$S_{2,1}$	$A_{2,1}$	$V(1), V(x)$	$S_{3,2}$
	$S_{2,2}$	$A_{2,1}$	$V(1), V(x) + aW(1)$, $a > 0$	$S_{3,2}$
	$S_{2,3}$	$A_{2,1}$	$V(1), W(e^x)$	$S_{3,3}$
	$S_{2,4}$	$2A_1$	$V(1), W(1)$	$S_{3,2}\dot{+}W(x)$
	$S_{2,5}$	$2A_1$	$W(1), W(x)$	$S_{2,1}\dot{+}N$
	$S_{2,6}$	$2A_1$	$W(x), W(f)$, $\ddot{f} \neq 0$	N(for$f$ generic*)
3	$S_{3,1}$	$s\ell(2, R)$	$V(1), V(x), V(x^2)$	$S_{3,1}$
	$S_{3,2}$	$A_{2,1} \oplus A_1$	$V(1), V(x), W(1)$	$S_{3,2}$
	$S_{3,3}$	$A_{2,1} \oplus A_1$	$V(1), W(e^x), W(1)$	$S_{3,3}\dot{+}W(x)$
	$S_{3,4}$	$A_{3,1}$	$V(1), W(x), W(1)$	$S_{3,4}\dot{+}V(x)\dot{+}W(x^2)$
	$S_{3,5}$	$A_{3,2}^a$	$V(1), W(e^x), W(e^{ax})$, $-1 \leq a < 1, a \neq 0$	$S_{3,5}\dot{+}W(1)$
	$S_{3,6}$	$A_{3,3}$	$V(1), W(xe^x), W(e^x)$	$S_{3,6}\dot{+}W(1)$
	$S_{3,7}$	$A_{3,4}^a$	$V(1), W(e^{ax}\sin x)$, $W(e^{ax}\cos x), a \geq 0$	$S_{3,7}\dot{+}W(1)$
	$S_{3,8}$	$3A_1$	$W(1), W(x), W(f)$, $\ddot{f} \neq 0$	N(for f generic*)
	$S_{3,9}$	$3A_1$	$W(x), W(f_1), W(f_2)$ $(1, x, f_1$ and f_2 linearly independent)	N(for f generic*)

C. *Subalgebras of the Entire Symmetry Algebra*

The nontwisted subalgebras of $L_1 \oplus L_2 \oplus L_3$ are obvious, the twisted ones require a little care. Let us consider one–dimensional subalgebras of the direct sum of two $\hat{u}(1)$ algebras. Three classes of $d = 1$ subalgebras of $\tilde{u}(1)$ exist, hence we can form 6 types of twisted subalgebras. Using elements of the appropriate normalizers, such as $\exp \lambda V_1(x_1)$, $\exp \lambda V_2(x_2)$, and discrete transformations (7.19), we obtain the following representative twisted subalgebras of $L_1 \oplus L_2$:

$$\begin{aligned} &V_1(1) + \kappa V_2(1),\ V_1(1) + W_2(1),\ V_1(1) + W_2(x_2), \\ &W_1(1) + aW_2(1),\ W_1(1) + W_2(x_2),\ W_1(x_1) + \varepsilon W_2(x_2) \\ &\kappa = \pm 1,\ a \in \mathbb{R}\ (a \neq 0). \end{aligned} \tag{7.49}$$

The next step is to twist together subalgebras (7.49) with those of L_3. The results are summed up in Table 3.

Table 3. Representative list of one-dimensional subalgebras of the symmetry algebra of the 3WRI equation.

We have $a \in \{R/0\}, \kappa^2 = 1, \kappa_1^2 = 1, \kappa_2^2 = 1$.

$L_{1,1}$	$= V_1(1)$
$L_{1,2}(\kappa)$	$= V_1(1) + \kappa V_2(1)$
$L_{1,3}(\kappa_1, \kappa_2)$	$= V_1(1) + \kappa_1 V_2(1) + \kappa_2 V_3(1)$
$L_{1,4}$	$= V_1(1) + W_2(1)$
$L_{1,5}$	$= V_1(1) + W_2(x_2)$
$L_{1,6}(\kappa)$	$= V_1(1) + W_2(x_2) + \kappa W_3(x_3)$
$L_{1,7}(\kappa)$	$= V_1(1) + \kappa V_2(1) + W_3(x_3)$
$L_{1,8}$	$= W_1(1)$
$L_{1,9}$	$= W_1(x_1)$
$L_{1,10}(a)$	$= W_1(1) + aW_2(1)$
$L_{1,11}$	$= W_1(1) + W_2(x_2)$
$L_{1,12}(\kappa)$	$= W_1(x_1) + \kappa W_2(x_2)$
$L_{1,13}(\kappa_1, \kappa_2)$	$= W_1(x_1) + \kappa_1 W_2(x_2) + \kappa_3 W_3(x_3)$

A list of all conjugacy classes of two–dimensional subalgebras of the symmetry algebra is quite long. In order to perform symmetry reduction to ordinary differential equations we need subalgebras, all elements of which generate an action on coordinate space. Subalgebras with elements lying entirely in the $W_1 \oplus W_2 \oplus W_3$ space are hence not needed. Moreover, one–dimensional subalgebras, involving only one space coordinate ($L_{1,1}$, $L_{1,4}$, $L_{1,5}$ and $L_{1,6}$ of Table 3) lead to PDE's in two variables that we can solve directly. We are hence only interested in two–dimensional subalgebras, all elements of which are conjugate to elements in $L_{1,2}$, $L_{1,3}$ or $L_{1,7}$ of Table 2.

A list of representatives of all two–dimensional subalgebras needed for symmetry reduction, is given in Table 4.

Table 4. Representative list of two-dimensional subalgebras all elements of which are conjugate to elements in $L_{1,2}$, $L_{1,3}$, or $L_{1,7}$.

We put $\kappa^2 = 1$, $\kappa_i^2 = 1$, $a \in R, b \in R$.

Type	No	Basis	
		X	Y
Abelian	$L_{2,1}(\kappa_1, \kappa_2)$	$V_1(1) + \kappa_1 V_2(1)$	$V_1(1) + \kappa_2 V_3(1)$
$[X, Y] = 0$	$L_{2,2}(\kappa_1, \kappa_2)$	$V_1(1) + \kappa_1 V_2(1)$	$V_1(1) + \kappa_2 V_3(1) + W_1(1)$
	$L_{2,3}(\kappa_1, \kappa_2, a)$ $a \geq 0$	$V_1(1) + \kappa_1 V_2(1)$ $+W_3(x_3)$	$V_1(1) + \kappa_2[V_3(1) - W_1(x_1)$ $-\kappa_1 W_2(x_2)] + aW_1(1)$
Nonabelian	$L_{2,4}(\kappa_1, \kappa_2, a)$ $a \geq 0$	$V_1(1) + \kappa_1 V_2(1)$	$V_1(x_1) + V_2(x_2) + \kappa_2 V_3(1)$ $+aW_1(1)$
$[X, Y] = X$	$L_{2,5}(\kappa_1, a, b)$ $a \geq 0$; if $a = 0$, then $b \geq 0$	$V_1(1) + \kappa_1 V_2(1)$	$V_1(x_1) + V_2(x_2) + aW_1(1)$ $+bW_3(1)$
	$L_{2,6}(\kappa_1, a)$ $a \geq 0$	$V_1(1) + \kappa_1 V_2(1)$	$V_1(x_1) + V_2(x_2) + aW_1(1)$ $+W_3(x_3)$
	$L_{2,7}(\kappa_1, a)$	$V_1(1) + \kappa_1 V_2(1)$ $+W_3(x_3)$	$V_1(x_1) + V_2(x_2) - V_3(x_3)$ $+aW_1(1)$
	$L_{2,8}(\kappa_1, \kappa_2, a, b)$ $a \geq 0$; if $a = 0$ then $b \geq 0$	$V_1(1) + \kappa_1 V_2(1)$ $+\kappa_2 V_3(1)$	$V_1(x_1) + V_2(x_2) + V_3(x_3)$ $+aW_1(1) - bW_2(1)$

7.6 SYMMETRY REDUCTION FOR THE 3WRI EQUATIONS

A. *Reductions by One-Dimensional Subgroups*

Subalgebras $L_{1,1}, \ldots, L_{1,7}(\kappa)$ of Table 3 all lead to reductions to PDE's in two variables. The remaining suablgebras, $L_{1,8}, \ldots, L_{1,13}(\kappa_1, \kappa_2)$ correspond to groups acting on solutions alone, not on the space–time variables. All reductions are treated in ref. [20]. Here we restrict ourselves to one nontrivial example, when we can solve the obtained PDE's completely, namely $L_{1,6}$ of Table 3. We have

$$X = V_1(1) + W_2(x_2) + \kappa W_3(x_3), \quad \kappa^2 = 1. \tag{7.50}$$

The equation

$$X\Phi(x_1, x_2, x_3,\ u_1, u_2, u_3,\ u_1^*, u_2^*, u_3^*) = 0 \tag{7.51}$$

leads to 8 invariants, namely

$$\{x_2, x_3, F_1 = e^{-i(x_2-\kappa x_3)x_1}u_1, F_2 = e^{-i\kappa x_3 x_1}u_2,$$
$$F_3 = e^{ix_1x_2}u_3, F_1^*, F_2^*, F_3^*\}. \tag{7.52}$$

We solve for u_i and put:

$$u_1 = F_1(x_2, x_3)e^{i(x_2-\kappa x_3)x_1}, \quad u_2 = F_2(x_2, x_3)e^{i\kappa x_3 x_1},$$
$$u_3 = F_3(x_2, x_3)e^{-ix_1x_2}. \tag{7.53}$$

Substituting into the 3WRI equations (7.1) we obtain the reduced equations

$$(x_2 - \kappa x_3)F_1 = \epsilon_1 F_2^* F_3^*, \quad (x_2 - \kappa x_3)F_{2,x_2} = i\varepsilon_1\varepsilon_2 F_2|F_3|^2$$
$$(x_2 - \kappa x_3)F_{3,x_3} = i\varepsilon_1\varepsilon_3 F_3|F_2|^2. \tag{7.54}$$

The first equation expresses F_1 algebraically in terms of F_2 and F_3. The remaining two complex equations can be decoupled and integrated. We obtain an explicit solution involving four arbitrary functions of one variable each:

$$u_2 = f(x_3)\ \exp i[\varepsilon_1\varepsilon_2 \int \tfrac{g^2(x_2)dx_2}{x_2-\kappa x_3} + \alpha(x_3) + \kappa x_3 x_1]$$
$$u_3 = g(x_2)\ \exp i[\varepsilon_1\varepsilon_3 \int \tfrac{f^2(x_3)dx_3}{x_2-\kappa x_3} + \beta(x_2) + x_1 x_2]$$
$$u_1 = \tfrac{\varepsilon_1}{x_2-\kappa x_3} f(x_3)g(x_2)\ \exp i[-\varepsilon_1\varepsilon_2 \int \tfrac{g^2 dx_2}{x_2-\kappa x_3} - \varepsilon_1\varepsilon_3 \int \tfrac{f^2 dx_3}{x_2-\kappa x_3}$$
$$-\alpha(x_3) - \beta(x_2) + (x_2 - \kappa x_3)x_1]. \tag{7.55}$$

The algebra $L_{1,1}$ of Table 3 leads to a noninteracting solution for which only one wave, say u_1 is nonvanishing. The algebras $L_{1,4}$ and $L_{1,5}$ lead to solutions similar to (7.55), depending on 4 arbitrary functions of one variable in each case.

More general solution can be obtained from these representative ones by applying a general symmetry group transformation.

The subalgebras $L_{1,2}$, $L_{1,3}$ and $L_{1,7}$ lead to PDE's in two variables for which we are not able to obtain an explicit general solution. In the $L_{1,2}$ and $L_{1,3}$ cases we obtain either a degenerate case, or the general case of the 3WRI equations in one space dimension, for which a large class of solutions has been obtained by the inverse scattering method [**52**,53]. The equations obtained using the $L_{1,7}$ reduction could also be solved by inverse scattering, though to our knowledge this has not been done. We shall inbed $L_{1,2}$, $L_{1,3}$ and $L_{1,7}$ into two-dimensional subalgebras of Table 4, in order to obtain new explicit.

B. *The Painlevé Property for Ordinary Differential Equations*

All subalgebras of Table 4 provide reductions to systems of 3 complex first order ODE's for 3 invariant functions $F_i(\xi)$ of one real variable. Putting

$$F_j(\xi) = \rho_j(\xi)e^{i\phi_j(\xi)}, \quad 0 \leq \rho_j < \infty, \quad 0 \leq \phi_j < 2\pi, \quad j = 1,2,3 \tag{7.56}$$

and separating out the real and imaginary parts of the equations, we obtain six real equations for the norms ρ_j and phases ϕ_j of the waves. These equations can in all cases be decoupled and we are left with one nonlinear ODE to solve. It turns out that *in all cases* we are able to solve this equation explicitly.

The reason for this is that the original 3WRI equations are integrable by inverse scattering techniques [**42–46**]. According to the Painlevé conjecture [**54**], we expect all ODE's obtained from an integrable PDE to have the Painlevé property, or to be transformable into an equation with the Painlevé property.

We recall that an ODE has the Painlevé property if none of its solutions have movable critical points. A "critical point" of a function is a singularity other than a pole (e.g. a branch point, or an essential singularity); "movable" means that the position of the singularity depends on the initial conditions (i.e. on the integration constants). If an ODE has the Painlevé property, then its general solution $u(t)$ can be expanded into a convergent Laurent series about an arbitrary point in the complex t–plane, with a finite number of negative powers in the expansion. This provides a test, the Painlevé test [**54**], that any equation having the Painlevé property must pass. Thus, if an n–th order ODE has the Painlevé property, the general solution can be written as

$$y(t) = \sum_{k=0}^{\infty} a_k (t - t_0)^{k+\alpha}, \tag{7.57}$$

where

(i) α is a negative integer.

(ii) A recursion relation of the form

$$P(k) a_k = F(t_0, a_0, a_1, \ldots, a_{k-1}) \tag{7.58}$$

exists for the expansion coefficients a_k and has $n - 1$ *resonances*. A resonance is a nonnegative integer k for which the function $P(k)$ vanishes. The coefficient a_k at a resonance is not determined and can hence be freely chosen.

(iii) The resonance condition

$$F(t_0, a_0, a_1, \ldots, a_{k-1}) = 0 \tag{7.59}$$

must be satisfied identically in t_0 (the position of the pole) and in a_r, $0 \le r \le k-1$, where a_r are the coefficients at the earlier resonances.

If an equation passes the Painlevé test, then it has at least a formal solution (7.57), having a pole of order α and depending on n arbitrary constants. The test is not sufficient, since the series may have a zero radius of convergence and also, because other solutions may exist, not contained in (7.57) and having movable essential singularities.

The above test has been implemented as a MACSYMA program [55]. The test and the concepts involved have been generalized to PDE's [56,57,23].

Nonlinear ODE's that have the Painlevé property are in general much easier to solve than those that do not. The Painlevé property is invariant under the transformation

$$u(x) = \frac{\alpha(x)y(t)+\beta(x)}{\gamma(x)y(t)+\delta(x)}, \quad t = \phi(x), \quad \dot{\phi} \neq 0, \quad \alpha\delta - \beta\gamma \neq 0 \tag{7.60}$$

where $\alpha, \beta, \gamma, \delta$ and ϕ are arbitrary smooth functions of x. This invariance under Möbius transformations of the dependent variable and arbitrary smooth transformations of the independent variable makes it possible to classify Painlevé type equations into equivalence classes.

In particular the following types of equations having the Painlevé property have been classified

$$\dot{y} = F(y,t) \tag{7.61}$$

$$\dot{y}^2 = F(y,t) \tag{7.62}$$

$$\ddot{y} = H(\dot{y}, y, t) \tag{7.63}$$

$$[\ddot{y} - P_1(y)\dot{y} - P_2(y)]^2 = a\dot{y}^3 + Q_2(y)\dot{y}^2 + Q_4(y)\dot{y} + Q_6(y). \tag{7.64}$$

Above F is rational in y and analytic in t, H is rational in $\dot{y}$ and y, analytic in t, P_k and Q_k are polynomials of order k in y, with coefficients analytic in t and a is an analytic function of t.

Equation (7.61) (if it has the Painlevé property) is equivalent [58] to the Riccati equation $\dot{u} = \alpha(x)u^2 + \beta(x)u + \gamma(x)$. Equation (7.62) can either be reduced to the Riccati equation, or is equivalent to the equation [58]

$$\dot{u}^2 = a_0 + a_1 u + a_2 u^2 + a_3 u^3 + a_4 u^4, \quad a_i = \text{const} \tag{7.65}$$

yielding elliptic functions, or elementary ones, if the polynomial on the right hand side has multiple roots.

Equations of the form (7.63) were classified by Painlevé and Gambier [59,60,58]. Fifty equivalence classes of such equations exist [58], some of them depending on parameters, or arbitrary functions. Six of them can be solved in terms of specifically introduced functions, the Painlevé transcendents $P_I, \ldots, P_{VI}$. The others can be solved in terms of elliptic functions, elementary functions, solutions of the Riccati equation, or linear equations.

Equations of the form (7.64) and also more general equations of the form

$$A(y, \dot{y}, t)\ddot{y}^2 + B(y, \dot{y}, t)\ddot{y} + C(y, \dot{y}, t) = 0, \quad A \neq 0 \tag{7.66}$$

with A, B and C polynomial in y and $\dot{y}$ and analytic in t were studied by Bureau and collaborators [**61**,62] (see also ref. [63]).

C. *Reductions by Two-Dimensional Subgroups*

The subalgebras $L_{2,1}$ and $L_{2,2}$ of Table 4 provide reductions to ODE's that can be solved in terms of Jacobi elliptic functions, or in special cases, in terms of elementary functions. The well-known soliton solutions belong to this class. As an example, consider the algebra $L_{2,2}(\kappa_1, \kappa_2)$. We obtain the following expressions for the wave amplitudes u_i in terms of invariants ρ_i, ϕ_i and ξ:

$$u_1 = \rho_1(\xi)e^{i\phi_1(\xi)}, \; u_2 = \rho_2(\xi)e^{i\phi_2(\xi)}e^{-i\kappa_2 x_2}, \; u_3 = \rho_3(\xi)e^{i\phi_3(\xi)}e^{i(x_1-\kappa_1 x_2)}$$
$$\xi = x_1 - \kappa_1 x_2 - \kappa_2 x_3. \tag{7.67}$$

The moduli and phases satisfy

$$\begin{aligned} \dot\rho_1 &= \varepsilon_1 \rho_2 \rho_3 \sin\phi & \rho_1\dot\phi_1 &= \varepsilon_1\rho_2\rho_3 \cos\phi \\ \dot\rho_2 &= -\kappa_1\varepsilon_2\rho_3\rho_1 \sin\phi & \rho_2\dot\phi_2 &= -\kappa_1\varepsilon_2\rho_3\rho_1 \cos\phi \\ \dot\rho_3 &= -\kappa_2\varepsilon_3\rho_1\rho_2 \sin\phi & \rho_3\dot\phi_3 &= -\kappa_2\varepsilon_3\rho_1\rho_2 \cos\phi \\ \phi &= \phi_1 + \phi_2 + \phi_3 + \xi. \end{aligned} \tag{7.68}$$

First integrals of (7.68) are

$$I_1 = \varepsilon_1\rho_1^2 + \kappa_1\varepsilon_2\rho_2^2, \quad I_2 = \varepsilon_1\rho_1^2 + \kappa_2\varepsilon_3\rho_3^2,$$
$$I_3 = \rho_1\rho_2\rho_3 \cos\phi + \tfrac{1}{2}\varepsilon_1\rho_1^2. \tag{7.69}$$

Solving for $\rho_2, \rho_3, \cos\phi$ and $\sin\phi$ in terms of ρ_1, and substituting into (7.68), we obtain the ODE

$$\begin{aligned} \dot Z_1^2 &= 4\kappa_1\kappa_2\varepsilon_2\varepsilon_3 Z_1^3 - [4\kappa_1\kappa_2\varepsilon_1\varepsilon_2\varepsilon_3(I_1+I_2)+1]Z_1^2 \\ &\quad + (4\kappa_1\kappa_2\varepsilon_2\varepsilon_3 I_1 I_2 + 4\varepsilon_1 I_3)Z_1 - 4I_3^2, \\ Z_1 &= \rho_1^2. \end{aligned} \tag{7.70}$$

Rewriting (7.70) as

$$\dot Z_1^2 = 4\kappa_1\kappa_2\varepsilon_2\varepsilon_3(Z_1-\beta_1)(Z_2-\beta_2)(Z_3-\beta_3), \quad \beta_i = \text{const} \tag{7.71}$$

we obtain solitary wave solutions for

$$0 \le \beta_1 < \beta_2 = \beta_3, \quad \kappa_1\kappa_2\varepsilon_2\varepsilon_3 = 1, \tag{7.72a}$$

or

$$0 \le \beta_1 = \beta_2 < \beta_3, \quad \kappa_1\kappa_2\varepsilon_2\varepsilon_3 = -1. \tag{7.72b}$$

The explicit formula in the case (7.72a) is

$$\rho_1 = \sqrt{Z_1} = [\beta_2 \ \sinh^2 \sqrt{(\beta_2 - \beta_1)}(\xi - \xi_0) + \beta_1]^{1/2}$$
$$\times [\cosh \sqrt{(\beta_2 - \beta_1)}(\xi - \xi_0)]^{-1}, \tag{7.73}$$

representing a soliton oriented downwards (a "well"). The formula in the case (7.72b) is similar, but represent a "bump".

Finite doubly periodic solutions are obtained for

$$0 \le \beta_1 \le Z_1 \le \beta_2 < \beta_3, \quad \kappa_1 \kappa_2 \varepsilon_2 \varepsilon_3 = 1, \tag{7.74a}$$

or

$$\beta_1 \le \beta_2 \le Z_1 \le \beta_3, \quad \kappa_1 \kappa_2 \varepsilon_2 \varepsilon_3 = -1, \quad 0 \le \beta_2 < \infty. \tag{7.74b}$$

The explicit formulas are quite similar in the two cases; for instance if (7.74b) is satisfied, we obtain

$$\rho_1 = \sqrt{Z_1} = [-(\beta_3 - \beta_2)\, sn^2(\sqrt{(\beta_3 - \beta_1)}(\xi - \xi_0), k) + \beta_3]^{1/2},$$
$$k^2 = \tfrac{\beta_3 - \beta_2}{\beta_3 - \beta_1} \tag{7.75}$$

where $sn(z, k)$ is a Jacobi elliptic function [**64**].

The subalgebra $L_{2,4}$ of Table 4 leads to an equation of the Painlevé–Gambier type (7.63). It can be solved [**20**] in terms of the Painlevé transcendent P_V, a solution that to our knowledge is new.

The subalgebras $L_{2,5}$ and $L_{2,6}$ lead to uninteresting solutions for which two of the waves, say u_1 and u_2 vanish:

$$u_1 = u_2 = 0, \quad u_3 = A \tfrac{1}{\eta} e^{ia\ell n \eta}, \quad \eta = x_1 - \kappa x_2. \tag{7.76}$$

Finally, the three remaining subalgebras $L_{2,3}$, $L_{2,7}$ and $L_{2,8}$ all lead to equations of the Bureau type (7.64). As an example, consider the subalgebra $L_{2,8}$.

The reduction formulas in this case are

$$\begin{aligned}
u_1 &= (-\kappa_1 x_2 + \kappa_2 x_3)^{-1} \rho_1(\xi) e^{i\phi_1(\xi)} \exp[ib\ell n(-\kappa_1 x_2 + \kappa_2 x_3)] \\
u_2 &= (x_1 - \kappa_2 x_3)^{-1} \rho_2(\xi) e^{i\phi_2(\xi)} \exp[-ia\ell n(x_1 - \kappa_2 x_3)] \\
u_3 &= (x_1 - \kappa_1 x_2)^{-1} \rho_3(\xi) e^{i\phi_3(\xi)} \exp[i(a-b)\ell n(x_1 - \kappa_1 x_2)] \\
\xi &= (x_1 - \kappa_1 x_2)(x_1 - \kappa_2 x_3)^{-1}.
\end{aligned} \tag{7.77}$$

The system of ODE's is

$$\begin{aligned}
\dot\rho_1 &= -\varepsilon_1 \tfrac{1}{\xi} \rho_2 \rho_3 \sin\phi & \rho_1 \dot\phi_1 &= -\varepsilon_1 \tfrac{1}{\xi} \rho_2 \rho_3 \cos\phi \\
\dot\rho_2 &= -\varepsilon_2 \kappa_1 \tfrac{1}{\xi(\xi-1)} \rho_3 \rho_1 \sin\phi & \rho_2 \dot\phi_2 &= -\varepsilon_2 \kappa_1 \tfrac{1}{\xi(\xi-1)} \rho_3 \rho_1 \cos\phi \\
\dot\rho_3 &= \varepsilon_3 \kappa_2 \tfrac{1}{\xi-1} \rho_1 \rho_2 \sin\phi & \rho_3 \dot\phi_3 &= \kappa_2 \varepsilon_3 \tfrac{1}{\xi-1} \rho_1 \rho_2 \cos\phi \\
\phi &= \phi_1 + \phi_2 + \phi_3 + (a-b)\ell n\xi + b\ell n(\xi - 1).
\end{aligned} \tag{7.78}$$

First integrals of this system are

$$\varepsilon_1\rho_1^2 + \varepsilon_2\kappa_1\rho_2^2 + \varepsilon_3\kappa_2\rho_3^2 = I_1,$$
$$\rho_1\rho_2\rho_3 \cos\phi - \tfrac{1}{2}\varepsilon_1(a-b)\rho_1^2 + \tfrac{1}{2}\varepsilon_2\kappa_2 b\rho_3^2 = I_2$$
$$\varepsilon_3\kappa_2\rho_3^2 + \varepsilon_2\kappa_1(\varepsilon\rho_2^2 - \textstyle\int \rho_2^2 d\xi) = I_3. \tag{7.79}$$

Using (7.79) and putting

$$Y(\xi) = \int \rho_2^2(\xi)d\xi, \tag{7.80}$$

we obtain the following ODE for $Y(\xi)$:

$$\begin{aligned}\ddot{Y}^2 = \tfrac{4}{\xi^2(\xi-1)^2}\{&-\kappa_2\varepsilon_1\varepsilon_3\xi(\xi-1)\dot{Y}^3 + [\kappa_1\kappa_2\varepsilon_1\varepsilon_2\varepsilon_3(\xi-1)(I_3+\kappa_1\varepsilon_2 Y)\\
&-\kappa_1\kappa_2\varepsilon_1\varepsilon_2\varepsilon_3\xi(I_1 - I_3 - \kappa_1\varepsilon_2 Y) - \tfrac{1}{4}(a\xi - a + b)^2]\dot{Y}^2\\
&+\{\kappa_2\varepsilon_1\varepsilon_3(I_1 - I_3 - \kappa_1\varepsilon_2 Y)(I_3 + \kappa_1\varepsilon_2 Y) - \tfrac{1}{2}\kappa_1\varepsilon_2[(a-b)I_1 + 2I_2\\
&- aI_3 - a\varepsilon_2\kappa_1 Y](a\xi - a + b)\}\dot{Y} - \tfrac{1}{2}[(a-b)I_1 + 2I_2\\
&- aI_3 - a\kappa_1\varepsilon_2 Y]^2\}.\end{aligned} \tag{7.81}$$

Equation (7.81) is transformed to a canonical form, integrated by Bureau et al.[**61**,62], by a linear transformation of the dependent variable

$$\begin{aligned}Y(\xi) &= \lambda(\xi)u(\eta) + \mu_1\xi + \mu_0\\
\eta(\xi) &= \tfrac{1}{\sqrt{3}}(\kappa_2\varepsilon_1\varepsilon_3)^{-2/3}\textstyle\int \xi^{-2/3}(\xi-1)^{-2/3}d\xi\\
\lambda(\xi) &= (\kappa_2\varepsilon_1\varepsilon_3)^{1/3}\tfrac{1}{\sqrt{3}}(\xi-1)^{1/3}\xi^{1/3}\end{aligned} \tag{7.82}$$

where μ_1 and μ_0 are constants (depending on I_1, I_2, I_3). The equation for $u(\eta)$ is

$$\begin{aligned}(\ddot{u} - 2\mathcal{P}u)^2 = &-4(\dot{u})^3 + \kappa\mathcal{P}(\dot{u})^2 + (12\mathcal{P}u^2 - \kappa\dot{\mathcal{P}}u + H)\dot{u}\\
&-4\dot{\mathcal{P}}u^3 + \kappa\mathcal{P}^2u^2 - \dot{H}u + g_0\end{aligned} \tag{7.83}$$

where $\mathcal{P} = \mathcal{P}(\eta;0,1)$ is a Weierstrass elliptic function, $H(\eta)$ is a solution of the Lamé equation

$$\ddot{H} - 2\mathcal{P}H = 0 \tag{7.84}$$

and K and g_0 are constants. Equation (7.83) can be solved [**62**] in terms of the sixth Painlevé transcendent P_{VI}.

The method of symmetry reduction has thus allowed us to obtain large classes of solutions of the 3WRI equations. It is worth mentioning that no new transcendents occur: the obtained Painlevé type equations can always be solved in terms of elementary functions, elliptic functions, or the known Painlevé transcendents.

8. Exact Analytic Solutions of Nonlinear Partially Integrable Classical Field Equations

8.1 INTRODUCTORY COMMENTS

We now turn to nonlinear PDE's that cannot be integrated by the linear techniques, mentioned at the beginning of Section 7. In particular, no Lax pair is available, and hence no inverse scattering techniques.

Prototypes of physically interesting equations of this type are nonlinear Schrödinger equations (NLSE) and nonlinear Klein–Gordon equations (NLKGE) with polynomial nonlinearities. Recent articles have been devoted to multidimensional NLKGE of the form [**26**]

$$\begin{aligned} \Box_\varepsilon \phi &= a_0\phi + a_1\phi^3 + a_2\phi^5, \quad \phi \in \mathbb{R},\ a_i \in \mathbb{R} \\ \Box_\varepsilon &= \frac{\partial^2}{\partial x_0^2} + \varepsilon\left(\frac{\partial^2}{\partial x_1^2} + \cdots + \frac{\partial^2}{\partial x_n^2}\right), \quad \varepsilon = \pm 1 \end{aligned} \tag{8.1}$$

and NLSE of the form [**24**,25]

$$\begin{aligned} i\psi_t + \Delta\psi &= a_0\psi + a_1|\psi|^2\psi + a_2|\psi|^4\psi, \quad \psi \in \mathbb{C},\ a_i \in \mathbb{R} \\ \Delta &= \frac{\partial^2}{\partial x_1^2} + \cdots + \frac{\partial^2}{\partial x_n^2}. \end{aligned} \tag{8.2}$$

In this Section, we shall concentrate on the NLSE (8.2) and use group theory to reduce it to an ODE. Equation (8.2) for $a_2 \neq 0$ is not integrable, even for $n = 1$; for $n \geq 2$ it is not integrable, even for $a_2 = 0$, as long as we have $a_1 \neq 0$. We do hence not expect all reduced equations to have the Painlevé property and the PDE (8.2), for all cases except the integrable one with $a_2 = 0$, $n = 1$, does not have the Painlevé property. We shall use Painlevé analysis to pick out specific ODEs among those obtained by symmetry reduction, namely those that do have the Painlevé property. These specific equations we then reduce to their standard form and then integrate.

It turns out that quite a few of the interesting reductions do have the Painlevé property; in this sense we shall call equations like (8.1) and (8.2) "partially integrable". The NLKGE (8.1) plays an important role in relativistic nonlinear field theories. It is also important in condensed matter theory, for instance in Landau–Ginzburg type theories of phase transitions. The NLSE (8.2) occurs in applications as diverse as nonlinear optics, wave propagation in water, interactions of laser beams with plasmas, turbulence, and many others. To convince oneself of its importance, it suffices to look at how often it occurs in other lectures in this volume.

8.2 SYMMETRIES OF THE CUBIC AND QUINTIC NONLINEAR SCHRÖDINGER EQUATION

Applying the MACSYMA program [**28**] and solving the determining equations, we find that the NLSE (8.2) in $n = 3$ space dimensions is invariant, for $a_2 \neq 0$, $a_1 \neq 0$, only under the extended Galilei group G. A convenient basis for its Lie algebra is given by the vector fields

$$\begin{aligned}
&P_i = \partial_{x_i}, && T = \partial_t - a_0\,\partial_\phi \\
&J_i = -\varepsilon_{ik\ell}x_k\partial_{x_\ell}, && K_i = t\,\partial_{x_i} + \tfrac{1}{2}x_i\,\partial_\phi, \quad M = -\partial_\phi, \\
&\Psi = Ne^{i\phi}, && N,\phi \in \mathbb{R}, \quad i,k,\ell = 1,2,3, \quad 0 \leq N < \infty, \quad 0 \leq \phi < 2\pi.
\end{aligned} \tag{8.3}$$

The operators P_i, J_i and K_i generate space translations, rotations and Galilei transformations, respectively, M generates changes of phase of the wave function ψ and T generates time translations, accompanied by a change of phase of ψ.

If either a_1 or a_2 vanishes (but not both simultaneously), the symmetry group is a larger one, namely the Galilei–similitude group G^d, including dilations. The Lie algebra of G^d contains one more basis element, in addition to (8.3), namely the dilation operator

$$\begin{aligned}
&D = 2t\,\partial_t + \sum_{i=1}^{3} x_i\,\partial_{x_i} - \sigma N\,\partial_N - 2a_0t\,\partial_\phi, \\
&\sigma = \begin{cases} 1 & \text{if} \quad a_2 = 0, \ a_1 \neq 0 \\ \frac{1}{2} & \text{if} \quad a_1 = 0, \ a_2 \neq 0 \end{cases}.
\end{aligned} \tag{8.4}$$

We see that the symmetry group is finite–dimensional, but quite large, 11– or 12–dimensional, as the case may be. The Lie algebra of the Galilei group G allows the Levi decomposition [**34**]

$$\mathfrak{g} \approx \{J_1, J_2, J_3\} \ni \{T, K_1, K_2, K_3, P_1, P_2, P_3, M\}, \tag{8.5}$$

where the radical (maximal solvable ideal) is $\{T, K_i, P_i, M\}$ and is actually nilpotent. The radical contains the Heisenberg algebra $\{K_i, P_i, M\}$. The Levi decomposition of the Lie algebra of the Galilei–similitude group is

$$\mathfrak{g}^d \approx \{J_1, J_2, J_3\} \ni \{D, T, K_1, K_2, K_3, P_1, P_2, P_3, M\}. \tag{8.6}$$

The nonzero commutation relations are

$$\begin{aligned}
&[J_i, J_k] = \varepsilon_{ik\ell}J_\ell, \quad [J_i, P_k] = \varepsilon_{ik\ell}P_\ell, \ [J_i, K_k] = \varepsilon_{ik\ell}K_\ell \\
&[T, K_i] = P_i, \quad [K_i, P_k] = \tfrac{1}{2}\delta_{ik}M \\
&[D, P_i] = -P_i, \quad [D, K_i] = K_i, \quad [D, T] = -2T.
\end{aligned} \tag{8.7}$$

The Galilei and Galilei–similitude group transformations are obtained by integrating the vector fields (8.3) and (8.4) and are quite obvious. Thus, if $\psi(\vec{x},t)$ is a solution of the NLSE then so is

$$\psi'(\vec{x}',t') = \psi(\vec{x},t)e^{-\sigma d} \exp \tfrac{i}{2}[\vec{v}(\vec{x}+\vec{b}+\tfrac{1}{2}\vec{v}t) - 2a_0 te^{2d} + \phi_0] \tag{8.8}$$

with

$$\begin{aligned} t &= e^{-2d}t' - t_0 \\ \vec{x} &= R(e^{-2d}\vec{x}') - \vec{b} - \vec{v}e^{-2d}t' \\ &\vec{b},\vec{v} \in \mathbb{R}^3, \quad R \in SO(3), \quad t_0,\phi_0 \in \mathbb{R}, \quad 0 \le \phi_0 < 2\pi \\ &d \in \mathbb{R} \quad \text{if} \quad a_1 a_2 = 0, \quad (a_1,a_2) \neq (0,0) \\ &d = 0 \quad \text{if} \quad a_1 a_2 \neq 0. \end{aligned} \tag{8.9}$$

In addition, the NLSE is invariant under reflections in the coordinate planes and under time reversal:

$$\begin{aligned} \Pi_i &: x_i \to -x_i, \; t \to t, \quad \psi \to \psi, \quad i = 1,2,3 \\ TRI &: x_i \to x_i, \quad t \to -t, \quad \psi \to \psi^*. \end{aligned} \tag{8.10}$$

The classification algorithm of Section 5 has been applied to find all subalgebras of both the Galilei and Galilei–similitude algebras [**24**]. The decomposition used was not the Levi decomposition but rather

$$\mathfrak{g} \approx \{J_1,J_2,J_3,K_1,K_2,K_3\} \ni \{T,P_1,P_2,P_3,M\} \tag{8.11}$$

$$\mathfrak{g}^d \approx \{D\} \ni \mathfrak{g}. \tag{8.12}$$

The advantage of (8.11) is that the ideal $n \approx \{T,P_1,P_2,P_3,M\}$ is abelian. That of (8.12) is that the action of the dilation operator D on the Galilei algebra $\mathfrak{g}$ is very simple, and that full use can be made of the knowledge of the subalgebra structure of $\mathfrak{g}$.

The lists of representative subalgebras of $\mathfrak{g}$ and $\mathfrak{g}^d$ are quite long. The classification of subalgebras has many possible applications. Here we shall restrict ourselves to one application only: symmetry reduction of the NLSE to an ODE. We are considering the case of a PDE with two real dependent variables (the norm and phase of ψ) and four independent ones ($\vec{x}$ and t). The subgroups of interest are hence those that have generic orbits of codimension 3 in the space of independent and dependent variables. The corresponding subgroup action must admit three real invariants

$$\xi = I_1(\vec{x},t), \quad I_2(\vec{x},t,N,\phi), \quad I_3(\vec{x},t,N,\phi)$$

$$\begin{vmatrix} \frac{\partial I_2}{\partial N}, & \frac{\partial I_2}{\partial \phi} \\ \frac{\partial I_3}{\partial N}, & \frac{\partial I_3}{\partial \phi} \end{vmatrix} \neq 0. \tag{8.13}$$

This in itself requires that the corresponding Lie algebras $\mathcal{L}_0$ have dimension $\dim \mathcal{L}_0 \geq 3$ and that the element M alone should not be in $\mathcal{L}_0$ ($M \notin \mathcal{L}_0$). Indeed, if $M \in \mathcal{L}_0$, then the invariants in (8.13) are independent of ϕ and the Jacobian in (8.13) will be singular.

The above requirements restrict the number of pertinent subgroups of the symmetry group to a managable one. All of them have been analyzed and the obtained ODE's solved, whenever possible [**24**,25].

Each reduction and each subgroup is adapted to a specific type of boundary conditions, imposed on the NLSE. Indeed, consistent boundary conditions can only be imposed, if the boundary itself is invariant under the reduction group. Boundary conditions will always break the original symmetry of the equation. A subgroup $G_0 \subset G$ can be used to perform symmetry reduction, if the equations, *and the boundary conditions*, are invariant under G_0. In these lectures, we shall consider just one example, namely boundary conditions imposed on a cylinder.

8.3 SOLUTIONS OF THE NLSE COMPATIBLE WITH A CYLINDRIC GEOMETRY

We shall consider the case when the function ψ and the squared gradient $(\nabla\psi)^2$ are given for some time $t = t_0$ on a cylinder $\rho = \rho_0$:

$$\begin{aligned} \psi(x,y,z,t)|_{(t,\rho)=(t_0,\rho_0)} &= f_1(\theta,z) \\ (\nabla\psi)^2|_{(t,\rho)=(t_0,\rho_0)} &= f_2(\theta,z) \\ x_1 = \rho \cos\theta, \; x_2 = \rho &\sin\theta, \; x_3 = z. \end{aligned} \tag{8.14}$$

Conditions of the type (8.14) can in any case only be imposed for specific choices of the functions f_1 and f_2. The conditions (8.14) restrict the reduction algebras to one of the following four classes:

$$\begin{aligned} L_1^{a,b} &= \{J_3 + aM, \; P_3, \; T + (b - a_0)M\}, & a \geq 0, \; b \in \mathbb{R} \\ L_2^{a,b} &= \{J_3 + aM, \; P_3, \; D + bM\}, & a \geq 0, \; b \geq 0 \\ L_3^{a,b} &= \{J_3 + aM, \; K_3, \; D + bM\}, & a \geq 0, \; b \geq 0 \\ L_4^{a,b} &= \{J_3 + aM, \; P_3, \; T, \; D + bM\}, & a \geq 0, \; b \geq 0. \end{aligned} \tag{8.15}$$

The subgroup corresponding to $L_4^{a,b}$ acts transitively on space–time with the z–axis cut out and has as its only invariants

$$N_0 = \rho^\sigma N, \quad \phi_0 = \phi - a\theta + a_0 t + b\ell n\rho, \tag{8.16}$$

where ρ and ϕ are polar coordinates σ is the dilation weight in (8.4) and N_0 and ϕ_0 are constants. We express the solution ψ in terms of invariants as

$$\psi = N_0 \rho^{-\sigma} \ \exp\, i(a\theta - a_0 t - b\ell n\rho + \chi_0). \tag{8.17}$$

Substituting (8.17) back into the NLSE (8.2) we find that χ_0 remains an arbitrary constant and N_0 satisfies an algebraic equation. The reality of N_0 implies $b = 0$ and we obtain

$$N_0 = \begin{cases} \left(\frac{1-a^2}{a_1}\right)^{1/2} & \text{for} \quad a_1 \neq 0,\ a_2 = 0,\ \sigma = 1 \\ \left(\frac{1-4a^2}{4a_2}\right)^{1/2} & \text{for} \quad a_1 = 0,\ a_2 \neq 0,\ \sigma = \frac{1}{2} \end{cases}, \tag{8.18}$$

where a must be so chosen that $N_0 \in \mathbb{R}$.

All other subalgebras in (8.15) provide reductions leading to second order ODE's. Let us consider the subalgebra $L_1^{a,b}$. The three invariants are

$$I_1 = \rho, \quad \chi = I_2 = \phi - a\theta + bt, \quad I_3 = N. \tag{8.19}$$

In terms of these invariants, we have

$$\psi(\vec{r}, t) = f(\rho)e^{i(a\theta - bt)}, \quad f(\rho) = N(\rho)e^{i\chi(\rho)}. \tag{8.20}$$

We substitute (8.20) into the NLSE (8.2) and obtain

$$\ddot{f} + \frac{1}{\rho}\dot{f} - \frac{a^2}{\rho^2} f = (a_0 - b)f + a_1 f|f|^2 + a_2 f|f|^4. \tag{8.21}$$

We separate out the real and imaginary parts of (8.21) and solve for the phase χ:

$$\chi = S_0 \int \frac{d\rho}{\rho N^2} + \chi_0, \tag{8.22}$$

where S_0 and χ_0 are real integration constants. The norm $N(\rho)$ then satisfies

$$\ddot{N} - \frac{S_0^2}{\rho^2 N^3} + \frac{1}{\rho}\dot{N} - \frac{a^2}{\rho^2} N = (a_0 - b)N + a_1 N^3 + a_2 N^5. \tag{8.23}$$

This second order nonlinear ODE is quite complicated and we cannot integrate it simply by inspection. We subject it to the Painlevé test described in Section 7.6, making use of a MACSYMA program [**55**]. Comparing the leading terms we find that we have $\alpha = -\frac{1}{2}$ in the expansion (7.57), i.e. as it stands eq. (8.23) does not have the Painlevé property. We hence perform the substitution

$$N(\rho) = H(\rho)^{1/2} \tag{8.24}$$

which turns the movable square root branch point of $N(\rho)$ at $\rho = \rho_0$ into a pole of $H(\rho)$. The function $H(\rho)$ satisfies

$$\ddot{H} = \frac{\dot{H}^2}{2H} - \frac{1}{\rho}\dot{H} + \frac{2a^2}{\rho^2} H + \frac{2S_0^2}{\rho^2 H} + 2(a_0 - b)H + 2a_1 H^2 + 2a_2 H^3 \tag{8.25}$$

and this equation passes the Painlevé test if and only if one of the two following conditions are satisfied

$$\text{(i)}\ a_2 = 0, \quad a_1 \neq 0, \quad a^2 = \tfrac{1}{9}. \tag{8.26a}$$

$$\text{(ii)}\ a_2 \neq 0, \quad a^2 = \tfrac{1}{16}, \quad a_0 - b = \tfrac{3}{16}\tfrac{a_1^2}{a_2}. \tag{8.26b}$$

The result (8.26) clearly illustrates the importance of a careful subgroup analysis. Had we simply assumed that cylindrical invariance implies that J_3 must be an element of the corresponding subalgebra, we would have concluded that we have $a = 0$ in (8.15). Hence conditions (8.26) would not be satisfied and the reduced ODE's would never have the Painlevé property.

Let us now consider the quintic NLSE and assume that condition (8.26b) is satisfied. We shall transform eq. (8.25) under the conditions (8.26b) to its standard form. This will prove that the equation does indeed have the Painlevé property and moreover, we will simultaneously obtain the solution. Several cases need be considered.

1. $a_2 \neq 0,\ a_1 \neq 0.$

We put

$$H(\rho) = \lambda(\rho)W(\eta), \quad \eta = \eta(\rho) \tag{8.27}$$

and look for functions λ and η such that $W(\eta)$ satisfies one of the 50 standard Painlevé equations [**58**]. The appropriate choice is

$$\lambda = \tfrac{1}{2}\varepsilon\sqrt{\varepsilon a_1}\left(\tfrac{3}{4a_2}\right)^{3/4}\rho^{-1/2}, \quad \eta = \sqrt{\varepsilon a_1}\left(\tfrac{3}{4a_2}\right)^{1/4}\rho^{1/2}, \quad \varepsilon = \pm 1 \tag{8.28}$$

and we obtain the equation $PXXXI$ in Ince's list. This is one of the six irreducible cases, namely that of the fourth Painlevé transcendent:

$$\ddot{W} = \tfrac{1}{2W}\dot{W}^2 + \tfrac{3}{2}W^3 + 4\eta W^2 + 2\eta^2 W + \tfrac{2^9}{9}\left(\tfrac{S_0 a_2}{a_1}\right)^2 \tfrac{1}{W} \tag{8.29}$$

and the solution is

$$W(\eta) = P_{IV}(\eta), \quad N(\rho) = [\lambda(\rho)]^{1/2}[P_{IV}(\eta)]^{1/2}. \tag{8.30}$$

2. $a_2 \neq 0,\ a_1 = 0.$

We again perform the transformation (8.27) and put

$$\lambda = \lambda_0\rho^{-1/2}, \ \eta = 4\varepsilon\left(\tfrac{a_2}{3}\right)^{1/2}\lambda_0\rho^{1/2} + \eta_0, \ \varepsilon = \pm 1, \ \lambda_0, \eta_0 \in \mathbb{R}. \tag{8.31}$$

The function $W(\eta)$ will this time satisfy the reducible equation $PXXX$:

$$\ddot{W} = \tfrac{\dot{W}^2}{2W} + \tfrac{3}{2}W^3 + \tfrac{3S_0^2}{2a_2\lambda_0^4 W}. \tag{8.32}$$

The first integral of (8.32) is

$$\dot{W}^2 = W^4 + 4CW - \frac{3S_0^2}{a_2\lambda_0^2} \equiv P(W), \tag{8.33}$$

where the fourth order polynomial $P(W)$ can be rewritten as

$$P(W) = (W - W_1)(W - W_2)(W - W_3)(W - W_4). \tag{8.34}$$

If all 4 roots of $P(W)$ are different, then W can be expressed in terms of Jacobi elliptic functions. If any of the roots coincide, we obtain elementary solutions. Let us consider individual subcases

2a. $a_2 > 0,\ W_1 = W_2 = W_3 = W_4 = 0,\ c = 0,\ S_0 = 0.$
The solution of the NLSE in this case is

$$\begin{aligned}&\psi(\vec{r},t) = \frac{1}{2}\left(\frac{3}{a_2\rho}\right)^{1/4} (\varepsilon\sqrt{\rho} + c_0)^{-1/2}\ \exp i\left(\frac{\theta}{4} - a_0 t + \chi_0\right)\\ &\varepsilon = \pm 1,\ c_0 = \text{const},\ \varepsilon\sqrt{\rho} + c_0 > 0.\end{aligned} \tag{8.35}$$

We see that the norm $N(\rho)$ is singular for $\rho = 0$ and for $\sqrt{\rho} = -\varepsilon c_0$ (if $\varepsilon c_0 > 0$).

In all other cases with $a_2 > 0$ the 4 roots are distinct, 2 are real, 2 mutually complex conjugate.

2b. $a_2 > 0,\ W_1 = r,\ W_2 = s,\ W_{3,4} = -\frac{r+s}{2} \pm iq.$

$$\begin{aligned}q &= \tfrac{1}{2}(3r^2 + 3s^2 + 2rs)^{1/2} > 0,\ r_0 \le 0 \le s,\ r < s\\ S_0^2 &= -\frac{a_2\lambda_0^4}{3} rs(r^2 + s^2 + rs) > 0,\ C = -\tfrac{1}{4}(r+s)(r^2 + s^2).\end{aligned} \tag{8.36}$$

The solution is

$$\begin{aligned}\psi(\vec{r},t) &= \rho^{-1/4}[\lambda_0 W(\xi)]^{1/2} e^{i\chi(\xi)} e^{i\left(\frac{\theta}{4} - a_0 t + \chi_0\right)}\\ W(\xi) &= \frac{-rA + sB + (rA + sB)\,cn(\xi,k)}{(A+B)\,cn(\xi,k) - A + B}\\ \chi(\xi) &= 2\varepsilon\left(\frac{-rs(r^2+s^2+rs)}{AB}\right)^{1/2} \int \frac{(A+B)\,cn(\xi,k) - A + B}{-rA + sB + (rA+sB)\,cn(\xi,k)}\,d\xi\\ k^2 &= \frac{(A+B)^2 - (s-r)^2}{4AB},\ A = (r^2 + 2rs + 3s^2)^{1/2}\\ B &= (3r^2 + 2rs + s^2)^{1/2},\ \xi = 4\varepsilon\sqrt{AB}\sqrt{\frac{a_2}{3}}\lambda_0\rho^{1/2} + \xi_0.\end{aligned} \tag{8.37}$$

Here $cn(\xi, k)$ is a Jacobi elliptic function and we have obtained singular "cnoidal" waves. The singularities of $W(\xi)$ occur at

$$\begin{aligned}&\xi = \xi_a + 4nK,\ cn(\xi_a, k) = \frac{A-B}{A+B}\\ &a = 1, 2,\ -1 < \frac{A-B}{A+B} < 1,\end{aligned} \tag{8.38}$$

where $4K$ is the real period of $cn(\xi, k)$. Since we must have $\lambda_0 W(\xi) \ge 0$ we must choose

$$\lambda_0 > 0 \quad \text{for} \quad \xi_1 < \xi < \xi_2$$
$$\lambda_0 < 0 \quad \text{for} \quad \xi_2 < \xi < 4K \quad \text{or} \quad 0 < \xi < \xi_1$$
$$(\text{mod}\, 4K).$$

2c. $a_2 < 0$, 2 real roots, 2 complex ones.
We put

$$W(\eta) \equiv R(\zeta), \; \eta = i\zeta \tag{8.39}$$

and obtain

$$\dot{R}^2 = -\left[R^4 + 4CR + \frac{3S_0^2}{(-a_2)\lambda^2}\right]. \tag{8.40}$$

The 4 roots satisfy

$$R_1 = r, \; R_2 = s, \; R_{3,4} = -\frac{r+s}{2} \pm iq$$
$$q \;= \tfrac{1}{2}(3r^2 + 3s^2 + 2rs)^{1/2}, \; r < s \leq 0, \quad \text{or} \quad 0 \leq r < s.$$

We introduce A and B as in (8.37) and obtain *finite* cnoidal waves

$$W(\xi) = \frac{(rA - sB)\, cn(\xi, k) + rA + sB}{(A - B)\, cn(\xi, k) + A + B}$$
$$\chi(\xi) \;= \frac{\varepsilon}{2}\left[\frac{rs(r^2 + s^2 + rs)}{AB}\right]^{1/2} \int \frac{(A - B)\, cn(\xi, k) + A + B}{(rA - sB)\, cn(\xi, k) + rA + sB} d\xi$$
$$\xi \;= 4\varepsilon\sqrt{\frac{-a_2 AB}{3}}\lambda_0 \rho^{1/2} + \xi_0, \; k^2 = \frac{(r-s)^2 - (A-B)^2}{4AB}. \tag{8.41}$$

In (8.41) $W(\xi)$ is finite since we have $(A + B)/(A - B) > 1$, and the denominator in the expression for $W(\xi)$ cannot vanish for real ξ.

We conclude that the quintic NLSE with $a_2 \neq 0$, $a_1 = 0$ in a cylindrical geometry has elementary algebraic solutions and also periodic cnoidal wave solutions. No solitary waves or kinks occur. The same is true for the cubic NLSE ($a_2 = 0$, $a_1 \neq 0$) in a cylindrical geometry. The results for the NLSE with initial conditions imposed on a sphere are also similar [**24**]. For solutions of the NLSE equation obtained using the subalgebra $L_2^{a,b}$ and $L_3^{a,b}$ of (8.15), we refer to the original articles [**24**].

8.4 EXAMPLES OF TRANSLATIONALLY INVARIANT SOLUTIONS OF THE QUINTIC NLSE

Let us consider a different three–dimensional subalgebra, namely

$$\{T + (a - a_0)M, \; P_1, \; P_2\}.$$

The reduction formulas are

$$\psi(\vec{r}, t) = N(z)e^{i\chi(z)}e^{-iat + i\chi_0} \tag{8.42}$$

where

$$\chi(z) = S_0 \int \frac{dz}{N^2}, \tag{8.43}$$

and N satisfies the ODE

$$\ddot{N} = \frac{S_0^2}{N^3} + (a_0 - a)N + a_1 N^3 + a_2 N^5. \tag{8.44}$$

The Painlevé test suggests the substitution

$$N = [H(z)]^{1/2} \tag{8.45}$$

where $H(z)$ satisfies

$$\ddot{H} = \frac{1}{2H}\dot{H}^2 + \frac{2S_0^2}{H} + 2(a_0 - a)H + 2a_1 H^2 + 2a_2 H^3. \tag{8.46}$$

Equation (8.46) passes the Painlevé test for all values of a, a_0, a_1 and a_2. It does indeed have the Painlevé property and we reduce it to standard form by putting

$$H(z) = \lambda_0 W(\eta), \;\; \eta = \varepsilon \left(\frac{4a_2}{3}\right)^{1/2} \lambda_0 z + \eta_0. \tag{8.47}$$

The function $W(\eta)$ satisfies the $PXXX$ equation

$$\ddot{W} = \frac{1}{2W}\dot{W}^2 + \frac{3}{2}W^3 + \frac{3}{2}\frac{a_1}{a_2\lambda_0}W^2 + \frac{3}{2}\frac{a_0 - a}{a_2\lambda_0^2}W + \frac{3}{2}\frac{S_0^2}{a_2\lambda_0^4}\frac{1}{W}. \tag{8.48}$$

A first integral is

$$\dot{W}^2 = W^4 + \frac{3a_1}{2a_2\lambda_0}W^3 + \frac{3(a_0 - a)}{a_2\lambda_0^2}W^2 + 4CW - \frac{3S_0^2}{a_2\lambda_0^4}. \tag{8.49}$$

We again obtain the standard equation for elliptic functions

$$\dot{W}^2 = (W - W_1)(W - W_2)(W - W_3)(W - W_4). \tag{8.50}$$

In this case all possibilities occur, namely:

(i) $W_i \in \mathbb{R}, \quad i = 1, 2, 3, 4$
- (i1) $W_1 = W_2 = W_3 = W_4$
- (i2) $W_1 \neq W_2 = W_3 = W_4$
- (i3) $W_1 = W_2 \neq W_3 = W_4$
- (i4) $W_1 \neq W_2 \neq W_3 = W_4 \neq W_1$
- (i5) W_i all different.

(ii) $W_1, W_2 \in \mathbb{R}, \; W_{3,4} = p \pm iq, \; p, q \in \mathbb{R}, \; q > 0.$
- (ii1) $W_1 = W_2$
- (ii2) $W_1 \neq W_2$.

(iii) $W_{1,2} = p \pm iq, \; W_{3,4} = r \pm is, \; p, q, r, s \in \mathbb{R}, \; q > 0, \; s > 0.$
- (iii1) $p + iq = r + is$
- (iii2) $p + iq \neq r + is$.

Without going into details, we give two examples of physically important finite elementary solutions.

Case (i3): $W_1 = W_2 = 0,\ W_3 = W_4 = -\frac{\alpha}{2},\ \alpha \in \mathbb{R}$.
The solution is

$$\psi(\vec{r}, t) = \left[-\tfrac{3a_1}{4a_2}\left(1 + \exp\sqrt{\tfrac{3}{a_2}\tfrac{a_1}{2}}(z - z_0)\right)^{-1}\right]^{1/2}$$

$$\exp i\left[\chi_0 - \left(a_0 - \tfrac{3a_1^2}{16a_2}\right)t\right]$$

$$a_2 > 0,\ a_1 < 0. \tag{8.51}$$

This is a "kink" solution, satisfying

$$|\psi| \to 0, \qquad |\psi| \to \sqrt{-\tfrac{3a_1}{4a_2}}$$

$$z \quad \to -\infty \quad z \quad \to +\infty.$$

Case (i4): $W_i \in \mathbb{R},\ W_1 = W_2 > W_3 > W_4$.

$$W(z) = \tfrac{\alpha+\beta\ \cosh\tau}{\gamma+\delta\ \cosh\tau}$$

$$\tau = [(W_1 - W_3)(W_1 - W_4)]^{1/2}\left(\tfrac{4a_2}{3}\right)^{1/2}\lambda_0(z - z_0), \quad a_2 > 0. \tag{8.52}$$

The constants $\alpha, \beta, \gamma, \delta$ depend on W_i, $\gamma/\delta < 1$. This solution is a solitary wave (a bump) since it satisfies

$$W \to \tfrac{\beta}{\delta} = W_1, \quad W_1 \le W \le W_3$$

$$z \quad \to \pm\infty.$$

For a systematic study of all reductions of the NLSE to ODE's, see refs. 24, 25.

Acknowledgements

The authors research is partially supported by research grants from NSERC of Canada and FCAR du Québec.

References

1. Olver, P.J. (1986), 'Applications of Lie Groups to Differential Equations', Springer, New York.
2. Ovsiannikov, L.V. (1982), 'Group Analysis of Differential Equations', Academic Press, New York.
3. Ibragimov, N.H. (1985), 'Transformation Groups Applied to Mathematical Physics', Reidel, Boston.
4. Bluman, G.W. and Cole, J.D. (1974), 'Similarity Methods for Differential Equations', Appl. Math. Sci., no 13, Springer, New York.

5. Anderson, R.L. and Ibragimov, N.H. (1979), 'Lie–Bäcklund Transformations in Applications', SIAM, Philadelphia.
6. Fushchich, V.I. and Nikitin, A.G.(1983), 'Symmetries of Maxwell Equations', Naukova Dunka, Kiev, (in Russian).
7. Ames, W.F. (1972), 'Nonlinear Partial Differential Equations in Engineering', Academic Press, New York.
8. Levi, D. and Winternitz, P. (Editors) (1988), 'Symmetries and Nonlinear Phenomena', World Scientific, Singapore.
9. Kosmann-Schwarzbach, Y. (1989), Lectures at this ASI.
10. Mikhaïlov, A.V. (1989), Lectures at this ASI.
11. Kumei, S. and Bluman G. (1982), SIAM J. Appl. Math. 42, 1157.
12. Kac, V.G. (1983), 'Infinite–Dimensional Lie Algebras', Birkhauser, Boston.
13. Goddard, P. and Olive, D. (1986), Int. J. Mod. Phys. A1, 303.
14. Patera, J., Winternitz, P. and Zassenhaus, H. (1975), J. Math. Phys. 16, 1597; (1975), ibid 16, 1615; (1976), ibid 17, 717.
15. Patera, J., Winternitz, P., Sharp, R.T. and Zassenhaus, H. (1977), J. Math. Phys. 18, 2259.
16. del Olmo, M.A., Rodriguez, M.A., Winternitz, P. and Zassenhaus, H. (1987), Preprint CRM-1480, Lin. Alg. Appl. (to appear).
17. Hussin, V., Winternitz, P. and Zassenhaus, H. (1988), Preprint CRM-1567.
18. David, D., Kamran, N., Levi, D. and Winternitz, P. (1985), Phys. Rev. Lett. 55, 2111; (1986), J. Math. Phys. 27, 1225.
19. Champagne, B. and Winternitz, P. (1988), J. Math. Phys. 29, 1.
20. Martina, L. and Winternitz, P. (1989), Preprint CRM-1594.
21. Kruskal, M.D., Grammaticos, B. and Ramani, A. (1989), Lectures at this ASI, Les Houches.
22. Bessis, D. (1989), Lectures at this ASI, Les Houches.
23. Weiss, J. (1989), Lectures at this ASI, Les Houches.
24. Gagnon, L. and Winternitz, P. (1988), J. Phys. A21, 1493; (1989), ibid A22, 469; (1989), Phys. Rev. A39, 296; (1989), Phys. Lett. A134, 276.
25. Gagnon, L., Grammaticos, B., Ramani, A. and Winternitz, P. (1989), J. Phys. A22, 499.
26. Grundland, A.M., Tuszyński J.A. and Winternitz, P. (1987), J. Math. Phys. 28, 2194; (1988), J. Phys. C21, 4931.
27. Schwarz, F. (1985), Computing 34, 91.
28. Champagne, B. and Winternitz, P. (1985), Preprint CRM-1278, Montréal.
29. Rand, D.W., Winternitz, P. and Zassenhaus, H. (1988), Lin. Alg. Appl. 109, 197.
30. Rand, D.W. (1986), Comp. Phys. Commun. 41, 105, (1987), ibid 46, 311.

31. Rand, D.W., Winternitz, P. and Zassenhaus, H. (1987), Comp. Phys. Commun. 46, 297.
32. Goursat, E. (1889), Ann. Sci. Ec. Normale Sup. (3), 6, 9.
33. Du Val, P. (1964), 'Homographies, Quaternions and Rotations', Clarendon Press, Oxford.
34. Jacobson, N. (1979), 'Lie Algebras', Dover, New York.
35. Patera, J., Sharp, R.T., Winternitz, P. and Zassenhaus, H. (1976), J. Math. Phys. 17, 986.
36. Boyer, C.P., Sharp, R.T. and Winternitz, P. (1976), J. Math. Phys. 17, 1439.
37. Ablowitz, M.J. and Segur, H. (1981), 'Solitons and the Inverse Scattering Transform', SIAM, Philadelphia.
38. Novikov, S., Manakov, S.V., Pitaevskii, L.P. and Zakharov, V.E. (1983), 'Theory of Solitons', Consultants Bureau, New York.
39. Kadomtsev, V.V. and Petviashvili, V.I. (1970), Sov. Phys. Dokl. 15, 539.
40. Davey, A. and Stewartson, K. (1974), Proc. R. Soc. London, Ser. A338, 101.
41. Anker, D. and Freeman, N.C. (1978), Proc. R. Soc. London, Ser. A360, 539.
42. Zakharov, V.E. and Shabat, A.B. (1974), Funkts. Anal. Prilozh. 8, 43.
43. Ablowitz, M.J. and Haberman, R. (1975), J. Math. Phys. 16, 2301.
44. Cornille, H. (1979), J. Math. Phys. 20, 1653.
45. Kaup, D.J. (1980), Physica 1D, 5.
46. Kaup, D.J. (1981), in Proc. NATO ASI 'Nonlinear Phenomena in Physics and Biology', Plenum Press, New York.
47. Cartan, E. (1909), Ann. Ec. Norm. Sup. 26, 93.
48. Teichmann, J. and Winternitz, P. (to appear).
49. Jimbo, M. and Miwa, T. (1983), Publ. RIMS Kyoto 19, 943.
50. Winternitz, P. (1988), in 'Symmetries and Nonlinear Phenomena', World Scientific, Singapore.
51. Neuman, F. (1985), Proc 23rd International Symposium on Functional Equations, Gagnano, Italy.
52. Zakharov, V.E. and Manakov, S.V. (1976), Sov. Phys. JETP 42, 842.
53. Kaup, D.J., Reiman, A.H. and Bers, A. (1979), Rev. Mod. Phys. 51, 275.
54. Ablowitz, M.J., Ramani, A. and Segur, H. (1980), J. Math. Phys. 21, 715; (1980), ibid 21, 1006.
55. Rand, D. and Winternitz, P. (1986), Comp. Phys. Commun. 42, 359.
56. Weiss, J., Tabor, M. and Carnevale, G. (1983), J. Math. Phys. 24, 522.
57. Weiss, J. (1983), J. Math. Phys. 24, 1405.

58. Ince, E.L. (1956), 'Ordinary Differential Equations', Dover, New York.
59. Painlevé, P. (1902), Acta Math. 25, 1.
60. Gambier, B. (1910), Acta Math. 33, 1.
61. Bureau, F.J. (1972), Ann. Mat. Pura Appl. (IV) 91, 163.
62. Bureau, F.J., Garcet, A. and Goffar, J. (1972), Ann. Mat. Pura Appl. (IV) 92, 177.
63. Bureau, F.J. (1964), Ann. Mat. Pura Appl. IV 64, 229; (1964), ibid 66, 1; (1972), ibid 94, 345.
64. Byrd, P.F. and Friedman, M.D. (1971), Handbook of Elliptic Integrals for Engineers and Scientists, Springer, Berlin.

THE NONLINEAR EVOLUTION EQUATION FOR THE ORDER PARAMETER IN SUPERFLUID HELIUM-FOUR

RADHA BALAKRISHNAN
The Institute of Mathematical Sciences
Madras 600 113
India

ABSTRACT. Using the procedure of Calogero and Eckhaus it is shown that the nonlinear evolution equation for the order parameter in superfluid helium-four is not C-integrable. It is likely to be S-integrable.

Starting with a microscopic model Hamiltonian for a system of hard-core bosons (helium-four atoms) with an attractive nearest neighbour (n-n) interaction of magnitude V_o, we have recently derived [1] the nonlinear evolution equation for the order parameter η in superfluid helium-four. $\phi = 2\eta$ satisfies (in 1+1 dimensions),

$$i\hbar\phi_t = \{b\,[1-(1-|\phi|^2)^{\frac{1}{2}}]-\mu\}\,\phi - \frac{\hbar^2}{2m}(1-|\phi|^2)^{\frac{1}{2}}\,\phi_{xx} - \frac{V_o a^2}{4}(1-|\phi|^2)^{-\frac{1}{2}}\,\phi\left[(|\phi|^2)_{xx} + \frac{1}{2}(1-|\phi|^2)^{-1}(|\phi|^2_x)^2\right], \qquad (1)$$

Here m and μ denote the mass and chemical potential of a helium-four atom, a = (average) n-n distance and $b = 3(\hbar^2/ma^2 - V_o)$.

Calogero-Eckhaus limiting procedure [2]: In Eq.(1), set $\phi = \varepsilon\,\psi(\xi,\tau)\exp(ikx-\omega t)$, where $\xi = \varepsilon^p(x-vt)$ and $\tau = \varepsilon^q t$. ($p,q > 0$ and $\varepsilon \to 0$). Here v is found as $v = d\omega/dk$ by linearizing Eq.(1) to yield $\hbar\omega = (\hbar^2k^2/2m) - \mu$ and $v = \hbar k/m$ (ω,v real). Substituting for ϕ in Eq.(1) and equating leading order terms in ε gives p = 1, q = 2 and

$$i\psi_\tau + B\,\psi_{\xi\xi} - N(k)\,|\psi|^2\psi = 0\ . \qquad (2)$$

Since $B = \hbar/2m \neq 0$(real) and $N(k) = 1/2\hbar\,(b - \hbar^2k^2/2m)$ is not identically zero, Eq.(1) is not C-integrable (i.e., not linearizable by a change of variables). Further, since N(k) is real, Eq.(2) is essentially the nonlinear

R. Conte and N. Boccara (eds.), Partially Integrable Evolution Equations in Physics, 569–570.

Schrodinger equation which is S-integrable (i.e., linearizable by the inverse scattering transform method). If one conjectures [2] that integrability properties are inherited through such limiting procedures (as confirmed by many examples) then Eq.(1) must also be S-integrable. In view of this, it would be interesting to investigate the existence of a Lax Pair for this equation.

References

[1] Balakrishnan Radha, Sridhar.R. and Vasudevan.R (1989) 'Hydrodynamics of superfluid helium-four in a pseudospin model', Phys. Rev. B 39, 174; See also 'Nonlinear dynamics in superfluid helium-four' Phys. Lett. A 125, 489 (1987)

[2] Calogero F. and Eckhaus W (1987) 'Nonlinear evolution equations, rescalings, model PDEs and their integrability: I', Inverse Problems 3, 229.

Quasimonomial Transformations and Integrability

L. Brenig and A. Goriely.
Université Libre de Bruxelles. CP 231. Boulevard du Triomphe. 1050 Bruxelles. Belgium

In a recent paper,[1] we have introduced the so-called "Lotka-Volterra representation" for systems of nonlinear polynomial differential equations:

$$\dot{x}_i = \lambda_i x_i + x_i \sum_{j=1}^{m} A_{ij} \prod_{k=1}^{n} x_k^{B_{jk}} \qquad i = 1, \ldots, n \qquad m \geq n \tag{1}$$

where A and B are rectangular real matrices (respectively $(n \times m)$ and $(m \times n)$). The x_i's are real functions of the time t.

Any system of O.D.E.'s which contains a finite number of polynomial terms can be written in this representation with an appropriate choice of matrices. Therefore, almost any physical system modelized by O.D.E's can be cast into form (1) and, as we shall see, their analysis can be made through the study of the fundamental matrices A and B.

Equation (1) appears to be form-invariant under the "quasimonomial" transformations :[2] .

$$T_n(C) \ : \ x_i = \prod_{k=1}^{n} x_k'^{C_{ik}} \tag{2}$$

where C is a $(n \times n)$ invertible matrix.

Indeed, system (1) is mapped by transformation (2) on

$$\dot{x}'_i = \lambda'_i x'_i + x'_i \sum_{j=1}^{m} A'_{ij} \prod_{k=1}^{n} x_k'^{B'_{jk}} \qquad i = 1, \ldots, n \qquad m \geq n \tag{3}$$

with

$$\lambda'_i = (C^{-1}\lambda)_i, \quad A'_{ij} = (C^{-1}A)_{ij}, \quad B'_{ij} = (BC)_{ij} \tag{4}$$

In paper,[1] we have used representation (1) and transformations (2) in order to define *equivalence classes* among O.D.E.'s systems. Each class is characterized by the matricial products BA and $B\lambda$ (which are invariant under $T_n(C)$, see eq. (5)). Furthermore, if matrices B and A are regular, there always exist universal *canonical forms* on which system (1) can be mapped. In the case of B regular (*i.e.* of rank n), one can obtain the first canonical form by mapping system (1) on a m-dimensional quadratic Lotka-Volterra system (in ref,[2] we have shown that this form can always be obtained, even if B is not a matrix of rank n). If matrix A is regular, a second canonical form can be reached by putting $C = A$ in $T_n(C)$. In this form, the system (4) contains at most one nonlinear term per equation. These canonical forms can be used to exhibit integrability conditions and to construct the corresponding solutions.[3,4]

Other classes of integrals can be found by directly using transformations (2).[5] Indeed, if matrix A is of rank r $(r \leq n)$, there exist $(n - r)$ first integrals of the following form :

$$I = I(\{x\}, t) = (x_1^{E_1} x_2^{E_2} \ldots x_n^{E_n}) e^{\rho t} \tag{5}$$

R. Conte and N. Boccara (eds.), Partially Integrable Evolution Equations in Physics, 571–572.

where E_i and ρ are real parameters and at least one of the E_i is non-vanishing.

Indeed, we know from matricial algebra that there always exists an invertible matrix C such that

$$A' = C^{-1}A = \begin{pmatrix} 0 & 0 & \dots & 0 \\ \vdots & \vdots & \ddots & \vdots \\ 0 & 0 & \dots & 0 \\ A'_{n-r+1,1} & A'_{n-r+1,2} & \cdots & A'_{n-r+1,m} \\ A'_{n-r+2,1} & A'_{n-r+2,2} & \cdots & A'_{n-r+2,m} \\ \vdots & \vdots & \ddots & \vdots \\ A'_{n1} & A'_{n2} & \dots & A'_{n,m} \end{pmatrix} \tag{6}$$

By applying on system (1) a transformation $T_n(C)$ with a matrix C satisfying (6), system (3) reads now

$$\begin{aligned} \dot{x}'_l &= \lambda'_l x'_l & l &= 1, \dots, n-r \\ \dot{x}'_i &= \lambda'_i x'_i + x'_i \sum_{j=1}^{m} A'_{ij} \prod_{k=1}^{n} x_k'^{B'_{jk}} & i &= n-r+1, \dots, n \end{aligned} \tag{7}$$

with B' and λ' defined in (4).

In this new set of variables, the $(n-r)$ first equations of system (7) are linear and can be immediatly integrated and in term of the original variables, we obtain the $(n-r)$ quasimonomial integrals :
where the K_l's are arbitrary constants.

$$x_1^{C^{-1}_{l1}} x_2^{C^{-1}_{l2}} \dots x_n^{C^{-1}_{ln}} = e^{\lambda'_l t} K_l \qquad l = 1, \dots, n-r \tag{8}$$

where the K_l's are arbitrary constants.

Conclusions

These new methods (using both the representation (1) and transformations (2)) have been applied to well-studied systems (such as the Lotka-Volterra equations, the Lorenz system,...). It has then been shown that one can obtain all the known cases of integrability as well as new integrals. Furthermore, we hope that the universal reduction to simple canonical forms will lead to a unified approach of nonlinear systems.

References

[1] Brenig, L. and Goriely, A. 1989. Phys. Rev. A 4, 4119 (1989).
[2] Brenig, L. 1988. Phys. Lett. A **133**,378.
[3] Goriely, A. 1989 Mémoire de licence "*Transformations quasi-monomiales et intégrabilité*"
[4] Goriely A., Brenig, L. 1989. *"A New Picture of Dynamical Systems and Integrability"* (submitted to Physica D).
[5] Goriely A., Brenig, L. 1989. *"Algebraic Degeneracy and Partial Integrability for systems of ODE's"* (submitted to Phys. Lett. A).

PARTIAL INTEGRABILITY OF THE DAMPED KINK EQUATION

José M. CERVERO and P.G. ESTEVEZ
Departamento de Física Teórica. Facultad de Ciencias
Universidad de Salamanca
37008 - SALAMANCA
Spain

ABSTRACT. A generalization of the damped kink equation with a space-time dependent damping factor is considered. The exact general solutions are obtained and some particular features concerning an infinite set of multikink solutions arising from the general solution are described. The physical significance of such a multikink solution is also discussed.

1. INTRODUCTION

In a recent paper we have explored the physical consequences arising from the presence of a damping term in the non-linear kink-equation (Ref. 1). By exactly integrating such a differential equation we were lead to discuss some striking physical properties which greatly differ from usual linear wave theory. Firstly, the conservation of the topological charge in spite of the presence of damping. Secondly, the quantization of velocity for the damped kink as a consequence of the break down of (1+1)-dimensional Lorentz invariance.

Encouraged by these novel results we present in this paper a generalization of the damped kink equation which contains as a particular case the previously studied kink (Ref. 1) while generalizes it in a noticiable manner. The idea is to allow for a damping term which is a space-time function. In the case of such a function being a constant we recover the case of Ref. 1. Indeed, we have also found that this new equation is only integrable for a very small set of functional forms. It is also worth to point out that the damped kink equation with space-time dependent damped functions has also appeared in other contexts such as Yang-Mills classical solutions with finite energy and action (Ref. 2). In this case the equation was shown to be of quasi-integrable type due to the functional form of the damping function. In the present analysis we shall study the one parameter set of functional forms that give rise to a fully integrable case and we shall also explore the physical consequences.

We shall start with the general damped cubic kink equation:

R. Conte and N. Boccara (eds.), Partially Integrable Evolution Equations in Physics, 573–574.

$$\frac{d^2f}{dx_o^2} - \frac{d^2f}{dx_1^2} + \lambda(x_1,x_o)\,\frac{df}{dx_o} - f + f^3 = 0 \qquad (1)$$

In Ref. 1 the dimensionless function was treated as a constant λ_o. Now $\lambda(x_1,x_o) = \lambda_o F(x_1,x_o)$. Obviously $F(x_1,x_o) = 1$ in Ref. 1. Since we are searching for travelling waves is reasonable to call:

$$z = \frac{x_1-(v/c)x_o}{[1-(v/c)^2]^{1/2}} \quad ; \qquad \frac{v}{c} = \frac{3}{(9+2\lambda_o^2)^{1/2}} \qquad (2)$$

Here, (2) expresses the general form of a travelling wave for a relativistic evolution equation. The velocity quantization condition was extensively discussed in Ref. 1. Using (2) in (1) we obtain:

$$\frac{d^2f}{dz^2} + \left[\frac{3\lambda(x_1,x_o)}{2^{1/2}\,\lambda_o}\right]\frac{df}{dz} + f - f^3 = 0 \qquad (3)$$

Let us now suppose that $\lambda(x_1,x_o) = \lambda_o F(x_1, x_o)$. Only the following functional form give rise to an integrable type equation through standard Painlevé analysis (Ref. 3):

$$F(x_1,x_o) = \frac{\sinh\{2^{1/2}(z+z_o)\}}{[\cosh\{2^{1/2}(z+z_o)\} - \alpha_o]} \qquad (4)$$

Where z and v/c are given by (2) and z_o and α_o are arbitrary constants. In order to avoid singularities $\alpha_o < 1$. If $z_o \longrightarrow \infty$ then the function $F(x_1,x_o) = 1$. The case of Ref. 1. Using now (4) in (3) we obtain an integrable ODE. The general solution can be given in terms of elliptic functions (Ref. 4) as:

$$f(z) = 2^{1/2}A_o[\cosh\{2^{1/2}(z+z_o)\}-\alpha_o]^{-1/2}\,\mathrm{ds}[A_o(u+\vartheta_o)\mid (A_o^2+\alpha_o)/2A_o^2] \qquad (5)$$

where u is given by:

$$u = \mathrm{cn}^{-1}[\{1-\alpha_o\}^{1/2}[\cosh\{2^{1/2}(z+z_o)\}-\alpha_o]^{-1/2} \mid (1+\alpha_o)/2] \qquad (6)$$

One can easily see that for integer and rational values of A_o, special solutions can be found using addition theorems for the elliptic functions. These solutions correspond to multikink configuration, to be discussed extensively elsewhere.

REFERENCES

1. Cerveró, J.M. and Estévez, P.G. (1986) Physics Letters **114A**, 435.
2. Cerveró, J.M. (1984) Letters in Mathematical Physics **8**, 233.
3. Ince, E.L. (1956) "Ordinary Differential Equations". Dover. Chp. XIV.
4. Byrd, P.F. and Friedman, M.D. (1971) "Handbook of elliptic integrals for Engineers and Scientists". Springer Verlag. Berlin.

NEW SIMILARITY REDUCTIONS OF BOUSSINESQ-TYPE EQUATIONS

PETER A. CLARKSON,
Department of Mathematics,
University of Exeter,
Exeter, EX4 4QE,
ENGLAND

In this paper new similarity reductions are presented for several physically interesting partial differential equations, including the Boussinesq and modified Boussinesq equations

$$u_{tt} + uu_{xx} + u_x^2 + u_{xxxx} = 0, \tag{1}$$

$$u_{tt} - u_t u_{xx} - \tfrac{1}{2}u_x^2 u_{xx} + u_{xxxx} = 0, \tag{2}$$

which are completely integrable soliton equations [1,2], and the SRLW equation

$$u_{tt} + u_{xx} + uu_{xt} + u_x u_t + u_{xxtt} = 0, \tag{3}$$

which is thought to be non-integrable [3]. These similarity reductions are obtained using a direct method developed with Martin Kruskal [4], which involves no group theoretical techniques.

The classical method for determining similarity reductions uses Lie groups of infinitesimal transformations [5], for which symbolic manipulation programs have been developed [6]. Applying this method to the Boussinesq equation (1) gives the travelling wave solution $u(x,t) = w_1(x - ct)$ and the scaling reduction $u(x,t) = t^{-1}w_2(xt^{-1/2})$, where $w_1(z)$ and $w_2(z)$ are solvable in terms of the first and fourth Painlevé transcendents, respectively [7,8]. However the Boussinesq equation also possesses the similarity reduction $u(x,t) = w_3(x + \lambda t^2) - 4\lambda^2 t^2$, where $w_3(z)$ is solvable in terms of the second Painlevé transcendent [2,7,8,9], which is *not* obtained using the classical Lie group method.

The method developed in [4] (hereafter called the *Direct method*), involves seeking solutions in the form $u(x,t) = U(x,t,w(z(x,t)))$ and demanding that the result be an ordinary differential equation for $w(z)$. This imposes conditions upon U, z and their derivatives in the form of an overdetermined system of equations, whose solution yields the similarity reductions. It is shown in [4] that the general similarity reduction of the Boussinesq equation (1) is given by

$$u(x,t) = \theta^2(t)w(z) - \frac{1}{\theta^2(t)}\left(x\frac{d\theta}{dt} + \frac{d\sigma}{dt}\right)^2, \qquad z = x\theta(t) + \sigma(t), \tag{4a}$$

where $\theta(t)$, $\sigma(t)$ are any solutions of

$$\frac{d^2\theta}{dt^2} = A\theta^5, \qquad \frac{d^2\sigma}{dt^2} = (A\sigma + B)\theta^4, \tag{4b}$$

with A, B constants, and $w(z)$ satisfies

$$w'''' + ww'' + (w')^2 + (Az + B)w' + 2Aw = 2(Az + B)^2, \tag{5}$$

R. Conte and N. Boccara (eds.), Partially Integrable Evolution Equations in Physics, 575–576.

with $' := \mathrm{d}/\mathrm{d}z$. This is solvable in terms of (i), the first Painlevé transcendent, if $A = 0$, $B = 0$; (ii), the second Painlevé transcendent, if $A = 0$, $B \neq 0$; and (iii), the fourth Painlevé transcendent, if $A \neq 0$. Examples of new similarity reductions include:

$$u(x,t) = t^2 w_1(z) - (x + 6\lambda_1 t^5)^2/t^2, \qquad z = xt + \lambda_1 t^6,$$
$$u(x,t) = t^{-1} w_2(z) - \tfrac{1}{4}(x - 3\lambda_2 t^2)^2/t^2, \qquad z = xt^{-1/2} + \lambda_2 t^{3/2},$$
$$u(x,t) = \wp^{-1}(t)\left[w_3(z) - \tfrac{1}{4}\left(z\mathrm{d}\wp/\mathrm{d}t + 2\lambda_3 \wp^{3/2}(t)\right)^2\right], \qquad z = \wp^{-1/2}(t)\left[x + \lambda_3 \zeta(t)\right],$$

where λ_1, λ_2, λ_3 are constants, $\wp(t;0,g_3)$, $\zeta(t;0,g_3)$ are the Weierstrass elliptic and zeta functions [10], and $w_1(z)$, $w_2(z)$, $w_3(z)$ satisfy (5) with $A = 0$, $B = 30\lambda_1$; $A = 3/4$, $B = 0$; $A = -3k/4$, $B = 0$, respectively. Subsequently Levi and Winternitz [11] have shown that all the similarity reductions found in [4] can be obtained using the "nonclassical method" due to Bluman and Cole [12].

For the modified Boussinesq equation (2) the Direct method yields the similarity reductions $u(x,t) = w_1(x + \lambda_1 t + \mu_1 t^2) - x(\lambda_1 + 2\mu_1 t)$ and $u(x,t) = w_2(xt^{-1/2} + \mu_2 t^{3/2}) - \lambda_2 \ln t - 2\mu_2 xt$, where λ_1, λ_2, μ_1, μ_2 are constants and $w_1(z)$ and $w_2(z)$ are solvable in terms of the second and fourth Painlevé transcendents, respectively [13]. These similarity reductions are not obtainable using the classical Lie group method (except in the special cases $\mu_1 = 0$ and $\mu_2 = 0$). For the SRLW equation (3) the Direct method yields the similarity reductions $u(x,t) = w_1(x + \lambda_1 t)$, and $u(x,t) = t^{-1} w_2(x + \lambda_2 \ln t) - \lambda_2^{-1} t$, where λ_1, λ_2 are constants, $w_1(z)$ is solvable in terms of the first Painlevé transcendent and $w_2(z)$ in terms of an equation which is not of Painlevé type since its general solution has movable logarithmic singularities [14]. The first similarity reduction is obtainable using the classical Lie group method [6], whereas the second is not.

The difference between similarity reductions obtained by the classical Lie group method and those not (which are usually called "nonclassical" similarity reductions), is that the group associated with a classical similarity reduction of a partial differential equation maps *all* solutions of the equation into other solutions of the equation, whereas for a nonclassical similarity reduction, the associated group only maps a subset of solutions of the equation into other solutions of the equation (generally, the group maps solutions of the equation into solutions of a different equation — see [4,11,13,14]).

References

[1] V.E. Zakharov, *Sov. Phys. JETP*, **38** (1974) 108–110.
[2] G.R.W. Quispel, F.W. Nijhoff and H.W. Capel, *Phys. Lett.*, **91A** (1982) 143–145.
[3] C.E. Seyler and D.L. Fenstermacher, *Phys. Fluids*, **27** (1984) 4–7.
[4] P.A. Clarkson and M.D. Kruskal, *J. Math. Phys.*, **30** (1989) 2201–2213.
[5] G.W. Bluman and J.D. Cole, "*Similarity Methods for Differential Equations*," (Springer-Verlag, Berlin, 1974).
[6] F. Schwarz, *Computing*, **34** (1985) 91–106; *SIAM Review*, **30** (1988) 450–481.
[7] T. Nishitani and M. Tajiri, *Phys. Lett.*, **89A** (1982) 379–380.
[8] P. Rosenau and J.L. Schwarzmeier, *Phys. Lett.*, **115A** (1986) 75–77.
[9] P.J. Olver and P. Rosenau, *SIAM J. Appl. Math.*, **47** (1987) 263–278.
[10] E.E. Whittaker and G.M. Watson, "*Modern Analysis*," 4th Ed. (C.U.P., Cambridge, 1927).
[11] D. Levi and P. Winternitz, *J. Phys. A: Math. Gen.*, **22** (1989) 2915–2924.
[12] G.W. Bluman and J.D. Cole, *J. Math. Mech.*, **18** (1969) 1025–1042.
[13] P.A. Clarkson, *J. Phys. A: Math. Gen.*, **22** (1989) 2355–2367.
[14] P.A. Clarkson, *J. Phys. A: Math. Gen.*, **22** (1989) 3821–3848.

THE HOMOGRAPHIC INVARIANCE OF PDE PAINLEVÉ ANALYSIS

Robert CONTE
Service de physique du solide et de résonance magnétique
Centre d'études nucléaires de Saclay
F-91191 Gif-sur-Yvette Cedex, France

ABSTRACT. The whole Painlevé analysis of PDE is shown to be invariant under an arbitrary homographic transformation of the function φ defining the singular manifold.

The Painlevé analysis of partial differential equations (PDE) [1] consists in looking for the general solution under the form of a series expansion in a neighborhood of a singular manifold $\varphi(x,t) = 0$:

$$u = \sum_{j=0}^{+\infty} u_j \chi^{j+p} \tag{1}$$

where p is usually a constant negative integer and χ an expansion variable vanishing on the singular manifold. Our goal is to optimize the choice of χ, classically taken as $\chi = \varphi$.

As is well known, the only relevant feature of the Painlevé property of ordinary differential equations (ODE) is the singularity structure of the solutions, which is obviously invariant under an arbitrary homographic transformation of the independent variable z:

$$z \longrightarrow \frac{az+b}{cz+d}, \quad (a,b,c,d) \in \mathcal{C}, \quad ad - bc = 1 \ \ (\text{group } \mathcal{H}) \tag{2}$$

In the PDE case, Painlevé property must also be invariant under $\mathcal{H}$, now acting on φ. We therefore require χ to be a homographic function of φ with coefficients depending on its derivatives $D\varphi$, and we choose these coefficients so that every first derivative of χ, which is a polynomial of degree 2 in χ, has its 3 coefficients invariant under $\mathcal{H}$. A particular solution[2] is:

$$\chi = \left(\frac{\varphi_x}{\varphi - \varphi_0} - \frac{\varphi_{xx}}{2\varphi_x} \right)^{-1} \tag{3}$$

where φ_0 is an arbitrary complex constant and x any independent variable such that $\varphi_x \neq 0$. The associated Riccati equations read

$$\chi_x = 1 + \frac{S}{2}\chi^2 \tag{4a}$$

R. Conte and N. Boccara (eds.), Partially Integrable Evolution Equations in Physics, 577–578.

$$\chi_t = -C + C_x\chi - \frac{1}{2}(CS + C_{xx})\chi^2 \tag{4b}$$

where S and C are two elementary differential homographic invariants:

$$S = \{\varphi; x\} = \frac{\varphi_{xxx}}{\varphi_x} - \frac{3}{2}\left(\frac{\varphi_{xx}}{\varphi_x}\right)^2, \quad C = -\frac{\varphi_t}{\varphi_x} \tag{5}$$

linked by the cross-derivative condition

$$\varphi_x^{-1}\left((\varphi_{xxx})_t - (\varphi_t)_{xxx}\right) \equiv S_t + C_{xxx} + 2C_xS + CS_x = 0 \tag{6}$$

One proves that, with the above choice of χ as an expansion variable, all results of the PDE Painlevé analysis (sequence of coefficients u_j, compatibility conditions, conditions for a truncation, etc) are invariant under action of the Möbius group $\mathcal{H}$ on φ, and therefore depend on the derivatives of φ only through the elementary invariants. This result, much stronger than the one for ODE, considerably simplifies the whole analysis. It is valid for any number of dependent variables, independent variables and equations, the equations being polynomial in the dependent variables and their partial derivatives.

Let us simply mention two of the important consequences:

1) if the series converges at one point of the φ complex plane, it converges everywhere;

2) in the Darboux transformation obtained by the Weiss truncation procedure, the two different solutions of the PDE have the same expression, evaluated at two different φ_0 values, namely 0 and ∞.

Moreover, if ψ denotes the scalar wave vector of the unknown underlying Lax pair, then $\frac{\psi}{\psi_x}$ is also[3] an invariant choice of χ, and the function Y defined by

$$\frac{\psi_x}{\psi} = \frac{\varphi_x}{\varphi - \varphi_0} - \frac{\varphi_{xx}}{2\varphi_x} + \frac{Y}{2} \tag{7}$$

is invariant under $\mathcal{H}$. It can be interpreted as one of the components of a nonlinear pseudopotential, by construction of Riccati type.

These invariant Painlevé analyses have been implemented in a computer algebra program written in the AMP language[4].

[1] Weiss, J., Tabor, M. and Carnevale, G. (1983) "The Painlevé property for partial differential equations", J. Math. Phys. 24, 522-526.
[2] Conte, R. (1989) "Invariant Painlevé analysis of partial differential equations", Phys. Lett. A140, 383-390.
[3] Musette, M. (1990) "Painlevé-Darboux transformations in nonlinear partial differential equations" in P.C. Sabatier (ed.), Inverse problems in action, Springer Verlag, Berlin.
[4] Drouffe, J.-M. (1986) "AMP reference manual (version 6.6)" , SPhT, Centre d'études nucléaires de Saclay, F-91191 Gif-sur-Yvette Cedex.

THE NONLOCAL AMPLITUDE EQUATION[1]

F.J. Elmer and T. Christen
Institut für Physik, Universität Basel
Klingelbergstr. 82
CH-4056 Basel
Switzerland

ABSTRACT: Stationary states, stability, and bifurcation scenarios are presented for a spatially nonlocal extension of the Newell-Whitehead-Segel amplitude equation.

In the weakly nonlinear regime a stationary finite wavelength ($k_c \neq 0$) instability breaking a translation symmetry will lead to the Newell-Whitehead-Segel amplitude equation or time-dependent Ginzburg-Landau equation (see e.g. P. Coullet's lecture in this volume). A general nonlocal extension of this equation is

$$\partial_t A = (\epsilon + b_3 \int_{-\infty}^{\infty} G(x - x')|A(x')|^2 dx' + b_4|A|^2)A + \partial_x^2 A, \tag{1}$$

where A is the complex amplitude of a standing wave with wave number k_c and where ϵ is the control parameter. We assume that the Fourier transform $F_G(q) \equiv \int G(x)e^{-iqx}dx$ is an even function which decreases monotonically to zero. Further we assume that a scaling form: $F_G(q) \equiv F_g((q/a)^2)$ with $-F_g'(0) = F_g(0) = 1$ and $F_g'' > 0$, where $1/a$ defines the width of the kernel G. For $a \gg k_c$ the kernel is approximated by a δ-function and the nonlocal amplitude equation becomes a local one. If $1/a$ is much larger then the length of the system L we get a strong nonlocality $\frac{1}{L}\int_0^L |A|^2 dx$. Such strongly nonlocal amplitude equations are important for several physical systems (e.g. ferromagnetic resonance [1], Taylor-Couette flow [2]). This work generalizes results for the strongly nonlocal amplitude equation [3].

The nonlocal amplitude equation has three different types of stationary solutions:

1. uniform solution, $A = 0$. It becomes unstable for $\epsilon > 0$ and bifurcates at the neutral curve $\epsilon = (k - k_c)^2$ into a
2. spatially periodic solution, $A = Re^{i(k-k_c)x}$, with $R^2 = (\epsilon - (k - k_c)^2)/(-b_3 - b_4)$. For $b_3 + b_4 < 0$ the bifurcation is supercritical. This kind of solution becomes unstable either through a long wavelength ($q_c \to 0$) phase instability (Eckhaus instability) or through a stationary finite wavelength ($q_c \neq 0$) amplitude instability. At these instability curves (see fig. 1) the periodic solution bifurcates into a
3. spatially modulated solution, $A = R(q_c x)e^{i(k-k_c)x+\Phi(q_c x)}$. In the strongly nonlocal limit of (1) the 2π-periodic functions R and Φ are elliptic functions [3].

[1]This work was supported by the Swiss National Science Foundation.

R. Conte and N. Boccara (eds.), Partially Integrable Evolution Equations in Physics, 579–580.

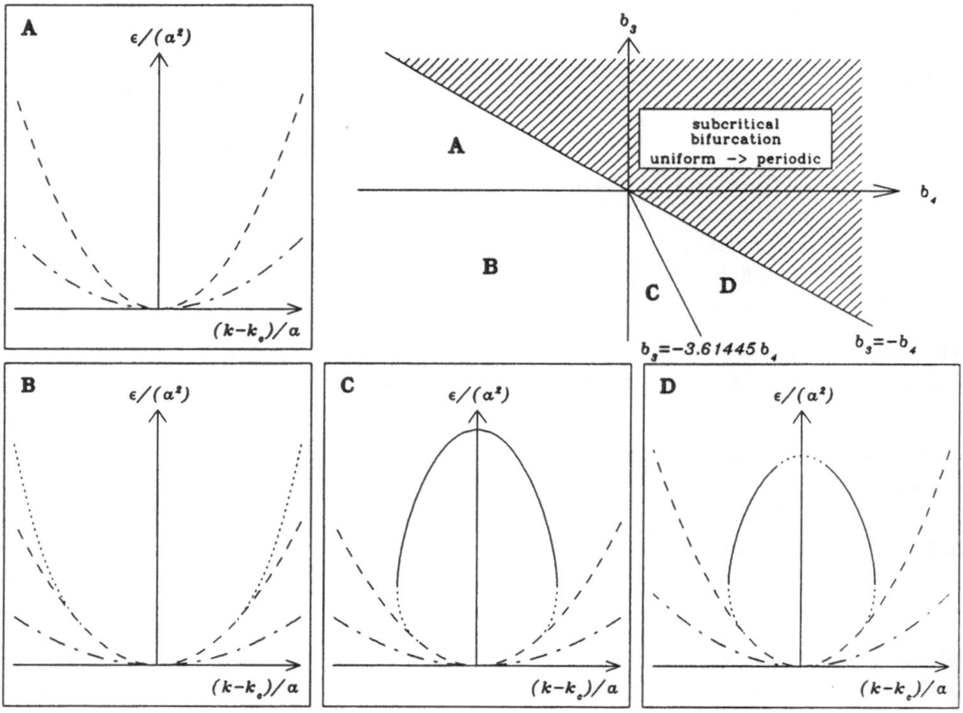

Figure 1: Bifurcation scenarios for a gaussian kernel. Instability curves: – · – · – neutral curve, - - - - - Eckhaus instability, ······ (———) amplitude instability with sub(super)critical bifurcation.

The bifurcation type (i.e. sub- or supercritical) at the amplitude instability depends on the kernel G.

We emphasize that in some cases (C and D in fig.1) already in the weakly nonlinear regime nonlocality leads to a secondary instability for an increasing control parameter.

References

[1] Elmer, F.J. (1988) 'Spatial pattern formation in FMR - An example for nonlocal dynamics', J. de Phys. Coll. C8, Suppl. to No. 12, 49, 1597-1598.

[2] Hall, P. (1984) 'Evolution equations for Taylor vortices in the small-gap limit', Phys. Rev. A 29, 2921-2923.

[3] Elmer, F.J. (1988) 'Nonlinear and nonlocal dynamics of spatially extended systems: Stationary states, bifurcations and stability', Physica D 30, 321-342.

PRESSURE WAVES IN FLUID-FILLED NONLINEAR VISCOELASTIC TUBES

H. A. ERBAY and S. ERBAY
Research Institute for Basic Sciences,
Department of Applied Mathematics,
P.O. Box 74, 41401 Gebze-Kocaeli, TURKEY

ABSTRACT. Using the reductive perturbation method, wave propagation in fluid-filled nonlinear viscoelastic thin tubes is investigated in the long wave approximation. For different scales of the relaxation constant τ, the Korteweg-de Vries (KdV), Burgers and Korteweg-de Vries-Burgers (KdVB) equations are obtained.

1. BASIC EQUATIONS

Many authors have studied the propagation of small but finite amplitude waves corresponding to the long wave approximation in fluid-filled tubes [1],[2],[3]. The most significant differences in these studies arise from the assumed behaviour of the vessel wall. Here, a generalization of the Mooney-Rivlin elastic material to nonlinear viscoelasticity given in [4] is taken as a constitutive equation for the tube and the resulting pressure-area relation is extended to include inertial terms. The dimensionless form of governing equations of the tube filled with an incompressible, inviscid fluid are given in [5] as follows

$$\frac{\partial B}{\partial t} + \frac{\partial (Bv)}{\partial z} = 0 \qquad \text{(Conservation of mass equation)} \qquad (1)$$

$$\frac{\partial v}{\partial t} + v\frac{\partial v}{\partial z} + \frac{\partial P}{\partial z} = 0 \qquad \text{(Conservation of momentum equation)} \qquad (2)$$

$$\left(\frac{\partial}{\partial t} + \frac{1}{\tau}\right)\left\{2BP-\delta\gamma\left[\frac{\partial^2 B}{\partial t^2} - \frac{1}{2B}\left(\frac{\partial B}{\partial t}\right)^2\right]\right\}=\left(\frac{\partial}{\partial t} + \frac{m}{\tau}\right)\Phi(B) \qquad \text{(Pressure-Area relation)} \qquad (3)$$

where t is time, z is the axial coordinate of the tube, B is the internal area of the tube, v is the fluid velocity in the z direction, P is the pressure difference between the inside and the outside of the tube, τ is the relaxation parameter, δ is the ratio of solid and fluid densities and γ is the ratio of the thickness to the inner radius of the tube.

R. Conte and N. Boccara (eds.), Partially Integrable Evolution Equations in Physics, 581–582.

2. EVOLUTION EQUATIONS IN THE LONG WAVE APPROXIMATION

We apply the reductive perturbation method to the above model. If the following coordinate stretching is made

$$\xi=\varepsilon^{\alpha}(z-ct) \quad , \quad \eta=\varepsilon^{\alpha+1}t \qquad \alpha>0 \tag{4}$$

where $c=[m\Phi'(B_0)/2-P_0]^{1/2}$ is linear wave speed and dependent variables are expanded about a constant state $(B_0,P_0,0)$ as integral power series of ε

$$B=B_0+B_1\varepsilon+B_2\varepsilon^2+\ldots,\ P=P_0+P_1\varepsilon+P_2\varepsilon^2+\ldots,\ v=v_1\varepsilon+v_2\varepsilon^2+\ldots \tag{5}$$

we readily get for ε^0: $2P_0B_0=m\Phi(B_0)$, ε^1: $v_1=cB_1/B_0$, $P_1=c^2B_1/B_0$ and for ε^2

$$\frac{\partial B_1}{\partial\eta}+\left[\frac{c}{2B_0}+\frac{m}{4c}\Phi''(B_0)\right]B_1\frac{\partial B_1}{\partial\xi}-\frac{\tau}{4}(1-m)\varepsilon^{\alpha-1}\Phi'(B_0)\frac{\partial^2 B_1}{\partial\xi^2}$$
$$+\varepsilon^{2\alpha-1}\frac{\delta\gamma c}{4}\frac{\partial^3 B_1}{\partial\xi^3}=0 \tag{6}$$

a) For $\alpha=1$ and $\tau=O(1)$, we get the Burgers equation

$$\frac{\partial B_1}{\partial\eta}+\kappa B_1\frac{\partial B_1}{\partial\xi}-\frac{\tau}{4}(1-m)\Phi'(B_0)\frac{\partial^2 B_1}{\partial\xi^2}=0\ ;\ \kappa=\frac{c}{2B_0}+\frac{m}{4c}\Phi''(B_0) \tag{7}$$

b) For $\alpha=1/2$ and $\tau=\tilde{\tau}\varepsilon^{1/2}$ where $\tilde{\tau}=O(1)$, we find the KdVB equation

$$\frac{\partial B_1}{\partial\eta}+\kappa B_1\frac{\partial B_1}{\partial\xi}-\frac{\tilde{\tau}}{4}(1-m)\Phi'(B_0)\frac{\partial^2 B_1}{\partial\xi^2}+\frac{\delta\gamma c}{4}\frac{\partial^3 B_1}{\partial\xi^3}=0 \tag{8}$$

c) For $\alpha=1/2$ and $\tau=\tilde{\tau}\varepsilon^{3/2}$ where $\tilde{\tau}=O(1)$, we get the KdV equation

$$\frac{\partial B_1}{\partial\eta}+\kappa B_1\frac{\partial B_1}{\partial\xi}+\frac{\delta\gamma c}{4}\frac{\partial^3 B_1}{\partial\xi^3}=0 \tag{9}$$

References

[1] Hashizume, Y. (1985) "Nonlinear pressure waves in a fluid-filled elastic tube", J.Phys.Soc. Japan 54, 3305-3312.

[2] Johnson, R.S. (1971) Ph.D. dissertation, University of London.

[3] Ravindran, R. and Prasad P. (1979) "A mathematical analysis of nonlinear waves in a fluid-filled viscoelastic tube", Acta Mech. 31, 253-280.

[4] Tait, R.J. and Moodie, T.B. (1984), "Waves in nonlinear fluid-filled tubes", Wave Motion 6, 197-203.

[5] Erbay, H.A. and Erbay,S. (1988) "Wave propagation in fluid-filled nonlinear viscoelastic tubes", (submitted).

Elliptic Function Solutions for Landau-Ginzburg Equation

A.M. Grundland
Dép. de mathématiques
UQTR
Trois-Riv., QC, G9A 5S7

J.A. Tuszyński
Dept of Physics
Univ. of Alberta
Edmonton, ALTA, T6G 2J1

P. Winternitz
Centre de rech. math.
Univ. de Montréal
Mtl, QC, H3C 3J7

Abstract

The symmetry reduction method is used to obtain a number of new exact solutions in terms of Jacobi elliptic functions.

The symmetry group reduction method **[1-2]** for a quintic N.L.K.G.E. is applied to a Landau-Ginzburg model in $4D$,

$$\Box\phi + 2(a_2\phi + 2a_4\phi^3 + 3a_6\phi^5) = 0, \quad a_6 \neq 0 \tag{1}$$

with real constants a_2, a_4 and a_6. In the Minkowski space, (1) is invariant w.r.t. the Poincaré group $P(3,1)$ ($a_2 \neq 0 \neq a_4$) and w.r.t. the similitude group sim (3,1) ($a_1 = 0 = a_4$). We consider only the case when symmetry variables are invariants of subgroup of sim (3,1) having generic orbits of codimension one. In [1] we show that the introduction of a symmetry variable ξ, in a reduction formula of the form $\phi(x) = \sigma(x)F(\xi)$, $\xi = \xi(x)$, reduces (1) to one of several O.D.E.s. Symmetry reduction by subgroups of $P(3,1)$ leads to

$$F_{\xi\xi} + \frac{k}{\xi}F_\xi + 2\lambda(a_2F + 2a_4F^3 + 3a_6F^5) = 0, \quad \lambda = \pm 1, \; k = 0,1,2,3 \tag{2}$$

and all other reductions lead to

$$(\alpha_1 + \alpha_2\xi + \alpha_3\xi^2)\,F_{\xi\xi} + (\beta_1 + \beta_2\xi)\,F_\xi + \gamma F + (\mu_1 + \mu_2\xi)\,F^5 = 0 \tag{3}$$

where $\alpha_i, \beta_i, \gamma, \mu_i \in R$. After the transformation $F = H^{1/2}(\xi)$. (2) has the Painlevé property (P.P.). For $k = 0$, we get

$$\dot{w}^2 = w(w^3 + pw^2 + qw + \gamma) =: P(w) \tag{4}$$

where $F(\xi) = (-8\lambda a_6)^{-1/2}w$, $p = \left(\frac{-8\lambda}{a_6}\right)^{1/2} a_4$ and $q = -8\lambda a_2$. For $k = 2$, $a_2 = a_4 = 0$, with $F(\xi) = \varepsilon^{1/2}\left(\frac{-\lambda}{8a_6}\right)^{1/4} \cdot (\xi^{-1}w(\ell n\xi))^{1/2}$, $\varepsilon = \pm 1$, we find that $w(\ell n\xi)$ satisfies (4) for $p = 1$ and where q is an integration constant. We put $F = H^{1/2}(\xi)$ into (3) and the resulting equation is tested for the P.P. Next, the substitution $H(\xi) = \lambda(\xi)\,w(\eta)$ with $\eta = \eta(\xi)$ transforms (3) into P-type equations PXXIX or PXXX; both with simple first integrals. In the examples in Table 1, we obtain real and finite doubly periodic solutions in terms of Jacobi elliptic functions (for distinct roots of $P(w)$), with moduli k ($0 < k^2 < 1$). This ensures that the elliptic solutions possess one real and one purely imaginary period, and that their arguments are real.

R. Conte and N. Boccara (eds.), Partially Integrable Evolution Equations in Physics, 583–584.

References

1. Winternitz, P., Grundland, A.M. and Tuszyński, J.A. (1987), "Exact Solutions of the Multidimensional Classical ϕ^6–Field Equations Obtained by Symmetry Reduction", J. Math. Phys. **28**, 9, 2194-2212.
2. Winternitz, P., Grundland, A.M. and Tuszyński, J.A. (1988), "Exact Results in the Three-Dimensional Landau-Ginzburg Model", J. Phys. C, Solid State Phys. **21**, 4931-4953.

Table 1

Elliptic function solutions of $\dot{H}^2 = AH(H - H_1)(H - H_2)(H - H_3)$. Four real distinct roots. – Finite solutions.

No	Roots	F	β	k
1	$A > 0$ $H_1 < 0 < H_2 < H_3$	$\left[\frac{H_1 H_2}{H_2 sn^2(\beta\xi,k) - H_2 + H_1}\right]^{1/2} sn(\beta\xi, k)$	$\frac{1}{2}[AH_3(H_2 - H_1)]^{1/2}$	$\left[\frac{H_2(H_3-H_1)}{H_3(H_2-H_1)}\right]^{1/2}$
2	$A > 0$ $0 < H_1 < H_2 < H_3$	$\left[\frac{H_1 H_3}{H_3 - H_1 cn^2(\beta\xi,k)}\right]^{1/2} sn(\beta\xi, k)$	$\frac{1}{2}[AH_2(H_1 - H_3)]^{1/2}$	$\left[\frac{H_1(H_3-H_2)}{H_2(H_3-H_1)}\right]^{1/2}$
3	$A < 0$ $0 < H_1 < H_2 < H_3$	$\varepsilon\left[\frac{H_1(H_3-H_2)\, sn^2(\beta\xi,k) - H_2(H_3-H_1)}{(H_3-H_2)\, sn^2(\beta\xi,k) - (H_3-H_1)}\right]^{1/2}$	$\frac{1}{2}[-AH_2(H_3 - H_1)]^{1/2}$	$\left[\frac{H_1(H_3-H_2)}{H_2(H_3-H_1)}\right]^{1/2}$
4	$A < 0$ $H_1 < 0 < H_2 < H_3$	$\varepsilon\left[\frac{H_2 H_3}{(H_2-H_3)\, sn^2(\beta\xi,k) + H_3}\right]^{1/2}$	$\frac{1}{2}[AH_3(H_2 - H_1)]^{1/2}$	$\left[\frac{H_1(H_2-H_3)}{H_3(H_2-H_1)}\right]^{1/2}$
5	$A < 0$ $H_1 < H_2 < 0 < H_3$	$\left[\frac{H_2 H_3}{H_3\, sn^2(\beta\xi,k) - H_3 + H_2}\right]^{1/2} sn(\beta\xi, k)$	$\frac{1}{2}[AH_1(H_3 - H_2)]^{1/2}$	$\left[\frac{H_3(H_1-H_2)}{H_1(H_3-H_2)}\right]^{1/2}$

APPLICATION OF A MACSYMA PROGRAM FOR THE PAINLEVÉ TEST TO THE FITZHUGH-NAGUMO EQUATION

W. HEREMAN
Mathematics Department
University of Wisconsin at Madison
480 Lincoln Drive
Madison, WI 53706, USA

ABSTRACT. *A Macsyma program for the Painlevé test is applied to the Fitzhugh-Nagumo equation. An auto-Bäcklund transformation and solitary wave solutions are derived.*

1. SCOPE OF THE PROGRAM

A Macsyma program [1] has been developped to determine whether a single nonlinear ODE or PDE with (real) polynomial terms fulfills the necessary conditions for having the Painlevé property. For a PDE (say, in two variables) to pass the *Painlevé Test*, it is necessary that its solution $u(x,t)$, expressed as a Laurent series in the complex plane,

$$u(x,t) = g^{\alpha} \sum_{k=0}^{\infty} u_k \, g^k, \tag{1}$$

has no worse singularities than movable poles. In (1), $u_0(x,t) \neq 0$, α is a negative integer, and $u_k(x,t)$ are analytic functions in a neighborhood of the singular, non-characteristic manifold $g(x,t) = 0$, with $g_x(x,t) \neq 0$. For an ODE, $g = x - x_0$; x_0 being the initial value for x. Specifics of the algorithm for the Painlevé test may be found elsewhere in these Proceedings. Details about the program are given in [1].

2. APPLICATION OF THE PROGRAM

We will use the Painlevé program to derive solitary wave solutions of the Fitzhugh-Nagumo (FHN) equation [2],

$$u_t - u_{xx} + u(1-u)(a-u) = 0, \tag{2}$$

where $-1 \leq a < 1$ is a constant parameter. To determine the speeds of traveling wave solutions of (2), it suffices to test the ODE,

$$c\,\phi_z + \sqrt{2}\,\phi_{zz} - \sqrt{2}\,\phi\,(1 - \sqrt{2}\,\phi)(a - \sqrt{2}\,\phi) = 0, \tag{3}$$

where $\phi(z) = \phi(x - \frac{ct}{\sqrt{2}}) = \frac{1}{\sqrt{2}}\,u(x,t)$. Observe that $\frac{c}{\sqrt{2}}$ is the physical speed of the wave. The program determines $\alpha = -1$, $\phi_0 = \pm 1$, and establishes there is a resonance that

R. Conte and N. Boccara (eds.), Partially Integrable Evolution Equations in Physics, 585–586.

at $k = 4$, thus u_4 is arbitrary. The compatibility condition at the resonance is satisfied provided,

$$c(c - 2a + 1)(c + a - 2)(c + a + 1) = 0. \tag{4}$$

The each of these c, for which thus the FHN equation passes the Painlevé test, one can construct an exact solution to (2). We will do so using a truncation of the series in (1). Running the program for the PDE (2), yields $\alpha = -1$ and $u_0(x,t) = \frac{\sqrt{2}\, g_x}{g}$. Kawahara and Tanaka [2] showed that the Laurent series, truncated at the constant level term,

$$u(x,t) = \sqrt{2}\, \frac{g_x(x,t)}{g(x,t)} + u_1(x,t), \tag{5}$$

allows to 'bilinearize' the FHN equation. Formula (5) also serves as an auto-Bäcklund tranformation. Indeed, substituting (5) into (2) and equating power terms in g, leads to an overdetermined system of equations for g and u_1:

$$g_t - 3g_{xx} - \sqrt{2}\, g_x(3u_1 - 1 - a) = 0, \tag{6}$$

$$g_{tx} - g_{xxx} + g_x\, [3{u_1}^2 - 2(1+a)u_1 + a] = 0, \tag{7}$$

$$(u_1)_t - (u_1)_{xx} + u_1(1 - u_1)(a - u_1) = 0. \tag{8}$$

Starting from a trivial solution u_1 to the FHN, (5)-(7) allow to construct a nontrivial solution u. For $u_1 = 0$ (the cases $u_1 = 1$ and $u_1 = a$ being similar), (6) and (7) are readily integrated:

$$g = A\, exp[\frac{1}{\sqrt{2}}(x - \frac{1}{\sqrt{2}}(2a-1)t)] + B\, exp[\frac{a}{\sqrt{2}}(x - \frac{1}{\sqrt{2}}(2-a)t)] + C, \tag{9}$$

where A, B and C are integration constants. Substitution of g into (5) then gives

$$u = \frac{A\, exp[\frac{1}{\sqrt{2}}(x - \frac{1}{\sqrt{2}}(2a-1)t)] + B\, exp[\frac{a}{\sqrt{2}}(x - \frac{1}{\sqrt{2}}(2-a)t)]}{A\, exp[\frac{1}{\sqrt{2}}(x - \frac{1}{\sqrt{2}}(2a-1)t)] + B\, exp[\frac{a}{\sqrt{2}}(x - \frac{1}{\sqrt{2}}(2-a)t)] + C}. \tag{10}$$

This exact solution, describing the coalescence of two wave fronts, was obtained by Hirota's bilinear method in [2]. Single solitary wave solutions follow as special cases of (10). For example, with $A = 0$, (10) simplifies to

$$u(x,t) = \frac{a}{2}(1 + \tanh[\frac{a}{2\sqrt{2}}(x - \frac{(2-a)}{\sqrt{2}}t) + \delta_1]), \tag{11}$$

where $\delta_1 = \frac{1}{2}\ln(\frac{B}{C})$ is the arbitrary constant phase. This is the solution corresponding to $c = 2 - a$ in (4). The other traveling wave solutions are constructed analogously.

References

[1] Hereman, W. (1988) 'Macsyma program for the Painlevé test of nonlinear ordinary and partial differential equations' in P.G.L. Leach and W.-H. Steeb (eds.), *Finite Dimensional Integrable Nonlinear Dynamical Systems*, World Scientific, Singapore, pp. 117-129.

[2] Kawahara, T. and Tanaka, M. (1983) 'Interactions of traveling fronts: an exact solution of a nonlinear diffusion equation', Phys. Lett. 97A, 311-314.

THE STRONGLY DISSIPATIVE TODA LATTICE

T. KUUSELA and J. HIETARINTA*
Wihuri Physical Laboratory and
Department of Physical Sciences
University of Turku
20500 Turku, Finland

The Toda lattice was proposed by M. Toda in 1967 [1]. It is a 1-dimensional system of equal masses connected by nonlinear springs with exponential restoring force: $m\ddot{y}_n = a\{\exp[-b(y_n - y_{n-1})] - \exp[-b(y_{n+1} - y_n)]\}$. In addition to this mass-spring interpretation there are other physical realizations; one of the most common is in terms of wave propagation in a nonlinear electrical transmission line [2,3,4]. The network equations describing this transmission line are

$$L\,\partial_t I_n = V_{n-1} - V_n, \qquad \partial_t Q_n = I_n - I_{n+1},$$

where V_n is the voltage, I_n the current and Q_n the nonlinear (differential) charge at a lattice point n. If the voltage dependency of the capacitance is $C(V) = Q_0/(F_0 - V_0 + V)$, then $Q_n = Q_0 \ln(1 + V_n/F_0)$. This system is then equivalent to the Toda lattice if we relate y_n and V_n by $\ln(1 + V_n/F_0) = -b(y_n - y_{n-1})$, and the parameters by $LQ_0/F_0 = m/(ab)$.

In real applications various perturbations are present and this leads to problems beyond the purely integrable ones. In a real transmission line we can have nonideal behaviour due to, e.g., inhomogenities in the lattice, stray capacitance and inductance between components of different lattice points, voltage dependence of the capacitance differing from the one given above, various types of losses in the components, etc. Some of these negative effects can be eliminated by careful construction, and in practice it turns out that the most important perturbation mechanism is dissipation. This is also the most common effect since every physical system is subject to some energy losses. If the dissipation is strong the methods of soliton theory are not even perturbatively applicable, but one can still observe some solitonic properties, as we have shown.

There are many ways to include dissipation in the Toda lattice, but if we use the transmission line analogue certain dissipation terms arise naturally: the (serial) resistance R is usually due to the losses in the thin wire used in the coils, while the (parallel) conductance G is normally caused by parasitic resistance in the capacitance diodes. The corresponding network equations are [3]

$$L\,\partial_t I_n + R\,I_n = V_{n-1} - V_n, \qquad \partial_t Q_n + G\,V_n = I_n - I_{n+1}.$$

*Supported by the Academy of Finland

R. Conte and N. Boccara (eds.), Partially Integrable Evolution Equations in Physics, 587–588.

After eliminating I_n this yields

$$\partial_\tau^2 \ln(1+v_n) + \widetilde{R}\partial_\tau \ln(1+v_n) + \widetilde{G}\partial_\tau v_n + \widetilde{R}\widetilde{G}v_n = v_{n+1} + v_{n-1} - 2v_n,$$

where $v_n = V_n/F_0$, $\tau = v_* t$, $\widetilde{R} = R/(Lv_*)$, $\widetilde{G} = G/(C_0 v_*)$, $C_0 = Q_0/F_0$, and $v_* = 1/\sqrt{LC_0}$ is the characteristic velocity (or frequency) of the lattice.

We have investigated the dissipative Toda lattice numerically, experimentally, and analytically, especially when $\widetilde{R}$ and $\widetilde{G}$ are not small [5,6]. In this system the solitary wave deforms markedly from the ideal soliton as it travels in the lattice: the amplitude of the soliton decays and the tail is created behind it. This might suggest that the system is far from the ideal Toda lattice, however, our numerical results show that certain properties of exactly solvable soliton systems persist even when we have to give up the idea of a permanent traveling wave.

Our main findings based on numerical simulations are the following:

1) The dissipative Toda lattice has some decreasing traveling wave solutions which attract all initial conditions even in the case of strong dissipation.
2) The soliton part of the solution seems to depend on only one parameter $f(\widetilde{R}, \widetilde{G}) \approx \widetilde{R} + \widetilde{G}$.
3) The form of the tail depends on both $\widetilde{R}$ and $\widetilde{G}$, and the tail disappears when $\widetilde{R} \approx \widetilde{G}$. The tail can be accurately described by a linear approximation using a conserved quantity.
4) The collision of decaying solitary waves is elastic within numerical accuracy.

Apparently there are two basic time scales involved in the strongly dissipative Toda lattice: Firstly there is the time needed for an initial pulse to obtain a soliton form, secondly there is the decay time constant. For large dissipation the traveling wave changes so fast that there is no time for the soliton to recover its original shape, thus the decay is not adiabatic. Nevertheless, the recovery time is short enough for all initial pulses to approach some traveling wave configuration(s) which takes the role of the soliton(s). The extension of analytical methods to describe this situation will be an important problem.

[1] M. Toda : J. Phys. Soc. Jpn. **22**, 431 (1967); *ibid* **23**, 501 (1967); Prog. Theor. Phys. Suppl. **45**, 174 (1970); Phys. Reports **18C**, 1 (1973).

[2] R. Hirota : Prog. Theor. Phys. Suppl. **59**, 64 (1976); R. Hirota, K. Suzuki : PIEEE **61**, 1483 (1973); H. Nagashima, Y. Amagishi : J. Phys. Soc. Jpn. **47**, 2021 (1979).

[3] H. Nagashima, Y. Amagishi : J. Phys. Soc. Jpn. **45**, 680 (1978).

[4] T. Kuusela, J. Hietarinta, K. Kokko, R. Laiho : Eur. J. Phys. **8**, 27 (1987).

[5] T. Kuusela, J. Hietarinta : RIMS Kokyoroku **650**, 29 (1987).

[6] T. Kuusela, J. Hietarinta : Phys. Rev. Lett. **62**, 700 (1989).

A PERTURBATIVE APPROACH TO HIROTA'S BILINEAR EQUATIONS OF KdV-TYPE.

F. LAMBERT and R. WILLOX.
Vrije Universiteit Brussel, Theoretische Natuurkunde,
Pleinlaan 2, 1050 Brussel, Belgium.

The occurrence of shock-like solitary wave solutions of the PKdV (or Burgers) type to a dispersive NLPDE is not exceptional. Starting from linear terms $\partial_x^{2m-1}V(x,t)$ or $\partial_t\partial_x^{2m-2}V(x,t)$, $m \geq 2$, it is easy to construct several nonlinearities which produce the "one-soliton balance"... The existence of a PKdV-like soliton, however, requires an exceptional balance between the dispersive terms of the linear part of the equation L(V) and the nonlinear part K(V).
A necessary (but not sufficient) condition is the "two-soliton balance": the equation must also have PKdV-like two-soliton (potential) solutions. This much stronger constraint constitutes an intermediate step towards KdV-like integrability of the equation. It happens to be automatically satisfied for those equations which are derived from Hirota's bilinear equations of KdV-type [1].
Yet, it is not clear which precise conditions allow for the bilinearization of a given NLPDE. Nor has there been any attempt, so far, to interpret the role of the D-operators from a direct point of view (their definition has to be accepted as a clever ansatz)...
The present approach aims at expressing the bilinear two-soliton balance in terms of explicit conditions on the NLPDE for the original variable. We adopt a direct perturbative procedure (it does not rely on any particular skill in choosing appropriate changes of variables) to relate the D-operators to lower order consistency conditions which are *necessary* for the existence of PKdV-type two-soliton solutions. Three necessary conditions, relating L(V) to the quadratic part $K^{(2)}(V)$ and the cubic part $K^{(3)}(V)$ of the nonlinearity, are readily obtained at the two lower orders of a Rosales perturbation [2] in the nonlinearity parameter. Two of these conditions involve only one set of soliton parameters and are automatically satisfied when the equation possesses the prescribed solitary wave solutions. The third condition is the lowest order *two-soliton consistency condition.* It can be used to check whether a PKdV-type solitary wave has any chance to qualify as a soliton potential.
More important for our purpose is the fact that the *two-soliton consistency condition* allows the construction, out of a given L(V), of the quadratic and cubic parts of nonlinearities which realize the two-soliton balance. We restrict our discussion, for simplicity, to L-operators of the form:

$$L = \sum_{\ell} d_{\ell}\, \partial_x^{2m_\ell - 1} + \sum_j \bar{d}_j \partial_t \partial_x^{2m_j - 2}, \qquad m_{\ell,j} = \text{integer} \tag{1}$$

and require that the two-soliton consistency condition be satisfied as a polynomial identity in such a way that each dispersive term be compensated by a specific nonlinearity, leading to a specific two-soliton coupling factor (phaseshift parameter), and that a linear combination of various dispersive

R. Conte and N. Boccara (eds.), Partially Integrable Evolution Equations in Physics, 589–590.

terms be compensated by the same combination of corresponding nonlinearities (linear superposition principle). The condition then decouples into two explicit constraints which associate one particular set of quadratric and cubic nonlinearities with each dispersive term of L(V). These *"standard nonlinearities"* are expressible by a simple *combinatorial rule for* distributing a given number of x-derivatives (one more than those contained in the linear term) among quadratic and cubic terms, which remains valid at any order of the dispersion. Thus, a dispersive term $\partial_x^{2m-1}V$ is found to be compensated by a nonlinearity $K_m(V)$ which contains as many quadratic (cubic) terms as there are different decompositions of the integer 2m as a sum of two (three) smaller *even* integers:

$$2m = 2r+(2m-2r): \qquad K_m^{(2)} = \sum c_{2m}(2r) V_{(2r-1)x} V_{(2m-2r-1)x} \tag{2}$$

$$2m = 2r+2s+(2m-2r-2s): \qquad K_m^{(3)} = -\sum c_{2m}(2r,2s) V_{(2r-1)x} V_{(2s-1)x} V_{(2m-2r-2s-1)x} \tag{3}$$

where $V_{px} = \partial_x^p V$ and where each coefficient represents the combinatorial weight of the corresponding decomposition (number of ways of dividing 2m distinct elements into as many boxes with as many elements as indicated by the decomposition if each box is only characterized by the number of elements it contains).
The standard quadratic and cubic nonlinearities associated with a term $\partial_t\partial_x^{2m-2}$ are similarly determined by the decompositions of 2m-1 as a sum of two and three integers of which only one is odd. The remarkable balance expressed by these rules suggests that the same principle should also hold at higher orders in the nonlinearity as a general recipe for the construction of standard two-soliton generating nonlinearities. It turns out that the Hirota D-operator is precisely the tool needed to formalize this construction. Thus, by collecting th r^{th} degree terms $P_{2m}^{(r)}$ of the polynomial:

$$P_{2m}(q_x, \ldots q_{2mx}) \equiv e^{-2q} D_x^{2m}(e^q . e^q) \tag{4}$$

it is found that $-\partial_x^{2m-1}(V = -2q_x) \equiv P_{2m}^{(1)}$ and that $K_m^{(r)}(V = -2q_x) \equiv P_{2m}^{(r)}, \quad r = 2,3.$ (5)

One can also verify that the same combinatorial rule applies at any higher value of r [3].
It is worth noticing that the presence of simple combinatorial rules, like those which underlie the bilinear machinery (thus providing a short cut of lengthy reductions of bilinear forms), is also a peculiarity of the Hopf-Burgers hierarchy [4] of equations which are linearized by the Cole-Hopf transformation. In this sense it is reasonable to consider the bilinear equations:

$$F(D_t,D_x) \equiv D_x L(D_t,D_x) = 0 \tag{6}$$

as being the simplest category of equations after the linearizable Hopf-Burgers class.

[1]: Hietarinta, J. (1987) "A search for bilinear equations passing Hirota's three-soliton condition. I. KdV-type bilinear equations" J. Math. Phys. 28 , 1732-1742.

[2]: Rosales, R. (1978) "Exact solutions of some nonlinear evolution equations" Stud. Appl. Math. 59 , 117-151.

[3]: Lambert, F., Willox, R. (1989) "On the balance between dispersion and nonlinearity for a class of bilinear equations". J. Phys. Soc. Japan 58, 1860-1861.

[4]: Choodnovsky D.V., Choodnovsky G.V. (1977) "Pole expansions of nonlinear partial differential equations" Nuovo Cim. 40 B, 339-353.

CONSTRUCTION OF TWO DIMENSIONAL SUPER POTENTIALS FOR CLASSICAL SUPER SYSTEMS

S.C. Mishra
Department of Physics and Astrophysics, University of Delhi, Delhi-11007, India

ABSTRACT : Construction of super potentials for two dimensional classical super systems (for $N \geq 2$) is carried out. Some interesting potentials have been studied in their super form.

In recent years there have been considerable interest in the study of supersymmetry which is one of the elegent creations in theoretical physics. There was an attempt (Bouquiaux et al. 1987) to discuss the general formalism of N=2 pseudomechanics in the superspace for one spatial dimensions, but not much effort (Tripathy and Tripathy, 1988) has been made to construct the superpotentials for two dimensional classical supersystems ($N \geq 2$, superspace). In this note we concentrate to construct only the superpotentials for two dimensional classical supersystems. Such a study could be profitably applied to the spining string or membrane theory having both bosonic and Grassmann degrees of freedom. Here we make use of "super" time variable which involves the usual t and the Grassmann time variables θ_i and $\bar{\theta}_i$. Correspondingly, the super position variable is given by

$$Z_i = Z_i(t, \underset{\sim}{\theta}, \underset{\sim}{\bar{\theta}}) = q_i(t) + i\bar{\theta}_i \psi_i(t) + i\theta_i \bar{\psi}_i(t) + \theta_i \bar{\theta}_i A_i(t) \quad i=1,2$$

where $q_i(t)$ position variables, ψ_i and $\bar{\psi}_i$ fermionic variables (which describe the spin degrees of freedom) and $A_i(t)$ is bosonic variables.

Following Bouquiaux et al. 1987, it is possible to get the set of equations of motion

$$\dot{q}_i = [q_i, H]^* = p_i, \ \dot{p}_i = [p_i, H]^* = \ddot{q}_i = -(\tfrac{1}{2}(W'_{q_i})^2 + W''_{q_i} \bar{\psi}_i \psi_i)$$

$$\dot{\bar{\psi}}_i = [\bar{\psi}_i, H]^* = iW''_{q_i} \bar{\psi}_i, \ \dot{\psi}_i = [\psi_i, H]^* = -iW''_{q_i} \psi_i$$

which led us to construct few potentials in their superform for two dimensional supersystems.

i. For the potential $V = \frac{1}{2}\lambda^2 q^2 = \frac{1}{2} W'^2(q)$ we get the superpotential $W(Z_i)$ as

$$W(Z_i) = \frac{1}{2} \lambda q^2 + i\lambda q_i (\theta_i \bar{\psi}_i + \bar{\theta}_i \psi_i) + \lambda^2 q_i^2 \theta_i \bar{\theta}_i + \lambda \theta_i \bar{\theta}_i \psi_i \bar{\psi}_i$$

ii. For the potential $V = \lambda^2 q^{-2/3}$, we get $W(Z_i)$ as

$$W(Z_i) = \frac{3}{2} \lambda q^{2/3} + i\lambda \frac{q_i}{q^{4/3}} (\theta_i \bar{\psi}_i + \bar{\theta}_i \psi_i) + \theta_i \bar{\theta}_i \frac{\lambda}{q^{4/3}} [\frac{\lambda q_i^2}{q^{4/3}}$$

$$+ (1 - \frac{4}{3} \frac{q_i^2}{q^2} \ \bar{\psi}_i \psi_i)]$$

. Conte and N. Boccara (eds.), Partially Integrable Evolution Equations in Physics, 591–592.

iii. For the potential $V = \frac{\lambda q^2}{1+gq^2}$, we get $W(Z_i)$ as

$$W(Z_i) = \frac{\sqrt{\lambda(1+gq^2)}}{g} + \sqrt{\frac{\lambda q_i}{(1+gq^2)}}\; i(\theta_i\bar{\psi}_i + \bar{\theta}_i\psi_i)$$

$$+ \theta_i\bar{\theta}_i \sqrt{\frac{\lambda}{(1+gq^2)}}\left[\frac{\sqrt{\lambda q_i^2}}{\sqrt{(1+gq^2)}} + \left(1 - \frac{gq_i^2}{(1+gq^2)}\right)\bar{\psi}_i\psi_i\right]$$

iv. For the potential $V = -\frac{\lambda^2}{q}$, we get $W(Z_i)$ as

$$W(Z_i) = 2i\lambda\sqrt{q} - \frac{\lambda}{q^{3/2}}(q_i\theta_i\bar{\psi}_i + \bar{\theta}_i q_i\psi_i) + \frac{i\lambda}{q^{3/2}}\theta_i\bar{\theta}_i\left[i\lambda\frac{q_i^2}{q^{3/2}}\right.$$

$$\left. + \frac{1}{2q^2}\{(2q_2^2 - q_1^2)\bar{\psi}_1\psi_1 + (2q_1^2 - q_2^2)\bar{\psi}_2\psi_2\}\right] .$$

Here we discuss the superpotentials in superspace with spatial dimensions ≥ 2. Invoking Dirac's constraints, the dynamical system under consideration reduces to a time–independent one. All these potentials have been discussed in their superform which contain bosonic and Fermionic part have some nice features. The fermionic part which effectively involves the fermionic number operator implies that the super symmetric formalism could be envisaged as dealing with "scalar" spinors. In some cases the potentials yield extra degeneracies in the corresponding quantum systems.

REFERENCES

Bouquiaux, L., Dauby, P. and Hussin, V., 1987, "Superspace formulation of N=2 pseudomechanics and superpotentials", J. Math. Phys. 28, 477-484.

Sudarshan, E.C.G. and Mukunda, N., 1974, Classical Dynamics a Modern Prospective (Wiley, New York).

Tripathy, K.C. and Tripathy, L.K., 1988, "Integrable two dimensional super system ICTP, Trieste, Preprint No. IC/88/233.

APPLICATIONS OF NONLINEAR PDE'S TO THE MODELLING OF FERROMAGNETIC INHOMOGENEITIES

J.A. TUSZYNSKI
Department of Physics, The University of Alberta,
Edmonton, Alberta, Canada, T6G 2J1

P. WINTERNITZ AND A.M. GRUNDLAND
C.R.M., Université de Montréal
Montréal, Québec, Canada, H3C 3J7

ABSTRACT. Several approaches lead to the Landau-Ginzburg free energy for magnetic inhomogeneities in ferromagnets. Minimization of this functional in both reorientational and spin-ordering processes yields a 3-D Nonlinear Klein-Gordon equation for the order parameter. Symmetry Reduction Method is used to obtain solutions of this PDE. Physical interpretation is given.

The Landau-Ginzburg Free Energy Functional

$$F = \int d^n x \; [A_2\eta^2 + A_4\eta^4 + A_6\eta^6 - h\eta + \tfrac{1}{2} D(\nabla\eta)^2] \qquad (1)$$

where n is spatial dimensionality, $A_2 = \alpha(T-T_c)$, A_4, A_6 are constants, η is the order parameter and h is the conjugate field, can be used to describe both field - ($h \neq 0$) and temperature-induced ($h = 0$), first ($A_4 < 0$) or second order ($A_4 > 0$) phase transitions. A more or less formal justification of its applicability to ferromagnetic materials near criticality can be demonstrated either through a continuum approximation applied to the Heisenberg Hamiltonian [1], or through an evaluation of the corresponding partition function [2]. It is quite common, however, to simply postulate (1) on the basis of symmetry arguments and elementary statements of the catastrophe theory [3].

In particular, for spin-ordering transitions ($P \rightarrow F$) occurring in the presence of an easy magnetization axis z, η represents the z-component of magnetization, D refers to the nearest-neighbours exchange constant and A_2, A_4, A_6 are "molecular field" constants due to higher coordination spheres. Minimization of F gives

$$D\nabla^2\eta = 2A_2\eta + 4A_4\eta^3 + 6A_6\eta^5 - h \qquad (2)$$

which is a 3-D time-independent nonlinear Klein-Gordon equation. Symmetry reduction method has been recently applied [4,5] to analyze Eq. (2) and its solutions. When $h = 0$ and the system is either at the critical or the tricritical point, the symmetry group is the similitude group. Then, the symmetry variables represent: 1,2 or 3-D hyperspheres, planes, spirals or cones. Otherwise, the symmetry group is Euclidean and only plane or hyperspherical symmetry variables are allowed. The effect of the external field h is, first of all, to

R. Conte and N. Boccara (eds.), Partially Integrable Evolution Equations in Physics, 593–594.

of the external field h is, first of all, to destroy the scale invariance if the system is at either the critical or the tricritical point. Secondly, h affects the form of the solutions. For translationally-invariant ones kinks are disallowed and are replaced by bumps. We believe that this plays a crucial role in producing a curvature of the Arrott plot (h/η vs η^2) for magnetization processes. Moreover, the amplitudes and frequencies of the periodic solutions are shifted as a result of switching on the field.

For reorientational processes occurring in the presence of an easy-magnetization plane, the order parameter η is two-component, i.e. consists of the two directional cosines α_1 and α_2. The expansion coefficients for the free energy analogous to eq. (1) correspond to the appropriate anisotropy constants. In 1-D approximation the minimization of the free energy results in a very similar type of problem as that described above. Substituting $(\alpha_1, \alpha_2) = (\sin\theta, \cos\theta)$ gives the first integral of the Euler-Lagrange equation in terms of θ which is identical to that obtained from eq. (2) in 1-D space. The resultant kink solution has been known as the so-called Néel domain wall when $(\alpha_1, \alpha_2) = (\alpha_x(z), \alpha_z(z))$ and the Bloch domain wall when $(\alpha_1,\alpha_2) = (\alpha_x(z), \alpha_y(z))$. Periodic solutions in terms of Jacobi eliptic functions can readily be found and correspond to the various ferromagnetic and antiferromagnetic spin waves.

Eq. (2) represents only magnetic structures $\eta(\underline{x})$ which minimize the free energy (1). In order to model thermodynamically stable fluctuations it is required of them to satisfy

$$A_2\eta^2 + A_4\eta^4 + A_6\eta^6 - h\eta + \frac{1}{2} D(\nabla\eta)^2 = E = \text{const} \tag{3}$$

Interestingly, Eq. (3) has the same symmetry group as Eq. (2). Moreover, in 1-D situations it takes the same form as the first integral of Eq. (2). The only difference there is the sign of D. This means that the functional extema of (1) for $D > 0$ correspond to equilibrium fluctuations of (1) for $D < 0$ and vice versa. Consequently, the entire spectrum of thermodynamically stable inhomogeneities can be found this way.

Work on the energies of the solutions, the influence of external fields and the modelling of equilibrium fluctuation is in progress and will be published elsewhere.

REFERENCES

[1] Anderson, P.W. (1984) Basic Notions of Condensed Matter Physics, Benjamin, Menlo Park,

[2] Amit, D.J. (1978) Field Theory, the Renormalization Group and Critical Phenomena, McGraw-Hill, New York.

[3] Poston, T. and Stewart (1978), Catastrophe Theory and Its Applications, Pitman, London.

[4] Winternitz, P., Grundland, A.M. and Tuszyński, J.A. (1987) J. Math. Phys. 28, 2194-2212.

[5] Winternitz, P., Grundland, A.M. and Tuszyński, J.A., (1988) J. Phys. C 21, 4931-4953.

AUTHOR INDEX

SUBJECT INDEX